TRANSACTIONS
OF THE
INTERNATIONAL
ASTRONOMICAL UNION

VOLUME XIVA – REPORTS

INTERNATIONAL COUNCIL OF SCIENTIFIC UNIONS
INTERNATIONAL ASTRONOMICAL UNION
UNION ASTRONOMIQUE INTERNATIONALE

TRANSACTIONS
OF THE
INTERNATIONAL ASTRONOMICAL UNION
VOLUME XIVA

REPORTS ON ASTRONOMY

Edited by

CORNELIS DE JAGER
Assistant General Secretary of the Union

PUBLISHED FOR THE INTERNATIONAL COUNCIL OF SCIENTIFIC UNIONS
WITH THE FINANCIAL ASSISTANCE OF UNESCO

D. REIDEL PUBLISHING COMPANY / DORDRECHT-HOLLAND
1970

Published on behalf of
the International Astronomical Union
by
D. Reidel Publishing Company
Dordrecht - Holland

Unesco subvention – 1969 – DG/2.1/414/30
Library of Congress Catalog Card Number 30-10103

SBN 90 277 0154 7

Printed in The Netherlands by D. Reidel, Dordrecht

PREFACE

The tri-annual preparation of the *Reports on Astronomy* is an impressive world-wide undertaking, involving not only the Presidents, Vice-Presidents and Organizing Committees of our thirty-eight Commissions, but also the greater part of the Commission members, and many other astronomers – certainly more than a thousand people. These large co-operative enterprises have led in the past to the successive *Draft Reports* and *Transactions* of the IAU. These constitute a series of monumental milestones characterizing in thirteen big steps the imposing progress of our science in the last fifty years.

An important change occurs in the present Reports. Financial reasons made it impossible for the Union to keep to the tradition of first printing the *Draft Reports* which, prior to the General Assembly were airmailed to all IAU members, and later printed again, finally amended and corrected. The Executive Committee of the Union had to decide that the reports should be printed directly, distributed to the participating IAU members at the General Assembly, and sent later by surface mail to the other members of the Union.

Furthermore it appeared necessary to reduce the size of these Reports to slightly more than one half of that of *Transactions* **XIIIA** – the previous *Reports on Astronomy*. It is fully realized that these constraints made the task of the Commission Presidents and their coworkers heavier than usual. First of all, the deadline for submitting the manuscripts had to be a fairly strict one since the Reports had to be printed and bound in time, and the subsequent addition of material was no longer possible.

However, the greatest difficulty was the reduction of the size. The Executive Committee suggested Commission Presidents to achieve this by the following measures:

1. By deleting the full lists of references; these took about one third of the previous Reports. This suggestion seems justified by the recent appearance of the first issue of *Astronomy and Astrophysics Abstracts*, the successor to the *Astronomischer Jahresbericht*. This new publication gives a complete review of the astronomical literature, and appears within six to twelve months after the publication of the abstracted papers; it is more complete and certainly more up to date than would have been possible in our Reports. (We should add, however, that these restrictions were allowed to be interpreted more liberally in the case of Commission 14, where the inclusion of fairly long lists of references from fields adjacent to our science was thought to be useful.)
2. By a reduction of the contents of the Reports to the really essential matter.
3. It was furthermore thought that the value of the Reports could increase by the introduction of previews about the further development of our science, and by an enumeration of future plans and projects.
4. It was further decided no longer to publish the detailed lists of Commission members since these lists are already published in *Transactions* **XIIIB** with the subsequent corrections listed in the various issues of the *Information Bulletin*.

These suggestions were found acceptable by many IAU members but not by all of them. The opinion was even expressed "that the new form of the Reports may make future volumes of less permanent value than Volume **XIIIA**, which will be the last one to attempt a full coverage of astronomy" (*Observatory*, **89**, 240, 1969). The reader is also referred to the introduction to the report of Commission 12.

Whether the value of the present Reports is less than that of previous ones is a matter to be decided by its users. Clearly, the precise form of the Reports needs further definition in certain respects, and some more experimenting may continue to be desirable. However, it is my personal belief, based on the reading of the present Commission Reports, that by intelligent reduction of the available material, and by carefully separating the essential from the trivial, the Commission Presidents have generally succeeded in presenting a more condensed, but yet complete coverage of the whole field of astronomy in all its fascinating aspects, a review that, in certain respects, may even be more balanced than some of the previous Reports, and certainly not less in value than would have been the case with the inclusion of more, but perhaps less important details and references. That each report reflects the personal style of its author(s) is one of their traditional and charming aspects.

We wish to express our very sincere thanks to the Commission Presidents and their co-workers for the way they have interpreted our suggestions, and for so strictly observing the deadline for submission of the manuscripts. Our thanks are also due to the printer and publisher of this Volume, Mr A. Reidel and his staff, for their excellent and always pleasant cooperation.

C. DE JAGER
Assistant General Secretary

Utrecht
January 1970

CONTENTS

4. EPHEMERIDES (EPHÉMÉRIDES)

PRESIDENT: G. A. Wilkins.
VICE-PRESIDENT: J. Kovalevsky.
ORGANIZING COMMITTEE: G. M. Clemence, G. A. Chebotarev, R. L. Duncombe, W. Fricke.

INTRODUCTION

The appended reports of the directors of the national ephemeris offices provide a record of the current activity in the publication and preparation of astronomical ephemerides, including those published for use by navigators and surveyors and those prepared for astronomers whose special needs are not met in the printed almanacs. A brief review of the relevant work of other institutions and individuals is also given in the appendix. The reports also cover to some extent activities in positional astronomy and celestial mechanics (i.e. in the fields of interest of Commissions 7, 8, 17, 19, 20 and 31). This report is therefore restricted to reviewing, firstly, the action that has been taken on matters raised at and since the 1967 meetings of the Commission and, secondly, the matters that will require attention at and after the 1970 meetings.

Before that review, it is appropriate to pay tribute to the memory of that almost legendary astronomer, Gaston Fayet, who died on 27 December 1967 at the advanced age of 93. As member of the Bureau des Longitudes responsible for the production of the *Connaissance des Temps* from 1930 to 1961, and President of Commission 4 from 1938 to 1952, he contributed greatly to our subject, in addition to his researches in other fields particularly in respect of the minor planets.

INTRODUCTION OF THE IAU SYSTEM OF ASTRONOMICAL CONSTANTS

The changes in the printed ephemerides resulting from the adoption (a) in 1964 of the new IAU system of astronomical constants and (b) in 1967 of new bases for the solar part of the lunar theory and for the minor planet ephemerides will be virtually completed in the almanacs for 1972. In all but a few cases the corrections have been incorporated into the main tabulations.

The problem of the epochs of the (ephemeris) time scales to which the solar and lunar ephemerides refer remains unresolved. The difficulties arise, firstly, from the discontinuity in the ephemeris of the apparent longitude of the Sun following the change in the constant of aberration and, secondly, from the suggestion that the solar and lunar ephemerides are referred to equinoxes that differ in right ascension by 1ˢ34. The question therefore arises whether it would be desirable to bring the solar ephemeris into accord with the lunar ephemeris by introducing another discontinuity or whether it is sufficient merely to adjust the values of $\Delta T =$ E.T. − U.T. that are obtained from the comparisons between theory and observation. The latter appears to be preferable, but a clear statement of the basis of the adjustments is obviously required before such a procedure could be formally recommended.

ASTRONOMICAL NAVIGATION

The use of conventional astronomical methods for navigation and surveying shows little sign of extinction in spite of the increasing use of new methods based on radio transmissions, artificial satellites, and inertial navigation systems. Minor changes in the content and format of the various almanacs and auxiliary tables have been introduced and the process of worldwide standardization continues slowly. The introduction of a new system of international time signals (see below) will entail additional effort by the user if the full accuracy of astronomical techniques is to be utilised.

UNIT OF TIME

The thirteenth General Conference on Weights and Measures adopted, on 13 October 1967, a resolution defining the unit of time in the International System (SI) of Units in terms of a transition frequency of caesium and abrogating the earlier resolutions defining the SI second in terms of ephemeris time. At present therefore, the responsibility for the definition and realization of the ephemeris time scale rests solely with the IAU, and Commission 4 should consider very carefully the future policy to be adopted in respect of the ephemeris time scale. In particular, consideration should be given to the possibility of adopting the SI second in place of the ephemeris second. This would once again make the coefficient of T in the expression for the Sun's mean longitude subject to determination by observation. Such an approach would still leave open the question of the identity of the gravitational and atomic time scales; any fundamental difference between the two scales would give rise to apparent secular accelerations that could not be explained in terms of other known forces. The term ephemeris time could still be retained to provide a meaningful name for the identification of epochs determined by the comparison of observations and ephemerides rather than by 'clocks' (e.g. A.T.) or the rotation of the Earth (U.T. and S.T.).

OTHER SI UNITS

In recent years there has been a marked increase in the use of SI units in scientific literature, especially as many of the leading journals have recommended their use. Even apart from the question of the unit of time, there are three principal reasons why SI units should not be introduced into all astronomical ephemerides, and their auxiliary tables and explanations. Firstly, the use of the astronomical units of distance and mass in planetary dynamics allows distances and masses to be specified to a higher precision than would be possible if metres and kilogrammes were to be used; the significance of data expressed in astronomical units is also more easily understood. Secondly, ephemerides for practical uses must be designed so as to minimize both the effort and the risk of error associated with their use; this applies particularly to ephemerides for navigational purposes where the reductions must be made quickly, and often under adverse conditions. The unit of tabulation must therefore correspond to the units in which the measurements are made. In particular, the continued use of sexagesimal units for angle and time appears to be inevitable. Thirdly, in some circumstances it may be desirable, or even necessary, to specify some quantities in terms of the legal units of the country of publication.

With these reservations it seems desirable that SI units and conventions (e.g. restricting multiple and sub-multiple units to steps of one-thousand and one-thousandth, respectively) should be used in preference to, or in addition to, other units. It would be useful, however, if the Commission were to adopt a list indicating recommended usages in the field of dynamical astronomy so that the attempt to introduce a higher degree of standardization does not, in fact, have the opposite effect.

INTERNATIONAL TIME SIGNALS

Responsibility for the international coordination of the transmissions of time signals rests with the Consultative Committee on International Radio (CCIR). Following the adoption of the new definition of the SI second an international working group has recently recommended a change from the present system in which the primary transmissions are maintained to within one-tenth of a second of U.T.2 by means of step adjustments in both frequency and epoch. If the recommendations are adopted, the primary transmissions will be based, as from the beginning of 1972, on a fixed frequency with respect to the SI second and on step adjustments of exactly 1 second. This implies that the departures from U.T.2 will exceed half-a-second and so will be much greater than the errors of human discrimination, such as in the timing of occultations and in the use of a sextant. The IAU has already expressed its willingness to accept such a change provided that certain conditions are met; these are intended to ensure that U.T.2 will be readily available, even under field conditions, to a precision of 0.1 without danger of ambiguity. Such a change has no direct effect on the publication of astronomical ephemerides except in so far as it will be necessary to warn users about the departures from U.T.2 and to provide appropriate instructions for the application of corrections.

INTERNATIONAL INFORMATION BUREAU ON ASTRONOMICAL EPHEMERIDES

It was agreed at the 1967 meeting that the membership of the Commission's Working Group on Space Ephemerides should be enhanced by the co-option of other interested scientists. As a result, and at the invitation of the President of COSPAR, the Group met during the course of the COSPAR meeting in Prague in May 1969. After reviewing the current activities and requirements in the field of astronomical ephemerides, the Group considered how best it could facilitate both the exchange of ephemerides between the institutions engaged in preparing them and the supply of ephemerides to those other organisations that required them for research purposes. It was agreed to recommend that IAU and COSPAR should cooperate in setting up an information service that would provide information to the international scientific community on the availability of astronomical ephemerides, particularly those that exist in machine-readable form. Subsequently on the recommendation of Dr J. Kovalevsky, who is Chairman of COSPAR's Working Group I, it was agreed that the bureau should be sponsored by the IAU only, but with COSPAR represented on the advisory committee. The Commission will have to decide whether to adopt such a proposal.

OBSERVATIONS OF THE POSITIONS OF THE MOON AND PLANETS

The further improvement in the fundamental ephemerides of the Sun, Moon and planets depends on both the development of better theories and the comparison of the theories with observations to determine their arbitrary parameters. The Working Group on Space Ephemerides considered that, for the purposes of the proposed information bureau, astronomical ephemerides should be deemed to include collections of observations of the positions of planets and satellites. The Commission should consider whether it should, and could, do more to encourage and assist astronomers to prepare and make available collections of past observations and to make and publish new observations. (At the present time no Commission appears to cover adequately these activities for the Moon and major planets.) The task of locating, punching and, possibly, re-reducing past observations is so onerous that it would be a pity if there were to be unnecesary duplication in this valuable, but prosaic, activity. Further, the full rewards of the developments of new theories (often using much expensive computer time) will be lost if there is a lack of adequate observational material with which to compare the theories.

COLLOQUIUM ON THE IAU SYSTEM OF ASTRONOMICAL CONSTANTS

As a result of the discussions at the 1967 meeting and of subsequent correspondence it was decided to hold a Colloquium on the IAU System of Astronomical Constants at Heidelberg from 12 to 15 August 1970. The primary purpose of the meeting is to provide the bases for subsequent formal proposals for the adoption or otherwise of new values of those astronomical constants that were left unchanged when the IAU system was adopted in 1964. The principal constants concerned are the precessional constants and the masses of the planets. It is necessary to consider not only the strength of the evidence for more accurate values, but also the possible procedures for, and consequences of, the introduction of new values. At the meetings of the Commission at Brighton it will be necessary to decide what action should be taken on the recommendations of the Colloquium. The present consensus of opinion seems to be that, if it is decided to adopt new values, the formal resolution should be considered at the General Assembly in 1973.

The question of the year of introduction into the almanacs of the use of the equinox of 2000·0 has been raised by Dr J. Kovalevsky. Only a small number of comments have so far been received, but the majority of them are strongly in favour of making no decision until the question of the change in the precessional constants has been settled, with the possibility of introducing both changes into the ephemerides for 1980. It has also been suggested (a) that the preparation of high-precision star catalogues referred to the new equinox should similarly be deferred, and (b) that the E-terms of aberration should be removed from the mean places so as to simplify rigorous reductions to and from apparent places.

FUTURE PUBLICATION OF FUNDAMENTAL EPHEMERIDES

As previous Presidents have implied the national almanacs are becoming less suitable for the publication of fundamental ephemerides, whose primary purpose is to provide internationally recognized standards with which observations and new theories can be compared. This arises partly from the increasing use of computers in the preparation of ephemerides and in the reduction of observations, and partly from the introduction of new techniques of observation in which quantities other than apparent right ascension and declination are measured. Further, the present system necessarily entails a time lag of between five and eight years before an improved ephemeris can be introduced into the main tabulations printed in the almanacs. This is now quite unsatisfactory since it means that in practice other ephemerides are used; the Commission should therefore consider alternative procedures.

The principal desiderata for a fundamental ephemeris are: firstly, that it should cover a period of many years and be available in both printed and machine-readable form; and, secondly, that it should be precisely computed on a recognized and clearly-defined basis. These conditions are often better met by an independent publication (such as the ephemerides in the *Astronomical Papers of the American Ephemeris*) than by inclusion in an annual publication, which has also to provide ephemerides for other purposes.

It is therefore suggested that in future the fundamental ephemerides should first be published in independent form covering many years, using computer-based techniques so as to reduce the cost and effort required. The task of the Commission would then be to decide when to recognize a new ephemeris as an international standard. The final decision need be taken only after the ephemeris has been published and scrutinized to establish the quality of its numerical values and documentation, although naturally there would usually be prior consultation. The Commission could assign a suitable reference code (such as is now adopted for the various forms of the lunar ephemeris) so as to facilitate the specification of the particular ephemeris used in a comparison between observation and theory. The Commission could recognize both heliocentric and geocentric ephemerides as well as special ephemerides, such as an ephemeris of the rectangular coordinates of the geocentre with respect to the centre of mass of the solar system for use in the reduction of observations of pulsars, etc. The adoption of such a system would allow greater freedom in the design of the almanacs to meet the day-to-day requirements of the users.

CONCLUSIONS

The period of change that started with the proposal in 1950 for the introduction of ephemeris time is not yet over, but in the next few years we may hope to consolidate the progress that has been made since 1950 and provide a firm foundation for dynamical astronomy during the remainder of the twentieth century.

G. A. WILKINS
President of the Commission

APPENDIX: REPORTS OF THE DIRECTORS OF THE NATIONAL EPHEMERIDES

Ephemeris Department of the Institute for Theoretical Astronomy, Leningrad, U.S.S.R.

During the period under review two issues of the *Astronomical Yearbook of the U.S.S.R.* for the years 1971, 1972 have been published, and that for 1973 is in preparation. The positions and proper motions of the stars are in the FK4 system as recommended by Commission 4 in 1961. Further, as recommended at the meeting in Hamburg in 1964 (*Trans. IAU*, **12 B** (1964), 105, 1966) the IAU System of Astronomical Constants has been introduced into the tabulations of the day numbers, solar and lunar eclipses, ephemerides for physical observations of planets, and the factors *S* & *C*. The corresponding differential corrections to the Moon's ephemeris as well as re-calculated eclipse data for 1971 have been tabulated separately in the Appendix to the *Yearbook* for 1971. As from the edition of the *Yearbook* for 1972, the lunar ephemeris $j = 2$ will be published in accordance with the recommendations adopted in Prague in 1967.

The regular publication of the ephemeris of the lunar crater Mösting A has been continued in the editions of the *Yearbook* mentioned above.

Fundamental ephemerides of the Sun, Moon, and planets have been printed in full conformity with the first part of the *Astronomical Ephemeris* circulated in advance by H. M. Nautical Almanac Office, Royal Greenwich Observatory.

The navigational almanacs have regularly been produced as before. The *Nautical Almanac* has been issued for the years 1970, 1971, and the *Air Almanac* for 1969, 1970.

The predictions of lunar occultations for 1968, 1969, 1970, computed by and received from H. M. Nautical Almanac Office, Royal Greenwich Observatory, were distributed, as usual, to the Soviet observatories.

A new electronic computer BESM-4 was installed in January–March 1969 making practicable the systematic calculation of ephemerides.

A series of papers on different topics of celestial mechanics and computational procedures were published in the *Bulletin* of this Institute.

V. K. ABALAKIN

Bureau des Longitudes, Paris, France

Le travail de calcul et de rédaction des publications du Bureau des Longitudes a été poursuivi à un rythme accéléré au cours des trois dernières années sans toutefois qu'il ait encore été possible d'avancer les dates de parution des volumes.

Le contenu de la Connaissance des Temps n'a pas été modifié dans son ensemble. Le système UAI de constantes astronomiques a été introduit depuis le volume pour 1968. Toutefois, on n'a pas tenu compte du changement de constantes dans le calcul des coordonnées vraies du Soleil et des planètes. Pour la Lune, on a continué à publier les éphémérides avec les anciennes constantes mais on donne un tableau de corrections pour passer au nouveau système.

La totalité des tableaux variables de la Connaissance des Temps à l'exception de ceux relatifs aux satellites de Jupiter, les éclipses et quelques courtes tables sont composés de manière automatique grâce à un convertisseur de bandes 7 canaux en bandes Monotype.

Les recherches concernant le mouvement des satellites galiléens de Jupiter se poursuivent. A court terme, une amélioration dans la présentation des éphémérides permettant le calcul facile des coordonnées rectangulaires tout en introduisant les termes antérieurement négligés de la théorie de Sampson a été étudiée. Elle sera introduite dans la Connaissance des Temps pour 1972, en même temps qu'une méthode rigoureuse de calcul des phénomènes sera utilisée. A plus long terme, on a commencé à refaire les calculs de la théorie de Sampson avec les constantes d'intégration tirées des travaux de De Sitter. Enfin, en liaison avec l'Instituto Tecnológico de Aeronáutica (São José dos Campos, Brésil) les études pour la construction d'une théorie nouvelle sont activement menées.

La totalité des Ephémérides Nautiques est calculée sur ordinateur et composée automatiquement.

Les Ephémérides aéronautiques, dont des épreuves reproductibles sont désormais envoyées directement par le U.S. Nautical Almanac Office et le H. M. Nautical Almanac Office, continuent à être publiées sous la même forme qu'auparavant.

J. KOVALEVSKY

H. M. Nautical Almanac Office, Royal Greenwich Observatory, Herstmonceux, England

The Office has continued, without substantial change, the production of the four ephemerides: *The Astronomical Ephemeris* – for astronomers; *The Nautical Almanac* – for surface navigation; *The Air Almanac* – for air navigation; and *The Star Almanac* – for land surveyors. The first three of these ephemerides are completely unified with the corresponding U.S. editions, which are printed separately from identical reproducible material; this material can be made available, at a small charge, to the official almanac-producing agency in any country that desires to publish similar ephemerides in this form.

The fundamental ephemerides of the Sun, Moon and planets contained in the first part of *The Astronomical Ephemeris* for 1972 and 1973 have been circulated to ensure that the tabulated data are available to all agencies responsible for the preparation of navigational and other almanacs.

As from the year 1972 there will no longer be in *The Astronomical Ephemeris* an appendix containing the differential corrections, or the formulae from which they can be derived, necessary to convert the printed ephemerides to those based on the IAU System of Astronomical Constants. However, in strict accord with the recommendations of the Commission at the thirteenth General Assembly (*Trans. IAU*, **13 B**, 47–49, 1968) the printed ephemerides of the Sun, Mercury, Venus and Mars do not incorporate these corrections, they are in fact strictly continuous with the ephemerides as printed up to the end of 1971. The corrections are generally small and are often negligible; formulae (similar to those previously given in the appendices) will be included in the Explanation for 1972 and future years. It should, however, be noted that the Sun's longitude is now tabulated to $0''.01$, so that the corrections to be applied are now significant in that unit, and may reach $0''.04$ in the apparent longitude.

The only substantial changes in the tabulations for 1972 are: the incorporation of the changes, necessary to convert to the IAU System, in the ephemerides of the Moon (which is now in accord with the specification for $j=2$), the outer planets (Jupiter to Pluto) and the four minor planets; the tabulation of an additional figure in the Moon's horizontal parallax for 0^h and 12^h on each day, and the inclusion of hourly values to $0''.01$; the introduction of a new ephemeris, for 0^h and 12^h on each day, of the distance of the Moon in units of the Earth's equatorial radius; and the extension of the ephemerides of the minor planets throughout the year.

The fundamental ephemerides for 1972 have been set by a photocomposing system controlled from the output of the ICL 1909 computer, on which all the calculations have been made. Due to limitations in the available equipment, the process has not been entirely satisfactory in practice, but it has served to demonstrate that the editing programs possess all the flexibility required to produce copy to a high standard – both of accuracy and presentation. Improved equipment will be used for future years.

A start has been made on the major project of recalculating, to a considerably higher precision than that aimed at by E. W. Brown, the effect of planetary perturbations on the motion of the Moon. Little real progress has been made with the theoretical work, other than to recognize the considerable difficulties that arise in using Brown's method to higher precision. However, a comprehensive program of numerical integration designed to provide a basis for the comparison of the theory with values obtained by direct numerical integration of the differential equations is now in operation.

Co-ordinates of the Earth with respect to the centre of gravity of the Sun and planets (S9) have been prepared, and made available, for the analysis of current observations of pulsars.

The availability of the ICL 1909 computer has made practicable the reduction of observations of lunar occultations almost as and when they are received, and observers can thus be informed of the results of their observations within a short time. This has led to a considerable encouragement to observers, as well as to improvements in methods of observation.

The limb corrections given in *The Marginal Zone of the Moon* (*A.P.A.E.*, Vol. XVII) by C. B. Watts have now been put into machine-readable form, so that in future these need no longer be laboriously taken out by hand. The systematic reduction of observed occultations, including limb corrections, is now complete from 1958 to 1968, and will be extended backwards in due course to 1943. The observations from 1960–1966 have been subjected to a detailed analysis (*Mon. Not. R. Astr. Soc.*, **144**, 129, 1969), the main result being that the duration of the second (the SI second, as now adopted internationally) differs very little from that of the ephemeris second as determined from the ephemeris of the Moon, $j=1$.

The programme of predictions of occultations by the Moon, including those of radio sources, has been somewhat extended; a computer programme for the systematic prediction of grazing occultations, is almost complete. A comprehensive programme for the prediction of occultations of stars by the planets (including minor planets) has just been completed. The earlier programme, by inspection, brought to notice the occultation of a star by Neptune on 1968 April 7, the collected observations of which have made possible (*Mon. Not. R. Astr. Soc.*, in press) an extremely accurate relative position of the planet and star, and a new determination of the planet's diameter.

D. H. SADLER

Astronomisches Rechen-Institut, Heidelberg, Germany

Astronomical Constants. When the IAU System of Astronomical Constants was adopted in 1964, the value of the astronomical unit in kilometers was based on radar measurements of interplanetary distances. The values known from determinations by the dynamical method deviated from the radar value. The discrepancy has been clarified and eliminated in the past three years. A determination of the mass of the Earth-Moon system has been carried out by Zech (Heidelberg, 1969) by numerical integration of the orbit of Eros. The result $1/m_{E+M} = 328894 \pm 30$ (m.e.) agrees very well with recent determinations by Rabe (Cincinnati, 1967) and Lieske (New Haven, 1968), and all three are in very satisfactory agreement with the mass resulting from the inter-planetary radar distance measurements.

The correction to Newcomb's value of lunisolar precession determined by Fricke (Heidelberg, 1967) has been the object of further investigations. A re-determination on the basis of Newcomb's data has cleared up the origin of Newcomb's error, and an uncertainty arising from an unexplained part in the variation of obliquity of the ecliptic has been explained and eliminated. The results of these latter investigations will soon be published.

Fundamental Catalogues. The systematic differences between 71 observational catalogues and FK3, which had been determined and applied in the construction of FK4, were published by Nowacki and Strobel (Heidelberg, 1968). Under the supervision of Lederle the documentation of star catalogues on punched cards has been continued; these data are made available, in accordance with the recommendations of the Working Group on Space Ephemerides. Investigations have been made with the aim of developing improved methods for the construction of fundamental and general catalogues of stars.

APFS and exchange of ephemerides. Within the period under review four volumes of *Apparent Places of Fundamental Stars* were published (1969–1972). Apart from the communication of apparent and mean places for the FK4 and FK4 Sup stars in listed or printed form, the material is also available in machine-readable form (punched cards or 5-channel paper tape) on request.

W. FRICKE

Hydrographic Department, Tokyo, Japan

The Japanese Ephemeris has continued to be published for the years 1969 to 1971. The ephemerides of the Sun, Moon and planets have been reproduced from the advanced proofs of the *Astronomical*

Ephemeris provided by the H.M. Nautical Almanac Office and the remaining parts have been computed in the Department. Corrections to the Moon's ephemerides for conversion to the IAU system are given for every half day.

The Nautical Almanac, *The Abridged Nautical Almanac*, and *The Polaris Almanac for Surveying* for the years 1968 to 1970 have continued to be published without substantial change. *The Altitude-Azimuth Almanac for the Antarctic Expedition* and *The Abstract from the Japanese Ephemeris* have been compiled annually at the request of National Defence Agency.

In consultation with the Nautical Society of Japan, the Department re-examined the contents of the *Astronomical Navigation Tables* (H.O. No. 601), which had been employed as standard for sight reduction in Japan, and the revised edition was published in 1969.

The value of ΔT for use in predictions of eclipses and compilations of almanacs and the national calendar, which is compiled at the Tokyo Astronomical Observatory, has been settled by arrangement with the Tokyo Astronomical Observatory from results of occultation observations made at the Department, taking the tendency of A.T. – U.T. into account.

A. M. SINZI

Nautical Almanac Unit, Regional Meteorological Centre,
India Meteorological Department, Calcutta, India

During the period under review, the volumes of the *Indian Ephemeris and Nautical Almanac* for 1969 and 1970 were published. The issue for 1971 is expected to be published by February, 1970. No substantial change in respect of the contents of the publication has been introduced during these years. As regards the IAU System of Astronomical Constants, some of the revised values have already been introduced in the 1968-issue. From 1969 the revision of the horizontal parallax of the Moon has been indicated in the relevant tables, while those of the planets have been fully revised.

In addition to the above main publication, the following two other publications continue to be issued annually:

(i) *Tables of Sunrise, Sunset and Moonrise, Moonset* – a reprint from *the Indian Ephemeris and Nautical Almanac.*

(ii) *Rashtriya Panchang* in twelve languages, giving details of Indian calendar and festival dates. The issues for 1891 Saka Era (1969–70 A.D.) were published in 1969.

The publication of the English translation of the Marathi book on the history of Indian astronomy the *Bharatiya Jyotish Sastra* by late S. B. Dixit has been taken up. The first part of the book covering the Vedic and Vedanga periods was edited and published. The remaining parts covering the Siddhantic and modern periods will be published soon.

N. C. LAHIRI

Instituto y Observatorio de Marina, San Fernando (Cádiz), Spain

The work in this Ephemeris Office has been to continue the publication of the *Efemerides Astronomicas*, the *Almanaque Nautico*, and the *Almanaque Aeronautico*, with only minor modifications. During the period January 1967 to January 1969 we have carried out the mechanization of most of the calculations of this Office.

M. RODRIGUEZ

Nautical Almanac Office, U.S. Naval Observatory, Washington, D.C. 20390, U.S.A.

The Nautical Almanac Office is responsible for the publication of four annual ephemerides: *The American Ephemeris and Nautical Almanac* – for astronomers; *The Nautical Almanac* – for surface navigation; *the Air Almanac* – for air navigation; and *The Ephemeris* – for land surveyors, as well as the pamphlet *Astronomical Phenomena.*

The *American Ephemeris*, unified with the *Astronomical Ephemeris* issued by H. M. Nautical

Almanac Office, Royal Greenwich Observatory commencing with the volume for 1960, has continued to be published. The volumes for 1968, 1969, 1970 and 1971 include appendices which contain corrections to convert the tabular data to the IAU System of Astronomical Constants (*Trans. IAU*, **12B**, 105, 1966). Beginning with the *American Ephemeris* for 1972, the IAU System of Astronomical Constants has been included in the tabular material.

A computer-controlled photo-composition method utilizing paper tape has been used to prepare the *Air Almanac* for 1969 and 1970. Starting with the *Air Almanac* for 1971 a faster photo-composition system using magnetic tape has been introduced.

Preparation of the circumstances of solar eclipses with argument in U.T. has been continued (see *U.S. Naval Observatory Circulars*, nos. 116, 122, 123 and 125). The cooperative project to put star catalogs in machine-readable form has also been continued (see *U.S.N.O. Circ.* no. 114). A special ephemeris of the radio longitude of the central meridian of Jupiter, System III, has been computed for the years 1968–71 (see *U.S.N.O. Circ.* no. 117). Additional astronomical information has been prepared and published in *U.S.N.O. Circulars* as follows: 'Phases of the Moon 1960–2003' in no. 119 and 'Sunlight, Moonlight and Twilight for Antarctica, 1969–1971' in no. 120.

The research program on the motions and masses of the planets has been continued (see report of Commission 7). Reprints of Volumes VI (Newcomb's Tables of the four inner planets) and X, Part II (Robertson's Zodiacal catalog) of the *Astronomical Papers of the American Ephemeris* have been issued.

R. L. DUNCOMBE

Brief reports and notes

J. Arias-de Greiff reports that the publication of the *Anuario de Observatorio Astronómico Nacional*, Bogotá, Colombia, has continued regularly since 1966 with only minor changes. Most of the material has been provided by H.M. Nautical Almanac Office, the U.S. Naval Observatory and the Astronomisches Rechen-Institut: the remainder has been computed in the Centro de Cálculo Electronico at the University of Columbia.

The Jet Propulsion Laboratory, Pasadena, U.S.A., has undertaken an extensive programme of ephemeris computation in connection with space missions. Descriptions of much of this work are given in the *Technical Reports* of the Laboratory and copies of the ephemerides on magnetic tape can normally be made available on a reciprocal basis. A summary of the work on the lunar ephemeris has been given by J. D. Mulholland (*Nature*, **223**, 247–9, 1969).

I. I. Shapiro reports that at the Massachusetts Institute of Technology, Cambridge, U.S.A., over 150000 optical observations of the right ascension and declination of the Sun, Moon and planets for the period 1750 to 1968 have already been put into a machine-readable form. It is hoped that arrangements can be made to make these data available to others.

Several large-scale comparisons between collected observational data and planetary orbits obtained by numerical integration are in progress and preliminary results have been reported by Melbourne and O'Handley (Jet Propulsion Laboratory) and by Oesterwinter and Cohen (U.S. Naval Weapons Laboratory, Dahlgren, Va., U.S.A.). These will form a valuable supplement to the recent determinations of astronomical constants from analyses of the orbits of individual natural and artificial bodies.

G. A. WILKINS
President of the Commission

5. DOCUMENTATION

(Committee of the Executive Committee)

PRESIDENT: J. B. Sykes.
VICE-PRESIDENT: K. F. Ogorodnikov.
No Organizing Committee.

The title of the Commission was changed from Abstracts and Bibliography to Documentation in 1968 (*IAU Inf. Bull.* no. 21, p. 6).

Primary publications

Five European astronomy journals (*Annales d'astrophysique, Bulletin astronomique, Bulletin of the Astronomical Institutes of The Netherlands, Journal des Observateurs, Zeitschrift für Astrophysik*) ceased separate publication and joined forces as a new, mainly English-language, journal *Astronomy and Astrophysics* at the beginning of 1969.

The change was noted in *Babel*, the journal of the Fédération Internationale des Traducteurs, with the comment that the use of English would facilitate communication between scientists, but might on the other hand make more difficult the establishment of correct and uniform terminology in the various European languages at the level of teaching or industrial application.

Kemp proposes a study of the citations in the 1969 and 1974 volumes of *Astronomy and Astrophysics*, to see whether the new form of publication has had any influence on habits of choosing journals to write in and to read in.

Secondary publications

(1) *Astronomy and Astrophysics Abstracts.* Members of the Commission were sent copies of announcements concerning the changeover from the annual *Astronomischer Jahresbericht* in German to the twice-yearly *Astronomy and Astrophysics Abstracts*, mainly in English, beginning with the literature of 1969. (See also *IAU Inf. Bull.* no. 20, p. 17; no. 22, p. 23.) Authors' abstracts are printed whenever available; a good response has been obtained by direct application to authors when abstracts are not published.

(2) *Referativnyj Žurnal Astronomija* has included, since January 1969, for all papers originally published in Russian, titles in both Russian and English, and authors' names in both Russian and Western transliteration (*IAU Inf. Bull.* no. 21, p. 6). Lack of facilities for translation into English prevented continuation of the project to publish an edition entirely in English (*Trans. IAU*, **XIIIB**, p. 54). This difficulty was alleviated by the development of *Astronomy and Astrophysics Abstracts* (see above).

(3) *Astrophysical Abstracts* has been published quarterly by Gordon and Breach (New York) since August 1969.

ICSU abstracting board

A description of the work of the International Council of Scientific Unions Abstracting Board (IAB) has been given in the previous Report (*Trans. IAU*, **XIIIA**, p. lxxvi). I attended, as representative of the IAU, the General Assemblies of IAB in July 1968, January 1969, and September 1969. I also served on the IAB Planning and Steering Committee from January 1969. At the September 1969 meeting I presented on behalf of the Union some comments on probable developments of abstracting services in the next few years.

IAB has set up Working Groups in five main subject fields; the one for Astronomy comprises representatives of the IAU and of the abstracting services in astronomy. The function of these groups is to promote collaboration and investigate problems arising in their respective fields.

Member Unions of IAB are to pay an annual subscription of $ 100, beginning in 1970. (Member Services pay $ 1000 or more.)

In May 1969 the Planning and Steering Committee of IAB resolved to recommend to the Scientific Unions "to reinforce their action for the purpose of obtaining abstracts prepared by authors and editors, to be published with all scientific papers".

IAB is also assisting in the project for a world science information system (UNISIST) and is conducting an extensive study of the needs of users of scientific information.

Classification

Staff of *Bulletin Signalétique*, under the auspices of IAB, have carried out a comparison of the latest revised subject classification schemes used by *Astronomy and Astrophysics Abstracts*, *Bulletin Signalétique* (corresponding to the classification in use at the Paris Observatory), *Referativnyj Žurnal*, and the Universal Decimal Classification. It appears possible to arrive quickly at a common scheme which will facilitate exchange of abstracts and the specification of each service's coverage.

The revision of the Universal Decimal Classification for astronomy, proposed by VINITI* (*Trans. IAU*, **XIIIB**, p. 54) was circulated on a small scale by the Féderation Internationale de la Documentation in August 1969 as a "draft P note" (1969: 17) for comment within three months. This will probably be followed by wider distribution as a "P note" with six months for comment, and the final stage is inclusion of the revised classification in the UDC itself.

Kemp has plans for a system linking three levels of operations in subject classification: book arrangement, the classified arrangement of card indexes, and a coding system for machine retrieval. He would be glad to hear from anyone who is considering the use of machine retrieval systems for astronomical literature.

Current-awareness publications

VINITI is to publish in Russian and English a twice-monthly current-awareness journal in astronomy (*Signal'naya informacija astronomija*), beginning in 1971, instead of the proposed printed cards (*Trans. IAU*, **XIIIB**, p. 56), which would have been too expensive. Items will appear about two months earlier than in *Referativnyj Žurnal.* Photocopies of originals will be available from VINITI.

The ESRO/ELDO Space Documentation Service has introduced a series of 111 Standard Profiles (including about 15 astronomical subjects), providing a subscription service of monthly listings of recent literature.

Information use

Meadows and O'Connor (University of Leicester, U.K.) have published a small number of copies of a report giving the results of an inquiry into sources of information and information retrieval by U.K. astronomers and space scientists in 1967. This is part of a continuing study of information needs for astronomy in the U.K. and elsewhere.

Bibliography of Astronomy 1881–1898

The proposed publication in book form (*IAU Inf. Bull.* no. 19, p. 25) had to be abandoned owing to insufficient support (ibid. no. 22, p. 15).

A British microfilm company has agreed to produce a 35 mm microfilm version of the 60000 slips forming the Bibliography. This will be sold at $ 180 per copy, with 10% of the charge paid as a royalty to the IAU. After being microfilmed, the slips will remain available for consultation by visitors to the Brussels Observatory.

Collinder has published the Swedish contribution separately (*Ark. Astr.*, **4**, No. 19, 1966).

* Vsesojuznyj Institut Naučnoj i Tehničeskoj Informacii, the Soviet Institute of Scientific and Technical Information.

Other bibliographies

Lavrova has compiled and published a bibliography of Russian astronomical literature 1800–1900 (*Trudy gos. astr. Inst. Šternberga*, **37**, 1968). The 4250 titles cover all publications by Russian astronomers, including those published in other countries. She also assisted P. G. Kulikovskij in preparing the annual bibliography of the history of astronomy.

K. Maichel's *Guide to Russian reference books*, vol. V, *Science, technology and medicine* (Stanford, 1967) deals with astronomy on pp. 82–88, including 22 bibliographies and 11 dictionaries.

Transliteration

Agreement has been reached on a standard transliteration of the Russian alphabet to be used by *Astronomy and Astrophysics Abstracts* and *Referativnyj žurnal*. The agreed system is as follows:

а	a	к	k	х	kh
б	b	л	l	ц	ts
в	v	м	m	ч	ch
г	g	н	n	ш	sh
д	d	о	o	щ	shch
е	e	п	p	ъ	″
ё	e	р	r	ы	y
ж	zh	с	s	ь	′
з	z	т	t	э	eh
и	i	у	u	ю	yu
й	j	ф	f	я	ya

It is for consideration by Commission 5 at the IAU General Assembly whether this system should now be recommended for adoption by the Union in place of that given in *Trans. IAU*, **XIIC**, pp. 69, 115.

Terminology

Hirose reports that a draft list of Japanese and English astronomical terms has been completed and will probably be published in 1970 after correlation with the terminology in other branches of science.

Catalogues

Alter reports that the second edition of the *Catalogue of Star Clusters and Associations* by Alter, Ruprecht and Vanýsek is to be published by the Hungarian Academy of Sciences under the editorship of Alter, Ruprecht and Balázs. It will include the contents of the annual supplements published since 1958 in the *Bulletin of the Astronomical Institutes of Czechoslovakia*.

List of non-commercial publications and catalogues from observatories

This work (*Trans. IAU*, **XIIA**, p. lxxix) is being completed by Mr J. Rosenberg of the Utrecht Astronomical Institute. It will list serials published (current and ceased) and availability, probably as a looseleaf volume. Lavrova has collated the information concerning U.S.S.R. observatories. The International Federation of Library Associations has granted the sum of 1000 Swiss francs to cover printing costs.

Library holdings

The Special Libraries Association (U.S.A.) group of astronomical librarians is investigating means of constructing a union list of the contents of astronomical libraries. Lavrova is conducting a similar inquiry in the U.S.S.R.

Dissertations

The Special Libraries Association group is endeavouring to ensure that all astronomy dissertations at U.S. universities are included in *International Aerospace Abstracts*.

Ancient books

Mme G. Grassi Conti (Osservatorio Astronomico di Roma) is collecting data for a union catalogue of manuscripts, incunabula and rare books. This work also has had financial support from the International Federation of Library Associations.

J. B. SYKES
President of the Commission

6. ASTRONOMICAL TELEGRAMS
(TÉLÉGRAMMES ASTRONOMIQUES)

(Committee of the Executive Committee)

PRESIDENT: Dr F. L. Whipple.
VICE-PRESIDENT: Dr P. Simon.
DIRECTOR OF THE BUREAU: Dr B. G. Marsden, Smithsonian Astrophysical Observatory, 60 Garden Street, Cambridge, Massachusetts 02138, U.S.A.
ASSOCIATE-DIRECTORS OF THE BUREAU: Dr O. J. Gingerich and Dr R. B. Southworth.

INTRODUCTION

The operation of the Central Bureau for Astronomical Telegrams, thanks to the diligence of its Director and his collaborators, has continued to remain relatively tranquil and quite effective since January 1968, following the methodology established by the previous Director, Dr Owen J. Gingerich. Financial problems, temporarily at least, seem not to be of major concern. The effective subsidy by the Smithsonian Astrophysical Observatory has been reduced almost to a tolerable level. Policy with regard to the content of the *Circulars*, however, deserves continuing surveillance by the subscribers.

F. L. WHIPPLE
President of the Commission

REPORT OF THE CENTRAL BUREAU FOR ASTRONOMICAL TELEGRAMS

As anticipated in the last report, there was an increase in the activity in the Central Bureau during 1967 because of the unprecedented number of periodic comets predicted to return: ten of the 17 were in fact recovered, and with four new comets discovered, as well as one located on plates taken in 1963, the total number of comets announced was a record. During 1968 the activity in the Bureau was greater still, for as many as seven *new* comets were discovered, and 21 supernovae were announced by the Bureau. In both years the British amateur G. E. D. Alcock succeeded in discovering a naked-eye nova, and in each case rapid announcement made it possible for spectroscopic observations to be secured before the nova reached its maximum. The Bureau also helped disseminate the first post-conjunction observations of Icarus in 1968, enabling last-minute orbit corrections to be made and ensuring the success of the radar experiments conducted when the planet passed by the earth three days later. Just half of the first 26 pulsars to be discovered were first reported on the Bureau's *Circulars*, while the startling news of the optical identification of the Crab pulsar NP 0532 was originally transmitted by telegram on 17 January 1969. Several notes appeared on the *Circulars* during the ensuing months as information was obtained about X-radiation from the pulsar as well as changes in the period. Although the Bureau's activity in 1969 was at a lower level than in 1968, it remained significantly higher than in 1967.

It is debatable whether the Bureau should be involved in distributing information about pulsars, or indeed any topic where scientists wish merely to establish priority for some particular observation, as opposed to reporting news about some transient celestial object that needs to be observed as quickly as possible. Other media are certainly available for reporting pulsar news, even when spectacular discoveries are being made every week. For a similar reason some subscribers may object to the appearance of extensive predicted ephemerides (often calculated long in advance) for faint periodic comets, or for positional observations of new comets long after they are required for providing satisfactory ephemerides. Ever since the IAU was established there has been an "informal agreement" with Commission 20 for publishing ephemerides and positional observations on the IAU *Circulars*, however, and in the absence of any more suitable medium (with the possible exception, in the case of the ephemerides, of the *Handbook of the British Astronomical Association*), this practice will be continued. An effort is made to ensure that reliable ephemerides are available on the *Circulars* for *all* comets, new and periodic, that might be observable at any particular time.

The number of "telegram books" and the number of occasions on which *Circulars* have been issued during the triennium were as follows:

	Telegrams	*Circulars*
1967	28	42 (Nos. 1987–2045)
1968	58	59 (Nos. 2046–2123)
1969	42	51 (Nos. 2124–2196)

The total number of subscribers to the *Circulars* is now approximately 750. When a subscription is shortly to expire, an invoice is generally prepared by the computer at the time a batch of *Circulars* is addressed, and it is necessary to repeat here that *failure to renew promptly automatically results in interruption of service*. Following the suggestion made at the Prague meeting, we investigated the possibility of billing subscribers on a yearly basis, rather than for a fixed number of *Circulars*. There is a distinct possibility that by the time this report is published the billing system will have been changed over to a yearly basis.

The Communications department of the Smithsonian Astrophysical Observatory has been of tremendous assistance in both the receipt and transmission of telegrams. Subscribers in North America, more than 30 for some categories of messages, receive telegrams collect. Most of the European distribution, involving up to 70 subscribers, continues to be handled very dependably via Dr P. Simon at Meudon, although we have found it more efficient to send the messages to him directly by telex, rather than over the AGIWARN network. Further distribution takes place from Stockholm and Prague. As before, deposits for these European subscribers are collected in Cambridge, and it is gratifying to note that since 1967 everyone has been most cooperative in this matter. Astronomers in The Netherlands, the U.S.S.R., and Australia and New Zealand are served via the AGIWARN network – operating since June 1968 from the Environmental Science Services Administration in Boulder, Colorado. The messages are sent directly from the Smithsonian over U.S. government lines for distribution in Chile, Argentina and Japan. Until recently this arrangement existed also for South Africa, but it has now become necessary for the messages to be sent there by commercial cable. It has been difficult to find a rapid yet inexpensive way of sending messages to Brazil, but we have recently been able to set up a route via the Johns Hopkins University and their TRANET station in São José dos Campos.

Not all of the activity of the Central Bureau comes to the notice of subscribers to telegrams and *Circulars*. Before and at the time of the close approach of Icarus to the earth the Bureau received letters, telegrams and telephone calls from the popular press and a frightened public; but it was encouraging to note that Geographos passed by a year later without a single communication of this type. The Bureau receives a considerable number of reports of suspected comets, particularly when the press has given publicity to some successful comet hunter. Many of these reports can be quickly dismissed, but for the more plausible ones we request confirmations from some of the Smithsonian's tracking stations. Sometimes we ask instead interested amateur astronomers in North America: this has proven to be a particularly rapid and reliable way of securing confirmations, and we are most grateful to those who have participated in this so-often unrewarding business.

The increased rates for *Circulars*, introduced in December 1967, have improved the financial situation of the Central Bureau, although the effect was somewhat offset by the reduction in the IAU subvention. The Smithsonian Astrophysical Observatory continues to subsidize the operation of the Bureau in the form of personnel, but provided that the IAU subvention is not reduced below the present amount of $ 533 per year, it may be possible for us to operate for the next triennium without a further increase in the rates for the *Circulars*.

B. G. MARSDEN
Director of the Bureau

APPENDIX

The following changes have been made in the code for astronomical telegrams (see *Trans. IAU*, **XIIC**, pp. 34–38, **XIIIA**, pp. lxxxii–lxxxiii):

1. The letter X, used for a suppressed digit, has been replaced by a hyphen (-). It counts as zero in checksums.

2. In messages giving ephemerides, the letter D has been replaced by the figure 9 and the letter R by the figure 8. This change causes no ambiguity, but it should be noted that these figures must be included in the checksum.

3. Following the usual checksum, there is an additional checksum, formed in the normal manner, but by adding *only* the groups containing right ascension, declination and magnitude information. Dates, daily motions, differential offsets, and geocentric and heliocentric distances are *not* included in this. When orbital elements are being transmitted, the second checksum is limited to the groups giving the angular elements ω, Ω and i.

Examples (for transcriptions see *Circulars* Nos. 2156 and 2079):

Old: KOHOUTEK 1969B COMET KOHOUTEK 19501 90724 0013X 19446 22648 X1148 10182 20010 83789 BERGEDORF

New: KOHOUTEK 1969B COMET KOHOUTEK 19501 90724 0013- 19446 22648 -1148 10182 20010 83789 43242 BERGEDORF

Old: 1968B COMET MILBOURN 19504 80623 15205 22016 D0399 R1267 15215 22320 15226 22554 D0468 R1282 15238 22804 15251 22955 D0539 R1299 15264 23129 80703 13261 CANDY

New: 1968B COMET MILBOURN 19504 80623 15205 22016 90399 81267 15215 22320 15226 22554 90468 81282 15238 22804 15251 22955 90539 81299 15264 23129 80703 23261 27177 CANDY

7. CELESTIAL MECHANICS (MÉCANIQUE CÉLESTE)

PRESIDENT: W. J. Eckert.
VICE-PRESIDENT: G. N. Duboshin
ORGANIZING COMMITTEE: G. A. Chebotarev, Y. Hagihara, J. Kovalevsky, Y. Kozai, P. J. Message, K. Stumpff, V. G. Szebehely, F. Zagar.

INTRODUCTION

The most exciting feature of contemporary celestial mechanics is the close interaction with three of the most dramatic engineering achievements of our time: the electronic computer, artificial celestial objects, and the precise measurement of distances in the solar system. The last two provide new information for and make new demands on celestial mechanics and the former provides an effective means of response. These developments have also stimulated people with backgrounds other than celestial mechanics to make contributions in the field.

Since the last meeting of the Union, the use of the computer for literal theoretical developments has become effective and widespread.

The creation of artificial celestial objects not only requires the services of celestial mechanics but provides the means of obtaining new measurements that may throw light on the physical laws that govern the motion of celestial objects. In other words, we can now perform experiments as well as observe.

Until the present decade celestial mechanics was based on the measurement of the directions of other objects from the observer, and distances were found by triangulation from the angular measures. Even in the case of the Moon there was a loss in accuracy of nearly two orders of magnitude in the distances. During the past few years important results have been obtained by radar and other electronic methods and the laser method, which promises an improvement in accuracy of several orders of magnitude over the best angular measurements for the Moon, has been started. We in celestial mechanics are looking forward to the availability of extensive series of uniform observations of distance made in the same spirit as the angular measurements on which we now rely.

At the last meeting of the Union, Commission 7 held a colloquium on the Use of Electronic Computers for Analytical Developments in Celestial Mechanics. For the coming meeting Professor Duboshin is now organizing a colloquium on Analytical Methods for the Orbits of Artificial Celestial Objects, and Professor Clemence is organizing a discussion of The Impact of Precise Measurements of Distances on Celestial Mechanics.

An important function of this report is to assist workers with different backgrounds to find in what areas of celestial mechanics various people are working and where they publish their results. This requirement has influenced the arrangement of the report and especially that of the bibliography.

The subject matter has been classified according to the scheme in Table A. This is a somewhat modified form of that used by Hagihara in the previous report. The discussion of the material in the ensuing text follows in general the order of Table A, and each reference in the biobligraphy carries classification designations for the article. Section N contains general questions as well as several specific topics.

The names of the publications to which reference is made in the bibliography are not printed in the bibliography but separately in Table B. The term "periodical" is used in the same sense as in page 76 of the *IAU Astronomer's Handbook* (vol. XIIC) and the periodicals marked with an asterisk are contained in the list starting on the same page of the Handbook. The bibliography is in alphabetical order by author and the references are in coded form, for example the reference

FcJ68 42,72,382

indicates that the article was classified Fc and J according to Table A, and was published in 1968 in the journal listed as number 42 in Table B (*Astronomical Journal*), volume 72, page 382. For articles in press as of November 1969 the year is replaced by asterisks. In references by joint authors the year is followed by "a" for the first author and "b" for the others. Obvious variations of the code are used for publications such as reports and dissertations. This form of bibliography appears to be more convenient than the conventional ones and it occupies about 60 % of the space.

In writing the report and in compiling the bibliography, we have made no attempt at completeness nor strict consistency with a concise set of rules; we have tried to give a representative picture. The report was developed in the following manner. The author made a quick inspection of fourteen of the journals used most frequently in the previous report to prepare a preliminary bibliography, listed by author. Each author who was either a member of Commission 7 or who had at least three titles (aside from joint articles with members of the Commission) was requested to correct, enlarge, and classify his portion. A second bibliography incorporating these corrections was arranged by classification for review.

Sections of the material were reviewed by members of the Commission as follows: Clemence (sections E and K), Garfinkel (F and I), Herget (A and L), Kovalevsky (B and C), and Szebehely (G and H). The reviewers' reports have received only general editing. I am most grateful to these gentlemen for reviewing the material and doing it on very short notice. I am also indebted to Professor Duboshin for a report on work in the U.S.S.R. which facilitated the work of the reviewers. Jose Fortoul and Sara Bellesheim of the IBM Watson Laboratory were largely responsible for the preparation of the bibliography.

The following review of the subject matter is presented in the order indicated in Table A. References in the Bibliography are referred to in the text by the name of the author and the number of the reference as it appears in his list. For joint authors the number given is that of the first.

Table A. Classification scheme

A. Two-Body Problem and Orbital Improvement
B. Perturbation Theory and General Dynamics
C. Planetary and Satellite Theory
D. Lunar Theory
E. Relativity and Other Non-Newtonian Effects
F. Artificial Celestial Objects
  a. General
  b. Air Drag, Light Pressure, and Rotation
  c. Lunar Orbiters and Artificial Satellites of Planets
  d. Optimization Problem
G. Three-Body Problem
  a. Regularization
  b. Numerical Survey
  c. Analytical Theory
H. Periodicity, Ergodicity, Stability
I. Resonance
J. *N*-Body Problem and Galactic Structure
K. Astronomical Constants
L. Minor Planets, Comets, and Meteors
M. Computing Methods
N. Other Topics

A. TWO-BODY PROBLEM AND ORBITAL IMPROVEMENT

The review of many articles is confined to their listing in the Bibliography. Benima (1) *et al.*, have given polynomials to replace the tables formerly used in the Gauss-Marth method of finding the position in a nearly parabolic orbit. Izvekov (1) concludes that the integration of the Variational

Equation is not necessary for the correction of minor planet orbits, except when they are subjected to extreme perturbations. Kustaanheimo (1) has prepared an introductory text which is developed in terms of vector and spinors, leading to relativity theory. Mulholland (2) attempts to ameliorate the differential correction of highly perturbed satellite orbits, where the integration of the Variational Equation is required to achieve a satisfactory result.

B. PERTURBATION THEORY AND GENERAL DYNAMICS

Hori (1) attacked the problem of general perturbations by the use of Lie-series in such a way that the new procedure is now valid for any set of canonical variables. The formulae are presented in the form of canonical invariance. The perturbations of the elements or coordinates are given in an explicit form and the theory is applicable whatever variables appear in the undisturbed part of the Hamiltonian. Deprit (13) extended the concept of these transforms in order to include cases when the generating function depends upon a small parameter and he gave a recursive algorithm for the solution. Canonical transformations are generated explicitly as well as their inverses as explicit chains of Poisson's brackets without inversions or substitutions.

Kevorkian (1) proposed a procedure, called the "two variable method" in which he introduces a fast variable and a slow variable whose ratio is of the order of the small parameter and the final series are derived as simultaneous functions of these two variables. This method was compared with Von Zeipel's method by Kevorkian (2). Applications of averaging methods to celestial mechanics were studied by Grebenikov (2) who evaluated the errors of these methods for intervals of time of the order of μ^{-1}.

Boigey (1) studied parametric Lagrangian systems and applied her results to the reduction of equations, regularization and separation of variables. Gustavson (1) constructed formal integrals of a Hamiltonian system near an equilibrium point while Kikuchi (2) worked on the complete system of integrals in celestial mechanics. The ultimate behaviour of orbits with respect to an unstable critical point was investigated by Conley (2).

Losco (2) investigated relationship existing between integrals and invariant curves or surfaces of a dynamical system and properties of solutions associated with an invariant surface in equations in involution (3). Henon (5), Contopoulos (7) and Hadjidemetriou, and Roels (2) and Henon made systematic studies of invariant curves in various area conserving mappings.

Practical methods in obtaining formal solutions of various perturbation problems with computers were given by Deprit (8, 9) and Rom. Grebenikov (1) investigated the effects of small variations of initial conditions on the respective magnitudes of secular and long-periodic perturbations, while Kholshevnikov (2) studied the speed of formal convergence of the solution of a Hamiltonian system by means of successive approximations. Much effort was directed towards the description of the motion of a particle in various gravitational fields.

The problem of the motion of a point attracted by two fixed masses has been further investigated because it provides a good intermediate orbit for an artificial satellite and even, as proposed by Lukashevich (1) for natural satellites. Most results obtained up to now in this field are presented in the book by Demin (see Table B). Aksenov (3) and Marchal (1) derived general theories of the motion in such a field and Aksenov (2) put this problem in a canonicalform. Kiryushenkov (1) expanded solutions around circular orbits in small parameters. Degtjarow (1) and Evdokimova studied the stability of these orbits. Timoshkova (1) investigated elliptic orbits in the problem and Chepurova (1) analyzed the hyperbolic case. A. H. Cook (1) showed that the exact solution of this problem obtained in spheroidal coordinates cannot be extended to ellipsoidal coordinates.

A generalized axisymmetrical Earth-like field depending upon J_2 alone provides a simple example of a dynamical system with two degrees of freedom. This is why Danby (2) studied numerically the phase space in the regions where non-integrability very probably occurs and where the orbits become *wild*. Mangeney-Ghertzman (1) made a more restricted study using osculating elements and found a critical value of J_2 after which, the osculating perigee circulates while the mean anomaly oscillates. Separable potentials in triaxial ellipsoidal coordinates were studied by Madden (1) and A. H. Cook (1). Kholshevnikov (1) studied the potential of a body of an arbitrary shape.

Less 'normal' force fields are often studied because of their evident applications to the motion of stars in clusters. As Kovalevsky (1) has shown, some problems in stellar dynamics can be solved by methods pertaining to celestial mechanics. So, Hori (3) computed orbits in the plane of symmetry of the Galaxy while Woolley (1) and Candy studied the perturbations of such orbits by the local field irregularities. Barbanis (2) and Prendergast represented the gravitational field of a disc galaxy by Legendre functions multiplied by Fourier series and Barbanis (4) gave orbits and integrals of motion in a spiral field.

A similar type of problem arises in the study of the motion of planets or satellites, when the perturbation methods do not apply, as shown by Morando (3). A general investigation on such strongly perturbed systems was made by Jefferys (6). It is the case of long range behavior of satellites of high inclinations and large eccentricity studied analytically by Kozai (4) and R. S. Harrington (1) or numerically by Chebotarev (1, 2). It is also the case of a close approach of a planetoid to a planet examined by Stellmacher-Amilhat (1). Numerical methods are often used in these investigations. They can be improved by regularizing the equations as shown by Stiefel (2) and Scheifele. Leimanis (1) applied Sonneschein summation methods to get the analytical continuation of the solution.

Kyner (1, 2) and Bennett improved the numerical Encke's method in introducing part of the first-order perturbations in the reference orbit. Stiefel (3) and Bettis investigated the stability of Cowell's method of numerical integration.

Radar measurements of the rotation of Mercury and Venus and photometric observations of artificial satellites raised considerable interest in investigations of the rotational motion of celestial bodies.

Goldreich (2, 5) and Peale studied the spin-orbit coupling in the solar system and applied their theory to the resonant rotation of Venus (3). More generally, Blitzer (2) considered the rotational resonances of a rigid body in a Keplerian orbit. Vinti (1, 3) discussed the various types of Liapounov stability in the case of a free rotation as function of the ratio of the moments of inertia. Brumberg (4) wrote the equations for the rotational motion of the planets when corrections for general relativity are introduced.

Colombo (1) showed that the second and third laws of Cassini are independent from the first and could be satisfied if the Moon's inertial ellipsoids were rotationally symmetric. Peale (4) generalized these laws; he computed relations between the moments of inertia for stable commensurabilities between the orbital mean motion and the spin-angular velocity. Habibulin (1) solved the equations for the physical libration of the Moon for the non-linear terms, using Krylov-Bogolioubov method.

A complete review of the rotational motion of an artificial satellite was given by the book by Beletzky (see Table B), who also (1) gave the recent advances in this field. Recently, Holland (1) and Sperling derived a first order theory for a triaxial rigid body orbiting an oblate planet.

The general dynamical problem of the motion of two finite rigid bodies also includes cross-terms giving the effect of the rotation on the motion of the center of mass. This effect was studied in particular by Schinkarik (1) in a central Newtonian field, Osipov (1) in a generalized Hill problem and Johnson (1) and Kane in a special case of the relative motion of two rigid ellipsoids of revolution. The equilibrium of a rotating non-homogeneous fluid body was investigated by Marchal (2) and Volkov (2) who introduced a spherical core in the model, while Miyamoto (1) considered a possible influence of an extended halo and Aizenman (1) studied this problem in a system of two bodies. Yabushita (1) investigated the stability of the rotation of Saturn's rings in function of the density distribution and the total mass.

C. PLANETARY AND SATELLITE THEORY

The planetary disturbing function was constructed by Meffroy (5) and by Brumberg (3) who gave an iterative formulation while Petrovskaya (1, 3) investigated the accuracy with which it can be represented by trigonometric polynomials.

Musen (2) has investigated Hansen's planetary theory and improved the convergence by introducing a new method of computing the function dw/dt. Seidelmann (1) has programmed an iterative procedure for determining a planetary theory based on Hansen's method and continues doing research on the technique. Nacozy (2) has tested Hansen's method of partial anomalies and applied it to the motion of Comet Encke.

Meffroy investigated the applicability of Von Zeipel methods to a planetary theory in order to eliminate short period terms of the first and second order (1, 2, 3, 8).

Brouwer's method in rectangular coordinates was analyzed and programmed by Hamid (1) who gave the solution to the first order, and more recently, some second order terms. Musen (1) proposed some improvements to the original Brouwer's formulation. Broucke (2) gave an iterative procedure for the computation of planetary rectangular coordinates.

General methods of literal developments on a computer described by Kovalevsky (2) were applied by Chapront (3) who applied a first order term and all second order terms factored by the square of the external mass of a purely literal "LeVerrier type" planetary theory. The coefficients are literal functions of the ratio of the semi-major axes put under special explicit forms involving polynomials and more complicated divisors valid for any value of the ratio between zero and 0.7. Ferraz-Mello (1) has introduced a new approach to the study of the motion of quasi-circular and quasi-resonant orbits in Hill's normalized coordinates, and extended this method to second order terms (3) and to the case when the mutual inclination is not negligible (4). Along a somewhat similar line, Brumberg (6) combined Hill's method with the principle of elimination of short period variables, gave a solution of the variational equations obtained and presented a method to obtain a general solution of the equations of planetary motion.

Morando (2) has completed a generalized semi-numerical theory of the motion of Vesta giving a first approximation orbit in terms of trigonometric functions of four independent arguments. The same method is now being applied to the Jupiter-Saturn system, some of the long periodic arguments being computed in a purely literal form and compared with results obtained using Krylov-Bogoliubov method.

Musen (5) proposed the use of Hill's method of secular perturbations in obtaining the zero order perturbations. Skripnichenko (1) applied Weil's method to study secular perturbations of some planets of the solar system.

A long-range study of the elements of the Earth motion over 30 million years was made by Sharaf (2) and Boudnikova. A long-range numerical integration of the motions of Pluto, Neptune, and Uranus was made by C. J. Cohen (1, 2) *et al.* and the orbits of these planets were investigated also by Duncombe (1) *et al.* and Seidelmann (3) *et al.*

Duncombe (3) *et al.* derived new numerical theories of the motion of Ceres, Pallas, Juno, and Vesta, based on a series of observations collected by E. S. Jackson. Clemence's theory of Mars was studied by Böhme (1) and is currently compared with observations by Laubscher.

The disturbing function of the motion of a satellite was studied and iterative formulations were given by Brumberg (1) and by Challe (1) and Laclaverie.

Giacaglia (3, 8) applied to the solution of Hill's equation a method of transformation of generalized Mathieu equations. Elmabsout (1) gave a semi-numerical method of solving equations in the Hill-Brown method and applied it to the motion of Phoebe. Various determinations of luni-solar perturbations of an artificial satellite, as those by Kozai (1, 2) pertain as well to the study of the motion of a natural satellite.

Wilkins (1) obtained a new satsfactory representation of the orbits of Phobos and Deimos free of acceleration, while Vashkovyak (1), and Liakh constructed two other independent theories of the motion of these satellites. Sudbury (1) worked on Jupiter's fifth satellite while Charnow (1) *et al.* applied Hansen's method to the tenth satellite of Jupiter. Gill (1) and Gault redetermined a new orbit of Triton.

Ferraz-Mello (1) derived a theory of the motion of the Galilean satellites of Jupiter that includes most of the second order terms. This task is being continued by Sagnier for other higher order terms and terms depending upon the inclination while D. T. Vu is applying Sampson's method to the same problem.

D. LUNAR THEORY

The dramatic interaction between celestial mechanics and engineering mentioned in the Introduction is most striking in the case of the Moon. The motion of the Moon has always been a most effective means of testing the laws governing the motions in the solar system and now, in a comparatively few years, we are achieving a new overall order of accuracy. This increase in accuracy cannot fail to have important consequences.

The comparison of the theory of the motion of the Moon with observation requires not only the measurement of the direction or distance of the Moon, but also of the many other quantities necessary to refer the position of the observer to the center of gravity of the Earth and of the point observed to the center of gravity of the Moon, and the distribution of mass in the Earth and in the Moon. For many years the effects of several of the components on the overall accuracy of the comparison have been roughly of the same order of magnitude [W. J. Eckert (1)]. Fortunately the new technology is advancing in all these areas by means of earth satellites, lunar orbiters [Mulholland (1) and Sjogren], and lasers.

The most dramatic of the new observing techniques is of course the use of the laser in conjunction with the reflector array placed on the Moon by the Apollo astronauts. This method not only gives direct measures of distances with a new order of accuracy but it uses a sharp point on the lunar surface. A long series of observations from several stations on the Earth and hopefully with one or two additional reflectors should give excellent resolution of the necessary parameters. The following communication concerning the experiment has been received from Mulholland.

Present plans of the experimenters (Alley, Bender, Dicke, Faller, Kaula, MacDonald, Mulholland, Plotkin), approved by NASA, call for an observing program to be carried on for at least 9 years from the primary observatory (McDonald Obs., Ft. Davis, Texas). The experimenters are encouraging other observatories to establish observing programs, providing widespread coverage. The data, which are expected to yield a 1-nanosecond (15 cm) resolution, will be used to investigate a variety of minute effects, as well as providing for significant improvement to our knowledge of lunar motion. The phenomena upon which information will be obtained include the polar motion of Earth, the physical librations of Moon, a test of relativistic concepts, and several matters of geophysical interest.

In the meantime progress is being made in improving the angular measures (see Commission 8); this includes improved reductions of old observations and new techniques for observing occultations.

There is not yet a complete lunar theory covering all phases of the problem with the precision demanded for the near future, but results already achieved and work in progress indicate that no observational accuracy will be compromised for lack of accuracy in the theory. Parts of the problem are being attacked by several independent methods and this should not only assure precision for practical purposes but should throw light on mathematical developments when pushed to a dozen significant figures.

The electronic computer can now be applied to all the traditional methods of celestial mechanics including numerical integration and development in harmonic series. The series development for the main problem includes Delaunay's method where all the parameters m, e, e', γ, α are literal, that of Hill-Brown where m is numerical, and that of Airy where all are numerical. Brown's theory had a precision of 2×10^{-8} for most of the main problem and about an order of magnitude less precision in the perturbations.

During the past three years much progress has been made. The solution of the main problem by the method of Airy (precision of 1×10^{-12} in most terms) was completed by Eckert (4) and Smith in 1967. The solution was made both with and without the effect of the factor $(E+M)/m'$. Since then a great deal of effort has been devoted to the detailed publication to facilitate critical analysis of the work. The stimulating paper "Lunar Disturbing Function" by Barton (1) has been followed by Barton (2). A new formalism has been devised and checked by Chapront (2) and Mangeney-Ghertzman and is now being programmed for computation up to high orders of the parameters as well as to permit a study of the planetary perturbations.

The solution of the main problem by the Hill-Brown method with 18 decimals and for several values of m and of $(E+M)/m'$ (including zero) is under way; the terms of zero order in e, e', k, α have been published [W. J. Eckert (1) and D. A. Eckert], and the first and second order terms have been completed. Deprit is engaged on a literal solution of the Main Problem. Musen (3) has re-examined Hansen's theory and Thiry (1) discussed action variables and Delaunay's theory.

Mulholland (3, 5) and his associates are using a combination of numerical integration and analytic techniques to prepare precise ephemerides for comparison with observations.

E. RELATIVITY AND OTHER NON-NEWTONIAN EFFECTS

Probably the most important development in this field has been the measurement of the oblateness of the photosphere of the Sun by R. H. Dicke and his collaborators, from which they deduce a motion of the perihelion of Mercury of some 3″ per century, which destroys the previously supposed excellent agreement with Einstein's prediction, and if substantiated, may require profound modifications in the theory of relativity. Dicke postulates the existence of a scalar field superimposed on Einstein's tensor field. Those results have appeared in the literature of physics rather than astronomy, chiefly the *Physical Review* and *Physical Review Letters*; e.g., Dicke (1, 4), Dicke (2, 3) and Goldenberg.

Several papers, including Carstoiu (2, 3) and Dăngvu (1), have appeared on the propagation of gravitational waves, which still awaits definite observational confirmation.

At least two papers, Eisenstaedt (1) and Kurmakaev (1), purport to give solutions of the relativistic two-body problem, which, according to the textbooks, is impossible.

F. ARTIFICIAL CELESTIAL OBJECTS

(a) *General.* The use of intermediate orbits for artificial earth's satellites has been the subject of numerous papers. A novel spherical-coordinate intermediary, introduced by Aksnes (2) has drastically simplified the calculation of perturbations. The author used this intermediary to construct a first- (1) and then a second-order theory (4). The latter paper is the first complete second-order theory for a geopotential including the J_2, J_3, and J_4 spherical harmonics. The work is noteworthy for its use of the method of Lie-series as formulated by Hori, and for the use of Hill variables to preclude the appearance of mixed secular terms. The Aksnes intermediary and its close relation to that of Garfinkel (1958) are discussed in Garfinkel (3) and Aksnes. The troublesome question of mixed-secular terms and their elimination by the use of action-angle variables is investigated in Garfinkel (5), Hori, and Aksnes.

The Vinti intermediary, separable in spheroidal coordinates has been adapted to the calculation of nearly polar orbits [Vinti (2)]. Practical applications of the Vinti potential have been investigated by Getchel (1) and O'Mathuna (2). An extension of the Vinti-type potential to *triaxial* ellipsoid has been studied by Madden (1). A closely related intermediary, corresponding to the solution of the problem of two fixed centers, has been used by Marchal (1).

The use of electronic machines to construct a formal artificial satellite theory is illustrated by the work of Deprit (19) and Rom. There the author solves the Main Problem, extending the Brouwer solution to the third order in J_2. The paper is noteworthy for the programming of the Lie-series method. However, not being in closed form, the solution is restricted to small eccentricities.

The luni-solar perturbations of an artificial earth's satellite have been studied by Kopal (1) and Roy (3). The effect of these perturbations on the 24-h satellite has been investigated by Martynenko (1) and Morando (1).

The problem of orbit determination was studied by Senjalow (1), Bazhenow (1), and Lundquist (1);

the problem of position determination from simultaneous angular observations was treated by Zhongolovitch (1).

That the critical inclination is, indeed, real was reiterated by Garfinkel (2) in his comment on a controversial paper by Lubowe (1).

(b) *Air Drag, Light Pressure, and Rotation.* Second-order perturbations in a and i, due to the combined effect of drag and oblateness, were calculated by Fominov (1). The lifetime of satellites of large eccentricity was investigated by G. E. Cook (2) and Scott. The drag effects on a cone-shaped satellite were studied by Po (1). Sehnal (1) and Mills discussed the short-periodic drag perturbations. Further studies of contraction of satellite orbits due to drag were carried out by G. E. Cook (3) and King-Hele. The "shadow equation", governing the radiation pressure of a satellite, was solved by Batrakov (1) for the case of small eccentricity.

(c) *Lunar Orbiters and Artificial Satellites of Planets.* The theory of a lunar orbiter was studied by Giacaglia (9), Oesterwinter (1), Forga (1), Roy (1, 2), Evdokimova (2), and Kirpichnikov (1). The latter included the effects of lunar and solar radiation pressure. Classification of orbits was discussed by Felsentreger.

(d) *Optimization Problem.* Hiller (1) considered impulsive transfer between non-coplanar elliptic orbits having collinear major axes. Tapley (1, 2) *et al.* investigated the advantages of regularization in optimal trajectories. J. Breedlove is using the Lie-series method to obtain a first-order solution of the Lawden Problem for the case of low thrust.

G. THE PROBLEM OF THREE BODIES

The restricted problem of three bodies dominates the field. The main areas of interest are periodic orbits, regularization, motion around the libration points and stability. The general problem of three bodies received less attention. The principal approaches are analytic and numerical.

Various aspects of periodic orbits and their stability in the restricted problem are discussed by Bozis (3, 4, 5), L. H. Carpenter (2) and Stumpff, Colombo (2) *et al.*, Conley (1, 2), Deprit (3) and Henrard, Giacaglia (1), Guillaume (2), Henon, (2, 3, 4) and Guyot, Jefferys (2, 4) and Standish, Kevorkian (5), Kozai (6), Lanzano (2), Meffroy (6), and Szebehely (6, 9) and Nazocy. In the general problem of three bodies periodic and quasi-periodic orbits are treated by Duboshin, Jefferys (3) and Moser, Moser (1, 4, 5), and Szebehely (16, 22) and Peters. Motion and stability around the equilibrium points of the restricted problem have been studied in considerable detail by Deprit (1, 5, 7, 12, 15, 16, 17, 20, 22) *et al.*, França (1, 3), Giacaglia (12) and França, Hadjidemetriou (3), Henrard (11), Katsis (1), Junqueira (1), Rabe (1, 2, 5), Roels (1), Schanzle (1), and Szebehely (1, 8, 11, 21) *et al.*

The elliptic restricted problem received attention from Choudhry (1), Contopoulos (3), Guillaume (1), Lanzano (1), Lukjanow (1), Rabe (7, 8), and Szebehely (2).

The regularization of the three dimensional problem was discussed by Deprit (10), Kustaanheimo (4), Peters (3, 4), Stiefel (1) *et al.*, and Waldvogel (1). Perturbation studies using regularized variables were performed by Pierce (1, 2), Szebehely (5), and C. Williams (1). General comments on and applications of regularization were made by Broucke (1), Giacaglia (6), and Szebehely (3, 10) and Pierce.

Applications to orbits of interest in space explorations have been made by Giacaglia (7), Grebenikov, Jarov-Jarovi, Kirpichnikov, Noroselov, Petrovskaya (U.S.S.R), Kevorkian (6) and Brachet, Lancaster (1), Shi (1, 3) and Eckstein, and Szebehely (7, 4).

Asymptotic orbits were investigated by Danby (1) and Deprit (14) and Henrard. Capture in the restricted and general problems of three bodies was discussed by Sung (1) and by Kotsakis (2) *et al.* who also discussed drag effects in the restricted problem (3).

Orbits in the general problem of three bodies were treated by Agekyan (1, 2) and Anosova, Alekseev (1), Sconzo (2), and Szebehely (12, 14, 19) and Peters.

Integrals of motion in the restricted problem have been investigated by Bozis (1, 2) and Contopoulos (2, 3).

A book containing the major accomplishments and listing references on the restricted problem up to 1967 is *Theory of Orbits* by Szebehely (see Table B).

H. PERIODICITY, ERGODICITY, STABILITY

The articles which discuss the above subjects with applications to the problem of three bodies have been reviewed under Class G. The main trends in this category are quasi-periodic orbits, stability, and the study of integrals of motion. This last subject is discussed by Contopoulos (1, 5, 6, 7) *et al.*, and Deprit (21) and Henrard, in addition to Bozis (see Section G). Stability questions are treated by Contopoulos (9), Losco (1), Moustakhichev (1), Nahon (2), and Pius (1). Normalization is treated by Deprit (18) *et al.* and Krassinsky (1). Quasi-periodic orbits are discussed by Moser (1, 3, 4, 5), see also under G.

I. RESONANCE

The many papers on the subject of resonance dealt with general theories of resonance and with a variety of special problems found in the solar system. The first group includes Lectures on Hamiltonian Systems, Moser (5); spin-orbit coupling and dynamics of planetary rotations, Goldreich (2, 3) *et al.*; resonant structure of the solar system, Molchanov (1), comment by Henon (6); the relation of the third integral to resonance, Barbanis (1), Contopoulos (1), Hori (2); evolution of commensurabilities, Dermott (1, 2); and the ideal resonance problem, Garfinkel (4), Jupp (1). The group of special problems is illustrated by the tesseral harmonics resonance, Allan (1, 2), Blitzer (1); rotational resonance in a Kepler orbit, Blitzer (2); resonance in the restricted three-body, L. H. Carpenter (2) and Stumpff, Colombo (2), Giacaglia (11); and in the restricted four-body problem, Kolenkiewicz (1) *et al.*; resonance in the neighborhood of the Lagrange points, Roels (3, 4); asteroids commensurable with Jupiter, Sinclair (1); periodic Trojans, Deprit (15) and Rabe; Kirkwood gaps, W. H. Jefferys (1), Schweizer (1); Cassini laws, S. J. Peale (4); the Hilda-type asteroids, Schubart (1); periodic orbits emanating from resonant equilibrium, Henrard (10); and the Mimas-Thetys commensurability, R. R. Allan (3).

A novel method of treating resonance problems was proposed by Shi (2) *et al.* In a class by itself is the paper on the radar determination of the rotation of Venus and Mercury, Dyce (2), Pettengill and Shapiro, which furnished observational data for much of the subsequent theorizing.

J. N-BODY PROBLEM AND GALACTIC STRUCTURE

Many papers in this area of importance to celestial mechanics are reviewed in the report of Commission 33 and others in Section B of this report.

The proceedings of the IAU Colloquium on the Gravitational and *N*-Body Problem, held in Paris, in 1967 were published in *Bulletin Astronomique* (many of the references marked J68 7,... were presented at the Colloquium). A new IAU Colloquium on the Gravitational *N*-Body Problem will take place, in August 1970 at the Institute of Theoretical Astronomy, Cambridge, England.

K. ASTRONOMICAL CONSTANTS

The application of radar to the direct measurement of lunar and interplanetary distances has provided a new dimension in celestial mechanics (at least for the terrestrial planets), and has already resulted in several important contributions to the system of astronomical constants, and to the values of the masses of the planets. Indeed, the work of the past five years has added as much to our knowledge of these subjects as that of the fifty years preceding. Observations of the planets by means of radar include Dyce (3) *et al.*, Evans (1, 2) *et al.*, Goldstein (1) Pettengill (1) *et al.* Determinations of masses and other astronomical constants include Ash (1) *et al.*, Bec (1), Duncombe (1) *et al.*, Klepczynski (2), Lieske (2, 4), Lieske (3) and Null, Null (1), O'Handley (1), Rabe (3), Rabe (4) and Francis, Seidelmann (3) *et al.*, and Zielenbach (1).

In addition, with the aid of radar the gravitational field of the near side of the Moon has been mapped for the first time [Bray (2) *et al.*, Chuikova (1), Derr (1), Goudas (3), Koziel (1), Safranov (1), Volkov (3) and Schober], and our knowledge of the gravitational field of the Earth has been greatly increased with the aid of artificial satellites [Bivas (1), Kaula (1), King-Hele (2, 5) *et al.*, Kozai (3, 5), Murphy (1)].

Coherent light pulses have been sent to, and reflected from, the Moon. A systematic series of such observations, continued over a number of years, would provide a direct test of the Theory of the Motion of the Moon, which can be obtained in no other way (see Section D).

L. MINOR PLANETS, COMETS, AND METEORS

Everhart (1, 2) has made a very interesting statistical study of energy changes and captures of comets in a hypothetical solar system. His results are not in agreement with the known short-period comets. Herget (2) derived satisfactory orbits for J-VIII to J-XII, including the integration of the Variational Equation. The mass of Jupiter is not well determined, in contradiction to the results of Miss A. Bec (1). Hunter (1, 2) has made a statistical study of carefully selected hypothetical satellites of Jupiter and minor planets, to find their evolution and stability. His supposed region of minor planets between Jupiter and Saturn is not substantiated by the Palomar-Leiden survey, which reached magnitude 20·5 and discovered 15 new Trojans. Marsden (4, 7) has varying degrees of success in representing short-period comet observations in three or more oppositions by including an unspecified non-gravitational term in the equations of motion. The effect is mainly an outward radial force, in agreement with Whipple's comet model. Polozova has computed the first order general perturbations of (11) Parthenope by Venus to Neptune, using Hill's method. The residuals reach 300″ at dates 60 years from the epoch. Schubart (2) used the observations from five oppositions of (1221) Amor (1932–1964) to derive improved elements, but he found it necessary to adjust the mass of Earth + Moon (in agreement with the radar value for the astronomical unit) and he encountered the difficulties usually associated with the correction of highly perturbed orbits. The detailed study of the motions of comets in the solar system by Kazimirchak-Polonskaya (1, 3, 4, 5, 6) has been completed.

M. COMPUTING METHODS

A review of computing methods as such is not a part of this report, but many papers in celestial mechanics contain contributions to the subject. We have attached 'M' to a number of references and commented on a few of them.

The principal development has been in the use of computers for the manipulation of large harmonic series with literal coefficients. The exciting paper of Barton (1) has been followed by the sequel (2). The report of the colloquium on the Use of Computers for Analytical Developments in Celestial Mechanics held at the meeting in Prague has been published, Eckert (3), with papers by Davis (1), Kovalevsky (2), Deprit (8) and Rom, Chapront (1) and Mangeney Ghertzman, and LeShack (1) and Sconzo. Other papers include those by Deprit (2) and Rom, Chapront (2) and Mangeney-Ghertzman, and Glebova (1).

Section D of this report contains examples of large literal and numerical calculations. Many of the references marked B, Ga, and J involve extensive numerical work; the paper by Prendergast (2) and Miller is of particular interest. Propagation of errors in numerical methods is discussed by Kinoshita (1), by Miachine (1), and by Stiefel (3) and Bettis.

N. OTHER TOPICS

Minor changes were made in the classification scheme of Table A during the course of collecting the bibliography, and some references have retained classifications that were assigned before the scheme was in final form. This explains some of the anomalies of classification, particularly in this section.

General and historical questions were discussed by Duboshin (1), Levy (1) and Neugebauer (1). A conference on the general problems of celestial mechanics and of astrodynamics was held in Moscow in March 1967; a symposium, "The General Questions of Celestial Mechanics", in Leningrad in May 1969; and a conference on the qualitative methods of celestial mechanics in October 1969.

Discussions of the internal structure of the Earth, Moon and planets include Goldreich (4), Jeffreys (1), Khan (1), Kozlovskaya (1), Mikhailov (1), Sconzo (4), Shimazu (1), and Volkov (1, 2). A symposium, "The Theory of the Figures of the Earth and Moon", was held in Lvov in October 1968 and another. "The Figures and Rotation of Celestial Objects", in Tiraspol in October 1969.

Secular effects, cosmic dust, etc., were treated by Sharaf (3, 4), and Boudnikova, Divari (1), Gerstenkorn (1, 2), Giuli (1), Lebedinets (1) and Kashcheev, McCord (1, 2), Roosen (1) and Ruskol (1).

W. J. ECKERT
President of the Commission

PERIODICALS

35. *A. Rev. Astr. Astrophys.**
36. *Acta. astr.**
37. *Adv. Astr. Astrophys.**
38. *AIAA J.*
39. *Akad. Nauk., Arm. USSR., Astrophys.*
40. *An. Acad. brasil. Cienc.*
41. *Astr. Astrophys.*
42. *Astr. J.**
43. *Astr. Nachr.**
44. *Astr. Pap. Washington**
45. *Astr. Zu.**
46. *Astrophys. J.**
47. *Astrophys. Norw.**
48. *Astrophys. Space Sci.*
49. *Bjull. Inst. teor. Astr.**
50. *Bull. Am. astr. Soc.*
51. *Bull. astr., Paris**
52. *C. r. Acad. Sci. Paris**
53. *Cel. Mech.*
54. *Commun. Sternberg astr. Inst.*
55. Dissertation, ETH, Zurich
56. Dissertation, Yale Univ.
57. *Dokl. Akad. Nauk. SSSR**
58. ESRO Publ.
59. *Icarus**
60. *J. astronaut. Sci.**
61. *J. diff. Eq.*
62. *J. geophys. Res.**
63. JPL Publ. (Cal. Inst. Tech.)
64. *Math. Annln*
65. *Math. USSR Sbornik*
66. *Mem. Am. math. Soc.*
67. *Mem. Soc. astr. ital.**
68. *Mon. Not. R. astr. Soc.**
69. NASA Publ.
70. *Nature**
71. *Num. Math.*
72. *Observatory**
73. Ph.D. Thesis, Univ. Conn.
74. *Phys. Rev.**
75. *Phys. Rev. Lett.**
76. *Planet. Space Sci.**
77. *Proc. nat. Acad. Sci. Am.**
78. *Proc. R. Soc. London Ser. A**
79. *Publ. astr. Obs. Helsinki**
80. *Publ. astr. Soc. Japan**
81. *Publ. astr. Soc. Pacific**
82. *Q. appl. Math.*
83. *Res. cel. Mech. diff. Eq., Univ. Sao Paulo*
84. *Science**
85. *SIAM J. appl. Math*
86. *SIAM Rev.*
87. *Smithsonian Inst. Obs., Spec. Rep.**
88. *Tech. Ann.*
89. *Trudy Inst. teor. Astr.**
90. *Vest. Leningrad. gos. Univ.**
91. *Vest. Mosk. gos. Univ.**
92. *Vistas Astr.*
93. *Z. angew. Math. Phys.*
94. *Z. Astrophys.**

Table B. Publications

BOOKS

1. Abalakin, V. K., Aksenov, E. P., Grebenikov, E. A., Riabov, J. A. *Reference Book of Celestial Mechanics* (in Russian), Moscow, in press.
2. Baker, R. M. L. 1967, *Astrodynamics: Applications and Advanced Topics*, Academic Press, New York.

3. Baker, R. M. L., Makemson, M. M. 1967, *An Introduction to Astrodynamics*, 2nd. ed., Academic Press, New York.
4. Beletzky, V. V. 1965, *The Motion of an Artificial Satellite About Its Center of Mass* (in Russian), 'Nauka', Moscow.
5. Captuo, M. 1967, *Gravity Field of the Earth from Classical and Modern Methods*, Academic Press, New York.
6. Cherbotarev, G. A. 1967, *Analytical and Numerical Methods of Celestial Mechanics*, American Elsevier, New York.
7. Contopoulos, G. 1966, *The Theory of Orbits in the Solar System and Stellar Systems*, IAU.
8. Demin, V. G. 1968, *The Motion of an Artificial Satellite in a Non-Central Field of Gravitation* (in Russian), 'Nauka', Moscow.
9. Duboshin, G. N. 1968, *Celestial Mechanics, Fundamental Problems and Methods*, 2nd ed., *Manual* (in Russian), 'Nauka', Moscow.
10. Duncombe, R. L., Szbehely, V. G., Eds. 1966, *Methods in Astrodynamics and Celestial Mechanics*, Academic Press, New York.
11. El'yasberg, P. E. 1967, *Introduction to the Theory of Flight of Artificial Earth Satellites*, Israel Program Sci. Transl.
12. Escabol, P. R. 1968, *Methods of Astrodynamics*, Wiley, New York.
13. Giacaglia, G. E. O., Ed. 1969, *Proceedings of the Symposium on Periodic Orbits, Stability and Resonances*, University of São Paulo, Brazil, in press.
14. Grodzovskii, G. L., Ivanov, Y. N., Tokarev, V. V. 1969, *Mechanics of Low-Thrust Spaceflight*, Israel Program Sci. Transl.
15. Hagihara, Y. *Celestial Mechanics*, vols. I, II, MIT Press, in press, vols. III, IV, V, in preparation.
16. Herrick, S. *Astrodynamics*, vol. I, van Nostrand Reinhold, New York, in press, vol. II, in preparation.
17. Kaula, W. M. 1966, *Theory of Satellite Geodesy*, Blaisdell, Waltham, Mass.
18. Kovalevsky, J. 1967, *Introduction to Celestial Mechanics*, Reidel, Dordrecht, The Netherlands.
19. Kovalevsky, J. 1966, *Trajectories of Artificial Celestial Bodies as Determined by Observations*, Springer-Verlag, New York.
20. Markowitz, W., Guinot, B. 1968, *Continental Drift, Secular Motion of the Pole and Rotation of the Earth*, IAU.
21. Marsden, B. G., Cameron, A. G. W. 1966, *The Earth-Moon System*, Plenum Press, New York.
22. Melchior, P. J. 1966, *Earth Tides*, Pergamon Press, Elmsford, N.Y.
23. *Modern Questions of Celestial Mechanics*, Internazionale Matematico Estivo, Edizioni Cremonese, Roma 1967.
24. Morando, B., Ed. 1970, *Dynamics of Satellites*, Springer-Verlag, New York.
25. Mueller, I. 1969, *Introduction to Satellite Geodesy*, Ungar, New York.
26. Mueller, I., Rockie, J. D. 1966, *Gravimetric and Celestial Geodesy*, Ungar, New York.
27. Pollard, H. 1966, *Mathematical Introduction to Celestial Mechanics*, Prentice-Hall, Englewood Cliffs, N.J.
28. Rosser, J. B., Ed. 1966, *Lectures in Applied Mathematics*, vols. 5–7, Space Mathematics, Am. Math. Soc., Providence, R.I.
29. Runcorn, S. K., Ed. 1967, *Mantles of the Earth and Terrestrial Planets*, Interscience, London.
30. Sedov, L. I. 1968, *Analytical Mechanics, Stability of Motion, Celestial Ballistics*, Part I, Israel Program Sci. Transl.
31. Sternberg, Shlomo 1969, *Celestial Mechanics*, Part I and Part II, W. A. Benjamin, New York.
32. Stiefel, E., Ed. 1966, *Mathematische Methoden der Himmelsmechanik und Astronautik*, Bibliographisches Inst., Mannheim.
33. Subbotin, M. F. 1968, *Introduction to Theoretical Astronomy* (in Russian), 'Nauka', Moscow.
34. Szebehely, B. G. 1967, *Theory of Orbits*, Academic Press, New York.

BIBLIOGRAPHY

Aarseth, S. J.	J68 51,3,105	Aizenman, M. L.	B68 46,153,511
Abalakin, V. K.	K68a49,11,481	Aksenov, E. P.	B68 45,45,858
Abitt, M. W. Jr.	I69b53,1,31		C68 45,45,1284
Agekyan, T. A.	Gb67a45,44,1261		B68 54,no.155, p. 3
	Gc68a39,4,31	Aksnes, K.	Fa67 47,10,149

	Fa67 47,10,69
	L67b42,72,952
	Fa69 56
	Fa**b42
	Fa**b42
Alekseev, V. M.	Gc68 65,5,no.1
Allan, R. R.	I67 76,15,53
	FI67 76,15,1829
	IN69 42,74,497
Allen, W. A.	Fc69 38,7,890
Anand, S. P. S.	E68a59,8,492
Anosova, Z. P.	Gb67b45,44,1261
	Gc68b39,4,31
Aoki, S.	K67 80,19,585
	B69 42,74,284
Ash, M. E.	K67a42,72,338
Astapovich, I. S.	L67b45,44,614
Aver'yanova, T. V.	E66a45,43,1301
Baierlain, R.	E67 74,162,1275
Barbanis, B.	I66 42,71,415
	B67a42,72,215
	H68b48,2,134
	B68 42,73,784
Barton, D.	DM66 42,71,438
	DM67 42,72,1281
Batrakov, Y. V.	Fb67 49,11,14
Bazhenow, G. M.	Fa67 49,11,116
Bec, A.	K69 41,2,381
Beletzky, V. V.	B** 24
Belorizky, D.	Gc66 52,262,1133
Belous, L. M.	L66 49,10,543
Belyaev, N. A.	L67 45,44,461
	L67b45,44,614
Benima, B.	A69a81,81,121
Bennett, M. M.	B66b42,71,579
Bernstein, I. S.	H68b59,9,281
Bettis, D. G.	CM69b71,13,154
Betz, H. R.	Fa69b53,1,91
Bivas, R.	K68 51,3,517
Blitzer, L.	I66 62,71,3557
	IB67 42,72,988
	I68 62,73,2304
Böhme, S.	C68 43,290,249
	E** 43
Boigey, F.	B66 51,1,265
Boudnikova, N. A.	C66b89,11,89
	BC67b49,11,231
	CN68b57,182,291
	CN69b89,14,48
Bourke, R. D.	K67b62,72,1265
Bozis, G.	GbH66 42,71,404
	GcH67 42,72,380
	Gb68 42,73,616
	Gb69 72,89,75
	Gb** 13
Brachet, G.	Gc69b38,7,885
Brady, J. L.	L67a42,72,365
	L67 42,72,1184
Branham, R. L. Jr.	L68 42,73,97
Bray, T. A.	Gc67a42,72,202
	K67a59,7,76
Brockelman, R. A.	K66b42,71,897
Broucke, R.	Ga67 59,7,221
	C69 53,1,110
Brouwer, D.	I66a42,71,543
Brumberg, V. A.	C66 89,11,3
	C67 49,11,73
	C67 45,44,204
	EB68 45,45,828
	E68 45,45,1110
	C** 13
Burdet, C. A.	BGa67b69,N67-29387
Campbell, D. B.	KB67b42,72,330
Campbell, J. W.	J68b42,73,611
Candy, M. P.	B68b68,139,231
	J68b68,141,277
Carpenter, E.	L67b42,72,365
Carpenter, G. C.	M69b53,1,72
Carpenter, L. H.	I67b42,72,180
	GcI68a43,291,25
Carstoiu, J.	E68 52,266,754
	E69 52,268,201
	E69 52,268,261
Cary, C. N.	Fc68a84,160,875
Castaing, C.	H67 52,264,333
Challe, A.	DFaC69a41,3,15
Chandrasekhar, S.	J68a68,139,135
Chapront, J.	D68a42,73,214
	D69a41,2,425
	C** 41
Charnow, M.	DC68a60,15,303
Chebotarev, G. A.	B68 49,11,341
	B69 49,11,625
Chepurova, V. M.	B68 54,no.154,p.14
Cherniack, J. R.	A69b81,81,121
Choudhry, R. K.	Gc66 49,10,523
Chuikova, N. A.	K68 45,45,1293
Clemence, G. M.	E67a42,72,1324
Cochram, J. E.	Gc**b42
Cohen, C. J.	C67a42,72,973
	C68a42,73,290
Cohen, L.	J68a51,3,213
Colombo, G.	B66 42,71,891
	GcIL68a42,73,111
Conley, C.	Gc68 85,16,620
	B69 61,5,136
Contopoulos, G.	HI67 51,2,223
	GcH67 42,72,191
	GcH67 42,72,669
	HI68 46,153,83
	H68a48,2,134
	D68 70,220,1018
	BH68a42,73,86
	H** 42
	H** 42
Cook, A. H.	B66 68,134,253

Cook, G. E.	K67b76,15,741
	Fb67a76,15,1549
	Fb68a78,303,17
	K69b76,17,629
Cronin, J.	H68a59,9,281
Danby, J. M. A.	Gb67 42,72,198
	C68 42,73,1031
Dăngvu, H.	E69 52,268,297
Davies, R. D.	K68b68,140,537
Davis, M. S.	M68 42,73,195
DeBra, D. B.	K67b62,72,1265
Degtjarow, W. G.	BH66a49,10,516
Deprit, A.	Gb67a68,137,311
	M67a51,2,425
	Gc67a42,72,158
	Gc67a42,72,173
	Gb67a59,6,381
	A68a51,3,315
	Gb68a37,6,2
	BM68a42,73,210
	B68a38,6,1234
	Ga68 93,19,369
	Gb68a59,9,336
	Gb68 42,73,730
	B69 53,1,12
	Gc69a42,74,308
	GbI69a42,74,317
	Gb69a41,1,427
	Gb69a41,3,88
	H**a53
	Fa**a53
	Gb**a13
	H**a48
Deprit-Bartholomé, A.	Gc67b42,72,173
	A68b51,3,315
Dermott, S. F.	I68 68,141,349
	I68 68,141,363
Derr, J. S.	K67 59,7,261
Devine, C. J.	D68b84,160,874
Dicke, R. H.	E64 70,202,432
	E67a75,18,313
	E67a70,214,1294
	E67 84,157,960
Ditto, F. H.	N69 53,1,130
Divari, N. B.	N66 45,43,1273
Douglas, B. C.	FI69b76,17,1505
Duboshin, G. N.	N67 49,11,1
	J69a45,46,895
Duncombe, R. L.	CK68a42,73,830
	CK69b42,74,776
	C** 44
Dupont, E. N.	K66b42,71,897
Dyce, R. B.	KB67b42,72,330
	IK67a42,72,351
	K67a42,72,771
Eckert, D. A.	D67b42,72,1299
Eckert, W. J.	D65 42,70,787
	D67a42,72,1299

	M68 42,73,195
	D**a44
Eckstein, M. C.	GcJ67b42,72,685
	I68b42,73,275
	GcJ68b42,73,653
	I69a42,74,551
Eisenstaedt, J.	E66 52,263,149
Elmabsout, B.	C** 41
Evans, J. V.	K66a42,71,897
	K66a42,71,902
Evdokimova, L. S.	BH66b49,10,516
	Fc68 49,11,523
Everhart, E.	L68 42,73,1039
	L69 42,74,735
Fahlman, G. G.	E68b59,8,492
Feix, M. R.	J67b46,147,1164
	JM68a51,3,289
Felsentreger, T. L.	Fa68 76,16,285
Ferraz-Mello, S.	C66 51,1,287
	C68 40,40,447
	BC69 52,268,198
	BC69 52,268,985
	C** 13
Finzi, A.	K68 59,9,191
Fominov, A. M.	F68 49,11,507
Forga, R.	F69 41,2,75
Foursenko, M. A.	N67 49,11,127
	K68b49,11,481
França, L. N. F.	Gc68 83,1,42
	Gc**b13
	Gc** 83
Francis, M. P.	K67b42,72,856
Franklin, F. A.	GcIL68b42,73,111
Freeman, K. C.	J66 68,133,47
	J66 68,134,1
	J66 68,134,15
Freeman, K. G.	J68a51,3,269
Froeschlé, C.	N68 52,266,747
	N68 52,266,846
Gaposchkin, E. M.	E68a87,no.283
Garfinkel, B.	I66 42,71,657
	FaI69 53,1,11
	Fa**a42
	I** 13
	Fa**a42
Gault, B. L.	C68b42,73,95
Gehrels, T.	L67 42,72,929
Gerstenkorn, H.	N67 59,6,292
	N67 59,7,160
Getchell, B. C.	Fc** 38
Giacaglia, G. E. O.	Ga67 42,72,386
	GNJ67 42,72,674
	BC67 42,72,998
	J68 51,3,63
	I68 87,no.278
	GaB68 83,1,1
	Gc68 83,1,3
	BC68 42,73,379

	Fc69 69,N69-28050
	N** 53
	I** 42
	Gc**a13
	I** 13
Gill, J. R.	C68a42,73,95
Giuli, R. T.	N68 59,9,186
Glebova, N. I.	M67 49,11,177
Goldenberg, H. M.	E67b75,18,313
	E67b70,214,1294
Goldreich, P.	K66a59,5,375
	ICH66a42,71,425
	ICH67a42,72,662
	N67 62,72,3135
	ICH68a35,6,287
Goldstein, R. M.	K68 42,73,829
Gonzalez, C.	J68a51,3,209
Goudas, C. L.	Gc67b42,72,202
	K67b59,7,76
	K67 42,72,955
Graham, R.	J67 68,137,25
Grebenikov, E. A.	B66 45,43,882
	B68 49,11,293
Guillaume, P.	Ga67 51,2,533
	GBGc69 41,3,57
Guinot, B.	N67 51,2,449
Gustavson, F. G.	B66 42,71,670
Guyot, M.	Gb69b13
Hadjidemetriou, J. D.	Gc67 42,72,865
	BH68b42,73,86
	Gb68 42,73,104
	A69 48,3,330
	A69 48,3,31
	GH69 42,74,789
Habibulin, S. T.	B68 36,18,20
Hamid, S. E.	C68 87,no.285
	L68a42,73,727
Hanson, L. B.	K66b42,71,897
Harrington, R.	J68b51,3,269
Harrington, R. S.	B68 42,73,190
Hayli, A.	J68 51,3,189
Henon, M.	BH67b51,2,267
	Gb68 51,3,377
	Gb69 41,1,223
	Gb69a13
	B** 82
	I** 59
Henrard, J.	Gb67b68,137,311
	Gc67b42,72,158
	Gb67b59,6,381
	Gb68b37,6,2
	Gc69b42,74,308
	Gb69b41,3,88
	H**b53
	Gb**b13
	H**b48
	I** 53
	Gb** 41

Herget, P.	L68 42,73,729
	L68 42,73,737
Hiller, H.	Fd66 76,14,773
Hirst, W. P.	A67 87,no.253
Hockney, R. W.	J68 51,3,199
	J69a42,74,1102
Hohl, F.	J67a46,147,1164
	J68 51,3,227
	JM68b51,3,289
	J68a42,73,611
	J69b42,74,1102
Holland, R. L.	B69a42,74,490
Hori, G.	B66 80,18,287
	I67 80,19,229
	NB68 80,20,204
	Fa**b42
Hubbard, E. C.	C67b42,72,973
	C68b42,73,290
Hunter, R. B.	L67 68,136,245
	L67 68,136,267
Iannini, G. M.	N68 42,73,743
Ingalls, R. P.	K66b42,71,902
Ingram, D. S.	Gc**b42
Izvekov, V. A.	A67 49,11,48
	C67 49,11,130
	K68 49,11,366
Jackson, E. S.	N67 42,72,360
Jefferys, W. H.	I66b42,71,543
	G66 42,71,566
	GcH66a42,71,568
	G66a42,71,982
	I67 42,72,872
	B68 42,73,522
Jeffreys, H.	N67 68,136,311
Johnson, D. B.	B69a42,74,563
Junqueira, J. L. A. N.	Gc69 83,2,42
Jupp, A. H.	I69 42,74,35
Kane, T. R.	B69b42,74,563
Kashcheev, B. L.	N66b45,43,854
Katsis, D.	Fb66 59,5,577
Kaula, W. M.	K66 62,71,4377
Kazimirchak-Polonskaya	L67 45,44,439
	L67a45,44,614
	L67 89,12,3
	L67 89,12,24
	L67 89,12,63
	L67 89,12,86
Kent, J. T.	Fa69a53,1,91
Kevorkian, J.	B66 28,p.206
	B66 42,71,878
	JG67b42,72,959
	Gc68b38,6,1986
	Gc68a42,73,791
	Gc69a38,7,885
Khan, M. A.	N68 62,73,5335
Kholshevnikov, K. V.	B68 90,no.7,p.128
	B68 90,no.1,p.149
Kiang, T.	L66 59,5,437

Kikuchi, S.	I67 43,289,241
	B68 43,291,9
King-Hele, D. G.	Fb66 78,294,261
	K67a76,15,741
	Fb68b78,303,17
	Fb69a76,17,217
	K69a76,17,629
Kinoshita, H.	M68 80,20,1
Kirpichnikov, S. N.	Fc68 45,45,675
Kiryushenkov, V. N.	B68 91,no.4,p.70
Klepczynski, W. J.	CK68b42,73,830
	K69 42,74,774
	CK69b42,74,776
Knothe, H.	N69 53,1,36
Knowles, S. H.	K69 42,74,291
Kolenkiewicz, R.	I67a42,72,180
Kopal, Z.	Fa67 59,6,298
	K67b59,7,76
Kotsakis, D.	G68 94,68,130
	G68 94,69,337
	G69a88,May
Kovalevsky, J.	B67 51,2,245
	CM68 42,73,203
Kozai, Y.	C66 87,no.234
	C66 87,no.235
	K68 80,20,24
	B68 92,11,103
	K69 87,no.295
	Gc69 80,21,267
Koziel, K.	K67 59,7,1
Kozlovskaya, S. V.	N66 45,43,1081
Krassinsky, G. A.	H68 49,11,411
	B69 89,13,105
Krefetz, E.	E67 42,72,471
	E67 42,72,1242
Kurmakaev, Z. K.	E66 45,43,1025
Kustaanheimo, P. E.	AB66 79,no.117
	E66 79,no.112
	E67 79,no.128
	A67 79,no.130
Kyner, W. T.	B66a42,71,579
	B66 42,71,583
Laclaverie, J. J.	DFaC69b41,3,15
Lancaster, J. E.	Gc68a38,6,1986
	Gc68b42,73,791
Lanzano, P.	G67 59,6,114
	G67 59,7,105
Lausberg, A.	E69 41,3,150
Lazović, J. P.	N67 49,11,57
Lebedinets, V. N.	N66a45,43,854
Lecar, M.	J66 42,71,706
	J68 51,3,91
	J68b51,3,209
	J68b51,3,213
Lee, E. S.	J68b68,139,135
Leimanis, E.	BGA68 51,3,21
LeSchack, A. R.	M68a42,73,217
Levy, J.	N68 51,3,13
Lewallen, J. M.	Fd69b38,7,1010
Lieske, J. H.	J67 63,32-1206
	K68 42,73,628
	KE69a42,74,297
	K69 42,74,572
	K** 41
Liu, H. S.	I67 62,72,4759
Loomis, A. A.	K67a62,72,1265
Losco, L.	H68 51,3,433
	B68 52,267,50
	B69 52,268,431
Lubowe, A. G.	I69 53,1,6
Lukashevich, E. L.	C68 45,45,850
Lukjanow, L. G.	G69 49,11,693
Lundquist, C. A.	FN67 87,no.248
Lutze, F. H. Jr.	I69a53,1,31
Lynden-Bell, D.	E67 68,135,413
	J67 68,136,101
	J68 51,3,305
Madden, S. J. Jr.	BFc** 53
Madore, J.	E66 52,263,746
Mamedov, M. A.	A66 49,10,549
Mangad, M.	Ga67 42,72,467
Mangeney-Ghertzman, L.	B66 51,1,223
	B67 51,2,243
	D68b42,73,214
	D69b41,2,425
Marchal, C.	FB66 51,1,189
	NB68 51,3,341
Marochnik, L. S.	J68a45,45,516
Marsden, B. G.	E66a87,no.236
	L67a42,72,952
	L67 42,72,1170
	L68 42,73,367
	L68b42,73,727
	A69b81,81,121
	L69 42,74,720
Martynenko, B.	FaI67 49,11,33
Maury, J.	DC68b60,15,303
McCord, T. B.	N66 42,71,585
	N68 62,73,1497
Meffroy, J.	BC66 52,263,145
	BC66 87,no.229
	BC67 51,2,343
	BC67 69,N67-33278
	BC67 69,N67-39923
	BC68 69,N68-28687
	Gc68 59,8,166
	BC69 52,269,309
Merman, G. A.	HI69 89,13,3
	HG** 49
Message, P. J.	B67 51,2,253
	H** 53
	HI** 13
Miachine, V.	M68 51,3,81
Michelson, I.	N66 68,133,17
	N68 59,8,265
Mikhailov, A. A.	N66 45,43,1313

Safronov, V. S.	K67 59,7,275
Sagitov, M. U.	K69 45,46,907
Sanchez, A. D.	K67b42,72,771
Sarris, E.	G69b88,May
Schanzle, A. F.	GL67 42,72,149
Scheifele, G.	BGa68 55,no.4212
	B68b52,267,950
Schinkarik, T. K.	B68 49,11,533
Schober, T. I.	F68 49,11,530
	K69b49,11,565
Schubart, J.	I68 42,73,99
	K69 41,2,173
	KL69 41,3,256
Schweizer, F.	I69 42,74,779
Sconzo, P.	BGa67 67,vol.38
	Gc67 43,290,163
	M68b42,73,217
	N69 63,32-1386
	B69 58,SN-97
Scott, D. W.	K67b76,15,741
	Fb67b76,15,1549
	Fb69b76,17,217
	K69b76,17,629
Sehnal, L.	Fb66a87,no.223
Seidelmann, P. K.	C68 73
	CK68b42,73,830
	CK69a42,74,776
Sejnalow, R. A.	FM66 49,10,537
Sekiguchi, N.	K67 80,19,596
Severne, G.	J68b51,3,289
Shapiro, I. I.	K67b42,72,338
	IK67b42,72,351
	N67 42,72,1309
Sharaf, S. G.	C66a89,11,89
	BC67a49,11,231
	CN68a57,182,291
	CN69a89,14,48
Sheffield, C.	M69 53,1,46
Sherman, N. W.	N68 51,3,61
Shi, Y. Y.	GcJ67a42,72,685
	I68a42,73,275
	GcJ68a42,73,653
	I69b42,74,551
Shimazu, Y.	N66 59,5,455
Shkodrov, V. G.	K69 45,46,446
Silva, R. R.	K66b42,71,902
Sinclair, A. T.	I69 68,142,289
Sjogren, W. L.	DFc67b84,155,74
	Fc68b84,160,875
Skripnichenko, W. I.	C67 49,11,441
Smalley, V. G.	N66 59,5,491
Smith, H. F. Jr.	D**b44
Smith, W. B.	K67b42,72,338
Sochilina, A. S.	Fa67 49,11,18
Soter, S.	K66b59,5,375
Sperling, H. J.	J68 59,9,305
	B69b42,74,490
Standish, E. M. Jr.	G66b42,71,982
	J68 51,3,135
Stanyukovich, K. P.	E66b45,43,1301
Stellmacher-Amilhat, I.	B66 51,1,215
Stiefel, E.	BGa67a69,N67-29387
	B68a52,267,950
	CM69a71,13,154
Stoyko, A.	N67 51,2,411
Stumpff, K.	J68a60,15,257
	GcI68b43,291,25
Sudbury, P. V.	C69 59,10,116
Sufiyanova, A. S.	A68 49,11,377
Sung, C.	Gc69 56
Szebehely, V.	Gc66 10,p.3
	Gc66 32,p.21
	Ga66a42,71,886
	GcG66, 28,p.150
	GcGa66 42,71,968
	Gb66 77,56,1640
	G67 23,p.173
	Gc67 42,72,7
	Gb67a42,72,184
	Ga67 42,72,370
	Gb67a42,72,373
	Gb67 77,58,60
	Ga67a38,5,1520
	Gb67a42,72,876
	Fd67b63,37-47v.4
	Gb67a42,72,1187
	E67b42,72,1324
	J68 51,3,33
	GbJ69 50,1,263
	Fd69b38,7,1010
	Gc**a42
	Gb** 13
	NGc** 53
Tapley, B. D.	Fd67a63,37-47v.4
	Fd69a38,7,1010
Terent'eva, A. K.	L67b45,44,614
Thiry, Y.	D67 52,264,1109
Thompson, I. H.	E68a68,139,499
Thüring, B.	N68 51,3,177
Timoshkova, E. I.	B68 49,11,465
Triplet, J. M.	N67 51,2,465
Tupper, B. O. J.	E68b68,138,67
Ulam, S.	J68 51,3,265
Van Flandern, T.	Gb67b42,72,373
Vashkovyak, S. N.	C69 91,no.1,p.87
Victor, E. L.	K68b76,16,195
Vinti, J. P.	B69 53,1,59
	Fc69 42,74,25
	B** 13
Volkov, M. S.	N67 49,11,157
	NB67 49,11,262
	K69a49,11,565
von Hoerner, S.	J68 51,3,147
Wagner, C. A.	K68 62,73,4661
	FI69a76,17,1505
Waldvogel, J.	Ga67 51,2,295

	BGa67b69,N67-29387		J68a68,141,277
Walter, H. G.	Fa67 42,72,994	Worrall, G.	G67 68,135,83
Weiss, E. H.	J68b60,15,257	Wright, J. P.	E66b87,no.236
Whipple, F. L.	L68b42,73,727		E68b87,no.283
Whitrow, G. J.	E68b68,139,499	Yabushita, S.	B66 68,133,247
Wielen, R.	J68 51,3,127	Zeinalov, R. A.	A68 45,45,1275
Wilkins, G. A.	C67 29,p.77	Zhongolovitch, L. D.	Fa66 49,10,509
Williams, C.	Ga66 42,71,976	Zielenbach, J. W.	K69 42,74,567
Williams, D.	K68a68,140,537	Zikides, M.	G69b88,May
Woolley, R.	B68a68,139,231		

8. POSITIONAL ASTRONOMY (ASTRONOMIE DE POSITION)

PRESIDENT: A. A. Nemiro.
VICE-PRESIDENT: W. Fricke.
ORGANIZING COMMITTEE: A. N. Adams, P. Lacroute, R. H. Stoy, R. H. Tucker, J. E. B. von der Heide, H. Yasuda, M. S. Zverev.

It is with deep regret that we mention the death, on 15 July 1969, of Prof. Dr G. Demetrescu who has done so much for the development of Astronomy in Roumania.

This report is based on letters received from members of the Commission, whom I wish to thank most cordially.

The most important events in the development of positional astronomy during the three past years were:

(1) The construction and use of new types of astronomical instruments based on new techniques, and

(2) The collective observations of large numbers of stars, especially in the Southern hemisphere on the basis of international cooperation.

I. THE CONSTRUCTION OF NEW INSTRUMENTS

At the USNO a design study for a ten-inch automatic meridian circle (ATC) was completed in 1968. Construction and installation should be completed by the end of 1970. A modified Cassegrainian optical system with a Cer-Vit primary mirror is being used. For some other important parts of the instrument new materials will also be used. At each end of the horizontal axis of the instrument Inductosyns will be mounted.

An Inductosyn system for digital readout of the pointing angle of the telescope was installed on the six-inch transit circle of the USNO in 1967 for testing and calibration. An IBM-1800 data acquisition and control system was on line with the same instrument.

At the Pulkovo Observatory Suharev's mirror transit circle was tested with photoelectrical observations in right ascension of bright FK 4 stars and faint stars of KSZ up to the magnitude 8·5. From April 1967 more than 2500 observations have been made. The results of these observations are promising. The mean error of one determination of n is $\pm 0\overset{s}{.}010$. The mean pointing error with photoelectrical micrometers equals $\pm 0\overset{''}{.}06$. The stability of the optical axis of the tubes is good; the variations of their azimuths are less than $0\overset{s}{.}003$ per degree of temperature. The investigation of the instrument and the observations with it are in progress.

The new axis of the mirror transit circle of the Porto Observatory has been installed. The strong dependence of the flexure upon the reaction of the pivots was removed.

Star observations in right ascension have started. Studies are being made for digitizing the circle readings.

On the Canadian mirror circle Dr E. G. Woolsey writes:

"The Ottawa mirror transit project has been abandoned owing to economy measures. The principal difficulties were:

1. A thermal instability of the piers supporting the central section evidenced by a large nightly drift in azimuth.

2. Poor seeing within the long horizontal optical paths, which have been only partially controlled.

To overcome these defects would require expenditure of time and resources which are not available. The instrument and building are on a temporary site and must be dismantled."

The transit circle of the Hamburg Observatory was equipped with a photoelectrical device for observations in both coordinates. The instrument has been installed in Perth (Western Australia).

Regular observations of the SRS and BSstars and the planets have been started intensively from the end of 1967. The Tokyo meridian circle has been equipped with a new digital micrometer during August 1969. It makes it possible to record observational data in computed readable form. All 2160 diameter corrections were determined by the Bruns-Fayet method with an accuracy of $\pm 0\overset{''}{.}05$ (m.e.)

An automatic editing and "marking off" system is being developed at the Royal Greenwich Observatory (Herstmonceux), so that the progress of an observing programme can be monitored by a computer.

The first results Obtained with the tracking photoelectric micrometer of the meridian circle of the Bordeaux Observatory have been confirmed: the m.e. of a single observation is ± 0.012 for equatorial stars as faint as $9\overset{m}{.}0$. This error is independent of magnitude and spectral type.

Since May 1968 the photoelectric micrometer has been in use on the meridian circle of the Liège Observatory at the experimental station of Dorres (near Font-Romeu, Eastern Pyrenees). Unfortunately, an instability in the azimuth of the meridian circle prevented one obtaining the accuracy expected in this mountain site (height +1677 m).

At the Bordeaux Observatory a new tracking photoelectric micrometer for the simultaneous determination of right ascension and declination is under development.

The micrometer of the meridian circle of the Besançon Observatory is under reconstruction. Photocomputing techniques with a Høg's grating will be introduced.

A new measuring-machine for the declination film is constructed at the Lund observatory. It is hoped to carry out a total determination of the division errors of the new circle in the immediate future.

At the Golosseyevo Observatory a similar photoelectric semi-automatic machine was built. The construction of a new type device for photographing the vertical circle divisions is nearing completion.

An automatic photoelectric machine for measuring the films with circle divisions of the meridian circle of the Odessa Observatory was built.

An analogous machine but of different construction was built at Pulkovo for measuring the films of circle divisions of the meridian circle of the Santiago Observatory (Chile). This machine is already used for reductions of SRS and BS star observations made in Chile in previous years.

A special machine for measuring star tracks on the plates of the Pulkovo photographic vertical circle was also built. It is in the testing phase.

At the Observatory of the Dresden Technical University an astrometric application of the TV-technique was developed and partly realized. The first results with the new circumzenithal (90, 1350 mm) telescope are promising. A new type of pendulum level with electronic equipment was built at the same observatory. The mean error of one level-reading is about $\pm 0\overset{''}{.}06$ to $\pm 0\overset{''}{.}09$.

A proposal has been made by Dr D. V. Thomas (Royal Greenwich Observatory) for a reflecting astrolabe, using the aluminized surfaces of a Cer-Vit prism to ensure a stable altitude of observation. An experimental modification of the Danjon astrolabe is envisaged after its return from the Cape.

II. INTERNATIONAL REFERENCE STAR PROGRAMME

The President of the AGK3R and SRS Working Groups Dr F. P. Scott reports:

A. *AGK3R programme*

In accordance with Resolution 5 adopted by Commission 8 at the Berkeley Meeting, all individual observations of AGK3R stars and associated fundamental stars collected at the U.S. Naval Observatory during the programme were sent to the Pulkovo Observatory.

Copies of the catalogue of final positions of 21499 AGK3 stars in machine readable form were sent to the Bergedorf, Pulkovo and Strasbourg Observatories. Copies of this catalogue on magnetic tape may now be obtained on a tape exchange basis. Requests should be addressed to the U.S. Naval Observatory, Washington, D.C. 20390, U.S.A.

It is hoped that proper motions for the majority of the AGK3R stars may be derived in the not too distant future. A pilot proper motion programme by T. E. Corbin, U. S. Naval Observatory, led to proper motions with mean errors of $\pm 0\overset{''}{.}35$/century for 2679 AGK3R stars with favorable observational histories in the region $+0°$ to $+30°$. The principal difficulty at the moment is the lack of a sufficient number of earlier catalogues in machine readable form.

About the state of AGK 3 as of December 1, 1969 Prof. W. Dieckvoss and Dr. H. Kox report:

"Least squares solutions are finished for the 1939 pairs of plates with reference positions from AGK 2 A (reduced to FK 4) and the final AGK 3 R using proper motions from AGK 3 preliminary for reducing to individual epochs of plates. The zones from the pole down to $+36°$ declination of AGK 2 (revised) and AGK 3 are on magnetic tapes and on punched cards. Duplicates of the tapes have been sent to the U.S. Naval Observatory Washington and to the Royal Greenwich Observatory. The remaining zones are in preparation. Remeasuring of stars found to be of inferior quality during the first calculations as well as editing magnitudes, spectral types and BD-numbers as auxiliary data is in progress.

Data from individual plates, zones from the pole down to $+35°$ so far, were sent to P. Lacroute of Strasbourg, France. Prof. Lacroute uses these data for extensive calculations by modern overlap-methods, including terms depending on magnitude, and applying individual weights to different plates."

Harley Wood of Sydney, Australia, is continuing his photographic work of covering the whole southern hemisphere by a modern wide angle objective aiming at a fourfold overlap. Four zones of 5° width are finished, three zones are in progress.

Dr. Dorrit Hoffleit of Yale University Observatory reported progress on the Yale Zone Catalog work.

"Zone −30° to −35°	Published, *Trans. astr. Obs. Yale Univ.*, **28**, 1967.
−35 to −40	Published, *Trans. astr. Obs. Yale Univ.*, **29**, 1968.
−40 to −50	In preparation for publication
−60 to −70	Measurements completed at Yale have been delivered to the U.S. Army Map Service for reduction
−70 to −90	Measurements nearing completion. Preliminary reductions in progress at Yale. Final reductions will be carried out at the Army Map Service under the guidance of Prof. H. K. Eichhorn von Wurmb.

For the future we are contemplating partially remeasuring the plates on the −40° to −50° zone and adding all of the stars in the *Cape Catalogue of Faint Stars* which were not included in the *Zone Catalogue* now being prepared for printing. In addition we plan to measure all possible stars in both of the *Cape Catalogues* of this zone together with selected galaxies on the first epoch plates that are being taken at the Yale-Columbia Southern Station at El Leoncito. (The older Yale plates, *ca.* 1940, do not show many galaxies.)"

B. *SRS programme*

Dr. F.P. Scott writes: "Table 1 shows the state of the SRS Programme on 1 July 1969.

It is a pleasure to note that the Felix Aguilar Observatory of the National University of Cuyo, San Juan, Argentina commenced its commitment of 2 observations of each of 7190 stars in the zone −40° to −90° during June 1969.

The Abbadia commitment has been completed and the results of observation were sent to the U.S. Naval Observatory during October 1968. Approximately 70% of the Bordeaux results arrived at the U.S. Naval Observatory during July 1967. The remainder is expected during the early months of 1970. The results for the Cape zone −40° to −52° were published in the *Third Cape Catalogue for 1950·0*.

The U.S. Naval Observatory seven-inch transit circle went into operation at El Leoncito, Argentina during May 1967. A few months were spent in testing the performance of the instrument and

Table 1. State of SRS Programme – 1 July 1969

Observatory	Zone	Commitment Stars	Commitment Obs'ns	Date Started	Completed Obs'ns	Completed Red'ns
Abbadia	+ 5° to − 15°	1560	4	62·3	100%	100%
Bordeaux	+ 5 to − 15	1560	4	62·5	89	100
Bucharest	+ 5 to − 10	1176	4	62·6	100	45
Nicolaiev	0 to − 20	5984	2	64·3	100	90
San Fernando	− 10 to − 30	3709	4	63·3	80	83
Tokyo	− 10 to − 30	3560	4	63·3	100	80
USNO 6-in	+ 5 to − 30	{ 8706 1233 }	{ 2 4 }	66·5	60	60
Cape	{ − 30 to − 40 − 40 to − 50 − 50 to − 90 }	10082	4	61·3	{ 90 100 0 }	{ 90 100 0 }
Santiago-Pulkovo	{ − 25 to − 47 − 47 to − 90 }	11496	4	63·1	{ 90 100 }	{ 35 50 }
Bergedorf (Bickley)	+ 5 to − 90	20495	4	67·9	43	70
USNO 7-in (El Leoncito)	{ + 5 to − 20 − 20 to − 75 − 75 to − 75SP }	{ 7683 12121 1382 }	{ 2 4 4 }	68·7	25	70
San Juan	− 40 to − 90	7190	2	69·5	2	2

in training observers before undertaking work on the SRS Programme. In addition to differential zone observations of the reference stars and fundamental stars, special all-night tours of observations are taken about once a week in which all FK 4 and FK 4 Supp stars between +30° and −70° S.P. are included. The latter series of observations are intended for improvement of the fundamental system. The programme at El Leoncito also includes the Mount Stromlo and La Plata PZT star list, 156 stars in the Scorpio-Centaurus association, all close double stars whose quota of observations had not been completed while the instrument was in Washington, and the four brightest minor planets."

Prof. Slaucitajs reports:

"The meridian circle loaned to La Plata by the Lick Observatory has not been installed in La Plata as yet. This instrument is intended to make the observations corresponding to the SRS Programme. Modernization of the instrument (change of pivots and supports, and overhauling of the eye-piece micrometer) is being done by Carl Zeiss Oberkochen, who have promised to finish the work towards the end of next November. In the meantime the existing pavilion will be reconditioned (slit enlargement) and the rebuilding and modernization of other parts of the instrument will be finished here at the Observatory's shops. The above-mentioned observations with this meridian-circle will start in 1970."

Dr. F. P. Scott concludes his report with the words:

"It is difficult at this time to predict the date of completion of the SRS Programme. However, from the information supplied by the observers, all commitments should be completed by 1973.

The U.S. Naval Observatory has made over 196000 apparent place, 119000 refraction, and 103000 m, n, and c computations for the SRS Programme."

Simultaneously with the SRS Programme five major and four bright minor planets were observed at Bickley with the meridian circle of the Bergedorf Observatory. The visual magnitudes of all stars were determined with the photoelectrical device.

With the SRS Programme are connected the works on the photographic catalogues.

S. V. M. Clube reported from the Cape Observatory:

"*Photographic Survey of the Southern Hemisphere*

This report describes progress as of September 1969 on the programme outlined briefly in *Highlights of Astronomy* IAU (1967).

1. *Observations with the Cape camera*

Coverage of the zone $0^{\circ} > \delta > -52^{\circ}$ is virtually complete (4000 plates) and that of $-52^{\circ} > \delta > -90^{\circ}$ well advanced. Arrangements have been made to obtain a series of sub-polar overlapping plates in order to examine the local refraction constant.

2. *Measurements*

All stars in the SRS programme have been measured on the Cape Zone plates (approximately 700 plates in the region $-40^{\circ} > \delta > -52^{\circ}$), in order to test analytical procedures.

3. *Analysis*

A reduction procedure using the Herstmonceux ICL 1909 has been developed for these measures, and has so far revealed a radially symmetric distortion pattern, partly dependent on magnitude, which is produced by the objective, and a differential equation between the photographic and meridian positions, mostly in right ascension, of unexplained origin. Eliminating these systematic effects, relative positions of individual stars on the overlapped photographic system are found to be good to at least $0''05$ (s.e.). The material used by Hough in producing the *Cape Zone Catalogue for 1900* (Gill and Hough, HMSO 1923) has now been punched ($\approx$ 30000 cards) at Herstmonceux, and is being analysed by the procedure used on the modern plates in order to test the production of a photographic proper motion system. Relative proper motions promise to be as good as $0''002$ p.a. (s.e.) over large areas of the sky."

The sixth and final volume of the *Cape Photographic Catalogue for 1950* has been published in 1968 (*Cape Ann.*, vol. XXII). It covers the zone between -80° and -90° for an epoch which is approximately 1954.

Dr. K. Aa. Strand reports from the U.S. Naval Observatory:

"Work on an astrographic catalogue is under way at the U.S. Naval Observatory. The catalogue will cover the sky south of declination $+5^{\circ}$ and will be nearly complete to visual magnitude 9. Blue and yellow plates will be taken with a new twin astrograph with 2 meter focal length, $f/10$ cameras (scale: 100″ arc/mm). An automatic engine will be used to measure positions and approximate B and V magnitudes. Both the astrograph and the measuring engine are currently under construction. Dr J. F. Wanner is in charge of this programme."

C. *BS programme*

The following observatories are observing the Bright Stars in their SRS zones: Abbadia, Bordeaux, Bucharest, Nicolahev, Tokyo, Cape, Bergedorf (Bickley) and Santiago-Pulkovo. In addition, Bucharest is observing 586 Bright Stars in the zone $+5^{\circ}$ to $+20^{\circ}$ and the USNO 6-in. all Bright Stars from -30° to $+90^{\circ}$. The state of completion of the observations of the Bright Stars at each observatory is approximately the same as it is for the Southern Reference Stars. Observations of Bright Stars in the zone -10° $+90^{\circ}$ have been finished at Kiev University Observatory. At the Tashkent Observatory the Bright Stars in the declination zone -20° to $+90^{\circ}$ were observed in right ascension only. Nine zone catalogues have been published and one is in press. The Bright Stars in the zone -15° to $+90^{\circ}$ are also being observed (right ascensions only) at the Kharkov University Observatory. Thus, as a result of a generous response to a Resolution adopted by Commission 8 at the Moscow Meeting, the Bright Star list is being well observed at this epoch.

D. *DS programme*

The list of double stars which might be troublesome to measure on photographic plates (Draft

Report Commission 8' Prague Meeting) is being observed as a whole from $-25°$ to $-90°$ by the Russian and Chilean observers at Santiago. Bucarest is observing 680 double stars, of which 360 are from the above list, in the zone 0° to +20°.

Unless other observatories have these stars under observation, there is great danger that they may not be adequately observed at this epoch.

The *Double Star List* may still be obtained from the U.S. Naval Observatory.

E. *KSZ programme*

At Odessa the meridian circle is being used for differential observations in the FK 4 system of 645 FKSZ stars in the zone $-30°$ to $+90°$. The catalogue will be published at the end of 1970. Before these observations the instrument was tested and partially modernized (photographing of the δ micrometer readings, motor-driven travelling thread). The catalogues of KSZ stars (declinations only) in the FK 4 system have been compiled at this observatory (973 stars in the zone $-10°$ to $-5°$ and 422 stars in the zone $-9°$ to $+3°$ of declination); the catalogues will be published in 1970.

At the Bucharest Observatory 61 selected KSZ stars surrounding galaxies were observed simultaneously with the observations of the BS and DS stars. About 90 % of the observations are accomplished. The *Bucharest FKSZ catalogue* (643 stars in the zone $-30°$ to 90°) published in 1957 was revised and improved in the FK 4 system. The results obtained by M. Tudor and E. Toma were published in *Studii Cerc. Astr.*, **13**, no. 1, 1968. The catalogue of 3913 KSZ stars in the zone $-10°$ to $+10°$ was also improved in the FK 4 system. The manuscript is in preparation for print.

At the Engelhardt Observatory the same improvement was made for the catalogue of 537 KSZ stars, observed at this observatory. The main results will be published in *Trudy 18 astrometr. conf. SSSR.*"

In Moscow (Sternberg Institute) a catalogue of KSZ stars in the zone $+30°$ to $+45°$ was published by A. P. Guliaev, Y. A. Korobova and M. P. Homiakova (*Trudy gos. astr. Inst. Sternberga* v. XXXIX). A catalogue of KSZ stars in the zone $+60°$ to $+90°$ is in press.

The Moscow catalogues of FKSZ stars 1941 and 1956 were improved and reduced to the FK 4 system by L. M. Hommik using an original method worked out by the author (*Trudy gos. astr. Inst. Sternberga*, v. XXXIX, *Soobšč. gos. astr. Inst. P. K. Sternberga*, no. 156).

Observations of the FKSZ stars made in 1948–54 with the meridian circle of the Kiev (University) Observatory were re-reduced in the FK 4 system. Two catalogues were published (N. A. Tchernega, *Vest. Kievskogo Univ., ser. astronomii*, no. 10, 1968 and no. 12, in press).

Reductions for FKSZ star declinations taking also into account the relations between the "Instrumental System and Observer" and "Fundamental System" deduced from the "Küstner Series" were finished at the La Plata observatory.

7500 positions of selected minor planets and the data for reference stars have been received at the Institute for Theoretical Astronomy (Leningrad) from 18 observatories of different countries (from 1949).

Thus the programme of observations of selected minor planets is nearing completion.

For determining the ΔA and $\Delta\delta$ it is necessary to reduce the received Yale positions to a fundamental system, which evidently will be possible when the AGK3 and Southern photographical catalogues referred to the SRS positions are available.

At the Pulkovo Observatory determinations of absolute proper motions of stars relative to 250 galaxies in 85 areas continued. The preliminary results were reported at the 13th IAU General Assembly in Prague. An additional investigation was made of the magnitude equation. After taking the corrections into account a change was found in the previously given systematic corrections of the proper motions of AGK3R stars; in particular there is a considerable decrease in $(\Delta\mu_\delta)_\delta$.

A meniscus astrometric astrograph with a field of 25 sq. deg. was mounted on Mount El Roble (near Santiago, Chile) by the common efforts of Chilean and Soviet astronomers. A start was made on photographing the first epoch plates of 160 areas with galaxies in the declination zone $-90°$ to 0°.

At the Astronomical Institute of the Academy of Sciences of Uzbekistan (Tashkent Astronomical Observatory) the determination of absolute proper motions of stars relative to galaxies was continued according to the Pulkovo plan (Prof. Deutsch's plan). Plates taken more than 20 years ago with the Tashkent normal astrograph were used.

At the Sternberg Astronomical Institute absolute proper motions of 670 stars relative to the galaxies were determined. Plates taken with the 15-inch astrograph were used. Photographing of the first epoch plates of the regions containing selected FKSZ stars with the 15-inch astrograph of the Sternberg Institute has been completed.

The photographic observations (first epoch) according to the Pulkovo programme (galaxies) has been accomplished at the Golosseyevo observatory.

The astrographic observations of the galaxies (first epoch) with the Carte du Ciel Astrograph in the declination zone $-5°$ to $-25°$ of the Pulkovo programme has also been completed at the San-Fernando (Cadiz) Observatory. A catalogue will be published in the near future. Photographic observations in the zone $-3°$ to $-9°$ for a second San Fernando catalogue are scheduled to begin in 1970.

F. *PZT and latitude star programmes*

The Askania Transit Circle of the Belgrade Observatory is being used for differential observations in the FK 4 system of 3941 stars from the zenith-telescope programmes. More than eleven thousand declination observations have been made. The observational programme will be finished in 1970. The division errors were investigated using three methods (Heufelink, Bruns and a new method worked out by L. Nikolič – a collaborator of the Belgrade Observatory).

At the Sternberg Institute (Moscow) the zenith-telescope ZTL-180 is used for observations to obtain differential declinations of stars from all latitude programmes. The method used consists of micrometric measurements in narrow zones of 1°30. The reference catalogues used are FK 4 and AGK3R.

Observations in the FK 4 system (δ only) with the new meridian circle of the Sternberg Institute of latitude and PZT stars, and observations in R. A. of the stars from the PZT programmes are being continued.

At the Odessa Observatory a catalogue of declinations of 194 stars from the Pulkovo ZTL-180 programme was compiled. The catalogue will be published in *Trudy glavn. astr. Obs. Pulkovo.*

A catalogue of declinations of the 192 stars from the new Moscow programme for the determination of the micrometer scale (wide pairs) was compiled at the same Observatory. The catalogue is in the press.

Corrections of the proper motions of the Washington zenith catalogue of scalar star pairs were determined at the Milan Observatory (Proverbio, *Contr. Oss. astr. Milano-Merate*, no. 317). At the Uccle Observatory the declinations and proper motions of all the stars from the ILS programme were deduced from discussions of 11501 positions from 463 star catalogues (P. Melchior et R. Dejaiffe, *Ann. Obs. r. Belgique, 3rd series*, vol. X, no. 3, 1969).

The Askania transit circle at Uccle was equipped with a device for photographing micrometer readings. The observations of 1102 stars (all latitude stars included) were begun at the end of 1968.

Proper motions for 22 stars in the Herstmonceux-Calgary zenith zone were provided for PZT work at the Royal Greenwich Observatory.

Observations of 3941 stars in the latitude and PZT programmes have been continued with the vertical circle at Pulkovo. R.A. of PZT stars were observed with the meridian circle.

At the Engelhardt Observatory observations of these stars have been finished in the "clamp west" position of the meridian circle. Observations in the other position of the instrument have been begun.

G. *Astrolabe programme*

A general card-index of the star positions deduced from astrolabe observations made at different places was compiled at the Paris Observatory as preliminary work for compiling a general catalogue of stars observed with astrolabes.

The astrolabe observations made at Paris in 1956–69 have been studied preliminary with a view to use them for proper-motion determinations.

Observations made with the astrolabe in Haute-Provence for the improvement of FK 4 positions have been finished in December 1968. They are under dicussion.

Observations of the planets with astrolabes of some observatories have been organized by Dr Débarbat. She worked out a strict method for correcting these observations for the phase effect. (*Bull. astr. ser. 3*, **2**, no. 4, 569, 1967). Results of the Moon observations received from instruments permitting observations at constant altitudes (astrolabes; circumzenithals) made it clear that such observations will be very useful and must be organized. (*Bull. Soc. r. Sci. Liège*, **38**, No. 5–6, 205, 1969.)

Dr T. Okuda reports:

"At Mizusawa observations with the Danjon astrolabe have been made regularly since January 1966. The results of the observations have been published annually in the Bulletins, Time Service of Mizusawa Observatory. Provisional individual corrections $\Delta\alpha$ to 117 FK 4 and 24 FK 4 Sup stars and $\Delta\delta$ to 72 FK 4 and 7 FK 4 Sup stars were computed, all of which are included in the declination zone between $+9°$ and $+69°$. The mean errors $\varepsilon_{\Delta\alpha}$ and $\varepsilon_{\Delta\delta}$ are comparatively large, and affected by those of group corrections" (*Publ. int. Latit. Obs. Mizusawa*, **6**, no. 2, 185).

According to Dr D. V. Thomas more than 36000 transits were observed at the Cape Observatory with the Danjon astrolabe transferred from Herstmonceux. The observations were obtained during four years of operation of this programme, which was completed in April, 1969. A detailed analysis of the observations has now begun. Improved positions should eventually result for about 1000 stars brighter than 5.6 in the declination zone $-4°$ to $-64°$.

It has been shown that absolute declinations may be obtained in principle from observations made with a series of Danjon astrolabes suitably distributed in latitude or with a single reflecting instrument observing at two alternative altitudes (Herstmonceux Conference, 1969, *Mon. Notes astr. Soc. Sth. Afr.* **26**, 2).

Dr L. Quijano reports from San Fernando (Cadiz):

"In April 1968 observations of 11 fundamental groups of stars were started mainly for the study of time and latitude. A selection of 250 FK 4 and FK 4 Supplement stars was made from these groups, 58 of which are observed in both E and W transits. From April 1969 on, small groups of catalogue stars were included in the observations. A rate of some 200 observations of fundamental groups per year is foreseen in which about 3000 transits can be inserted. The reductions are made in the Paris Observatory."

25000 transits have been observed at Pulkovo with the Danjon astrolabe during the last 7 years. Six annual cycles of observations have been discussed. Provisional corrections $\Delta\alpha$ and $\Delta\delta$ of about 300 stars in the declination zone $+30°$ to $+80°$ were thus obtained.

At Kriukovo (near Moscow), in Irkutsk and in Novosibirsk observations for time determinations were made. At Poltava the observations were produced for latitude determinations. All these observations may be used for the determination of star coordinate corrections.

III. OBSERVATORIES AND INSTITUTES PROGRAMMES AND INVESTIGATIONS

Improvement of the Fundamental Catalogue and revision of the Albany General Catalogue

Prof. W. Fricke reports from the Astronomisches Rechen-Institut at Heidelberg:

The work reported here is part of a current project of the Astronomisches Rechen-Institut, Heidelberg, to prepare future improvements of FK 4 and to revise the *Albany General Catalogue*.

Under the supervision of Lederle the documentation of star catalogues on punched cards has gone on steadily; these data are made available, according to the conception as started by IAU Working group on Space Ephemerides. Nowacki and Strobel (Heidelberg, 1968) have published the systematic differences between 71 observational catalogues and FK 3; these data formed part of the work on the construction of FK 4. By means of these systematic differences and the differences FK 4–FK 3 published in FK 4 any user may reduce the observational catalogues to the system of FK 4.

Work on new methods for deriving systematic differences between star catalogues was continued by Brosche. The differences are expanded in spherical harmonics, and all significant terms of the series development are taken into account. The number of terms in the series development is limited by the choice of $m=12$ and $n=22$ as the highest suffixes of the spherical harmonics. By means of the covariance matrix of the coefficients the mean error of the systematic differences is determined as a function of the position in the sky. The programme compiled for machine computation of systematic differences between FK 4 and important observational catalogues has the following advantages: optimal separation of systematic and individual parts of differences; consideration of the two-dimensional character of differences; the spectrum of coefficients forms a convenient basis for investigating sources of errors.

The determination of the correction to Newcomb's lunisolar precession has been the object of further investigations by Fricke: A redetermination on the basis of Newcomb's data has shown that Newcomb's error mainly originated from Auwers' declination system of proper motions of Bradley's stars; a rediscussion of equinox determinations has sufficiently cleared up the origin of the error in the motion of Newcomb's equinox; there is no variation of the obliquity of the ecliptic significantly different from Newcomb's value."

At the Pulkovo time service department a general catalogue of right ascensions of 807 stars to the 7th magnitude in the declination zone $-30°$ to $+90°$ was compiled by N. N. Pavlyv, P. M. Afanacieva and S. S. Staritsyn. The catalogue is based upon 185371 observations, made during the IGY and IYC by the 9 Soviet time services. The mean epoch of the catalogue is close to 1958·0. The mean errors of the catalogue positions are: $\pm 0^{s}.0025\ (-10° < \delta < +18°)$, $\pm 0{\cdot}0022\ (20° < \delta < 40°)$ and $\pm 0{\cdot}037\ (55° < \delta < 76°)$. The proper motions in α of 807 stars were obtained from a comparison of their right ascensions as given in the compiled general catalogue and those in FK 4, FK 3GC and N 30. Systematically these proper motions are completely independent from those of the fundamental systems.

A catalogue of α and δ in the FK 4 system of 972 stars in the visual magnitude range from 3·0 to 11·0 and in the declination interval from $-8°$ to $+65°$ based upon observations made at the Brørfelde (Copenhagen University) Observatory from August 1964 to January 1967 was published (*Publ. mind. Medd. Kbh. Obs.*, 197, 1968). The catalogue includes stars from 6 lists: FK 4 Supplement stars; O and B types stars nearer than 20 parsec, high velocity stars, brighter G and K stars in the photometric classification programme in progress at the Brørfelde and Lowell observatories, and variable and double stars selected by G. van Herk.

The automated photographic micrometer described by Lausten in 1967 was used. Most stars were observed four times, but all stars in the programme which were observed at least two times in right ascension and declination were included in the catalogue. The following mean errors of a single observation in right ascension and declination were found:

$$\varepsilon_\alpha \cos\delta = \pm\, 0.0152 \pm 0^{s}.0002; \qquad \varepsilon_\delta = \pm\, 0''.219 \pm 0''.003.$$

At the Golosseyevo Observatory (near Kiev) a catalogue of absolute declinations of 1181 bright (FK 3) stars in the zone $-31°$ to $+90°$ and 611 faint (FKSZ) stars in the zone $-26°$ to $+90°$ has been published (A. K. Korol, *Declinations of Bright and Faint Stars in a Common System*, Naukova Dumka, Kiev, 1969 (in Russian)). The catalogue is based on the observations (more than 12000) made with the Wanschaff vertical circle (D 19 cm, F 252 cm) in 1954–62. The instrument was carefully investigated, especially its flexure. The observations of stars were reduced with the values of flexure determined during the same nights. The mean error of one observation is represented by the formula

$$\delta(z)^2 = (0''.320)^2 + (0''.165\ \mathrm{tg}\, z)^2$$

The catalogue is close to the fundamental systems FK 4 and N 30 (a little nearer to the latter).

The results of observations made with the six-inch transit circle of the U.S. Naval Observatory between 1956–62 have been published (Adams, A. N. and Scott, D. K., *U.S. Naval Observatory Publ.*, 19, part 2, 1968). The six-inch transit circle observations of the Moon made in Washington in 1925–68 have been reduced with the revised system of astronomical constants adopted by the IAU in 1964 and adjusted to the FK 4 system. The observed positions were compared with the Improved Lunar Ephemeris. The results are published (*U.S. Naval Obs. Publ.*, 19 part 3). An analysis of these results by B. L. Klock and D. K. Scott for equator and equinox corrections and for corrections to the lunar orbital elements is published in the Astronomical Journal.

The current six-inch transit circle USNO programme of fundamental observations of the Sun, Moon, planets, four bright asteroids, FK 4 and FK 4 Supplement stars and "Bright Stars" north of −30° declination will be completed in 1970.

A catalogue of R.A., Declination and proper motions of 520 circumpolar stars in the declination zone from +84° to 90° was published by Telniuk-Adamchuk (*Astr. Obs. Kievsk. Univ.*, Kiev, 1969). The positions of stars are based on the observations made in 1876–82 by Fabritius with the Repsold meridian circle of the Kiev Univ. Obs.; the proper motions were determined using 9 catalogues of circumpolar stars from the catalogue of Carrington up to AGK3.

At the Sternberg Astron. Inst. (Moscow) a general catalogue of R. A. of circumpolar stars is being compiled by A. P. Guliaev. Observations made at several Soviet observatories in previous years have been used.

Reductions of the observations of circumpolar stars made in 1909–14 by L. O. Struve, N. N. Evdokimov and B. Y. Kudrevich with the meridian circle of the Kharkov Observatory are in progress. The first part of the work – the compilation of a catalogue of declinations of 1047 circumpolar stars – will be accomplished in the near future. Reductions of the semi-absolute determinations of the declinations of 270 stars made in 1924–33 by N. N. Evdokimov and B. P. Ostachenko-Kudriavcev are also in progress.

The first Greenwich catalogue for 1950, based on observations made with the Airy Transit Circle 1931–38 has been published (Woolley *et al.*, *Roy. Greenwich Obs.* Annals, No. 3, 1969). The contents of the catalogue were made available many years before publication and have been incorporated in the N 30 and FK 4 catalogues.

At Herstmonceux the programme of observing stars with good radial velocities and poor proper motions has been concluded. Lists of selected variable stars have been observed in various differential and fundamental programmes. Groups of reference stars in the vicinity of quasi-stellar sources were observed differentially with groups of FK 4 stars in order to provide a systematic framework for photographic positions of faint optical objects identified with the sources.

Proper motions for 182 semi-regular and RV Tauri variables have been published with an account of the methods employed (*Roy. Obs. Bull.*, 142). Proper motions of 19 Praesepe stars have also been determined.

Observations of absolute declinations of FK 4 stars, major planets, and of the four brightest minor planets are being continued with the vertical circle of the Munich Observatory in Munich.

The compilation of a catalogue of positions (δ and α) and proper motions of 130 eclipsing variable stars based on the observations with the meridian circle at the Odessa Observatory is almost completed. The catalogue will be published in 1970.

Observations of 9537 zodiacal stars up to the visual magnitude 8·0 (in β limits from −15° to +15°) were started by I. E. Gordon with the Repsold meridian circle of the Nikolaev Observatory. It would be of great importance that other observatories join this work.

An investigation of the systematic errors of some AG catalogues in equatorial zones was made by L. F. Gorel using the results of observations of AGK3R and KSZ stars made at the Nikolaev Observatory in previous years. She has compiled a catalogue of proper motions of 12590 AGK3R and KSZ stars and determined the correction to the precessional constant.

At the Besançon Observatory the meridian observations of the stars, the occultations of which had been observed at the observatory, were started. In the computing department of this observatory the proper motions of the 3rd catalogue of Besançon (List of Blaauw stars) were prepared for publication. The differential positions of the fundamental stars from the 2nd catalogue of Besançon were also prepared for print.

Regular observations of the Sun and major planets (especially Mars and Venus) and fundamental stars were continued also at many observatories (Bickley, Herstmonceux, Kazan (Engelhardt) Kharkov, Kiev (University), Milan, Munich, Moscow (Sternberg Institute), Nikolaev, Pulkovo, Tashkent, Tokyo and Washington). At some observatories the four brightest minor planets were observed too (Bickley, Herstmonceux, Milan, Munich, Tokyo and Washington).

I was also informed that photographic observations of the major and bright minor planets were made at Nikolaev, Golosseyevo, Moscow-Sternberg Institute and at Pulkovo.

The results of observations of bodies of the Solar System are partly published (*Vestnik Kievsk. Univ. Ser. Astronom.*, nos. 11 and 12, 1969 (in Ukrainian with a Russian résumé); *Circ. astr. Obs. Milan*, nos. 23 and 25, 1967; *Ann. Tokyo astr. Obs.* **5**, 1967; *Trudy Pulkovo Obs.*, **77**, 1969, *Izv. glavn. astr. Obs. Pulkovo*, no. 183, 1968; *Bjull. Inst. teor. astr.* (*Leningrad*) **5** (128)).

The *Second Cape Catalogue* for 1950·0 including all Sun, Moon and Planet observations made at the Cape between 1935 and 1960 was published in 1969.

A discussion of the absolute observations in δ made at the Melbourne Observatory in 1928–41 has been completed at Pulkovo. The corresponding catalogues of absolute declinations of 2466 stars have been published: (K. N. Tavastsherna, *The catalogue of Declinations of 2249 Backlund-Hough List Stars from +35 to −90° of Declination, observed at the Melbourne observatory during 1928–41.* Me_{50} I(δ), and K. N. Tavastsherna and N. A. Schacht – three small declination catalogues Me_{50} II(δ) Me_{50} III(δ) and Me_{50} IV(δ). *Edition of the Pulkovo Observatory*, Leningrad, 1968).

The discussion of the corresponding observations in right ascension is continued.

Observations of absolute declinations of the Southern Hemisphere stars of the FK4 catalogue are being continued at the Southern Astrometric Station in La Leona.

Dr M. S. Zverev reports: "In connexion with the observations of SRS, BS, DS and FK 4 stars with the meridian circle of the Cerro Calan Observatory (Chile) corrections $\Delta\alpha_\delta$ for 1044 southern FK 4 stars were determined, using a 'quasi absolute method' (M. S. Zverev, *Russ. astr. J.* **46**, 6, 1969) on the base of about seven thousand observations (211 fundamental series). It was shown that the FK 4 system needed corrections of the form $\Delta\alpha_\delta$ in the equatorial region and especially in the declination zone close to −35° and −75° (C. Anguita, G. Carrasoo, D. D. Polojentsev, M. S. Zverev, *Astr. Cirk. U.S.S.R.* **506**, 1969; *Inf. Bull. Southern Hemisphere*, **15**, 1969).

Large systematic errors of the Cape catalogue of circumpolar stars (W. Gliese, *R. Obs. Bull.* **106**, London, 1966) in the Declination zone −80° to −83° have been revealed.

All these results were confirmed on the basis of observations of 839 bright southern stars made at Cerro Calan observatory with the Zeiss transit by P. R. Loyola and V. N. Shishkina (Publ. no. 4, Obs. Cerro Calan, Santiago de Chile, 1968, V. N. Shishkina, M. S. Zverev, *Inf. Bull. Southern Hemisphere*, **16**, 1970, in print).

Dr von der Heide reports from the Perth Observatory on the determination of $\Delta\alpha_\delta$ and $\Delta\delta_\delta$ corrections from differential observations of the SRS and BS stars. He remarks that "the $\Delta\alpha_\delta \cos\delta$ seems to have a maximum of $-0\overset{''}{.}35$ at $\delta = -65°$; the $\Delta\delta_\delta$ go up to about $-0\overset{''}{.}5$ at $\delta = -35°$."

S. Slaucitajs reports from the La Plata Observatory. "The possibility is being considered for the determination by means of 'Küstner Series' of systematic errors $\Delta\alpha_\alpha$ and $\Delta\alpha_\delta$ that have been observed in previous years as well as those that are still in observation."

At the Yale Observatory in collaboration with the Numerical Analysis Group, Section 312 (U.S.A.) the final solution for the Yale Minor Planets Project was obtained. 6912 observations of 15 minor planets were used. Corrections to the six orbital elements for each planet and 108 corrections of $\Delta\alpha_\kappa$ and $\Delta\delta_\kappa$ for 54 regions in the declination zone from −30° to +30° for the *Yale Zone Catalog* and 120 corrections for 60 regions for GC have been obtained (198 (210) corrections in total).

To solve the system of 6812 equations with 198 or with 210 unknowns (for GC), computer programmes and techniques developed by the Numerical Analysis Group for the IBM 7040-7094 and modified for use on the UNIVAC 1108 were used. This solution corrects the internal inconsistencies within the system of right ascensions and declinations and in addition it adjusts the location of the equator ($\Delta\delta_0 = +0\overset{''}{.}17$). It makes no adjustment to the location of the equinox along the equator $-\Delta A$. The results will be published in the Astronomical Papers of the American Ephemeris in 1970.

H. Yasuda (Tokyo Observatory) has found the periodic errors $\Delta\alpha_\alpha$ and $\Delta\delta_\alpha$ of the FK 4 system in the equatorial zone from meridian observations of the four brightest minor planets (*Ann. Tokyo astr. Obs.*, **5**, 1967). K. Hurukawa showed the same trend of $\Delta\delta_\alpha$ in the declination zone of $+30°$ from floating zenith tube observations at the Mizusawa Latitude Observatory and found a reliable coincidence among three independent determinations, the determination from minor planets, with a floating zenith tube and with the astrolabe at the Paris Observatory (*Pub. int. Latit. Obs. Mizusawa*, **6**, 1968).

Dr G. van Herk reports from the Leiden Observatory:

"I have continued my investigations related to possible sources of systematic errors in meridian circle catalogues from

Besançon, *Ann. Nouv. Série*, **6**, 1964.
Brørfelde, *Publ. Københavns Obs.*, no. 197, 1968.
München, *Veröff.*, **5**, no. 15, 1963.
Paris, *J. Obs.*, **42**, 94, 1959.
Washington, *Publ. U.S.N.O.*, **19**, part 1, 1964.

It turns out that the right ascensions agree better amongst each other than the declinations, in all respects. The declinations for Washington show some dependency on spectral type, derived from the published differences between upper and lower culmination, which is understood according to the lines of *BAN*, **11**, 489, 494, 1952. The differences in right ascension of various catalogues show in some cases some dependency on magnitude, that is the nominal magnitudes. As long as the true, observed, magnitudes are not published, this parameter is only inadequately known. The $\Delta\alpha$ for Washington and FK 4 show also a dependency on magnitude. I refer to *BAN*, **19**, 355, 1968."

"A careful discussion of vertical circle observations made at Pulkovo, Odessa, Munich and Canberra has proved the existence of an annual variation of the constant of refraction with an amplitude of about $0\overset{''}{.}1$" writes Dr F. Schmeidler from the Munich observatory.

From the Goethe Link Observatory (Indiana University) Dr R. d'E. Atkinson writes: "I am engaged on a re-evaluation of the results obtained with the Cookson Floating Zenith Telescope, 1911–36, to study the possibility (suggested by Jeffreys) that the leading terms in the nutations in longitudes and in obliquity may not be related in the manner required by a rigid earth. I hope to complete this work before the General Assembly."

New computing techniques have been used at many observatories for the reduction of observations. At some observatories new methods of reduction have been discussed. At the Golosseyevo Observatory the theory of the accidental sequences and fields has been used to study the distribution of the errors of observations with the Wanschaff vertical circle (Intern. Lat. star programmes). It was shown that the empirical distribution departs substantially from the normal one. Thus it is recommended to use weights in accordance with S. Newcomb's and G. Jeffreys' method.

D. P. Duma at this observatory has discussed problems connected with the determination of equator and equinox corrections from observations of major and minor planets and artificial satellites.

At the Sternberg Astron. Inst. (Moscow) a strict method for reductions of the differential observations to the fundamental system was worked out (L. M. Chommik).

A. P. Gulaiev from the same Institute has proved that it is necessary to investigate the systematical errors of the observations of the programme stars before using them for determinations of the parameters of the meridian circle.

A project for the determination of more definitive proper motions of the early type and Cepheid stars of Blaauw's list is being carried out under the supervision of R. W. Rhynsburger (USNO).

At the same observatory the project for placing star catalogues on punched cards has been continued by the *Nautical Almanac Office*.

Improved reconstructions and investigations of the meridian instruments and determinations of the instrumental and personal errors have been made at many observatories.

Information on work of this kind has been received from the observatories: Belgrade, Besançon, Brussels, Bucharest, Golosseyevo, Greenwich-Herstmonceux, Kharkov, Kazan-Engelhardt, Lund, Milan, Moscow-Sternberg Inst., Nikolaev-Odessa, Potsdam, Pulkovo.

Some work of this kind was already mentioned in previous parts of this report.

I would like only to give some details on the investigation of the division errors of the circles manufactured by Askania and Repsold made by P. Melchior using spectral analysis. The results will be published in the near future. The same method was used by K. G. Gnevysheva (Pulkovo) for investigation of some Askania circles.

A detailed investigation of the flexure of the Wanschaff vertical circle (Golosseyevo) was made by A. S. Kharin. It was shown that the flexure of this instrument is correlated with the temperature gradients (not with the temperature itself). The same method was used for the investigation of flexure at the vertical circle of the Belgrade Observatory (A. S. Kharin, G. K. Teleki).

A. I. Nefed'eva (Kazan-Engelhardt Observatory) has shown that systematical errors of the zenith distances measured with the meridian circle of this observatory remained unchanged during 120 years.

In the Vsesousny Nauchno-Issledovatelsky Institut Fisiko-technich. i Radiotechnich. Izmerenij (VNIIFRTRI (Moscow)) B. I. Vlasov proposed a method to improve fundamental catalogues by using space observations. The results were reported at the 18th Astrometrical Conference of the U.S.S.R. (Pulkovo 31 May–5 June 1969) and will be published in the *Trudy* of this Conference.

The second edition of the book by V. V. Podobed, *Fundamentalnaja Astrometria* (Fundamental Astrometry) was published in 1968.

IV. SUGGESTED TOPICS FOR DISCUSSIONS AT THE GENERAL ASSEMBLY

1. Recent progress in instruments and their accessories.
2. Progress with the existing mirror transit circles.
3. Results obtained with new types of meridian instruments.
4. The organization of absolute observations of stars at as many observatories as possible.
5. The international collaboration for the automation of reductions of observations.
6. The correction of the FK 4 for the southern hemisphere on the basis of modern meridian observations.
7. The reduction of differential observations to a fundamental system.
8. The discussion of the results of the SRS observations.
9. Theory of astronomical refraction based on new meteorological and aerological data.
10. The problem of site-testing for astrometry with Danjon astrolabes.
11. The refraction in pavilions.
12. The possibility to organize Symposia in 1971–72 on these problems: Cosmical and lunar astrometry; applicability of radio and laser astronomy to position astronomy; the project to refer motions to the system of the galaxies; the use of observations of bright minor planets in fundamental astronomy; distribution of data in machine-readable form and bringing existing data (e.g. large catalogues of positions and proper motions, BD or HD included) into machine-readable form by reasonable international cooperation.
13. The name of our commission: Dr Dieckvoss proposes "Basic Astrometry of Position".

A. A. NEMIRO
President of the Commission

9. INSTRUMENTS ET TECHNIQUES
(INSTRUMENTS AND TECHNIQUES)

PRÉSIDENT: J. Rösch.
VICE-PRÉSIDENT: V. B. Nikonov.
COMITÉ D'ORGANISATION: H. W. Babcock, W. A. Baum, R. G. Giovanelli, J. D. McGee, A. B. Meinel, O. A. Mel'nikov, B. Valniček.

INTRODUCTION

Le présent rapport ressemblera peu à ceux présentés aux précédentes Assemblées Générales de l'Union; d'abord en raison des prescriptions du Comité Exécutif concernant la limitation du texte, la suppression des références bibliographiques et la latitude laissée au Président dans la conception même du Rapport; ensuite parce qu'il est apparu que la Commission 9 avait un rôle et une position assez différents de ceux de la plupart des autres Commissions et que l'un des points auxquels il convenait de s'attacher était précisément le re-modelage de son activité.

Toutefois, un certain nombre de membres de l'Union ayant bien voulu résumer l'activité de leurs Institutions ou de leurs pays dans le domaine de l'instrumentation astronomique, leurs contributions trouveront place dans le présent Rapport et fourniront ainsi aux membres de l'Union une information non exhaustive mais cependant des plus utiles, en attendant que soit mise au point une nouvelle forme de documentation.

Enfin, l'un des groupes de travail de la Commission – sur les Tubes à images – ayant poursuivi son travail de coordination et d'information comme par le passé, un rapport spécial sur son activité figurera à la suite du Rapport Général de la Commission 9.

CONSIDÉRATIONS GÉNÉRALES SUR L'ÉVOLUTION DES INSTRUMENTS ET DES TECHNIQUES ASTRONOMIQUES

Avec les ouvertures vers les grandes longueurs d'onde par la Radioastronomie et vers les courtes longueurs d'onde par les moyens spatiaux, la connaissance de l'Univers a pris appui sur un domaine spectral de plusieurs dizaines d'octaves, alors qu'elle était limitée, il y a vingt-cinq ans, aux deux octaves s'étendant du proche ultraviolet au proche infra-rouge. Sur le plan des résultats scientifiques, on a cessé de borner ses ambitions aux objets dont le rayonnement thermique est maximum, ou au moins important, dans le domaine optique (encore est-il heureux qu'une bonne partie des étoiles soient dans ce cas!) et on a pu reconnaître des mécanismes d'émission beaucoup plus variés. Sur le plan des instruments et des méthodes, l'Astronomie a pu bénéficier de l'extraordinaire floraison des réussites technologiques les plus avancées – quand elle ne les a pas elle-même stimulées. Ce bénéfice n'est d'ailleurs pas limité aux domaines spectraux "nouveaux" et il suffit de citer, dans le classique domaine "optique" les diverses générations de récepteurs photoélectriques, ou l'apparition du laser qui, avec le radar – chacun dans son domaine – permet de parler maintenant d'une "astronomie active" inégalable dans la mesure précise des distances.

Parallèlement, l'abondance de l'information recueillie a imposé des méthodes nouvelles de traitement, que justement permettaient aussi les progrès généraux dans ce domaine, indépendamment de l'Astronomie. Par contre-coup, l'acquisition même de l'information est apparue comme limitative en certains cas, et un effort se dessine pour la rendre plus objective, plus automatique et plus rapide. Il n'est pas jusqu'à l'Astrométrie qui n'éprouve le besoin d'une révision et d'un perfectionnement de ses moyens.

Il est clair que la Commission 9 de l'UAI ne peut rester à l'écart de cette évolution, en limitant son objet aux télescopes classiques et aux récepteurs photoélectriques; ou, du moins, si des raisons pratiques obligent à une division des attributions, que ce soit en pleine conscience et avec tous les échanges souhaitables entre les parties intéressées.

Le problème revêt d'ailleurs des aspects différents selon les domaines en cause. Deux commissions, en effet, ont été créées à la suite des "ouvertures spectrales" mentionnées plus haut: celle de Radioastronomie (40) et celle des Observations Extra-Terrestres (44) qui ont chacune considéré les techniques, soit radioastronomiques, soit spatiales, comme relevant de leur domaine d'intérêt propre, sans d'ailleurs que la Commission 9, aux époques de leurs créations, ait soulevé le problème des attributions respectives. D'autres Commissions, par contre, plus anciennes, traitent de certains objets astronomiques par des moyens classiques qu'elles laissent aux spécialistes des instruments le soin d'étudier et de perfectionner. On peut citer comme telles, à des degrés divers, et qu'elles en soient conscientes ou non, les Commissions 8 (Astronomie de Position); 10 et 12 (Soleil); 14 et 29 (Spectroscopie); 15, 16 et 17 (Comètes, Planètes, Lune); 25, 27 et 42 (Photométrie et Variables); 30 (Vitesses Radiales); etc.... Il arrive que des membres de ces Commissions se plaignent de voir une partie de leurs séances, lors des Assemblées Générales, occupée par des discussions techniques qui, à leur sens, trouveraient mieux leur place dans les travaux de la Commission 9.

On peut donc concevoir que, d'une part, soient précisées les positions des Commissions 40 et 44 en face de la Commission 9 et, d'autre part, que soient envisagés des groupes de liaison avec celles des autres Commissions qui souhaiteraient l'aide de la Commission 9.

Pour ouvrir cette voie, et en même temps pour établir un bilan des domaines d'activité des membres de la Commission 9, un questionnaire leur a été envoyé.

ENQUÊTE AUPRÈS DES MEMBRES DE LA COMMISSION 9

Quatre-vingt-cinq formulaires ont été envoyés. Soixante-et-une réponses ont été reçues, soit une proportion de plus des deux tiers. Encore, sur les vingt-quatre membres qui n'ont pas répondu, s'en trouve-t-il une bonne douzaine dont l'activité dans le domaine de la Commission est assez notoire pour que leur silence ne soit pas dû à l'indifférence, et une dizaine seulement dont on peut penser que leur radiation passerait inaperçue. On doit certainement se féliciter d'un tel écho, et y voir la garantie d'un avenir fécond pour la Commission.

La première partie du questionnaire demandait une réponse sur les trois attitudes possibles ci-dessous; le nombre de membres se déclarant en faveur de chacun des points à l'exclusion des deux autres est indiqué entre parenthèses.

(1) *Ne pas* s'occuper des problèmes instrumentaux et techniques qui sont actuellement pris en charge par d'autres Commissions (3);

(2) S'occuper de *tous* les problèmes instrumentaux et techniques, y compris ceux ci-dessus (8);

(3) Entretenir avec les autres Commissions traitant de problèmes techniques des contacts étroits, par exemple en formant des Groupes de Travail communs (27).

Personne ne donne à la fois les points 1 et 2 (qui sont incompatibles). Mais cinq membres adoptent à la fois 1 et 3, et neuf, à la fois 2 et 3. Enfin, deux membres répondent affirmativement à 1, 2 et 3 ensemble. Ce qui prouve peut-être qu'en toute rigueur les questions n'étaient pas parfaitement bien posées. Sept membres n'ont pas répondu directement à ces questions. La conclusion la plus manifeste est qu'une large majorité (43 membres) est en faveur de groupes communs avec d'autres Commissions.

Par ailleurs, le Président de la Commission 40 pense que les membres de sa Commission ne désireront pas renoncer à traiter de leurs propres instruments mais seront favorables à des réunions mixtes avec la Commission 9 sur des sujets d'intérêt commun.

La deuxième partie du questionnaire demandait aux membres d'indiquer leurs domaines d'intérêt principal ou secondaire dans la liste suivante, où l'on a fait suivre chaque ligne du nombre de réponses affirmatives (sans détailler ici à l'intérieur des lignes):

(5) *Collecteurs optiques*: Mécanique, Optique (calculs), Optique (réalisation), Asservissements, Guidages (38);

(6) *Collecteurs radioélectriques* (en liaison avec la Commission 40): Mécanique, Aériens, Collecteurs Composites (4);

(7) *Collecteurs pour IR, Collecteurs pour UV et X* (en liaison avec la Commission 44) (8);

(8) *Analyseurs domaine optique*: Spectrographes, Filtres, Polarimètres, Analyse de Fourier (36);

(9) *Récepteurs* pour domaines X, UV, optique (photographie, photoélectricité, électronographie), Infra-rouge, radio; Amplificateurs divers (35);

(10) *Émetteurs*, techniques radar et laser (4);

(11) *Effets de l'atmosphère*, qualité des images, choix des sites (24);

(12) *Traitement de l'information*, automatisation des mesures, ordinateurs (28);

(13) *Conception générale des instruments*, optimisation, automatisation (26);

(14) *Normalisation*, rapports avec l'industrie (7).

La concentration de l'intérêt sur les instruments du domaine classique (points 5 et 8) est certaine, comme on pouvait s'y attendre, ainsi que pour les récepteurs. Mais le grand nombre de réponses sur les points 11, 12 et 13 démontre le souci, dont il faut être très heureux, d'améliorer l'efficacité des instruments terrestres, tant dans la collecte de l'information que dans son traitement.

On notera, au surplus, *qu'aucun* des points mentionnés ne s'est trouvé sans répondant, et l'on ajoutera qu'une rubrique 15, "Autres Sujets", a fait apparaître les spécialités suivantes:

Photométrie par balayage;

Application des techniques de télévision;

Sources de calibration;

Instruments pour l'astrométrie;

Magnétographes;

On est donc certain que la Commission peut déjà trouver chez ses membres actuels un très large éventail de compétences; certains sont même en mesure d'assurer déjà un contact avec les Commissions 40 et 44.

RÔLE ET MODES D'ACTION POSSIBLES DE LA COMMISSION 9

Il apparaît, d'ores et déjà, que l'idée d'établir la liaison avec d'autres Commissions par l'intermédiaire de *groupes mixtes* est assez largement admise par les membres de la Commission 9 pour ne pas appeler une discussion lors de la XIVème Assemblée Générale, qui devrait, par contre, être mise à profit pour une prise de contact avec les autres Commissions.

Il convient de bien préciser la justification de tels groupes: l'idée fondamentale doit être qu'un *instrument astronomique n'est pas une fin en soi* mais un moyen de parvenir à des observations aussi sûres et abondantes que possible; le but à atteindre est donc d'assurer la meilleure association possible entre "producteurs" et "consommateurs" d'instruments; les premiers doivent pousser leur "art" aussi loin que possible *dans le sens qui répond aux besoins des utilisateurs*, sans s'égarer dans des exploits techniques qui n'auraient pas leur justification astronomique; les seconds doivent leur indiquer les voies dans lesquelles des progrès techniques produiraient des progrès astronomiques, mais aussi prendre conscience des contraintes techniques inévitables. Fort heureusement, un bon nombre des membres de la Commission 9 sont déjà membres d'autres Commissions, et sont tout désignés pour former les noyaux de ces groupes mixtes. Mais il importe que les autres Commissions, dans leur ensemble et à partir de leurs Présidents, et non quelques-uns de leurs membres seulement, soient averties des vues de la Commission 9 et manifestent leur position devant cette suggestion qui n'est autre qu'une offre d'aide technique.

Ces groupes mixtes ne rempliront pas tous les rôles dévolus à la Commission 9 et doivent être en partie recoupés orthogonalement par des *groupes internes*, par spécialité, suggérés par de nombreux membres dans leurs réponses: les groupes "Tubes à images" et "Qualité des Images et choix des sites" en sont des exemples (mais les seuls) qu'il y aurait intérêt à multiplier.

Sur le plan pratique, on est en droit de se demander si la création de tels groupes internes ou mixtes ira au délà de l'élaboration de listes de noms lors de l'Assemblée Générale.

On peut répondre que les chances pour que des liens s'établissent et qu'un travail se fasse réellement, par correspondance (à défaut de rencontres), augmentent très vite lorsque le nombre de membres concernés diminue, et encore plus s'ils ont des points d'intérêt commun. La division de l'activité de la Commission 9 entre des groupes plus limités dans leurs préoccupations ne peut qu'aller dans le bon sens.

Bien entendu, il est toujours souhaitable de pouvoir organiser des colloques ou symposia sur des problèmes bien définis. Mais, là encore, la constitution de groupes restreints pourrait mettre fin à la dispersion des intérêts entre un grand nombre de membres, et conduire, justement, à l'organisation de réunions fécondes.

Une autre tâche s'offre parallèlement, à la Commission 9, si elle réussit à trouver les moyens matériels (relativement modestes) et surtout les bonnes volontés nécessaires: l'amélioration des sources d'information sur les instruments astronomiques (principaux ou auxiliaires) en service.

Parmi ces équipements, certains, quoiqu'en état de marche, sont peu ou mal employés, pour des raisons de programmes ou d'implantation géographique, et pourraient être utiles à d'autres observateurs, soit sur place, soit, éventuellement, après transfert. Le problème se pose différemment pour les instruments les plus récents et les plus puissants, qui sont plutôt trop demandés; encore ne serait-il pas mauvais de disposer d'un moyen commode de savoir qu'un instrument de telles caractéristiques existe en un lieu de telles coordonnées géographiques et pourrait permettre de faire telle observation à tel moment. Mais, surtout, les innombrables variantes dans leur construction, si elles étaient connues, éviteraient à d'autres des tâtonnements, des essais, voire des erreurs.

Il devrait être possible, en général, d'obtenir des astronomes qu'ils fournissent, de façon assez détaillée, les caractéristiques de leurs instruments, non seulement classiques, mais même nouveaux: il est rare, en effet, que la divulgation de ces détails puisse priver un chercheur de l'avance que lui garantit une disposition nouvelle, qu'un autre mettra un certain temps à mettre en oeuvre.

Des répertoires existent déjà, tels que "Les Observatoires Astronomiques et les Astronomes" ou "Observatories in the World". Mais ils se bornent à des indications très succinctes sur chaque instrument. D'autre part, on trouve peu de descriptions complètes d'instruments dans la littérature et en général seulement pour des instruments assez exceptionnels. En effet, les revues astronomiques de haut niveau sélectionnent les articles proprement astronomiques et feraient difficilement passer avant eux des articles de pure description d'instruments; au contraire, dans les revues consacrées aux instruments scientifiques en général et à l'optique, l'appareillage astronomique est noyé parmi d'autres. Par ailleurs, une publication exclusivement consacrée à cet appareillage ne serait sans doute pas justifiée.

La seule formule possible paraît donc être celle d'un *fichier* qui rassemblerait sur chacun des instruments (principaux on auxiliaires) en service quelques lignes, ou éventuellement plus, et dont le maintien à jour reposerait sur un secrétariat modeste et ... sur la bonne volonté des astronomes. En un mot, ce serait l'extension de ce qui a été présenté jusqu'ici dans les "Draft Reports" de la Commission, mais sous une forme plus souple, plus complète, et non liée à la périodicité des Assemblées Générales.

En attendant qu'un tel projet voie le jour, on trouvera ci-après, pour les trois dernières années, quelques informations données par un certain nombre de membres de la Commission ou responsables d'Institutions ou Groupes de Recherches.

DÉVELOPPEMENTS INSTRUMENTAUX DANS LA PÉRIODE 1967–1969

Les informations ci-après, simplement classées dans l'ordre alphabétique des noms de ceux qui ont pris le peine de les communiquer, sont, soit des citations intégrales ou partielles, soit des résumés:

L. Arbey (Observatoire de Besançon)

"Des essais concernant l'acquisition de données photométriques, à partir d'une grille de Høg, sur l'astrolabe impersonnel de Danjon, ont été effectués en Mai et en Octobre 1968 avec les moyens existants à l'Observatoire, et un photomultiplicateur Radio-Technique."

. . . .

"Il apparaît, d'ores et déjà, que la forme de la grille, son pas, la largeur des fentes claires, le choix d'un seuil optimal de discrimination des impulsions en amplitude, la fréquence de l'échantillonnage dans le temps, dépendent de la scintillation, de l'agitation, de l'altération, de la qualité des images, du type spectral de l'étoile observée, etc."

. . . .

A. Behr (Hamburger Sternwarte)

"Since 1967, Rudolph and Schmahl from the Göttingen Observatory have developed a method to produce spectroscopic high precision reflectance and transmission gratings by laser light interference on photoresist layers. Until now, blazed gratings up to a ruled area of about 100×100 mm^2 with more than 90% theoretical resolution and 50% efficiency have been produced. Such gratings have low straylight and are completely ghostfree. Very large gratings which are needed for large telescopes can be made soon.
(Rudolph, D., Schmahl, G. 1969, *Optik*, in press)"

P. Connes (Centre National de la Recherche Scientifique et Observatoire de Meudon)

Prototype de télescope composite de 420 cm de l'Observatoire de Meudon:

"Il s'agit d'un télescope Cassegrain, à monture azimutale et muni d'un foyer coudé. Le primaire est sphérique et constitué par une mosaïque de 36 éléments carrés en verre de 60×60 cm, rayon de courbure 12 m et donnant une tache d'aberration moyenne de 10″; ils ont été réalisés par polissage de glace (épaisseur 42 mm). Le miroir Cassegrain convexe, de diamètre 60 cm, est très asphérique (écart à la sphère 1 mm); il est en cours de polissage à l'Institut d'Optique (MM. Demarcq puis Marioge). Ces éléments seront asservis individuellement en orientation sur l'étoile visée elle-même; des essais en laboratoire sur étoile mobile artificielle laissent espérer une précision de l'ordre de 1″. Le guidage de l'ensemble sera également effectué par asservissements.

"A l'heure actuelle (Août 1969) l'ensemble de la mécanique, réalisé au Laboratoire de Bellevue, est terminé et installé sous toit mobile à Meudon. Les miroirs aluminés seront mis en place en Septembre et les premiers essais sur le ciel effectués en 1969. Les asservissements seront complétés au cours du premier semestre 1970.

"Le but de l'opération est double:

(1) Réaliser le plus rapidement possible avec des moyens relativement modestes" (...), "un collecteur de lumière entièrement consacré à la spectroscopie par transformée de Fourier dans l'infrarouge proche (1 à 12 μ) des étoiles et planètes, avec son laboratoire coudé. Cet instrument devrait permettre d'augmenter considérablement la production de spectres infrarouges obtenus jusqu'ici avec les téléscopes de 150 et 193 cm de Saint Michel" (...).

"(2) Constituer un prototype permettant d'essayer sur le ciel un certain nombre de solutions optiques et mécaniques extrapolables ultérieurement à un collecteur de plus grandes dimensions; la plus originale étant naturellement la réalisation d'un miroir primaire par une mosaïque d'éléments asservis."

R. B. Dunn (Sacramento Peak Observatory)

"An evacuated Tower Telescope"

Une tour solaire de type nouveau a été mise en service à l'Observatoire de Sacramento Peak en Octobre 1969. L'auteur définit ainsi les objectifs qui ont guidé la conception:

"(1) Reduce the deleterious effects on image quality caused by solar heating of the interior and exterior surfaces of the telescope, (2) provide a simple optical system with the fewest possible reflecting surfaces, with minimal heating of the mirrors, and with no change in illumination of the aperture during the day, (3) provide a flexible guiding arrangement so that the image can be scanned and rotated either about the center of the sun or the center of the field of view, and (4) provide for rapid and accurate switching of the solar image between a number of instruments so that different types of observations of rapidly changing solar phenomena can be obtained."

La tour quasi-conique, en béton armé, mesure 17 mètres de diamètre à la base et 1·5 m à 40 m au-dessus du sol. Elle porte, en son sommet, une tourelle altazimutale à deux miroirs à 45° dont le premier renvoie le faisceau solaire horizontalement, et le second verticalement. L'ouverture libre, de 76 cm de diamètre, est fermée par une lame de quartz fondu de 10 cm d'épaisseur. L'objectif principal est un miroir sphérique de 55 m de longueur focale, placé dans un puits de même axe que la tour à une profondeur telle que l'image solaire se forme au voisinage du niveau du sol et tombe sur divers spectrographes disposés verticalement. Un tube métallique porté par flottaison sur mercure dans la partie supérieure de la tour descend dans le puits et est fermé à sa partie inférieure par le barillet du miroir. La tourelle altazimutale repose également sur un flotteur à mercure et l'étanchéité est ainsi assurée depuis la lame d'entrée jusqu'à la lame de sortie disposée au voisinage de l'image solaire. On peut donc faire le vide dans tout l'espace parcouru par la lumière (450 m^3). Sur le tube sont fixés une plateforme de 12 m de diamètre, au niveau du sol, pour les observateurs, et, dans la partie comprise entre la plateforme et le flotteur, les spectrographes. L'ensemble tourne, indépendamment de la tourelle altazimutale, pour compenser la rotation de l'image solaire.

L'auteur écrivait en 1964 (*Applied Optics*, **3**, 1357):

"The design provides a telescope with a simple optical system, with environmental control, with complete control of image motion, and with a multiplicity of auxiliary instruments. The disadvantages of the design include the following: the turret is altazimuth and must be servo-controlled; the turret mirrors are placed at 45° and will introduce polarization during the daily observation that must be compensated if magnetic field observations are made; any bending or heating of the mirrors will introduce astigmatism; high-quality windows are required; rotation of a large mass is required to eliminate image rotation; part of the light beam is lost to the guider, the longer instruments must be vertical, and a sophisticated control system is required.

"Most of these problems have been solved in the design. None of them seriously limits the effectiveness of the instrument."

L'instrument est maintenant achevé et il sera extrêmement intéressant de connaître les résultats réels des solutions originales qui ont été mises en oeuvre.

H. Elsässer (Max-Planck-Institut für Astronomie)

"A Max-Planck-Institut für Astronomie has been founded in 1969 which will consist of a central institute in Heidelberg and two observatories, one in the Mediterranean (area) and one in the southern hemisphere. Both observatories will be equipped with a 2·2 m Ritchey Chrétien telescope which are under construction at C. Zeiss, Oberkochen. In addition a 3·5 m telescope will be constructed. It is not yet decided if it goes to the northern or southern hemisphere station. A 3·6 m disk as well as the mirrors for the 2·2 m telescopes are ordered from Schott-Mainz and are made of glasceramics which is more or less the same as Cervit."

Ch. Fehrenbach (Observatoire de Haute-Provence et Observatoire Austral Européen)

Les instruments mis en service depuis la dernière Assemblée Générale sont les suivants:

O.H.P.:

1. Télescope de 152 cm en 1968.
   Spectrographe Coudé avec des réseaux de 20 × 30 cm – octobre 1969.

Seront achevés en 1969:

2. Télescope de Schmidt (en liaison avec l'Université de Liège).

3. Télescope photométrique automatique de 40 cm de diamètre.

E.S.O.:

1. Télescope photométrique de 1 m (1967).
2. Télescope spectrographique de 1 m 52 (1969) avec spectrographe.
3. Prisme Objectif de 40 cm de diamètre (1968).
4. Télescope de 60 cm de l'Université de Bochum.
5. Télescope de 1 m de l'Université de Copenhague.

G. Godoli (Osservatorio Astrofisico di Catania)

"At the Catania Astrophysical Observatory four main instruments are now available for stellar work:

1. A Newtonian-Cassegrain reflector (91 cm of aperture and 400 cm and 1400 cm of focal length respectively).

2. A Schmidt-quasi Cassegrain reflector (41 cm of aperture and 119 cm of focal length in the Schmidt combination and 61 cm of aperture and 600 cm of focal length in the quasi-cassegrain combination).

3. Two Cassegrain reflectors (31 cm of aperture and 500 cm of focal length) which we individuate conventionally by the position (N or S) in their common dome.

All these instruments are equipped for photoelectric observations. Two photometers are equipped with FET amplifiers.

A photometer for simultaneous stellar photometry in three colours has been constructed in the laboratories of the Catania Observatory.

An apparatus for the automatic conversion of analogic data, given by the photometers, into digital data on punched cards is under construction."

"An equatorial spar is under construction at the Marchiori firm of Rome.

The spar is equipped with six optical trains (4 m length) for:

(a) visual and photographic observations of the photosphere;
(b) $H\alpha$ observations of the chromosphere;
(c) $H\alpha$ wing observations of the structure of the chromosphere;
(d) K observations of the chromosphere;
(e) $H\alpha$ observations of prominences;
(f) experiments."

J. S. Hall (Lowell Observatory)

"A 42-inch fork-mounted reflector, to be used in the Cassegrain, Nasmyth, and Coudé configuration, has been installed in the vicinity of the Perkins Reflector at the Lowell Observatory. Mechanical design and construction was carried out by Astro Mechanics, Inc., the optics were furnished by Davidson Optronics, Inc.

In conjunction with the International Planetary Patrol Program, established with NASA support and under the supervision of Baum, new Boller and Chivens 24-inch $f/75$ Cassegrain reflectors were installed both on Cerro Tololo and on Mauna Kea; planetary patrol cameras were designed and built especially for this program.

An electron-optical image stabilizer ("image tranquilizer") to be used primarily for planetary photography was completed under the supervision of Baum.

Boyce designed a dual-channel area scanning photometer primarily for planetary observations.

A photoelectric area scanner for photometry and astrometry of double stars was designed by O. G. Franz (*Lowell Obs. Bull.*, No. 152, 1969).

J. S. Hall designed and supervised the construction of a dual-beam scanning polarimeter that can also be used as a scanning colorimeter (*Lowell Obs. Bull.* **7** (1968), 61)."

G. Newkirk, Jr. (High Altitude Observatory, Boulder, Colorado)

"(a) *The Apollo Telescope Mount Coronagraph*

Although the ultimate goal of continuous observation of the outer solar corona has still not been realized, progress has been made toward improving the quality and coverage of coronal observations. Successful observation of the outer corona by our balloon borne coronagraph in 1964 and 1965 demonstrated the feasibility of extra-atmospheric coronal observations as well as contributed a unique sequence of data from which the three-dimensional structure of the corona could be inferred. However, only a satellite borne coronagraph can give the long term observations which are required. The Apollo Telescope Mount (ATM) coronagraph is designed to furnish these observations. The instrument itself, which is nearly an exact duplicate to the Coronascope II system, consists of a Lyot coronagraph of aperture 3·2 cm and effective focal length 44 cm. A set of three external occulting disks in front of the objective serves to reduce the instrumentally scattered light in the field of view to a maximum $\sim 3 \times 10^{-10} B_{\odot}$. Since the ATM is to be operated by astronauts at a space station, the image of the corona from 1·5 $R_{\odot}$ to 6 $R_{\odot}$ can be recorded photographically with all the advantages of the high information content of film."

. . . .

"Since the quantity of sunlight which will be scattered in the instrument remains one of the uncertainties in the design of a satellite coronagraph, two extensive photometric tests are being performed. The first test will measure light scattered in the 100-ft vacuum tunnel at HAO. This tunnel contains a synthetic sun of the appropriate angular diameter, at one end, and a "laboratory" chamber 10 ft in diameter at the other. The coronagraph can be placed in the "laboratory" and the scattered light measured, without contamination from light diffused by the air between the "Sun" and the instrument. The influence of various misalignments can also be determined, so that the astronauts, who will be operating the coronagraph for 56 days, can be supplied with a compilation of diagnostic photographs to help them analyze any misalignments which may occur.

Upon completion of the scattered-light tests in the vacuum tunnel, the flight verification unit will be mounted on the old Stratoscope-Coronascope gondola, and will be carried to about 115000 ft by balloon. This near-space flight will secure observations of the corona, from 1·5–6 $R_{\odot}$ continuously for a period of about 10 h, and will permit a realistic evaluation of the measurements of instrumentally scattered light previously made in the tunnel.

Although the Apollo capsule fire and other vagaries have delayed launch of the first ATM to mid-1972, we look forward to nearly two months of photographic coverage of the corona out to 6 $R_{\odot}$. This coverage will enable us to detect rapidly developing, transient phenomena as well as the slow evolution and changing aspects of the corona.

A unique problem in operating a coronagraph in a manned satellite is that the optical environment may be contaminated by a life-support system. Newkirk has examined the dynamics of particulate material ejected from manned spacecraft. He found that débris thrown from the vehicle will accompany it in orbit as a surrounding cloud for nearly an hour. Moreover, if most of the water vapor present in the spacecraft cabin condenses to form ice crystals as it escapes, the radiance of the sunlit cloud of debris may be so high as to preclude observations of the solar corona and other faint astronomical objects."

"(b) *Climax Observing Station*

In January 1968 the magnetograph was converted to a digitized mode of operation following the installation of a small computer (PDP-8) and associated equipment. Observations can now be collected on magnetic tape and the data reduction handled on the Control Data 6600. The digitized mode of operation does away with the calibration phase of the old procedure and allows for a more efficient use of available good weather. As a result the actual data output in 1968 was increased three to four times over that in previous years."

. . . .

"(c) *Mauna Loa Observing Station*

During 1968 we greatly enlarged our capability for study of the white light corona with the completion and installation at Mauna Loa of a new coronagraph, the Coronal Activity Monitor. It is a photoelectric polarimeter, similar in many respects to the coronameter, but with the special ability to scan at a selected range of position angles above the solar limb. Its purpose is to monitor the region above disturbed chromospheric regions in order to detect interactions between such events as flares and the overlying corona. By the end of the year, debugging of the instrument was nearly finished and many high quality survey observations had been made. However, we have not yet directed the coronagraph at the right place in the corona at the right time to monitor a major transient event."

"(d) *Solar Observations from Aircraft*"

....

"An experiment to study the two-dimensional form of $\lambda 10747$ emission and polarization was carried out by Eddy and Firor at the eclipse of 12 November 1966. The Observatory was granted a window for this purpose on the Convair 990 jet aircraft of the NASA Ames Research Center, operated for this expedition from Porto Alegre, Brazil. An 8-inch infrared telescope was gyroscopically guided on the corona through a 10×14 inch window of optical quality. A coronal image was formed through a $\lambda 10747$ interference filter of 14 Å half-width on the face of a two-stage image tube, furnished by the Department of Terrestrial Magnetism of the Carnegie Institution. The color-converted image was recorded on blue sensitive film at the image tube phosphor. The result is a series of unusual coronal photographs, taken in the light of the $\lambda 10747$ emission in three states of polarization. A similar series was taken for comparison in the nearby continuum at $1{\cdot}06\ \mu$ through a second filter. The pictures are of high quality, with resolution limited by the image tube to about 15 arc seconds; they show the intricate form of the emission corona extending as far as $1{\cdot}5\ R_{\odot}$ in some regions."

....

"Equally encouraging in this experiment was the performance of the gyroscopic stabilization unit, developed and built in the Observatory electronic and mechanical facility by Lee and Lacey. Beginning with a concept developed by Richard Dunn at Sacramento Peak for his 1965 eclipse coronal spectrograph, Lee and Lacey built a completely new system around the infrared telescope. In the improved design the center of system mass was located near the floor of the aircraft; a more sophisticated servo and torque motor system was used also. A low-frequency booster system to correct the dominant mode of aircraft motion, and an auxiliary roll-rate gyro, gave demonstrable improvement in guiding. Two-axis pointing accuracy achieved was about 2 arc seconds, rms, on the best flights. This engineering achievement demonstrates to us the promising future of the airborne platform in making eclipse and other observations."

....

"In 1967 Eddy and Firor, in collaboration with H. A. Gebbie of the National Physical Laboratory of England, succeeded in a pilot attempt to record photometric spectra of the Sun from about 100–700 μ from a high-altitude jet aircraft. Gebbie, whose far infrared spectrum of the Sun in 1957 was the first ever published, made available to HAO an interferometric spectrometer of the Michelson type fitted with a Golay cell. This was joined to the gyroscopically guided infrared telescope which Eddy and Firor had used in the 1966 airborne eclipse experiment. Observing time on the Convair 990 scientific aircraft was made available by the Airborne Science Office of the NASA Ames Research Center. Observations were made in August and October of 1967 at altitudes of about 40000 ft. Separate spectra were taken of the Sun and of the sky through special aircraft windows of high-density polyethylene and polypropylene. Spectral resolution was about 2 cm^{-1}. These spectra were calibrated with reference to a thermal source of known temperature and emissivity. Solar spectra are of the integrated solar disk, with no attempt at angular resoluto n."

V. B. Nikonov (pour l'ensemble de l'U.R.S.S.)

"The following works on astronomical instrumentation are being carried out in the U.S.S.R.

Telescopes for general use

The construction of the 6-m altazimuth telescope (designer in chief Dr B. K. Ioannisiani) has been completed except the main mirror. After the preliminary assembly of the telescope at the factory in 1967 it was removed to the Special Astrophysical Observatory of the Academy of Sciences of the U.S.S.R. located at the altitude of 2 km near Zelenčukskaya in the Northern Caucasus. The dome of the telescope has been erected too. The assembly of the telescope with a simulated main mirror, will begin in the nearest future. The work on figuring the mirror is in progress.

A modernized version of the well known 2·6-m Shajn telescope of the Crimean Astrophysical Observatory is under construction for the Byurakan Astrophysical Observatory of the Armenian Academy of Sciences.

The adjustment and testing of the 2-m telescope of the Shemacha Astrophysical Observatory of the Azerbaidjan Academy of Sciences has been completed. This telescope is a duplicate of that of the Ondřejov Observatory in Czechoslovakia. It was supplied by C. Zeiss, Jena, D.D.R. The telescope is equipped with spectrographs having dispersions from 75 to 25 Å/mm (Cassegrain focus) and from 24 to 4 Å/mm (coudé focus). A photoelectric photometer has been supplied too.

For the Struve Astrophysical Observatory of the Estonian Academy of Sciences at Tõravere near Tartu, a 1·5-m telescope is under construction. Being equipped with a computer, it will give the possibility to attain a high degree of automation in observations and their reductions.

Specialized Telescopes

The construction of two completely automated 1·25 m photoelectric telescopes is being continued. These telescopes will be supplied to the Crimean Astrophysical Observatory of the Academy of Sciences of the U.S.S.R. and the Abastumani Astrophysical Observatory of the Georgian Academy of Sciences.

We should like to notice the realisation of the system of two separate telescopes, controlled from the same desk. This system is used for photoelectric observations; photoelectric signals, coming from photoelectric photometers of both telescopes, are led to a common measuring device and treated simultaneously. Such a system of the two 0·5-m telescopes is working now at the Struve Observatory at Tõravere. Another system is being constructed at the Radioastrophysical Observatory of the Latvian Academy of Sciences at Baldone.

The Odessa Astronomical Observatory started its work on a series of coma-free telescopes (up to 1·5 m in diameter) according to the optical scheme designed by P. P. Argunov. In general these telescopes will be used for variable star observations.

Some new telescopes are to be used for lunar and planetary observations. At the Main Astronomical Observatory of the Academy of Sciences of the U.S.S.R. (Pulkovo Observatory) the construction of a 0·7-m planetary telescope is coming to the end. It was designed by the late D. D. Maksutov. The adjustment of a horizontal telescope ($D = 22$ cm, $F = 22$ m) has been finished at the Engelhardt Astronomical Observatory (near Kazan). This telescope will be used for large scale photography of the Moon.

A special astrometric double-meniscus telescope has been installed by the Chilean astrometric expedition of the Pulkovo observatory at Cerro Robles near Santiago, Chile. The diameter of meniscus is 0·7 m, the diameter of the field 5°, focal ratio 1/2·95, the diameters of images of faintest stars are 15–20 μ. The dome and all other facilities were supplied by the National University of Chile. The telescope is used both by Chilean and Soviet astronomers.

At the Kislovodsk mountain station of the Pulkovo Observatory and at the Astronomical Station of Sibizmir (Siberia) the two largest Lyot-type coronographs have been put in operation. Their object glasses have diameters of 53·5 cm and focal ratio 1/13. Coronographs are equipped with spectrographs having dispersion 1 Å/mm.

Special cameras for photographic observations of sputniks have been installed at some observatories including Pulkovo and the Astronomical Station of the Astronomical Council of the Academy of Sciences of the U.S.S.R. at Zvenigorod. Some of these cameras designed by Abelle and Lapushka have original four-axis mountings.

Domes for the Telescopes

An interesting experiment in the construction of rotating domes made of plastics have been carried out at the Struve Observatory, Tõravere. A 6·5-m dome of such type is being constructed at the Radioastrophysical Observatory, Baldone.

In response to an order from the Crimean Astrophysical Observatory, different kinds of white titanium dioxide paints for astronomical domes have been supplied by an industrial institution. The detailed investigation of more than 20 samples of these paints carried out at the Crimean Observatory for the last years, gave the possibility to find out the best samples as to their thermal properties and stability in the conditions of normal astronomical exploitation. One of the best types has been used for painting the dome of the Shajn 2·6-m telescope.

Auxiliary Equipment

The work on application of television methods in astrophysics is being carried out at the Crimean, Pulkovo and Abastumani Observatories.

At the Crimean Astrophysical Observatory the application of new types of very sensitive image orthicons, recently developed in the U.S.S.R. has made it possible to take photographs of very faint objects with very short exposures. Thus using a 0·5-m telescope it is possible to practice television photographic photometry of variable stars up to 17^m–18^m at exposures of few tens of seconds with a precision comparable to that of ordinary photography. It also has made it possible to obtain photographs of planets through light filters during the short moments of good seeing. A visual television survey for the search of supernovae in isolated galaxies has been organized too.

At the Pulkovo Observatory a two-channel television telescope is being constructed. The possibility of compensating for atmospherical disturbances is considered. The methods for obtaining television images of stellar objects in the infrared (up to 3.4 μ) and of laboratory television spectrophotometry are under development.

At the Abastumani Observatory a television device for lunar researches, such as luminescence effects, changes on the lunar surface, etc., is constructed.

At the Crimean and Abastumani observatories photoelectric scanners using photon counting are designed and constructed. Printing output is provided.

A three-channel photoelectric photometer is being constructed at the Radioastrophysical Observatory, Baldone. For the separation of spectral regions dichroic filters are used. The photometer is automated and photon counting is used.

The instrumentation for photoelectric polarimetry is being developed at the Crimean Observatory and at the Sternberg Astronomical Institute (Gaish). Polarimeters constructed at Gaish allow the immediate determination of Stokes's parameters and linear and elliptical polarization.

Photoelectric equipment, for the observations of twilight and meteors in the near infrared ($\sim 0.96\ \mu$) is being constructed at the Engelhardt Astronomical Observatory. Magnetic tape is used at the Main Astronomical Observatory of the Ukrainian Academy of Sciences (Kiev) for recording the solar spectrum as well as for the registration of other types of information.

Instrumentation for Reduction of Observations

The construction of new types of iris photometers, as well as the automation of existing ones is being carried out at Gaish, Pulkovo and Engelhardt observatories and at the Astronomical Observatory of the Leningrad University.

Fast recording automated microphotometers with the output on perforated tape are under construction at the Pulkovo and Byurakan observatories.

At the Odessa Astronomical Observatory a microphotometer with an oscilloscope output has been constructed and put into operation.

At the Pulkovo Observatory an experimental model of a completely automated coordinate measuring device is being adjusted.

A special device for automated measurements of photographic material obtained for the studies of the Moon and the planets is under construction at the Main Astronomical Observatory of the Ukrainian Academy of Sciences."

J. Rösch (Observatoire du Pic du Midi)

"La construction du télescope de 2 m (qui portera le nom de Bernard Lyot), décidée depuis plusieurs années et retardée pour diverses raisons, commencera effectivement en 1970, et les études sont très avancées. Le but essentiel est d'obtenir *la plus haute résolution spatiale possible* à partir du sol. A cet effet, une étude soignée du "microsite" a été faite (voir ci-après), et un certain nombre de dispositions particulières sont prévues. L'instrument sera au sommet d'une tour de 14 m de diamètre et 25 m de haut. La coupole n'aura qu'une ouverture circulaire du diamètre du faisceau, fermée par une lame à faces parallèles. L'optique comportera un miroir parabolique (en Cervit) de 10 m de longueur focale, un miroir plan amovible de 110 cm de diamètre donnant un "foyer primaire replié" au travers du trou central du parabolique (comme dans le télescope astrométrique de l'USNO), un miroir Cassegrain $f/25$, et un montage "coudé".

Dans un but tout autre que la haute résolution, a été entreprise la construction d'un collecteur de lumière composite de 10 m^2 de surface utile pour 10″ de pouvoir résolvant, destiné en premier lieu à la réception de signaux laser diffusés par la surface lunaire ou réfléchis par les cataphotes mis en place sur la Lune. Mais cet instrument sera aussi employé pour la photométrie photoélectrique d'objets faibles, pour laquelle ses performances ne doivent pas être très inférieures à celles d'un télescope classique, malgré son prix de revient, à surface égale, considérablement plus bas. La conception de l'appareil est entièrement différente de celle du collecteur de Connes décrit plus haut. Il sera, en effet, composé de 210 lentilles simples de 280 cm de longueur focale montées sur la face avant d'un tambour hexagonal montée en équatorial, la lumière étant collectée en chaque foyer par une fibre (dont l'entrée forme diaphragme) et ramenée sur un photomultiplicateur unique."

E. H. Schröter (Universitäts-Sternwarte, Göttingen)

"In March 1968 a new Gregory-Coudé solar telescope = 45 cm aperture, $f_{eff} = 24$ m) was installed at our Locarno solar station. It had been designed and constructed in our workshop. It is a considerably improved replica of the telescope delivered 1961 by Cox and Hargreaves, which had been found to be insufficiently accurate and stable. For the optical design see Brückner *et al.* (*Solar Phys.*, **1** (1967), 487). A short-focus refractor mounted parallel to the main instrument allows either for observations of the entire solar disc in Hα (and ± 1 Å) on a television monotor, or to expose time sequences of the Hα solar image. A Zeiss photoelectric guider mounted on the main instrument tube serves for a precise guiding. Its inclination with respect to the main optics can be changed by a paper tape-controlled electronic device. With various paper tape programs any solar region can be scanned automatically with different speeds and patterns (Schröter and Wiehr, *Mitt. Astron. Ges.* **25** (1968), 187). A speciqal optical device in front of the slit of the large spectrograph allows for very precise guiding and control of the solar feature under study. White light, Hα and K slit jaw pictures can be exposed. For spectrographic problems we designed a semi-automatic scanner for the high-dispersion (5–10 mm/Å) solar spectra. The design of the Locarno vector magnetograph as described by Brückner (*Z. Astrophys.*, **68** (1968) 48) has been changed completely as far as polarization optics and electronics are concerned. The new magnetograph avoids the disadvantages of the old one (see Beckers, *Z. Astrophys.*, **68** (1968), 406) and allows to measure all four Stokes's parameters simultaneously (see Wiehr, *Solar Phys.*, **9** (1969) 225).

S. D. Sinvhal (Uttar Pradesh Observatory)

"(a) *Stellar Telescopes*

1. A 22-inch reflector (Primary $f/4$, Nasmyth $f/15$) has been installed in April 1969. The telescope also has a Baker-Corrector ($f/3$) and a pair of objective prisms (Schott UK-50 glass) of angles 6° and 10°, capable of being used singly or in combination. Photoelectric photometer is provided.
2. A 40-inch reflector (Cassegrain $f/13$, Coudé $f/31$, Meinel $f/2$) is due to arrive from Zeiss (Jena) in October 1969 and will be installed by mid 1970. Suitable Cassegrain and Coudé spectrographs will be provided by 1971. The telescope has automatic guiding and programmed star changing device for photometry of clusters.

(b) *Solar Spectrograph*

1. A horizontal solar unit employing an 18-inch coelostat a 10-1/2 inch $f/66$ Skew Cassegrain objective (7″ solar image) and a double pass spectrograph capable of giving a dispersion of 1·3 Å/mm. in the first order is expected to be installed early in 1970. Photoelectric scanning and photography of the spectrum have been provided for."

J. Tremko (Astronomicky Ustav, Skalnaté Pleso)

"The new automatic photoelectric photometer at the 60-cm reflector of the Skalnaté Pleso Observatory has been placed into operation. Many operations are fully automatized: change of the input and output sensitivities, change of filters, etc. The photometer is equipped with an electronic clock and typewriter. Following data are typed: time, star number, filter, focal diaphragm, interval of the integration, output of the integrator and its gain. Development of the instrumentation for the punched tape recording is under way."

K. O. Wright (ainsi que D. H. Andrews, G. A. Brealey, J. B. Hutchings, G. J. Odgers, E. H. Richardson) (Dominion Astrophysical Observatory)

"(a) *The 1·8-m (72-inch) telescope*

In 1967 an all mirror spectrograph with an hyperboloidal, off-axis collimator constructed at the Observatory began operation at the Cassegrain focus. It is suitable for future use with an image intensifier. The camera has a focal ratio $f/5$ with 53 cm focal length. The exposure time at 15 Å/mm dispersion for a B star magnitude 5·0 at the zenith, seeing disc 2·8 seconds of arc is 3 min without an image slicer. The spectrogram is broadened to 0·4 mm; with the projected slit width 0·022 mm the practical resolution is 0·026 mm. Three dual-width slits have been constructed for use with the Cassegrain spectrograph. The narrow portion of each slit is for the comparison spectrum, the wide portion for the star and a sliding mask covers the portion of the slit not required for use."

. . . .

"(b) *The 1·2-m (48-inch) telescope*

(1) In 1967 a mosaic grating began operation in the coudé spectrograph. It is made up of four 154 mm × 180 mm Bausch and Lomb gratings having 830·8 grooves per mm. The mosaic is used with a camera of 2·4 m focal length. The mounting of the gratings is based on the Mount Palomar design of Bruce Rule, and a periscope arrangement permits alignment by one person.

(2) Also in 1967 superpositioning-type image slicers went into operation as part of the coudé spectrograph. In this design the stellar image is elongated to cover the desired length of slit, and each subsequent slice is focussed by a multiple-reflection mirror system and superimposed on the

previous slices. The system in use, which was constructed at the Observatory, has four useable slices and in practice, increased the spectrograph speed by more than a factor of 2.

(3) In 1969 a high reflectance coudé mirror train was installed in the telescope for the last four of the five reflections to the focus. A three position turret is situated at each reflection point. In order that small mirrors can be used the secondaries 15 cm in diameter change the focal ratio to *f*/145 from the *f*/4 primary; at the 5th reflection a totally-reflecting prism-lens combination changes the focal ratio to *f*/30. The obscuration of the primary mirror by the old coudé arrangement covered an area 50 cm in diameter whereas it is only 22 cm in diameter with the new (smaller) mirror system. Together with the greater efficiency of the multicoated mirrors, the total gain in light flux is 40% in the blue and more in the near infrared. The three mirrors on each turret are coated for use in the regions UV, blue, and infrared respectively. The exposure time at dispersion 2·4 Å/mm is 50 min for a 5·0 magnitude B star at the zenith with seeing disc 2·8 seconds of arc diameter."

....

Voir E. H. Richardson, "The Spectrographs of the Dominion Astrophysical Observatory," *J. r. astr. Soc. Can.*, **62** (1968), 313.

(c) *4-m Telescope*

Work on the design and construction of a 4-m telescope has continued at the Observatory although severely hampered by a shortage of funds. The 4-m fused silica mirror blank was completed in October 1967 and a 100-inch aluminium sphere for use in testing the secondary mirrors was ready for grinding in 1968. A polishing machine capable of finishing mirrors up to 4 m diameter was completed in October 1968.

Some of the design features of this telescope have been incorporated into large telescopes in advanced stages of construction. For Kitt Peak National Observatory's two 150-inch telescopes a design for the mirror-support system was evolved; this system incorporates two air-operated controllers with a feed-back device which increases the accuracy and sensitivity of the system. For the Anglo-Australian 150-inch telescope the concept of exchangeable upper ends of the telescope tube developed for the Canadian 4-m telescope has been adopted.

Refinements of the design of support systems for large mirrors have been made which reduce the weight of mirror cell to a minimum, and an extensive set of testing of bonding agents has been completed.

....

"An account of progress up to April 1968 is given *Mt Kobau Nat. Obs. Bull.*, No. 3."

....

Des développements importants ont été également donnés aux instruments de dépouillement (microphotomètre, analyse des spectres etc.).

W. H. Wehlau (University of Western Ontario)

"A 48-inch (122 cm) reflecting telescope of the classical Ritchey-Chrétien type, has been installed at the new University Observatory. For direct photography at the Cassegrain focus a doublet corrector lens and a field flattener provide a field of 1·1 diameter at *f*/8.26. The Nasmyth focus at *f*/9 and coudé focus at *f*/30.9 are also available."

....

(Spectrographe à faible dispersion au foyer Cassegrain; au "coudé", spectrographes à chambre de Schmidt et à correction par le collimateur.)

GROUPE DE TRAVAIL SUR LA QUALITÉ DES IMAGES ET LE CHOIX DES SITES

Bien que ce groupe n'est pas été formellement dissous, on doit constater qu'à part une séance lors de l'Assemblée Générale de Hambourg, aucune activité dans le cadre de l'UAI ne s'est manifestée dans ce domaine depuis le Symposium No. 19, tenu à Rome en 1962. Cette réunion répondait à un besoin et le volume de ses Comptes-Rendus a suscité un intérêt certain. Mais au-delà, chacun des groupes intéressés à une prospection a organisé son travail selon ses propres conceptions, en fonction des conditions géographiques qui lui étaient fixées et des moyens matériels mis à sa disposition. Il en résulte que si les méthodes d'évaluation de la qualité des images ont été quelque peu passées au crible de la critique internationale, peu de progrès ont été faits dans le sens d'une comparaison significative à l'échelle mondiale entre des sites inclus dans des campagnes de prospection différentes. Force est bien de regretter ce non-aboutissement, tout en accordant aux responsables des divers opérations l'excuse du délai généralement limité qui leur est imparti, et l'impossibilité d'attendre la mise au point d'une normalisation internationale. Un "groupe interne" de la Commission 9, au sens défini dans la première partie de ce Rapport, pourrait se consacrer à cette importante question.

En ce qui concerne les opérations de prospection conduites depuis 1967, l'auteur signalera celles dont il a eu connaissance, pleinement conscient qu'elles ne représentent qu'un échantillonnage arbitraire.

On a vu se dessiner – et c'est heureux – la tendance à spécialiser les modalités de prospection en fonction du type même de l'observatoire projeté.

C'est ainsi que s'est déroulée sous les auspices de Caltech une recherche de site d'observatoire *solaire* en Californie du Sud, qui a abouti au choix de *Big Bear Lake*, et qu'un groupe d'astronomes européens étudie, en vue de la création d'une station commune, dans la région méditerranéenne, (JOSO, Joint Organisation for Solar Observations), les sites supposés *a priori* devoir fournir de bonnes images pendant la journée.

C'est ainsi également qu'en France a été dissociée de la prospection pour un *observatoire astrophysique* celle d'un *observatoire astrométrique*, cette dernière étant conduite au moyen d'un astrolabe de Danjon, qui a l'avantage de fournir directement la qualité des observations futures.

Les autres prospections visent à l'implantation de grands télescopes et se fixent, en général, une altitude de 2000 m au minimum. Celle conduite sur une vaste échelle en U.R.S.S. a abouti au choix du site de Zelenčukskaya, sur le versant Nord du Caucase, pour le télescope de 6 m. Une campagne systématique a été menée par les astronomes français, en vue de l'implantation de divers télescopes, le meilleur site trouvé devant être réservé, en priorité, à un réflecteur de 360 cm. Elle a porté sur quatre sites dans le Sud de la France continentale, un site en Corse, un en Espagne du Sud (région d'Alméria) et a été étendue plus récemment à la Sierra de San Pedro Martir (Basse Californie) où la prospection avait déjà été amorcée par les travaux de Johnson et des astronomes américains. Les résultats sont en cours de dépouillement. Des prospections sont faites également par les astronomes allemands (Péloponèse et station française d'Alméria), italiens, brésiliens. De son côté, Walker a fait une étude approfondie de divers sites californiens et a cherché à analyser les raisons des différences de qualité des uns et des autres.

Signalons que pour le télescope de 2 m du Pic-du-Midi, le *site* étant déjà connu dans son ensemble et le but étant la "haute résolution", on a procédé à une étude du *microsite* par des essais en soufflerie sur maquette à 1/750 pour connaître les hauteurs des sillages pour tous les points possibles et toutes les directions de vent (avec vérification sur le terrain pour quelques points). On a conclu en tenant compte des vents dominants, des sources locales de chaleur et d'un certain nombre de contingences.

On est certes loin encore du but défini en 1961 par le groupe de travail *ad-hoc* de la sous-commission 9b, à savoir la détermination, sur la base de principes généraux, de la qualité d'images à attendre en tel ou tel point du globe. Cependant, l'accumulation et l'analyse des remarques faites au cours de toutes ces prospections conduisent progressivement à une meilleure compréhension des phénomènes, et si le progrès dans ce domaine n'est pas spectaculaire, il est néanmoins certain.

J. RÖSCH
Président de la Commission

GROUPE DE TRAVAIL SUR L'EMPLOI EN ASTRONOMIE DES RÉCEPTEURS PHOTOÉLECTRIQUES D'IMAGES

Introduction

Un article de synthèse a été publié par Kent Ford en 1968 (**1**). De plus, le 4ème Symposium international sur les récepteurs photoélectriques d'images s'est tenu à Londres en Septembre 1968 et les comptes rendus des communications paraissent actuellement (**2**); il y sera fait référence par la mention (AE, page).

Il est espéré que les travaux omis involontairement et les développements nouveaux pourront être présentés lors des séances de la Commission 9 à Brighton.

Le rapport considère les trois types principaux de récepteurs: ceux qui emploient l'électronographie, ceux qui utilisent un écran fluorescent (convertisseurs d'images) et enfin les appareils de télévision.

1. *Électronographie*

(a) *Centres de développement:* A l'Observatoire de Paris, A. Lallemand et ses collaborateurs essaient de mettre au point une caméra électronique à focalisation magnétique ayant une photocathode de 9 cm tandis que M. Duchesne fait porter son effort sur une caméra électronique à focalisation électrostatique ayant une photocathode de diamètre supérieur ou égal à 5 cm. A Meudon, Combes, Felenbok et leurs collègues développent des caméras électroniques pour applications spatiales (AE, 39). A l'Imperial College, à Londres, McGee, McMullan et leurs collaborateurs continuent l'amélioration du *spectracon* (AE, 61). A la station de Flagstaff de l'Observatoire naval des Etats-Unis, Kron et ses associés poursuivent le travail sur une caméra électronique dans laquelle la cathode et les plaques sont séparées par une vanne (AE, 1). Au département d'astronomie de l'université du Texas, un nouveau laboratoire a été créé et Griboval signale le travail entrepris pour réaliser une caméra électronique à focalisation magnétique, en visant une haute définition et une haute sensibilité. Enfin, dans un laboratoire industriel travaillant sous contrat de l'Armée américaine, Decker et Mestwerdt travaillent à une caméra électronique ayant une grande cathode et pouvant prendre beaucoup d'images (AE, 19).

(b) *Propriétés photométriques:* La caractéristique "éclairement-densité" a été étudiée notamment par Duchesne *et al.* (**3**; AE, 27), Kron *et al.* (AE, 1) et par Smyth et Brand (AE, 737). Les auteurs sont d'accord que le début de la caractéristique est linéaire (densité proportionnelle à l'éclairement). Au-delà d'une certaine valeur de la densité (par exemple $D = 1{\cdot}5$ pour l'émulsion Ilford G5) des désaccords apparaissent. Pour essayer de trancher cette question, Walker a proposé une action coopérative: un certain nombre de groupes reçoivent des émulsions provenant d'un même lot, les utilisent dans des conditions aussi comparables que les divers récepteurs électronographiques le permettent et les clichés obtenus font l'objet de mesures avec plusieurs types de microphotomètres. McGee et Kahan ont bien voulu se charger de fournir les émulsions et de centraliser les résultats qu'il pourra être intéressant de présenter et de discuter à Brighton. La comparaison entre la photographie et l'électronographie continue de faire l'objet de publications parmi lesquelles celle de Kron *et al.* (AE, 1) et celle de Chopinet et Duflot (**4**). On peut citer, d'autre part, l'essai réalisé par A. Labeyrie (AE, 899) de modifier une caméra électronique en un spectrographe utilisant la Transformée de Fourier. Enfin, la photométrie stellaire par électronographie continue à être étudiée par Ables *et al.* (**5**).

(c) *Emplois en astronomie:* A l'Observatoire de Haute-Provence, la caméra électronique est utilisée en spectrographie et en photographie directe. Un spectrographe à très haute résolution a été construit au foyer Coudé du télescope de 193 cm; le montage et les résultats sont indiqués dans trois publications: Baranne *et al.* (**6**), Duchesne *et al.* (**7**), Bied-Charreton *et al.* (AE, 27). Une bonnette pour photographie électronique est associée au foyer Newton du même télescope et a été utilisée par Wlérick pour l'étude photométrique de noyaux de galaxies et pour l'identification et la mesure de radiosources très faibles (ce dernier travail en collaboration avec P. Véron) (AE,

787); Bijaoui l'a employée pour la photométrie des régions centrales d'amas globulaires (AE, 27). D'autre part, à l'issue d'une série d'observations de la vitesse de Mercure, étalées sur plusieurs années, Duchesne, Feissel et Guinot ont pu obtenir une valeur précise de la constante de l'aberration. A l'Observatoire Lick, Walker, associé à divers collègues, a continué les travaux de spectroscopie électronographique sur des étoiles telles que SS Cyg, sur des nébuleuses gazeuses et sur des galaxies (Voir *Ap. J.*).

L'utilisation du spectracon (McGee, AE, 61) est prévue au foyer Coudé du télescope Isaac Newton de l'Observatoire Royal de Greenwich à Herstmonceux (**8**); il a été employé très largement par Walter pendant un séjour d'une année à l'Observatoire de Cerro-Tololo (Chili) pour la photométrie d'amas globulaires.

La caméra développée par Kron (AE, 1) est utilisée pour la photographie directe et la photométrie de champs stellaires, de nébuleuses diffuses (par H. D. Ables), d'amas globulaires, et la photographie en bande étroite de nébuleuses planétaires (par Kron et Walker). En photographie directe à bande large, une pose d'une heure permet d'atteindre la magnitude 23,8.

2. *Convertisseurs d'images*

(a) *Centres de développement*: Tandis que les récepteurs électronographiques sont principalement élaborés par les astronomes, les convertisseurs d'images, dont les applications sont étendues (militaires, rayons X, ...), sont généralement produits par des firmes industrielles notamment allemandes, américaines, anglaises, françaises, hollandaises, soviétiques. Une des exceptions notables est le travail qui continue à l'Imperial College dans le groupe de McGee (AE).

(b) *Installation dans les observatoires*: Elle s'est poursuivie dans de nombreux pays et principalement aux Etats-Unis grâce, entre autres, au "Comité Carnegie"; on peut dire qu'au cours des trois dernières années, les patients efforts de ce comité ont porté leurs fruits dans le domaine de la spectrographie. En effet, de nombreux grands instruments ont été équipés de spectrographes à convertisseurs, par exemple le télescope de 5 m du Mont Palomar, au foyer Cassegrain (Dennison, AE, 767) et au foyer Coudé, le télescope de 3 m de Lick où les petites dimensions du tube ont permis de l'utiliser au 1er foyer, et le télescope de 2·13 m de Kitt Peak au foyer Cassegrain.

(c) *Observations*: L'usage spectroscopique des convertisseurs est devenu suffisamment courant pour que les astronomes ne le mentionnent que très brièvement dans leurs publications; d'ailleurs, fréquemment ils emploient concurremment les convertisseurs et d'autres récepteurs. Il serait difficile d'établir une liste exhaustive des travaux exécutés avec les convertisseurs.

3. *Télévision*

L'ère spatiale est aussi celle de la télévision. On se souvient des découvertes faites avec les appareils de prises de vues placés à bord de satellites météorologiques (Tiros, Nimbus) ou de sondes lunaires telles que les "Rangers". Ces appareils sont, en fait, en compétition avec la photographie puisque, dans les cas des sondes sovietiques MAS et des satellites "Lunar Orbiter", de véritables stations photographiques existent à bord, avec retransmission ultérieure des clichés.

Mars s'est révélé un objet de choix pour la télévision; après le succès, il y a plusieurs années, de Mariner IV, deux sondes spatiales Mariner VI et Mariner VII, équipées de vidicon, se sont approchées de la planète et ont pris une série de remarquables "clichés". Ceux-ci ont révélé l'existence de deux types de cratères et d'un dépôt blanc très fin dans la calotte polaire Sud (travail d'un groupe dirigé par R. Leighton).

Si les vidicons conviennent bien pour des objets brillants, tels que la Lune et Mars, des récepteurs plus sensibles sont nécessaires pour les étoiles; c'est ainsi que l'OAO II (Orbiting Astronomical Observatory), le projet Célescope, réalisé par le Smithsonian Astrophysical Observatory, emploie 4 tubes "uvicons" qui fonctionnent dans l'ultra-violet et utilisent un étage amplificateur du type convertisseur, suivi d'un étage du type télévision (Goetze, AE, 159; Hansen, AE, 807).

Au sol, la télévision est utilisée en Union Soviétique notamment par Kuprevitch (**9**), Kerinbekov et Efendiev (études solaires, **10**), Abramenko *et al.* (photométrie stellaire, **11**). Elle a permis à Miller et Wampler d'identifier l'astre optique associé au pulsar de la nébuleuse du Crabe. Elle sert enfin d'instrument auxiliaire pour pointer un télescope vers un champ donné (**8**).

On peut mentionner pour terminer que c'est grâce à des caméras de télévision que 800 millions d'hommes ont pu voir les premiers pas de l'un des leurs sur le sol poussiéreux de la Lune.

Conclusion

L'astronomie a épousé son temps et il semble qu'elle a pris résolument au cours des trois dernières années le "tournant" électronique. Les récepteurs photoélectriques d'images ne sont que des récepteurs parmi d'autres mais les astronomes n'hésitent pas à les employer chaque fois qu'ils apparaissent les plus adaptés.

RÉFÉRENCES

1. Kent Ford, Jr. W. 1968, Electronic image intensification, *Ann. Rev. Astr. Astrophys.*, **6**, 1.
2. 1969, *Photoelectronic Image devices*, dans *Adv. electronic Electron Phys.*, Eds. J. D. McGee, D. McMullan, E. Kahan, B. L. Morgan, Academic Press, London et New York.
3. Duchesne, M. 1967, *J. Observateurs*, **50**, 123.
4. Chopinet, M., Duflot, R. 1967, *J. Observateurs*, **50**, 415.
5. Ables, H. D., Kron, G. E., Hewitt, A. V. 1969, *Bull. Am. astr. Soc.*, **1**, 231.
6. Baranne, Bastie, Bijaoui, A., Duchesne, M., Lecontel, J. M. 1967, *J. Observateurs*, **50**, fasc. 3.
7. Duchesne, M., Herman, R., Lecontel, J. M. 1967, *C. R.*, **256**, 1213.
8. Woolley, R., Stoy, R. H., Cousins, A. W. J. 1969, *Q. J. R. astr. Soc.*, **10**, 127.
9. Kuprevitch, N. F. 1968, *Izv. glav. astr. Obs. Pulkovo*, No. 184, p. 156.
10. Kerimbekov, M. B., Efendiev, Ch. A. 1968, *Soln. Dann. Bjull. SSSR.* No. 11, p. 93.
11. Abramenko, A. N., Petrov, P. P., Prokofeva, V. V. 1968, *Astr. Cirk. Izdav. Bjuro astr. Soobšč.*, No. 491 et 497.

G. WLERICK
Président du Groupe de Travail

10. SOLAR ACTIVITY (ACTIVITÉ SOLAIRE)

President: Z. Švestka.
Vice-President: J. T. Jefferies.
Organizing Committee: K. O. Kiepenheuer, R. Michard, S. Nagasawa, G. Newkirk, J. Rösch, A. B. Severnyj, M. Waldmeier.
Secretary: A. D. Fokker.

The reader is kindly asked to consider this report for some sort of greatly unbalanced equilibrium between the pressure exercised by the steadily increasing flow of papers discussing the vastly extending subject of solar activity, and the equally steady pressure exercised by the editor who insists on having this report twice as short as three years ago. Therefore, some selection of the subjects and papers had to be made, a fairly severe one, since only about one third of all published papers could have been mentioned in the references, and I am well aware that almost everyone who will read the report, could suggest some improvement of the selection scheme. Nevertheless, I believe that the report as it stands, can fulfill its basic purpose, i.e. to give to the readers a general picture on the progress, which has been achieved in the field of solar activity research since the Prague meeting and on some of the research activities which are in progress at the present time.

It is a great pleasure to acknowledge the enormous help I have received from the authors of the individual sections of the Draft Report, as well as from many other members of the Commission, who have very kindly provided me, or the co-authors, with useful information on the recent progress of their work. It is necessary to add that some highly valuable comments made by V. Bumba, J. T. Jefferies, C. Sawyer, E. v. P. Smith, B. Valníček and J. M. Wilcox, have also been included in other chapters than those prepared by themselves.

1. INTRODUCTION

One of the most remarkable characteristics of the recent years is an increasing effort to improve the resolving power both of ground-base and space research equipment. The excellent white-light and Hα photographs obtained at Culgoora, Athens, Sacramento Peak and many other stations, several ingenious improvements of solar magnetography, the new Australian and Japanese compound radio interferometers, and X-ray photographs of the Sun, going down to the order of 1″ in resolving power, are outstanding manifestations of this general effort. New solar stations built at water-surrounded sites and the ATM project indicate that this tendency will continue in the coming years. Thus the fine structure of solar active phenomena, detected by the theoreticians in the past years both from spectroscopic and energetic studies, is being attacked now by the solar observers.

Another great progress has been achieved in extending our knowledge of the solar spectrum towards both its ends. Very low frequency radio telescopes on satellites, and X-ray and EUV spectrometers and spectroheliographs, on the OSO satellites in particular, have contributed to this progress. In some research fields, and what I have in mind, are the magnetographic and X-ray studies, in particular, routine studies are being replaced by ingenious improvements in the quality of the obtained results. The continuously increasing sensitivity of receptors has enabled us to detect very small particle events in space and permanent particle emission from specific active regions. Well organized international cooperation has brought some new aspects into the study of complex solar phenomena which, in their consequences, reach far behind the orbit of the Earth.

For other summarizing considerations and all the details the reader is referred to the individual chapters of the Report. Before concluding this very brief introduction, however, I would like to mention two periodicals, which have extensively developed in the past three years and which, according to my opinion, have been of great assistance to research workers in the field of solar activity: *Solar Physics*, the journal starting in January, 1967 (edited by C. de Jager and Z. Švestka, published by D. Reidel Publ. Comp., Dordrecht-Holland), and *Solar Geophysical Data*, from 1969 divided into two parts, comprising prompt and comprehensive reports respectively (edited by J. Virginia Lincoln, published by ESSA at Boulder, U.S.A.). This second publication, jointly with the *Quarterly Bulletin on Solar Activity*, provides solar physicists, geophysicists, and space scientists with a very comprehensive selection of data from the field of solar-terrestrial physics, and the work of Miss Lincoln and her co-workers is to be greatly appreciated.

2. EQUIPMENT AND METHODS

(The reader is also referred to the "Progress Reports from Solar Institutes", published in the past three years in the journal *Solar Physics*).

2.1. *Optical observations*

(by B. Valníček, Ondřejov)

The effort to find best seeing conditions for solar work has resulted in the establishment of new solar stations at water-surrounded sites. The great advantage of such establishments has been well demonstrated by Becker (1969) for four different sites in California, and the place for the Aerospace Corporation Observatory was selected on a narrow neck of land that extended into the San Fernando reservoir. Results in Hα and in integrated light are convincing. Similar effect has been reached by the Mt Wilson and Palomar Observatory as a result of a 3-year site survey carried out by Leighton, Howard and Zirin. The new solar observatory is constructed in the middle of Big Bear Lake in the San Bernardino Mountains. In Canada, a peninsula extending into a wide bay of the Ottawa River is under development as a telescope site (Gaizauskas, 1969).

New solar towers are prepared for observation at Kandilli Observatory, with a 60/45 cm coelostat, and at Meudon Observatory, where the work starts in 1970. At Sacramento Peak a new solar tower started operation in autumn 1969, equipped with many innovations well known in space technology, but not previously incorporated into ground-based solar instruments. The domeless refractor at Capri evoked a similar instrument for the Aerospace Corporation Observatory at San Fernando, where a 24-inch Cassegrain-Gregorian telescope with spectroheliograph has been installed (Becker, 1969). At the Kiev-Goloseevo Observatory a horizontal telescope with 44-cm mirror and 17-m focal length is now in function, with 7-m spectrograph. In Utrecht, a total revision of the horizontal solar spectrograph has been completed by Namba. At Stockholm Observatory a new echelle spectrograph for flare studies is under construction, and at the McMath-Hulbert Observatory the vacuum spectrograph has been digitalized and prepared for realization of a high-dispersion digital atlas of solar spectrum. At Ondřejov the great horizontal telescope is now finished, with 35-m focal length and 10-m spectrograph.

Very characteristic for this epoch of solar research are multiple instruments on solar spars, which make complex work in solar activity studies easily possible. Equipment of this kind is now at the new Solar Station of the Dominion Observatory (equipped to monitor the disk and limb with different filters, and a selected active region in Hα and in integrated light), at the new Big Bear Solar Observatory (with two 10-inch refractors, a 15-inch off-axis Cassegrain system and 9-inch reflecting coronagraph) and at Hawaii (with a 25-cm coronagraph, two 10-cm coronagraphs for work in two different lines and a 15-cm chromospheric telescope). Also the Catania Observatory will be equipped from 1970 with a spar of the same type as at Monte Mario Observatory in Rome. The six-side equatorial spar supports a white light refractor, Hα disc patrol, Hα great size for special events, Hα prominence coronagraph and K monochromator.

In general, refractors are in renascence for solar work today. At Culgoora a 30-cm refractor equipped with 1/8 Å Hα filter is applied to fine chromospheric studies in different parts of the Hα profile. At the National Observatory in Athens a 40-cm refractor with 505-cm focus is working with practical resolution of 0·6″ in the Hα line. At Ondřejov, a Clark objective, 20-cm aperture and 285 focus, is applied for fine structure studies of the photosphere, in connection with a 0·8 Å Šolc filter. Refractors equipped with 0·25 Å filters are also working at Hawaii and San Fernando.

Several new patrol instruments have been built. For the Hα line at Tel Aviv, Big Bear Lake, Uttar-Pradesh and Kandilli Observatories, and for the Ca II K-line in Athens, Uccle and at the Kandilli Observatory again.

Coronagraphs are still in development and many special instruments are now in use for the coronal work. A coronal activity monitor, a branch of coronal polarimeter similar to the coronameter, has been established at Mauna Loa in Hawaii. A balloon-borne coronagraph, for observations of the F-corona in the infrared region near 2·2μ has been described by McQueen (1968). Also at the High Altitude Observatory, Boulder, a coronagraph for the Apollo telescope mount has been constructed, with 3·2-cm aperture and 44-cm focus. The corona can be photographed from 1·5 R to 6 R with the aid of three occulting disks in front of the objective (Newkirk, 1967*a*). New coronagraphs have been built at the Kislovodsk station of Pulkovo Observatory, with 53-cm objective (Gnevyshev *et al.*, 1967), and at the Norikura Corona Station in Japan, equipped with Littrov spectrograph, focus 7 m, and a 2·5-m echelle-spectrograph. An interesting theoretical study of the coronagraph system has been made by Demarq and Rösch (1967).

A polarization-interference filter with thermo-optically compensated parts, practically eliminating the need for a precise thermostat, has been put into use for coronal observation and investigated by Joffe and Smirnova (1967). A very sophisticated conception of a birefringent filter applying optical rotation of the birefringent plates is proposed by Katzenstein (1968), but its practical use has not been documented.

2.2. *Magnetic field measurements*
(by V. Bumba, Ondřejov)

Because of the urgent need for increased spatial and time resolution in solar observations, a considerable effort in the past three years has gone into the modernization of existing photoelectric magnetographs, improvement of measuring and evaluation methods, and invention of new types of observational devices. At Mt Wilson a more accurate guider, an electro-optic crystal yielding an improved signal-to-noise ratio, an improved Doppler compensator, and a completely automatic data-recording system were introduced (Howard *et al.*, 1968). A new electronic scheme of a magnetograph for simultaneous recording of the longitudinal and vector components of the transversal magnetic field at Crimea was described by Nikulin (1967). Regular observations have been started at Pulkovo with the aid of a six-channel two-slit magnetograph with one photomultiplier and rotating $\lambda/2$ plate measuring longitudinal as well as transversal fields (Ikhsanov and Platonov, 1967). Deubner and Liedler (1969) have described the new vectormagnetograph of the Fraunhofer Institute at Capri, which measures all components of the magnetic field vector strictly simultaneously. A photographic method is applied which presents longitudinal and transversal field components, vector, azimuth, Doppler shift and intensity in the form of maps. Late in 1967 the Kitt Peak magnetograph expanded to allow transverse measurements, and since the fall of 1969 a 40-channel magnetograph with both longitudinal and transverse output is in operation at the same observatory. It uses advances in fiber optics and micro-circuit technology, with spatial resolution 0·5″–2″ for making photoelectric recordings comparable in resolution with spectroheliograms. A new technique for displaying magnetograph observations has been developed by Frazier and Sherrer (1969) and applied to the 12-channel magnetograph at Kitt Peak. Using the data from a raster-scan, a digital "spectroheliogram" is constructed on the face of a cathode ray tube and photographed. The new Gregory-coudé telescope in Locarno (Göttingen station) obtained several improvements with special regard to the magnetic field measurements (Schröter and Wiehr, 1968). At the modified vector magnetograph all four Stokes parameters are measured simultaneously, using a $\lambda/13{\cdot}7$ plate and four photomultipliers (Wiehr, 1969). A detailed description of the lambdameter was published by Rayrole (1967) demonstrating some of the advantages of this photographic method.

At the Rome Observatory a new kind of a very narrow pass-band filter using the atomic-beam techniques in absorption is used as a "Babcock window" for fitting of a magnetograph (Cimino *et al.*, 1968). Giovanelli has announced first preliminary tests of a new type of magnetograph, which uses a high-resolution filter ($\Delta\lambda = 0{\cdot}05$ Å) and incorporates three automatically controlled Fabry-Pérot interferometers which can be adjusted to transmit any desired wavelength. The monochromatic solar image is thrown through a $\lambda/4$-wave plate and a polarizing system and photographed in oppositely circularly polarized components with a short ($\sim 0{\cdot}3$ s) exposure time. Beckers (1968*a*) and Ramsey *et al.* (1968) have described two similar modifications to a birefringent filter, which allow high spatial resolution observations of magnetic and velocity fields in rapid sequences of exposures. A new magnetograph, which is now under construction for the Naval Research Laboratory, Marshall Space Flight Center, and Göttingen Observatory, will be able to produce magnetograms nearly in real time, showing the complete magnetic vector with a resolution of 1″ (Brückner, 1968). It will use a birefringent filter and a new SEC vidicon tube which converts the polarization information into electrical currents simultaneously for all picture elements. Digitized electrical signals will be processed by a computer which will prepare the magnetic maps. A real-time video magnetograph using digital enhancement is being constructed by Janssens at the Aerospace Observatory (Becker, 1969).

A solar magnetograph without electronics or any moving parts has been designed by Stenflo (1968*b*). Its principle is the use of a subtracting double-dispersion optical system. Stenflo (1969*c*) also described another method which modulates the light entering the spectroheliograph in the wing of a magnetic-sensitive line by putting alternately a + and − $\lambda/4$ plate in front of the slit and placing a fixed polarizer after the modulation $\lambda/4$ plate. In the resulting spectroheliogram the magnetic regions show a fringe pattern where the contrast between the fringes is a direct measure of the field strength. A similar result can be obtained by replacing the varying $\lambda/4$ plate with a stationary Babinet compensator whose fringes are parallel to the spectrograph dispersion (Beckers and Stenflo, 1969).

Severnyj (1967) was the first who pointed out the problem of large discrepancies between the theoretical calibration curves of a vector magnetograph calculated for an optically thin spectral line in a homogeneous field and those obtained photographically or from visual measurements of Zeeman splitting or from the field strength of a laboratory magnet. The same question was discussed theoretically by Rachkovskij (1967*b*) who demonstrated that with an assumption of (a) only true line absorption, (b) only line scattering, (c) thin layer line formation, it is not possible to get a good agreement between the theoretical calibration curves and the measured transversal signal. The fact that the use of any theory of spectral line formation in a homogeneous magnetic field cannot explain the discrepancies, has also been demonstrated by Obridko (1968*a*). In a critical paper, Beckers (1968*b*) has pointed at an erroneous signal for the Doppler shift which may also lead to incorrect values of the magnetic field strength by several existing magnetographs. Wiehr (1968, 1969) has discussed the miscentering by the Doppler compensator, and uncertainties that are associated with the calibration of the response of a vector magnetograph in terms of H and γ using Unno's theory, especially for the line $\lambda 5250$ which is rather sensitive to temperature changes. According to Harvey and Livingston (1969) magnetograph measurements with temperature sensitive lines (such as Fe I $\lambda 5250$ Å) lead to an underestimation of magnetic intensities by a factor of 2 to 4 outside of sunspots.

Staude (1969) has demonstrated that the discrepancies between the theoretical and experimental calibration curves may be explained by unresolved horizontal inhomogeneities in the magnetic field region and a similar problem has also been studied by Wittmann. A filamentary model of the magnetic field and magnetic microturbulence with optically thin magnetic turbulent elements have been studied by Stenflo (1968*a*), who shows that pure microturbulence does not falsify the observed value of the average field, but inhomogeneous nature of the magnetic field may make the measured values too low.

2.3. *Observations from space vehicles*

(by E. v. P. Smith, University of Maryland)

Probably the most important advance of the last few years has been the increased spatial and spectral resolution of observations made both from satellites and from rockets. For example, Vaiana *et al.* (1968) have obtained remarkable photographs of flares in the soft X-ray region with a resolution of 1″ during rocket flights, with instrumentation relying on image forming grazing-incidence optics. A spectral resolution ranging from 0·01 Å at 2 Å to 0·6 Å at 300 Å was obtained by the Goddard Space Flight Center spectrometer flown on OSO 3 (Neupert *et al.*, 1969). This instrumentation included both crystal spectrometers and a grating spectrometer. OSO 4 included the spectrometer experiments from NRL which covered the wavelength range from 0·5 to 8·5 Å. The instrument integrates over the entire solar disk but the wavelength determinations have an error of only about 0·03 Å (Meekins *et al.*, 1968). The Harvard Observatory photoelectric spectrometer experiments have been flown both on OSO 4 and OSO 6. OSO 4 included the wavelength region 300 to 1400 Å with a spatial resolution of 1′ while on OSO 6 the wavelength range is from 280 to 1360 Å and a resolution of 35″. The instrument has the capability of several modes of operation: Spectroheliograms of the entire disk of the Sun in any pre-selected line in this wavelength range can be made every 8 min, and alternatively in the time scan mode, spectroheliograms 7·0′ by 7·5′ are possible every 30 s. The third mode records spectra at any position within 23′ of the center of the solar disk. In addition to the Harvard spectroheliometer, other solar experiments on OSO 6 include the NRL X-ray spectrometer extending from 0·14 to 25 Å, the hard X-ray detector for five energy bands between 20 to 200 keV of the University of Bologna, and the University of London EUV monitor (180 Å to $L\alpha$).

Excellent time resolution was attained in the soft X-ray region by the Leicester experiment on OSO 5 which includes a grazing incidence telescope that maps the Sun in 7 bands between 8 and 15 Å with a spatial resolution of 1·6′ and time resolution of 5 s. A second Leicester experiment on the same satellite scans the Sun with an aperture of 2″ by 2° in the 3–9 Å band every 7 s (Evans and Pounds, 1968). Leicester is also preparing a new crystal spectrometer to be flown, initially on rockets, which will use field collimators to select and limit the area of the solar disk to be observed rather than integrating over the entire disk of the Sun as in the past. At the Physical Institute of the Soviet Academy of Sciences the group of Mandelshtam, Zhitnik and Tindo has flown X-ray experiments on three satellites of the COSMOS series during 1967–1969 and on the satellite Intercosmos 1. These experiments include X-ray photometers for the energy range 2–50 keV and spectroheliograms yielding X-ray spectroheliograms within 2–5 keV with resolving power of about 1′. The Astrophysics Research Unit at Culham Laboratory has flown in a rocket and will fly again in collaboration with the Stockholm Observatory an echelle spectrograph for the wavelength range λ1200–2200 Å. This group is also planning to fly again the two EUV and soft X-ray grazing incidence spectrographs ($\lambda\lambda$15–120 Å and 450–800 Å).

Dijkstra at the Space Research Laboratory at Utrecht is developing fresnel-type zone plates for imaging monochromatically at X-ray wavelengths. One fresnel system has already flown in a NASA rocket in October 1967, and improved instruments are expected to be flown in ESRO rockets in 1970 and 1971. Van Beek and Van Gils of the same group are also developing an equipment for monitoring hard X-ray bursts between 20 and 600 keV in 12 channels, aboard the ESRO-TD 1 satellite in 1972. Prior to that time balloon flights will test the instrument. Other X-ray spectral observations being planned are those of the group at Paris University under the leadership of Michard, which will include spectroscopic observations from rockets in the spectral range 1·8 to 9 Å and 40 to 80 Å, with 2 to 3′ spatial resolution. The first flights are anticipated toward the end of 1970. Valníček's group at the Ondřejov Observatory has flown equipment monitoring solar X-rays in three channels, one of them being for energies > 50 keV, on the Soviet satellite Intercosmos 1 in September 1969, and the experiment is planned to be repeated in 1970. In the past, hard X-rays have been observed from OSO 3 by the University of California group under the leadership of Peterson, and on OGO 3 by the Goddard group (Cline *et al.*, 1968).

The Apollo telescope mount (ATM) for 1972 will carry five solar oriented experiments. These will include ultraviolet spectrographs and spectroheliographs, and X-ray imaging telescope, and a remote occulting telescope for white-light coronagraph. The participating institutions are Harvard Observatory, Naval Research Laboratory, American Science and Engineering, Goddard Space Flight Center and the High Altitude Observatory. The California Institute of Technology, The Jet Propulsion Laboratory and Mt Wilson and Palomar Observatories have participated in the development of a 65-cm solar telescope to be flown on a satellite to obtain extremely high resolution films in various wavelengths. The American Science and Engineering imaging X-ray telescope and a scaled-down version of the Naval Research Laboratory ATM spectroheliograph will be flown on a rocket as part of a flare-watch program late in 1969. Similarly Novick at Columbia is preparing a flare rocket for determining the polarization of flares in the soft X-ray region for 1970. A balloon-borne 30-cm telescope with a compact high resolution spectrograph has been developed by Fraunhofer Institute (Freiburg). With a resolution of 0".4 and 11 mÅ and a guiding stability of 0·1 simultaneous observations of spectra and white-light or Hα pictures are aimed for. The first results of testing of a solar stratoscope were described by Krat (1968). A "solar" space probe going up to 0·3 A.U. is under consideration in cooperation of GFR and USA (Porsche, 1968).

Both Michigan Observatory and Goddard Space Flight Center have flown radio low-frequency dynamic spectrographs to study solar bursts. The Michigan instrument flown on OGO 3 has the range from 2 to 4 MHz while the Goddard instrument has a range from 0·45 to 3·0 MHz.

Narrow band filters for optical wavelengths suitable for prolonged satellite operation are being developed by Harvard Observatory in conjunction with Perkin-Elmer Corporation and Thin Film Industries. These are solid etalon Fabry-Pérot interferometers combined with multilayer dielectric filters, with high stability over an extended period. Filters with ½Å bandwidth have been made, and wavelengths include $\lambda\lambda$6563 (Hα), 6103 (Ca I), 5250 (Fe I), 3933 (Ca II), and 3968 Å (Ca II).

3. CYCLIC VARIATIONS OF SOLAR ACTIVITY

(based on a draft prepared by J. Xanthakis, Athens)

From a study of the annual values of the 5303 Å coronal line intensity during 1954–1965 Gnevyshev and Chistjakov (1968) have confirmed the existence of two activity maxima in the 11-year solar cycle. The application of autocorrelation methods by Vitinskij (1967) and Vitinskij and Kopecký (1968) has shown the existence of possible periodic variations of the mean importance of spot groups with periods smaller than 11 years, and in some cases the presence of Gnevyshev's two peaks in the 11-year cycle. Xanthakis' (1969) analysis has shown that these two maxima also appear in the distribution of proton flares and as a function of the spot areas index. Kopecký and Kuklin (1969*a*) have concluded that the double maximum of the 11-year cycle is determined by the importance of spot groups and not by the frequency of the spot group formation.

Gleissberg (1967) has studied the mean curves of the 80-year solar cycle, which generally show a symmetry of the ascending and descending branch. Zhukov and Muzalevskij (1969), however, have carried out a correlational-spectral analysis of the solar activity periodicities which casts a serious doubt on the reality of the 80-year solar cycle. Their analysis has detected a cycle with a period of about 200 yr, with power of the harmonic exceeding powers of all the other long-period components. Vitinskij (1969*b*) has found that a secular cycle of sunspots, unlike the 11-year cycles, is of the character of free pseudo-harmonic oscillations. An increasing attention has been paid to the hypothetical influence exerted by planets upon the Sun (Memeth, 1967; Bigg and Mulhall, 1967; Takahashi, 1968; Blizard, 1969).

Leighton (1969) has presented a kinematical model of the solar cycle, based upon field amplification by the solar differential rotation. The rather broad quantitative agreement of his model with the known features of the solar cycle led him to a conclusion that other factors which have been suggested as contributing to the solar cycle, such as meridional circulatory currents and torsional oscillations, are unlikely to be important. This model is now being extended to include variation of field strength and sunspot eruption with longitude as well as latitude in the hope of elucidating the well-known spatial correlations in sunspot eruptions. At Munich Observatory, Born and Wellmann have derived the diffusion coefficient for magnetic fields from study of the development of Ca II plages, which is in good agreement with Leighton's theory. Other models of the solar cycle have been proposed by Nakagawa and Swarztrauber (1969), who have solved the hydromagnetic equations for a differentially rotating Sun and found that the solutions obtained show several interesting topological features of the general magnetic field, and by Steenbeck and Krause (1969) who presented models of alternating field dynamos, which are in good agreement with the real situation on the Sun.

4. ACTIVE REGIONS ON THE SUN

(by K. O. Kiepenheuer, Freiburg)

At the 1967 IAU Symposium *Structure and Development of Solar Active Regions* (Kiepenheuer, 1968*a*) a number of important results have been presented. Important contributions also came from the CSSAR Project, organized by Michard (1968), and the PFP Project, organized by Švestka (1968*b*). The synoptic presentation of solar magnetic fields for the period August 1959 to June 1966 prepared by Howard *et al.* (1967) represents a basic and indispensable contribution for future investigations of the solar magnetic cycle and the production and evolution of active regions. Also the magnetic classification of active regions, introduced by Smith and Howard (1968) and by Martres (1968), has become an efficient tool in this context.

In this section we shall understand active regions as physical units and not as an assembly of single phenomena. From this point of view, the concept of an active region has undergone considerable changes during the last years. New types of observations as well as the systematic presentation and digital evaluation of solar magnetic data have brought a wealth of new and significant information. The mechanism of an active region is intimately connected with the model of the solar activity cycle. It becomes increasingly clear, however, that the well-known model proposed by Babcock, which is able to explain a number of characteristic features of the cycle quite adequately and which represents, because of its perspicuity and clearness, an enormous stimulus for research, is no more in position to account for these complex observations (Kiepenheuer, 1968*b*; Bumba *et al.*, 1968). Therefore, it appears most important to work out a consistent global mechanism for solar activity as, e.g., Leighton (cf. Section 3) is attempting to do in his recent theory.

An active region can no more be regarded as an independent organism produced only by general field amplification which is brought to the surface by magnetic buoyancy. The recent studies have shown that the significant factors influencing the birth and the evolution of active regions seem to be mainly the following:

(a) Deep-seated magnetic subsurface sources – probably at the base of the convection layer – which produce active regions over a longitude zone of several tens of degrees. These sources rotate with a synodic period of about 27·0 days (Bumba and Howard, 1969; Bumba *et al.*, 1969). These long-lived components in the solar magnetic field distribution, which are obscured by rapidly changing structures of smaller dimensions, are clearly connected with the interplanetary magnetic field structure (Wilcox and Howard, 1969; Schatten *et al.*, 1968) and with the so-called "Bartels Active Longitudes" (Bumba and Howard, 1969). This also seems to have some bearing on the concentration of faculae in longitude (Dodson and Hedeman, 1968). The effect of active longitudes is found most pronounced for the most active phenomena (Levickij, 1968; Vitinskij, 1969; Švestka, 1969), including the proton-flare regions on the Sun (Sakurai, 1967*a*; Švestka, 1968*a*; Haurwitz, 1968; Bumba and Obridko, 1969).

(b) Continuous interaction of old background fields with the field of new active regions, the former being responsible for the observed recurrence of active regions. The locations of the centers of recurrence rotate with the 27·0 day period (Bumba and Howard, 1969).

(c) Active regions, once formed usually do not rotate any more with the 27·0 day period. They somehow withstand, however, the shearing effects of differential rotation as long as the magnetic fields involved are strong enough (Bumba and Howard, 1969; Dodson and Hedeman, 1968).

(d) The formation of a significant number of active regions was followed in numerous instances within a solar rotation by the development of another active region on the opposite site of the Sun, about 180° away (Dodson and Hedeman, 1968*a*).

(e) The concentration, dilution and redistribution of magnetic fields is done by a system of convective motions occurring in the convection layer of the Sun: granulation, supergranulation (Bumba, 1967) and the "giant cells", all three interacting with differential rotation of the Sun in a not yet quite perceptible way (Schmidt, 1969).

The concept of giant cells has received support from theory: Simon and Weiss (1968) have shown that the observed concentration of magnetic field between supergranules can be produced only by subphotospheric convection in cells of about 30000 km, which must extend over several scale heights. From such a model a third characteristic scale of motion follows with cells of about 300000 km in diameter.

5. FINE-SCALE MAGNETIC FIELDS

(by V. Bumba, Ondřejov)

Spectrograms obtained during moments of good seeing, making use of the 80-cm solar image at Kitt Peak, demonstrate magnetic fields of several hundred gauss in tiny areas as small as 500 km in extent in regions sometimes well removed from areas of sunspot activity. They can be observed as weakenings or gaps in temperature-sensitive lines and as local darkening in the continuum (Sheeley, 1967) and are similar to Beckers and Schröter's (1968*a*) magnetic knots (cf. subsection 7.2). Time-lapse sequences of spectroheliograms using Leighton's photographic technique allowed to be found a steady flow of these tiny field concentrations laterally outward of sunspots and the birth of new elements of this network (Sheeley, 1969). Such photographic observations seem to attack one of the most important problems of the dynamics of active region development. On the other hand, however, line scans with a photoelectric magnetograph using the same instrument did not show magnetic elements below about 1500 km in size (Livingston, 1968). The influence of the resolving power of a magnetograph on statistical results concerning the mean field strength, the average size and the number of magnetic elements of different polarity was examined by Severnyj (1968*a*).

The reality of tiny concentrations of magnetic lines of force and their relations to flares and bright points has been confirmed by the photographic technique using a birefringent filter (Ramsey *et al.*, 1968). The importance of such small-scale fields in the field diffusion was considered theoretically by Kassinskij (1969) who showed that these fine-scale field elements are transported through the large-scale motions during a very long time without any considerable damping.

At Capri (Deubner, 1968) and Crimea (Severnyj, 1968*c*) short periodic oscillations, with periods from 7 to 9 min, of magnetic field strength have been found on repeated scans along a fixed line, or on one place on the disk, outside of active regions. These measurements were carried out photoelectrically with relatively high resolution, the aperture slit of the magnetograph being of the order of 10 square seconds of arc. At Capri close correlation has been found between these oscillations and simultaneously measured contrast changes. The physical nature of this phenomenon is unknown, although the seeing cannot yet be fully excluded as its possible source.

6. FACULAE AND PLAGES

(extracted from a draft* prepared by G. Godoli, Catania)

Daily maps of Hα and K plages observed at Catania have been made since 1967, and characteristic figures are regularly published in *Osservatorio Astrofisico di Catania Pubblicazioni*. The McMath-Hulbert Observatory has continued its daily measurement of position, area, and intensity of K plages and has distributed these measurements through the World Data Center A, ESSA, Boulder. Regular observations of faculae and plages also have been made at Kislovodsk. Definitive

* Available on request.

characteristic figures of K plages for IGY and IQSY and maps of K plages for the second year of IQSY were published by Godoli (1967) and Godoli *et al.* (1967). Starting January 1968 the Rome Observatory regularly publishes films with reproductions of daily Hα and K_{232} images of the Sun as *The Photographic Journal of the Sun.*

Loughhead *et al.* (1968) have shown that chromospheric faculae regions seen in Hα may be partially resolved into a granular structure. According to Bray (1968) the correlation between photospheric faculae and plages is weaker than has hitherto been supposed. A detailed study on the chromospheric and photospheric network is in progress at Kitt Peak, carried out by Chapman, Engvold and Sheeley. Spectroheliograms obtained by Fredga (1969) in the Mg II λ2795·5 Å line have been found very similar to K spectroheliograms, and a detailed comparison between Mg II and Ca II is in progress.

A correlation has been found between the K plage intensity and the $L\alpha$ (Monsignori Fossi *et al.*, 1968), and the X-ray flux (Landini *et al.*, 1968), respectively. A correlation between enhanced radio emission at 9·1 cm and X-rays found by Reidy *et al.* (1968) indicated a common physical origin of both these types of radiation in the plage region.

The newly constructed double-beam-splitter magnetograph at Kitt Peak is currently used by Sheeley to study the relation between intensity and magnetic fields of plages. Magnetic fields in plage areas have also been discussed by Tandberg-Hanssen, Porter, Bumba and Godoli at the *IAU Symposium* no 35, as well as by Steshenko (1968) and Livingston (1968). Photospheric and chromospheric longitudinal magnetic fields and the brightness of plages have been compared by Tsap (1967), who also studied the correlation between longitudinal and transversal magnetic fields on different heights (Tsap, 1969). The correlation between velocities at different levels has been investigated by Gopasjuk and Tsap (1969*b*). Gopasjuk (1967) has shown that the brightest portions of a plage as well as the brightest points of the chromospheric network coincide with the neutral line of the radial velocity or lie close to it. The gases move both toward and from the neutral line and do not correspond to the pattern of supergranulation (Gopasjuk and Tsap, 1969*a*).

At IZMIRAN in Moscow, Ioshpa, Mogilevskij, Obridko and Shelting are using simultaneous measurements at two different levels in the photosphere and chromosphere for constructing the spatial structure of the magnetic field for the whole active region, and for this purpose active prominences have also been used by Ioshpa. Mogilevskij and Shelting also use simultaneous magnetic and velocity observations in order to reveal three-dimensional interconnection of the large-scale elements in the magnetic and velocity fields. Alexandrovich and Stepanov (1969) have detected regions of compressed and rarefied plasma in the chromosphere, and found that the brightest parts of plages are situated in the regions of compressed gas.

Weart and Zirin (1969) have shown some characteristic features, that occur in plages prior to sunspot occurrences, some of them confirming discoveries made three decades ago by Waldmeier. Other work of these authors is in progress. They find an unequal development of the Hα plage brightness in association with the preceding and following part of the active region, respectively, while this brightness asymmetry does not appear in the K-line. They also have attempted to determine weak magnetic fields by examination of on-band and off-band high-resolution Hα pictures and report that 5-G contours of plages can be distinguished in this way. At the Cambridge solar magnetograph operating in Malta, observations of weak magnetic fields have been carried out since early 1966 (Von Klüber, 1968), which may contribute to the understanding of plage formation. At McMath-Hulbert Observatory Mohler is investigating the correlation between magnetic measurements and line profiles in plages.

Heights of plages have been studied by Öhman at Kitt Peak comparing K_2 and K_3 components in limb phenomena and work of this type is now under consideration at the Swedish station at Anacapri. Structure of plages has been discussed by Anichini and Godoli (1967), Livshitz (1968) and Bachmann (1969). Zirker (1967) has found larger velocities in a plage than reported for spicules in quiet regions. Sitnik and Mitropolskaja (1968) have concluded that turbulent velocities in faculae exceed the photospheric values and the same result is obtained by Bodajan and Livshitz. Current investigations carried out by Mitropolskaja and Sitnik show that the turbulent velocities decrease with the depth. According to Howard *et al.* (1968) the 5-min oscillations are damped in plages relative to the quiet photosphere, which also has been confirmed by E. Smith (1969). Mazzucconi and A. Righini are studying area and intensity fluctuations of plages in order to establish whether these fluctuations are preferentially present before or after flares, and photometric investigations on the flare-associated plages are in progress at the Kandilli Observatory by Soytürk.

At Harestua Observatory observations of photospheric faculae in several wavelength regions are at present carried out. From photoelectric measurements Schmahl (1967) has found that the intensity differences facula/photosphere are non-zero even for weak faculae in the disk-centre, and therefore it is not necessary to suppose a temperature deficit in deep layers of faculae. In the infrared, the facula contrast has been studied by Kuzminykh and Sitnik (1967). A comparative study of hydrogen lines in spectra of faculae and photosphere has been carried out by Vojhanskaja (1967). Pande *et al.* (1969) have calculated the dependence of facula/photosphere contrast in molecular lines on dissociation energy and the CO molecule in faculae has been studied by Polonskij (1967).

Some other papers on the structure and development of plages can be found in *IAU Symposium* no. 35 and *Annals of IQSY*, Vol. 3.

7. SUNSPOTS

(by W. Mattig, Freiburg)

7.1. *Sunspot spectrum and abundances*

For the last thirty years the Utrecht photometric atlas of the solar spectrum was extremely helpful for the whole solar research. Now two similar publications for sunspots are in preparation. At the Kitt Peak Observatory maps are made for a photographic atlas of the umbral spectrum taken through a Babinet compensator, and photochromic glass is utilized to reduce the photospheric to umbra intensity gradient. At the Locarno station of the Göttingen Observatory photoelectric records have been made by Schröter and Wöhl in the wavelength region 3900–8000 Å in order to produce a photometric atlas of an umbra spectrum.

The abundance of some elements has been determined from high dispersion spot spectra. Li, In, Rb and Cs have been studied by Wiehr *et al.* (1968). A special program to determine the lithium abundance very accurately from the 6707 Å line is under way by Beckers and Hoyt. The appearance of the elements Eu and La in sunspots has been investigated by Bachmann, Pflug and Staude at Potsdam. Kumar at the McMath-Hulbert Observatory and Branch (1970) have determined the abundance ratio of Mg isotopes, which significantly differ from the terrestrial values.

7.2. *Sunspot fine structure*

A substantial progress has been made in the field of the fine structure in spots and their surroundings. Beckers and Schröter (1968*b*) who observed the umbral fine structure (umbral dots) in two continuum wavelengths, came to the conclusion that, after correcting their measurements for the influence of finite resolution, the umbral dot's intensity is nearly the same as the photosphere, and that the dots have a true diameter of 150 to 200 km. These umbral dots are concentrated mostly in the outer part of the umbra and have a lifetime of $\sim$ 25 min. Wilson (1969*b*) has discussed several models with these bright structures on the basis of a three-dimensional radiative transfer analysis.

A very rapidly changing fine structure in the umbral chromosphere was detected by Beckers and Tallant (1969) and confirmed by Wittmann (1969). These bright features, observable in the K-line by filters and with the spectrograph, and called umbral flashes, have diameters smaller than 2700 km, magnetic field strength of about 2000 G, and there are usually about 3–5 flashes in an umbra at the same time. The life-time of such a flash is about 50 s and there is a tendency for their repetition after about 150 s. The umbral flashes show a proper motion of 40 km/s in the penumbra direction and Doppler velocities of 6 km/s. Havnes in Oslo has examined the hypothesis that an umbral flash may be caused by a magnetoacoustic wave phenomenon and, in general, the observations agree with the wave hypothesis.

The chromospheric fine structures in Hα in the surroundings of spots have been studied by Loughhead (1968). The dark filaments, fibrils, were found to be the continuation of the filamentary structure of the penumbra. In the case of this "superpenumbra", most fibrils end within the penumbra and some penetrate right to the border of the umbra. The fine structure of the intensity variation and magnetic field around sunspots has been investigated by Beckers and Schröter (1968*a*). The so-called magnetic knots coincide with dark intergranular space. More than 2000 knots can be found around a unipolar spot. The diameter of a typical knot is about 1100 km, the magnetic field amounts up to 1400 G, it generally shows a downward motion and has a lifetime of more than 30 min. It seems that there exists a critical level of the magnetic field ($\sim$ 1500 G) for the smallest sunspot to appear (Bumba, 1967*b*; Steshenko, 1967). Similar to the magnetic knots are Sheeley's (1967) "gaps" (cf. Section 5). Both knots and gaps coincide with regions of bright Ca II emission. An attempt for an explanation of these phenomena has been presented by Zwaan (1967).

7.3. *Sunspot magnetic fields and motions*

The magnetic field distribution over an isolated spot has been investigated again by Beckers and Schröter (1969*b*). The distribution of the total magnetic field across a spot is given by $H(r) = -H(0)\,(1 + r/R)^{-1}$ and the inclination of the field vector is nearly the same as found by Nicholson. Deubner and Göhring at Freiburg are in agreement concerning the inclination angle, but the field distribution differs remarkably. They found a lower field strength in the outer part of the spot. Kjeldseth Moe (1968*b*) has found a field distribution in good agreement with that of Bumba of 1962. As Beckers and Schröter have confirmed, the magnetic field does not drop to zero at the boundary penumbra-photosphere, which is in accordance with Kassinskij's (1967) remark that the magnetic radius of a spot is larger than that of the penumbra. Recordings of the magnetic field at different photospheric levels made by Kotov (1969) show the presence of a strong transversal field within the umbra and large (up to 90°) changes in the orientation of H with depth in certain parts of the spot. He also obtained the distribution of the vertical and horizontal components of the electric current ($\sim 10^3 - 10^4$ CGSE).

The time variation of the maximum magnetic field strength observed by Kolpakov (1968) shows a wide range of $10^{-4} - 5 \times 10^{-2}$ G/s for different spot types. Ikhsanov (1967) has found that the magnetic field strength varies with time in parallel to the change of spot area, but the rate of change is smaller for larger spots. Jakimiec is investigating the distribution of the magnetic force and electric currents in the surface layers of sunspots. Künzel (1969) has demonstrated that the shape of the penumbra is very closely correlated with the distribution of the magnetic field polarity in sunspot groups, which means that the shape of the penumbra is influenced by the magnetic field of spots in the neighbourhood.

The fine structure of the magnetic field in the penumbra has been studied by Beckers and Schröter (1969*a*) and by Mattig and Mehltretter (1968). The results are contradictory: Beckers and Schröter find the stronger and more horizontal magnetic field in the interfilamentary dark regions, while the result given by Mattig and Mehltretter is vice versa. In accordance to each other are the results on the velocity fine structure by Beckers (1969) and Mattig and Mehltretter. The Evershed effect is restricted to the dark interfilamentary regions in the penumbra. Steshenko and Shcherbakov (1969) have discovered motions of the type of pulsations in the radial direction, along the penumbra filaments. Obviously, more observations are needed to determine the physical parameters in the penumbral fine structure elements.

At Oslo, Maltby and Bönes have made observations to study the Evershed velocity fine structure. Haugen's (1969) studies of the chromospheric velocity field have revealed that the velocities of different elements vary in a wide range of -35 to $+50$ km/s. The mean velocity vector is directed towards the spot and downward, the inclination being about 20° and the mean velocity amounts to 7 km/s, but the rms deviation of line-of-sight velocity is of the same order. The depth gradient of line-of-sight velocity near the penumbral radius has been determined by Sattarov (1969) as about 0·01 km/sec/km. Chitre (1968) has discussed the Evershed flow from a general point of view. Assuming that the Evershed flow starts from the base of the spot with a velocity that is adequate to carry the convective flux at that level, the velocity at the surface turns out to be of the order of 1 km/s, in agreement with the observations.

Strong efforts have been made to explain the measured polarization and line contours of magnetically split lines by solution of the transfer equations of the Stokes parameters under different conditions, as for example by Rachkovskij (1967*a*) and Kjeldseth Moe (1968*a*), who has obtained very complete solution of the equations with a homogeneous field but without any special assumption regarding the model atmosphere. Kai (1968*b*), Beckers (1969) and Göhring have looked for the influence of the Faraday rotation. Numerical integrations of the transfer equations of the four Stokes parameters have been carried out by Kjeldseth Moe, Beckers (1969), and Göhring to investigate the accuracy of the simple Unno theory and the influence of the curvature of the magnetic field. Stenflo (1969*b*) has shown that the reradiated light in the case of longitudinal Zeeman effect will not be completely circularly polarized, if the life time of an atom in an excited state is comparable or shorter than the period of Larmor precession. Most recently, the problem of line formation in a magnetic field has also been discussed by Rees (1969).

An interesting discussion has developed around the problem of the appearance of a π-component of a normal Zeeman triplet in sunspot umbrae, which many a time shows circular polarization opposite to that of the main Zeeman splitting. Recently, e.g., Deubner (1969) has measured in the "π-component" an opposite circular polarization corresponding to a magnetic field of 600 G. Henoux (1968*a*) has interpreted the existence of this π-component as a saturation effect and Göhring (1969), applying Unno's theory of line formation in a magnetic field for the case of true absorption arrived at the same conclusion. But, as Beckers and Schröter (1969*a*) have shown, the Stokes parameter V cannot be explained by the saturation effect described above. Possibly magnetic field inhomogeneities caused by umbral dots are responsible for the existence of the π-component. This is in agreement with the results of Mehltretter (1969) who inspected weak and medium strong magnetically affected lines. Interesting observational results to the solution of this problem have been added by Surkov (1969) who demonstrated that not only the presence and behaviour of the π-component, but also the asymmetry of the whole split line, depends on the age, size, and heliographic position of the studied sunspot.

7.4. *Physical state of sunspots*

A review paper on the structure of sunspots has been written by Zwaan (1968). Many observations of the sunspot spectrum have been made and lots of results are available to derive umbral and penumbral models. According to Rossbach's observations at Göttingen, the spot intensity is nearly independent of the area, even for spots or pores with diameter down to 2″. Henoux (1968*b*) has also assumed that all spots have the same intensity. These results are in contradiction to measurements carried out by Kneer and Mattig (1968). From far infrared observations ($\lambda 24\mu$) Léna (1969) has derived umbral temperatures near 3000 K and penumbral temperatures of about 4000 K. At the Oslo Observatory several pinhole photometers have been used to determine the umbral intensities in a wide wavelength region $4040 \text{Å} < \lambda < 40\mu$. A very small temperature gradient in the umbra at $0{\cdot}2 < \tau_0 < 1$ was called for by Mattig (1969*a*) from the reduction of older limb darkening observations, and by Wittmann and Schröter (1969) from new observations. Wilson (1969*a*) has demonstrated that a low temperature gradient is a consequence of the influence of umbral inhomogeneities on the center-to-limb variation.

Teplitskaja and Turclina (1969) have discussed once more the equivalent width measurement by Elsässer and Fricke and came to the result that the lines are formed in a dense sunspot with a hydrostatic pressure distribution. Ruhm's (1969) rediscussion, however, establishes the former results of Fricke and Elsässer. For two different sunspots Henoux (1968*a*, *b*) has derived homogeneous sunspot models for the depth range $10^{-5} < \tau_0 < 8$. His results include umbra in hydrostatic equilibrium, transparency of the spot, microturbulent velocity equal to that of the photosphere, and no radiative equilibrium in layers deeper than $\tau_0 = 1$. Obridko (1968*b*) has shown that the magnetic field determined from ionized lines is only half of that measured in neutral lines and proposed a two component sunspot model, with 5–10% bright elements covering the umbra.

Elste at McMath-Hulbert Observatory has obtained high dispersion spot spectra, which provide the basis for a number of current research projects. Hagyard and Harvey at Kitt Peak have employed photoelectric and photographic techniques for a comparison of theory and observation of the line profile of Fe I λ5250 in sunspots. Wiehr at Göttingen and Kneer at Freiburg are using magnetically unaffected lines to establish sunspot models. At Utrecht, first preliminary results by Zwaan show a microturbulence velocity smaller than 1 km/s for non-split lines and a temperature gradient somewhat steeper at $\tau_0 > 3$ than that of Henoux. As molecular lines are concerned, Makita (1968) obtained a rotational temperature from Ti I lines of 3000 K while Webber (1969) found remarkably higher temperatures using MgH (3900 K) and CaH (3500 K), respectively. Measurements of molecular line strengths have been made by Sotirovski at Meudon and by Wöhl (1969), who could identify H_2O and NiH in sunspots.

Kjeldseth Moe and Maltby (1969) have derived a homogeneous penumbra model in hydrostatic equilibrium which is able to explain the observations of the continuum and the equivalent widths. From highly resolved white light photographs and spectra Beckers, Schröter and Wittman have determined intensity fluctuations, curves of growth and line wings of strong lines. The aim is to deduce an either homogeneous or inhomogeneous model of the penumbra.

The height of formation of several lines was determined by Mattig (1969) from sunspot observations very close to the limb. He deduced the same scale height in the umbra as in the photosphere and therefore hydrostatic equilibrium. Ruhm (1969) has rediscussed these observations and found a remarkably larger height-scale in the umbra. Engvold (1969) has found height differences between the Hα core and the continuum in the range 1000 to 2300 km. From the displacement of the sunspot border in the lines of Fe I and Ti I Baranovskij (1969) has found that the umbra is 10 times more transparent than the photosphere, and the displacement of the Hα line leads to the height of formation of this line equal to 740 km above the photosphere. He has also found P_g and P_e in the umbra to be 1·5 to 2 times lower, and the gradient of gas pressure somewhat higher than that computed for the undisturbed photosphere. Hence, one is induced to conclude that the pressure and height scales are indeed different from spot to spot. There are probably spots in hydrostatic equilibrium and other spots which differ remarkably from this condition.

The Wilson effect has been studied by Wittman and Schröter (1969), who derived an umbral depression of about 650 km and a density scale height of about 450 km, in good agreement with the results given by Mattig (1969*b*). By analyzing the three-dimensional equation of radiative transfer, Wilson (1968*b*) has shown that by assuming a spot depression of 700 km near unit optical depth it is possible to reproduce qualitatively all features of the observed intensity profiles. According to Jensen *et al.* (1969), hydrostatic models for the penumbra and umbra and a depression of 600 km give a good fit for both the Wilson effect and the intensity profiles near the limb.

7.5. *Theory*

Erikson and Maltby (1967) have suggested longitudinal compression waves propagating parallel to the solar surface from the umbra to the penumbra, and could show that the resulting line profile is asymmetrical in the same sense as observed in the penumbra. With the Evershed effect interpreted as an acoustical wave phenomenon, the propagation may also be perpendicular to the magnetic lines of force. The effect of magnetosonic waves on special lines was considered by Maltby (1968*a*) with the result that magnetically active lines show a larger wavelength shift than inactive lines. He believes to have found this effect in the Evershed observations by Servajean. Maltby (1968*b*) has also shown that the effect of Alfvén waves, in terms of line broadening, should be largest in regions where the magnetic field is perpendicular to the line of sight.

Deinzer's sunspot theory has been improved by Yun (1968), introducing new boundary conditions. A refinement of an earlier sunspot model has been given by Chitre and Shaviv (1967). Wilson

(1968*a*) has brought up new aspects concerning the structure and development of a sunspot. He starts with an inhibition of the convection by the magnetic field in the deep layers, resulting in the formation of a "cool zone" above that region, and the development of the penumbra is described as due to the interaction of strong magnetic fields with thermal forces. Altschuler (1967), taking into consideration that the conductivity gradient near the spot boundary may be relatively large, finds that the photosphere is current-free and that ring currents might surround the sunspots. The development in time of a single sunspot has been described by Altschuler *et al.* (1968). With the initial condition $v=0$ everywhere and a magnetic field distribution of a ring current embedded in the photosphere, it can be shown that Evershed motion will develop within a wide range of initial parameters assumed. Convective instability and overstability in the presence of a uniform vertical magnetic field has been studied by Saito and Kato (1968). The overstable convection may be a cause of some dynamical features (umbral dots) but cannot be the main contributor to mechanical energy transport. Getling and Tverskij (1967) have described a possible mechanism for producing sunspot magnetic fields.

Kopecký and Obridko (1968) and M. Kopecký (1969) have drawn attention to the fact that Joule dissipation of magnetic fields might play a rather important role in some active processes on the Sun. The dissipation time of the spot magnetic field has been discussed by V. Kopecký (1968). Assuming a number of realistic models for the magnetic field, he shows that the dissipation time comes close to the time determined according to Chandrasekhar's formula. The conductivity coefficient for a sunspot model has been derived by Oster (1968) and a recomputation of the electric conductivity was carried out by Kopecký and Kuklin (1969), within a wide range of temperature and gas pressure. The conductivity, however, becomes lower by several orders of magnitude if the interaction between particles and ionic-sound waves is taken into account, which is an effect currently studied by Kuklin and Ribakov.

8. SOLAR FLARES

In 1968, two symposia were held on the subject of solar flares: in Tokyo, on *Solar Flares and Space Research*, the Proceedings of which contains 13 review papers on various aspects of the flare and flare-associated phenomena (de Jager and Švestka, 1969), and at Anacapri, on *Mass Motions in Flares and Related Phenomena* (Öhman, 1968). Other review papers on flares were presented by Bruzek at the *Conference on Plasma Instabilities in Astrophysics*, held at Pacific Grove in October 1968, and by Axford and de Jager at the *Third ESLAB/ESRIN Conference* held in Noordwijk in September 1969.

When compared with the years preceding the last Draft Report, one observes that after 1966 the main attention has shifted from studies of the common chromospheric flare phenomenon to space investigations of the high-temperature flare regions in the corona, and to flares associated with particle events in space. This obviously has been a natural consequence of the fast progress in space research.

8.1. *Spectrophotometry of flares*

Only relatively small progress has been achieved in spectrophotometry of flares. Isophotal photometry of flares has been carried out at the Kandilli Observatory (Dizer, 1969), and a photometric analysis of high-resolution pictures of flares, photographed at Athens, is being made by Falciani, Macris and Rigutti. A new photographic technique of constructing flare isophotes has been suggested by Falciani *et al.* (1967), and a special photometric analyzer called "videometer" has been developed by Tallant (1970) and is now being used at the NASA Manned Spacecraft Center in Houston.

Central intensities and equivalent widths of 918 emission spectral lines in a flare spectrum have been published by Polupan (1968). Electron density, determined from Hydrogen Balmer series in 16 flares, has been discussed by Fritzová and Švestka (1967) in dependence on the flare importance, phase of development, and on the height in the solar atmosphere, and spectrum of the flare of September 2, 1966, has been analyzed by Stepanjan (1969) who has confirmed the existence of "flare knots" and "interflare matter" in the flare region. Babin (1969) has discussed striking differences in the Hα and K_{232} emission in some knots of the limb flare of 11 July, 1966. Severnyj (1968*b*) has shown that the line asymmetry observed in spectra of flares and moustaches, can be explained by incoherent scattering in contracting and expanding shells, with a gradient of velocity inside. He has also found that in many cases the effect of asymmetry in limb flares is connected with a splitting of the flare emission into two distinct features, separated slightly along the dispersion, as if we had two emitting and partly overlapping layers, moving in opposite directions along the line of sight.

Steshenko (1970) finds that in most cases the continuum emission of flares appears in other regions of the flare than the line emission and supposes that this difference depends mainly on the separation of the emission sources in the horizontal direction. The mechanism for the continuous emission, proposed by him, is a plasma jet moving with a velocity of about 100 km/s from the flare, which hits the photosphere and increases its temperature up to 6700 K.

8.2. *Flare association with active regions and magnetic fields*

Zirin and Werner (1967) and Moreton and Severnyj (1968) published detailed analyses of the large spot group of September 13–26, 1963, during its transit over the solar disk. Simultaneous occurrence of flare brightenings in regions of opposite magnetic polarity, mostly very closely adjacent to the line $H_{\parallel} = 0$, close association of flares with "bifurcated" regions, and well expressed coincidence of flare knots with the strongest electric currents are among the principal results obtained by Moreton and Severnyj. Zirin and Werner found that most flares in the studied region were homologous and some were triggered by disturbances elsewhere in the region. This seems to be in support of Gold's (1968) opinion that it is unlikely that one simple instability is responsible for the flare event within its whole extent. If, however, due to some disturbance, a limited volume is taken out of the region of stability, the energy release in this volume causes a disturbance, which is responsible for triggering all the other accessible instabilities in the region.

As the homology is concerned, an astonishingly good example of two extensive homologous flares occurring nearly 54 apart, has been presented by White and Janssens (1970). The homology included flare shape, filament agitation and other associated prominence activities, as well as close similarity in the microwave radio bursts. Convincing examples of radio homology were earlier presented by Fokker (1967), and examples of X-ray homology were published by Pintér (1969). On the other hand, McKenna-Lawlor (1968) in her analysis of the active region of 1959, July 7–21, has cast some doubt on the actual "homology" of flares. McKenna's studies of this region are still in progress.

The relations of flares to magnetic fields were reviewed by Severnyj (1969*a*). Severnyj and Zvereva (1969) have found that the change of the net magnetic flux before flares leads to the e.m.f. $\sim 10^8$–10^9 V in the fixed circuit containing the active region, and for important flares this emf is about 10 times larger than the emf due to inductance as it can be evaluated from the measured velocities. In the case of the proton flare of 1966 July 7, Severnyj (1969*b*) has found a striking variation of the longitudinal field gradients and total magnetic flux, which dropped from highest values before the flare to low values after its occurrence. The flare appeared in the area of the junction of oppositely directed electric currents (Severnyj and Zvereva, 1969) and the brightest flare portions have been identified with the areas of strongest currents. Smith and Ramsey (1967) have found a strong correlation between the shape and intensity of the Hα fine structure and $H_{\parallel}$, except in the vicinity of the $H_{\parallel} = 0$ line.

Bruzek (1968) has concluded that changes in the chromospheric fine structure suggest changes in the magnetic field associated with the flare occurrence. Malville and Tandberg-Hanssen (1969) have discovered field fluctuations in the flare of 1967 May 21, with average period 7·5 min close to the period observed in photospheric fields (cf. Section 5). At Culgoora, the magnetograph designed for moderately high angular resolution and rapid frame repetition rates is being used to look for rapid changes in magnetic fields around times of solar flares.

According to Fortini and Torelli (1968), the most energetic solar regions are due to a fusion of two or more active centers. Martres (1968) has shown that the magnetically complex active regions, in which the most energetic flares occur, only are born at positions where an old active center has existed before, and the complexity is greater if the old active center is younger. This might explain why the non-random distribution of activity in solar longitude is most pronounced for the most important solar phenomena, and for the proton-active regions, in particular (cf. Section 4). Martres *et al.* (1968) have also found that flares are often associated with two spotted evolutionary features evolving in opposite senses, one growing, the other declining.

8.3. *Flare motions*

At the Anacapri Nobel Symposium (Öhman, 1968), our rather limited knowledge on motions in flares has been summarized by Švestka (1968*c*) and Dodson and Hedeman (1968*b*). Attention has been drawn, in particular, to ascending motions observed at some limb events during the first 2–5 min of their development, which may be missed in the disk flare spectra, and to the expansion of two-ribbon flares. Švestka pointed out the difference between the behaviour of flares of this type when occurring inside sunspot groups and outside of them, and Dodson and Hedeman are now currently studying the properties of well observed, major Hα flares in regions with either very small or no detectable spots. The expansion of two-ribbon flares associated with type IV radio bursts has been studied in detail by Ogir (1967, 1969) who finds an increase of the ribbon velocity with the increasing magnetic field gradient, and generally, the flare motions directed along the vectors of the transversal magnetic field. On the other hand, the rate of drift of a separating ribbon has been found increasing with the increasing distance from strong magnetic fields by Smith and Ramsey (1967).

Bruzek's study of fine structure of flares has provided evidence that brightening and expansion of small flare elements are associated with or preceded by enhanced mass motion and fast changes in the surrounding small-scale chromospheric structures and that inflow of material is prevailing. This may be considered as a support for Hyder's (1967) impact theory. A study by Janssens and White (1970), which shows a proximity of the falling material to the flare brightenings, also indicates that an impact-brightening process plays a role in the energy release. At Culgoora, observations of the flare fine structure lead to the conclusion that apparent motions in flares are due not to velocity of the flare material, but to successive brightening of new flare fibrils, which appear to have precisely the same pattern as the pre-existing chromospheric structures in the same region.

8.4. *Proton flares*

Great attention has been paid to the particle-producing flares on the Sun. The results of the first Proton Flare Project (PFP) concerning the July 1966 event have been published in *Annals of the IQSY*, Vol. 3. This book contains 60 original contributions on the proton-active region, the proton flare itself, and its consequences in space and in the magnetosphere and ionosphere of the Earth. A summary on the development and activity of the region has been prepared by Banin *et al.* (1969), on energetic particles by Krimigis (1969), and general summary and conclusions have been written by Simon and Švestka (1969). Solar aspects of the July PFP period have been summarized by Švestka (1968*b*). Summary on the results of the second August/September 1966 PFP period, has been prepared by Švestka and Simon (1969). Fairly extensive PFP results obtained in GDR were published by Krüger and Lauter (1969).

Something like a retrospective PFP was organized by Virginia Lincoln (1969), who prepared a valuable compilation of data on the solar particle event of May 23, 1967 and its geophysical effects. There are contained, among others, contributions by De Mastus and Stover (also see 1968) on the white-light flare emission, Dodson and Hedeman on the Hα flare, McIntosh on the spot group, and several authors contributed on the radio burst. Van Allen, and Kane and Winckler contributed on the X-ray burst, and Lanzerotti (also see 1969) on the particle event. A similar study is now in progress for other selected periods. Another enterprise of somewhat similar type was the Third

ESRIN/ESLAB Conference, held at Noordwijk in September 1969, aimed in particular at the cosmic-ray flare of 1969 February 25. A second Proton Flare Project organized by Švestka and Simon from May to July 1969, unfortunately did not meet with success, due to an unexpectedly low solar activity during that period.

Due to the limited extent of the Draft Report, we cannot review here all the numerous results obtained during the 1966 Proton Flare Project, and the readers are referred to the abovementioned summaries, and to the extensive review paper by Švestka and Simon (1969), in particular. From the solar point of view, perhaps the most valuable results, hardly obtainable without the organized PFP enterprise, concern the changes in the active regions prior to the proton flare occurrence, which start to occur about three days in advance (Severnyj, 1969*b*; Friedman and Kreplin, 1969; Tanaka *et al.*, 1969*a*) and become, at least in some cases, very pronounced during the last 12 h preceding the flare event (Popovici and Dimitriu, 1969; Dodson, 1969; McIntosh, 1969; Friedman and Kreplin, 1969). A uniquely low elevation condensation in the corona was detected following the proton flares of 1966 July 7 and September 2 (Newkirk *et al.*, 1969). Eliseeva (1967) and Křivský and Nestorov (1968) have found that SID effects are associated with substantially more flares in proton-active regions than in other regions on the Sun. An association of high-energy flares with transient umbrae has been suggested by McKenna-Lawlor (1969).

In the flare of 1966 August 28, Zirin and Russo-Lackner (1969) have observed a very fast separation of the two bright ribbons during the initial phase of the flare development, which slowed down after a few minutes. The fast expansion coincided in time with the hard X-ray burst, and this observation confirms the earlier conclusion made by Valníček (1967) that the fast flare expansion, the first maximum of the microwave radio burst, and the maximum of hard X-ray burst, all coincide in time. One can suspect, therefore, that this is the flare phase when acceleration process takes place in the proton flare region (Švestka, 1969). At the proton flare of 28 August, 1966, also the first evidence of a strong ultraviolet flash was recorded by McClinton and Kreplin, coincident again with the hard X-ray burst. A possible relationship of this UV enhancement to the continuum of the white-light flare has been suggested (Friedman, 1969).

Problems related to short-term forecasts of flares and proton flares, in particular, have been discussed and reviewed by Simon *et al.* (1969) and Severnyj and Steshenko (1969). The problems associated with long-term forecasts of proton flares have been discussed by Švestka (1968, 1969). Attempts to work out a method of long-term regional forecasts of flares are also made at Pulkovo by Vitinskij and his co-workers. A classification system of solar proton events is being prepared by Smart and Shea.

9. FLARE THEORIES

The efforts of theoreticians in trying to find an explanation for the flare process have not diminished during the past three years, and many new or modified theoretical considerations on this problem have appeared. Some of them, including older theories, have been critically reviewed by Sweet (1969).

Unsöld (1968) has proposed that flares should be considered as non-stationary analogues of plages, caused by sudden splashes of the waves that originate in the turbulent convection zone. The produced fast heating in the chromosphere produces shock waves giving rise to the flare-associated high-energetic phenomena. Pneuman (1967) has suggested that flares may be caused by chocking off of the normal energy flux in the corona by the closed magnetic fields of a plage. In this model, the flare is produced by the sudden increase in energy density of trapped kinetic and magnetic energy, which initially might be in the form of Alfvén waves and later could be partitioned into a wide variety of other forms including the acceleration of high-energy particles. Menzel (1968) has tried to show that the flare can be simply due to an existing natural conflict between mechanical and thermodynamical processes, which prevents the development of either a static or a steady-state equilibrium in a sunspot group. He has shown that a dynamic outward flow pumps gas in between the sunspot pair in a bipolar group, thus giving rise to a pressure and temperature increase in that region. After breaking the restraining magnetic barriers the gas explodes and a shock wave develops.

Schatzman (1967) has suggested that concentration of matter in filamentary structures aligned by the magnetic field, exists in the active region before the flare occurrence and becomes visible as a flare if there is a change of heat balance. This may appear when a decompression wave reaches the region, and such a wave can originate through an opening of the magnetic bottle above the active region. Particles can be accelerated to lower energies by a field of random Alfvén waves inside the bottle, and to high energies by the bottle opening. Elliot (1969) has supposed that the energy released in flares comes from energetic protons trapped in sunspot fields. Those particles may acquire their energy over a period of hours or days and the flare phenomenon corresponds to the catastrophic dumping of the particles into the chromosphere.

Hyder (1967) tries to explain the chromospheric flare brightenings, and those of two-ribbon flares in particular, by prominence material lifted out at disparitions brusques and falling down along the field lines into the chromosphere. He supposes that the kinetic energy of fall is then dissipated in the flaring areas. Although the original arguments favouring this model were largely circumstantial, in more recent data on 85 flares Hyder finds 75% to show rapidly falling material which is intimately associated with bright flare regions. As mentioned in the preceding Chapter, Hyder's ideas also receive support from studies carried out by Bruzek and Janssens and White (1970). Support for this theory also can be found in the studies published by Slonim (1968, 1969), who has found many spatial coincidences of flares and quiescent filaments, both in active regions and away from them. Nakagawa and Hyder have considered the way in which the infalling material modifies the temperature and density structure of the chromosphere, and their results are being used by Athay and Hyder to predict the observed line profiles. More spectroscopic Doppler shift data and polarization measurements are needed to confirm the infall-impact theory.

Other theoretical considerations are aimed at the problem of instabilities and the fast rate of energy dissipation. Sturrock's (1968) model explains the flare by reconnection of magnetic field in a bipolar magnetic-field structure by a tearing-mode resistive instability. The energy stored in the pre-flare closed-field region is derived from photospheric motion. The time-scale of the reconnection process is of the order 10^2–10^3 s if reconnection proceeds by the Petschek mechanism, with the modification that resistive diffusion is replaced by the more rapid Bohm diffusion. The plasma affected by the reconnection process is believed to be highly turbulent, which would lead to stochastic acceleration of both electrons and protons. Highly turbulent plasma, rendered by current-driven electrostatic instability, is also supposed by Friedman and Hamberger (1969) in Petschek's model of magnetic field annihilation, with plasma flowing through the boundary layer, where its magnetic energy is released. The same authors (Friedman and Hamberger, 1968) have also shown that the region near the neutral point, whose dimensions are governed only by diffusion processes, would be subject to current induced micro-instabilities, which would lead locally to a fast rate of magnetic field diffusion and hence to physically reasonable dimensions of this region. Syrovatskij (1969) supposes that the process of dynamic dissipation of magnetic energy near the zero lines of the magnetic field is the base of the solar flare mechanism. Křivský (1968) has proposed a magnetic model of two-ribbon flares in the form of an interaction of an emerging magnetic channel from the photosphere, entering a field of an already existing interaction of two systems of magnetic fields of a complex active region. According to Schmidt (1969) who summarized the theoretical aspects of the flare phenomenon, the annihilation of magnetic flux in a current sheet combined with a hypothetical hydromagnetic wave mode seems to give the fastest rate of dissipation. The explosive onset of this dissipation may be possible if the energy is stored in the sheared magnetic fields supporting a gradually condensing filament against gravity or in force-free currents induced gradually by the action of convective rolls in the penumbra.

Severnyj's (1969*a*, *b*; Moreton and Severnyj, 1968; Severnyj and Zvereva, 1969) measurements of the magnetic fields in flare regions have given strong support to Alfvén and Carlqvist's (1967) theory that a solar flare is caused by a sudden interruption of electric currents. The energy of the flare is supposed to be stored as magnetic energy in the current system. If, due to a local constriction at a certain point, the current density exceeds the maximum value which the plasma can carry in a stable way, a local decrease in plasma density is produced and an over-voltage arises in the circuit which will give rise to a rapid release of the stored energy. In the created high impedance region ions and electrons can be accelerated to very high energies within a reasonably short time comparable with the flash phase duration of the flare (Carlqvist, 1968).

Acceleration processes in flares have been discussed by De Jager (1968). He has suggested as the most probable case a two-step acceleration, with the Alfvén-Carlqvist mechanism as the first, and Fermi acceleration as the second step. Particle acceleration by nonlinear waves (in chromospheric ion solitons) has been suggested by Ginzburg (1967). By means of comparing spectral diagrams of type IV bursts with the particle production, Sakurai (1967*c*) has concluded that there are two rates of space-time change in sunspot magnetic fields, one of an explosive nature and the other gradual, on which the efficiency of particle acceleration depends. Boischot and Clavelier (1968*a*) suggest that electrons in the corona can be accelerated by the mechanism described by Tsytovich, if an ascending perturbation coming from the chromosphere attains some critical value of the Mach number.

As the ultimate flare energy is concerned, Anzer (1968) has proposed that the energy of flares comes from the convective motion in the penumbrae of sunspots. Stenflo (1969*a*) has shown how the kinetic energy of the rotational motion of a sunspot can be transferred to electromagnetic energy in filamentary currents, thus building up flare energies within reasonable time. Sakurai (1967*b*) has suggested that a solar flare is the result of a twisting of the magnetic lines of force due to a rotational motion of sunspot groups driven by the upward convective motion near the photospheric surface.

10. FLARE-LIKE AND FLARE-ASSOCIATED OPTICAL PHENOMENA

The fine broad spectral features known as moustaches have been discussed by several authors. Bruzek reports a continuing study in which he has measured sizes of over 700 such features, finding angular sizes of 0·5 arc seconds under best seeing conditions and he believes that the true size is, in fact, somewhat smaller. This is not inconsistent with Engvold and Maltby's (1968) study, or the earlier results of Severnyj. Bruzek has also confirmed that there is no conspicuous association between the moustaches and flares. From a study of metallic line emission in moustaches, Koval (1967) has deduced a velocity increase with height. The blue asymmetry, particularly pronounced in lines formed in greater depths, decreases with the moustache development. Asymmetry in the far wings has been studied more in detail by Koval and Severnyj (1970), and Severnyj (1968) explains the asymmetrical line profile as due to incoherent scattering in contracting shells (cf. Subsection 8.1). The continuous emission accompanying moustaches mostly shows rapid growth towards the violet (Koval and Severnyj, 1970).

Flare-like "dashes", first discovered by Prokofjeva more than 10 yr ago, have been studied by Krat (1968), who considers them for a separate class of flare-like objects closely connected with spicules. Kiepenheuer's (1968*c*) suspicion that surges originate in the photosphere or very close to it, in close connection with the granulation, might indicate, that there is some connection among all the three phenomena, spicules, dashes, and surges.

A review paper summarizing the flare-associated optical phenomena was written by Bruzek (1969). A number of expanding coronal phenomena associated with flares have been found by Bruzek in recent Sacramento Peak coronal movies taken in $\lambda 5303$ Å. In particular: a slow and decelerated expansion of concentric coronal arches ($v < 10$ km/s), and a fast and accelerated expansion of coronal arches ($\geqslant 150$ km/s) which terminates by the explosion of the arch. The nature of flare sprays and their association with other solar phenomena has been discussed by Elske Smith (1968). Ogir (1969) has shown that sprays from two flare knots situated in regions of opposite polarity are directed towards each other and may lead to the formation of a loop. Smith, however, has concluded that sprays by themselves do not instigate other interesting phenomena, which perhaps is a disappointing conclusion when one considers how spectacular some of these events are. They may be related to surges, but are almost an order of magnitude more energetic. This recalls McKenna's (1967) remark, pointing out that the various active phenomena associated with an active region appear to require very different energies, so that these events may be considered as monitors of the magnitudes of the accelerating disturbances which initiate them.

Dodson and Hedeman (1968*e*) and Zirin and Russo Lackner (1969) have described a large expanding feature which moved from the site of the proton flare of 1966 August 28, with a velocity of about 700 km/s. A similar phenomenon also was observed in association with the flare of 1967 May 23 (Dodson and Hedeman, 1969). The velocity was ~ 1000 km/s. These flare waves seem to be the visible manifestation of disturbances which cause the disparitions brusques. Models for such wave disturbances following flares have been presented by Uchida (1968) and Meyer (1968). Labrum and Smerd (1968) have discussed two events, in each of which it appears that shock waves arising from a flare produced distant prominence activity associated with type IV continuum. A similar event was also discussed by Wild *et al.* (1968) and Wild (1969). On the other hand, Křivský (1969) has interpreted these results as due to a moving radio type IV condensation, having its origin in the flare itself.

11. PROMINENCES

(by J. T. Jefferies, University of Hawaii)

Observations and analysis of spectral line intensities and shapes in prominences have been reported by Geichenko, Morozhenko, Sergejeva, Smotrelkina and Zeldina in several papers published in the Volume *Voprosy Astrofysiki* (Porfirov, 1967). Characteristic of much of the newer work is the recognition that a filamentary model may be needed to explain the spectral data and that a satisfactory model may need also to include temperature and density variations. An exhaustive study of 8000 emission line widths of four quiescent prominences whose spectra were obtained at the 1966 total eclipse is reported by Hirayama who concluded that kinetic temperature is around 6000 K inside and 12000 K near the edge of the prominence; the turbulent velocity varies from 8 km/s inside, to 20 km/s on the edge, the He I emission originates in the low temperature regions, and the electron density is about 4×10^{10} cm^{-3}. Further work is aimed at obtaining a two-dimensional model of the filament. These results, where comparable, appear to be characteristic of the results obtained by Stellmacher (1969).

Leroy (1967, 1969) has obtained monochromatic filter (10 Å bandpass) images of prominence in the light of He I, He II, Hα and Fe X, exposed so as to allow particular study of the faint peripheral prominence regions. The similarity of the He I and He II images gives support to the idea that λ4686 (He II) is excited by coronal radiation as Falciani and Rigutti (1967) have tentatively suggested. Leroy has been able to photograph extremely faint emissions extending far above the limb using this coronagraph technique. Future studies of the motion and disappearance of such faint wisps, which he plans, should be very interesting. Similar work is in progress in Hawaii, where McCabe and Fisher have completed a detailed study of the motion of the eruptive prominence of 1969 March 1, where Hα emission was seen extending beyond a solar radius. Similar work is also reported to be in progress at Kandilli by Soytürk using isophotal contours of Hα photographs of a surge to study morphological changes.

The microstructure of prominences is still not satisfactorily understood either in terms of the range of temperatures and densities present or the degree to which prominences exhibit subtelescopic filamentary structure. Much more refined studies of the excitation of the prominence gas are needed, as are spectroscopic observations at high spatial resolution at coronagraph stations. Measurements of the wing shapes of Balmer and other lines could give valuable information as to the mechanism of excitation. Such studies should also consider the influence of motion on the visibility of Hα prominence filtergrams, as Rompolt (1967) has done in a preliminary way.

Öhman *et al.* (1968) have drawn attention to the frequent occurrence in spectra of quiescent prominences of slanted spectral lines, and of "lines" with a circular appearance, and with other peculiar forms. Similar phenomena are well known in flares, moustaches and spicules. The standard interpretation in terms of a rotating structure has been broadened by Öhman to a self-twisting toroid (like a smoke ring). In this connection it is interesting that observations of eruptive material in prominences and surges are frequently strongly suggestive of an untwisting motion (Valníček, 1968; Kleczek, 1968; Dizer, 1968; Valníček *et al.*, 1969). Detailed spectroscopic studies of such twisting and untwisting give promise of elucidating the general questions of factors underlying prominence stability and instability. A model for quiescent filaments with helical structure has recently been completed by Anzer and Tandberg-Hanssen (1969). The structure arises from the combination of a pair of fields, one having a photospheric source, the other deriving from currents along the cylindrical filaments. This model is similar in nature to one suggested earlier by Ioshpa (1968).

Suggestive in this context, too, is the tentative association by Liszka (1968) between the amplitude of small scale prominence motions and the degree of activity in the neighboring photosphere. A more refined study which accounts roughly for his observed relation between field strength and random velocity amplitude is given by Malville (1968) who suggests that these persistent motions derive their source in oscillations of field lines which are, in turn, driven by motions in the denser photospheric layers. Large amplitude horizontal oscillations, giving rise to the "winking" filaments, are studied by Kleczek and Kuperus (1969). They attribute the dissipation of the oscillations to the generation of acoustic waves in the surrounding coronal plasma.

Bruzek (1967, 1969) has studied in detail the Hα arch filament systems closely associated with bipolar spot groups. Material is found to be moving at high speed (50 km/s) down both sides of the arches and upward (at 10 km/s) at the tops. Bruzek suggests that these systems are produced by expansion of the spot field with a consequent elevation of material into the chromosphere where it slides down field lines under gravity. They are not related to flare occurrence.

Some very interesting measurements of the line-of-sight component of the magnetic field in prominences above the limb have been reported by a number of investigators: Malville (1968), Ioshpa (1968), Rust (1967), Malville and Tandberg-Hanssen (1969), Harvey and Tandberg-Hanssen (1968), and Lee *et al.* (1969). The results seem generally consistent, giving fields characteristically of about 6 G for quiescent and about 30 G for active prominences with wide variations around these values – fields up to 200 G have been reported. It is found that lines of various elements all give essentially consistent values. Smolkov and Bashkirev (1969) have studied line profile parameters in spectra of quiescent prominences in order to derive the magnetic field configuration in the prominence region, and Smolkov (1969) also followed the time evolution of the longitudinal component of the prominence magnetic field and its relation to the redistribution of brightness. Von Klüber (1967) has measured the field variation across filaments confirming their preferred occurrence along neutral lines. According to Harvey (1969), the typical prominence field is characterized as having a non-zero component normal to its axis and a monodirectional field along the entire length.

The formation of solar prominences from coronal gas has been considered by Raju (1968) for a uniform magnetic field distribution. He finds that, although the instability leads to a strong temperature decrease, with a time scale of order 10^6 s, it would result in prominence densities not significantly greater than those in the surrounding corona. Kuperus and Tandberg-Hanssen (1967) have discussed the condensation problem also, but have adopted a field geometry more in keeping with observation, in which the prominence is formed along a neutral sheet. They showed that this is a thermally unstable region which, through cooling of the matter, results in the field lines being drawn together with consequent compression and radiative cooling in a characteristic time of 10^5 s. In order for this situation to be preserved against heat conduction from the corona, however, these authors suggest that the magnetic field in the region of the neutral plane must become unstable (a tearing mode instability) and break up into tubular filaments of characteristic size ~ 300 km. Of very great interest to such theoretical studies would be observational studies aimed at following the formative period of prominence material in condensing from the hot coronal gas.

Other theoretical work of interest in prominence formation has been carried out by Anzer (1969) who has obtained necessary and sufficient conditions on the magnetic field gradients for the Kippenhahn-Schlüter mechanism to apply. His stability analysis leads to the early conclusion of Kippenhahn and Schlüter that the horizontal component of field strength increases with height, which has been confirmed by Rust (1967) for many prominences, and that the vertical component of the field at the filament boundary should decrease with height. The nature of prominences is also being further investigated by Sturrock, with the aim of determining the magnetic field structures of prominences and the role of thermal instabilities in these objects.

The reader's attention also is drawn to the proceedings of the 9th Nobel Symposium (Öhman,

1968), which contains a valuable collection of articles dealing with the measurement and analysis of motions in prominences, sprays and surges.

12. CORONAL CONDENSATIONS

(by G. Newkirk, Boulder*)

The past three years have seen extensive work on the study of coronal condensations, especially in the UV and X-ray regions (Goldberg, 1967; Neupert, 1969). Not only have these new observations permitted the construction of detailed models of the electron density and temperature structure, but they have also yielded information on the temporal evolution of a "typical" such region (Neupert, 1969). These details generally confirm the ideas derived from earlier, optical observations of the corona.

According to the UV and X-ray observations, the typical condensation exists at a temperature of from 1·8 to $2{\cdot}4 \times 10^6$ K, has an electron density of approximately 3 or 4×10^9 cm^{-3}, a volume of about 10^{29} cm^3, and an emission measure, $N_e^2 V \sim 10^{48}$. Within such a region, temperatures up to 4×10^6 K and densities up to 7×10^9 may occur. Such a region appears to correspond well to the permanent and sporadic condensations visible in the green and yellow coronal lines and described in the earlier literature.

The evolution of a typical such condensation shows a rapid increase in the electron density during the first two weeks of development of the underlying active region (Neupert, 1967). At about the time when the region reaches maximum activity the temperature also reaches its maximum of approximately $2{\cdot}5 \times 10^6$ K. That the temperature declines more rapidly than the emission measure is demonstrated by the fact that the lines of cooler coronal ions (Fe XIV and Fe XII) reach a maximum a week or two after the maximum in such ions as Fe XVI.

The systematic investigation of a wide variety of lines originating from active regions (Noyes *et al.*, 1969) has allowed not only the determination of the average conditions in the corona and chromosphere-corona transition region, but also a measurement of the conductive flux in such regions. These observations show a typical active region with a temperature of approximately $2{\cdot}5 \times 10^6$ K and enhancement of about 5 times in the pressure in the chromosphere-corona transition region, and electron density enhancement of approximately 4 times, and a conductive flux from the corona back into the chromosphere increased by about 5 times.

There appears to be some indication of a disagreement in the conditions in coronal condensations inferred from the spacecraft observations and inferred from optical emission line observations in the direction of the short wavelength observations requiring somewhat higher electron densities and a rather higher prevalence of temperatures of $3\text{–}4 \times 10^6$ K than would be suggested by the optical observations. In the absence of the detailed study of the consistency of the optical, radio, UV, and X-ray observations of particular regions in the corona, it is unknown whether this apparent disagreement is real or not. It may well reflect the fact that the short wavelength observations preferentially see the densest and hottest regions and that the optical observations seldom give reliable information concerning the lowest portions of the corona.

Another disagreement which appears to have arisen is that of the maximum temperature which is present in coronal condensations. Optical observations would most certainly place this maximum at $4\text{–}5 \times 10^6$ K. However, UV and X-ray measurements (Hudson *et al.*, 1969; Neupert, 1967; Neupert, 1969) which show the spectral lines up to Fe XXV require the presence of material with temperatures of up to 3×10^7 K even in the absence of reported solar flares. At the present time it is not known whether this radiation originates in the corona or whether it represents the impact of non-thermalized particles of energies of 10–15 keV into the lower solar atmosphere. A strong case has been made that, for solar flares at least (Acton, 1968), the observations are best described by a corona which has a maximum temperature of $4\text{–}5 \times 10^6$ K and a substantial nonthermal component.

* A more extensive draft available on request.

In the future we can expect spacecraft observations with good angular resolution to play an increasing role in the investigation of coronal condensations. Because the regions can be followed completely across the solar disk, the evolution and response of the condensation to such cataclysmic effects such as flares can be documented in a manner hitherto impossible. Since these measurements also allow conditions in the crucial chromosphere-corona transition region to be examined, we can expect detailed knowledge of the change in energy transport in these regions as well. Hopefully, satellite and ground-based observations of particular regions can be combined to further elucidate the details of the temperature density structure in such regions.

The problem of explaining the high temperature of the solar corona has received considerable attention in the past several years. Most of the theories (for a review see Kopp, 1968; Kuperus, 1969) have been concerned with the details of the generation of mechanical energy in the lower solar atmosphere and the deposition of this energy by shock waves in the chromosphere and corona. Recently the role of ion acoustic waves, which dissipate their energy by Landau damping in the lower corona, has been pointed out by d'Angelo (1969). Perhaps, the most important aspect of these theoretical investigations has been their demonstration of the fact that the magnetic field in the lower atmosphere can be expected to play a very large role in determining the rate of energy deposition in the corona as well as the structure of the chromosphere-corona transition region (Kopp and Kuperus, 1968). As mentioned in the section on coronal condensations, there now appears to be substantial evidence (Noyes *et al.*, 1969) that the energy deposition into the corona is, in fact, increased substantially above active regions.

Temperatures in the outer corona above active regions have been inferred from the duration of type III radio bursts for several years. The advent of space-borne radio receivers operating at frequencies down to 0·4 MHz has extended these observations out to what appears to be 30 $R_\odot$ (Alexander *et al.*, 1969; Hartz, 1969; Slysh, 1967). The usual interpretation of these observations is that the type III burst radiates at the local plasma frequency, that the duration of the burst is determined by the local collision frequency and that the local temperature can thus be obtained. The relation of temperature to distance from the sun is then derived by assuming a density model. If the density over the active region is taken to be some constant times the ordinary coronal models or even includes a simple evaluation of the role of the magnetic pressure in influencing the density structure, the temperature is found to fall off much more rapidly than would occur in the conduction model of the corona. Such observations would then suggest that the streamers over active regions are dense and cold compared to the rest of the corona. In fact, they would appear to be in subsonic expansion (Newkirk, 1967*b*). An alternate hypothesis (Alexander *et al.*, 1969) is that the temperature in the active region streamer is the same as in the coronal background, which is taken as that of the solar wind heated by conduction from below. The question of subsonic streamers above active regions is then avoided. Such a streamer would have a maximum density enhancement of about 5 times the background at about 30 $R_\odot$. It should be noted that this is approximately the density and temperature behaviour which is predicted by the theoretical streamer model of Pneuman and Kopp (1970). Recent evidence (Warwick, 1967), suggests that local heating by strings of type III bursts far out in the corona is required to explain the appearance of type V radio bursts.

An examination of the rotation of coronal condensations during the period 1964 to 1967 (Hansen *et al.*, 1969) demonstrates that the rotation rate varies from year to year and in any given year may be different between the two hemispheres. At lower latitudes, the rate appears to agree with that determined from filaments and sunspots while at high latitudes the rate appears to be similar to that of filaments, which is over 2° per day faster than that displayed by polar faculae. These data show a dependence of angular velocity with height which appears to depend upon both latitude and the phase of the solar cycle. Both increasing and decreasing angular velocity with height appear. The details of this phenomenon are poorly understood. These is some indication that the high latitude coronal features are being influenced by the more rapidly moving magnetic centers at lower latitude.

In spite of the fact that the determination of magnetic fields in the corona is extremely difficult, the question of such fields and their influence on the solar corona has been one of tremendous activity in the past several years. Although, in theory, capable of yielding information on these fields, the measurements of coronal emission line polarization (Eddy and Malville, 1968; Hyder *et al.*, 1968) have proved difficult to interpret in terms of the magnetic field in any detailed manner. Faraday rotation of artificial and natural radio sources (Levy *et al.*, 1969) is also a potentially useful tool although no quantitative measurements of the coronal field have yet been determined. The Razin effect in solar radio emission (Boischot and Clavelier, 1968*b*; Ramaty and Lingenfelter, 1968; Bohlin and Simon, 1969) appears to have given us the most unambiguous determination of the field at a value of 0·3 to 0·5 G at 2 $R_{\odot}$ above an active region. All these researches refer to the observation of a single type IV burst and similar observation of other events is urgently required. A similar value has been inferred from the weak polarization of type III bursts (Kai, 1970). At the end, one has to emphasize that the observation of radio bursts at 80 MHz per second with high angular resolution with the Australian radioheliograph has opened an entirely new vista in the study of magnetic fields and high energy particles in the corona (Wild, 1968*a*, 1969*a*, *b*; Kai, 1969, 1970).

13. RADIO ASPECTS OF SOLAR ACTIVITY

(by A. D. Fokker, Utrecht)

(See also the section Solar Radio Emission of the Commission 40 report).

The past three years are distinguished by a further improvement of observational instrumentation and by some coordinated efforts to collect the largest possible material for some selected solar events. Progress has been made in the treatment of gyro-synchrotron emission and a new element in theoretical discussion was the attention paid to the possible influence of the Razin-effect on the shape of burst spectra. Certain notions about the radiation mechanism of type II and type III bursts have become rather generally accepted, but the detailed treatments of processes cannot yet be summarized in a textbook kind of presentation which could be understood by everybody. The occurrence of type I storms is so common as to make people forget that these, in particular, are still ill-explained. Two review papers on the theory of radio bursts and on the generation of radio waves in astrophysical plasmas, respectively, have been written by Takakura (1967) and by Smerd (1968).

13.1 *Progress in equipment*

The main instrumental achievement in interferometry has been the completion of the 80 MHz Culgoora radio heliograph (Wild, 1967), which consists of 96 dishes each of 13 m diameter and equally spaced around a circle of 3 km diameter. Sidelobes are suppressed by connecting the aerials with each other via different phase lags according to a fast computer-controlled sequence. The pencil beam is made to scan the Sun two-dimensionally along a grid of 60×48 points and a two-dimensional picture of the Sun with a resolving power of 3·5′ is displayed twice a second on a cathode ray tube.

The 16-element array of the 169 MHz Nancay interferometer has been provided with a system of coaxial cables which permits simultaneous reception of solar radiation in 31 equidistant fan beams. With this multibeam system instantaneous one-dimensional pictures are obtained (Vinokur, 1968). At Toyokawa, the $(32+2)$-element compound interferometer for 3750 MHz with a half power beam width of 23″, completed in 1967, has been provided with rotary phase shifters which permit the Sun to be scanned once every 10 s with a 1$'$.1 beam (Tanaka *et al.*, 1969*b*). Two-dimensional mapping of the Sun at λ3 cm with a 32×16 EW/NS array takes 20 min and the pencil beam measures 1$'$.6 × (1$'$.6–3$'$.2). At Algonquin Observatory, an array of 32 reflectors for operation at 2800 MHz has been completed and yields a half power width of 1$'$.5 (Covington *et al.*, 1967). Four additional reflectors will be added to form a compound interferometer with a beam of 0$'$.25 width.

High-resolution (35″) fan-beam scans at λ21 cm are obtained at Fleurs, Australia with a 32-element array. It is intended to develop there a fast-scanning device which permits a continuous display to be obtained of the Sun's radio profile. A new (8 + 3)-element compound interferometer operating at 160 MHz with a resolving power of 1'.8 in the E–W direction was completed by Tokyo Astronomical Observatory in 1969. The fan beam is made to sweep the Sun once in 0·1 s. At Stanford University, a new radio telescope is under construction, with a resolution of 20 seconds of arc and with provision for polarization measurement.

A novel type of radiospectrograph was put into operation by the Utrecht Observatory. The octave from 160 to 320 MHz is covered by 60 equidistant single-frequency channels. The radiospectrograph displays the instantaneous intensities as referred to a level which is the intensity averaged over the past three or ten seconds, so that relatively small fluctuations are reproduced with considerable contrast (De Groot and Van Nieuwkoop, 1968). At Fort Davis, a new sweep-frequency receiving equipment, covering the band 50–2000 MHz is being put into operation with an 85-ft paraboloid antenna. A 64-channel spectrometer with a time resolution of 0·001 s for the range 400–1420 MHz is being developed by the radio observatory of Kiel University. A swept receiver from about 310 to 340 MHz has been in use since 1967 at Harestua, and from 50 to 200 MHz since 1968 at Ondřejov and Irkutsk. For low-frequency spectrographic work aboard satellites cf. Subsection 2.3.

At Harestua, a receiver at 159 MHz, working since summer 1968, has been specially constructed for the recording of extremely short-lasting solar bursts. At the Heinrich Hertz Institute in Berlin, improved polarization measurements are now available at 23, 30, 40 and 115 MHz, and in 1971 polarization measurements also will be carried out at 793, 1470 and 9500 MHz.

13.2 *Observations of radio events*

More detailed information on radio events has become available thanks to the diversification of world-wide observations and to the progress of instrumental sophistication.

High resolution observational results on microwave bursts obtained at Toyokawa were reported by Énomé *et al.* (1969). A change in polarization was observed by Tanaka and Énomé during a fairly strong simple-structured outburst and several cases of multi-source bursts have been recorded both by drift-scanning and quick-scanning. The source size of the impulsive burst was found of the order of 0'.5.

Observational results with the Culgoora radio heliograph were reported for several events. The heliograph has revealed interactions between distant centers around the Sun (Wild *et al.*, 1968) and established a new kind of phenomenon in which two distant sources appear within a few seconds or even within one second of one another (Wild, 1968). The magnetic structure of a type IV event could be inferred from total radiation and polarization images (Wild, 1969*b*). An expanding arch structure of a great solar radio outburst has been described by Kai (1970). Several reports on other observations made with this equipment are contained in Vol. 1, nos. 4 and 5, of the *Proceedings of the Astronomical Society of Australia.*

Detailed observations of spectrum and position of a weak moving type IV event on 1966 September 14 were reported by Boischot and Clavelier (1968*b*). This event has received a lot of discussion, by Boischot and Daigne (1968), Lacombe and Mangeney (1969) and Ramaty and Lingenfelter (1968). The shape of its spectrum seems to be strongly influenced by the Razin effect. A strong stimulus for making multifixed frequency observations of solar radio events is provided by the widespread interest for solar particle events (Švestka and Simon, 1969). Castelli *et al.* (1967) noted that U-shaped spectra of type IV bursts (with minimum flux around 2000 MHz) are indicative for production of high-energy particles. Extensive observational material of the Heinrich Hertz Institute (Berlin) on several big radio events was presented and discussed by Böhme and Krüger (1968).

Burst observations in the very low frequency range, made with satellites, have been reported and discussed by Slysh (1967), Hartz (1969) and Alexander *et al.* (1969). On the other hand, solar radio observations have been pushed through to higher frequencies so as to encompass the mm wavelength range. Some bursts at $\lambda 3{\cdot}3$ mm were reported by Shimabukuro (1968), who found good correlation with solar flares. These bursts reached relatively low intensities, in contrast with an outstanding solar radio burst at $\lambda 4{\cdot}2$ mm reported by Croom and Powell (1969), which reached a peak intensity of 47×10^{-19} Wm^{-2} Hz^{-1}. This outburst was associated with a proton flare. Infrared techniques were employed by Clark and Park (1968) to map the sun at $\lambda 1{\cdot}2$ mm. In their 3′ beam they observed transient enhancements of brightness temperature of the order of 100 K, which apparently were associated with certain filaments.

13.3. *Application of the theory of gyro-synchrotron emission to microwaves and type IV bursts*

Computations of the spectrum and time variation of microwave and type IV bursts have been carried through in more and more detail. Gopasjuk *et al.* (1967) discussed the role of compression and expansion of the region that generates synchrotron radiation, in order to explain microwave bursts of the explosive type. Takakura *et al.* (1968) dealt with the time variation of the spectrum of gyro-synchrotron emission, assuming a power-law distribution of electron energies. In a paper by Holt and Ramaty (1969) detailed theories of gyro-synchrotron emission, the Razin effect, gyro-resonance absorption and gyro-synchrotron reabsorption were applied, and for a given set of parameters the overall spectrum of the type IV event associated with the proton flare of 1966 July 7, could be reproduced.

Whereas the above authors based their computations on fixed values of the magnetic field and the electron density in the source region, Yip (1967) interpreted the properties of type IV bursts on centimetre and decimetre wavelengths on the basis of a model for the electron density distribution and the magnetic field configuration, using theories of Doppler-shifted cyclotron radiation, its excitation in a stream-plasma system and of gyro-resonance absorption. Kai (1968*a*) discussed evolutional features of microwave type IV bursts and made estimates of the strength of the magnetic field and the energy distribution function of electrons in the source regions where subsequent peaks in one and the same event originate. Ramaty and Lingenfelter (1968) interpreted the spectrum of the type IV burst of 1966 September 14, which originated high in the corona, in terms of gyro-synchrotron emission with due allowance for the Razin effect. A value of the magnetic field and the range of electron energies were derived.

14. SOLAR EUV RADIATION, X-RAYS AND GAMMA RAYS

(by E. v. P. Smith, University of Maryland)

This subject was recently reviewed in an excellent article by Neupert (1969). Since space is limited, we will largely avoid recapitulating material covered in that review paper. A vast amount of data has accumulated from OSO 3 through OSO 6 spacecraft, and in fact is continuing to be accumulated especially by OSO 6. This material is supplemented by data obtained from rocket flights, which usually have the advantage over satellite data in having far better spatial resolution. Among the outstanding results one should mention the localized nature of the flare X-ray emission and the large number of lines in the soft X-ray region of the spectrum which become prominent at times of flares.

The group at American Science and Engineering (Vaiana *et al.*, 1968) obtained high resolution X-ray photographs of the Sun at the time of the 1 N flare of 1968 June 8, showing considerable fine structure in both the flare and in active regions. A very close correspondence has been found between Hα and X-ray features, but the X-ray regions extend well above the Hα counterparts at the limb. The brightest part of the X-ray flare was located at the position of a filament that had disappeared minutes earlier.

Many soft X-ray flare lines are observed in the region from 1 to 20 Å (Neupert *et al.*, 1967, 1969; Meekins *et al.*, 1968; Evans and Pounds, 1968; Evans *et al.*, 1969; Jones *et al.*, 1968). A large number of these have been identified by Gabriel and Jordan (1969) with the resonance, intercombination and forbidden lines of helium-like ions such as O VII, Ne IX, Mg XI, and Si XIII. Neupert and Swartz (1969) have added to the identification other transitions of this type along the helium isoelectric sequence up to Ni XXVII. Gabriel and Jordan showed how the relative intensities of these triplet lines can be used to calculate electron densities of the emitting region, while the temperatures obtain from the presence of a particular stage of ionization. They find that temperature and density increase together, and the densest regions tend to be the most localized.

The dominant spectral line below 3 Å is that near 1·9 Å. This has been observed both by Goddard and NRL groups at the times of flares but they disagree on its identification. Neupert (1969) argues that it is Fe XXV while Meekins *et al.* (1968) prefer to identify the line with inner-shell transitions. The fact that this line reaches maximum intensity earlier than lower ionization lines has led Neupert to suggest that a very hot plasma exists or is ejected into the corona at the start of a flare and the lower ionization lines arise as the plasma cools and recombination occurs. Since a large range in ionization stages is observed simultaneously, it is evident that a wide range of temperatures must exist at a given time. Relative intensities of continuum to line energy show that the continuum emission is about a factor 10 greater than the integrated line emission during flares (Meekins *et al.*, 1968; Neupert *et al.*, 1969). This continuum radiation is "hardest" near the maximum of the X-ray flare.

The University of Leicester group under K. Pounds (Evans *et al.*, 1969) has observed X-ray flares in the 1 to 18 Å range on OSO 4 with a time resolution of 16 s. The bursts range from being very impulsive with 1 to 2 min duration to long enduring events of several hours. Many events have a precursor in the form of a weak gradual increase in X-radiation, corresponding to a temperature about 8×10^6 K some minutes before the flare event. Kane and Winckler (1969) also find that X-ray bursts have two components, an impulsive and a more slowly varying one. A catalog of 2–12 Å solar X-ray flux is being produced and has been completed for the period from July 1966 to September 1968 by Drake *et al.* (1969), based on the measurements with the University of Iowa equipment on Explorers 33 and 35.

In contrast to the thermal or quasi-thermal character of soft X-rays, the higher energy or hard X-ray bursts are nonthermal (Cline *et al.*, 1968; Arnoldy *et al.*, 1968a; Hudson *et al.*, 1969*a*). Frost reports on the OSO 5 observations in the 14 to 250 keV energy range of the 1969 March 1 X-ray burst. Remarkable fine structure in the intensity-time profile was revealed by all 9 channels within this energy range, but especially at the lower energies. The intensity fluctuations closely followed the microwave burst structure. On the other hand in the flare of 1968 July 8, Van Allen and Wende (1969) have found only a general association of soft X-ray and centimeter radio flux, with lack of detailed similarity. Frost (1969) also has found that both X-rays and radio bursts were quasi periodic with a period of about 35 s. Similar, 10-s periodic pulsations, with X-ray spectrum hardening during the peaks, which also correlated well with the microwave data, have been observed by Parks and Winckler (1969) at the flare of 1968 August 8.

An Atlas of 10–50 keV Solar Flare X-rays observed by the OGO satellites from September 1964 to the end of 1967 has been published by Arnoldy *et al.* (1968*b*) and Kane and Winckler (1969*b*). Another catalog of X-ray bursts with energies in excess of 7·7 keV is being prepared by Petersen and his colleagues, based on OSO 3 observations. They find that nearly all Hα flares and subflares produce X-rays above 7·7 keV, but a significant number of bursts do not seem to have corresponding Hα events. Generally, while the time development of soft X-rays closely follows the Hα flare variations, the duration of harder X-ray bursts becomes progressively shorter and peaks earlier, before the Hα maximum, as the X-ray energy increases (see, e.g. Švestka and Simon, 1969). Good correlation is found between these bursts and first maxima of microwave radio bursts, while X-ray and radio bursts late in the flare development do not correlate, and the correlation with type III bursts is much lower as if not all X-ray bursts affect the upper corona (Arnoldy *et al.*, 1967; Hudson *et al.*, 1969*a*). As to the height of the X-ray emitting region, both Beigman *et al.* (1969) in an X-ray flare and Broadfoot (1967) for X-ray emission from three coronal condensations find $20–25 \times 10^3$ km above the photosphere.

Active regions were observed over a wide range of wavelengths ($\lambda\lambda$280–1360) by the OSO 4 Harvard spectrometer-spectroheliometer (Noyes *et al.*, 1968; Hearn *et al.*, 1969). Because of the wide wavelength coverage it was possible to obtain data pertaining to heights in the solar atmosphere from the low chromosphere to the corona. The enhancement of emission lines in non-flaring regions increases with the ionization temperature. Analysis of spectroheliograms in the λ537 and

λ584 lines of neutral helium comparing quiet and active regions shows that increased intensities can be explained purely in terms of increased electron density up to a factor 2·5. During flares many lines show an intensity increase of a factor 5 or more at maximum. As a rule, pre-flare intensities prevail within 20 min, for the flare duration is shorter in the UV lines than in Hα. Hall and Hinteregger (1969) have used an extreme ultraviolet spectrometer aboard the OSO 3 satellite to measure flare-associated enhancements in the wavelength range from 260 to 1300 Å over a six months period. They have found very complex phenomena, the elaboration of which is still in progress. Kreplin and McClinton (Friedman, 1969) and Monsignori Fossi *et al.* (1969) have observed short-lived enhancements of the UV continuum close to, but outside of the $L\alpha$ line, in the flash phases of the flares of 1966 August 28 and March 20, respectively (also see subsection 8.4). Lemaire (1969) has obtained profiles of the ultraviolet Mg II lines with good spatial and spectral resolution, including active and spot regions, while Fredga (1969) obtained spectroheliograms in these lines. Allen (1969) has discussed systematic photometry of XUV solar images. A review paper on twenty years of extreme ultraviolet solar research has been written by Tousey (1967).

The importance of polarization measurements of X-ray bursts below 8 Å in the initial phase of a flare, which can distinguish bremsstrahlung from synchrotron and Compton radiation, has been emphasized by Korchak (1967) and Elwert (1968). A model of hard X-ray and microwave bursts has been proposed by Holt and Ramaty (1969), mostly derived from observations of the 1966 July 7 event (Holt and Cline, 1968). Both kind of bursts are shown to be consistent with a single source of non-thermal electrons.

15. HIGH ENERGY PARTICLES

The limited extent of this Report does not allow us to touch all the interesting observations carried out in space and in the Earth environment and all the conclusions drawn from them. From the point of view of solar physics, however, three kinds of observation are of particular interest: the discovery of permanent fluxes of ~1 MeV protons from particular active regions on the Sun, records of electrons in space, up to relativistic energies, and an evidence of particle flux arriving to the Earth from the invisible solar hemisphere.

Enhanced proton fluxes in the energy range 0·6 to 13 MeV have been observed to come from the Sun continuously over heliocentric longitude ranges as great as 180° in association with specific active regions (Fan *et al.*, 1968). An explanation of these extremely wide cones have been suggested by Jokipii and Parker (1969) in term of random fluctuations of magnetic lines of force which lead to rapid diffusion of energetic particles across the average magnetic field. The enhanced fluxes exhibit definitive onsets, when the active region is 60° to 70° East of the central meridian, and cut-offs, when the center reaches 100° to 130° West (Pioneer 6 and 7 measurements). Superposed on this permanent flux level are occasional large and discrete flare-produced intensity increases from these centers of activity. McCracken (1969), in his very instructive review paper on solar particle events, has expressed an opinion that the generation of energetic particles is a normal feature of all solar flares; nevertheless, as Švestka and Simon (1969) have shown, some necessary conditions have to be fulfilled in the active region to make it capable of producing significant particle events, the permanent particle emission probably being one of them.

Measurements of protons in the energy range of the order of 1 MeV on Soviet space probes Zond 3 and Venus 2, 3, 4 from 1965 to 1967, as well as preliminary results from Venus 5 and 6 have been summarized and presented by Vernov *et al.* (1969). An extensive study of long-lived streams of low-energy solar electrons and protons has been made by Anderson (1969). He has suggested that the energetic storm particles, delayed flare events and fluxes associated with active regions, are all probably the same basic phenomenon viewed in somewhat different ways depending on the age of the region and its location on the solar disk.

Electrons with kinetic energy > 40 keV in space were studied by Lin and Anderson (1967). They have found that such electrons follow many solar flares, promptly or delayed, and distinguish two types of the prompt events: those coming from flares beneath large radio noise regions, when electrons are emitted into fairly wide cones of propagation, and all the other events, when the cone of propagation is very narrow, with only about 30° opening angle. Solar flare electron events do not show filamentary structure even at times when protons from the same flare do. And these electrons also are only mildly anisotropic, while low-energy protons coming later from the same flare usually show a sharp anisotropy (Rao *et al.*, 1969).

Arnoldy *et al.* (1968*a*) have found a strong correlation between interplanetary solar flare electrons and X-ray bursts, while this correlation is weaker for solar proton events. They also have inferred a strong propagation asymmetry for flare electrons along the spiral interplanetary magnetic field, but the most unusual propagation feature is that the electron flux in delayed particle events is centered several hours before the corresponding proton flux, as has been discovered by Anderson (1969). Jokipii (1969) has shown that at least a part of this separation may be due to difference in drift velocities of protons and electrons in space. Grigorev *et al.* (1969) and Lin (1969) have reviewed the most up-to-date results on the electron events. They incline to the conclusion that the acceleration and emission of ~ 40 keV electrons is a phenomenon distinct from the production of solar protons and relativistic electrons.

Relativistic electrons with energy > 3MeV were detected in space aboard the IMP satellites by Cline and McDonald (1968). The first clear event was associated with the proton flare of 1966 July 7. The authors conclude that these interplanetary electrons are created simultaneously with the flare electrons which cause the microwave and X-ray burst. Probably only a small fraction of flare electrons escape into interplanetary space. In 1968 and 1969, Datlowe *et al.* (1969) with equipment on board OGO-5, and Koechlin *et al.* (1969) on the ESRO satellite HEOS A 1 measured on several occasions an excess flux of > 10 MeV electrons in association with solar flares.

Evidence has been given that some fairly strong particle events recorded close to the Earth were due to flares situated on the invisible solar hemisphere. This seems to be the case with particles recorded on 1966 July 16–19, which Dodson *et al.* (1969) ascribe to a flare that occurred near the central meridian on the opposite solar hemisphere, in the active region which produced the July 7 cosmic ray flare; the most extraordinary event of this type, however, occurred on 28 January 1967 (Baird *et al.*, 1967). On that day, a GLE was recorded by neutron monitors, without any obvious source on the visible solar hemisphere. This fact, as well as combined measurements with space probes in different distances from the Earth-Sun line, and the unusual isotropy of the particle flux, give strong support to the origin of the particles far behind the solar limb (Innanen *et al.*, 1968; Bukata *et al.*, 1969); a position 60° behind the western limb seems to be the most probable one (Dodson and Hedeman, 1969*b*; Lockwood, 1968). An interesting feature of the event also was a low-energy precursor of the cosmic-ray enhancement (Blake *et al.*, 1969; Masley, 1969).

Many papers have been published on proton measurements during individual particle events in space, and the most detailed studies were presented by Lindgren (1968) and Lanzerotti (1969) on the 1967 May 24 and 28, events. Several selected events and periods were studied in an international cooperation, which has greatly improved the possibility to get a fairly complex picture of the solar event and its consequences in space. These were the events of 1966 July 7, (results published in *Annals IQSY*, Vol. 3), 1966 August 28 and September 2 (Švestka and Simon, 1969), 1967 May 23 (Lincoln, 1969), and 1969 February 25 (*Third ESRIN/ESLAB Symposium* at Noordwijk). The strongest event during this period was that of 1966 September 2. Dosage of ~ 60 *R*/h behind a 0·22 g/cm^2 thick aluminium wall was measured during its maximum phase (Kane and Winckler, 1969), and it also was the fifth case, the first one after 1961, when nuclei with charges greater than 2 were detected in the particle flux (Durgaprasad *et al.*, 1968).

The acceleration of charged particles to high energies on the Sun has stimulated the search for high energy neutrons also associated with flares. OSO 1 satellite results (Hess and Kaifer, 1967) have failed to show any neutron effects associated with specific flare activity. A positive flux at the Earth of $\sim 10^{-1}$ neutrons/cm^2 s for 50–500 MeV range has been reported by Daniel *et al.* (1967) in association with a subflare, but Holt (1967), on the base of negative OGO 1 measurements, has raised serious doubt as to whether such a high flux was possible without the detection of decay protons in interplanetary space. No increase higher than 10^{-2} neutrons/cm^2 s has been found by Zych and Frye (1969) for periods after two flares of importance 1, and by Daniel *et al.* (1969)

after a 2B-class flare. Of course, no strong particle events have been investigated in this way, and theoretical considerations indicate that a measurable neutron flux could still be detected after such events as that of 1966 September 2 (Lingenfelter and Ramaty, 1967; Lingenfelter, 1969).

16. SOLAR WIND DISTURBANCES

(by J. M. Wilcox, Berkeley)

Review papers on solar wind plasma and magnetic fields were written by Dessler (1968), Wilcox, (1968), Axford (1968), Hundhausen (1968), and Parker (1969). Solar wind plasma observations included electrons and several ions not previously observed. Solar wind electrons (Montgomery *et al.*, 1968; Formisane, 1969) were observed to have temperatures always greater than or equal to proton temperature, with the range between 7×10^4 and 2×10^5 K. A two fluid model of the solar wind (Hartle and Sturrock, 1968) showed that the collision energy exchange rate between electrons and protons is sufficiently slow that electrons and protons can have quite different temperatures. The average helium abundance in the solar wind (Ogilvie and Wilkerson, 1969) was observed to be approximately 0·05, suggesting that the solar abundance may be about half the previously quoted estimates, while helium abundances in the range of 0·15 observed for a few hours at a time may have been associated with the driver gas of a solar flare-induced shock wave. Hundhausen *et al.* (1968) compared the observed flux of various solar wind oxygen species and determined the temperature of the coronal region from which the solar wind came, since the ionization state of the expanding coronal plasma is established deep in the corona and remains unchanged in transit through interplanetary space.

Hundhausen *et al.* (1967) observed the quiet solar wind velocity to be near 320 km/s with variations between 280 and 750 km/s. Proton temperatures ranged from 8×10^5 to 6×10^3 K. An east-west asymmetry in the solar wind velocity caused by fast stream-slow stream interaction was observed by Siscoe *et al.* (1969), and a sector-dependent asymmetry in the north-south component of the solar wind flow was inferred by Siscoe and Coleman (1969). Siscoe and Finley (1969) used steady-state hydrodynamic equations to determine what pressure perturbations near the Sun are required to produce at Earth the observed north-south deviations from radial flow. Heating of collisionless solar wind plasma by damping of hydromagnetic waves was discussed by Barnes (1969), and heating by long wavelength turbulence was discussed by Jokipii and Davis (1969).

The large-scale sector structure continued to be a prominent feature of the interplanetary magnetic field (Wilcox and Colburn, 1969). The formation of a new interplanetary sector was related to development of a solar active region that produced magnetic loops in the interplanetary medium by Schatten *et al.* (1968). Wilcox and Howard (1968) compared interplanetary and photospheric magnetic fields and postulated a solar sector pattern with sector boundaries at almost constant longitude in a latitude range from 40° N to 40° S. A correlation between right-handed polarization of the solar continuum radial emission and away polarity interplanetary magnetic sectors was observed by Krüger *et al.* (1968). A model for the transition between coronal and interplanetary magnetic fields was proposed by Schatten *et al.* (1969), and the solar origin of the interplanetary magnetic field was also studied by Wilcox and Ness (1967), Winters *et al.* (1969), and Wilcox *et al.* (1969). Carovillano and Siscoe (1969) with approximate solutions to the hydromagnetic equations were able to find reasonable agreement with sector structure observations. Bumba and Obridko (1969) have shown that flares are most probable close to solar sector boundaries. Švestka (1968*d*) and Lanzerotti (1969) have found that in association with some flare events energetic particles are stored within the range of a sector boundary, and with decreasing energy tend to accumulate toward and behind the eastern boundary limit.

Solar flares produce interplanetary shock waves that are decelerated as they expand and propagate through the quiescent solar wind plasma (Dryer and Jones, 1968; Hirshberg, 1968; Ogilvie *et al.*, 1968; Hundhausen and Gentry, 1969; Ness and Taylor, 1969; Ogilvie and Burlaga, 1969). All clearly defined *ssc*'s at the Earth are caused by hydromagnetic shocks (Burlaga and Ogilvie, 1969), but generally a *ssc* or *si* is often the result of tangential discontinuities in the solar wind, and does not imply the existence of a shock wave (Gosling *et al.*, 1967). Levy *et al.* (1969) observed Faraday rotation of the telemetry signal from Pioneer 6 when it was occulted by the Sun. Transient magnetic effects were related to limb flares (Schatten, 1970) and the large-scale Faraday rotation was consistent with the interplanetary magnetic sector structure.

17. SOLAR-TERRESTRIAL RELATIONS

(extracted from a draft* prepared by C. Sawyer, Boulder)

The relative dearth of very energetic particle events (GLE) at the maxima of cycles 18 and 19 has not been observed in cycle 20 (Dodson and Hedeman, 1969*b*). This identifies the inhibition of these events with a critical level of activity, rather than with a particular phase of the cycle (Švestka, 1969). Another difference in the new cycle, perhaps due to the relatively low solar activity and unusually stable interplanetary conditions, is the identification of particle emission with flares on the invisible hemisphere of the Sun (cf. Section 15).

Several authors have continued studies on small-scale statistical increases in cosmic ray intensity after flares of smaller importance. Increases of about 0·1% at balloon altitudes have been found, larger for flares located west of central meridian and at low heliographic latitude (Adagarova *et al.*, 1968; Vladimirskij, 1968). Evidence for weaker solar and interplanetary magnetic fields at high heliographic latitudes comes from the semi-diurnal wave in cosmic-ray intensity (Lietti and Quenby, 1968; Patel *et al.*, 1968). An index of cosmic ray intensity and so of solar activity over periods $> 10^5$ yr may be provided by measurement of ^{59}Ni in dated sedimentary layers (Hasegawa *et al.*, 1968).

Geomagnetic disturbance most often occurs 2 or 3 days after a sector boundary sweeps past the earth, while the arrival at the Earth of the boundary follows 4 or 5 days after central meridian passage of the corresponding photospheric boundary (Wilcox, 1968). Weak and transient disturbances, however, sometimes can be directly associated with the boundary crossing at the Earth (Švestka, 1968*d*). Roosen (1969) has shown that recurrent high values of the geomagnetic character figure C9 follow 4 days after CMP of solar regions with low radio emission in the microwave region. These "coronal suppressions" are supposed to be associated in a solar-wind model with enhanced temperature and velocity, diminished density, and low polytropic index; the low density leads to the observed low radio brightness temperature.

The time sequence and interrelations among variations of the indexes D_{st}, A_E, and PCA are consistent with the hypothesis that the solar "storm cloud" may contain a large flux of protons with energy in the 10 to 100 keV range, and that geomagnetic effects are explained by these solar particles, with no necessity for acceleration or storage in the magnetosphere (Haurwitz, 1969*a*). This hypothesis is consistent with enhancement of polar-glow (optical) aurora at the time of *ssc* (Sanford, 1967), and with subsequent direct observations of large and variable flux of particles in the energy range outside and within the magnetosphere (Frank, 1967). Arguments are advanced for interpreting the observed north-south asymmetry of solar activity and its terrestrial consequences entirely in terms of solar radiation, with no asymmetry in the propagation of the particles that cause *ssc*, PCA or GLE (Popovici and Maris, 1968; Maris and Dinulescu, 1969). Ballario (1967) has studied the history of active regions in relation to geomagnetic disturbances and distinguishes different types of plages, which are associated with disturbed and quiet days, respectively. Haurwitz (1969*b*) has shown that the period of Pi2 micropulsations may offer a means of devising terrestrial indices that would separate the magnitude of the solar stimulus from the sensitivity of the Earth's response. Extensive lists of well-defined geomagnetic storms of sporadic and of recurrent type have been compiled by Mustel (1968).

From studies of ionospheric effects of flares, especially SFD which shows F-region effects (Donnelly, 1967), or from incoherent-scatter measurements of electron density as a function of height and time (Garriott, 1967; Garriott *et al.*, 1969), the relative enhancement within several broad wave-

* Available on request.

length ranges can be estimated. Line and continuum radiation in the 500 to 1000 Å range is found to be the major contributer to the SFD. Ohshio is using the fluctuation of signal strength at 25 kHz to derive a daily and monthly index presumed to measure 0–3 Å radiation. At the Max-Planck Institute for Aeronomy at Lindau, the solar Lyman $L\alpha$ and X-radiation are compared with D-layer absorption and E-layer critical frequency; 10-cm flux, sunspot number and UV flux are compared with F-layer electron-density distribution and with temperature and density in the thermosphere. The Ionospheric Group at Leicester University collaborates with the X-ray group in comparing recorded SID effects with X-ray bursts. Organized efforts to observe ionospheric flare effects are sponsored by IUCSTP, under the leadership of J. W. King.

A very thorough treatment of the influence of solar activity on the weather has been given by Schuurmans (1969). One can agree with W. O. Roberts' opinion that work in this field is going to become very important in the decade ahead. It is a field that has long suffered from a good deal of sloppy and uncritical work, and it is therefore good to have high-quality papers, as this one by Schuurmans, on the subject. This problem is also well treated in three review papers on solar-terrestrial relations published by Mustel (1968), which are concerned with the quiet and disturbed solar wind and the influence of solar particle streams on the lower layers of the Earth's atmosphere. Mustel's studies of dependence of barometric pressure on solar activity, as described by geomagnetic disturbances, suggest that solar particles may contribute to heating the stratosphere and influence the circulation of the lower atmosphere.

Proceedings of a Symposium on Solar-Terrestrial Physics organized by the Sovet Solnce-Zemlja of the Soviet Academy of Sciences at Crimea in October 1968 have been edited by Obridko (1969). Gnevyshev is preparing a compilation of Soviet studies of solar activity effects in the biological sphere.

BIBLIOGRAPHY

Acton, L. W. 1968, *Astrophys. J.*, **152**, 305.
Adagarova, Kh., Ivanov, V., Kolomeets, E. 1968, *Can. J. Phys.*, **46**, 5745.
Alexander, J. K., Malitson, H. H., Stone, R. G. 1969, *Solar Phys.*, **8**, 388.
Alexandrovich, I. I., Stepanov, V. E. 1969, *Izv. SibIZMIR*, no. 2, in press.
Alfvén, H., Carlqvist, P. 1967, *Solar Phys.*, **1**, 220.
Allen, C. W. 1969, *Solar Phys.* **8**, 72.
Altschuler, M. D. 1967, *Solar Phys.*, **1**, 377.
Altschuler, M. D., Nakagawa, Y., Lillienquist, C. G. 1968, *Solar Phys.*, **3**, 466.
Anderson, K. A. 1969, *Solar Phys.*, **6**, 111.
Anichini, M., Godoli, G. 1967, *Mem. Soc. astr. ital.*, **38**, 259.
Anzer, U. 1968, *Solar Phys.*, **2**, 298.
Anzer, U. 1969, *Solar Phys.*, **8**, 37.
Anzer, U., Tandberg-Hanssen, E. 1970, *Solar Phys.*, **11**, 61.
Arnoldy, R. L., Kane, S. R., Winckler, J. R. 1967, *Solar Phys.*, **2**, 171.
Arnoldy, R. L., Kane, S. R., Winckler, J.R. 1968a, *IAU Symposium* no. 35, p. 490; *Astrophys. J.*, **151**, 711.
Arnoldy, R. L., Kane, S. R., Winckler, J. R. 1968*b*, School of Physics and Astronomy, University of Minnesota, CR-108.
Axford, W. I. 1968, *Space Sci. Rev.*, **8**, 331.
Babin, A. N. 1969, *Izv. Krym. astrofiz. Obs.*, **41–42**, in press.
Bachmann, G. 1969, *Astr. Nachr.*, **291**, 131.
Baird, G. A., Bell, G. G., Duggal, S. P., Pomerantz, M. A. 1967, *Solar Phys.*, **2**, 491.
Ballario, M. C. 1967, *IQSY Notes* 21, 36; 1969, *Mem. Soc. astr. ital.*, in press.
Banin, V., de Feiter, L. D., Fokker, A. D., Martres, M. J., Pick, M. 1969, *Ann. IQSY*, **3**, 229.
Baranovskij, E. A. 1969, *Izv. Krym. astrofiz. Obs.*, **40**, **41**, in press.
Barnes, A. 1969, *Astrophys. J.*, **154**, 751; **155**, 311.
Becker, R. A. 1969, *Space Physics at the Aerospace Corporation*, El Segundo, California.
Beckers, J. M. 1968*a*, *Solar Phys.*, **3**, 258.
Beckers, J. M. 1968*b* *Solar Phys.*, **5**, 15.
Beckers, J. M. 1969, *Solar Phys.*, **9**, 372.

Beckers, J. M., Schröter, E. H. 1968*a*, *Solar Phys.*, **4**, 142.
Beckers, J. M., Schröter, E. H. 1968*b*, *Solar Phys.*, **4**, 303.
Beckers, J. M., Schröter, E. H. 1969*a*, *Solar Phys.*, **7**, 22.
Beckers, J. M., Schröter, E. H. 1969*b*, *Solar Phys.*, in press.
Beckers, J. M., Stenflo, J. O. 1969, *Solar Phys.*, **6**, 480.
Beckers, J. M., Tallant, P. E. 1969, *Solar Phys.*, **7**, 351.
Beigman, I. L., Grineva, Yu. I., Mandelshtam, S. L., Vainstein, L. A., Zhitnik, I. A. 1969, *Solar Phys.*, **9**, 160.
Bigg, E. K., Mulhali, P. S. 1967, *Proc. astr. Soc. Australia*, **1**, 53.
Blake, J. B., Paulikas, G. A., Freden, S. C. 1969, Symposium on *Solar Flares and Space Research*, p. 258.
Blizard, J. B. 1969, Paper presented at the *15th Annual AAS Meeting*, Denver.
Bohlin, J. D., Simon, M. 1969, *Solar Phys.*, **9**, 183.
Böhme, A., Krüger, A. 1968, *HHI Suppl. Ser. Solar Data*, **1**, 51.
Boischot, A., Clavelier, B. 1968*a*, *IAU Symposium* no. 35, p. 565.
Boischot, A., Clavelier, B. 1968*b*, *Ann. Astrophys.*, **31**, 445.
Boischot, A., Daigne, G. 1968, *Ann. Astrophys.*, **31**, 531.
Branch, D. 1970, *Astrophys. J.* **159**, in press.
Bray, R. J. 1968, *Solar Phys.* **4**, 318.
Broadfoot, A. L. 1967, *Astrophys. J.*, **149**, 675.
Brückner, G. 1968, *NRL Space Res. Seminar*, NRL Washington, D.C., p. 27.
Bruzek, A. 1967, *Solar Phys.*, **2**, 451.
Bruzek, A. 1968, *Nobel Symposium* no. 9, p. 67.
Bruzek, A. 1969*a*, Symposium on *Solar Flares and Space Research*, p. 61.
Bruzek, A. 1969*b*, *Solar Phys.*, **8**, 29.
Bukata, R. P., Gronstal, P. T., Palmeira, R. A. R., McCracken, K. G., Rao, U. R. 1969, *Solar Phys.*, **10**, 198.
Bumba, V. 1967*a*, *Plasma Astrophysics*, Intern. School of Physics E. Fermi, 39, p. 77.
Bumba, V. 1967*b*, *Solar Phys.*, **1**, 371.
Bumba, V., Howard, R. F. 1969, *Solar Phys.*, **7**, 28.
Bumba, V., Obridko, V. N. 1969, *Solar Phys.*, **6**, 104.
Bumba, V., Howard, R. F., Martres, M. J., Soru-Iscovici, I. 1968, *IAU Symposium* no. 35, p. 13.
Bumba, V., Howard, R. F., Kopecký, M., Kuklin, G. 1969, *Bull. astr. Inst. Csl.*, **20**, 18.
Burlaga, L. F., Ogilvie, K. W. 1969, *J. geophys. Res.*, **74**, 2815.
Carlqvist, P. 1968, *Nobel Symposium* no. 9, p. 193, and 1969, *Solar Phys.*, **7**, 377.
Carovillano, R. V., Siscoe, G. L. 1969, *Solar Phys.*, **8**, 401.
Castelli, J. P., Aarons, J., Michael, G. A. 1967, *J. geophys. Res.*, **72**, 5491.
Chitre, S. M. 1968, *Solar Phys.*, **4**, 168.
Chitre, S. M., Shaviv, G. 1967, *Solar Phys.*, **2**, 150.
Cimino, M., Cacciani, A., Sopranzi, N. 1968, *Solar Phys.*, **3**, 618.
Clark, C. D., Park, W. M. 1968, *Nature*, **219**, 922.
Cline, T. L., McDonald, F. B. 1968, *Solar Phys.*, **5**, 507; 1969, *Ann. IQSY*, **3**, 205.
Cline, T. L., Holt, S. S., Hones, E. W. 1968, *J. geophys. Res.*, **73**, 433.
Covington, A. E., Legg, T. H., Bell, M. B. 1967, *Solar Phys.*, **1**, 465.
Croom, D. L., Powell, R. J. 1969, *Nature*, **221**, 945.
D'Angelo, N. 1969, *Solar Phys.*, **7**, 321.
Daniel, R. R., Joseph, G., Lavakare, P. J., Sunderrajan, R. 1967, *Nature*, **213**, 21.
Daniel, R. R., Gokhale, G. S., Joseph, G., Lavakare, P. J., Sekhon, B. S. 1969, *Solar Phys.*, **10**, 465.
Datlowe, D. W., L'Heroux, J., Meyer, P. 1969, Paper presented at the *11th Int. Conf. on Cosmic Rays* Budapest.
De Groot, T., Van Nieuwkoop, J. 1968, *Solar Phys.*, **4**, 332.
De Jager, C. 1968, *Nobel Symposium* no. 9, p. 171.
De Jager, C., Švestka, Z. (eds.) 1969, *Solar Flares and Space Research*, 11th Plenary Meeting of COSPAR Symposium, North-Holland Publ. Comp., Amsterdam.
Demarq, J., Rösch, J. 1967, *Adv. opt. Techn.*, **7**.
De Mastus, H. L., Stover, R. R. 1967, *Publ. astr. Soc. Pacific*, **79**, 615.
Dessler, A. J. 1967, *Rev. Geophys.*, **5**, 1.
Deubner, F. L. 1968, *IAU Symposium* no. 35, p. 230.

Deubner, F. L. 1969, *Solar Phys.*, **7**, 87.
Deubner, F. L., Liedler, R. 1969, *Solar Phys.*, **7**, 87.
Dizer, M. 1968, *Solar Phys.*, **4**, 99.
Dizer, M. 1969, *Solar Phys.*, **10**, 416.
Dodson, H. W. 1969, *Ann. IQSY*, **3**, 21.
Dodson, H. W., Hedeman, E. R. 1968*a*, *IAU Symposium* no. 35, p. 56.
Dodson, H. W., Hedeman, E. R. 1968*b*, *Nobel Symposium* no. 9, p. 37.
Dodson, H. W., Hedeman, E. R. 1968*c*, *Solar Phys.*, **4**, 229.
Dodson, H. W., Hedeman, E. R. 1969*a*, World Data Center A – Report UAG – 5, 7.
Dodson, H. W., Hedeman, E. R. 1969*b*, *Solar Phys.*, **9**, 278.
Dodson, H. W., Hedeman, E. R., Kahler, S. W., Lin, R. P. 1969, *Solar Phys.*, **6**, 294.
Donnelly, R. 1967, *J. geophys. Res.*, **72**, 5247; 1969, ibid. **74**, 1873.
Drake, J. F., Gibson, J., Van Allen, J. A. 1969, *Solar Phys.*, **10**, 433.
Dryer, M., Jones, D. L. 1968, *J. geophys. Res.*, **73**, 4875.
Durgaprasad, N., Fichtel, C. E., Guss, D. E., Reames, D. V. 1968, *Astrophys. J.*, **154**, 307.
Eddy, J. A., Malville, J. M. 1968, *Astrophys. J.* **150**, 289.
Eliseeva, L. A. 1967, *Izv. Krym. astrofiz. Obs.*, **37**, 153.
Elliot, H. 1969, Symposium on *Solar Flares and Space Research*, p. 356.
Elwert, G. 1968, *Mitt. astron. Ges.*, **25**, 201.
Engvold, O. 1969, *Solar Phys.*, **8**, 284.
Engvold, O., Maltby, P. 1968, *Nobel Symposium* no. 8, p. 109.
Énomé, S., Kakinuma, T., Tanaka, H. 1969, *Solar Phys.* **6**, 428.
Eriksen, G., Maltby, P. 1967, *Solar Phys.*, **2**, 249.
Evans, K., Pounds, K. A. 1968, *Astrophys. J.*, **152**, 319.
Evans, K., Balstrone, R., Parkinson, J., Pounds, K. A. 1969, *Solar Phys.*, in press.
Falciani, R., Rigutti, M. 1967, *Observatory*, **87**, 170.
Falciani, R., Righini, A., Rigutti, M. 1967, *Z. Astrophys.*, **67**, 481.
Fan, C. Y., Pick, M., Pyle, R., Simpson, J. A., Smith, D. R. 1968, *J. geophys. Res.*, **73**, 1555.
Fokker, A. D. 1967, *Solar Phys.*, **2**, 316.
Formisano, V. 1969, *J. geophys. Res.*, **74**, 355.
Fortini, T., Torelli, M. 1968, *IAU Symposium* no. 35, p. 50.
Frank, L. A. 1967, *J. geophys. Res.*, **72**, 3753.
Frazier, E. N., Sherrer, P. H. 1969, *Solar Phys.*, **10**, 297.
Fredga, K. 1969, *Solar Phys.*, **9**, 358.
Friedman, H. 1969, Symposium on *Solar Flares and Space Research*, p. 87.
Friedman, H., Kreplin, R. W. 1969, *Ann. IQSY* **3**, 78, 144.
Friedman, M., Hamberger, S. M. 1968, *Astrophys. J.*, **152**, 667.
Friedman, M., Hamberger, S. M. 1969, *Solar Phys.*, **8**, 104.
Fritzová-Švestková, L., Švestka, Z. 1967, *Solar Phys.*, **2**, 87.
Frost, K. 1969, *Astrophys. J. Letters*, **158**, in press.
Gabriel, A. H., Jordan, C. 1969, *Nature*, **221**, 947, and *Mon. Not. R. astr. Soc.*, in press.
Gaizauskas, V. 1969, Report presented to the Pasadena *Meeting of the A.A.S.* in February 1969.
Garriott, O. 1967, *J. geophys. Res.*, **72**, 6099.
Garriott, O., da Rosa, A., Davis, M., Wagner, L., Thome, G. 1969, *Solar Phys.*, **8**, 226.
Getling, A. V., Tverskij, B. A. 1967, *Astr. Zu.*, **45**, 606.
Gintzburg, M. A. 1967, *J. geophys. Res.*, **72**, 2749; 1968, *Astr. Zu.*, **45**, 610.
Gleissberg, W. 1967, *Astr. Nachr.*, **290**, 11.
Gnevyshev, M. N., Chistjakov, V. F. 1968, *Soviet Astr.* **12**, no. 5.
Gnevyshev, M. N., Nikolskij, G. M., Sasanov, A. A. 1967, *Solar Phys.*, **2**, 223.
Godoli, G. 1967, *Oss. Mem. Oss. astrofis. Arcetri*, **87**.
Godoli, G., Mazzucconi, F., Monsignori Fossi, B. C. 1967, *Ann. Geophys.*, **19**, 95; **20**, 183.
Göhring, R. 1969, *Solar Phys.*, **8**, 271.
Gold, T. 1968, *Nobel Symposium* no. 9, 205.
Goldberg, L. 1967, *A. Rev. Astr. Astrophys.*, **5**, 279.
Gopasjuk, S. I. 1967, *Izv. Krym. astrofiz. Obs.*, **38**, 65; 1969, ibid., **40**, 111.
Gopasjuk, S. I., Tsap, T. T. 1969*a*, *Astr. J.*, **46**; 1970, *Izv. Krym. astrofiz. Obs.*, **43**, in press.
Gopasjuk, S. I., Tsap, T. T. 1969*b*, *Izv. Krym. astrofiz. Obs.*, **41–42**, in press.
Gopasjuk, S.I., Erjushev, N. N., Neshpor, Ju. I. 1967, *Izv. Krym. astrofiz. Obs.*, **37**, 109; 1969, ibid. **39**, 298.

Gosling, J. T., Asbridge, J. R., Bame, S. J., Hundhausen, A. J., Strong, I. B. 1967, *J. geophys. Res.*, **72**, 3357; 1968, ibid. **73**, 43.
Grigorev, N. L., Lutsenko, V. N., Pisarenko, N. F., Savenko, I. A. 1969, Paper presented at the *11th Conf. on Cosmic Rays*, Budapest.
Hall, L. A., Hinteregger, H. E. 1969, Symposium on *Solar Flares and Space Research*, p. 81.
Hansen, R., Hansen, S., Loomis, H. 1969, *Solar Phys.*, **10**, 135.
Hartle, R. E., Sturrock, P. A. 1968, *Astrophys. J.*, **151**, 1155.
Hartz, T. R. 1969, *Planet. Space Sci.*, **17**, 267.
Harvey, J. W. 1969, Thesis, University of Colorado, Boulder.
Harvey, J., Livingston, W. 1969, *Solar Phys.*, **10**, 283.
Harvey, J. W., Tandberg-Hanssen, E. 1968, *Solar Phys.*, **3**, 316.
Hasegawa, H., Yamakoshi, K., Noma, M., Maihara, T. 1968, *Canad. J. Phys.*, **46**, 5930.
Haugen, E. 1969, *Solar Phys.*, **9**, 88.
Haurwitz, M. W. 1968, *Astrophys. J.*, **151**, 351.
Haurwitz, M. W. 1969*a*, *J. geophys. Res.*, **74**, 2348.
Haurwitz, M. W. 1969*b*, *Planet. Space Sci.*, in press.
Hearn, A. G., Noyes, R. W., Withbroe, G. L. 1969, *Mon. Not. R. astr. Soc.*, **144**, 351.
Henoux, J. C. 1968*a*, *Solar Phys.*, **4**, 315.
Henoux, J. C. 1968*b*, *Ann. Astrophys.*, **31** 511.
Hirshberg, J. 1968, *Planet. Space Sci.*, **16**, 309.
Holt, S. S. 1967, *J. geophys. Res.*, **72**, 3507.
Holt, S. S., Cline, T. L. 1968, *Astrophys. J.*, **154**, 1027.
Holt, S. S., Ramaty, R. 1969, *Solar Phys.*, **8**, 119.
Howard, R., Bumba, V., Smith, S. 1967, *Atlas of Solar Magnetic Fields*, Carnegie Inst. Washington, D.C. Publ. 626.
Howard, R., Tanenbaum, A. S., Wilcox, J. 1968, *Solar Phys.*, **4**, 286.
Hudson, H. S., Peterson, L. E., Schwartz, D. A. 1969*a*, *Solar Phys.*, **6**, 205; *Astrophys. J.*, **157**, 389.
Hudson, H., McKenzie, D., Zirin, H., Ingham, W. 1969*b*, 18–21 February 1969 *AAS Meeting*, Pasadena.
Hundhausen, A. J. 1968, *Space Sci. Rev.*, **8**, 690.
Hundhausen, A. J., Gentry, R. A. 1969, *J. geophys. Res.*, **74**, 2908.
Hundhausen, A. J., Asbridge, J. R., Bame, S. J., Strong, I. B. 1967, *J. geophys. Res.*, **72**, 1979.
Hundhausen, A. J., Gilbert, H. E., Bame, S. J. 1968, *J. geophys. Res.*, **73**, 5485.
Hyder, Ch. L. 1967, *Solar Phys.*, **2**, 49, 267.
Hyder, Ch. L., Mauter, H. S., Shutt, R. L. 1968, *Astrophys. J.*, **154**, 1039.
Ikhsanov, R. N. 1967, *Astr. Zu.*, **44**, 1048, 1211.
Ikhsanov, R. N., Platonov, Y. P. 1967, *Soln. Dann.*, **11**, 78.
Innanen, W. G., Krimigis, S. M., Armstrong, T. P. 1968, paper presented at the *Midwest Cosmic Ray Conference*, Iowa City.
Ioshpa, B. A. 1968, *Soln. Aktivn.*, **3**, 44.
Janssens, T. J., White, K. P. 1970, *Solar Phys.*, in press.
Jensen, E., Brahde, R., Ofstad, P. 1969, *Solar Phys.*, **9**, 397.
Joffe, S. B., Smirnova, T. A. 1967, *Astr. Zu.*, **45**, 1037.
Jokipii, J. R. 1969, Paper presented at the *11th Conf. on Cosmic Rays*, Budapest.
Jokipii, J. R., Davis, L., Jr. 1969, *Astrophys. J.*, **156**, 1101.
Jokipii, J. R., Parker, E. N. 1969, *Astrophys. J.*, **155**, 777.
Jones, B. B. Freeman, F. F., Wilson, R. 1968, *Nature*, **219**, 252.
Kai, K. 1968*a*, *Publ. astr. Soc. Japan*, **20**, 140.
Kai, K. 1968*b*, *Publ. astr. Soc. Japan*, **20**, 154.
Kai, K. 1969, *Proc. astr. Soc. Austr.*, **1**, 186.
Kai, K. 1970, *Solar Phys.*, in press.
Kane, S. R., Winckler, J. R. 1969*a*, *Solar Phys.*, **6**, 304.
Kane, S. R., Winckler, J. R. 1969*b*, School of Physics and Astronomy, University of Minnesota, CR-134.
Kassinskij, V. V. 1967, *Soln. Dann.*, **2**, 59.
Kassinskij, V. V. 1969, *Soln. Dann.*, **4**, 76; **5**, 82.
Katzenstein, J. 1968, *J. opt. Soc. Am.*, **58**, 1348.
Kiepenheuer, K. O. (ed.) 1968*a*, *IAU Symposium* no. 35, *Structure and Development of Solar Active Regions*, D. Reidel, Dordrecht, The Netherlands.

Kiepenheuer, K. O. 1968*b*, *IAU Symposium* no. 35, p. 3.
Kiepenheuer, K. O. 1968*c*, *Nobel Symposium* no. 9, p. 123.
Kjeldseth Moe, O. 1968*a*, *Solar Phys.*, **4**, 267.
Kjeldseth Moe, O. 1968*b*, *IAU Symposium* no. 35, p. 202.
Kjeldseth Moe, O., Maltby, P. 1969, *Solar Phys.*, **8**, 275.
Kleczek, J. 1968, *IAU Symposium* no. 35, p. 280.
Kleczek, J., Kuperus, M. 1969, *Solar Phys.*, **6**, 72.
Kneer, F., Mattig, W. 1968, *Solar Phys.*, **5**, 42.
Koechlin, Y., Raviart, A., Tregner, L., Blaud, C. J., Degli Antoni, G., Dilworth, C., Maccagni, D., Tanzi, E. G. 1969, Paper presented at the *11th Conf. on Cosmic Rays*, Budapest.
Kolpakov, P. E. 1968, *Izv. Krym. astrofiz. Obs.*, **38**, 59.
Kopecký, M. 1969, *Bull. astr. Inst. Csl.*, **20**, 296.
Kopecký, M., Kuklin, G. V. 1969*a*, *Bull. astr. Inst. Csl.*, **20**, 22.
Kopecký, M., Kuklin, G. V. 1969*b*, *Solar Phys.*, **6**, 241.
Kopecký, M., Obridko, V. 1968, *Solar Phys.*, **5**, 362.
Kopecký, V. 1968, *Bull. astr. Inst. Csl.*, **19**, 362.
Kopp, R. 1968, Doctoral Thesis, Harvard University, Cambridge, Mass.
Kopp, R., Kuperus, M. 1968, *Solar Phys.*, **4**, 212.
Korchak, A. A. 1967, *Dokl. A. N. SSSR*, **173**, 291; *Nature*, **213**, 1209.
Kotov, V. A. 1969, *Izv. Krym. astrofiz. Obs.*, **41–52**, in press.
Koval, A. N. 1967, *Izv. Krym. astrofiz. Obs.*, **37**, 62.
Koval, A. N., Severnyj, A. B. 1970, *Solar Phys.*, in press.
Krat, V. A. 1968, *Nobel Symposium* no. 9, p. 89.
Krat, V. A. 1969, *Izv. glav. astr. Obs. Pulkovo*, **185**, in press.
Krimigis, S. M. 1969, *Ann. IQSY*, **3**, 457.
Křivský, L. 1968, *IAU Symposium* no. 35, p. 465.
Křivský, L. 1969, *Solar Phys.*, **9**, 194.
Křivský, L., Nestorov, G. 1968, *Bull. astr. Inst. Csl.*, **19**, 197.
Krüger, A., Lauter, E. A. (eds.) 1969, *Geod. geoph. Veröff.* R.II. H. 10.
Krüger, A., Bumba, V., Howard, R., Kleczek, J. 1968, *Bull. astr. Inst. Csl.*, **19**, 180.
Künzel, H. 1969, *Astr. Nachr.*, **291**, 265.
Kuperus, M. 1969, *Space Sci. Rev.*, **9**, 713.
Kuperus, M., Tandberg-Hanssen, E. 1967, *Solar Phys.*, **2**, 39.
Kusminykh, V. D., Sitnik, G. F. 1967, *Mitt. Staatl. astr. Sternberg Inst.*, **147**, 3 and 7.
Labrun, N. R., Smerd, S. F. 1968, *Proc. astron. Soc. Australia*, **1**, 140.
Lacombe, C., Mangency, A. 1969, *Astr. Astrophys.*, **1**, 325.
Landini, M., Monsignori Fossi, B. C., Poletto, G., Tagliaferi, G. L. 1968, *Solar Phys.*, **5**, 546.
Lanzerotti, L. J. 1969, *J. geophys. Res.*, **74**, 2851.
Lee, R. H., Harvey, J. W., Tandberg-Hanssen, E. 1969, *Appl. Opt.*, in press.
Leighton, R. B. 1969, *Astrophys. J.*, **156**, 1.
Lemaire, P. 1969, *Astrophys. Letters*, **3**, 43, also in *IAU Symposium* no. 36.
Léna, P. 1969, *Solar Phys.*, **7**, 217.
Leroy, J. L. 1967, *Ann. Astrophys.*, **30**, 249.
Leroy, J. L. 1969, *Solar Phys.*, **7**, 221.
Levickij, L. S. 1967, *Izv. Krym. astrofiz. Obs.*, **37**, 137.
Levy, G. S., Sato, T., Seidel, B. L., Stelzried, C. T., Ohlson, J. E., Rusch, W. V. T. 1969, *Science*, **166**, 596.
Lietti, B., Quenby, J. 1968, *Canad. J. Phys.*, **46**, 5942.
Lin, R. P. 1969, Paper presented at the *11th Int. Conf. on Cosmic Rays*, Budapest, and 1970, *Solar Phys.*, in press.
Lin, R. P., Anderson, K. A. 1967, *Solar Phys.*, **1**, 446.
Lincoln, J. V. (ed.) 1969, World Data Center A Report UAG-5.
Lindgren, S. T. 1968, *Solar Phys.*, **5**, 382.
Lingenfelter, R. E. 1969, *Solar Phys.*, **8**, 341.
Lingenfelter, R. E., Ramaty, R. 1967, Inst. Geophys. Planet. Phys., UCLA, Publ. 595, 34.
Liszka, L. 1968, *Nobel Symposium* no. 9, p. 117.
Livingston, W. C. 1968, *Astrophys. J.*, **153**, 929.
Livshitz, M. A. 1968, *Soln. Aktivn.* **3**, 78.

Lockwood, J. A. 1968, *J. geophys. Res.*, **73**, 4247.
Loughhead, R. E. 1968, *Solar Phys.*, **5**, 489.
Loughhead, R. E., Bray, R. J., Tappere, E. J., Winter, J. G. 1968, *Solar Phys.*, **4**, 185.
Makita, M. 1968, *Solar Phys.*, **3**, 557.
Maltby, P. 1968*a*, *Solar Phys.*, **4**, 96.
Maltby, P. 1968*b*, *Solar Phys.*, **5**, 3.
Malville, J. M. 1968, *Solar Phys.*, **4**, 323; **5**, 236.
Malville, J. M., Tandberg-Hanssen, E. 1969, *Solar Phys.*, **6**, 278.
Maris, G., Dinulescu, V. 1969, *Stud. si Cerci. Astron.*, **14**, 45.
Martres, M. J. 1968, *IAU Symposium* no. 35, p. 25.
Martres, M. J., Michard, R., Soru-Iscovici, I., Tsap, T. 1968, *IAU Symposium* no. 35, p. 318; *Solar Phys.*, **5**, 187.
Masley, A. J. 1969, Symposium on *Solar Flares and Space Research*, p. 279.
Mattig, W. 1969*a*, *Solar Phys.*, **6**, 413.
Mattig, W. 1969*b*, *Solar Phys.*, **8**, 291.
Mattig, W., Mehltretter, J. P. 1968, *IAU Symposium* no. 35, p. 187.
McCracken, K. G. 1969, Symposium on *Solar Flares and Space Research*, 202.
McIntosh, P. S. 1969, *Ann. IQSY*, **3**, 40.
McKenna, S. M. P. 1967, *Astrophys. J.*, **150**, 1087.
McKenna-Lawlor, S. M. P. 1968, *Astrophys. J.*, **153**, 367.
McKenna-Lawlor, S. M. P. 1969, *Astrophys. J.*, in press.
McQueen, R. M. 1968, *Appl. Opt.*, **7**, 1149.
Meekins, J. F., Kreplin, R. W., Chubb, T. A., Friedman, H. 1968, *Science*, **162**, 891.
Mehltretter, J. P. 1969, *Solar Phys.*, **9**, 387.
Memeth, T. 1967, *Geofiz. Közl. Magyar*, **16**, No. 1–2.
Menzel, D. H. 1968, *Nobel Symposium* no. 9, p. 183; 1969, Symposium on *Solar Flares and Space Research*, p. 363.
Meyer, F. 1968, *IAU Symposium* no. 35, p. 485.
Michard, R. 1968, *IAU Symposium* no. 35, p. 307.
Monsignori Fossi, B. C., Poletto, G., Tagliaferri, G. L. 1968, *Mem. Soc. astr. ital.*, **39**, 201; *Space Research* vol. 10, p. 196.
Monsignori Fossi, B. C., Poletto, G., Tagliaferri, G. L. 1969, *Solar Phys.*, **10**, 196.
Montgomery, M. D., Bame, S. J., Hundhausen, A. J. 1968, *J. geophys. Res.*, **73**, 4999.
Moreton, G. E., Severnyj, A. B. 1968, *Solar Phys.*, **3**, 282.
Mustel, E. R. 1968, *Astr. Sov. Akad. Nauk. SSSR Sci. Inf.*, **10**.
Nakagawa, Y., Swartztrauber, P. 1969, *Astrophys. J.*, **155**, 295.
Ness, N. F., Taylor, H. E. 1969, *Ann. IQSY*, **3**, 336.
Neupert, W. M. 1967, *Solar Phys.*, **2**, 294.
Neupert, W. M. 1969, *A. Rev. Astr. Astrophys.*, **7**, 121.
Neupert, W. M., Swartz, M. 1969, *Phys. Rev. Letters*, in press.
Neupert, W. M., Gates, W., Swartz, M., Young, R. 1967, *Astrophys. J.*, **149**, L79.
Neupert, W. M., White, W. A., Gates, W. J., Swartz, M., Young, R. M. 1969, *Solar Phys.*, **6**, 183.
Newkirk, G. 1967a, *Planet. Space Sci.*, **15**, 1267.
Newkirk, G. 1967*b*, *A. Rev. Astr. Astrophys.*, **5**, 213.
Newkirk, G., Hansen, R. T., Hansen, S. 1969, *Ann. IQSY*, **3**, 49.
Nikonov, O. V., Nikonova, E. S. 1969, *Soln. Dann.*, **4**, 97; **5**, 86.
Nikulin, N. S. 1967, *Izv. Krym. astrofiz. Obs.*, **36**, 76.
Noyes, R. W., Goldberg, L., Parkinson, W. H., Reeves, E. M., Withbroe, G. L. 1968, *Astr. J.*, **73**, 573.
Noyes, R. W., Withbroe, G. L., Kirshner, R. P. 1969, *Conf. on the Chromosphere-Corona Transition Region*, HAO Boulder.
Obridko, V. N. 1968*a*, *Soln. Dann.*, **4**, 68.
Obridko, V. N. 1968*b*, *Bull. astr. Inst. Csl.*, **19**, 183, 186.
Obridko, V. N. (ed.) 1969, *Solnechno-zemnaja fizika* 1, Akad. Nauk. SSSR, Sovet Solnce-Zemlja.
Ogilvie, K. W., Burlaga, L. F. 1969, *Solar Phys.*, **8**, 422.
Ogilvie, K. W., Burlaga, L. F., Wilkerson, T. D. 1968, *J. geophys. Res.*, **73**, 6809.
Ogilvie, K. W., Wilkerson, T. D. 1969, *Solar Phys.*, **8**, 435.
Ogir, M. B. 1967, *Izv. Krym. astrofiz. Obs.*, **36**, 69; **37**, 94.
Ogir, M. B. 1969, *Izv. Krym. astrofiz. Obs.* **41–42**, and **43**, in press.

Öhman, Y. (ed.) 1968, *Mass Motions in Solar Flares and Related Phenomena, Nobel Symposium* no. 9.
Öhman, Y., Hosinsky, G., Kusoffsky, U. 1968, *Nobel Symposium* no. 9, p. 95.
Oster, L. 1968, *Solar Phys.*, **3**, 543.
Pande, M. C., Gaur, V. P., Tripathi, B. M. 1969, *Solar Phys.*, **7**, 370.
Parker, E. N. 1969, *Space Sci. Rev.*, **9**, 325.
Parks, G. K., Winckler, J. R. 1969, *Astrophys. J.*, **155**, L117.
Patel, D., Sarabhai, V., Subrahmanian, G. 1968, *Can. J. Phys.*, **46**, 5981.
Pintér, Š. 1969, *Solar Phys.*, **8**, 149.
Pneuman, G. W. 1967, *Solar Phys.*, **2**, 462.
Pneuman, G. W., Kopp, R. 1970, *Solar Phys.*, in press.
Polonskij, W. W. 1967, *Astr. Cirk.*, **416**, 1.
Polupan, P. N. 1968, *Soln. Aktivn.*, **3**, 125.
Popovici, C., Maris, G. 1968, *Stud. si Cerc. de Astron.*, **13**, 3.
Popovici, C., Dimitriu, A. 1969, *Ann. IQSY*, **3**, 31.
Porfirov, V. (ed.) *Voprosy Astrofiziky*, Izdatelstvo Naukove Dumka Kiev.
Porsche, H. 1968, *Mitt. astron. Ges.*, **25**, 55.
Rachkovskij, D. N. 1967*a*, *Izv. Krym. astrofiz. Obs.*, **36**, 9, 56.
Rachkovskij, D. N. 1967*b*, *Izv. Krym. astrofiz. Obs.*, **36**, 51.
Raju, P. K. 1968, *Mon. Not. R. astr. Soc.*, **139**, 479.
Ramaty, R., Lingenfelter, R. E. 1968, *Solar Phys.*, **5**, 531.
Ramsey, H. E., Smith, S. F., Angle, K. L. 1968, Final Report High Resolution Solar Photography, LMSC-681495.
Rao, U. R., Allum, F. R., Bartley, W. C., Palmeira, R. A. R., Harries, J. A., McCracken, K. G. 1969, Symposium on *Solar Flares and Space Research*, p. 267.
Rayrole, J. 1967, *Ann. Astrophys.* **30**, 257.
Rees, D. E. 1969, *Solar Phys.*, **10**, 268.
Reidy, W. P., Vaiana, G. S., Zehnpfennig, T., Giaconni, R. 1968, *Astrophys. J.*, **151**, 333.
Rompolt, B. 1967, *Acta Astr.*, **17**, 329.
Roosen, J. 1969, *Solar Phys.*, **7**, 448.
Ruhm, H. 1969, *Solar Phys.*, **10**, 104.
Rust, D. M. 1967, *Astrophys. J.*, **150**, 313.
Saito, M., Kato, S. 1968, *Solar Phys.*, **3**, 531.
Sakurai, K. 1967*a*, *Rep. Iono. Space Res. Japan*, **21**, 29.
Sakurai, K. 1967*b*, *Rep. Iono. Space Res. Japan*, **21**, 113.
Sakurai, K. 1967*c*, *Rep. Iono. Space Res. Japan*, **21**, 213.
Sanford, B. 1967, *Space Res.*, vol. **7**, p. 836.
Sattarov, I. 1968, *Soln. Dann.*, **8**, 76.
Schatten K. H. 1970, *Solar Phys.*, in press.
Schatten K. H., Ness, N. F., Wilcox, J. M. 1968, *Solar Phys.*, **5**, 240.
Schatten K. H., Wilcox, J. M., Ness, N. F. 1969, *Solar Phys.*, **6**, 442.
Schatzman, E. 1967, *Solar Phys.*, **1**, 411.
Schmidt, H. U. 1968, *IAU Symposium* no. 35, p. 95.
Schmidt, H. U. 1969, Symposium on *Solar Flares and Space Research*, p. 331.
Schröter, E. H., Wiehr, E. 1968, *Mitt. astr. Ges.*, **25**, 187.
Schuurmans, C. J. E. 1969, *K. Ned. Meteorol. Inst. Med. Verhandl.* no. 92.
Severnyj, A. B. 1967, *Izv. Krym. astrofiz. Obs.*, **36**, 22.
Severnyj, A. B. 1968*a*, *Izv. Krym. astrofiz. Obs.*, **38**, **3**.
Severnyj, A. B. 1968*b*, *Nobel Symposium* no. 9, p. 71.
Severnyj, A. B. 1968*c*, *IAU Symposium* no. 35, p. 233.
Severnyj, A. B. 1969*a*, Symposium on *Solar Flares and Space Research*, p. 38.
Severnyj, A. B. 1969*b*, *Ann. IQSY*, **3**, 11.
Severnyj, A. B., Steshenko, N. V. 1969, *STP Notes*, in press.
Severnyj, A. B., Zvereva, A. M. 1969, *Izv. Krym. astrofiz. Obs.*, **41–42**, in press.
Sheeley, N. R. 1967, *Solar Phys.*, **1**, 171.
Sheeley, N. R. 1969, *Solar Phys.*, **9**, 347.
Shimabukuro, F. I. 1968, *Solar Phys.*, **5**, 498.
Simon, G. W., Weiss, N. O. 1968, *IAU Symposium* no. 35, p. 108.
Simon, P., Švestka, Z. 1969, *Ann. IQSY*, **3** , 469.

Simon, P., Martres, M. J., Legrand, J. P. 1969, Symposium on *Solar Flares and Space Research*, p. 405.
Siscoe, G. L., Coleman, P. J., Jr. 1969, *Solar Phys.*, **8**, 415.
Siscoe, G. L., Finley, L. T. 1969, *Solar Phys.*, **9**, 452.
Siscoe, G. L., Goldstein, B., Lazarus, A. J. 1969, *J. geophys. Res.*, **74**, 1759.
Sitnik, G. F., Mitropolskaja, O. N. 1968, *Astr. Zu.*, **45**, 456.
Slonim, J. M. 1968, *Astr. Zu.*, **45**, 286, 726.
Slonim, J. M. 1969, *Astr. Zu.*, **46**, 570, 697.
Slysh, V. I. 1967, *Astr. Zu.*, **44**, 487; *Kosm. Issled.*, **5**, 897.
Smerd, S. F. 1968, *Proc. astr. Soc. Austr.*, **1**, 124.
Smith, E. v . P. 1968, *Nobel Symposium* no. 9, 137.
Smith, E. v. P. 1969, *Bull. Am. astr. Soc.*, **1**, 262.
Smith, S. F., Howard, R. 1968, *IAU Symposium* no. 35, p. 33.
Smith, S. F., Ramsey, E. 1967, *Solar Phys.*, **2**, 158.
Smolkov, G. J. 1969, *Issl. geomagn. aeron., fizika Solnca*, in press.
Smolkov, G. J., Bashkirtzev, V. S. 1969, *Issl. geomagn. aeron., fizika Solnca*, in press.
Staude, J. 1969, *Solar Phys.*, **8**, 264.
Steenbeck, M., Krause, F. 1969, *Astr. Nachr.*, **291**, 49.
Stellmacher, G. 1969, *Astr. Astrophys.*, **1**, 62.
Stenflo, J. O. 1968*a*, *Medd. Lunds astr. Obs. Ser.* I, no. 233 and Ser. II, nos. 152 and 153.
Stenflo, J. O. 1968*b*, *Solar Phys.*, **3**, 482.
Stenflo, J. O. 1969*a*, *Solar Phys.*, **8**, 115.
Stenflo, J. O. 1969*b*, *Solar Phys.*, **8**, 260.
Stenflo, J. O. 1969*c*, *Solar Phys.*, **6**, 476.
Stepanjan, N. N. 1969, *Astr. Zu.*, **46**, 580.
Steshenko, N. V. 1968, *Izv. Krym. astrofiz., Obs.*, **38**, 52.
Steshenko, N. V. 1970, *Izv. Krym. astrofiz. Obs.*, **43**, in press.
Steshenko, N. V., Shcherbakov, A. G. 1969, *Izv. Krym. astrofiz. Obs.*, **41–42**, in press.
Sturrock, P. A. 1968, Proc. E. Fermi Int. Summer School of Physics, 39th Course: *Plasma Astrophysics*, p. 168, and *IAU Symposium* no. 35, p. 471.
Surkov, E. P. 1969, *Soln. Dann.*, **5**, 91.
Švestka, Z. 1968*a*, *Solar Phys.*, **4**, 18.
Švestka, Z. 1968*b*, *IAU Symposium* no. 35, p. 513.
Švestka, Z. 1968*c*, *Nobel Symposium* no. 9, p. 17.
Švestka, Z. 1968*d*, *Solar Phys.*, **4**, 361.
Švestka, Z. 1969, *Space Research*, in press; Proceedings of the *3rd ESLAB/ESRIN Symp.*, in press.
Švestka, Z., Simon, P. 1969, *Solar Phys.*, **10**, 3.
Sweet, P. 1969, *A. Rev. Astr. Astrophys.*, **7**, 149.
Syrovatskij, S. I. 1969, Symposium on *Solar Flares and Space Research*, p. 346.
Takahashi, K. 1968, *Solar Phys.*, **3**, 598.
Takakura, T. 1967, *Solar Phys.*, **1**, 304.
Takakura, T., Uchida, Y., Kai, K. 1968, *Solar Phys.*, **4**, 45.
Tallant, P. E. 1970, *Solar Phys.*, in press.
Tanaka, H., Kakinuma, T., Énomé, S. 1969*a*, *Ann. IQSY*, **3**, 63.
Tanaka, H., Kakinuma, T., Énomé, S., Torii, C., Tsukiji, Y., Kobayashi, S. 1969*b*, *Proc. Res. Inst. atm. Nagoya Univ.*, **16**, 113.
Teplitskaja, R. B., Turchina, V. D. 1969, *Astr. Zu.*, **46**, 74.
Tousey, R. 1967, *Astrophys. J.*, **149**, 239.
Tsap, T. T. 1967, *Izv. Krym. astrofiz. Obs.*, **39**, 265.
Tsap, T. T. 1969, *Izv. Krym. astrofiz. Obs.*, **41–42**, in press.
Uchida, Y. 1968, *Solar Phys.*, **4**, 30.
Unsöld, A. 1968, *Q. J. R. astr. Soc.*, **9**, 294.
Vaiana, G. S., Reidy, W. P., Zehnpfennig, T., Van Speybroek, L. 1968, *AScience*, **161**, 564.
Valníček, B. 1967, *Bull. astr. Inst. Csl.*, **18**, 249.
Valníček, B. 1968, *IAU Symposium* no. 35, p. 282.
Valníček, B., Godoli, G., Mazzucconi, F. 1969, *Ann. IQSY*, **3**, 113.
Van Allen, J. A., Wende, C. D. 1969, *J. geophys. Res.*, **74**, 3046.
Vernov, S. N., Chudakov, A. E., Vakulov, P. V., Gorchakov, E. V., Kontor, N. N., Legachev, Yu, I.,

Ljubimov, G. P., Pereslegina, N. V., Timofeev, G. A. 1969, Paper presented at the *3rd ESRIN/ESLAB Symp.* at Noordwijk, Sept. 1969.
Vinokur, M. 1968, *Ann. Astrophys.*, **31**, 457.
Vitinskij, Ju. L. 1967, *Izv. glavn. astr. Obs. Pulkovo*, **24**, no. 6.
Vitinskij, Ju. L. 1969*a*, *Solar Phys.*, **7**, 210.
Vitinskij, Ju. L. 1969*b*, *Izv. glavn. astr. Obs. Pulkovo*, no. 186, in press.
Vitinskij, Ju. L., Kopecký, M. 1968, *Izv. glavn. astr. Obs. Pulkovo*, no. 184.
Vladimirskij, B. 1968, *Izv. Krym. astrofiz. Obs.*, **38**, 132.
Von Klüber, H. 1967, *Mon. Not. R. astr. Soc.*, **137**, 297.
Von Klüber, H. 1968, *Observatory*, **88**, 45; *Mon. Not. R. astr. Soc.*, **141**, 469.
Vojhanskaja, N. F. 1967, *Astr. Obs. Trudy*, **24**, 73.
Warwick, J. W. 1967, *Astrophys. J.*, **150**, 1081.
Weart, S. R., Zirin, H. 1969, *Publ. astr. Soc. Pacific*, **81**, 270.
Webber, J. C. 1969, Thesis, Pasadena.
White, K. P., Janssens, J. T. 1970, *Solar Phys.*, in press.
Wiehr, E. 1968, *IAU Symposium* no. 35, p. 259.
Wiehr, E. 1969, *Solar Phys.*, **9**, 225.
Wiehr, E., Stellmacher, G., Schröter, E. H. 1968, *Astrophys. Letters*, **1**, 181; 1969, ibid. **3**, 91.
Wilcox, J. M. 1968, *Space Sci. Rev.*, **8**, 258; Symposium on *Solar Flares and Space Research*, p. 294.
Wilcox, J. M., Colburn, D. S. 1969, *J. geophys. Res.*, **74**, 2388.
Wilcox, J. M., Howard, R. 1968, *Solar Phys.*, **5**, 64.
Wilcox, J. M., Ness, N. F. 1967, *Solar Phys.*, **1**, 437.
Wilcox, J. M., Severnyj, A. B., Colburn, D. S. 1969, *Nature*, **224**, 353.
Wild, J. P. (ed.) 1967, The Culgoora radio heliograph, *Proc. Instn. Radio El. Eng. Austr.*, **28**, 277.
Wild, J. P. 1968*a*, *Proc. astr. Soc. Austr.*, **1**, 137.
Wild, J. P. 1968*b*, *Australian Physicist*, **5**, 117.
Wild, J. P. 1969*a*, *Proc. astr. Soc. Austr.*, **1**, 181.
Wild, J. P. 1969*b*, *Solar Phys.*, **9**, 260.
Wild, J. P., Sheridan, K. V., Kai, K. 1968, *Nature*, **218**, 536.
Wilson, P. R. 1968*a*, *Solar Phys.*, **3**, 243, 454.
Wilson, P. R. 1968*b*, *Solar Phys.*, **5**, 338.
Wilson, P. R. 1969*a*, *Solar Phys.*, **9**, 391.
Wilson, P. R. 1969*b*, *Solar Phys.*, **10**, 404.
Winters, J. V., Ballif, J. R., Jones, D. E. 1969, *Solar Phys.*, **7**, 278.
Wittmann, A. 1969, *Solar Phys.*, **7**, 366.
Wittmann, A., Schröter, E. H. 1969, *Solar Phys.*, **10**, 357.
Wöhl, H. 1969, *Solar Phys.*, **9**, 394.
Xanthakis, J. 1969, *Solar Phys.*, **10**, 168.
Yip, W. K. 1967, *Austr. J. Phys.*, **20**, 421.
Yun, H. S. 1968, Thesis, Indiana University.
Zhukov, L. V., Muzalevskij, Ju. S. 1969, *Astr. Zu.*, **46**, 600.
Zirin, H., Russo Lackner, D. 1969, *Solar Phys.*, **6**, 86.
Zirin, H., Werner, S. 1967, *Solar Phys.*, **1**, 66.
Zirker, J. B. 1967, *Solar Phys.*, **1**, 204.
Zwaan, C. 1967, *Solar Phys.*, **1**, 478.
Zwaan, C. 1968, *A. Rev. Astr. Astrophys.*, **6**, 135.
Zych, A. D., Frye, G. M. 1969, *J. geophys. Res.*, **74**, 3726.

Z. ŠVESTKA
President of the Commission

12. RADIATION AND STRUCTURE OF THE SOLAR ATMOSPHERE (RADIATION ET STRUCTURE DE L'ATMOSPHÈRE SOLAIRE)

PRESIDENT: M. N. Gnevyshev.
VICE-PRESIDENT: R. G. Athay.
ORGANIZING COMMITTEE: L. Delbouille, R. Michard, M. Minnaert, E. A. Müller, A. K. Pierce, Z. Suemoto.

INTRODUCTION

By action of the Executive Committee of IAU, the report for 1970 is restricted to half the length of the 1967 draft report; the bibliography and the abstracting of work already published are essentially eliminated. The report is no longer to serve primarily as a review of accomplishments between General Assemblies but rather as a report of work in progress. These changes were adopted by the Executive Committee in order to reduce the cost of publishing the reports.

Several commission members have voiced the opinion that the extensive bibliographies in past reports have been the most valuable contributions of the reports and they have objected to the elimination of the bibliography. Also, material from some observatories and from some national groups that have been submitted for inclusion in this draft report have contained only abstracts of work already published. The rules adopted by the Executive Committee do not permit the report to abstract published works except in evaluating general progress in solar astronomy. We express our regret to these commission members and to these institutions that we have not been able to include the abstracts and bibliography in the present report.

In preparing the present report, we have been forced to ask the obvious question, *"What function should the report serve?"* If it is not to contain a bibliography, its use is restricted effectively to the current General Assembly and perhaps the following year. Although the report may have some historical value, it obviously will not serve as a useful reference document for research purposes. What, then, is the function of the report? We have assumed for the present report that its purpose is to inform those engaged in research in solar physics of what is happening in solar physics in 1970. It is not obvious that this is the correct assumption to make, and we urge commission members to communicate their whishes with respect to the report to the commission officers.

If the reports continue in the future to serve primarily as a report of work in progress at the time of the General Assembly, it will be mandatory that more commission members contribute to the report. Requests for information for the current draft report received very limited response. As a result, we are not able to provide a well balanced report. A further consequence of a report that does not extensively survey the literature is that it will reflect inevitably the personality and interest of the person preparing the report and will tend to be more editorial in character. This adds to the need for each commission member to submit appropriate material for future reports.

Strong contributions to this report were made by L. Delbouille, M. N. Gnevyshev, R. G. Giovanelli, C. de Jager, M. Kopecký, G. Newkirk, A. K. Pierce, G. Righini, M. Rigutti, Z. Suemoto and M. Waldmeier. The final report was prepared by R. G. Athay, although most of the coronal section was taken verbatim from a report by G. Newkirk.

INSTRUMENT DEVELOPMENT

Instrumentation in solar astronomy is undergoing relatively rapid growth. Major new telescopes are under construction, new observatories are being established and new instrumental techniques are being developed.

The new 45 m tower telescope at Meudon is producing a good solar image and spectroscopic observations will begin during 1970.

The large new tower telescope at the Sacramento Peak Observatory is producing good images also, and will soon be ready for spectroscopic work. This instrument contains several new features including a rotatable telescope floating in a mercury bath. The tower extends 41·5 m above ground and sits over a pit approximately 60 m deep. The entire telescope and spectrograph assembly are evacuated. A turret mirror system of 76 cm aperture at the top of the tower feeds a primary mirror of 55 m focal length to produce a 51 cm solar image.

New horizontal telescopes include the Meudon 40 cm telescope with 20 m focal length and the Kiev 44 cm telescope with 17 m focal length. The latter telescope is operated by the Astronomical Observatory of the Ukrainian Academy of Sciences.

Coronagraph systems with 53 cm aperture designed for large angular and spectral resolution have been installed at Kislovodsk, Abastumani and Alma-Ata. Companion instruments are being prepared for installation at the Crimean Observatory and at Debrecen (Hungary).

We draw attention to the successful operation of the OSO VI satellite carrying an improved spectrometer capable of both spectral scanning at a fixed point on the sun and raster scanning fixed areas at a given wavelength. The raster scan mode of operation produces spectroheliograms of areas measuring approximately 7 arc min on a side with a time resolution of 30 sec or of areas measuring 46 arc min on a side with a time resolution of 8 min. The spectrometer covers the wavelength band 280–1360 Å and has a spatial resolution of approximately 35 arc sec.

We draw attention also to the development of successful X-ray telescopes that are producing high angular resolution solar X-ray images of excellent quality, and to the development of a zone-plate X-ray camera at Utrecht.

Narrow band filters utilizing two etalon Fabry-Pérot filters in series (Kononovich, Sternberg) or one or two etalon Fabry-Pérot filters in combination with multilayer dielectric filters (Goldberg, Harvard) are being successfully developed and tested. Tests indicate that the filters are stable if controlled to $\pm 0{\cdot}5\,^{\circ}$C and may give pass bands as narrow as 1/20 Å.

A particularly interesting experimental program is the stellar atmosphere simulator being developed at the Max-Planck Institut für Extraterrestriche Physik (Specht, Garching). This device will heat a H-He plasma to 3500–7000° and will permit the controlled introduction of other elements.

A Josephson-Junction detector capable of very fast time response and high sensitivity in mm and far infrared regions of the spectrum has been used successfully by Ulrich (University of Texas). Development of the detector and its application to solar astronomy are continuing.

THE CONTINUUM SPECTRUM

Renewed interest in the continuum spectrum of the sun has been stimulated by a desire to find more accurately the location and value of the temperature minimum and to determine the temperature structure in the low chromosphere and upper photosphere. Most of this effort is directed towards observations in the far infrared and submillimeter regions of the spectrum. Parallel, but less intensive, efforts are directed toward obtaining more accurate data in the far ultraviolet.

Present data indicate that the temperature minimum region is observed at wavelengths of about $\lambda 1650$ and at about $300\,\mu$, or greater. The exact location in the submillimeter region is difficult to determine from existing data and is still quite uncertain. At present, the submillimeter data of Eddy, Léna and MacQueen of the High Altitude Observatory and far ultraviolet data of Parkinson and Reeves of Harvard College Observatory each suggest that the minimum temperature is near 4400°. There are uncertainties in both sets of data, however, and more work remains to be done in determining reliable absolute intensities in these regions of the spectrum.

New infrared observations are in progress at Institute d'Astrophysique in Paris in the 0.3 to $30\,\mu$ band (Peyturaux), at Observatoire de Paris in Meudon in the 10–$1000\,\mu$ band (Gay and Léna), in Switzerland in the 15–$1000\,\mu$ band (Müller, Geneva, Knubuhl and Stettler, Zurich), at the Jungfraujoch in the $1{\cdot}2$–$12\,\mu$ band (Delbouille and Roland), in Liege (spectral band unknown,

Zander, Delbouille and Roland), and at High Altitude Observatory in Boulder, in the 300–1000μ band (Eddy, MacQueen and Mankin).

New observations of the ultraviolet continuum in the $\lambda\lambda$1400–1875 band are planned at Harvard (Parkinson and Reeves) and in the $\lambda\lambda$1800–3000 band at Verrires (Bonnet). These were the only ultraviolet continuum observing programs reported to the author. It is both somewhat surprising and regrettable that more effort is not being made in this important spectral region.

In the violet end of the photographic spectrum we are still plagued by uncertainties in the location of the continuum and by uncertainties in opacities. Houtgast and Namba (Utrecht) have attempted to locate the continuum more accurately between $\lambda\lambda$2988 and 4087. A number of workers report difficulty in fitting the violet continuum with theoretical models. Thus, additional opacity is required from $\lambda\lambda$1800–5000 by Heintze (Utrecht), for $\lambda < 5000$ by Sitnik (Sternberg) and $\lambda < 2000$ by Nishi (Japan). It is not clear at this time, however, just how much of the difficulty is due to unknown opacity sources and how much is due to either uncertainty in the continuum location or faulty model atmospheres.

The Lyman continuum observations from OSO IV and the newer data from OSO VI obtained by the Harvard group are highly valuable contributions for studying middle and upper chromospheric structure. Similar experiments in the far ultraviolet with improved spectral resolution could provide much badly needed information about the structure of the low chromosphere and upper photosphere.

There is still some dispute over the spectral distribution of energy observed at the center of the solar disk. The results of Labs and Neckel differ somewhat from those of Peyturaux, for example. Again, it seems more work is needed and we particularly need confirming results from different investigators.

THE LINE SPECTRUM

A number of factors are contributing to the current high level of interest in the line spectrum. These include: the opening of new spectral regions in the infrared and far ultraviolet, interest in the high photosphere and temperature minimum region, interest in improved chromospheric models, contradictory results on abundances (notably iron), the study of mass motions and their influences on line profiles, and the advent of improved techniques for treating non-LTE problems in line formation. The new tower telescopes with high resolution spectrographs at Meudon and Sacramento Peak Observatories will undoubtedly spur even greater interest in the line spectrum. Also, the high resolution vacuum telescope at the Jungfraujoch has been automated to obtain high resolution, low noise spectra.

The comprehensive *Second Revision of Rowland's Preliminary Table of Solar Spectrum Wavelengths* by Charlotte M. Sitterly, M. G. J. Minnaert and J. Houtgast (National Bureau of Standards Monograph, 61) represents another milestone in solar astronomy. Mrs Sitterly reports that section 4 of *The Analyses of Optical Atomic Spectra*, National Bureau of Standards Publication 306 was scheduled for publication in September 1969. Earlier sections were published in 1968 and 1969. Mrs Sitterly is currently preparing a revised table of ionization potentials together with the ionization limits for each spectrum and the energy levels of the terms measured from the ground configuration. The work has already been completed for atoms H through Cl.

Precision wavelength measurements of lines in the interval $\lambda\lambda$2900–10000 using interferometric wavelengths of Thorium lines as standards are being made by Brault, Brechinridge and Pierce (Kitt Peak Observatory). They report that a large number of previously unidentified faint solar lines have been identified with CN by using new CN laboratory spectra in the red and violet.

Identification of many faint lines in sunspot spectra with the molecules NiH, H_2O and O_2 has been reported by Wöhl (Göttingen). He further reports that lines of MgH are broadened by magnetic splitting and not by blending as supposed by some authors.

High resolution photoelectric scans of the spectrum in the range $\lambda\lambda$3000–3600 by Mohler and Mitchell (Michigan) have added many new lines, with equivalent widths, relocated wavelengths and suggested identifications. This work is currently being extended to λ2900. A similar atlas extending from $\lambda\lambda$3000–8000 is in preparation by Delbouille and coworkers at the Jungfraujoch .

Tables of wavelengths and identifications of the lines measured in the *Photometric Atlas of the Solar Spectrum from λλ7498–12016* (Delbouille and Roland) are in press.

Roddier (Nice) is developing an atomic beam spectrograph to obtain precision profiles of the Sr I 4607 and Ca I 4227 lines. It is planned to make the instrument light enough for balloon work as well as for use on refracting telescopes.

Observations of line profiles and precise wavelength measurements in the far ultraviolet and infrared spectral regions remain as two of the major unexplored observational problems in solar physics. New work is beginning in these directions, however. We note especially the observations of the profiles of lines of O I and C II by workers at the Laboratory of Atmospheric and Space Physics (Boulder) and the observations of Mg II and Lyman α profiles by the Verrieres group. Both groups plan to extend their observations to other lines in the near future.

ROTATION

The solar oblateness reported by Dicke (Princeton) together with the lingering problems of differential rotation are sustaining a relatively strong interest in solar rotation problems. Theoretical work on solar rotation is being carried out in Boulder independently by Nakagawa, Durney, Benton and Gilman, in Tübingen by Unz and Walter and in London by Roxburgh.

The observation that the angular velocity of the lower solar atmosphere varies with height (Livingston, Kitt Peak) raises the question as to whether a similar phenomenon exists in the solar corona. An examination of the rotation of coronal condensations during the period 1964 to 1967 (Hansen, Hansen and Loomis, HAO Hawaii) demonstrates that the rotation rate varies from year to year and in any given year may be different between the two hemispheres. At lower latitudes, the rate appears to agree with that determined from filaments and sunspots while at high latitudes the rate appears to be similar to that of filaments, which is over 2° per day faster than that displayed by polar faculae. These data show a dependence of angular velocity with height which appears to depend upon both latitude and the phase of the solar cycle. Both increasing and decreasing angular velocity with height appear. The details of this phenomenon are poorly understood. There is some indication that the high latitude coronal features are being influenced by the more rapidly moving magnetic centers at lower latitude.

A question of continued interest is that of the co-rotation of the corona and loss of angular momentum of the solar wind. Theoretical investigations have demonstrated that the magnetic field of the sun imposes effective co-rotation out to approximately 15 $R_\odot$. These models predict an azimuthal velocity of the solar wind at 1 AU of approximately 1 km s^{-1}, which is considerably lower than the approximately 10 km per second observed (Brandt, Goddard; Hundhausen, Los Alamos). These models predict, moreover, that even within 15 $R_\odot$ no strict, rigid rotation of the corona can be expected and that there should be a very slight spiraling of coronal structures toward the east. Empirical corroboration of these theoretical results appears impossible at the present time. Although rocket observations of the outer corona (Koomen, Seal, Purcell and Tousey, U.S. Naval Research Laboratory) show that streamers out to 10 $R_\odot$ appear straight, any realistic solar wind model would produce such a result even in the absence of a torque produced by the solar magnetic field.

A calculation by Weber (Kitt Peak) suggests that the angular velocity of *ions* in the corona may increase with height in special circumstances in closed magnetic loops. No empirical evidence for such a surprising phenomenon has been obtained.

Radar observations of the corona (James, El Campo) have produced detailed measures on macroscopic motions in the corona. This appears to be the only direct measurement of the solar wind, which has an average velocity of approximately 20 km s^{-1} at a distance of 1·5 $R_\odot$. Echoes also appear with velocities of up to 200 km s^{-2}. The lack of angular resolution in the radar observations makes their interpretation difficult so that it is presently largely a matter of speculation as to which echoes correspond to large blobs of moving coronal material and which may represent shock waves.

THE PHOTOSPHERE

The studies of phenomena related to super granules, of magnetic fields, and of mass motions in the photosphere currently play strong roles in photospheric physics. Further interest has been added through efforts to determine more accurately the location and value of the temperature minimum, and through recent controversies over the solar abundance of iron. The Bilderberg conference on the photosphere has added extra stimulus and has provided a trial model through which we can bring our different studies to a common standard of reference. As a result of these stimuli interest in the solar photosphere stands, perhaps, at a higher level than ever before.

A. *Granules, super-granules and their systematic motions*

Rösch and his coworkers at Pic du Midi are continuing their study of the evolution and structure of granules using cinematography. Some of the Pic du Midi observations together with cinematographic observations of granules made at Sacramento Peak are being studied by P. Mein and his coworkers at Meudon. Additional observations are planned with the new Meudon tower. Léna successfully observed granule structure in the infrared region of the spectrum at Kitt Peak and is planning to continue granule studies in the infrared. M. Rossbach (Göttingen) is studying sizes and lifetimes of granules near sunspots on high resolution photographs obtained by Beckers and Schroter at Sacramento Peak. Namba and coworkers (Utrecht) are studying granulation structure from stratoscope I photographs obtained at Princeton. They find large granules which darken in the center and divide into smaller cells giving the appearance of several granules each of shorter lifetimes than the lifetime of the large granule and its fragments. They are studying also velocity shifts and line profiles in granule spectra of the C I $\lambda 5052$ line obtained at Sacramento Peak Observatory.

Edmonds (University of Texas) reports that a statistical study of granule data by the Tukey fast-Fourier-transform method leads to power and coherence spectra that differ significantly from results obtained from the same data with the mean-lag-product method. The Tukey method is capable of higher resolution and is less sensitive to finite data sampling and smoothing.

Edmonds further reports that continuum brightness fluctuations (granules) have wavelengths concentrated near 1200 km whereas low frequency radial velocity oscillations have wavelengths concentrated near 4000 km. He concludes, therefore, that the two phenomena are distinct. The new results come from bi-dimensional power spectra.

H. Reiling (Göttingen) is studying the relationship between line asymmetry and intensity fluctuations using spectra obtained by Livingston at Kitt Peak. The objective of this study is to obtain a better idea of the true velocity of unresolved elements in granules and intergranular spaces.

V. A. Krat and Ch. Abusametov (Pulkovo) report that the average velocities of features in the photosphere decrease as the size of the element decreases.

From one dimensional scans obtained with the velocity servo of a magnetograph, Tanenbaum, Wilcox, Frazier and Howard (Mt. Wilson) have analyzed velocity fields for the five-minute oscillations and for the flow in super granules. The oscillating elements of average size less than 6000 km appear to be randomly distributed but can statistically combine into a large oscillating area. In the photosphere they find that the brightness fluctuation leads the velocity by less than 90° and conclude that traveling waves are present at this level. In the low chromosphere the phase lead is 90° and they conclude that standing waves are present.

The super granule flow has a diameter of about 10000 km, an amplitude of 100 m/s and a lifetime of over 3 h.

R. Howard and A. Bhatnagar report that they have obtained an excellent granulation spectrum which they are analyzing to obtain physical characteristics of granular and intergranular spaces.

A. Title (Harvard) is analyzing right and left circular polarized pairs of two-dimensional arrays of spectra. Regions one arc min square are scanned and spectra are obtained in each polarization at 0·5 arc sec intervals. Each spectrum is a 2Å band centered at $\lambda 5250$. The arrays provide information on the magnetic field, velocity field, intensity field or any other property of the spectra desired.

Observers measuring velocity fields should be cautioned by the theoretical profiles computed by Beckers and more recently by Athay that when velocity gradients are present in the region of line formation standard techniques for measuring velocities may give completely erroneous results.

B. *Magnetic fields*

Large scale mapping of the solar magnetic field continues to be carried out at Mt. Wilson, Kitt Peak and Sacramento Peak Observatories and at the Crimean Observatory. The new solar observatory at Aerospace Corporation (Los Angeles) is using Leighton's photographic subtraction technique with the 24″ vacuum telescope and spectrograph to study the fine structure of the magnetic field.

Title (Harvard) reports that one of the more interesting features of his pairs of polarized two-dimensional spectra (see granules and super granules) are the very large field gradients near sunspots, up to 2500 G/arc sec at the umbral boundary and up to 1500 G/arc sec at the outer penumbral boundary.

Frazier (Aerospace Corporation) has found a high correlation between longitudinal magnetic field strength and the presence of downward velocities.

Perhaps the most remarkable new discovery in solar magnetism is that the sun observed as a star by Severny (Crimea) has a net magnetic field that varies roughly periodically with half the solar rotation period and that the sign of the residual solar field correlates with the sign of the interplanetary magnetic field (Severny, Wilcox and Colburn).

Staude (Heinrich-Hertz Institute) and Rachkovsky (Crimea) are studying the effect of magnetic field inhomogeneities on line profiles and the reverse problem of inferring magnetic fields from line profiles when unresolved inhomogeneities are present. Again, observers should make note that when gradients are present either in the magnetic field or in velocity standard techniques may give incorrect magnetic field strengths.

Krat (Pulkovo) reports that "magnetic knots" are small pores of approximately 0·3 to 0·4 arc sec in size and are located at the boundaries between radial velocities of opposite sign.

C. *Chemical composition and microturbulence*

Photoelectric scanning of the spectrum with double pass spectrometers can yield precision profiles of Fraunhofer lines. As line profiles become more reliable there is a turning from the curve of growth method of abundance analysis to profile representation or spectrum synthesis. One should now insist that profiles as well as equivalent widths should be explained. For those elements for which only a few lines are observable, often badly blended, profile fitting (spectrum synthesis) is the only method acceptable. For others it is a major improvement in analytical technique, utilizing more data and capable of yielding correspondingly more information.

Still further information may be extracted from the lines by using center-limb data. One often finds that an adequate explanation of line profiles and equivalent widths at disk center is no longer adequate for the same lines observed near the limb. Similarly, an adequate explanation for lines of a given range of excitation potentials may fail when the range of excitation potentials is expanded. Some authors are finding that when many lines are studied or when lines are studied at different disk positions the chemical abundance deduced for an element depends upon excitation potential or upon position on the solar disk. Such results may be indicative of poor *f*-values, of non-LTE effects or of poor models for the photosphere. We can never have great confidence in abundances, however, until all such difficulties are either eliminated or properly explained.

Aside from problems with *f*-values (line spectrum), the outstanding difficulties in properly explaining the lines and obtaining proper abundances are in the effects of non-LTE and mass motions. It is clear that non-LTE effects are present in many line profiles and that the microturbulence is not homogeneous. We are still in need of analytical techniques that allow adequately for these effects. These are difficult problems, to be sure, but it is time to face them.

Studies of microturbulence in photospheric molecular lines are reported by Sitnik (Sternberg).

Studies using precision profiles are reported as being carried out by the following workers or by groups associated with them: Aller (UCLA), Altrock (Sacramento Peak), Gurtovenko *et al.* (Kiev), de Jager (Utrecht), Jefferies (Hawaii), Müller (Geneva), Pecker (Paris, Nice), White (High Altitude Observatory) and J.-P. Swings (Liege). Much of the observational work for such studies has been obtained at Kitt Peak, Sacramento Peak and Jungfraujoch Observatories in collaboration with personnel at these observatories.

New Fe I *f*-values recently obtained by such workers as Garz and Kock (Kiel) and by King and collaborators (California Institute of Technology) have cast serious doubt on the reliability of the Fe I *f*-values published in recent years by the U.S. National Bureau of Standards. This is both an encouraging and disturbing development; encouraging because of the new interest in the problem, the new techniques developed, and the promise of better data in the future; disturbing because of the suspicion it casts on other *f*-values as well. Goldberg calls attention to the fact that although the new Fe I *f*-values give a photospheric iron abundance that agrees with that obtained from [Fe II] lines the new abundance disagrees by an order of magnitude with the abundance obtained from Fe II lines by Warner. Thus, we still have a discrepancy and we still must hold *f*-values as suspect.

D. *Models*

The Bilderberg continuum model has served usefully as a reference model, whether used as a working model or as a target for criticism. It seems, at this time, to be most deficient at $\tau_c > 1$ where it gives temperatures that are too low and at $\tau_c < 10^{-2}$ where the model was uncertain from the beginning. A revised model is in preparation by Gingerich (Harvard).

The author believes that the usefulness of the Bilderberg model as a reference model owes its success in large measure to the fact that many people contributed to the discussion leading to the model and that they, and their colleagues, therefore felt some allegiance to it. Had the same model been produced by one person there seems little doubt that it would have received much less attention than has been accorded the Bilderberg model. It seems in the best interest of solar physics, therefore, that the Bilderberg model be revised by a process similar to that which gave birth to it.

Recent ultraviolet data obtained by Parkinson and Reeves (Harvard) and infrared data obtained by Eddy, Léna and MacQueen (High Altitude Observatory) indicate that the minimum temperature in the upper photosphere is near 4400°. Both the infrared opacity and the ultraviolet opacity (Cuny, Paris) indicate that $T_{\min}$ occurs near $\tau_5 = 10^{-4}$. The proper evaluation and location of $T_{\min}$ is of crucial importance for comparison with theoretical models and for evaluating the mechanical energy flux required to sustain the observed temperature distribution.

Feutrier (Meudon), Jordon (Goddard) and Thomas and Gebbie (Joint Institute for Laboratory Astrophysics, Boulder) have studied the rise in temperature at small optical depths resulting from the Cayrel mechanism. All of these studies suggest that the rise in temperature in the low chromosphere or upper photosphere is largely due to the Cayrel mechanism rather than mechanical energy dissipation. However, they find temperatures in excess of 4600°, particularly when $\tau_5 \leqslant 10^{-3}$.

The only known mechanism for cooling the external layers of the photosphere below 4600° is line blanketing. Athay (High Altitude Observatory) has computed a model photosphere using non-LTE blanketing but omitting the Cayrel mechanism. He finds a radiative equilibrium temperature of $4340° \pm 150$ at $\tau_5 = 10^{-4}$. The surface cooling effect increases rapidly if the surface temperature is raised and would strongly resist a warming tendency by the Cayrel mechanism.

Studies of strong Fraunhofer lines with the intent of determining the solar model more accurately in the region of $T_{\min}$ and the low chromosphere are being carried out at a number of places. Some of the groups involved in this work are listed in the preceding section on the discussion of profile fitting. Others include Avrett (Harvard), Athay and Skumanich (HAO, Boulder), Johnson (Indiana),

Kostik and Orlova (Kiev), Linsky (JILA, Boulder), Mugglestone (Queensland), Sitnik (Sternberg), Suemoto (Japan), Cuny and Dumont (Paris), and Tanaka (Japan).

New techniques for handling non-LTE line transfer problems in inhomogeneous atmospheres and in moving atmospheres are being studied by Hummer (JILA, Boulder), Jones (Goddard), Kalkofen and Rybicki (Harvard), Skumanich and House (HAO, Boulder) and Wilson (Sydney).

THE CHROMOSPHERE

It is no longer possible to distinguish sharply between some chromospheric and photospheric studies. A single strong Fraunhofer line may be formed in regions extending from the low photosphere to the middle chromosphere. The continuum spectrum also has been extended into spectral regions where the emission is chromospheric.

The temperature minimum region is a phenomenon of both the photosphere and the chromosphere, and the current lively interest in this region is unifying much of the photospheric and chromospheric research. Similarly the current emphasis on the interpretation of the profiles of strong Fraunhofer lines is unifying research in these two areas. Accordingly much of the discussion in the section of this report dealing with the photosphere pertains to the chromosphere as well.

The generally accepted belief that the mechanical energy heating the chromosphere and corona passes through the photosphere further unites photospheric and chromospheric research. There is widespread interest in the nature of the mechanical energy. How much is required? How does it propagate? Where and how is it dissipated? These questions are both "photospheric" and "chromospheric" and should not be divided by an artificial line such as the solar limb. It appears that much of the recent effort in chromospheric research has been "dissipated" in the region of temperature minimum.

A. *Eclipse spectra*

The tabulation of line intensities observed at the 1962 eclipse by Dunn *et al.* (*Ap. J. Suppl.*, 1968) represents a valuable and long awaited addition to the literature. Other reports of spectra from the 1962 and 1966 eclipses already published or to be published soon come from Hurokawa *et al.* (Japan, 1962 eclipse), Hiei and Hirayama (Japan, 1966 eclipse) and the Utrecht group (1966 eclipse). The 1966 data obtained by Hiei and Hirayama have sufficient spectral resolution to provide very good line profiles and represent an especially valuable contribution.

Plans to observe the chromosphere at the March 7, 1970 eclipse include groups from the Kwasan and Utrecht Observatories.

B. *Models*

Analysis of continuum data for the low chromosphere from the 1966 eclipse is being continued by Gebbie and Thomas (JILA). Linsky (JILA) and Gingerich, Noyes and Avrett (Harvard) are combining studies of the ultraviolet, infrared, radio and visual continuum in an effort to obtain a better model of the low and middle chromosphere.

Analyses of Lyman continuum data obtained with OSO IV and OSO VI are continuing by Noyes and Kalkofen (Harvard).

Although considerable effort has been devoted to chromospheric models in the past three years there is little to report in the way of real progress. The chromospheric part of the Bilderberg model was largely a step backward since it ignored well established eclipse data giving electron densities both in the chromosphere and corona. Much of the model evolution since the Bilderberg model has tended to re-emphasize the electron density models and to restructure temperature models to fit the various other data as well as possible. It has been known for many years that eclipse data give an electron density at 500 km above the limb of approximately 2×10^{11} cm^{-3} and a neutral hydrogen density of approximately 10^{14} cm^{-3} (cf. Henze, *Solar Phys.*, 1969, **9**, 65). The only known source for such a high electron density is hydrogen ionization and there seems no way to avoid the conclusion that the temperature is high enough to produce a fractional ionization of hydrogen of about 0·1 %. This requires a temperature near 5700° if proper allowance is made for non-LTE effects. In any case, it seems clear that the temperature at 500 km above the limb is some 1000° to 1500° above the temperature at the solar minimum.

The eclipse values of hydrogen and electron densities at 500 km lead to an optical depth at this height of $\tau_5 \approx 10^{-5}$. Recent trends to set $T_{\min} \approx 4300$ and to place it near $\tau_5 \approx 10^{-4}$ require that the temperature gradient between $\tau_5 \approx 10^{-4}$ and $\tau_5 \approx 10^{-5}$ be large. This feature was absent from the Bilderberg model.

A second obvious defect of the Bilderberg model is that at very small τ_5 it gives densities that are below coronal densities. The only way proposed thus far (Athay, HAO) for avoiding this difficulty is to markedly reduce the thickness of the low temperature regions of the chromosphere and to require a second steep rise in temperature in the general range of 1000 to 2000 km above the limb. Newer models are tending in this direction (Noyes and Kalkofen, Harvard; Avrett and Linsky, Harvard and JILA).

There is no doubt that the chromosphere has lateral structure of significance and that single component models must finally give way to multicomponent models. As a point of departure for such multicomponent models, however, it is important that we establish the best possible single component model.

Some of the studies of line data in progress have been mentioned in the photospheric discussion. Efforts to properly explain the K_2 emission phenomena are still in progress. The trend here seems to be in the direction of favoring the interpretation based on an effectively thick chromosphere, which is not inconsistent with the trend in models of the low chromosphere. Some authors (e.g. Tanaka, Japan) still favor the effectively thin interpretation, however.

Tandberg-Hanssen and Smythe (HAO) are studying the excitation of resonance doublets in the chromospheric spectrum. They report that observed doublet ratios in Ba II of 2·5 and greater can be accounted for if interterm collision rates are sufficiently small.

Efforts to obtain better chromospheric models during the past few years have concentrated on the low chromosphere and the chromosphere-corona transition region. New attempts are being made to arrive at a better understanding of the middle chromosphere, however. Avrett (Harvard) and Poland (HAO) are studying helium excitation using both XUV and eclipse data. Studies of the hydrogen Lyman lines (Athay, HAO) and of the C II lines (Berger, LASP, Boulder) are in progress. Athay and Canfield (HAO) find from their study of the profile of the $\lambda 1304$ O I line in the XUV that a microturbulent velocity of 7–8 km/s is needed in the middle chromosphere. Existing chromospheric models give the correct line flux, but they give too much self-reversal in the center of the line.

Pecker (Paris) calls attention to the need in chromospheric research for coordinated studies incorporating radio, IR, XUV and visual data, both continuum and lines. Too often, in the past, models have been based upon overly restricted data, only to be replaced by other models based upon equally restricted data.

C. *Structure and spicules*

Although we have accumulated considerable data on the nature of the structure observed in different spectral lines, we have made little progress in relating the structure in one line to that in another, and we have made even less progress towards relating the structure observed in the lines to physical structure in the chromosphere. These are difficult problems, of course, and we cannot expect that a given structure will produce the same changes in brightness or have even the same size in two different lines even though the different lines may be formed at the same height. The problem is further compounded by the ambiguities introduced by Doppler shifts and, in some cases, magnetic splitting.

Bray and Loughhead (Sydney) are continuing their high resolution studies. Loughhead has suggested that spicules are seen as dark mottles on the disk. He identifies the bright mottles with bright features seen above the limb, but reports that these features are distinct from spicules.

Spicule studies at the limb are being continued at Pic du Midi and at eclipse by Mouradian (Paris) and at Kislovodsk by Nikolsky and Sazanov. The latter find that the line-of-sight velocity can change sign for a given spicule and may reach 20 km/s. Similar results were reported by Pasachoff, Noyes and Beckers using data from Sacramento Peak.

Pasachoff (Harvard) notes in his study of the emission structure observed in the K line that:
(1) The emission is often confined to one side of the line, usually the violet side.
(2) The maximum intensity of the violet peak exceeds that of the red.
(3) The $\Delta\lambda$ positions of K_2 show an rms deviation of 0·04 Å.
(4) K_3 tends to be displaced from line center away from the most intense K_2 emission.
(5) The structure of K_2 emission is not necessarily related to continuum threads. Suemoto (Japan) is conducting a similar study of the K line disk structures.

Studies of oscillatory motions in the chromosphere are reported by Orrall (Hawaii) and by Zhugzhda (ITMIP, Moscow).

D. *Chromosphere-corona transition region and chromospheric heating*

Far ultraviolet emission line data obtained from satellites and rockets have made it possible to study the chromosphere-corona transition region in some detail. A conference on the transition region was held in Boulder in August, 1969, with about 40 participants. A limited number of the proceedings of this conference are still available. They can be obtained on request by writing the High Altitude Observatory, Boulder.

New data from the Harvard experiment on OSO VI and from rocket experiments at Culham are being used by those groups to obtain improved models of the transition region and new information about transition region phenomena.

The problems of heating the chromosphere and of the energy mechanisms in the chromosphere-corona transition region are being studied by a number of authors including Delache (Nice), Dubov (Crimea), Kopp (HAO), Kuperus (Utrecht), Mäckle (Tübingen), Moore (Stanford), Stein (Brandeis) and Ulmschneider (Heidelberg). Although there are still major questions to be resolved, it appears that heating by shock waves is adequate. The magnetic field structure and the period of the waves are known to play important roles in the propagation and dissipation of the waves. Exactly how to take these phenomena into account in constructing models is not well understood, however. The energy dissipated by shocks must be disposed of mainly by radiation in the chromosphere and by radiation, conduction and mass motion in the transition region. The radiation losses and mass motions are difficult to take into account properly and are still limiting the usefulness of such models. Much improvement is being made, however, and we can confidently expect significant improvement in the theoretical models in the near future.

E. *Abundances*

Pecker and Pottasch have developed a method for obtaining relative abundances in the chromosphere that avoids, in some cases, major uncertainties from non-LTE effects. The method combines the use of central intensities of disk lines with the total intensities of lines observed at eclipse and is applied wherever possible to transitions from the ground state. Abundances derived using this technique show better agreement with coronal and meteoritic abundances than with classical photospheric abundances. This is particularly true of the iron abundance, which agrees with the newer photospheric values from Fe I and Fe II and with the coronal abundances from forbidden lines.

THE CORONA

A. *The coronal density structure*

Continued observation of the corona has confirmed the validity of the *form* of the classical coronal density models. However, all observations point to the fact that the previous sunspot maximum (1957–1958) and minimum (1962) were characterized by electron densities about twice those of the Van de Hulst models for corresponding periods of solar activity. A detailed documentation of the variation of coronal electron densities from 1964 through 1967 has recently been published by Hansen, Hansen, Garcia and Loomis (HAO, Hawaii).

The density structure of the corona has often been treated as a spherically symmetric component with the superposition of one or more streamers. Researchers now have at their disposal several new models to describe "typical" coronal streamers as well as ellipsoidal distributions attempting to reproduce the dependence of density upon longitude in an average corona.

One aspect of the density structure of the corona which requires considerable attention in the future is that of the fine scale structure. It is now clear that such structure extends from the upper chromosphere out into the chromosphere-corona transition region (Withbroe, Harvard) and into the corona (James, El Campo). The scale of these fluctuations in the corona varies from the 3×10^4 km which some authors (Newkirk and Harvey, HAO, Boulder; Ivanchuk, Kiev) identify with the supergranulation down to the few hundred kilometers which are responsible for interplanetary scintillation. The origin of such fluctuations is largely unknown.

Looking to the future, we can hope for the resolution of several problems concerning the density structure of the corona. Synoptic observations of the outer corona are going to be an essential observation in order to be able to determine the true three-dimensional distribution of material. The relation of such structures as streamers and rays to solar surface features is also presently largely unknown. Even less is understood about the evolution of such structures although there is some indication that streamers endure for several solar rotations (Bohlin, Mt. Wilson). Even in the area of the gross, global structure of the corona much remains to be done. Many of the present models employ convenient but unproved assumptions about the symmetry of the corona along the line-of-sight. Clearly, a systematic investigation of the uniqueness of the density models used to fit the observations is required.

Observations of the white light corona from ground stations are being continued by Hansen (HAO, Hawaii) and by Dollfus (Meudon). Eclipse observations (1970) are planned by Saito and Hata (Japan). Balloon observations are in progress at Meudon (Dollfus).

The Apollo 11 astronauts obtained several photographs of the *F*-corona, which are yielding useful results on brightness near the sun (Kovar, Manned Spacecraft Center, Houston). A white light coronagraph is being prepared for flight on a future Apollo mission by Newkirk (HAO, Boulder).

B. *Heating and temperature of the solar corona*

The problem of explaining the high temperature of the solar corona has received considerable attention in the past several years. Most of the theories have been concerned with the details of the generation of mechanical energy in the lower solar atmosphere and the deposition of this energy by shock waves in the chromosphere and corona. Recently the role of ion acoustic waves, which dissipate their energy by Landau damping in the lower corona, has been pointed out by d'Angelo (European Space Research Institute, Rome). Perhaps, the most important aspect of these theoretical investigations has been their demonstration of the fact that the magnetic field in the lower atmosphere can be expected to play a very large role in determining the rate of energy deposition in the corona as well as the structure of the chromosphere-corona transition region (Kopp and Kuperus, HAO and Utrecht and Mäckle, Tübingen). As mentioned in the section on coronal condensations, there now appears to be substantial evidence (Noyes, Withbroe and Kirshner, Harvard) that the energy deposition into the corona is, in fact, increased substantially above active regions.

Little progress has been made in gathering empirical information which adds measurably to our knowledge of the temperature structure of the corona with a view to distinguishing between competing theories of coronal heating. The classical measurements of coronal line widths generally have insufficient spectroscopic resolution for a definitive measure of the profile and are often confused by macroscopic motions in the corona (Delone and Makarova, Sternberg). The identification of new coronal lines (Edlén, Lund; Wagner, Sacramento Peak) of Ni XI and Fe IX suggests that even the quiet corona contains a temperature fine structure.

About the only empirical evidence bearing on the problem of coronal heating is that obtained by solar radar echoes (James, El Campo) at a radar frequency of 38 MHz, which is reflected at approximately 1·5 $R_{\odot}$, where some echoes are found moving with a velocity of approximately 150 km s^{-1}. If these echoes are identified as the reflection off shock waves moving through the corona, the slowing down of these waves at approximately 1·6 $R_{\odot}$ can be interpreted (Brandt, Goddard) as delineating the extent of coronal heating out to 1·6 $R_{\odot}$. Since the radar observations do not have any appreciable angular resolution and the exact location and amplitude of the disturbances are unknown, it is impossible at this time to judge whether such waves are truly sufficient to account for the required energy flux in the corona. Another piece of information favoring the heating of the corona over a substantial depth comes from the measurement of the relation between the velocity and temperature of protons in the interplanetary medium (Burlaga and Ogilvie, Goddard). The present solar wind models can adequately account for the quiet period temperature and velocities but appear to require the hypothesis of an extended region of coronal heating near the sun to account for the higher range of temperature and velocity which is observed during disturbed periods.

Spacecraft observations have also provided (Hundhausen, Gilbert and Bame, Los Alamos; Kozlovsky, Israel) the first direct evidence for the high temperature of the corona. The presence of ions of O IV to O VII in the solar wind at 1 AU demonstrates that even though the local temperature of the wind is quite low (the most probable temperature is $4{\cdot}8 \times 10^4$ K while the average temperature is $1{\cdot}4 \times 10^5$ K), the state of ionization is effectively frozen into the solar wind rather low in the corona. The O VII to O VI ratio implies that the ionization is fixed at a level where the temperature is approximately $2{\cdot}8 \times 10^6$ K. During periods of activity in the solar wind the ion ratios imply a temperature of approximately 3×10^6 K suggesting that small, hot coronal condensations may, in fact, have a large contribution to the solar wind. The observation of He I in the solar wind has been suggested as evidence for the presence of a cold, perhaps subsonic component, to the wind; however, it appears likely that charge exchange in the neighborhood of the earth can equally well account for the observed abundance of singly ionized He (Hundhausen, Los Alamos).

X-ray and XUV spectrograms and spectroheliograms of the corona also, in principal, provide the basis for improved information on both the density and temperature structure of the corona. A number of groups are now involved in X-ray and XUV programs. These include: Michard (Meudon), Dijkstra (Utrecht), Blake (Chicago), Vaiana (American Science and Engineering), Goldberg *et al.* (Harvard), Peterson (University of California, San Diego), Neupert (Goddard), Friedman (U.S. Naval Research Laboratory), Teske (Michigan), Pounds (Leicester), Burton *et al.* (Culham) and groups in the U.S.S.R.

A number of researchers report new theoretical work on collision cross-sections, *f*-values and ionization equilibria for coronal ions. Van Rensbergen (Holland) is studying the influence of excited levels and of autoionization and dielectronic recombination on the equilibrium of coronal ions. Studies of collision cross-sections and *f*-values are reported by Blaha (Goddard), Lexa (Czechoslovakia) and Belý (Nice).

Coronal magnetic fields

In spite of the fact that the determination of magnetic fields in the corona is extremely difficult, the question of such fields and their influence on the solar corona has been one of tremendous activity in the past several years. Although, in principal, capable of yielding information on these fields the measurement of coronal emission line polarization (Eddy and Malville, HAO, Boulder; Hyder, Mauter and Shutt, Sacramento Peak; Nikolsky and Sazanov, Abastumani) have proved difficult to interpret in terms of the magnetic field in any detailed manner. Other researchers reporting studies of coronal line polarization are Charvin (Pic du Midi), Delone and Makarova (Sternberg) and Nikolsky and Tetruashvili (Abastumani).

Faraday rotation of artificial and natural radio sources (Levy *et al.*, Jet Propulsion Laboratory,

Pasadena) is also a potentially useful tool although no quantitative measurements of the coronal field have yet been determined. The Razin effect in solar radio emission (Boischot and Clavelier, Paris; Ramaty and Lingenfelter, Goddard; Bohlin and Simon, California Institute of Technology) appears to have given us the most unambiguous determination of the field at a value of 0·3 to 0·5 G at $2R_{\odot}$ above an active region. All these researches refer to the observation of a single Type IV burst and similar observations of other events is urgently required.

The connection of coronal magnetic fields to the density structure of the corona has been attacked from two directions. In the first, the surface distribution of the fields over the entire sun is used to calculate the fields in the corona; the shapes of the field lines are then compared to the visual appearance of the corona (Newkirk, Altschuler and Harvey, HAO, Boulder; Schatten, University of California). Although there is always some question about the validity of such current-free calculations in the corona, the correspondence between the shapes of the field lines and the small scale density structures in the corona such as arches and rays suggests that much of the small scale structure is due to the filling of magnetic tubes by various amounts of coronal material. These same analyses suggest that coronal streamers, indeed, form over extended regions of opposite polarity on the surface of the sun. Such a picture makes the apparent association of streamers with filaments a quite natural consequence. A similar conclusion was reached by Sturrock and Smith (Stanford) but without the benefit of a quantitative model of the coronal magnetic fields.

A second approach to the role played by magnetic fields in forming coronal streamers has involved the detailed solution of the hydromagnetic equations for the solar wind with a simple distribution of magnetic fields on the surface of the sun (Pneuman and Kopp, HAO, Boulder). These models predict rather well the shape of the streamers, and the density distribution within and outside of a streamer. The calculations may be extended to 1 AU where it is found that the streamer represents a small increase in the local density and velocity over the background corona.

The question of the extension of the corona and its magnetic fields out into the interplanetary medium has also received considerable attention. Although it is found that a correlation exists between the fields calculated for the corona at 1·6 $R_{\odot}$ and those at 1 AU (Schatten, Wilcox and Ness, University of California and Goddard) there still exists considerable disagreement as to whether the "mapping" or the "nozzle" hypothesis is a better interpretation of the data. At the present time, the connection between such visible structures in the corona as streamers and the features seen in the particles and fields at 1 AU is largely unknown.

R. G. ATHAY
Vice-President of the Commission

WORKING GROUP ON CENTRAL LINE INTENSITIES

Of a select list of about 40 Fraunhofer lines distributed between $\lambda\lambda$3083–7122, central intensities corrected for scattered light, ghosts and instrumental profile, has been obtained for 18 lines. Uncorrected values of central intensity for a number of other lines are available. A complete list of the results of all observers will be given at the General Assembly in Brighton.

A. KEITH PIERCE
Chairman of the Working Group

WORKING GROUP ON SOLAR ECLIPSES

The working Group on Solar Eclipses regrets Dr J. Houtgast's resignation as Chairman of the Group. However, Dr Houtgast will maintain his membership and continue to give his experience to the group.

The total eclipse of September 22, 1968 was actively observed by many groups in spite of its short duration. Reports on the observations are expected to be given during the 1970 IAU General Assembly.

A great deal of work has been done everywhere in view of the total eclipse of March 7, 1970.

The National Science Foundation has issued a Solar Eclipse 1970 Bulletin which has been widely distributed in order to help the observers in making their plans and to coordinate the observations. Dr Albert E. Belon, U.S. Coordinator for the 1970 Solar Eclipse, has devoted to this heavy task all his attention and care.

The U.S. Naval Observatory published a circular with details on the circumstances and relevant data on the 1970 total eclipse, which has regularly distributed.

The Working Group on Solar Eclipse has been in contact with the U.S. Coordinator and with Dr Edward R. Dyer, Jr., Secretary of the IUCSTP for a mutual cooperation in exchanging information about the organization of the researches and observations to be performed during the 1970 solar eclipse.

A comprehensive report about the activity of the group and about the work performed during the solar eclipses of 1968 and 1970 will be prepared after the coming IAU General Assembly.

M. RIGUTTI
Chairman of the Working Group

14. DONNÉES SPECTROSCOPIQUES FONDAMENTALES (FUNDAMENTAL SPECTROSCOPIC DATA)

PRESIDENT: M. Migeotte.
VICE-PRESIDENT: A. H. Cook.
ORGANIZING COMMITTEE: R. H. Garstang, G. Herzberg, W. Lochte-Holtgreven, Mrs Ch. Moore-Sitterly, J. G. Phillips, R. Tousey.

REPORTS

Mrs Ch. Moore-Sitterly has represented the IAU at the Fourth Annual Meeting of CODATA (June 26–27, 1969, Rome, Italy). She also agreed to represent the IAU at the Second International CODATA Conference to be held at St. Andrews (Scotland) from 7 to 11 September 1970 and at the Fifth Annual Meeting of CODATA which follows on September 11–12, 1970.

COMMITTEE 1: STANDARDS OF WAVELENGTH

The primary standard

Experiments on Lamb-dip stabilization of the 6328 Å He-Ne laser line made at NBS, NPL and PTB and reported in a joint paper (**10**) show that the wavelength from different lasers may differ by as much as one part in 10^7. Similar work on a CO_2 laser line at 10·6 μm and a xenon line at 3·5 μm is reported from NPL, the 1·15 μm line in a pure Ne 20 laser has been investigated at JILA (**5**), and several lines in a xenon ion laser are being studied at PTB. However, the method that holds the greatest promise of accuracy and stability is the use of saturated absorption to lock a laser wavelength to some molecular transition having a sharply defined and stable wavelength. Thus the 6328 Å He-Ne laser line can be fixed on a line in the molecule $^{127}I_2$ (**6**, **7**), and the 3·3913 μm line of a He-Ne laser has been connected to a line of CH_4, promising a unique precision and stability according to experiments made at JILA (**5**) and NRC. This technique will evidently yield excellent substitutes for the Kr 86 line, and may, when the time is ripe, lead to the adoption of an improved primary standard.

The wavelength of the primary standard line as emitted by an electrodeless microwave-excited Kr 86 lamp at room temperature (cf. report from the 1967 meeting) has been measured at BIPM (**11**), NPL (**14**) and NRC. The results are somewhat discordant, probably owing partly to differences in the power supplied to the discharge, giving values for the red-shift from 0·0001 to 0·0003 Å. For a provisional value, accurate enough for most applications, we may adopt $\Delta\lambda = +0{\cdot}0002 \pm 0{\cdot}0001$ Å. The Doppler width is as expected about twice that in the standard lamp (**14**). When part of the electrodeless lamp is cooled to 63 K, as in the work by Kaufman and Humphreys (**8**), it was found at NRC that the line broadening is very slight and the red-shift is 0·0001 Å. This shift is equal in magnitude and of opposite sign to the Doppler shift in the standard lamp, showing that the effects of pressure and current density are the same in the two lamps.

Secondary standards

Krypton 86: From the averages of all available precision measurements of 233 lines in ^{86}Kr I, Kaufman and Humphreys (**8**) have made a least-squares derivation of 45 even and 66 odd levels. From these level values they have then calculated and tabulated the wavelengths of 530 lines from

3345 to 40685 Å, which may be recommended for adoption as secondary standards. Subsequently, measurements made at BIPM by R. Czerwonka on nine infrared ^{86}Kr I lines have been reported and found to be in perfect agreement with the calculations (**8**). In addition, Humphreys and Paul have measured about 40 wavelengths in ^{86}Kr II.

Xenon 136: According to Humphreys, the advantages derived from the use of xenon 136 are that the large atomic mass permits lines of extreme sharpness to be produced at moderately low temperatures, and that the spectrum includes several lines of high intensity between 2·0 and 3·5 μm, thus providing an essentially unique example among sources of standards. Humphreys is now preparing a comprehensive report on his measurements in ^{136}Xe I. He has also developed a method for producing the ^{136}Xe II spectrum with very sharp lines, and has determined the wavelengths of 108 lines in the range 3800–6500 Å.

Thorium: The results of the interferometric measurements of thorium lines that have accumulated during the last decade (latest contributions to be found in **4** and **15**) have been compiled and critically evaluated in a paper by Giacchetti *et al.* (**3**). From weighted averages of all published wavelengths they derive a set of energy levels for Th I by means of which they calculate, with an average standard deviation of 0·0014 cm^{-1}, the wavenumbers of 1375 lines covering the range from 2650 to 12381 Å. By adding 181 accurately measured Th I and Th II wavelengths they obtain a total of 1556 lines that may be adopted as standards. This synthesis of all the measurements in the thorium spectrum made to date is a great step forward. At the same time it is evident, as emphasized by the authors, that the present results still come short of a full exploitation of the potential advantages of thorium as a source of standards. They recommend that further measurements be made by using cooled hollow-cathode sources and by direct reference to the primary standard.

Iron: Concerning the iron standards we note that H. M. Crosswhite is preparing a line list for publication as an NBS Monograph under the title *The Iron-Neon Hollow-Cathode Spectrum.*

Vacuum-UV Standards

The present situation regarding wavelength standards in the vacuum UV was briefly surveyed in a recent review (**1**). The latest contributions concern Cl I, Cu II and Zn II.

In a comprehensive work on Cl I, Radziemski and Kaufman (**12**) have calculated the wavelengths of 108 Cl I lines in the region 978–1397 Å to ±0·0015 Å or better, relying on some precision wavelength measurements made at NBS by L. Minnhagen.

The second spectrum of copper, Cu II, has been the subject of an exhaustive reinvestigation by Ross (**13**). He gives what should be the final list of Cu II Ritz standards, comprising 432 recommended wavelengths down to 675 Å, with uncertainties of 0·0004 Å to 0·0001 Å. There is a close agreement with the values earlier reported to this commission (**2**).

Martin and Kaufman (**9**) give wavelengths accurate to 0·002 or 0·003 Å for 40 lines of Zn II from 2105 to 1400 Å.

In this report we have used the following abbreviations:

BIPM = Bureau International des Poids et Mesures, Sèvres.
JILA = Joint Institute for Laboratory Astrophysics, Boulder, Colorado.
NBS = National Bureau of Standards, Washington, D.C.
NPL = National Physical Laboratory, Teddington.
NRC = National Research Council, Ottawa.
PTB = Physikalisch-Technische Bundesanstalt, Braunschweig.

REFERENCES

1. Edlén, B. 1969, in *Polarisation, Matière et Rayonnement*. Presses universitaires de France, Paris, pp. 219–227.
2. Edlén, B. 1965, *Trans. IAU,* **12A**, 137.

3. Giachetti, A., Stanley, R. W., Zalubas, R. 1970, *J. opt. Soc. Am.*, in press.
4. Goorvitch, D., Valero, F. P. J., Clúa, A. L. 1969, *J. opt. Soc. Am.*, **59**, 971.
5. Hall, J. L. 1968, *IEEE J. Quantum Electronics*, **QE4**, 638.
6. Hanes, G. R., Baird, K. M. 1969, *Metrologia*, **5**, 32.
7. Hanes, G. R., Dahlstrom, C. E. 1969, *Appl. Phys. Lett.* **14**, 362.
8. Kaufman, V., Humphreys, C. J. 1969, *J. opt. Soc. Am.*, **59**, in press.
9. Martin, W. C., Kaufman, V. 1970, *J. Res. NBS*, in press.
10. Mielenz, K. D., Nefflen, K. F., Rowley, W. R. C., Wilson, D. C., Engelhard, E. 1968, *Appl. Opt.*, **7**, 289.
11. Norlén, G. 1970, *Metrologia*, **6**, 19.
12. Radziemski, Jr., L. J., Kaufman, V. 1969, *J. opt. Soc. Am.*, **59**, 424.
13. Ross, Jr., C. B. 1969, Ph.D. thesis, Purdue University.
14. Rowley, W. R. C., Ward, L. 1968, *Metrologia*, **4**, 146.
15. Valero, F. P. J. 1968, *J. opt. Soc. Am.*, **58**, 484, 1048.

B. EDLÉN
Chairman of the Committee

COMMITTEE 2: TRANSITION PROBABILITIES

A very great deal of work has been done during the last three years on the determination of atomic transition probabilities, and complete coverage here is impossible. Wiese and his staff at the National Bureau of Standards, Washington, have continued to collect all numerical results, and they can supply information as to what is available for any particular atom. They have published (**1**) a critical compilation of atomic transition probabilities for the atoms sodium to calcium (inclusive) in essentially the same way as an earlier compilation covered hydrogen to neon. At the time of writing a complete new bibliography is also in preparation (**2**), which will be published as NBS Special Publication 320 and which is complete up to June 1969. Review articles published include that of Layzer and Garstang (**3**) on theoretical allowed and forbidden transition probabilities, and the two volume conference report (**4**) covers beam foil spectroscopy (including lifetime measurements) in some detail. A fairly complete review of forbidden line transition probabilities was given by Garstang (**5**), where references to many original papers may be found. In the following we shall confine ourselves to mentioning a few areas of particular interest; detailed references can be traced through the bibliographies, reviews and abstracting journals.

On the experimental side much attention has been given to the direct measurement of atomic lifetimes, and many individual cases have been studied. Outstanding has been the work by the phase shift technique by Lawrence, Hesser, Cunningham, and others. Atoms studied include O I, N II, Ne I, Ne II, Ne III and Al I. The direct decay of atomic states has been observed in other cases after pulsed electron excitation, after direct optical excitation and after beam-foil excitation. A major problem in these methods is cascading from higher states, and this limits the applicability of the methods, but there are many important states which have been, and could be, studied. As examples of what has been done we quote the work of Lawrence by pulsed electron excitation on Ar I, Ar II, Ar III, Ar IV, S II, Cl II and Xe I, of Wares and his group on various stages of ionization of oxygen by beam-foil spectroscopy, and of Wolff and Davis on decay following optical excitation in Cs I and Na I. The Hanle effect has been used very successfully by Gallagher for the resonance lines and infrared triplets in Ca II, Sr II and Ba II, and by deZafra and Marshall for Sn I and Pb I. Absorption spectroscopy has been used in many investigations, for example the work of Huber and Tobey and of Grasdalen, Huber and Parkinson on Fe I, Fe II, Cr I and Cr II and the work of Parkes, Keyser and Kaufman on the resonance triplet of O I. Emission spectroscopy of stabilised arcs has been used by Bridges and Wiese and by Foster on the S I and S II spectra. Particularly important are the papers by the Kiel group (**6**) on Fe I and Fe II, using a well stabilized argon arc with a small admixture of iron chloride. They derive solar abundances of 7·6 (on the scale with log H = 12) from their results, in agreement with coronal values.

On the theoretical side an important trend has been the inclusion of configuration interaction in calculations by Hartree-Fock techniques. Work by Fischer on the Al I isoelectronic sequence is of particular interest; she explained some unexpectedly low oscillator strengths in Si II. For highly ionized atoms intermediate coupling calculations must be performed, and configuration interaction must often be included as well. Intermediate coupling calculations were reviewed in (**3**); among more recent papers we mention the work of Mendlowitz on Ti II. Examples of configuration interaction calculations include papers by Froese on various stages of ionization of iron and silicon and of Weiss on C I and C II and on Al I. A number of calculations of oscillator strengths including electron correlation effects have been made by Sinanoglu and his students.

Many problems remain to be studied. Perhaps the most important is in the area of the oscillator strengths of high-excitation-potential lines in atoms such as Fe I. Independent measurements of oscillator strengths for some such lines by methods which do not involve a knowledge of Boltzmann factors are greatly desired to check on excitation potential effects in the existing measurements, and to place existing measurements on a more reliable absolute scale.

Recent work on forbidden lines was reviewed in (**5**). Confidence in the reliability of the theoretical results has been increased to some extent by Thackeray's comparisons of [Fe II] and [Ni II] intensities in η Carinae with theoretical intensities. Laboratory experiments by Hults on [Pb I] and [Pb II] lines in which there is a mixture of magnetic dipole and electric quadrupole radiation tend to confirm the theoretical ratios of these two types of radiation. Measurements of some ratios of lines in [O I] and in [S I] by LeBlanc, Oldenberg and Carleton, and by McConkey, Burns, Moran, Kernahan and Emeleus are in excellent agreement with the theory. Garstang has calculated the transition probabilities for a number of magnetic quadrupole transitions, and showed that one line in each of Fe IX, Fe XVII and Fe XXV may be a significant contributor to the de-excitation process in these ions in the solar corona. The $2\ {}^3S_1$–$1\ {}^1S_0$ line in the helium isoelectronic sequence has been shown by Griem to be due to magnetic dipole radiation made possible in a relativistic approximation.

REFERENCES

1. Wiese, W. L., Smith, M. W., Miles, B. M. 1969, Nat. Stand. Ref. Data System – NBS 22. Atomic transition probabilities, Vol. II.
2. Miles, B. M., Wiese, W. L. 1970, Nat. Bur. Stand. Special Pub. 320.
3. Layzer, D., Garstang, R. H. 1968, *A. Rev. Astr. Astrophys.*, **6**, 449.
4. Bashkin, S. 1968, *Beam-foil Spectroscopy*, 2 vols., Gordon and Breach, New York.
5. Garstang, R. H. 1969, *Mém. Soc. r. Sci. Liège*, **17**, 35–44.
6. Baschek, B., Garz, T., Holweger, H., Richter, J. 1969, *Astr. Astrophys.* **2**, 274–279 and 446–450; 1970, *Astr. Astrophys.* **4**, 229–233.

R. H. GARSTANG
Chairman of the Committee

COMMITTEE 3: COLLISION CROSS SECTIONS AND LINE BROADENING

1. LINE BROADENING

Two review articles have been published (**1, 2**).

1.1. *Collision broadening by neutral atoms.* The continued study of widths and shifts produced by neutral atom impacts at low densities has shown that the Van der Waals interaction ($-C_6 r^{-6}$) as used, for example, in (**3**), is an inadequate representation of the interatomic potential. The classical theory for a Leonard-Jones potential ($C_{12} r^{-12} - C_6 r^{-6}$) has been developed (**4**) and subjected to experimental test (**5**). Further improvements result from introducing a term $C_8 r^{-8}$ in the potential (**6**) and from allowance for quasi-static effects (**7**). Applications to broadening by H—atom impact are discussed in (**8**).

Progress has been made in the study of resonance broadening (collisions between atoms of the same kind). The theory has been subjected to further experimental test (**9**), and a study has been made of combined resonance interaction and dispersion interaction (**10**). Calculations (**11**) and experiment (**12**) for alkali atoms give information applicable to hydrogen (**12**).

1.2. *Line broadening by charged particles*. New and more general theoretical approaches have been developed (**13, 14**). The theory of Stark broadening of hydrogen lines has been improved (**15, 21**) and extended to lower densities and higher quantum numbers (**17**), and semi-empirical formulae have been given for line wings (**18**). The Lyα line has been studied, with special emphasis on the contribution from strong impacts (**19, 20**). Further calculations have been made for HeI lines, with forbidden components, using a theory similar to that used for HI (**22, 23, 24**).

Semi-classical impact theory has been applied to many lines of atoms and ions, and it has been shown that the domain of validity of the theory is greater than has been previously supposed (**25**). The method has been critically reviewed (**26**). Refinements in the theory for isolated lines of neutral atoms do not give significant changes in calculated results (**27**). For positive ions the approximation of rectilinear paths has been abandoned; the use of hyperbolic paths gives substantially greater widths, in better agreement with experiment (**25**) and with quantum mechanical calculations using reactance matrices obtained from close-coupling calculations for MgII (**28**a) and CaII (**28**b). A semi-empirical formula may be used for approximate calculations (**29**).

In turbulent plasmas collective fields can be more important than Holtmark fields (**30**).

2. COLLISION CROSS SECTIONS

2.1. *Data services*. We again draw attention to the important services provided by two Information Centers; details are given in our last report (**31**). A new journal is devoted to publication of atomic data (**32**).

2.2. *General reviews and bibliographies*. A second edition of a standard work of reference is being published in four volumes (**33**). Complete abstracts have been published of papers given at two international conferences on atomic collisions (**34**). Another recent book is concerned with bibliographies on chemical kinetics and collision processes (**35**); further bibliographies are also available (**36**).

2.3. *Heavy particle collisions at high energies*. References are given only to the most recent papers. Impact parameter calculations have been made for excitation of H by H (**37, 38**) and by He (**39**), and it has been shown that a simple approximation is reliable over an extensive range (**39**). Born calculations have been made for excitation of He by H (**40**), and H by H^+ (**41, 42**). Excitation of H by H^+ has also been studied using other approximations (**43, 44, 45**). Ionization by fast H^+ has been studied for H (**46**), He (**47**a) and heavier atoms (**47**b). Calculations have been made for electron loss from fast H and He in H or He (**48**) and a simple approximate formula has been given (**49**). Excitation of He by H^+ has been studied in the Born approximation (**50**) and in a more refined treatment (**51**). Charge transfer has been studied for H^+ in ground-state H (**52**) and He (**53**), and excited-state H (**54**). Various other calculations have been made for fast $H-H^+$ collisions (**55, 56**). Studies of the stopping of fast H in H (**57**) are of importance for theories of the solar wind.

We mention an earlier review (**58**), not cited in our last report, on heavy particle collision processes of importance in astrophysics, at both high energies and low energies.

2.4. *Heavy particle collisions at low energies*. Excitation of forbidden lines by heavy particle impact has been reviewed (**59**). Further work has been done on excitation of fine-structure transitions by proton impact (**60**).

New calculations (**61**) for radiative attachment of C and H, and of C^+ and H, give rate coefficients much larger than earlier estimates. Further rate coefficient calculations have been made for associative detachment of $H+H^-$ (**62, 63**). A rate coefficient of $2\times10^{-5}/T\ \mathrm{cm^3\ sec^{-1}}$ (T in K) has been reported for $H^+ + H^-$ mutual neutralisation (**64**). Theoretical studies have been made of Penning ionization (**65, 66**). Further work has been done on rotational excitation of H_2 by H impact (**67, 68**) but serious uncertainties remain. Recent work (**67**; see also **59**) suggests that earlier estimates of the cross sections for $O^3P_J \rightarrow {}^3P_{J'}$ by H impact, are in error by a factor of about 3. Improved calculations have been made for transitions between hyperfine structure levels in the H ground state, by H impact (**69**). Calculations based on a shape-resonance model have been made for three-body association, $H + H + M \rightarrow H + M$ (**70**).

A bibliography of work on ion-molecule reaction rate data is available (**71**).

2.5. *Electron collisions with molecules.* Review articles have been published on experimental (**72**) and theoretical (**73, 74**) work concerned with excitation of rotation and vibration in simple molecules by slow electrons. In two recent calculations for rotational excitation of CN (**75, 76**) different values of the dipole moment have been assumed; an experimental value has now been obtained (**77**).

A bibliography of work on low energy electron collision cross section data is available (**78**). There are also bibliographies on electron swarm data (**79**), and on cross sections for hydrogen and for rare gases (**80**).

2.6. *Excitation of atoms and ions by electron impact.* A major review article deals with experimental and theoretical aspects of electron impact excitation of atoms (**81**); further reviews are concerned with electron collisions with atoms (**82**) and ions (**83**) and with classical theories (**84**). Many references in these reviews are not repeated in the present Report. A major advance has been the development of a semi-classical approach, based on the correspondence principle (**85**), which should give greatly improved cross sections for transitions between highly excited states.

Electron collision processes in gaseous nebulae have been discussed in two review papers (**86, 87**) and work has continued on excitation of forbidden lines in configurations $2p^q$ (**88, 89, 90, 91**) and $3p^q$ (**92, 93, 94**). Further calculations have been made on transitions between levels of P terms in these configurations (**95**).

The 'close-coupling' approximation has been used to calculate cross sections for excitation of Mg^+ and Ca^+ (**96**), and cross sections for $2s^2 \rightarrow 2s2p\ {}^3P$ and ${}^1P$ in C III and other ions (**97**). Some very elaborate calculations for e-H scattering (**98**) give the best estimates currently available for the near threshold excitation of the $n = 2$ levels. Similar calculations have been made for excitation of He^+ (**99**).

Large numbers of cross sections for excitation of positive ions are required for the interpretation of coronal and chromospheric spectra. A detailed study has been made of the semi-empirical $\bar{g}$ formula (**100**). General computer programs are being developed, for the more accurate calculation of large numbers of cross sections (**101**).

A general review of excitation and ionization by electron impact, with emphasis on processes of astrophysical importance, is being prepared for publication (**102**).

2.7. *Ionization by electron impact.* An earlier review of experimental work has been followed by a companion review of the theory of electron-impact ionization (**103**). Several other papers have been concerned with reviews of available data and calculation of rate coefficients (**104, 105, 106**).

The most important recent advances, so far as astrophysical applications are concerned, have been concerned with work on the ionization of positive ions. References to experimental work are given in (**107**) and to theoretical work in (**108**). It has been confirmed that the Coulomb-Born approximation for ionization of ions is more accurate than the Born approximation for ionization of neutrals. Using a general computer program (**109**), Coulomb-Born calculations can be made for many different ions.

2.8. *Recombination.* Some further work has been done on the theory of dielectronic recombination at low densities (**110, 111, 112**), and a detailed study has been made of the effects of additional collisional processes which become important at higher densities (**113**).

REFERENCES

1. Traving, G. 1968, *Plasma Diagnostics*, Ed. W. Lochte-Holtgreven. North-Holland Publ. Co., Amsterdam.
2. Ben-Reuven, A. 1969, *Adv. at. molec. Phys.*, **5**, 201.
3. Warner, B. 1968, *Z. Astrophys.*, **69**, 161.
4. Hindmarsh, W. R., Petford, A. D., Smith, G. 1967, *Proc. R. Soc.*, **A297**, 296.
5. Vaughan, J. M., Smith, G. 1968, *Phys. Rev.*, **166**, 17.
6. Hindmarsh, W. R., du Plessis, A. N., Farr, J. M. to be published.
7. Behmenburg, W. 1968, *Z. Astrophys.*, **69**, 368.
8. Roueff, E., van Regemorter, H. 1968, *Astr. Astrophys.*, **1**, 69.
9. Kuhn, H. G., Lewis, E. L. 1967, *Proc. R. Soc.*, **A299**, 423.
10. Lewis, E. L. 1967, *Proc. Phys. Soc.*, **92**, 817.
11. Stacey, D. N., Cooper, J. 1969, *Phys. Lett.*, **30A**, 49.
12. Gompertz, G., Hindmarsh, W. R. 1969, *Mon. Not. R. astr. Soc.*, **142**, 97.
13. Smith, E. W., Hooper, C. F. 1967, *Phys. Rev.*, **157**, 126.
14. Zaidi Haider, R. 1968, *Phys. Rev.*, **173**, 132.
15. Kepple, P., Griem, H. R. 1968, *Phys. Rev.*, **173**, 317.
16. Smith, E. W., Cooper, J., Vidal, C. R. 1969, *Phys. Rev.*, **185**, 140; see also D. Voslamber, 1969, *Z. Naturf.*, **24A**, 1458.
17. Bengston, R. D., Kepple, P., Tannich, J. D. *Phys. Rev.* in press.
18. Edmonds, F. N., Schlüter, H., Wells, D. C. 1967, *Mem. R. astr. Soc.*, **71**, 271.
19. Shen, K. Y., Cooper, J. 1969, *Astrophys. J.*, **155**, 37.
20. Pfenning, H. to be published.
21. van Regemorter, H. 1969, *Phys. Lett.*, **30A**, 365.
22. Griem, H. R. 1968, *Astrophys. J.*, **154**, 1111.
23. Gieske, H. A., Griem, H. R. 1969, *Astrophys. J.*, **157**, 963.
24. Barnard, A. J., Cooper, J., Shamey, L. J. 1969, *Astr. Astrophys.*, **1**, 28.
25. Sahal-Brechot, S. 1969, *Astr. Astrophys.*, **1**, 91 and **2**, 322.
26. Roberts, D. E. 1968, *J. quantit. Spectrosc. radiat. Transfer*, **8**, 1241.
27. Cooper, J., Oertel, G. K. 1969, *Phys. Rev.*, **180**, 286.
28. (a) Bely, O., Griem, H. R. 1970, *Phys. Rev.*, in press; (b) Barnes, K. S., Peach, G. 1970, *Nature*, **225**, 374.
29. Griem, H. R., 1968, *Phys. Rev.*, **165**, 258.
30. Kunze, H.-J., Griem, H. R. 1968, *Phys. Rev. Lett.*, **21**, 1048; and 1969, *ibid.*, **23**, 1279.
31. *Trans. IAU*, **13A**, 1967, 239.
32. *Atomic Data: A Journal Devoted to Compilations of Experimental and Theoretical Results in Atomic Physics*, Academic Press, New York and London.
33. *Electronic and Ionic Impact Phenomena*, Oxford Univ. Press: Vol. I: H. S. W. Massey, E. H. S. Burhop, *Electron Collisions with Atoms*, 1969; Vol. II: H. S. W. Massey, *Electron Collisions with Molecules – Photo-Ionization*, 1969; Vol. III: H. S. W. Massey, *Slow Collisions of Heavy Particles*, in press; Vol. IV: H. S. W. Massey, E. H. S. Burhop, H. B. Gilbody, *Fast Collisions of Heavy Particles – Recombination – Slow Collisions with Positrons and Muons*, in preparation.
34. (*a*) Abstracts of Papers, 5th Int. Conf. Phys. Electr. and Atomic Coll., Nauka Publ. House, Leningrad, 1967; – (*b*) 6th Int. Conf., M.I.T. Press, Cambridge, Mass., 1969.
35. *Bibliography of Chemical Kinetics and Collisions Processes.* Ed. A. R. Hockstim, Plenum Press, New York, 1969.
36. The following bibliographies on Atomic Collisions are available from Dr. K. Takayanagi, Institute of Space and Aeronautical Studies, University of Tokyo: No. 5 and No. 6., *Ionic Impact* (experimental); No. 7, *Heavy Particle Collisions with Electronic Transitions* (theoretical); No. 8, *Electron Collisions with Atoms and Molecules* (experimental).
37. Levy, H. 1969, *Phys. Rev.*, **184**, 97.
38. Flannery, M. R. 1969, *J. Phys. B: At. Molec. Phys.* **2**, 909; and *Phys. Rev.*, **183**, 231 and 241.
39. Flannery, M. R. 1969, *J. Phys. B: At. Molec. Phys.*, **2**, 913.
40. Levy, H. 1969, *Phys. Rev.*, **185**, 7.
41. Khandelwal, G. S., Choi, B.-H. 1968, *J. Phys. B: At. Molec. Phys.*, **1**, 1218 and 1220; 1969, *ibid.*, **2**, 308.
42. Khandelwal, G. S. 1969, *J. Phys. B: At. Molec. Phys.*, **2**, 151.
43. Coleman, J. P. 1968, *J. Phys. B: At. Molec. Phys.*, **1**, 567.

44. Holt, A. R., Moiseiwitsch, B. L. 1968, *J. Phys. B: At. Molec. Phys.*, **1**, 36.
45. Flannery, M. R. 1969, *J. Phys. B: At. Molec. Phys.*, **2**, 1044.
46. Salin, A. 1969, *J. Phys. B: At. Molec. Phys.*, **2**, 631.
47. (a) Bell, K. I., Kingston, A. E. 1969, *J. Phys. B: At. Molec. Phys.*, **2**, 653; (b) Peach, G., 1968, *J. Phys. B: At. Molec. Phys.*, **1**, 1088.
48. Bell, K. L., Dose, V., Kingston, A. E. 1969, *J. Phys. B: At. Molec. Phys.*, **2**, 831.
49. Victor, G. A. 1969, *Phys. Rev.*, **184**, 43.
50. Bell, K. L., Kennedy, D. J., Kingston, A. E. 1968, *J. Phys. B: At. Molec. Phys.*, **1**, 218, 1028 and 1037.
51. Van den Bos, J. 1969, *Phys. Rev.*, **181**, 191.
52. Mapleton, R. A. 1968, *J. Phys. B: At. Molec. Phys.*, **1**, 529.
53. Colegrave, R. K., Stephens, D. B. L. 1968, *J. Phys. B: At. Molec. Phys.*, **1**, 856.
54. Bates, D. R., Reid, R. H. G. 1969, *J. Phys. B: At. Molec. Phys.*, **2**, 851 and 857.
55. Gallaher, D. F., Willets, L. 1968, *Phys. Rev.*, **169**, 139; see also I. M. Cheshire, D. R. Gallaher, A. J. Taylor in (**34***b*).
56. Coleman, J. P., Trelease, S. 1968, *J. Phys. B: At. Molec. Phys.*, **1**, 172.
57. Schafer, J., Trefftz, E. to be published.
58. Dalgarno, A. 1967, *Rev. Mod. Phys.*, **39**, 850.
59. Dalgarno, A., Reid, R. H. G. 1969, *Mém Soc. r. Sci. Liège*, **17**, 57.
60. Reid, R. H. G., Schwartz, J. in (**34***b*).
61. Klemperer, W. to be published.
62. Chen, J. C. Y., Peacher, J. L. 1968, *Phys. Rev.*, **167**, 30.
63. Browne, J. C., Dalgarno, A. 1969, *J. Phys. B: At. Molec. Phys.* **2**,, 885.
64. Victor, G. A. to be published.
65. Nakamura, H. 1969, *J. phys. Soc. Japan*, **26**, 1473.
66. Mori, M. 1969, *J. phys. Soc. Japan*, **26**, 773.
67. Nishimura, S. 1968, *Ann. Tokyo astr. Obs.*, **11**, 33.
68. Tang, K. *J. Chem. Phys.*, in press.
69. Allison, A. C., Dalgarno, A. 1969, *Astrophys. J.*, **158**, 523.
70. Roberts, R. E., Bernstein, R. B., Curtis, C. F. 1969, *J. Chem. Phys.* **50**, 5163.
71. Sinnott, G. A. 1969, *Bibliography of Ion-Molecule Reaction Rate Data*, JILA Information Center Report No. 9 (Joint Inst. Lab. Astrophys., Boulder, Colo.).
72. Phelps, A. V. 1968, *Rev. Mod. Phys.*, **40**, 399.
73. Takayanagi, K. 1967, *Progress theor. Phys.* (Kyoto), Suppl. No. 40, 216.
74. Takayanagi, K., Itikawa, Y. 1970, *Adv. At. Molec. Phys.*, **6**.
75. Takayanagi, K., Itikawa, Y. 1968, *Publ. astr. Soc. Japan*, **20**, 376.
76. Allison, A. C., Crawford, O. H., Dalgarno, A., 1969, *Astr. Astrophys.*, **2**, 451.
77. Thomson, R., Dalby, F. W. 1968, *Can. J. Phys.*, **46**, 2815.
78. Kieffer, L. J. 1969, *Atomic Data*, **1**, 19 and 122.
79. Dutton, J. 1967, *Bibliography of Electron Swarm Data*, JILA Information Center Report, No. 4 (Joint Inst. Lab. Astrophys., Boulder, Colo.).
80. Laborie, P., Rocard, J.-M., Rees, J. A. 1968, *Tables des Sections Efficaces Electroniques et Coefficients Macroscopiques: I: Hydrogéne et gaz rares*. McGraw Hill, New York.
81. Moiseiwitsch, B. L., Smith, S. J. 1968, *Rev. Mod. Phys.*, **40**, 238.
82. Burke, P. G. 1969, *Atomic Physics*, Ed B. Bedersen, V. W. Cohen and F. M. J. Pichanick. Plenum Press, New York, p. 265.
83. Seaton, M. J. 1969, *Atomic Physics*, Ed. B. Bedersen, V. W. Cohen and F. M. J. Pichanick. Plenum Press, New York, p. 295.
84. Burgess, A., Percival, I. C. 1968, *Adv. At. Molec. Phys.*, **4**, 109.
85. Percival, I. C., Richards, D. 1969, *Astrophys. Lett.*, **4**, 235.
86. Seaton, M. J. 1968, *Adv. At. Molec. Phys.*, **4**, 331.
87. Seaton, M. J. 1969, *Mém. Soc. r. Sci. Liège*, **17**, 45.
88. Saraph, H. E., Seaton, M. J., Shemming, J. 1969, *Phil. Trans. R. Soc.*, **A264**, 77.
89. Smith, K., Conneely, M. J., Morgan, L. A. 1969, *Phys. Rev.*, **177**, 196.
90. Henry, R. J. W., Burke, P. G., Sinfailam, A.-L. 1969, *Phys. Rev.*, **178**, 218.
91. Eissner, W., Martins, P. de A. P., Nussbaumer, H., Saraph, H. E., Seaton, M. J. 1969, *Mon. Not. R. astr. Soc.* **146**, 63.
92. Czyzak, S. J., Krueger, T. K. 1967, *Proc. Phys. Soc.*, **90**, 623.

93. Conneely, M. J., Smith, K., Lipsky, L. submitted to *J. Phys. B: At. Molec. Phys.*
94. Czyzak, S. J., Krueger, T. K., Martins, P. de A. P., Saraph, H. E., Seaton, M. J. *Mon. Not. R. astr. Soc.*, in press.
95. Blaha, M. 1968, *Ann. Astrophys.*, **31**, 311; 1969, *Astr. Astrophys.*, **1**, 42.
96. Burke, P. G., Moores, D. L. 1968, *J. Phys. B: At. Molec. Phys.*, **1**, 575.
97. Osterbrock, D. E. *Astrophys. J.*, in press; and *J. Phys. B: At. Molec. Phys.*, in press.
98. Burke, P. G., Ormonde, S., Whitaker, W. 1967, *Proc. Phys. Soc.*, **92**, 319; A. J. Taylor and P. G. Burke, *ibid.*, 336; P. G. Burke, A. J. Taylor, S. Ormonde, *ibid.*, 345; J. Macek and P. G. Burke, *ibid.*, 351; J. Macek, *ibid.*, 365.
99. Burke, P. G., Taylor, A. J. 1969, *J. Phys. B: At. Molec. Phys.*, **2**, 44.
100. Blaha, M. 1969, *Astrophys. J.*, **157**, 473.
101. Eissner, W., Seaton, M. J. see **(34b)**.
102. Bely, O., Van Regemorter, H. 1970, to appear in *A. Rev. Astr. Astrophys.*, **8**.
103. Rudge, M. R. H. 1968, *Rev. Mod. Phys.*, **40**, 564.
104. Gulyaev, R. A. 1967, *Sov. Astr. – AJ.*, **10**, 761.
105. Lin, Sin-Shong, Stafford, F. E. 1968, *J. Chem. Phys.*, **48**, 3885.
106. Lotz, W. 1968, *Z. Phys.*, **216**, 241.
107. Peart, B., Dolder, K. T. 1969, *J. Phys. B: At. Molec. Phys.*, **2**, 1169; Peart, B., Martin, S. O., Dolder, K. T. *ibid.*, 1176.
108. Bely, O. and Schwartz, S. B. 1969, *Astr. Astrophys.*, **1**, 281; Sato, S., Takahashi, K., and Takebe, H., 1968, *Publ. astr. Soc. Jap.*, **20**, 149.
109. Eissner, W., Moores, D. L. see **(34b)**.
110. Trefftz, E. 1969, *Proc. Sommerfeld Memorial Meeting, Munich*, North-Holland Publ. Co., Amsterdam.
111. Davies, P. C. W., Seaton, M. J. 1969, *J. Phys. B: At. Molec. Phys.*, **2**, 757.
112. Trefftz, E. *J. Phys. B: At. Molec. Phys.*, in press.
113. Burgess, A., Summers, H. P. 1969, *Astrophys. J.*, **157**, 1007.

M. J. SEATON
Chairman of the Committee

COMMITTEE 4: STRUCTURE OF ATOMIC SPECTRA

General work on Atomic Spectra

A current *Bibliography on The Analyses of Optical Atomic Spectra* is in print **(43)**. It appears in four sections, three of which cover the same elements as the respective volumes on *Atomic Energy Levels.* Relevant literature references from the dates of publication of these Volumes are continued for each spectrum to the present, as follows:

Section 1	^{1}H – ^{23}V	issued September 1968.
Section 2	^{24}Cr – ^{41}Nb	issued February 1969.
Section 3	^{42}Mo – ^{57}La ^{72}Hf – ^{89}Ac	issued May 1969.

The last Section contains references to rare-earth spectra:

Section 4	^{57}La – ^{71}Lu ^{89}Ac – ^{99}Es	issued September 1969.

The present report will be limited to later work.

Monographs, extended analyses, work in progress

The active programs in Lund are providing data on atomic spectra that are urgently needed. In the Physics Department of the University, Edlén and his colleagues have published or submitted for publication analyses on the following spectra:

Be I, Be II (**30**); B III (**46**); F I (**49**), F II (**48**); Al I (**20**).

Analyses are nearly completed or ready for publication for B II (**47**); C V (**35**); N I (**21**); Ti III, Ti IV (**59**). Comprehensive work is in progress on O II (**61**) and on F III, F IV, F V (**50**). Edlén has also reported his progress on the complex spectrum Fe IV (**16**), which is one of the most important contributions to astrophysics.

At the Lund Institute of Technology Minnhagen and his associates have investigated Ca III (**5**) and are working on Ne II (**51**), K II (**41**), and Mg III (**2**). Bockasten and associates at Lund and Uppsala have published work on O V (**4a**) and O IV (**7**).

In Madrid at the Institute of Optics the analysis of V III is in course of preparation (**33**) and the analyses of V II (**60**), Mn III (**24**) and Ni IV (**25**) have been published. For Ni IV, see, also (**52**).

At the National Bureau of Standards the analyses of Cl I (**53**), Zn II (**38**) and Tc I (**6**) have been extended.

At the Corona Laboratory Humphreys (**32**) has recalculated the values of selected energy levels for ^{86}Kr II and ^{136}Xe II derived from new interferometric observations in the 4μ region.

Vacuum UV atomic spectra

In the He I sequence lines in the spectra Ar XVII through Zn XXIX have been observed by Cohen and his associates in the range 4–1.4 Å (**11**). Gabriel and Jordan have reported observations of long wave satellites to the resonance lines of He-like ions in both the laboratory and the Sun; they list laboratory wavelengths for the spectra C through Al (**23**).

New classifications have also been made in the spectra of the iron-period elements of the Ar I, Cl I, and S I sequences by Fawcett *et al.* (**22**). These authors have studied the configurations $3s^2\ 3p^n$–$3s^2\ 3p^{n-1}4$, $5s$, 4, $5d$, in the far vacuum UV region.

Lines of O VI have been observed in emission near 150 Å with a laser-produced plasma as source (**8**).

Blake and House (**4**) have reported observations of KαX-ray-type transitions $1s^2\ 2s^2\ 2p^{n-1}$–$1s\ 2s^2\ 2p^n$ in the solar atmosphere and in laboratory plasmas for: Ne III, Si VII, Cr XVIII, Fe XX.

M. Chaghtai has observed and analyzed the spectra: Zr V, Zr VI, Zr VII; Nb VI, Nb VII, Nb VIII; Mo VII, Mo VIII, Mo IX (**9**). This work has been carried on at the Lab. de Chimie physique, Orsay and also in Lund.

Garton and his associates have continued their valuable program on the observation of absorption series: Be I (**26**), $2s\ np(n\sim 20)$; Ca, Sr, Ba $<$ 500 Å, inner shell excitation (**28**); Sr I Singlets of Principal Series to $n=33$ (**27**); Cs I, Rb I (**13**).

New absorption series and ionization potentials have been reported by Huffman *et al.* (**31**) for O I and for F I, Br I, Cl I, I I in the range 600–1500 Å.

Autoionization is a subject that becomes increasingly important as observational techniques develop. In Ne, autoionizing levels have been determined from observations of 13 absorption series observed by ion impact (**19**). Madden and his staff have observed resonances in the photo-ionization continuum of Ar I (20–150eV) (**37**).

Edlén has extended his identifications of the coronal lines by a study of the transitions from metastable levels in the configurations $3s^2\ 3p^k\ 3d$ with $k=3$, 4, 5 of Fe XI, Fe X, Fe IX, and Ni XIII, Ni XII, and Ni XI (**17**). These transitions "can well account for all of the remaining unidentified coronal lines". In the same paper he confirms the identification of λ4412 as Ar XIV and of λ 5533·4 as Ar X, by a study of the Z-dependence of the ^2p intervals of the configurations $2s^2\ 2p$ and $2s^2\ 2p^5$. Edlén *et al.* have also published a paper entitled 'Spin-Forbidden Resonance Multiplets in Light Elements' (**18**).

A number of theoretical papers deal with highly-ionized spectra. In the He I isoelectronic sequence analytical wave functions have been calculated up to N VI for the $1s\ np\ ^{1,3}P$ states for $n=2$, 3, 4, 5 (**1**). Eigenvalues and orbital wave functions have been calculated for the ^1S states, $n=1$–3 and for the ^3S states, $n=2$–4 for He I, Li II and Be III (**12**). Close-coupling calculations of the positions and widths of the lowest-lying autoionizing D states in He converging on the $n=2$ level of He II, have also been made (**14**).

In the B I isoelectronic sequence Lexa (**34**) has determined by extrapolation the 12 lowest levels in the spectra S XII through Ca XVI.

Hartfree-Fock Wave Functions have been calculated for the highly-ionized spectra Fe XXII, Fe XXIII, Fe XXIV, Fe XXV (**10**).

Rare-earth spectra

At the National Bureau of Standards, active work on rare-earth spectra of the lanthanides is being emphasized in order to provide data for Volume IV of *Atomic Energy Levels*. The spectra being investigated are: Ce I; Pr I, Pr III (**58**); Pm I, Pm II; Ho I; Tm I, Tm II, Tm III; Yb I and in addition La IV and Th I. A brief summary has been published (**44**).

With the aid of Zeeman observations, Blaise and Van Kleef have extended the analyses of Gd II (**3**). Infrared observations of Gd have been reported by Spector and Held (**57**).

Theoretical work on rare-earth spectra is being continued at the Hebrew University in Israel by Z. B. Goldschmidt and others. In a paper on 'Magnetic Interactions in Heavy Atoms' the authors discuss the $4f^N$ configurations in Pr III, Pr IV and Er IV, as well as the $3d^N$ configurations in third spectra of the iron group (**29**).

Theory and miscellaneous

A number of theoretical papers should be cited. Lotz has calculated 'Subshell Binding Energies of Atoms and Ions, H to Zn' (**36**), giving preference to experimental values in all doubtful cases. Czyzak and others have calculated collision strengths for the N I($2p^3$), O I($2p^4$), Si I($3p^2$) and S I($3p^4$) isoelectronic sequences in a paper on 'Excitation of Forbidden Lines' (**15**). Minnhagen and Svensson have calculated the energy level distributions and eigenvectors for the $3p^4\ nf$ configurations of Ar II (**42**). R. D. Cowan is working on *ab initio* theoretical calculations of atomic energy levels and transition probabilities, primarily in the $p^m l$ configurations of ions isoelectronic with the third-row elements (Na-Ar). Shadmi and his associates have calculated levels and g-values in the second and third spectra of the iron group taking into account effective interactions between d-electrons (**55, 55a**). They have also made a theoretical study of the even levels in W I (**54**). C. Roth has made extensive theoretical calculations of levels and g-values for odd configurations in the first, second and third spectra of the iron group, by fitting parameters to the observed data (**53a**). The odd configurations of Ti I have been interpreted theoretically by Smith and Siddall (**56**). *Ab initio* calculations of the odd configurations of Cd I have been reported by Wilson (**62**).

Martin and Sugar at the National Bureau of Standards have published a paper on 'Perturbations and Coupling in the $d^9\ sp$ configurations of Cu I, Zn II, Ag I, Cd II and Tl III' (**39**). One on the calculations of the $d^9\ ns$ configurations in Zn II and Ag I, which contains some new levels in these spectra (**40**) has, also, been published by Martin and Kaufman.

In response to a steady demand for a revised table of ionization potentials, such a compilation is now in course of preparation (**45**). It is planned to include for each spectrum not only the ionization potential in eV, but also the limit in cm^{-1} used in the calculation. In addition, the energy levels from the ground configuration, above the ground state zero, will be tabulated. References to the source material used for the various spectra will be included.

REFERENCES

1. Ali, M. A., Dasgupta, N. K. 1967, *Proc. phys. Soc. (London)*, **92**, No. 575, 50–54.
2. Andersson, E., Johannesson, G. A. 1969, unpublished, Lund.
3. Blaise, J., Van Kleef, Th. A. M. 1969, *Compt. Rend. Acad. Sci. Paris*, **268**, 792–795.
4. Blake, R. L., House, L. L. 1967, *Astrophys. J.*, **149**, No. 1, part 2, L22–L35; 1968, **154**, No. 3, Part 2, L141–L143.
4a. Bockasten, K., Johansson, K. B. 1969, *Ark. Fys. (Stockholm)*, **38**, 563–584.

5. Borgström, A. 1968, *Ark. Fys. (Stockholm)*, **38**, No. 12, 243–260.

6. Bozman, W. R., Corliss, C. H., Tech, J. L. 1968, *J. Res. Nat. Bur. Stand.*, **72A**, No. 6, 559–608.

7. Bromander, J. 1969, *Ark. Fys. (Stockholm)*, **40**, 257–274.

8. Burgess, D. D., Fawcett, B. C., Peacock, N. J. 1967, *Proc. Phys. Soc. (London)*, **92**, No. 577, 805–816.

9. Chagtai, M. S.-Z. 1969, J. opt. Soc. Am. **59**, No. 8, 969–970.

10. Chapman, R. D. 1969, *Astrophys. J.*, **156**, Part 1, No. 1, 87–91.

11. Cohen, K., Feldman, U., Swartz, M., Underwood, J. H. 1968, *J. opt. Soc. Am.*, **58**, No. 6, 843–846.

12. Cohen, M., McEachran, R. P. 1967, *Proc. Phys. Soc. (London)*, **92**, No. 575, 37–41; **92**, No. 577, 539–542.

13. Connerade, J. P. 1969, *Astrophys. J.*, **159**, 685–702.

14. Cooper, J. W., Ormonde, S., Humphrey, C. H., Burke, P. G. 1967, *Proc. Phys. Soc. (London)*, **91**, No. 572, 285–287.

15. Czyzak, S. J., Krueger, T. K., Saraph, H. E., Shemming, J. 1967, *Proc. Phys. Soc. (London)*, **92**, No. 578, 1146–1147(L).

16. Edlén, B. 1969, *Mon. Not. R. astr. Soc.*, **144**, 391–396.

17. Edlén, B. 1969, *Solar Phys.*, **9**, 439–445.

18. Edlén, B., Palenius, H. P., Bockasten, K., Hassin, R., Bromander, J. 1969, *Solar Phys.*, **9**, 432–438.

19. Edwards, A. K., and Rudd, M. E. 1968, *Phys. Rev.*, **170** No. 1, 140–144.

20. Eriksson, K. B. S. 1969, *Ark. Fys. (Stockholm)*, **39**, No. 28, 421–424.

21. Eriksson, K. B. S., Petterson, J. 1969, unpublished, Lund.

22. Fawcett, B. C., Peacock, N. J., Cowan, R. D. 1968, *J. Phys. B. (Proc. Phys. Soc. London)*, (2) **1**, No. 2, 295–306.

23. Gabriel, A. H., Jordan, C. 1969, *Nature*, **221**, No. 5184, 947–949.

24. Garcia-Riquelme, O. 1968, *Rev. Opt. Pura y Apl. (Madrid)*, **1**, No. 2, 53–72.

25. Garcia-Riquelme, O. 1969, *Physica*, **40**, 27–29.

26. Garton, W. R. S. 1968, private communication, August.

27. Garton, W. R. S., Codling, K. 1968, *J. Phys. B (Proc. Phys. Soc. London)*, [2], **1**, No. 1, 106–113.

28. Garton, W. R. S., Newson, S. 1969, unpublished.

29. Goldschmitt, Z. B., Pasternak, A., Goldschmidt, Z. H. 1968, *Phys. Lett. (Amsterdam)*, **28A**, No. 4, 265–266.

30. Holmström, J. E., Johansson, L. 1969, *Ark. Fys. (Stockholm)*, **40**, No. 12, 133–138

31. Huffman, R. E., Larrabee, J. C., Tanaka, Y. 1967, *J. Chem. Phys.*, **47**, No. 2, 856–857; **47**, No. 11, 4462–4471.

32. Humphreys, C. J. 1969, private communication, August.

33. Iglesias, L. 1969, *Rev. Opt. Pura y Apl. (Madrid)*, unpublished.

34. Lexa, J. 1967, *Bull. astr. Inst. Csl.*, **18**, No. 5, 269–273.

35. Löfstrand, Edlén, B. 1969, unpublished, August.

36. Lotz, W. 1968, *J. opt. Soc. Am.*, **58**, No. 7, 915–921.

37. Madden, R. P., Ederer, D. L., Codling, K. 1969, *Phys. Rev.*, **177**, No. 1, 136–151.

38. Martin, W. C., Kaufman, V. 1970, *J. Res. Nat. Bur. Stand.*, **74A**, No. 1, 11–22.

39. Martin, W. C., Sugar, J. 1969, *J. opt. Soc. Am.*, **59**, No. 10, 1216–1280.

40. Martin, W. C., Sugar, J. 1970, *J. Res. Nat. Bur. Stand.*, **74A**, No. 1, 7–10.

41. Minnhagen, L., Hallén, K. 1969, unpublished, May.

42. Minnhagen, L., Svensson, H. 1967, *Ark. Fys. (Stockholm)*, **34**, No. 7, 71–79.

43. Moore, C. E. 1969, *Nat. Bur. Stand. Special Publ. 306*, 4 Sections.

44. Moore, C. E. 1969, *Bull. Am. astr. Soc.*, **1**, No. 1, 85–86.

45. Moore, C. E. 1969, unpublished, September.

46. Ölme, A. 1969, *Ark. Fys. (Stockholm)*, **40**, No. 2, 35–47.

47. Ölme, A. 1969, unpublished, August.

48. Palenius, H. P. 1968, *Ark. Fys. (Stockholm)*, **39**, No. 3, 15–64.

49. Palenius, H. P. 1969, *Ark. Fys. (Stockholm)*, **39**, No. 29, 425–427.

50. Palenius, H. P. 1969, analysis in progress, August.

51. Persson, W. 1969, *J. opt. Soc. Am.*, **59**, No. 3, 285–287.

52. Poppe, R. 1969, *Physica*, **40**, 17–26.

53. Radziemski, L. J., Kaufman, V. 1969, *J. opt. Soc. Am.*, **59**, No. 4, 424–443.

53a. Roth, C. 1968, *J. Res. Nat. Bur. Stand.*, **72A**, No. 5, 505–520; 1969, **73A**, No. 2, 125–157, 159–171; in press, 1969.

54. Shadmi, Y., Caspi, E. 1968, *J. Res. Nat. Bur. Stand.*, **72A**, No. 6, 757–760.
55. Shadmi, Y., Oreg, J., Stein, J. 1968, *J. opt. Soc. Am.*, **58**, No. 7, 909–914.
55a. Shadmi, Y., Caspi, F., Oreg, J 1969, *J. Res. Nat. Bur. Stand.*, **73A**, No. 2, 173–189.
56. Smith, G., Siddall, J. 1969, *J. opt. Soc. Am.*, **59**, No. 4, 419–423.
57. Spector, N., Held, S. 1969, *Astrophys. J.*, in press, September.
58. Sugar, J. 1969, *J. Res. Nat. Bur. Stand.*, **73A**, No. 3, 333–381.
59. Swensson, J. W., Edlén, B. 1969, unpublished, August.
60. Velasco, R., Gullon, N. 1968, *Rev. Opt. Pura y Apl.* **1**, No. 2, 93–102.
61. Wenåker, I. 1969, work in progress, Lund, August.
62. Wilson, M. 1968, *J. Phys. B (Proc. Phys. Soc. London)*, [2], **1**, No. 4, 736–741.

CH. MOORE-SITTERLY
Chairman of the Committee

COMMITTEE 5: MOLECULAR SPECTRA

The laboratory study of molecules of interest to astrophysicists has been progressing at a steady pace. A complete bibliography for the year September 1968 to September 1969, for instance, would show work on 68 molecules of possible interest; the diatomic and simpler polyatomic molecules. During this period 38 papers appeared on H_2 alone, notably on its polarizability and the character of its spectrum near the ionization limit. Twenty-one papers on CO included studies in the vacuum UV, its Franck-Condon factors, chemiluminescence and IR absorptances. In the interests of saving space in this report, the detailed discussion of published work on each molecule of interest will be deleted. Instead, attention might be called to the fact that a complete card file bibliography is maintained by Phillips at Berkeley, and inquiries are welcome. In addition, a bi-monthly *Newsletter* is distributed from Berkeley in collaboration with S. P. Davis of the Physics Department. This *Newsletter* includes information on work currently in progress at various laboratories. At the present time it is being distributed to over 250 spectroscopists in various countries. A further source of information is a bibliography of articles on 'Spectroscopy of Interest to High Temperature Chemistry', which is distributed at frequent intervals by Leo Brewer in the Chemistry Department at Berkeley. Of more limited scope are the annual *Technical Reports* of the Laboratory of Molecular Structure and Spectra of the University of Chicago, which is a collection of their papers. A bibliography of similar scope is distributed periodically by the Joint Institute for Laboratory Astrophysics at Boulder, Colo. Finally, a quite complete bibliography is to be found in *Current Papers in Physics*, published twice monthly by The Institution of Electrical Engineers, Savoy Place, London W.C.2.

Two papers from the National Research Council of Canada should be mentioned, since they are of special astrophysical interest. These are the paper by the late P. G. Wilkinson on H_2 (*Can. J. Phys.*, **46**, 1968, 1225) and one by S. K. Luke on H_2^+ (*Astrophys. J.*, **156**, 1969, 761). The first gives the precise wavelengths of the UV H_2 lines, which will surely be observed in interstellar absorption in the near future, and the second gives the expected radio spectrum of H_2^+, which may also be observed sooner or later.

In addition to work in progress at various laboratories reported in the *Newsletters*, G. Herzberg has submitted the following for inclusion in this Report:

One of the most interesting results at NRC has been the recent observation by J. W. C. Johns of a spectrum of ArH corresponding to transitions between two Rydberg states of this molecule. The work on CH reported on in previous IAU reports is now in press (G. Herzberg and J. W. C. Johns). An extension of the spectrum of SiH in the vacuum UV has been completed (G. Herzberg, A. Lagerqvist, B. J. McKenzie, *Can. J. Phys.*, in press).

G. Herzberg has carried out a good deal of work on the absorption spectrum of molecular hydrogen (including HD and D_2) in the region below 900 Å. The discrepancy between the theoretical and experimental value for the dissociation energy of hydrogen has been resolved in favor of the theoretical value. The previous experimental values by Monfils and Herzberg were severely affected by lack of resolution and overlapping lines. The new experimental values are based on spectra taken at liquid nitrogen temperature with the 10·5 m vacuum spectrograph. In addition the Rydberg series of H_2 and its isotopes are being studied.

Following the work on a new spectrum associated with C_2 in the visible region (G. Herzberg and A. Lagerqvist, *Can. J. Phys.*, **46**, 1968, 2363) Milligan and Jacox at NBS by means of matrix work at liquid helium temperature have confirmed the NRC suggestion that this spectrum is due to C_2^-. Three new electronic transitions of C_2 have been observed in the vacuum UV by A. Lagerqvist, C. Malmberg, and G. Herzberg (*Can. J. Phys.*, **47**, 1969, 2735). These spectra may be of interest when far UV spectra of carbon stars become available.

Dr. Ch. Jungen, now of NRC, in collaboration with E. Miescher, has produced a very convincing analysis of the f complex of the NO molecule based on high resolution plates in the vacuum UV taken at NRC.

Venkateswarlu, also using the 10·5 m vacuum spectrograph at NRC, has obtained and analysed very extensive spectra of Br_2 and I_2 in the vacuum UV.

B. L. Lutz and A. E. Douglas at NRC have submitted a short paper in the *Canadian Journal of Physics* on the observation, for the first time, of the spectrum of SiH^+. A $'\pi$–$'\Sigma$ band system, analogous to the CH^+ system, has been found with its 0–0 band origin at 25025·20 cm^{-1}. This band system may give discrete interstellar absorption lines at 3993·400 Å [R(O), 0–0 band] and 3932·347 Å [R(O), 1–0 band] and also the system may occur in the solar spectrum.

Aert Schadee (*JQSRT*, **7**, 1967, 169) has stressed that an electronic f-value for molecular transitions is not uniquely defined, but that a wavelength dependent $f_{el}(\lambda)$, multiplied with the Franck-Condon factor, is a fair approximation of the band oscillator strength $f_{v'v''}$ for the stronger bands of a system. He also discusses the normalization of Hönl-London factors in connection with the relation between individual line and band oscillator strengths.

The Berkeley analyses of molecular spectra have resulted in the publication by the University of California Press of Volume 2 comprising analyses of 36 bands of the C_2 Swan system. Also included in this volume is the analysis of the HgH spectrum, which is of less astrophysical interest. Volume 3 on the spectrum of TiO should be ready for publication before the end of 1970. At present in the region 4000–9000 Å TiO analyses are complete for (a) 12 bands of the α-system, (b) 4 bands of the β-system, (c) one band of the γ'-system, (d) 17 bands of the γ-system, (e) one band of the δ-system. Work is continuing on further TiO bands in this same spectral region, and attempts are being made to extend the observations farther into the photographic IR.

Nicholls has been continuing the publication of his very useful series of identification atlases of molecular spectra. In addition to the three atlases referred to in our previous report, he and his co-workers have published (a) the O_2 ($\beta^3\Sigma_u^-$–$X^3\Sigma_g^-$) Schumann-Runge system, and (b) the C_2 ($A^3\pi_g$–$X'^3\pi_u$) Swan system.

Recent extensions of astronomical observational capabilities into the UV and IR have created demands for laboratory data that are, at the present time, unavailable. Some of these needs are listed below, in the hope that, by calling attention to them, they will stimulate further laboratory work in these directions.

As an aid in the study of the IR spectra of late-type stars, Spinrad and Wing have called attention to the need of analysis of the IR system of VO. Furthermore, more laboratory work is needed on the diffuse bands of VO between 9530 Å and 9590 Å that were found by Lagerqvist and Selin to accompany the 1·06 μ bands in their laboratory emission spectrograms.

A higher resolution laboratory study is needed of the first overtone sequence of CH rotation-vibration bands in the 2·0–2·3 μ region.

The CN Red System should be observed with high resolution in the IR, notably the $\Delta v = -1$ and $\Delta v = -2$ sequences. The recent improvement of the rotational and vibrational constants of this sytem by Fay, Marenin and Van Citters (*Astrophys. J.*, in press) makes possible improved predictions of line locations in the IR, but even these predictions become uncertain for higher J.

Infrared ZrO bands are known that have not yet been classified. Afaf has suggested an analysis involving a system of five-headed bands, indicating that the system involves either triplet or quintet levels.

A feature attributed to HF has been seen at $2{\cdot}3\,\mu$ in the spectrum of α Ori. Data on the molecular equilibrium are lacking.

The polyatomic molecules HCN and C_2H_2 are being considered more and more frequently in connection with absorption features in stellar spectra in the 2 and $3{\cdot}1\,\mu$ regions. Such studies are hampered by lack of measures of these band profiles and intensities for lines with $J > 30$.

In the case of late-type stars, Vardya has called attention to the importance of certain negative molecular ions, notably OH^-, C^-, CN^-, SH^-, H_2O^-. The corresponding rotational and vibrational constants are largely unknown. Furthermore, the electron affinity of HS, C_2 and CN are quite uncertain. Various values in the literature differ by as much as 1 eV. The absorption X-sections for none of these ions are yet available.

Methane and ammonia are important in the spectrum of Jupiter, however, one still lacks data on the absorption profiles of these molecules at the temperatures, pressures and concentrations encountered in Jupiter. For instance, Gillett, Low, and Stein (*Astrophys. J.*, **157**, 1969, 926,) call attention to the need to know the amount of CH_4 required to reproduce the observed absorption in the $3{\cdot}1$–$4{\cdot}3\,\mu$ range. The existing laboratory measurements do not extend to sufficiently large CH_4 abundances. Recently, Margolis and Fox (*Astrophys. J.*, **15**, 1969, 935) and others have resolved and analysed the R-branch of the $3v_3$ band of CH_4 at 9050 cm^{-1}, and point out that "an important contribution to the determination of the Jovian methane temperature and abundance would be a measurement of the half-width of the singlet lines in the $3v_3$ band, and especially a determination of the line shape".

In the radio region of the spectrum there is still some uncertainty in the location of the CH interstellar line; see, e.g., W. Miller Goss (*Astrophys. J.*, **145**, 1966, 707). Fabry-Pérot interferometer investigations of the $A^2\Delta$–$X^2\pi$ band of CH at 4300 Å results in a interstellar line as consequence of ground-level Λ-doubling at 3030 Mc-s, while van Vleck theory leads to 3400 Mc/s.

In the far IR, the wings of bands of H_2O in its rotation spectrum appear to be too strong. For example, in the 1 mm to 1 cm region one must multiply by four the sum of all absorptions calculated according to the van Vleck formula. It is suggested that this may be caused by the formation of dimers. This is important in the estimation of H_2O abundance in Venus. More laboratory studies are needed of H_2O absorptions over a wider range of temperatures. Also important would be studies of line broadening of H_2O and NH_3 in presence of foreign atoms or molecules, e.g., N_2, CO_2, H_2, He.

The source of 26 diffuse features in the spectrum of interstellar matter is still unknown. The strongest is the well-known feature at 4430 Å. One of a number of suggestions is that they are produced by occluded molecular species on the surfaces of interstellar grains.

J. G. PHILLIPS
Chairman of the Committee

OTHER INFORMATION

Through Mrs Ch. Moore-Sitterly, we have received the following information:

From W. R. S. Garton:

At the Imperial College of Science and Technology, the main advance in the last few years has lain in the development of techniques enabling the extension of absorption spectroscopy of atomic vapors into the range 100–600 Å. The techniques developed have been reported in *Appl. Opt.*, **8** (1969), 919.

Analyses of the p-shell spectra of Na I, K I, Rb I are provisionally complete. The data on Ca I, Sr I and Ba I are being worked up. All the metal vapor spectra previously reported (back in the 1930's) by Beutler have been re-photographed using modern techniques and much greater dispersion. As a result, the spectra of Zn I, Cd I, and Hg I due to excitation of the d-shell and including transitions due to two-electron excitation have been extended. The Tl I spectrum has been greatly extended and the work of Beutler on Rb I and Cs I has been subject to particularly drastic revision. Jointly with F. S. Tomkins at Argonne, the diamagnetic Zeeman effect has been studied in Ba I (in press in *Astrophysical Journal*) and Sr I; extensive series spectra of Sc I and Y I have been observed in the Schuman UV. In collaboration with Goldberg, Parkinson and Reeves of the Harvard College Observatory, the absorption spectrum of an ion, La II, has been obtained from the middle UV to about 1200 Å. The analysis is in progress.

From R. Wilson:

The Astrophysics Research Unit of the Culham Laboratory, Abingdon, Berkshire (England) continued to be very active in spectroscopy. A report on 'Spectral Classification' prepared by R. Wilson, mainly deals with emission lines from Fe IX to Fe XVIII as well as isoelectronic sequences. Other information may also be found in a report to be published jointly by Science Research Council and the Royal Society, under the title 'Space Research in the United Kingdom 1968/69'.

From C. Newkirk, Jr:

Summaries of the work done at the High Altitude Observatory, Boulder, Colorado, U.S.A., have been published in the Annual Reports of this Institution. The papers of interest for Commission 14 appeared in *Astrophysical Journal*, except a publication by W. J. Wagner and L. L. House on 'A Survey of Current Coronal Visible Line Identifications' (*Solar Phys.*, **5**, 1968, 55–60). These authors have found that of the ~100 lines visible at eclipse only about 20 have undisputed identifications. Six new identifications are suggested, based on Wagner's combined laboratory and theoretical study.

At the suggestion of J. Phillips, A. Lagerqvist has reported to us thatat the Institute of Physics of the University of Stockholm more and more work is done in the UV vacuum region. Among other molecules, the spectra of metal hydrides and deuterides are investigated. One research group is studying diatomic spectra of the platinum metals, for example the spectra of PtH, PtD, PdD, PtC, RhC, RuC, IrC, PtO. Efforts are also made on the spectrum of CuO.

Concerning suggestions regarding areas of research activity requiring greater attention than is currently being provided, A. Lagerqvist mentioned (a) UV spectra, (b) Rydberg series, (c) spectra of ionized molecules, (d) intensity and life time measurements of spectral lines, especially in perturbed regions, (e) the mechanism of excitation, (f) hyperfine structure, (g) investigations of very strong perturbations, (h) transition probabilities.

At the Institut d'Astrophysique of the University of Liège (Belgium), B. Rosen and Mrs Denis-Gausset hope to publish in 1970 a new edition of the International Tables of Molecular Constants for diatomic molecules.

The determination of experimental potential functions of np $^1\Pi$ and np $^1\Sigma_u^+$ states in H_2, HD and D_2 has been the subject of a publication by A. Monfils (*Bull. Acad. roy. belg., Cl. Sc.*, **54**, 1968, 44).

Measurements on the C_3 4050 group excited in a pulsed low pressure discharge have furnished a first estimation of 300×10^{-9}s for the lifetime of the excited state of this transition (F. Remy, Dr. degree thesis, Univ. of Liège, Oct. 1967).

I. Dubois has studied the spectra of SO_2 (*J. molec. St.*, **3**, 1969, 269), SeO_2 (*Bull. Soc. r. Sci. Liège*, **11–12**, 1968, 562), TeO_2 (*Bull. Soc. r. Sci. Liège*, in press), SiH_2 (*J. Chem. Phys.*, **47**, 1967, 4262, in coll. with G. Herzberg and R. D. Verna, *Can. J. Phys.*, **46**, 1968, 2485).

M. V. MIGEOTTE

President of the Commission

15. PHYSICAL STUDY OF COMETS
(L'ÉTUDE PHYSIQUE DES COMÈTES)

PRESIDENT: L. Biermann.
VICE-PRESIDENT: V. Vanýsek.
ORGANIZING COMMITTEE: G. Herzberg, B. J. Levin, E. Roemer, F. L. Whipple, K. Wurm.

NARRATIVE REPORT

A. *Introduction*

The partition of the subject matter of this report follows that of earlier reports. Drafts for individual sections or parts of such were contributed by C. Arpigny, L. Biermann, A. H. Delsemme, O. V. Dobrovolsky, L. Haser, G. Herzberg, B. J. Levin, Rh. Lüst, E. Roemer, V. Vanýsek, and K. Wurm. The overall editing was done by Mrs Rh. Lüst, V. Vanýsek, and the undersigned.

Before mentioning some of the more important developments it is noted that E. Roemer has continued to publish every few months short notes on comets in the Publications of the A.S.P. The annual reports in *Quarterly Journal of the Royal Astronomical Society* have been continued by B. G. Marsden and J. G. Porter.

Several collections of Russian papers have been published which give a survey of the work done in the U.S.S.R. The description of physical characteristics of comets which appeared from 1961–1965 was continued by S. K. Vsehsvjatskij, who also completed a book on *Nature and Origin of Comets and Meteoritic Matter.*

Two new observatories, primarily for comet observations, are under construction in the U.S.S.R.; one at Lesniki near Kiev, the other, with a 70 cm classical paraboloid and a 40 cm refractor, at Hissar.

The Isophotometric Atlas of Comets, edited by W. Hoegner and N. Richter, appeared in 1969. Furthermore the *Atlas of Cometary Forms*, ed. by J. Rahe, B. Donn and K. Wurm, became available in early 1970. An excellent review on cometary tails was written by J. Brandt.

One of the more important developments in our knowledge of the physical parameters which characterize a comet concerns the determination of the gas output. The work of D. Malaise on pressure effects in the rotational band structure of CN in high dispersion spectra of several recent comets led to estimates of the total gas density at a distance from the nucleus of 10^4 km of the order of 10^7 to 10^8 molecules/cm^3, which corresponds to a total outflow of 10^{31} to 10^{32} molecules/s. This result is in substantial agreement with that of three earlier attempts to estimate the total gas output, by L. Biermann and E. Trefftz, by W. Huebner, and by F. Probstein and M. Finson, all of whom had by different methods arrived at similar figures. Though objections may be raised against each of these four methods, the order of magnitude agreement of the results leaves hardly room for any doubt about their essential correctness. This confirms the earlier suggestion that collisions may be important in the inner parts of the heads of comets. Furthermore, it will most probably be possible very soon to measure the hydrogen component of future comets spectroscopically by observing the $L\alpha$ picture by means of spacecraft or stabilized sounding rockets. If hydrogen, as had been proposed by F. Whipple already in 1950, is the most important atomic constituent (by numbers), the proposed measurement will measure also the total gas output.*

Another recent development concerns the lifetime of the parent molecules of CN and C_2. It seems now certain that these do not reach larger distances from the nucleus than approximately 10^4 km.

* **Note added in proof:** In January 1970, Comet Tago-Sato-Kosaka 1969 has been observed with OAO 2 and was found (as expected) to possess a very large hydrogen atmosphere.

B. *Photometry*

1. *Visual observations*

Visual estimation of cometary brightness is still a method which supplies relatively the largest amount of information on the behavior of the integral brightness of comets. Besides the observations which have been carried out for several decades by the well-known observer M. Beyer (1966, 1969), a considerable number of observations are recorded by groups of amateur observers, for instance, the Comet Section of the Association of Planetary and Lunar Observers in the U.S.A. Only a part of these observations are available through publications scattered in different journals and circulars.

The most comprehensive source of brightness and other physical data of comets was published recently by Konopleva *et al.* (1967, 1968).

A critical analysis of visual data of some recent comets was made by Meisel (1969b), who discussed different reduction methods. The Bobrovnikoff telescope-aperture correction, applied to the standard reduction formula, seems to be the most efficient simple reduction method. Meisel's (1969a) results for the periodic Comet Tempel 2 indicates no fast decrease of "absolute" magnitude and he concluded that all estimates of a secular decrease of cometary brightness may be suspect. The average "absolute" visual magnitude of P/Tempel 2 was 9.4^m in the years 1873–1894, in 1925 increased to $11{\cdot}3^m$, and in 1967 is again near $9{\cdot}4^m$.

Dobrovolskij and Osherov investigated the dependence of the visible brightness of comets on the geocentric distance and showed statistically that the inverse square dependence may be more nearly correct than an inverse linear dependence suggested recently by Öpik.

2. *Photographic photometry*

Integral photographic magnitudes were published by Garasdo-Lesnyh and Konopleva (1969) for Comet Kilston 1966 V, whose brightness was almost constant in the range $9{\cdot}6^m$ to $9{\cdot}9^m$, and the mean error of one measurement was $\pm 0{\cdot}2$ to $\pm 0{\cdot}3^m$.

The photographic methods are more widely used for the study of intensity distribution in the cometary head and tail.

The spectral gradient of the cometary continuum, according to a recent photographic spectrophotometric study by Pflug (1967) of Comet Arend-Roland 1957 III is close to that of K2V stars and depends slightly on phase angle.

Microphotometry of tails of Comet Ikeya-Seki 1965 VIII with accuracy of about 5% was carried out by Vsehsvjatskij *et al.* (1968) and utilized to study the tail kinematics. Nazarchuk (1969) published a similar analysis of the distribution of intensity in the tail of Comet Arend-Roland. The equidensity method for study of coma and tail structure based on the Sabbatier effect was successfully developed by Hoegner and Richter. Their *Isophotometric Atlas of Comets*, which was recommended by Commission 15 at the Prague Assembly in 1967, became available in late 1969.

3. *Photoelectric photometry*

Photometry, both in narrow bands as well as in the standard UBV system, was used more widely in recent years. In a routine photoelectric program using the UBV system, Mrkos *et al.* (1968) obtained photometric data for Comets 1966 V, 1967 XI, 1968 I, 1968 VI. The unusual intensity distribution in cometary spectra when emission bands are present has the result that the observed color of comets cannot be perfectly transformed to any conventional color system. The U filter covers only the CN $\Delta v = 0$ band, while the V filter includes the $\Delta v = 0$ and $\Delta v = +1$ band sequences of C_2. Generally, the $U-B$ color is more sensitive to the behavior of cometary spectra than the $B-V$.

Bappu and Sivaraman (1967) used five interference filters with a half-width of about 50Å for photoelectric measurements of emission bands as well as continuum. Two of these filters were for CN($\Delta v = 0$) and C_2($\Delta v = +1$) bands, and the continua regions were centered at $\lambda = 4310$ Å, 4860 Å (Hβ) and 5890Å. The measured fluxes were utilized to determine absolute abundances of CN and C_2 which were found to be $10^{29{\cdot}2}$ and $10^{30{\cdot}2}$ molecules in the observed region of the coma. Energy distribution in the cometary continuum was similar to that of G8V stars, which is in agreement with a reddening of the cometary continuum observed in almost every previous photometrically or spectrophotometrically studied comet.

Vanýsek (1969a) found an unusually low relative molecular abundance of C_2 with respect to CN in "dust rich" Comet Ikeya-Seki 1967n from monochromatic measurements in CN($\Delta v = 0$) and C_2($\Delta v = +1$) band sequences as well as in the continuum region near 4860Å. The paucity of accurate photoelectric observations of comets is due partly to the fact that it is very difficult to reconcile the needs of cometary photometry with those of stellar photometry, especially in the three-color UBV system.

Since the H_β and uvby system is more widely used in stellar photometry, the possibility of successful application of a "stellar" color system to cometary photometry increases. The b and H_β filters in the H_β + uvby system can be used without any change for measuring C_2($\Delta v = +1$) emission bands and continuum flux near the H_β wavelength. The region close to 4860 Å is not contaminated by molecular emissions. Therefore the H_β photometry and b color regions seem to be the best choice for the recommended basic "two color" system for future routine cometary photometry, i.e. the minimal number of spectral regions required for the photometric studies of comets. Such a simple two-color system could be very effective for observations of faint comets and especially of brightness outbursts when the contribution of dust and gas to a sudden increase of coma brightness cannot be determined from observation in only one spectral range.

C. *Spectroscopy*

No really bright comet, suitable for high-dispersion spectrographic observations, appeared during the period under review, so that the published work in this field as well as the present report are to a great extent concerned with the recent sun-grazing comet, Ikeya-Seki 1965 VIII.

At the Lick Observatory, spectra of this comet were taken by Preston (1967) before perihelion passage at a heliocentric distance $r \sim 0{\cdot}4$ AU and after perihelion passage at the closer distance $r \sim 0{\cdot}14$. The latter are, of course, daytime observations. The spectra for the larger distances contain the usual cometary emissions CN, C_2, CH, C_3. The spectra taken at $r \sim 0{\cdot}14$ are governed by many emission lines, the strongest ones belonging to Na I, K I, Fe I, Ni I, Cu I. CN and Ca II are also present. Because of the brightness of the night sky the cometary emissions are detectable on the plates only at and close to the cometary nucleus. The lengths of the lines of highest intensity (Na D-lines) do not extend beyond 25″ (= 17500 km) from the center of the nucleus. The numerous Fe lines are less than half as long. A curve-of-growth analysis of the Fe I spectrum leads to a formal excitation temperature of 4500 K, the line intensities of Na I and K I to only 2600 K. Attempts to determine relative abundances of the observed elements seem to indicate an anomalously low K/Na and a very high Cu/Fe ratio. However, these results are unfortunately not very significant because the fluorescence excitation of the observed emissions has not been properly taken into account. Both the relative and the absolute number densities will have to be revised.

A similar analysis has been carried out by Slaughter (1969), who gives a very extensive list of line intensities for several points in the head and tail of Comet Ikeya-Seki just prior to and after perihelion from 1·1 mmA^{-1} spectra taken with the Kitt Peak solar telescope. Spinrad and Miner (1968) studied the structure of the Na D-lines of Comet Ikeya-Seki on high dispersion spectrograms (5 mmA^{-1}) obtained at the same period and with the same instrument. Since the heliocentric radial velocity was −187 km/sec, the comet's resonance wavelengths were shifted into the full solar continuum near λ5890. The spectrograms were secured with the slit through the nucleus at several position angles. The Na D-line emission spreads in the tail direction to some 100000 km, and normal to the comet-sun line to about 40000 km from the nucleus. The emission was most intense at and on the sunward side of the nucleus, but decreased rapidly in this direction. The Na coma must have had a shape similar to the one obtained theoretically from a fountain type model for a particle suffering a high repulsive force. However, this simple Na fountain model cannot be correct, since the lifetime of the neutral Na atoms until ionization was certainly much too short to reach the observed distances from the nucleus. The apparent fountain model shape must have been brought about by a precursor, probably small grains, which spread from the nucleus and released Na atoms not earlier than after a few hours. This idea of the authors is very probably correct.

Measurements of the polarization of the Na D2-line in Comet Ikeya-Seki made by Hyder (1966) were originally interpreted by him as due to a magnetic field which is dragged along with the comet. A more detailed discussion by Chamberlain (1967) revealed, however, that cometary observations of this type should not be affected by magnetic fields of such a small strength as would be expected here.

On the basis of high dispersion CN spectra of several rather bright recent comets Malaise (1970) tried to arrive at a refinement of the physics of cometary head atmospheres for heliocentric distances between $r = 1$ and $r = 0{\cdot}5$ A.U. Synthetic model structures of the (0, 0) band of CN constructed by taking into account all conceivable physical factors influencing intensities and profiles of the lines are compared with structures observed for one and the same comet close to the nucleus and at larger distances from it. According to the author there exist, in addition to the heliocentric distance, five other variable physical parameters which determine the finer shape of the band branches; of these the total gas density and the gas temperature are the most important ones. The numerical values of the parameters of the eight cases reported in the paper, which are internally consistent, were determined by attempting the best fit between the synthetic and the observed spectra. The most important result of the analysis is that the collision zones of the neutral coma atmospheres, where the rotation of the molecules is influenced by particle collisions and in which we might therefore expect bimolecular reactions, are much further extended than has hitherto been assumed by most authors. Concerning the total densities the author estimates for a distance $D_0 = 10000$ km from the nucleus minimum values between 10^7 and 10^8 cm^{-3}. Despite the careful work of the author and the excellent spectra used doubts may, nevertheless, remain whether the results are actually correct. Cometary heads are no neutral atmospheres as assumed, and it may well be and it is even probable that the more or less continuous stream of visible and invisible ions from the head through the whole coma is more effective in producing collisions than that of the neutrals among themselves. (See also section on type I tails.) It is accepted that such collisions are in fact necessary to explain the spectra. However, if one admits minimum densities of 10^7 to 10^8 cm^{-3} at $D_0 = 10000$ km, it is difficult to understand how narrow ion streamers with an observed origin close to the nucleus remain undisturbed throughout the whole coma despite the moderate flow velocity within them.

D. *Structure and origin of the nucleus*

Seemingly fundamental criticisms by Lyttleton of the reality of the Oort cloud of comets as a maximum in the space density of orbital aphelia have been answered by Le Poole and Katgert (1968), who point out that this is an incorrect interpretation of Oort's ideas. Sekanina, in two papers in *Bulletin of the Astronomical Institute of Czechoslovakia* investigates the perturbations of nearby stars on the comet cloud, extending earlier work by Makover, Steins and Sture, and others. He concludes from detailed study of the sphere of action of the solar system that, due to perturbations by α Cen, the Oort cloud would not remain under the influence of the solar system for extended periods of time. Encounters of interstellar comets with fast-moving stars are considered in the second paper. Vsehsvjatskij (1968) also sees difficulties in persistence of the comet cloud because of many close stellar approaches since the origin of the solar system.

Everhart (1967) has attempted to derive the intrinsic distributions of several orbital elements of long-period comets and the distribution of their absolute brightnesses from a more detailed study than has been attempted previously of effects of observational selection on discoveries.

Hamid *et al.* (1968) have concluded from an investigation of secular perturbations of the orbits of a number of short-period comets that a belt of comets exceeding one earth mass within a distance of 50 A.U. from the sun is excluded.

Recent findings regarding nongravitational effects in the motion of comets appear to modify substantially the entire picture of hyperbolic original and future orbits (Sekanina, 1967a).

With definitely observable nongravitational effects on the motions of many periodic comets now convincingly established (Marsden, 1968; Sekanina, 1968b), investigation has turned to analysis of the character of such effects. Marsden (1969) finds that a continually acting force, decreasing in intensity with time, explains the greater part of the deviations from gravitational motion. A radial component, almost invariably outward from the sun, generally dominates. The transverse component is about an order of magnitude weaker, while there appears to be no significant force component perpendicular to the orbit plane. There is some evidence of additional effects, presumably random and impulsive in nature. Sekanina has developed the model of nongravitational effects associated with a rotating nucleus, an idea first suggested by Whipple in 1950, in a series of seven papers in *Bulletin of the Astronomical Institute of Czechoslovakia* under the title "Nongravitational Effects in Comet Motions and a Model of an Arbitrarily Rotating Comet Nucleus". Sekanina suggests that splitting of the nucleus can be caused by an explosive ejection of material from a fragile nucleus, the fracturing being simultaneous with an impulsive effect on the orbital motion. In a few cases nongravitational forces have been large enough to produce clearly observable effects during a single perihelion passage. An outstanding example is Comet Burnham 1960 II. There also appear to be anomalous nongravitational effects associated with close approaches of individual comets to Jupiter. Marsden and P. Stumpff are working on this problem.

The magnitude of nongravitational effects on individual comets seems to be correlated with their apparent diffuseness. The fact that the motion of comets of asteroidal appearance can be represented gravitationally has led Marsden (1969) to suggest that such objects represent a transition phase between a comet and an asteroid of the Apollo type. This suggestion is particularly interesting in view of the fact that Monte Carlo calculations by Wetherill, aimed at identification of the source(s) of meteorites, seem to be leading to similar ideas regarding a relationship between Jupiter-family comets and Apollo asteroids (Wetherill, 1968). Öpik (1968) also has further developed his ideas on the cometary origin of meteorites, suggesting that fragments resulting from planetary collisions have been embedded in the ices of cometary nuclei.

Studying the origin of the Kreutz group of sungrazers, Marsden (1967) has presented considerable evidence that the sungrazers fall into two very compact groups, defined by strikingly similar orbital elements. He suggests that each group is composed of fragments of a comet disrupted only one or two revolutions ago, probably by tidal effects near perihelion. These parent comets may derive from disruption of a very large comet in the more remote past. Sekanina (1967b) considers that collisions of a massive proto-comet with a "cosmic projectile" at some distance from perihelion is a preferable hypothesis, with subsequent evolution of the dynamical characteristics of the group of sungrazers through less violent means.

Harwit (1968) has pointed out that splitting of all the comets studied by Stefanik occurred close to the ecliptic plane. Sekanina has studied splitting and impulsive perturbations of orbital motion both for individual events and as part of his more general studies of nongravitational effects. Objects studied include the sungrazer Ikeya-Seki 1965 VIII, P/Biela, and Comet Wirtanen 1957 VI (Sekanina, 1968a).

Roemer (1968) has found that the populations according to nuclear radius of periodic and near-parabolic comets are similar, the cumulative number of objects with radius larger than R being represented in the form $N = N{\cdot}R^{-s}$, with $s = 1{\cdot}5$ in each case.

Markovich and Tulenkova (1968), solving the heat conduction equation numerically, conclude that in the 30 days including perihelion passage, the sungrazing comet Ikeya-Seki must have lost mass amounting to 10^{14} g. Sekanina's recent work also includes theoretical calculations of heat and mass transfer in model comet nuclei as well as an attempt to reconcile directly observed mass loss rates of C_2 with the total mass loss rates derived dynamically for representative short-period comets.

E. *The coma*

Study of the heads of comets in monochromatic light has led to new data determining the mean paths τv (lifetimes $\tau \times$ ejection velocities v) of the observed radicals, as well as their hypothetical precursors (Miller, 1967; Vanýsek and Zacek, 1967; Vanýsek, 1968a and 1969a). C_2 and CN were often measured, C_3 and CH more rarely, the other radicals practically never; this is a pity for it neglects major constituents whose oscillator strength is weak, like OH. The observations by and large confirm the former value of the lifetimes of the radicals decaying in the solar field. The dependence of the total brightness of a given radical on heliocentric distance is not very well known. A good determination for Comet Rudniczki 1967 II (Mayer and O'Dell, 1968) shows that the desorption heat as defined by Levin's law is lower for C_3 than for C_2 or CN. However, the behavior often seems erratic because of outbursts.

More attention has been given to the lifetimes of the assumed precursors. Some knowledge has been gained by trying to fit the brightness profiles with Haser's model and selecting the best ratio for the radical/precursor lifetimes. For Comet Everhart 1964 IX, the best fit for this ratio is 10, but 5 is almost as good (Vanýsek and Zacek, 1967). The ratio seems to be the same for C_2, CN, and C_3. Malaise (1968) uses a simplified procedure on Comets Burnham 1960 II and Ikeya 1963 I, because his observations exclude the outer coma; the dependence of the precursors' mean paths on the heliocentric distance suggests that photodissociation can produce C_2 but not CN nor C_3; his study of the population of the rotational levels emphasizes collisions in the coma.

Upper limits of the precursors' lifetimes, often near 10^4 sec whatever the neutral radical concerned, are reached by other methods (Wurm and Vanýsek, 1969; Vanýsek, 1969b). By contrast, the upper limit for the CO^+ precursors' lifetime, the only ion seriously discussed, is much shorter. Wurm and Mammano (1967), Wurm (1968), Rahe (1968), Rahe and Wurm (1969), Rahe and Donn (1969) have accumulated more evidence of this fact, which has not yet been satisfactorily explained by any process. Jackson and Donn (1968), in a comprehensive analysis, conclude that photodissociation is adequate to explain observed steady state densities of radicals observed near the nucleus of a comet. However, none of the identifications of any hypothetical precursor with actual molecules has yet been successful in explaining lifetimes. The discrepancy is still one full order of magnitude. New unstable molecules could of course still be identified with precursors.

Another way of avoiding this difficulty is to build a model of the coma where the precursors are large ice grains stripped from the nucleus and evaporating in the solar field (Delsemme and Miller, 1969). Contrary to an earlier approach by Huebner and Weigert, a large optical depth which drastically limits the particle size is not postulated here. The shape of photometric profiles of both inner coma and continuum can be explained by radicals evaporating from a clathrate lattice in the grain. The presence of solid grains in the coma is now receiving much attention. Finson and Probstein (1967) as well as Shulman (1969) applied a gasdynamic approach to a flow in the nuclear region and showed that dust particles will be accelerated to velocities up to several hundreds of meters per second.

The continuum in the head and its polarization have been compared with theoretical size distributions of particles (Donn *et al.*, 1967). The polarization observed is too low for iron spheres but can be explained by dielectric grains. The bulk of cometary grains cannot have sizes larger than a few microns without requiring excessive mass, but a fraction of larger particles is not ruled out. An increase in the ratio of the intensities of the continuum/molecular emissions has been observed near inferior conjunction; Vanýsek (1968b) suggests that it comes from a forward scattering effect of the dust halo, which originates from particles whose size is at least as large as λ. Krishna Swamy and Donn (1968) study grain temperatures; the results are not corrected for the latent heat of evaporation at small heliocentric distances. Kaimakov and Sharkov (1967a) note that the escape velocity of dust grains is governed by temperature only, and depends little on the grain size. Dolginov (1967) and Egybekov (1969) study the formation of grains in the coma. To explain the decrease of brightness after a perihelion passage, Huebner (1967) must use a reduction of the sublimation rate.

At small heliocentric distances, other grains less volatile than ices are likely. Huebner (1970) studies the vaporization of sodium grains at short heliocentric distances. The Kovars (1968) show that the anomalous ratio of the Na I D doublet cannot come from a fluorescent mechanism only. They study (1969) the problem of the lifetime of Na I. Cherednichenko (1968) shows that the [O I] red line might partly come from dissociative recharge of ions O_2^+ with Na; the green line from dissociative recombination of ions NO^+ with electrons. The asymmetry of the coma is too often neglected. Miller (1967) measures the oblate form of the C_2 coma of 1960 II; he shows (1969) its perfect symmetry in Comet Honda 1968 VI, as opposed to the asymmetry of dust ejection. This obvious non-interaction of gas and dust at larger distances from the nucleus is satisfactorily explained by Finson and Probstein's theory (1968); the drag forces occur only in the inner head.

The hydrogen atmosphere of comets to be expected on the basis of current concepts of their chemical constitution and gas production has been studied by Biermann (1968) in order to provide estimates for the extent of the hydrogen coma, which might be observed in the light of $L\alpha$ by means of stabilized sounding rockets or space probes (cf. sections A and J).

F. *Type I tails*

No new comets with conspicuous plasma tails were observed over the last three years. In the tail spectrum of Comet Ikeya-Seki 1965 VIII the atoms dominated as compared to ions including molecular ions. The interpretation of the ion features near perihelion was discussed by Lüst and Schmidt (1968).

Plates of Comet Morehouse were rediscussed by several authors with the aim to determine certain parameters with higher precision than before or to investigate the compatibility with recent theoretical work. Lüst (1967) has measured the motion of envelopes in close distance of the head and discussed the possibility that these features are identical with the contact surfaces between cometary and solar wind plasma postulated by theory. Also, velocities of tail particles were deduced. Stumpff (1967) compared these findings with values in greater distance of the head, made by Wolf, and deduced a velocity field of the streaming tail material. Also Schlosser (1968) derived velocities and accelerations of tail structures. He found a decreasing acceleration with increasing distance from the nucleus, and no dependence of the velocities on the distance from the tail axis in contrast to other authors. Several other papers by Schlosser (1966, 1967), and Schlosser and Hardorp (1968) deal with the evolution of the wavy tail rays observed in this comet. From the wavy structure, a directional fluctuation of the solar wind was postulated. The bulk velocity of the solar wind in high heliographic latitudes (+46°) deduced from the aberration angles of the tail axes agrees with satellite measurements near the ecliptic plane.

Brandt's work on the statistics of the direction of cometary plasma tails as indicators of the direction of the solar wind flow was continued. In his first paper (Brandt, 1967), he compared the values for the bulk velocity of the solar wind derived from a great number of observations of tail aberration angles from different comets with values from the Mariner II mission and from geomagnetic storms. All values are in good agreement. Further, evidence was found for a mean tangential component of the solar wind velocity near the earth of ~9km/s directed in the sense of solar rotation. – A curious dependence of the solar wind velocity on heliographic latitude was found by Pflug (1966). From similar statistical material he derived that the solar wind velocity has a minimum between heliographic latitudes 20° to 30°, i.e. above the solar activity belts. According to Brandt, these findings might, however, be influenced by a biased sample of comet observations. Dispersions in the tail orientation were interpreted as being caused by discontinuities in the interplanetary medium by Brandt and Heise (1969). The authors confirmed the existence of a mean azimuthal plasma velocity of 9 km/s in the sense of co-rotation; this velocity component was furthermore found to be largest at periods of high solar activity.

Theoretical work on the interaction of the tail plasma with the solar wind was continued by Ioffe (Tadzhik State University), by Biermann, Brosowski, Meyer and Schmidt (Munich), and by Wallis (Manchester). The approach in most cases was the use of magneto hydrodynamics without explicit introduction of the magnetic field. Ioffe (1967) represented the comet by a line source of plasma, and the observed accelerations were related to the field of flow assuming approximate equality (within a factor of 2) of the magnetic tension and the pressure at the stagnation point, Wallis (1967, 1968) gave particular attention to a pure CO comet, whereas the work done in Munich was based on the concept of distributed sources using the picture developed earlier by Biermann and Trefftz (Biermann *et al.*, 1967; Brosowski and Schmidt, 1967).

Nazarchuk and Shulman (1968) developed a diffusion model of a plasma tail which was checked observationally and used for the estimate of tail plasma characteristics on the basis of brightness distribution in the tails of Comets Arend-Roland 1957 III and Mrkos 1957 V.

The determinations of the total gas production of a comet discussed in the introduction are of course basic to future discussions of the comet – solar wind interaction.

A new line of approach has become possible from the measurements of the rate constant of the dissociative recombination of CO^+ (see the section on laboratory work). From this it became possible to give a lower limit for the distance from the nucleus, at which the CO^+ is being formed, and an upper limit for the electron collision rate which turns out to be somewhat smaller than that of molecular collisions (Biermann, 1970).

From a list of 89 comets observed photographically since 1880 with information about type and length of tail, brightness, spectrum etc., Vsehsvjatskij (1969) tried to correlate the maximum tail length to perihelion distance, from which he drew conclusions about the composition of the tails. New classification criteria were proposed.

A summary of our knowledge concerning the structure of type I tails was presented by Wurm (1968).

G. *Type II and III tails*

Since Bredikhin's time type II tails of comets were usually regarded as syndynames with $1-\mu\sim 1$ to 2, it was difficult to combine this notion with the assumption that these tails are composed of dust. Therefore a hypothesis that they are composed of neutral gas had been put forward by Levin. The discovery of the presence of atomic oxygen in comet heads increased by several orders of magnitude the estimates of gas densities in the heads and of the rate of evaporation of the nucleus (see Introduction and section on cometary heads).

Dobrovolskij *et al.* (1966) reported their measurements of the position of the tail axis of Comet Arend-Roland 1957 III, from which they concluded that the tail was not a syndyname or a synchrone, but was produced by a continuous emission of particles of various dimensions. Clearcut evidence was not presented in this note. Later Finson and Probstein (1968) came to the same conclusion from the analysis of isophotes of the tail of this comet. They definitely showed that this dust tail was produced by a continuous but peaked emission of dust particles with various values of $1-\mu$. Peaks of dust emission produce type III tails which in case of a small rate of continuous emission cannot be superimposed on or combined with type II tails.

Thus it became clear that for previous comets reported to have type II tails, there is no convincing evidence that these were syndynames with $1-\mu\sim 1$ to 2. The nature of so-called "synchronic bands" sometimes observed in type II tails remains to be explained.

Dolginov (1968) showed that interplanetary dust particles can acquire rapid rotation and orientation under the influence of the solar wind. Such orientation can also take place in dust tails of comets and is manifested in their light scattering properties.

H. *Laboratory work*

Physicochemical studies of the evolution of gases and dust from comet nuclei have been handicapped by deficient knowledge of the behavior of possible cometary materials at low temperatures (50–200 K) and pressures. Efforts to fill this gap experimentally are being made by Kaimakov and Sharkov (1967b), who have studied water ice, and by Delsemme and Miller who are measuring vapor pressures of CH_4, NH_3 and their hydrates (see section on coma).

Furthermore a good deal of work has been done in various laboratories on the spectra of diatomic molecules which are of interest in connection with cometary spectra. These include CH (Herzberg and Johns, 1969), CH^+ (Carré, 1969), NH (Murai and Shimauchi, 1966; Horani *et al.*, 1967), NH^+ (Colin and Douglas, 1968). OH (Czarny and Felenbok, 1968), OH^+ (Merer *et al.*,

1969), SiH (Herzberg *et al.*, 1969), SiH^+ (Douglas and Lutz, 1970), C_2 (Phillips and Davis, 1968; Lagerqvist *et al.*, 1970), AlO (Mahieu and Bécart, 1968; McDonald and Innes, 1970). The transition probabilities for the Swan and Mulliken C_2 bands were newly determined by Smith (1969). In addition, a good amount of photoionization work has been reported, for example, on CO, which is obviously of considerable interest in connection with the question of the formation of the ionized molecules in comets. The rate constant of dissociative recombination of CO^+ has been measured by Mentzoni and Donohoe (1969) (for preliminary conclusions see section on type I tails). The ionization potential of CH has been determined from a Rydberg series to be 10·64 eV. Finally, experiments have been carried out on the fluorescence of a number of polyatomic molecules when excited in the far ultraviolet. For example (see Judge *et al.*, 1970), CH_4 gives in fluorescence the spectrum of CH, while C_2H_2 gives both CH and C_2 spectra by absorption of light quanta in the region below 900Å.

Furthermore laboratory physics of comets is being carried out at the Ioffe Physical and Technical Institute at Leningrad under the guidance of Konstantinov. The sublimation rates, heats, and temperatures have been measured for pure H_2O when mixed with dust. The experiments show that an icy nucleus may remain solid at heliocentric distances much less than 1 A.U. At $T > -50\,°C$ a hotbed effect was observed. The formation of a dust matrix and the dynamics of the dust particle outflow have been investigated.

Simulation experiments to study the flow in the coma of dust comets have been carried out by Brunner and Michel (1968). Atomic resonance absorption spectroscopy in a density range where dipole interactions determine the line widths was experimentally shown to be a useful tool for gas dynamic studies (Brunner, 1969). The experiments by Danielsson (1966) for a simulation of the solar wind interaction with comets have been continued.

J. *Space experiments*

During the period covered by this report several theoretical studies were prepared concerning rendez-vous-missions to a comet. (NASA-reports by Brereton *et al.*, 1967; Friedlander, 1967; Eades, 1968). The feasibility of such a mission has been proved, but high costs and the lack of a suitable comet have so far prevented the execution of such projects. A review about plans for a cometary mission is given by Lüst (1969).

Sounding rocket experiments prepared for the observation of the UV spectrum of comets were launched during the appearance of Comet Ikeya-Seki 1965 VIII, but no results have yet been reported. In any case this object would not have represented a typical comet because of its sungrazing orbit.

In several institutions sounding rocket experiments are being prepared. Spectrographic equipment combined with stabilizing and pointing units are designed mainly to study the spectrum between 1100Å and 3500Å with resolutions of about 30Å. In at least one such experiment, attention is given to the hydrogen Lα emission (cf. sections A, note added in proof, and E).

A barium plasma cloud experiment was performed by the Institute for Extraterrestrial Physics, Garching, in the magnetosphere at a distance of 12·5 earth's radii to study the interaction with the dilute magnetospheric plasma (Haerendel and Lüst, 1970).

REFERENCES

A. *General references mentioned in the introduction*

Roemer, E. 1967–1969, *P.A.S.P.* **79**, 81, 187, 272, 372, 503, 632; **80**, 106, 239, 345, 487, 628, 752; **81**, 79, 172, 289, 448, 700.
Marsden, B. G., Porter, J. G. 1967–1969, *Q. J. RAS* **8**, 274; **9**, 304; **10**, 240.
Konopleva, W. P. 1967, *Active Processes in Comets*, Naukova Dumka, Kiev.
Konopleva, W. P. 1969, Astrometry and Astrophysics, Vol. IV: *The Physics of Comets*, Naukova Dumka, Kiev.

Dobrovolsky, O. V., Ed. 1967, *Comets and Meteors*, Acad. of Sciences Tadjik SSR, Dushanbe.
Vsehsvjatskij, S. K., Ed. 1968–69, *Problems of Cosmical Physics*, No. **3** and No. **4**, Isdat. Kiev. University.
Vsehsvhatskij, S. K. 1966, *Astr. Z. U.S.S.R.* **43**, 1292.
Vsehsvjatskij, S. K. 1967, *Nature and Origin of Comets and Meteoric Matter* .
Hoegner, W., Richter, N. B. 1969, *Isophotometric Atlas of Comets*, J. A. Barth, Leipzig.
Rahe, J., B. Donn, K. Wurm 1970, *Atlas of Cometary Forms*, NASA SP–198.
Brandt, J. C. 1967, *A. Rev. Astr. Astrophys.*, **6**, 267.

B. *Papers mentioned in the text*

Bappu, M. K. V., Sivaraman, K. R. 1967, *Mon. Not. R. astr. Soc.*, **137**, 151.
Beyer, M. 1966, *Astr. Nachr.*, **289**, 173.
Beyer, M. 1969, *Astr. Nachr.*, **291**, 239, 257.
Biermann, L. 1968, JILA Report No. 93.
Biermann, L. 1970, *Sber. Bayer. Akad. Wiss.*, in press.
Biermann, L., Brosowski, B., Schmidt, H. U. 1967, *Solar Phys.*, **1**, 254.
Brandt, J. C. 1967, *Ap. J.*, **147**, 201.
Brandt, J. C., Heise, J. 1969, Preprint, to be published.
Brereton, R. G., Newburn, R. L., Giffin, Ch. E., Neugebauer, M. M., Smith, E. J. *et al.* 1967, Mission to a Comet, NASA-Cr-83113.
Brosowski, B., Schmidt, H. U. 1967, *Z.A.M.M.*, **47**, T140.
Brunner, W., Michel, K. W. 1968, *Mitt. astr. Ges.*, **25**, 220.
Brunner, W. 1969, *Z. Naturforschg.*, **24a**, 1063.
Carré, M. 1969, *Physica*, **41**, 63.
Chamberlain, J. W. 1967, *Ap. J.*, **149**, 455.
Cherednichenko, V. I. 1968, *Probl. Cosm. Phys. Kiev*, **3**, 30.
Colin, R., Douglas, A. E. 1968, *Can. J. Phys.*, **46**, 61.
Czarny, J., Felenbok, P. 1968, *Ann. Astrophys.*, **31**, 141.
Danielsson, L. R. 1966, *Appl. Phys. Letters*, **9**, 339.
Delsemme, A. H., Miller, D. C. 1969, *Bull. Am. astr. Soc.*, **1**, 339.
Dobrovolskij, O. V., Faiziev, E., Ibadinov, K. H. 1966, *Dokl. Acad. Nauk. Tadjik. SSR*, **9**, 14.
Dolginov, A. Z. 1967, *Astr. Z. U.S.S.R.*, **44**, 434.
Dolginov, A. Z. 1968, *Dokl. Acad. Nauk. S.S.S.R.*, **179**, 1070.
Donn, B., Powell, R. S., Remy-Battiau, L. 1967, *Nature*, **213**, 379.
Douglas, A. E., Lutz, B. L. 1970, *Can. J. Phys.*, in press.
Eades, J. B. 1968, A Survey of Comet Missions, NASA-TM-X 63295.
Egybekov, P. 1969, *Probl. Cosm. Phys. Kiev*, **4**, 78.
Everhart, E. 1967, *Astr. J.*, **72**, 716, 1002.
Finson, M. L., Probstein, R. F. 1967, A Theory of Dust Comets, MIT Fluid Mech. Lab. Preprint No. 67-4 (see also Probstein, R. F. in *Collection Dedicated to the 60th Birthday of Academician L.I. Sedov*, Nauka Moscow, 1969, p. 397).
Finson, M. L., Probstein, R. F. 1968, *Ap. J.*, **154**, 327, 353.
Friedlander, A. L. 1967, On the Problem of Comet Orbit Determination for Space Craft Intercept Missions, NASA-CR-84848.
Garasdo-Lesnyh, G. A., Konopleva, V. P. 1969, *Astrometry Astrophys. Kiev*, **4**, 174.
Haerendel, G., Lüst, R. 1970, in *Particles and Fields in the Magnetosphere*, Ed. B. M. McCormac, D. Reidel Publ. Co., Dordrecht, Holland, p. 370.
Hamid, S. E., Marsden, B. G., Whipple, F. L. 1968, *Astr. J.*, **73**, 727.
Harwit, M. 1968, *Ap. J.*, **151**, 789.
Herzberg, G., Johns, J. W. C. 1969, *Ap. J.*, **158**, 399.
Herzberg, G., Lagerqvist, A., McKenzie, B. J. 1969, *Can. J. Phys.*, **47**, 1889.
Horani, M., Rostas, J., Lefebvre-Brion, H. 1967, *Can. J. Phys.*, **45**, 3319.
Huebner, W. F. 1967, *Z. Astrophys.*, **65**, 185.
Huebner, W. F. 1970, *Astr. Astrophys.*, in press.
Hyder, L. C. 1966, *Ap. J.*, **146**, 748.
Ioffe, Z. M. 1967, *Astr. Z. U.S.S.R.*, **44**, 835, 1304.
Jackson, W. M., Donn, B. 1968, *Icarus*, **8**, 270.

Judge, D. L. *et al.* 1970, to be published.
Kaimakov, E. A., Sharkov, V. S. 1967*a*, *Astr. Z. U.S.S.R.*, **44**, 682.
Kaimakov, E. A., Sharkov, V. S. 1967*b*, *Bjull. Kom. Komet. Meteor.*, no. 15, 16 and 21.
Konopleva, V. P., Garasdo-Lesnyh, G. A., Sereda, E. M. 1967, *Active Processes in Comets*, Kiev, p. 100.
Konopleva, V. P., Garasdo-Lesnyh, G. A., Sereda, E. M. 1968, *Prob. Cosm. Phys. Kiev*, **3**, 112.
Kovar, N. S., Kovar, R. P. 1968, *Solar Phys.* **3**, 611.
Kovar, N. S., Kovar, R. P. 1969, Amer. Astr. Soc., Spring Meeting.
Krishna Swamy, K. S., Donn, B. 1968, *Ap. J.*, **153**, 291.
Lagerqvist, A., Malmberg, C., Herzberg, G. 1970, *Can. J. Phys.*, **47**, in press.
LePoole, R. S., Katgert, P. 1968, *Observatory*, **88**, 164.
Lüst, Rh. 1967, *Z. Astrophys.*, **65**, 236.
Lüst, Rh. 1969, *Space Sci. Rev.*, **10**, 217.
Lüst, Rh., Schmidt, H. U. 1968, *Mitt. astr. Ges.*, **25**, 211.
McDonald, J. K., Innes, K. K. 1970, *J. Mol. Spec.*, in press.
Mahieu, J. M., Bécart, M. 1968, *Can. J. Spec.*, **13**, 95.
Malaise, D. 1968, Thesis, University of Liège.
Malaise, D. 1970, in press.
Mayer, P., O'Dell, C. R. 1968, *Ap. J.*, **153**, 951.
Markovich, M. Z., Tulenkova, L. N. 1968, *Probl. Cosm. Phys. Kiev*, **3**, 25.
Marsden, B. G. 1967, *Astr. J.*, **72**, 1171.
Marsden, B. G. 1968, *Astr. J.*, **73**, 367.
Marsden, B. G. 1969, *Astr. J.*, **74**, 720.
Meisel, D. D. 1969*a*, *P.A.S.P.*, **81**, 65.
Meisel, D. D. 1969*b*, in press.
Mentzoni, M. H., Donohoe, J. 1969, *Can. J. Phys.*, **47**, 1789.
Merer, A. J., Achter, E. K., Horani, M., Rostas, J. 1969, *Can. J. Phys.*, **47**, 1723.
Miller, F. D. 1967, *Astr. J.*, **72**, 487.
Miller, F. D. 1969, in preparation.
Mrkos, A., Tremko, J., Vanýsek, V. 1968, *Bull. astr. Inst. Csl.*, **19**, 43.
Murai, T., Shimauchi, M. 1966, *Sci. Light*, **15**, 48.
Nazarchuk, G. K. 1969, *Astrometry Astrophys. Kiev*, **4**, 77.
Nazarchuk, G. K., Shulman, L. M. 1968, *Probl. Cosm. Phys. Kiev*, **3**, 11.
Öpik, E. 1968, *Irish astr. J.*, **8**, 185.
Pflug, K. 1966, *Publ. Obs. Potsdam*, **31**, No. 4.
Pflug, K. 1967, *Astr. Nachr.*, **290**, 89.
Phillips, J. G., Davis, S. P. 1968, *The Swan System of the* C_2 *Molecule*, Univ. of California Press.
Preston, G. W. 1967, *Ap. J.*, **147**, 718.
Rahe, J. 1968, *Z. Astrophys.*, **68**, 208.
Rahe, J., Donn, B. 1969, *Astr. J.*, **74**, 256.
Rahe, J., Wurm, K. 1969, in preparation.
Roemer, E. 1968, *Astr. J.*, **73**, S33.
Schlosser, W. 1966, Thesis, Hamburg.
Schlosser, W. 1967, *Z. Astrophys.*, **66**, 421; **67**, 321.
Schlosser, W. 1968, *Astr. Nachr.*, **291**, 1.
Schlosser, W., Hardorp, J. 1968, *Z. Astrophys.*, **69**, 228.
Sekanina, Z. 1967*a*, *Bull. astr. Inst. Csl.*, **18**, 369.
Sekanina, Z. 1967*b*, *Publ. astr. Inst. Charles Univ.*, No. 51.
Sekanina, Z. 1968*a*, *Bull. astr. Inst. Csl.*, **19**, 153.
Sekanina, Z. 1968*b*, *Bull. astr. Inst. Csl.*, **19**, 351.
Shulman, L. M. 1969, *Astrometry Astrophys. Kiev*, **4**, 101.
Slaughter, C. D. 1969, *Astr. J.*, **74**, 929.
Smith, W. H. 1969, *Ap. J.*, **156**, 791.
Spinrad, H., Miner, E. D. 1968, *Ap. J.*, **153**, 355.
Stumpff, P. 1967, *Z. Astrophys.*, **67**, 99.
Vanýsek, V. 1968a, *Acta Univ. Carol.*, No. 1/1968, 53.
Vanýsek, V. 1968*b*, *Icarus*, **8**, 510.
Vanýsek, V. 1969*a*, *Bull. astr. Inst. Csl.*, **20**, 355.
Vanýsek, V. 1969*b*, *P.A.S.P.*, in press.

Vanýsek, V., Zacek, P. 1967, *Acta Univ. Carol.*, no. 2/1967, 85.
Vsehsvjatskij, S. K. 1968, *Probl. Cosm. Phys. Kiev*, **3**, 98.
Vsehsvjatskij, S. K. 1969, *Astr. Z. U.S.S.R.*, **46**, 435.
Vsehsvjatskij, S. K., Demenko, A. A., Harpach, W. T. 1968, *Probl. Cosm. Phys. Kiev*, **3**, 57.
Wallis, M. K. 1967, *Planet. Space Sci.*, **15**, 1407.
Wallis, M. K. 1968, *Planet. Space Sci.*, **16**, 1221.
Wetherill, G. W. 1968, in *Origin and Distribution of the Elements*, Ed. L. H. Ahrens, Pergamon Press, New York and London.
Wurm, K. 1968, *Icarus*, **8**, 287.
Wurm, K., Mammano, A. 1967, *Icarus*, **6**, 281.
Wurm, K., Vanýsek, V. 1969, *Icarus*, in press.

L. BIERMANN
President of the Commission

16. PHYSICAL STUDY OF PLANETS AND SATELLITES
(ÉTUDE PHYSIQUE DES PLANÈTES ET SATELLITES)

PRESIDENT: J. S. Hall.
VICE-PRESIDENT: B. J. Levin.
ORGANIZING COMMITTEE: A. Dollfus, G. Herzberg, G. P. Kuiper, C. H. Mayer, C. E. Sagan, R. Wildt.

INTRODUCTION

The ground-based research activity in the planetary field has increased very rapidly since the Prague meeting in 1967. The space probes into the atmosphere of Venus and the flybys of Mars have not only added directly to our knowledge of these planets but have stimulated further research.

Because of this rapid growth and the large expense which would be required to publish detailed descriptions of the very numerous research projects carried on in the last three years, this report will differ in important respects from the previous ones. Drs Levin, Pettengill, and McElroy have provided summaries of progress made in three important fields of planetary science, and Levin has also kindly contributed a summary of planetary research carried on in the U.S.S.R. since the Prague meeting. On behalf of its members, the President of Commission 16 gratefully acknowledges each of these important contributions.

The summaries will be followed by a résumé of current (unpublished) research being carried on elsewhere. Future projects will also occasionally be mentioned. The material presented here is that which has been sent to the President in response to requests contained in newsletters which have been sent at more or less regular intervals to all members of Commission 16.

In accordance with the new policy of the IAU Executive Committee, the number of detailed references given in commission reports must be drastically reduced. Bibliographies are now published in English, German, and Russian, and many are now more nearly up-to-date than any which could be given here. Table I lists a number of recent symposia, conferences, and meetings, and indicates where the papers presented at these gatherings have been or will be published.

An effort has been made to see to it that available reports and bibliographies were distributed to all members of Commission 16. The following reports or bibliographies have already (December 1969) been distributed:

Bibliography of Lunar and Planetary Research Supplement for 1966 and 1967 – Office of Aerospace Research, Air Force Cambridge Research Laboratories, L. G. Hanscom Field, Bedford, Mass., U.S.A. The volume for 1968 is in press and the bibliography for the first quarter of 1969 has appeared in the July, 1969 issue of *Icarus*. In the future, these are scheduled to appear quarterly.

Planetary Astronomy (*An Appraisal of Ground-Based Astronomy*) and *Planetary Exploration* 1968–1975 – two publications of the National Academy of Sciences, Washington, D.C.

Summaries of Science for the year 1964–1965 (Itogi Nauki) – Institute of Scientific and Technical Information, Moscow, U.S.S.R. The next summary is to appear in 1970.

INTERNAL CONSTITUTION OF PLANETS
(by B. J. Levin)

To obtain some insight into the internal constitution and composition of planets one must know the equations of state (pressure-density relations) for materials of which planets are composed, and one must study by numerical calculations the radial distributions of density. Only the conclusion that Jupiter and Saturn contain much hydrogen can be reached without detailed calculations

Table 1. Symposia and Conferences Since Prague IAU Meeting

Subject	Date Place	Sponsor(s)	Publication
Physics of Moon and Planets	October 1968 Kiev	Astron. Council U.S.S.R.	Publishing House „Nauka", Moscow, 1970–71
Jupiter and the Outer Planets	December 1968 Dallas, Texas	Section D, AAS; AGU and GSA	*Icarus*, May 1969, See also *Science*, **162**, 588, (1968)
Working Group VII	May 1969 Prague	COSPAR	COSPAR
Planetary Atmospheres and Surfaces	August 1969 Woods Hole, Mass., U.S.A.	IAU, URSI	*Radio Science* February 1970
Arizona Conferences on Planetary Atmospheres	February 1967 Tucson, Ariz.	KPNO	Gordon & Breach, Publishers, 1968
	March 1968 Tucson, Ariz.	KPNO	*Journal of the Atmospheric Sciences*, July 1968
	May 1969 Tucson, Ariz.	KPNO	*Journal of the Atmospheric Sciences*, September 1969
Planetary Atmospheres	October 1969 Marfa, Texas	*IAU Symposium*, no. 40, McDonald Obs.	D. Reidel Publishing Co., Dordrecht, The Netherlands. Summer of 1970

(De Marcus, 1967). Unfortunately, the equations of state are poorly known and the improvement of planetary models strongly depends on the improvement of these equations.

Terrestrial planets

Pressure-density relations necessary to calculate models of the terrestrial planets can be obtained either by using the Earth as a prototype, or from experimental data. In the latter case one can use, for mantles, data on compressibility of olivine and its high-pressure modifications (spinel and post-spinel) or of principal oxides (SiO_2, MgO, Al_2O_3, CaO), assuming the additivity of partial molar volumes.

Differences in equations of state, naturally, lead to somewhat varying models. But the main controversy is about the nature of cores of the Earth and Venus: whether or not they contain metallized silicates. The problem is very difficult indeed. Experimental shock-compressions of rock-samples failed to effect their transition into metallic phase at pressures up to 5 million atmospheres. Therefore, many astronomers and geophysicists regard the metallized-silicate core hypothesis as disproved. Moreover, there is no doubt that Mercury contains about 60% metallic iron or nickel-iron (this seems to be the only possibility to account for its high density) and this proves that planets *can* contain much metal.

Although the iron core hypothesis, modified to suggest that the core is composed of an iron-nickel-silicon alloy, is able to explain the present-day mechanical structure of the Earth and Venus, there are unsurmounted and perhaps even unsurmountable difficulties connected with the *formation* of the iron core. Why did much iron remain in metallic form in the zone of accretion of the Earth, although the temperature dropped much lower than in the zone of Mercury? How did the interiors of the Earth and Venus become molten to permit iron to sink into the core? How do these planets get rid of heat liberated by segregation of iron into a core? As long as these questions remain unanswered, it is legitimate to regard the iron core hypothesis as unproven and continue to analyze the hypothesis of metallized silicates.

As Mars has only a small iron core compared to the Earth and Venus, if iron cores are accepted for the Earth and Venus, one can say that the content of metallic iron decreases with the heliocentric distance of the planet (Reynolds and Summers, 1969). This statement neglects the facts that the relative mass of the Earth's core is larger than that of Venus and that the absence of an iron core in the Moon requires a special explanation.

On the other hand, if one is to accept that the outer core of the Earth is composed of metallized silicates, while the inner one is of iron, and that the core of Venus has a similar constitution, then differences in composition of Venus, Earth, Moon and Mars are rather small – approximately within the limits of variants of composition for different types of chondritic meteorites (Levin, in *Interiors and Surfaces of Planets* (ed. A. Dollfus) Academic Press, 1970).

Calculations of the thermal histories of terrestrial planets are being done now for rather complicated models which take into account the gradual character of the melting of rocks, the dependence of heat conductivity on temperature and pressure, the redistribution of heat sources after partial melting of planetary interiors, etc. (Majeva, 1966, 1969; Fricker and Reynolds, 1968; Lee, 1968). Unfortunately, the variations of melting temperatures with pressure are only approximately known.

Calculations for Mercury show that it was never molten and therefore is probably core-less. At the present time its interior must be slowly cooling. If the core of Venus is composed of an iron-rich alloy, the release of gravitational energy during its segregation would lead to a molten state of the mantle and core of Venus. (The same would hold true for the Earth and this would be contradictory to seismic evidence for the solid state of the mantle.) Probably this cannot be reconciled with the locked-in rotation of Venus which requires some three-axiality of its figure. But calculations for a model with a metallized core indicate that all or nearly all mantle must be solid.

Thermal histories of Venus and the Earth, and probably also of Mars, critically depend on the possibility of stationary convection in their interiors. Mathematical theories of thermal and non-thermal convection were developed for highly idealized planetary models (Ashworth, 1968, 1969; Danes, 1969). But their applications to real planets are scarcely possible. (The same is true for the study of thermoelastic deformation of planetary globes by Kopal, 1968.) For example, although the thermal gradient, at least in some layers, is superadiabatic, and thus appropriate for thermal convection, the latter is scarcely possible because it requires a homogeneity of material, while in real semimolten interiors chemical and gravitational differentiation must block thermal convection.

Giant planets

The modern approach to the study of Jupiter and Saturn begins with the work by DeMarcus (1958) who analyzed cold ($T = 0$ K) models of these planets and checked the density distributions using values of multipole moments J_2 and J_4 derived from observations. Peebles (1964) took the temperature into account and obtained more precise values for H and He content. Hubbard (1968, 1969) improved the estimates of temperature, especially in a core of metallized hydrogen, and showed that the observed energy emission from Jupiter can derive from the thermal energy of an interior heated up to 10000–15000 K. For both planets – Jupiter and Saturn – Hubbard calculated chemically homogeneous convective models, but noted the possibility that they have small dense cores.

On the other hand, Smoluchovski (1967) proposed a qualitative model for Jupiter in which he takes into account a low solubility of He in metallized hydrogen at $P \lesssim 10^7$ bars. His model contains a He core surrounded by a "mantle" of metallized hydrogen which contains in its lower part some dissolved He. He assumes a slow motion outwards of the boundary between molecular and metallized hydrogen which, (a) shifts He admixed with molecular hydrogen towards the surface, and (b) produces a slow contraction of the planet. The shift of the boundary at the rate of 0·1 cm/y is sufficient to explain the observed heat flow from Jupiter.

Further studies of the internal constitution of Jupiter and Saturn must include a qualitative check of Smoluchovski's ideas, a check of the role of nonvolatile elements and compounds, and calculations of the thermal history of Jupiter, including the heating during accretion. Such heating would be the cause of a high energy emission in the past which would explain the stony composition of Io and Europa.

New models of Uranus and Neptune calculated by Ramsey (1967) differ only slightly from that by Reynolds and Summers (1965). (Ramsey includes metallic ammonium, NH_4 instead of NH_3, and includes Ne not with H and He but with CH_4, NH_4 and H_2O, and neglects the compressibility of the heavy component which forms a core in some models.) For Neptune the observed value of J_2 was used to select better models.

The new value of Neptune's diameter determined from photoelectric observations of the occultation of a star on April 7, 1968, will remove a peculiar large difference in the hydrogen content of Uranus and Neptune.

The result obtained by Kieffer (1967) is important for planetary cosmogony: calculating mass-radius relations for planetary bodies of different compositions, he showed that Jupiter can be composed of solar material, while Saturn is slightly enriched in less volatile elements.

An important step forward was taken recently in the classical field of the study of figures of rotating planets (in hydrostatic equilibrium). The Darwin-De Sitter theory has taken into account the second order terms. Zharkov and Trubizin (*Dokl. Acad. Nauk S.S.S.R.*, **186**, 791, 1969; *Astr. Zu.*, N6, 1969) developed a theory including the third-order terms, thus allowing the use of the multipole moment J_6 (besides J_2 and J_4). To begin with, the new formulae can be applied to studying the figure and internal structure of Jupiter.

Magnetic fields

Among the terrestrial group, the Earth seems to be the only planet possessing a substantial magnetic field (Mercury's field remains unknown). Combined analysis of measurements carried out by Venera 4 and Mariner V permitted substantial lowering of the upper limit for the magnetic field of Venus: it is at least 10000–15000 times weaker than that of the Earth (Dolginov, Eroshenko and Davis, 1969). The absence of the magnetic field on Venus is probably due to too slow a rotation of the planet. The Martian period of rotation is similar to that of the Earth and it must have a liquid iron core. However, to produce a magnetic field there must be systematic motions in the core, most probably thermal convection. The latter is impossible in the Martian core because of a low proportion of radioactive elements with metallic iron.

Radio-observations of Jupiter reveal that it possesses an intense magnetic field and strong magnetosphere. Variations of the plane of polarization observed at decimeter wavelengths, which are due to the tilt of the magnetic axis 10° from the rotation axis, have a period of $9^h55^m29^s.37$ which coincides with the period of bursts observed at decameter wavelengths. This period may be the most fundamental for Jupiter, because it probably reflects the rotation of the interior. Decametric radiation from Jupiter is influenced by the position of Io and Europa in their orbits, possibly through some unexplained interaction with the Jovian magnetosphere.

The most important problems for further study are the origin of the Jovian magnetic field, which is tied to the dynamics of the interior and its nonuniform rotation (Runcorn, 1964; Hide, 1967), and the nature of the radiation bursts and their influence on the atmosphere.

PLANETARY SURFACES

(by G. H. Pettengill)

Significant measurements of the surface properties of the terrestrial planets have been made over the past 3 years using radar techniques at frequencies between 430 and 7840 MHz, radiometric techniques at radio and infrared wavelengths, and optical telescopic observations. The results obtained using the first two of these three techniques have been well summarized in the joint IAU-URSI Symposium on Planetary Atmospheres and Surfaces, held at Woods Hole, Massachusetts in August, 1969 (Proceedings published in February, 1970, issue of *Radio Science*).

Mercury

The rotation of Mercury has been confirmed at a rate precisely 1·5 times faster than the average orbital revolution, using both radar and optical observations. The repercussions of this higher-order synchronism (as compared with the one-to-one synchronism between rotation and revolution assumed in earlier years) have received considerable attention, both from the point of view of possible modes of evolution and from a consideration of the likely effects on the distribution of subsurface temperatures. The latter have been observed radiometrically by several groups and have been found to display a marginally significant double maximum along the equator, as expected from the new value of rotation which forces the planet to expose the same two antipodal regions of the surface alternately to the sun at perihelion.

Observations of the surface characteristics of Mercury are not yet available in sufficient quantity or quality to permit firm deductions concerning its nature. Existing radar and radiometric observations, however, suggest a surface whose composition, density and roughness are not far different from the corresponding values for the Moon. Both radar and optical observations find evidence for separated surface areas with differing scattering characteristics. The relative difficulty of observing this planet, however, as compared to the other terrestrial planets, has held back good mapping of these areas.

Venus

This planet received intensive study during the interval surrounding the inferior conjunctions of 1967 and 1969. Radar observations at a variety of wavelengths yielded good agreement on rotation (retrograde sidereal period of $243{\cdot}0 \pm 0{\cdot}1$ d, with a pole approximately normal to the orbit) and the positions of half a dozen anomalously scattering surface features. Optical observations in the near ultraviolet found relatively longlived cloud formations that appear to circle the planet in a retrograde sense with a period of about 4 d.

Several relatively small regions with heights of 1·5 to 2 km above the average surrounding surface have been localized on Venus using radar techniques. The general impression, however, is of a surface (in the equatorial belt, at least) devoid of large-scale topographic relief. Some evidence exists for a slight (1·5 km) displacement between the center of mass and the center of shape.

It appears as though the surface dielectric constant as determined from interferometric observation of the emission polarization at radio wavelengths can now be reconciled with the corresponding value as determined from the radar reflectivity at normal incidence, when depolarization associated with the substantial absorption in the atmosphere of Venus is properly accommodated. The best current value for the dielectric constant of the radio surface is 4·8, which is significantly higher than for the Moon and Mercury (2·7), or Mars (3·5). A variety of radiometric evidence now favors a near-constant equatorial surface temperature of about 700 K.

Mars

The increasingly favorable oppositions of Mars in 1967 and 1969 have stimulated a considerable observing effort. Direct radar observations have disclosed substantial large-scale topographic relief in the northern tropics of about 12 km peak-to-valley vertical extent, when averaged over lateral distances of several hundred kilometers. No significant correlation between optical surface markings and this relief was noted, although fair correlation does appear to exist between the locations of regions exhibiting high radar reflectivity and low visual albedo. To the radar, Mars appears to have far more gentle slopes in its average small-scale surface structure than either Venus or the Moon, despite its significant large-scale topographic variations.

The successful photographs resulting from the Mariner space-probe missions in 1969 should make it possible to compare directly the small-scale structure of selected regions of the Martian surface with various areas of the Moon.

Icarus

Considerable radar and optical observing effort was expended on studying the minor planet Icarus during its close approach to earth in June of 1968. Several radar and a large number of optical observations were successful, yielding a rotational period for the asteroid of 2^h16^m, a radius lying between 0·5 and 1 km and an unexceptional surface reflectivity. The absolute optical magnitude was about 17. From the small amount of rotational modulation, a highly spherical shape was inferred.

RESEARCH IN PLANETARY ATMOSPHERES 1967–1970

(by M. B. McElroy)

Scope of this Review

This paper is intended to be a brief résumé of current research on planetary atmospheres. Work reported in the period 1967–1970 is emphasized. The preceding three years were reviewed by T. Owen in the report of Commission 16 to the Thirteenth General Assembly of the International Astronomical Union. He included a detailed bibliography which is not repeated or updated here. I would like to draw particular attention to the annual Arizona conferences (see Table I) on planetary atmospheres which were sponsored by the Kitt Peak National Observatory and offered a forum for review of research in planetary atmospheres. They were well attended by scientists both from the United States and from the international community.

The first conference, in February 1967, was devoted to the atmospheres of Mars and Venus and included a detailed review of the Mariner IV results. The proceedings give a good historical account of uncertainties in our understanding of Venus at that time. These uncertainties prevailed until the highly successful Mariner V and Venera 4 flights. The second Arizona conference was devoted specifically to Venus and the papers presented then reflected the great advances in the intervening year. The third Arizona conference was directed at the major planets and it was clear that our level of understanding of physical processes even for Jupiter is rudimentary, comparable to our understanding of Venus prior to the 1967 space observations.

The most recent broadly based conference was held at Marfa, Texas on October 1969 under the auspices of Commission 16. This conference included discussions of the recent Soviet Venus experiments (Veneras 5 and 6) and U.S. Mariner experiments (Mariners 6 and 7).

Our summary of present knowledge is ordered on a planet by planet basis, beginning with Mercury.

Mercury

Considerable effort was devoted to a spectroscopic search for an atmosphere on Mercury stimulated by the tentative identification of CO_2 by V. I. Moroz. This observation was not confirmed and present work establishes an upper limit of 0·58 m atm on the abundance of CO_2 in the Mercury atmosphere. Based on polarization studies, A. Dollfus concluded that Mercury had an observable atmosphere and estimated a surface pressure $p_s \sim 1$ mb. His conclusions are disputed, however, by B. T. O'Leary and D. G. Rea.

Venus

The important developments here are directly related to the Venera probes and to the Mariner 5 flyby. The Soviet measurements were made in atmospheric layers about 36–38 km thick and covered a pressure range from 0·5 to 27 bar and a temperature range from about 300 K to 600 K. The observed temperature gradient is close to the adiabatic limit and extrapolation of measurements gives surface pressure and temperature values of 140 bar and 800 K with Venera 5 data, 60 bar and 670 K with Venera 6 data. The difference between these results is attributed to a discrepancy in altitude measurements presumably related to malfunctions in the on-board radar altimeters. All three Veneras were crushed by pressure before reaching the planetary surface. Composition measurements indicated that CO_2 is the major constituent (93–97%) whereas N_2 and noble gases account for less than 5% of the atmospheric density. An upper limit of 0·4% was placed on the O_2

mixing ratio and the water vapor content was measured to be 4–11 mg per liter at a pressure level of 0·6 bar. Water vapor was not saturated at this level.

Venera 4 and Mariner 5 also detected a hydrogen corona around Venus and interpretations of these data involve a number of interesting though still speculative hypotheses. Mariner 5 revealed the presence of two components to the observed $L\alpha$ emission. Their scale heights differ by an approximate factor of 2 and published theoretical models associate the less extensive component with either H_2 or D. In either case the implications are far reaching. The H_2 theory suggests unexpectedly large abundances for this gas in the lower atmosphere. The D theory, in its simplest form, requires an enhancement in the D/H ratio for Venus as compared with Earth.

Mariner 5 detected ionospheres on both day and night sides of Venus. The dayside measurements revealed a sharp cut-off in the ionization above about 500 km (termed the anemopause on the hypothesis that it is associated with a solar wind interaction). The physics of this cut-off are not yet understood although detailed models have been constructed for the ionization observed at lower altitudes. A number of ionospheric theories conclude that atomic oxygen is a minor constituent in the upper atmosphere, and several invoke the presence of He concentrations in the Venus atmosphere larger by several orders of magnitude than values observed for the Earth's atmosphere.

The composition of the Venus clouds remains uncertain although recent studies by G. P. Kuiper and co-workers represent considerable progress. They find that the albedo of Venus is closely reproduced by laboratory reflection from a fine partially hydrated $FeCl_2$ powder. However, the implications of this hypothesis with regard to physical and chemical processes in the atmosphere have not yet been subjected to detailed examination. A problem exists as to the correct abundance of H_2O. Kuiper estimates a mixing ratio of 10^{-6} from studies in the spectral region near 2μ. Observations at 8189Å have been interpreted in terms of somewhat larger H_2O mixing ratios ($\sim 10^{-4}$), in better agreement with the Venera results. The discrepancy may be due in part to complications introduced by effects of radiative transfer which must be included in any completely consistent model of line formation. It is possible also that the strength of the H_2O absorption line is indeed variable. More work is required to clarify these problems and this research is directly relevant to the question of Venus cloud composition.

Mars

The only gases spectroscopically detected in the atmosphere of Mars are CO_2, H_2O and CO. The CO_2 abundance is 78 ± 11 m atm and the spectroscopically determined surface pressure is in the range 5·7–10 mb. Water vapor is variable but definitely established by the Texas observations (R. Schorn, paper presented at the Marfa meeting). Its average concentration is about 30μ and typically more H_2O is observed in the northern than in the southern hemisphere. There is no obvious seasonal effect although there appears to be more water in the midday atmosphere than in either morning or evening atmospheres. The CO concentration is about 5·6 cm atm which corresponds to a mixing ratio of approximately 10^{-3}. Important upper limits of 20 cm atm and 2×10^{-4} cm atm have been placed on the concentrations of O_2 and O_3. The latter result (unpublished) was derived by L. Wallace from analysis of data obtained from the Orbiting Astronomical Observatory.

The ultraviolet experiment on Mariners 6 and 7 failed to observe emissions associated with N_2 in the upper Martian atmosphere. This result should place an important upper limit on the N_2 concentration. Present preliminary analyses suggest an upper limit for the N_2 to CO_2 mixing ratio of about 5×10^{-2}. Emissions associated with H, O, CO and CO_2^+ were observed and the experiment establishes that Mars has a detectable H corona, a surprisingly small concentration of O in its upper atmosphere and an ionosphere composed mainly of CO_2^+.

M. Belton, D. Hunten and R. Wells recently reported a study of Martian topography which deserves special note. They observed the CO_2 band at $1{\cdot}05\mu$ using a three-channel multislit spectrophotometer with high spatial resolution on the planet. The observed strength of absorptions gives a quantitative estimate of elevation differences. They concluded that Syrtis Major was very high but found little correlation generally between elevation and albedo. Elevated areas were observed both in desert and dark regions. In several cases, dark areas were associated with relatively steep slopes. The spectroscopic survey is in good agreement with the radar maps derived earlier. The technique shows great promise as a new tool for exploration of the Martian surface.*

Jupiter

The only gases spectroscopically identified in Jupiter's atmosphere are H_2, CH_4, and NH_3. As for Venus, interpretation of absorption lines in terms of abundances is complicated due to the probable importance of radiative transfer in the line formation process. Errors are minimized to some extent if results are quoted in terms of mixing ratios.

The best estimate of the H_2 abundance is derived from observations of the quadrupole lines and with reflecting layer theory these observations give an abundance of 68 km atm above a cloud deck whose effective temperature is about 150 K. The mixing ratio of CH_4 to H_2 is about 5×10^{-4} and the distribution of NH_3 above the visible clouds is apparently determined by the saturation vapor pressure relation. The abundance of He remains unknown although spectroscopic studies suggest that the mixing ratio He to H_2 should not exceed 0·5. From studies of thermochemical equilibrium, J. Lewis concluded that apart from H_2, He, CH_4 and NH_3, the most abundant gases in the visible layers of Jupiter's atmosphere should be H_2O and H_2S. Their mixing ratios were estimated to be of order 10^{-12}. By way of comparison the current spectroscopic limits for mixing ratios of these gases are $7{\cdot}5 \times 10^{-7}$ and 4×10^{-6} respectively.

Accurate values for geometrical albedo as a function of wavelength are most important for many planetary studies and important data in the visual and near infrared spectral regions have been reported by W. M. Irvine and associates for Venus, Mars, Jupiter and Saturn. Recent rocket and satellite observations have yielded good data in the ultraviolet for Venus, Mars and Jupiter. The UV Venus and Jupiter observations confirm the importance of scattering for these planets and demonstrate the necessity of including radiation transfer effects in the interpretation of absorption lines.

Spectra of Jupiter in the far infrared have offered an important tool for the study of that planet's atmosphere above the visible clouds. They suggest the possibility of a thermal inversion at a temperature level of about 130 K. A similar result is indicated by recent data in the radio region near 1·25 cm. The existence of important internal energy sources for both Jupiter and Saturn has been demonstrated by F. Low and associates.

Saturn

Only H_2 and CH_4 have been established as component gases in Saturn's atmosphere. A recent study by Kuiper and associates has identified solid H_2O in the infrared reflection spectrum of the rings. As yet NH_3 has not been observed as an atmospheric gas.

Uranus

A recent analysis of the H_2 quadrupole lines indicates that the strength of these absorption lines may be consistent with an optically semi-infinite H_2 atmosphere, i.e. the lines are as strong as they can be and penetration of visible and near infrared radiation is limited by Rayleigh scattering. The only gas observed other than H_2 is CH_4 and there have been no recent revisions of its abundance ($\sim 3{\cdot}5$ km atm). The brightness temperature at 20μ is 53 K, and does not require the postulate of an internal energy source.

Neptune

The most important developments here are directly related to the occultation of the star

* See below the report of the current work on this problem being carried on at the McDonald Observatory.

BD – 17°4388 which occurred April 7, 1968. Observations of this event yielded a new radius for the planet of approximately 25000 km and suggested that the temperature in the highest atmospheric layers may be somewhat higher than values expected on the basis of model studies.

SHORT REPORT ON THE STUDIES OF PHYSICS OF PLANETS AND SATELLITES IN THE USSR IN 1967–1969

(by B. J. Levin)

As mentioned in the previous summary, Venera 4 revealed the existence of a hydrogen corona around Venus. The first signs of it were observed at the distance of 25000 km from the centre of the planet, while at 10000 km a hydrogen density of ~100 atoms per cm³ was measured (the Venera 4 data gave $n_H = 600/R/R_♀ \text{ cm}^{-3}$). The measurements gave for the density of atomic oxygen at 300–350 km above the night side of Venus the upper limit $n_0 < 2 \times 10^3 \text{cm}^{-3}$.

From the absorption of $L\alpha$-radiation the CO_2 density at the height of about 100 km was estimated to be $n_{CO_2} \approx 3 \cdot 10^{11} \text{ cm}^{-3}$.

No radiation belts were observed.

Peculiar refraction in the atmosphere of Venus was studied by M. M. Skotnikov.

A model of the atmosphere of Venus based on data obtained by radioastronomy, radar, and Venera 4 was compiled by A. D. Kuzmin and Yu. N. Vetukhnovskaya. (Paper presented at COSPAR Meeting, Prague, 1969.)

Photometric studies of Venus by O. M. Starodubzeva (Kharkov Obs.) confirmed the existence of the quasispecular reflection effect discovered by N. P. Barabashov.

Photometric studies of Mars were continued by N. P. Barabashov and his colleagues in Kharkov and by I. K. Koval′ and his colleagues in Kiev. According to I. K. Koval′ the wavelength-dependence of contrast between the continents and maria on Mars speaks in favour of an elevation of the continents. For normal transparency the optical depth of the atmosphere over maria is 2½ times larger than over continents.

According to the suggestion by M. S. Bobrov, the first six frames taken by Mariner IV represent photographs of the cloud layer, not of the surface of Mars.

V. I. Moroz in 1967 observed Venus, Mars and Jupiter in IR (8–14μ) and found brightness temperatures: for Venus (the centre of the disk) 223^{+12}_{-3} K; for Mars (subsolar point) 263^{+13}_{-5} K; for Jupiter 107^{+5}_{-10} K. In 1969 he made spectral observations of these planets in the 1–2μ region. Some observations were done with high angular resolution (up to 1″) – with resolution 200 Å. Other observations of the whole disk were made with spectral resolution: 8 Å for Jupiter, 5 Å for Mars, 2 Å for Venus. The NH_3 band $\lambda 1{\cdot}53$ on Jupiter is more intense in a latitude near 50° than on the equator, probably because of lesser height of the cloud layer. Preliminary analysis indicates that on Mars the CO_2 bands are more intense in dark areas than in the bright ones. On Venus, near inferior conjunction, the intensity of these bands is reduced but, even so, near the horns, they are definitely more intense than on the intensity equator.

From spectroscopic observations in the red region, V. G. Teifel′ and his colleagues (Alma-Ata) confirmed that on Jupiter and Saturn the cloud layer has no definite boundary and that the visible surface is indicated by a layer of haze with a low density of aerosols ($\sigma_a \sim 10^{-7} \text{ cm}^{-3}$). From observations of the CH_4 band at $\lambda 6800$, the rotational temperature in the atmosphere of Uranus is found to be about 60–68 K at the level where the effective pressure is $2{\cdot}5 \leqslant P_e \leqslant 5$ atm. About 2000 polarimetric measurements of separate details of the surface of Jupiter and Saturn were carried out at the Abastumani Observatory (V. P. Djapiashvily *et al.*).

V. G. Teifel′ published a monograph, *The Atmosphere of the Planet Jupiter*, (in Russian) (Publ. House "Nauka", Moscow, 1969) in which he reviews the composition of the Jovian atmosphere, its constitution and thermal properties, optical properties, clouds, and atmospheric circulation.

M. S. Bobrov, from the analysis of worldwide observations of Saturn's rings done in 1966, concluded that their thickness is about 3 or 4 km. His monograph, *The Rings of Saturn* (in Russian) is to be published in 1970 by the Publishing House "Nauka" (Moscow).

Calculations of the internal constitution of Mercury made by S. V. Kozlovskaja indicate that it must contain about 60% of metallic nickel-iron. With such a composition the planet was never molten and therefore, probably, has no iron core (S. V. Majeva). For Venus, various models containing both metallized silicates and iron alloy are compatible with the available data (S. V. Kozlovskaja). Metallized silicates had to appear, not at the end of the accumulation phase of Venus, but much later when the heating up of its interior had decreased the critical pressure for phase transition. S. V. Majeva calculated the thermal history of Venus for two models: with the iron core and with a core of metallized silicates. In the first model the melting of the mantle is much more pronounced due to the release of gravitational energy during the segregation of iron into the core.

The connection between planetary cosmogeny and the problem of the origin of meteorites was discussed by B. J. Levin. He concluded that new data speak in favour of the common origin of the Sun and protoplanetary cloud, of the type suggested by F. Hoyle or E. Schatzman, but not of the type suggested by A. G. W. Cameron.

V. S. Safronov published a monograph, *Evolution of the Pre-Planetary Cloud and the Formation of the Earth and Planets* (in Russian) (Publ. House "Nauka", Moscow, 1969) in which he summarizes and develops further his studies on this problem.

CURRENT UNPUBLISHED RESEARCH

Venus

H. Camichel states that he is continuing his observations of Venus. Recent data confirmed the reality of a displacement of the visible surface at a velocity of approximately one rotation in four days in the retrograde direction. This does not contradict a value for the rotation of the surface in 243 d, if one finds a way to explain the persistence of motion of such great velocity in the high atmosphere with respect to a surface which almost does not turn.

C. H. Mayer reports that studies are being made at the U.S. Naval Research Laboratory of the brightness temperature of Venus over the phase cycle at 2·7 cm wavelength. These observations are being made to determine the difference in temperature between the sunlit and dark hemispheres and specifically to check the phase measurements of several years ago at a nearby wavelength. The new observations do not show a significant dependence of Venus' brightness temperature at this wavelength with the phase of solar illumination. This result is not in agreement with our previous observations at 3·15 cm wavelength and those of Drake at 10 cm which indicated a lower brightness temperature just after inferior conjunction than near quadrature. It is, however, in agreement with more recent measurements by others which have suggested little variation with time or phase. The reason for the discrepancy is not clear. The new measurements are made with a more sensitive radiometer, but the sensitivity for the 1961 measurements was adequate for the signals observed between inferior conjunction and quadrature where the measurements were made. One would prefer to speculate that both observations are real and perhaps there was a change in Venus, but as yet one has no basis for such speculation.

Spectroscopic programs at the McDonald Observatory have failed to reveal water vapor over the last two years. Apparent CO_2 variations are being studied as a function of position on the disk and phase of the planet with respect to the sun (Gray, Young, Schorn, Barker).

Laboratory studies at the University of Massachusetts of the infrared reflectivity for frosts and powders of a variety of minerals which have been suggested as possible Venus cloud constituents have been carried out by Plummer and Carson (1970*b*). The marked influence of particle sizes on spectral reflectivity has been shown, and it has been demonstrated that most of the compounds which have recently been suggested as possible cloud constituents have unacceptable spectra when compared with Venus. Of the compounds studied, only small ice particles provide a good match to the Venus data.

Mars

H. Camichel mentions the continuation of his observations of the planets, particularly Mars and Venus. He is presently calculating the positions of spots on Mars which will provide the final values for their coordinates for the oppositions from 1956 to 1969.

Harlan J. Smith reports that intensive planetary studies have been made at the McDonald Observatory with the 82-inch Struve reflector and also with the new 107-inch telescope and its large coudé spectrograph equipment. The principal observations have been of Mars in the 0·8–0·9 μ H_2O bands and the various CO_2 bands in the 0·7 to 1·2μ region. Water vapor has been found to be present on Mars especially in the hemisphere of the evaporating polar cap, with amounts ranging up to 40μ of precipitable water vapor, but declining at times to immeasurably low values, below 10μ. A systematic program of study of the water vapor by location on the planet throughout a full Martian year is in progress (Schorn, Little, Owen, Barker). The CO_2 programs have given improved data for the weak short-wavelength CO_2 bands (Gray, Owen), show clear correlation with radar determinations of relative elevation over different portions of the Martian disk (Barker, Woszczyk), and have given the first indications of varying total amount of CO_2 (and therefore that the atmospheric pressure on Mars may be a function of the Martian season) (Barker).

Further observations are in progress to determine the H_2O and CO_2 variations throughout a complete Martian year (Barker).

With a spectrum scanner on the 36-inch reflector, the reflectivity curve of Mars from 0·3 to 1·2μ has been determined (Tull).

G. de Vaucouleurs, in conjunction with J. Bergstrahl and C. Michelis, is about to publish a paper entitled "Martian Transits 1877–1969", which contains the reduction and analysis of aerographic longitudes derived from transits of surface markings at the central meridian of Mars (750 transit observations of 142 stations by 19 observers at 21 oppositions from 1877 to 1969).

He also states that final reports on the Mars Map Project 1958–1969 and on "Observations of Mars in 1969" are in preparation.

G. de Mottoni reports that the following work is in press: *Cartographie de la Planète Mars sur Documentation Photographique Internationale à Partir de 1907* – Memoire III Oppositions de 1954 à 1958 – Memoire IV: Oppositions de 1960 à 1967:

E. H. Collinson, Director of the Mars Section of the British Astronomical Association, reports the reappearance of the dark shading in the Aethiopis region, reported early in the apparition by de Vaucouleurs, has been observed and sketched; in addition to this, changes have been observed in the region of Lunae Palus. Lunae Palus was unusually large and dark, as was Nilokeras. A broad dark shading was observed proceeding from Lunae Palus to Aurorae Sinus and another shading proceeding from Lunae Palus towards Ceranius which are not recorded on the IAU Map. A darkening of the region between Niliacus Lacus and Margaritifer Sinus was also observed. The Thoth-Nepenthes was faint. Numerous intensity observations of the dark areas were made by J. H. Botham at Johannesburg and by Baron R. de Terwangne at Antwerp.

The motions of well-defined transient bright spots (often assumed to be clouds) on Mars have been investigated by W. A. Baum and L. J. Martin at the Planetary Research Center of the Lowell Observatory. A search of several thousand plates in the Lowell Observatory collection yielded 28 groups of plates on which the positions of bright spots could be followed on a nearly daily basis. These groups of plates were from fifteen different oppositions of Mars, starting from 1907 and ending with 1958. All but two of these spanned four nights or more, and the maximum interval covered was thirty nights. Clouds near the limb were ignored.

The 28 groups of plates yielded 95 cloud histories. More than half appeared to be relatively stationary. Others showed definite motion well in excess of observational error but sometimes followed paths that partly doubled back upon themselves. The mean velocity for non-stationary clouds was found to be 5·6 km per hour, and the most commonly occurring direction of motion was eastward, particularly at high latitudes. The range of velocities found by this mapping procedure is nearly an order of magnitude smaller than values that have been estimated earlier by others from visual observations. These earlier observations are evidently in error, unless there exist clouds at high elevation, visible only on the limb, that can move much faster than those that were mapped from this photographic survey. More clouds were found in the northern hemisphere than in the southern, and there seemed to be avoidance of the relatively darker areas of the Martian surface. Certain regions seem to be more favored than others. A few recurrences at identical positions suggest the existence of related topographic features.

Their study is being continued with the use of films obtained during the Planetary Patrol mentioned elsewhere in this report.

The seasonal behavior of the boundaries of the Martian polar caps has been investigated in detail by G. E. Fischbacher, L. J. Martin, and W. A. Baum at the Planetary Research Center of the Lowell Observatory. Measurements of the cap boundaries in aerographic coordinates were made on more than 3000 yellow and red photographs covering 60 y of observation from 1905 through 1965.

The receding phase of each cap was found to follow a well-defined curve that repeated itself very closely from one Martian year to another; the mean deviations of measurements from various plates were only about 1°, and the variation from one Martian year to another was of similar magnitude.

Laboratory investigations at the University of Massachusetts have shown that the reflectivity of carbon suboxide (C_3O_2) agrees extremely well with the reflectivity of Mars over the range from 0·3 to 1·2μ (Plummer and Carson (1970*a*)).

Jupiter

C. H. Mayer reports that the Naval Research Laboratory has made fairly frequent observations of Jupiter at 1·6 and 2·7 cm wavelengths, partly to see whether its emission varies with time as has been suggested in the past. They have seen no evidence for variations over the past two or three years.

At the McDonald Observatory spectroscopic studies are concentrating on H_2 and on the behavior of the 1·1μ methane bands as a function of position on the disk (Owen, Bergstralh).

W. E. Fox, Director of the Jupiter Section of the British Astronomical Association, states that the apparition of 1968–1969 was well observed and that during this period the Red Spot became exceedingly prominent, mainly due to the fading of a section of the south temperate belt and also of the south equatorial belt.

Saturn

P. Guérin reports his discovery of the fourth ring and a new dark division in the system of Saturn's rings. This discovery has been made by photography with the Pic du Midi reflector, with a focal length of 60 m and an exposure of one second only, using a 35-mm commercial Tri-X film.

This fourth ring, which he calls "Ring D", is weaker than Ring C. It is separated from the latter by a dark division, similar to Cassini's division, and it extends almost as far as the globe itself. Its light decreases as it approaches the globe. It is equal at the border of the ring to 1/18 of the maximum brightness of Ring B in the neighborhood of Cassini's division.

All the negatives show this structure, and the composite images show them even better. The resolution obtained was about 0·25 arc second.

Recent polarimetric measurements made in the ultraviolet by J. S. Hall and L. A. Riley confirm the suspicion, mentioned at the third Arizona Conference, that the anomalous polarization found in the east-west direction was spurious. The new data indicate radial polarization in the ultraviolet for the disk of Saturn.

Uranus and Neptune

C. H. Mayer reports radio observations of Uranus and Neptune at two wavelengths, 1·6 cm and 2·7 cm. These observations made with Jupiter and Mars as controls (along with radio sources) confirm the high brightness temperatures, near 200 K in this region of the spectrum, and along with published observations at other wavelengths suggest the possibility of a peak in the apparent temperature spectrum.

MISCELLANEOUS RECENT DEVELOPMENTS

Professor L. M. Barreto reports that micrometric measurements are being made of Venus, Jupiter and Saturn at the National Observatory of Brazil. Migon and Barreto are measuring the phase variation of Venus to study the difference (O–C) for the dichotomy. Measurements of Jupiter's belts for latitude determination are being made by Migon, Nogueira and Barreto. Further micrometric measurements of the belts and ring system of Saturn are being performed by Migon and Nogueira. The results are expected to be published in 1970.

At the McDonald Observatory, high resolution planetary photography is occasionally attempted (Marin) and also high-spatial-resolution spectral scanning is being carried on (Trafton). Furthermore, measurements of planetary and lunar fluxes at 1 mm are being initiated (Ulrich).

A design for an electronic system for stabilizing planetary images at the focus of a telescope (image tranquilizer) was described by Baum of Lowell Observatory at the IAU General Assembly in Prague. Preliminary tests of this system, as modified by Thomas Pettauer, indicate that it holds the centroid of a planetary image nearly stationary against atmospheric displacements ("seeing") with a frequency response up to about 200 cycles per second.

G. de Vaucouleurs reports an article now in press entitled, Photometrie des Surfaces Planetaires, in *Interiors and Surfaces of the Planets* (Ed. A. Dollfus, Academic Press, 1970).

S. Runcorn states that studies are being made at the University of Newcastle-upon-Tyne regarding the origin of magnetic fields of planets and on the topic of convection in the "solid" mantles of the Moon and planets. In general, the dynamo theory of the geomagnetic field enables one to understand the absence of magnetic fields in Mars and the Moon, which have no cores or very small ones, and the presence of a field, discovered by radio emissions, on Jupiter. The case of Venus is being studied.

Basic research on theoretical problems of radiative transfer in planetary atmospheres is being undertaken at the University of Massachusetts by W. M. Irvine, D. J. van Blerkom, and A. Uesugi (on leave from the Institute of Astrophysics, Kyoto). Irvine also reports that data from his multicolor photoelectric photometry of the brighter planets, which have been published, have more recently been analyzed to provide accurate phase curves and albedos. The observations of Mars, Jupiter, and Saturn are still being studied in order to separate the average behavior with phase angle from fluctuations due to rotational effects, the rings of Saturn, and meteorological conditions (Hopkins and Irvine; Irvine, Higdon, and Ehrlich: *IAU Symposium*, no. 40).

Results of research at the University of Massachusetts which are now in press are:

Plummer, W. T. 1970, Venus Clouds: Infrared Reflectivity of Several Suggested Materials, submitted to *J. Geophys. Res.*

Plummer, W. T., Carson, R. K. 1970*a*, Mars: The Red Surface, *Science.*

Plummer, W. T., Carson, R. K. 1970*b*, Venus Clouds: Test for Carbon Suboxide, *Astrophys. J.*

Uesugi, A., Irvine, W. M. 1969, Computation of Synthetic Spectra for a Semi-Infinite Atmosphere, *J. Atmosph. Sci.*

Uesugi, A., Irvine, W. M. 1970, Multiple Scattering in a Plane Parallel Atmosphere. I: Successive Scattering in a Semi-Infinite Medium, *Astrophys. J.*

PLANETARY DATA CENTERS

At the Berkeley meeting of the IAU in 1961, a resolution was proposed by Commission 16 to facilitate international collaboration on planetary studies by the eventual establishment of two data centers, one in the U.S. and one in Europe. Following the adoption of this resolution by the Executive Committee, the IAU centers were subsequently established at Meudon and at the Lowell Observatory in Flagstaff.

With NASA support, a suitable building for carrying out the functions of collecting, classifying, reproducing and studying planetary photographs was completed at Flagstaff in 1965.

Photographs provided by European observatories have almost all been reproduced at Meudon as positive "composites" resulting from the superimposition of several images on one photograph. Copies of 1183 such composites have been made and sent to the Data Center at Lowell.

Similarly, copies of 11214 photographs and composites obtained at the American observatories of Lowell, Lick, Yerkes, McDonald, and Mt. Wilson have been reproduced at Lowell and sent to the Meudon Center. A total of 627 of these are Lowell positive composites.

The process of classifying, copying, and exchanging (between the two centers) photographs obtained prior to 1968 can now be considered almost complete.

Many copies of photographs were supplied to users for illustrating books, periodicals, reports and for making educational films. Computer-printed catalogues of the collection have been sent from the Lowell Center to users who have requested them.

Both collections are accessible to all qualified scientists having a research program. The reproductions and positive composites are presented in a form easily accessible and reproducible. The original negatives are filed separately.

Meudon

Certain programs, such as cartography of the planet Mars since 1907, the study of the movements of ultraviolet clouds on Venus, and the atmospheric activity in the bands of Jupiter, have been the object of prolonged studies of photographs covering a number of years. Other short-term programs have been executed, such as on rotation, coordinates, and cartography of the planet Mercury, the movement of yellow clouds in the atmosphere of Mars, and the study of the rings of Saturn. Much of this material has already been published. Still further studies have been conducted by other visiting scientists who have utilized temporarily the facilities of the Meudon Center for special programs.

The scientific direction of the Center at Meudon was undertaken by J. H. Focas. After his regrettable death on the 3rd of January, 1969, he was replaced by Messrs Ch. Boyer and T. Weimer. In June, 1969, the Center was transferred to a new location in the Laboratoire "Physique de Systeme Solaire" at Meudon, where it has been reorganized.

Flagstaff

Studies of cloud motions on Mars and the Martian polar caps described above have been carried out by the staff of the Planetary Research Center. In addition to these, the plate and film collection at the Center has provided observational data for investigations being carried out by a number of other people. This has included four separate Martian cartography projects, two studies of secular changes in Martian surface markings, three projects concerning Martian clouds and polar caps, the compilation of data for ground-based support for the Mariner 1969 mission, and two investigations of Jovian belts. Photographs from the collection have also been used to supplement independent photoelectric data.

The total number of planetary film copies (not including duplicates) produced at, and contributed to, the Flagstaff Center by other NASA-supported observatories is presently as follows:

New Mexico State University composites	– 1663
Table Mountain Observatory composites,	– 282
and other sets	– 481

The Planetary Center at Flagstaff has developed three major pieces of equipment which have already improved, or will soon improve, the effectiveness of this facility.

One of these is a projection image reader which has been extensively used to measure the positions of Martian features. It is briefly mentioned in *IAU Trans.*, **XIII B**, p. 96) and in *Sky Telescope* (**35**, 7, 1968). A second instrument of this kind will be available in early 1970. The second instrumental development is a modular planetary time-sequence camera which has already been extensively used in connection with the Planetary Patrol and is described below. The third piece of equipment, described in another section of this report, is an image tranquilizer.

The scientific staff of the Center in Flagstaff includes W. A. Baum (director), P. B. Boyce, R. L.

Millis, and occasional temporary appointees. Patrol observations and film editing are carried out by a group under the supervision of L. J. Martin. The staff of the photographic laboratory is supervised by S. E. Jones.

INTERNATIONAL PLANETARY PATROL PROGRAM

The Planetary Patrol is a worldwide cooperative program of six observatories, coordinated by the Planetary Research Center at the Lowell Observatory and supported by NASA Headquarters, for the ground-based photographic surveillance of the planets. This Patrol provides more observations with less interruption than ever before available. It will therefore help greatly in pursuing some of the current studies of Mars and Jupiter described elsewhere in this report, and it thus constitutes a valuable body of scientific data in its own right. The Patrol is under the direction of W. A. Baum, assisted by R. L. Millis and S. E. Jones.

At present, the Planetary Patrol network includes the Mauna Kea Observatory of the University of Hawaii, the Mount Stromlo Observatory in eastern Australia, the Republic Observatory in South Africa, the Cerro Tololo Inter-American Observatory in northern Chile, the Magdalena Peak Station of New Mexico State University, and the Lowell Observatory at Flagstaff. Equipment is now being prepared for adding the Kavalur Station of the Kodaikanal Observatory in India, and it is possible that further additions may follow.

All of the telescopes of the network have been optically equipped to produce identical image scales with diffraction-limited resolution. The planet cameras are also identical. They are of modular construction and include provision for focusing, guiding, color-filter selection, and sensitometric calibration. Exposures are made with 35-mm film transports controlled by solid-state electronic timing units with which each sequence of exposures is programmed in advance. The date, time, place, and color are automatically recorded on the edge of every film frame. At some stations the telescopes have been equipped with new ULE-quartz Cassegrain mirrors to produce the desired image scale directly (4·5 arc-seconds per millimeter), while at other stations the camera is preceded by a Barlow lens module to accomplish the same thing.

The present planning period of the Patrol spans five years, from 1969 through 1973. It includes three apparitions of Mars and five of Jupiter, which are the two primary targets. There will also be an ultraviolet campaign on Venus in the middle of 1970. Every patrol station photographs each accessible planet hourly on a pre-calculated schedule with fourteen-frame sequences in each of three colors. Observing logs are airmailed weekly to the Planetary Research Center at Lowell Observatory. Rolls of film are also airmailed there regularly for calibrating and processing under controlled conditions. After a number of steps related to inspecting, editing, copying, cutting into strips, mounting, labeling, and cataloguing, the final product for each color and each observing hour is a strip of film mounted in a fourteen-hole aperture card that is relatively easy to handle for image projection, measurement, microphotometry, and filing.

Except for relatively recent years, a large percentage of the available planet photographs obtained at various observatories in the past eighty years are now housed (at least in copy form) in Flagstaff. It is interesting to note that in 1969 alone, the Planetary Patrol has already nearly doubled the Planetary Center collection.

COMMITTEE 16C

Committee 16 C was created at the Berkeley meeting to secure and to study continuous records of the evolution and of rapid changes in atmospheric and surface features of planets. In order to do this, the committee promoted and coordinated observations between stations at several longitudes. It was also given the task of encouraging the formation of data centers and the collection of planetary photographs.

The committee, headed by the late Focas, organized a number of valuable international campaigns. The most successful of these was concerned with the photographic studies of Venus clouds in the ultraviolet and of the rings of Saturn at the moment of the passage of the earth through their plane in 1966.

Because of the development and growth of the data centers and the undertaking by Lowell Observatory of the NASA International Planetary Patrol program, it appears that the principal aims of the committee have already been achieved. Consultation which the President of Commission 16 has had with the heads of the two data centers and with Dollfus has indicated that one or both of these centers can now best organize special international planetary programs.

J. S. HALL
President of the Commission

17. LA LUNE (THE MOON)

PRÉSIDENT: A. Dollfus.
VICE-PRÉSIDENTS: A. A. Mikhailov, G. H. Pettengill.
SECRÉTAIRE: Miss B. Middlehurst.
COMITÉ D'ORGANISATION: G. P. Kuiper, D. H. Menzel, M. G. J. Minnaert.

Selon les instructions reçues de la part du Secrétaire Général de l'UAI, l'esprit général du présent "Draft Report" de la Commission 17 "La Lune" ne doit plus représenter, comme par le passé, un résumé des travaux effectués dans la discipline au cours des trois dernières années. Son rôle sera de donner un aperçu général sur les aspects et les tendances des recherches sur la Lune, la contribution présente et future de la Commission 17 du l'UAI dans son développement, ainsi que les coopérations dont elle est l'objet ou devrait l'être dans l'avenir.

L'évènement le plus marquant de ces trois dernières années a été l'exploration directe de la Lune. Pour la première fois un homme, N. Armstrong, a posé le pied sur le sol lunaire, le 21 Juillet 1969, à 2^h56^m T.U. Il était suivi peu après par E. Aldrin, tandis qu'un troisième co-équipier M. Collins, contribuait à la manœuvre. L'exploit, réussi par l'Organisation Américaine de Recherches Spatiales NASA, dans le cadre du programme "Apollo" a été renouvelé le 19 Novembre 1969 par C. Conrad, R. Gordon, et A. Bean. Ces missions d'exploration ont permis de ramener de très nombreux documents photographiques, une grande quantité d'échantillons sélectionnés du sol lunaire, de laisser sur le terrain des sismographes, des réflecteurs laser, des instruments de mesure de la densité de l'atmosphère lunaire, de l'intensité du champ magnétique, de la composition du vent solaire, etc.

Les possibilités nouvelles offertes à l'étude de la Lune par l'exploration directe entraînent une réorganisation fondamentale de la recherche dans ce domaine, dont nous allons examiner les implications concernant l'UAI.

STRUCTURATION DE LA COMMISSION 17 DE L'UAI

La Commission 17 de l'UAI La Lune a eu le regret de perdre l'un de ses Vice-Présidents, le Professeur A. Lebedinsky, décédé brusquement en mission scientifique, quelques semaines après sa nomination aux fonctions de la Commission. Il a été remplacé par le Professeur A. A. Mikhailov.

J. H. Focas est décédé le 3 Janvier 1969; il a été remplacé par le Professeur S. Miyamoto.

La Commission 17 de l'UAI s'est vu préciser sa mission de coordination des recherches lunaires internationales par le Comité Exécutif de l'ICSU de la façon suivante (ICSU Bulletin, 1966): *"Commission 17 of the IAU, Commission of the Moon, is agreed to be the most appropriate body to coordinate studies of problems relating to the Moon."*

Afin d'être en mesure de remplir la tâche de coordination des études lunaires que lui a confiées spécifiquement l'ICSU, dans le cadre du développement quasi explosif des recherches sur la Lune ces dernières années, la présente Commission a jugé efficace de diviser ses activités en 3 groupes de travail.

1. *Groupe de Travail: Nomenclature Lunaire*

Ce groupe de Travail a été créé lors de l'Assemblée Générale de l'UAI à Prague en 1967, avec pour mission:

"... The purpose of this Committee is the continuation of the work on Lunar Nomenclature, according to Resolution No. 8 adopted during the General Assembly at Prague in 1967: "The assignment of names and permanent designations to features on the far side of the Moon will be postponed until the XIVth General Assembly. As an interim measure, a Working Group will assign numbers to about 500 major lunar formations." It has been agreed that the members of this Committee – none of whom is personally and directly engaged in the work of Lunar Topography –

should represent the work of each country through the National Astronomical Committees concerned. It is desirable that, in a general way, the work on Lunar Nomenclature be guided in conformity with the decisions already adopted by the IAU, defined by Resolution No. 2 of Commission 16, taken during the General Assembly at Berkeley in 1961 (*Trans. IAU* **11**, 1961, 235) and completed by the list of names adopted in 1964 (*Trans. IAU* **12**, 1964, 202)."

Ce travail avait été préparé les années précédentes par le Professeur Z. Kopal, chargé dès 1965 par la Commission d'explorer le domaine. La composition du groupe de travail en 1967 était: D. Menzel (President), A. A. Mikhailov, M. Minnaert, A. Dollfus (*ex-officio*).

Au début de 1969, le Professeur Mikhailov a demandé à se faire remplacer par B. Levin. Le Groupe de Travail s'est réuni quatre fois en 1968–69 à Cambridge (Mass.), à Paris (France), à Moscou (U.R.S.S.), et à Meudon (France). Ses activités ont fait l'objet d'un rapport particulier préparé par le Prof. Menzel. Grâce à la consultation des Académies des Sciences de tous les pays membres de l'UAI, le Groupe de Travail a recensé un vaste choix de propositions de noms susceptibles d'être attribués à des cratères; elle en a effectué une sélection attentive, établi les biographies justificatives, recensé les cratères de la face arrière de la Lune dont les dimensions justifient une désignation, et attribué les noms en préservant une répartition alphabétique mnémotechnique. Des aménagements de la règle alphabétique sont apparues souhaitables pour une quinzaine de grands bassins d'une importance topographique évidente. La proposition résultant de ce travail doit être soumise aux remarques et à l'approbation de tous les membres de la Commission.

2. *Groupe de Travail: Figures et mouvements de la Lune*

Le champ des activités dont la Commission est responsable étant trop vaste, il apparut nécessaire de subdivider les membres en groupes plus restreints et plus spécialisés. Le premier groupe est concerné par les activités qui étaient antérieurement celles de l'ancienne Commission 17 "Mouvements et figures de la Lune" avant que ces attributions ne se soient élargies en 1965. Th. Weimer (Observatoire de Paris, France) a accepté d'en assurer la mise en place en lui définissant des objectifs exploratoires initiaux limités:

"... The activity of this group should cover the following fields: cartography (except nomenclature), selenodesy, gravific potential, orbital movement (theory and observations), rotation (theory and observations).

"In a first state, the group would assume a mission of information and relationship. Each member should address to the chairman of the group reprints or summaries of all works he knows. The list and a short analysis of these publications will be sent by the chairman to all members."

Les personnalités suivantes ont accepté de participer à ces activités:

Membres de la Commission 17: Arthur, D. W. G., Eckert, W. J., Habibullin, S. T., Hall, R. G., Hopmann, J., Jakovkin, A. A., Kopal, Z., Koziel, K., Mikhailov, A. A., Nefediev, A. A., Potter, H. J., Weimer, Th.

Membres des autres commissions de l'UAI: Goudas, C. L. (Patras, Grèce); Shapiro, I. I. (Lexington, U.S.A.); Yaplee, B. S. (Washington, U.S.A.).

Membres extérieurs à l'UAI: Alley, C. N., Univ. of Maryland, U.S.A.; Carder, R., ACIC-USAF, 2nd and Arsenal Street, St. Louis, Missouri 63118, U.S.A.; Hunt, M., AFCRL, Lunar Laser Field Unit, Lunar and Planetary Lab., University of Arizona, Tucson, Ariz. 85721, U.S.A.; Michael, W. H. Jr., NASA Langley Research Center, Hampton, Virginia 23365, U.S.A.; Orszag, Labor. de Physique Vignal, École Polytechnique, 17, Rue Descartes, Paris, France.

Des échanges réguliers de bibliographie sont en cours.

3. *Groupe de Travail: Lunar Geology and Geophysics (Géologie et Géophysique lunaires)*

Les nouveaux aspects de la physique lunaire résultant de l'exploration directe ouvrent le champ des recherches à de nouvelles disciplines. La nécessité d'y adapter la Commission 17 est évidente. La Commission possède parmi ses membres des spécialistes de géologie et géophysique, mais en nombre insuffisant et ceux-ci couvrent incomplètement la discipline; de nombreuses personnalités, membres de la Commission ou extérieurs, ont insisté sur le besoin de les regrouper dans leur spécialité, selon la formule du Groupe de Travail (Working Group) qui possède l'avantage de pouvoir s'adjoindre des participants non membres de l'UAI pouvant appartenir à des disciplines autres que celles de l'Astronomie:

"The principal aim of the Working Group "Lunar Geology and Geophysics" is to assist with the Geological and Geophysical Interpretation of the Moon in close connection with astrophysical implications. The Working Group might also wish to make recommendations about the most efficient means of analysing the lunar surface and interior, including returned samples."

G. Fielder a accepté de mettre en place ce Groupe de Travail comprenant au départ les membres concernés de la présente commission 17 de l'UAI, et un représentant de chacune des autres Associations Internationales concernées. Ce groupe devra s'adjoindre des spécialistes, géologues et géophysiciens appropriés:

Membres appartenant à la Commission 17: Kuiper, Levin, Liddel, Martynov, Menzel, Middlehurst, Minnaert, Miyamoto, O'Keefe, Shoemaker, Strom, Urey.

Représentants des unions scientifiques: Dollfus (Union Astronomique Internationale); Elston (American Geophysical Union); Fielder (International Union of Geological Sciences); Green (International Association of Planetology); Kaiser (Inter-union Commission on solar terrestrial physics) Dept. of Physics, Univ. of Sheffield, England; Pettengill (Union Radio Scientifique Internationale); Runcorn (Union Géodésique et Géophysique Internationale); School of Physics, Univ. of Newcastle upon Tyne, England; Sagan (COSPAR); Tazieff (International Association on Volcanology and Chemistry of the Earth's Interior) 15 Quai Bourbon Paris 14, France.

L'une des premières activités dépendant de ce Groupe de travail se trouve être la préparation d'un "Catalogue des cratères terrestres et structures connexes", entrepris par J. Fryer et C. Titulaer, à l'Observatoire de Meudon, avec l'appui de "European Space Research Organization" (ESRO). Le but est défini ainsi:

"... Ce catalogue est destiné à permettre une étude des cratères de toutes dimensions, comprenant les petites cavités artificielles, ou d'origine volcanique, les dépressions formées par les effondrements, les impacts de météorites jusqu'aux plus grands types connus. Il devra comprendre également le plus grand nombre d'informations accessibles relatives aux dimensions, à la position, à l'histoire géologique de ces cratères, etc."

La première partie du catalogue concerne le Canada et doit être publiée au début de 1970 ("Catalogue Général des formations géologiques circumlunaires terrestres: Partie I, Canada.")

NOUVELLES ORGANISATIONS INTERNATIONALES

Deux nouvelles *Organisations Internationales* se sont trouvées concernées par l'extension du domaine des études sur la Lune ces trois dernières années: le COSPAR et l'IUGS. La présente Commission de l'UAI a eu à définir, avec les intéressés, les domaines respectifs et les modes de coopération entre ces organismes:

(*a*) *COSPAR:* A la suite des entretiens avec les responsables des Commissions 16 (Planètes) et 17 (Lune) de l'UAI, le COSPAR a créé un *Groupe de Travail W.G. VII: "Space-related studies of the Moon and Planets"*. Les attributions de ce Groupe de Travail se définissent ainsi:

"... *Working Group VII* is: – to organize meetings within COSPAR on the subject of its title – to coordinate its activities with interested Unions, particularly with Commissions 16 and 17 of the International Astronomical Union and Commission V of the URSI – and to function otherwise in a manner similar to the other Working Groups. COSPAR intends to leave such traditional matters as lunar and planetary nomenclature and cartography entirely to the IAU."

Les membres du Working Group VII du COSPAR sont les suivants:

Chairman: A. D. Kuzmin, Lebedev Physical Institute, Moscow, U.S.S.R.

Vice-Chairman: C. Sagan, Center for Radiophysics and Space Research Cornell University, Ithaca, N.Y., U.S.A.

Membres: Abysov, S. S.; Colombo, G.; Dollfus, A.; Drake, F. D.; Eshelman, V.; Florensky, C. P.; Hall, J.; Hayakawa, S.; Johnson, F. S.; Kopal, Z.; Kuiper, G. P.; Levin, B. J.; Link, F.; Marov, N.; Mayer, C.; Ness, N.; Pimentel, G.; Prokofiev, V. K.; Rea, D.; Shoemaker, E.; Troitsky, V. S.

(*b*) *IUGS:* L'"International Union for Geophysical Sciences" a accepté l'affiliation d'une organisation nouvellement créée, l'"International Association of Planetology (IAP). Cette Association définit ainsi ses buts:

"... Our charter specifically deals with the investigation of planets, their satellites, and interplanetary matter from a geoscientific point of view. The research methods employed are geological, geophysical, geochemical, meteorological and astronomical, supported by recently acquired astrophysical and astronautical data."

La composition initiale de cette Association est:

Président d'honneur: G. N. Katterfeld.

Président: J. Green.

Secrétaire Général: Madame N. Stovickova.

Vice-Présidents: S. Miyamoto, H. Tazieff, K. von Bülow, K. Rankama, P. Hedervari, K. Beneš.

Les attributions de l'IAP et son incorporation dans la coordination internationale ont résulté de nombreux échanges de lettres entre les membres concernés de la Commission 17, son Président, et les membres du bureau de l'IAP.

IMPORTANTES INITIATIVES DE COORDINATION

(a) La branche de l'organisation américaine de recherches spatiales NASA, *"Apollo Lunar Exploration Office"*, dirigé par L. R. Scherer, a joué un rôle d'une importance particulière pour les domaines couverts par notre Commission. Elle a assuré la conception et la coordination des opérations qui ont conduit à l'exploration directe de la Lune et à l'exploitation scientifique des observations effectuées.

Les travaux ont conduit, notamment, à la mise en place au Manned Space-flight Center, Houston, Texas du *"Lunar Receiving Laboratory"* dirigé par E. King, puis par A. Callio, dont les objectifs sont les suivants:

"Functions of this laboratory are as follows: (1) to quarantine the lunar material for an appropriate period of observation; (2) to perform a preliminary examination of this material (chemical, mineralogical, and other examinations); (3) to perform certain time-critical experiments which could not wait until the end of quarantine; and (4) to distribute the lunar samples to approved investigators and to be depository for the permanent collection of lunar materials."

Afin d'assurer une étude scientifique exhaustive des premiers échantillons du sol lunaire ramenés sur Terre par la mission spatiale Apollo XI, 4·2 kg de poudre lunaire, et 30 blocs rocheux d'un poids total de 4 kg ont été distribués en Septembre 1969 à 106 laboratoires américains et 36 laboratoires de tous autres pays, tous sélectionnés sur les critères de leurs compétences particulières. Les résultats des différentes études ont fait l'objet d'une confrontation lors d'une conférence internationale en Janvier 1970 à Houston, Texas, et feront l'objet d'un rapport final unique dans la revue *Science*.

(b) La nouvelle publication scientifique "*The Moon* – An International Journal of Lunar Studies", Éditeurs H. Alfvén, Z. Kopal, H. C. Urey, Éditions Reidel, Dordrecht, Hollande. Premier fascicule paru en Novembre 1969. Les éditeurs définissent ainsi le but du journal:

"... *The Moon* is a journal which provides an inter-disciplinary but monothematic medium for publication of the results of original investigations in all fields of lunar studies, including: astronomy (both optical and radio-astronomy), astronautics, chemistry, geology, space physics, and all other aspects of the scientific study of our satellite by spacecraft as well as by the more traditional approaches of ground-based astronomy. In addition, the editors of this journal expect to publish from time to time invited review papers in the form of a critical discussion on problems of current interest; and keep a column open for book reviews, as well as reports on international meetings concerned with any aspect of lunar studies. An integral part of each issue of our journal should be a current critical bibliography of the entire field of lunar studies – an essential and truly indispensable service for a rapidly growing interdisciplinary domain of our science – containing titles, abstracts, or critical comments, on work published concurrently in all other journals or more informal sources."

(c) *Le ' National Space Science Data Center"* de la NASA (Goddard Space-flight Center – Code 601 – Greenbelt, Maryland 20771 – USA) définit ses responsabilités ainsi:

"... The National Space Science Data Center (NSSDC) was established by NASA to further the widest practicable use of reduced data obtained from space science investigations and to provide investigators with an active repository for such data. As such, it is responsible for the active collection, organization, storage, announcement, retrieval, dissemination, and exchange of data received from satellite experiments, sounding-rocket probes, and high altitude aeronautical and balloon investigations."

Ce Centre de Documentation regroupe en particulier tous les clichés et les mesures recueillis par les sondes spatiales Ranger, Surveyor, Orbiter, Explorer, Apollo, etc., et les tient disponibles pour des programmes de recherches scientifiques. Chaque nouvelle série de documents fait l'objet d'un "Data Announcement Bulletin."

(d) Le *"Lunar International Observers' Network"* (LION) of the Smithsonian Institution Center for short-lived phenomena" dirigé par R. Citron (Cambridge, Mass., U.S.A.) et décrit comme il suit:

"The Lunar International Observers' Network was organized by Miss Barbara Middlehurst during the past several years and now includes 176 astronomers in 31 countries around the world. The network is composed of professional and amateur astronomers who have particular interest in transient lunar phenomena .... The main objectives of these observing programs are to record the nature, magnitude, frequency, and distribution of transient lunar phenomena, and possibly obtain independent correlation of events from widely separated groundbased observatories."

A l'occasion de la mission de survol circumlunaire Apollo X, les observations ont permis de maintenir la Lune sous surveillance 24 heures par jour pendant 6 journées consécutives; 23 phénomènes lumineux ont été enregistrés dans 8 régions lunaires différentes. (Cf. "Communications Support for Transient Lunar Phenomena during the Apollo X Mission 28 May 1969, Smithsonian Institution Publications".)

COMMENTAIRES SUR LE RÔLE DE LA COMMISSION 17

La conduite des responsabilités de la Commission 17 a nécessité des consultations et échanges de nombreuses correspondances avec les membres de la Commission et avec d'autres personnes qualifiées. Voici quelques extraits de lettres choisis pour exprimer qualques-uns des besoins de la Commission:

Observations par le laser

– "... Pour la mise en oeuvre et l'exploitation de ces nouvelles possibilités, la coopération internationale est plus nécessaire que jamais. Mais cette technique n'étant plus, en général, du ressort des astronomes et des observatoires astronomiques au sens classique, il est urgent de compléter la Commission 17 par la nomination de nouveaux membres faisant des recherches dans ce domaine. Le Groupe de Travail "Figures et Mouvements de la Lune" créé au cours de l'été 1968 est un premier pas vers cette coopération internationale active."

– "... Des résultats ont été obtenus aux États-Unis par le groupe de Faller (Observatoire Lick); le groupe de Alley (Observatoire McDonald); par le groupe d'Eckhardt (Tucson) et par Orszag (Pic-du-Midi). Actuellement, en plus d'une coopération étroite entre les groupes de J. Faller et de C. O. Alley, il existe une coopération entre le second de ces groupes et le Pic-du-Midi. Certains contacts, d'autre part, étaient pris dans le cadre de la coopération franco-soviétique avec le groupe Kokurin. Un rôle possible pour la Commission 17 de l'UAI pourrait être d'encourager ces coopérations, en particulier dans la mesure où la mise en commun des résultats émanant de stations assez uniformément réparties en latitude accroît notablement la précision des déterminations lunaires."

Observations par le radar

– "At the present, all radar astronomy, both lunar and planetary, appears to be concentrated in the U.S.A. For this reason, the part played by the IAU in coordinating radar observations is minimal and will likely remain so."

Champ gravitationnel et concentrations de masse

– "... Commission 17 could attempt to encourage an international effort in application and interpretation of these results. Scientists not involved in particular experiments will, nevertheless, have access to the results, and they could conduct research investigations applying the results at a different point of view from those initially involved. Also, there will probably be numerous interesting areas of research in which multi-disciplinary information should be combined for particular objectives."

Luminescence du sol lunaire

– "... Parallèlement aux observations du sol lunaire, d'excellentes études de simulation du phénomène en laboratoire ont été menées ces dernières années. Particulièrement, l'analyse spectrale de la luminescence créée par bombardement de protons sur différentes roches (Geake et Alley, 1967; Nash, 1966), et l'étude de la thermo-luminescence (Blair et Edgington, 1968; Sun et Gonzalez, 1966). On voit l'intérêt de premier ordre de ces travaux de laboratoire lorsque les matériaux utilisés seront des roches lunaires. Un programme de collaboration s'impose entre les expériences en laboratoire et les observations de la luminescence lunaire à partir du sol."

Échantillons du sol lunaire

– "... The recent search of interest in the analysis of returned lunar material had led to the application of many specialists techniques to lunar studies. The investigators using these techniques should be invited to participate in a Commission 17 meeting: the importance of their producing results and of their analysis within the medium of the IAU should be stressed. Only if Commission 17 can be fed with all these pertinent data, will it be in a position to make balanced recommendations for the future of Lunar research."

Groupe de Travail "Géologie et Géophysique Lunaires"

– "... Exchange of views between those schooled in the disciplines Astronomy, Geology, Geomorphology, Meteorology, Geophysics, would be especially desirable within the forum of Commission 17. Therefore, members of the new Commission 17's structure should be selected from these disciplines as a minimum requirement. Furthermore, there should be representatives of all international and national bodies concerned with lunar studies.... I propose that, in the first instance, this Working Group should (a) nominate representatives for the main body of Commission 17, following the above scheme; (b) advise Commission 17 of the most useful and efficient way to organize and run meetings. It may be desirable for the Working Group to make recommendations about the most efficient means of analyzing the lunar surface (including returned samples) and interior; but the principal aim of the Working Group would be to assist in general terms with the geological and geophysical interpretation of the Moon in close connection with the relevant astrophysical topics."

– "... Pour le moment, il me semble en effet que la Commission 17 de l'Union Astronomique Internationale est l'organisme le plus indiqué pour constituer un tel groupe de travail. Je ne suis pas sûr qu'à l'avenir ce groupe restera le point de rencontre central."

– "... It must be assured that the quantitative astronomical, astrophysical, geophysical and geochemical approach will dominate the qualitative, descriptive, geological approach."

– "... I would not endorse any external policy-making recommendations of this Working Group in the area of Geosciences. This responsibility should probably be within the International Union of Geological Sciences, or within the International Union of Geodesy and Geophysics. Traditionally, the IUGS and the IUGG have had the authority to pass judgment on geoscientific recommendations made by any recognized national or international groups, when this recommendation was of international policy-making significance in the fields of geology, geophysics, and geochemistry."

– "... The problems of lunar geology involve the bringing together of a wide range of scientists. The assimilation and understanding of the flood of data which is pouring in on us will be a task not only for the astronomers in the countries who possess facilities for experiment and observation, but also for the countries where theoretical research dominates. It often happens that a theoretical man will hesitate to attack a problem because he is not sure what the exact observational situation is. Hopefully, this Working Group can add the isolated workers to contribute to the overall understanding of the Moon."

Rôle général de l'UAI

"... Commission 17 should recommend the release of national funds to activate personal exchanges on a regular footing. This is a most efficient way of disseminating knowledge and giving birth to new ideas. The way in which the IAU can assist in this aim is by drawing attention to the work done by various national groups, individuals, and by disseminating this knowledge sufficiently widely and with some direction, to assist research workers throughout the world."

– "... Commission 17 must respond to the modern trends and rapidly adapt itself to advise on the coordination of the various inter-disciplinary activities in different countries. The first requirement is to adjust the membership of Commission 17 to take in interests previously absent from its ranks.... Selected abstracts should be circulated to both professional and amateur lunar workers. Of necessity, a list of such workers should be prepared."

– "... I have thought about the future relations between Commissions 16 and 17 of the IAU.... It seems clear that there is much to be said for combining the functions of the two Commissions ultimately since the IAU's role in lunar science will not differ from that in planetary science (problems of nomenclature, international organization of worthy scientific projects, definition of coordinate systems, etc.). These functions were already combined before in Commission 16. Nevertheless, we also felt that with the nomenclature questions concerning the Moon and Mars now pending, it would be destructive to have the organization changed already at the next meeting (1970). Rather, a period of adjustment would seem indicated so that reunification would probably be made more appropriately in 1973..."

– "... Recent experience has made me feel that coordination of the many groups in such diffuse fields as Moon and planets would be impossible, that we are already witnessing too much competition for limited attention."

A. DOLLFUS
Président de la Commission

19. ROTATION DE LA TERRE (ROTATION OF THE EARTH)

PRÉSIDENT: P. Melchior.
VICE-PRÉSIDENT: H. M. Smith.
COMITÉ D'ORGANISATION: E. P. Fedorov, W. Markowitz, S. Yumi.

La commission déplore le décès de Madame Aksentieva, directeur de l'Observatoire de Poltava et du Professeur Demetrescu, ancien directeur de l'Observatoire de Bucarest.

1. INTRODUCTION

L'étude de la rotation de la Terre poursuit plusieurs objectifs scientifiques fondamentaux:

(1) Fournir des données d'observations continues d'où l'on pourra tirer un ensemble de paramètres servant de base à l'élaboration des théories sur la constitution interne de la Terre et sur la dynamique du système Terre-Lune.

(2) Fournir à tout moment la position d'un trièdre de référence attaché au manteau terrestre par rapport à un trièdre fixé dans l'espace, travail d'une nécessité primordiale pour l'astronomie fondamentale, la géodésie et la navigation spatiale.

(3) Permettre de nouvelles recherches sur certaines constantes fondamentales de l'astronomie (nutation, aberration) grâce à la constitution de très longues séries d'observations homogènes.

Définition des références utilisées

(a) Il nous paraît nécessaire de revenir sur le sens à donner à la définition de OCI.

Ce point est choisi de manière purement conventionnelle. Sa position résulte de la résolution par moindres carrés des équations traditionnelles dans lesquelles on introduit cinq latitudes adoptées conventionnellement comme latitudes moyennes des cinq stations du SIL: Mizusawa, Ukiah, Gaithersburg, Carloforte et Kitab.

Le point OCI n'est donc pas un "pôle" ni un "pôle moyen" et n'en possède aucune des propriétés auxquelles les géophysiciens sont intéressés. En outre, il n'est nullement rapporté à une époque préférentielle et c'est par exemple une absurdité que de le rapporter à l'intervalle "1900–1905" puisque la station de Kitab qui sert à le définir n'est entrée en fonction qu'en 1932...

Les termes choisis à Prague ont été soigneusement pesés: ce point est une origine conventionnelle internationale et rien de plus.

(b) En ce qui concerne le T.U.$_1$ il convenait également de choisir un point origine sur l'équateur de OCI.

Avant 1968 ce point O était défini par le BIH sur l'équateur du pôle moyen P_M adopté et à partir de la longitude de l'observatoire moyen (pondération des longitudes de tous les observatoires participant au travail du BIH).

A partir de 1968, pour éviter d'introduire une discontinuité dans T.U.$_1$, le BIH a adopté comme nouvelle origine l'intersection du méridien P_MO avec l'équateur de OCI. Ceci implique une révision de toutes les longitudes y compris celle de Greenwich qui n'est plus nulle.

(c) Plusieurs membres de la Commission préfèrent utiliser comme origine du mouvement un pôle moyen de la date. Celui-ci dérivant lentement serait le pôle d'inertie et aurait une signification géophysique.

D'autres membres estiment qu'une origine mobile est un inconvénient.

L'opinion du Président est qu'il est bien facile de satisfaire les uns et les autres en publiant les deux systèmes de coordonnées (x, y) comme cela avait été proposé lors de l'Assemblée de l'UGGI à Helsinki.

2. MESURES DE LA LATITUDE ET DU TEMPS UNIVERSEL

(a) *Modifications à la liste des stations figurant dans le Rapport présenté à Prague (1967)*

Calgary	Le PZT d'Ottawa a été transféré en cette station située sur le parallèle de Herstmonceux et des observations conjuguées sont commencées.
Ottawa	a été équipé d'un nouveau PZT.
La Plata	équipée d'un PZT à Punta Indio.
Poltava	ajouter LZV de 180 mm.
San Fernando	ajouter Astrolabe OPL.
Saint Michel	fermée fin 1968.
Uccle	supprimer 2 IP.

(b) *Chaîne de PZT dans l'hémisphère Sud*

Des efforts avaient été entrepris pour coordonner l'installation de trois PZT sur un même parallèle de l'hémisphère Sud. L'Australie et l'Argentine disposent maintenant d'un tel instrument et l'Afrique du Sud aurait sans doute pris la décision d'en acquérir un si le parallèle choisi par les deux autres pays avait traversé le continent africain (Le Cap $\varphi = -34°$).

Malheureusement il n'a pas été possible jusqu'ici d'obtenir le déplacement du PZT de Mount Stromlo ($\varphi = -35°$) vers le Nord et en conséquence l'Argentine a placé son PZT à Punta Indio (La Plata, $\varphi = -35°$). Ces deux stations travaillent en étroite coopération.

3. ÉTUDES ET RECHERCHES SUR LES INSTRUMENTS

A Mizusawa, Sugawa, Hurukawa, Ooe et Abe ont étudié la digitisation par servomécanisme du micromètre de la lunette zénithale. Ils ont également développé un servo-niveau pour maintenir automatiquement l'orientation du télescope.

Guinot met au point un astrolabe photoélectrique.

Pavlov a construit deux nouveaux instruments de passages photoélectriques.

Deux nouveaux PZT ont été construits dans les ateliers de l'U.S. Naval Observatory à Washington. Ils ont été installées à Ottawa et La Plata (Punta Indio).

4. PROBLÈMES DE RÉDUCTION LIÉS AUX CATALOGUES D'ÉTOILES

A Uccle, P. Melchior et R. Dejaiffe ont déterminé, dans le système FK4, les déclinaisons et mouvements propres en déclinaison de toutes les étoiles utilisées par le SIL depuis 1900. Pour ce faire, ils ont discuté 11 501 positions figurant dans 463 catalogues méridiens différents.

Y. Wako a montré que ce catalogue est en excellent accord avec les corrections de déclinaisons déduites aussi bien par la méthode en chaîne que par la méthode du terme z pour tous les programmes du SIL depuis 1900.

A Hambourg, les positions des étoiles déterminées dans le système du PZT de cet observatoire ont été comparées à celles rattachées au FK4 par l'U.S. Naval Observatory. On a constaté que si ce dernier système était utilisé le terme z local annuel disparaîtrait.

A Poulkovo, N. N. Pavlov a établi un catalogue des 807 étoiles utilisées pour la détermination de l'heure en U.R.S.S. basé sur 185 000 observations faites dans 9 observatoires.

Iijima et Fujii ont comparé les positions moyennes des étoiles du PZT de Tokyo avec le catalogue SAO et avec les résultats du cercle méridien de Tokyo (α, δ et mouvements propres).

A Herstmonceux on a procédé à une analyse des observations faites de 1958 à 1967 par la méthode en chaîne en vue de déterminer les erreurs des positions d'étoiles. Une correction de $+0''40$, déduite d'observations méridiennes et d'une comparaison PZT-astrolabe, a été appliquée aux déclinaisons de toutes les étoiles.

Shepunov utilise 10 années d'observations de latitude à Blagoveshchensk pour corriger déclinaisons et mouvements propres.

Takagi a abordé ces problèmes sous l'angle théorique.

Yokoyama a discuté les corrections aux positions des étoiles FK4 utilisées à l'astrolabe de Mizusawa.

A Washington également les observations du PZT (en latitude: 1915–1969 et en temps: 1935–1969) ont été réduites en utilisant de nouvelles positions d'étoiles.

Des calculs analogues sont en cours pour le PZT de Richmond.

5. NOUVELLES RÉDUCTIONS DE LONGUES SÉRIES D'OBSERVATIONS

L'accumulation progressive des observations faites suivant un même programme permet la constitution de séries temporelles que l'on soumet maintenant à l'analyse statistique suivant des critères modernes plus approfondis que par le passé.

Nous citerons en particulier plusieurs études intéressantes dues à

Billaud (Paris)	sur la série de Washington	1915–1969
Jorge-Silva (Paris)	sur la série de Greenwich	1958–1967
Kostina et Sakharov	sur la série de Pulkovo	1948–1968
Yusupov	sur les observations de Kazan	
Slavinskaya	sur la série d'observations à l'astrolabe de Poltava	1961–1964.

Dans cet ordre d'idées les astronomes de Kiev, sous la direction de E. P. Fedorov, se consacrent à une étude statistique de toutes les séries d'observations (SIL et stations indépendantes) en leur assignant des poids différents en fonction de critères rigoureux. Ils se proposent de déterminer ainsi les coordonnées du pôle pour la période 1897–1970 rapportées à OCI aussi bien qu'au pôle d'inertie.

Enfin Mrs Rykhlova (Sternberg A. I.) a discuté les observations faites de 1846 à 1891 à Washington, Greenwich et Poulkovo pour en déduire les coordonnées du pôle pour cet intervalle de temps.

Billaud (Paris) recherche les corrélations entre les observations simultanées faites avec deux astrolabes voisins.

Persianinova (Poulkovo) a recherché la composante non polaire des variations de latitude pour trois télescopes zénithaux, le PZT de Mizusawa et l'astrolabe de Paris et constaté que le terme z est différent pour chaque instrument et corrélé aux différences de température dans les salles d'observation.

6. RECHERCHE DES PARAMÈTRES CARACTÉRISTIQUES DES VARIATIONS DE LATITUDE

(a) *Mouvement chandlérien*

La période de Chandler τ son amplitude A et son temps de relaxation T ont fait l'objet de nouvelles recherches.

Sakharov a déduit des 20 ans d'observations de Poulkovo (1948–1967): $\tau = 1{\cdot}188$ an.

Dans son étude de l'intervalle 1846–1891, Mrs Rykhlova a obtenu:

$$\tau = 1{\cdot}208 \text{ an} \qquad T = 9 \text{ ans}.$$

Enslin déduit des observations effectuées de 1962·0 à 1968·0 au PZT de Hambourg:

$$\tau = 1{\cdot}19 \text{ an}$$
$$A = 0''15.$$

Takagi met en évidence un saut de valeur de τ dans l'intervalle 1918–1923.

Vanicek a fait une analyse spectrale du mouvement du pôle 1951–1966 et a trouvé $\tau = 434{\cdot}64\,j$ ou $435{\cdot}08\,j$ avec une amplitude A variant de 0″134 à 0″278 avec une période de 23·75 ans.

(b) *Nutation diurne libre*

N. A. Popov a étudié les longues séries d'observations de deux étoiles brillantes effectuées à Poltava depuis la fondation de cet observatoire. Il trouve une variation d'amplitude de la nutation diurne dont la période serait comprise entre 10 et 12 ans.

Iatskiv confirme ce résultat par l'analyse de la série 1915–1928 de Poulkovo.

S. Débarbat a fait une analyse détaillée des observations faites à l'astrolabe de Paris et obtient une période propre de $23^h57^m7^s$.

(c) *Variations diurnes*

Panchenko a analysé les séries d'observations des astrolabes de Paris, Herstmonceux et Poltava.

Kostina a analysé les données de Poulkovo de 1955 à 1961 en éliminant les variations forcées, les déviations de la verticale dues aux marées et en utilisant la nouvelle valeur de la constante d'aberration. Dans ces conditions l'amplitude de la variation diurne est réduite de moitié tandis que la phase reste inchangée: $\Delta\varphi_{\odot} = 0''015 \cos(t_{\odot} + 7^h9)$.

(d) *Marées terrestres et effets du noyau liquide de la Terre*

Comme nous l'avons montré à l'Assemblée de Prague, il existe un lien fondamental entre les marées et les nutations, chaque nutation correspondant à un couple d'ondes de marées.

C'est dire l'intérêt de la recherche des caractéristiques des marées terrestres par voie astronomique.

Les marées terrestre à longue période (15 jours, 6 mois) sont des marées zonales qui provoquent des déformations suivant l'axe polaire et par conséquent des variations concomitantes de la vitesse de rotation de la Terre.

Pilnik (Sternberg) a fait une analyse approfondie de ces variations et a trouvé pour le nombre de Love k (paramètre rhéologique de la Terre) la valeur:

$$k = 0{\cdot}301 \pm 0{\cdot}004$$

en bon accord avec les résultats les plus récents des mesures directes de marées terrestres (Melchior: $k = 0{\cdot}328$).

Les déviations de la verticale, qui perturbent évidemment la position du plan méridien, ont été recherchées dans les séries d'observations à l'astrolabe de Paris par S. Débarbat qui a déterminé des valeurs numériques possibles pour la combinaison des nombres de Love h et l: $\Lambda = 1 + k - l$.

Proverbio (Milano) étudie le même phénomène.

Gubanov (Poulkovo) a recherché l'amplitude de la nutation bi-mensuelle et les termes d'Oppolzer dans le mouvement du pôle.

Takagi a déduit les équations différentielles stochastiques de la rotation de la Terre par la théorie des fonctions aléatoires. Il a étudié aussi un modèle de l'intérieur de la Terre.

Kakatu (Mizusawa) a discuté l'effet de la compressibilité du noyau liquide sur les conditions à la frontière manteau-noyau exprimées dans la théorie de Molodensky de la nutation et des marées terrestres. Il conclut à un effet de quelques pour cents donc non négligeable.

Il examine également les oscillations forcées hydromagnétiques induites dans le noyau par le mouvement du pôle et conclut à une corrélation théorique entre la dérive du champ magnétique et un mouvement du pôle à longue période associé à de petites variations du géopotentiel à la surface de la Terre.

Sugawa a tenté de déduire l'inégalité des moments équatoriaux A et B de la Terre de l'ellipticité du mouvement chandlérien de 1900 à 1965. Il résulte de ses calculs une différence des axes équatoriaux a, b dix fois plus forte que celle donnée par les perturbations d'orbites de satellites artificiels. Sugawa invoque la possibilité d'un effet de mouvements dans le noyau liquide.

Signalons encore que, dans le but d'étudier l'effet du noyau liquide de la Terre dans les marées tesserales associées aux nutations astronomiques, Melchior, Bonatz et Blankenburgh ont installé des appareils au Spitzberg (mesure directe par pendules horizontaux du couple de précession-nutation appliqué à la Terre, couple qui est maximum au pôle).

(e) *Constantes fondamentales de l'Astronomie*

Les longues séries d'observations de latitude constituent, en raison de leur homogénéité, un matériel de choix pour l'étude des constantes fondamentales de nutation et d'aberration.

A Washington on utilise les observations de 1915 à 1969 corrigées des erreurs de catalogue.

Une série de 20 années d'observations (1948–1967) exécutées à Poulkovo et comportant 48000 déterminations de latitude est à l'étude.

Wako attribue le terme z annuel aux erreurs commises sur les coefficients adoptés pour la nutation semi-annuelle en longitude et en obliquité (effets du noyau liquide).

Fleckenstein poursuit des recherches sur la constante de l'aberration.

7. RELATIONS AVEC LES PHÉNOMÈNES GÉOPHYSIQUES

(a) *Effets météorologiques*

L'étude des effets du vent a été reprise à Potsdam où l'on a trouvé une corrélation avec le mouvement du pôle et la vitesse de rotation.

Kirita étudie la variation saisonnière du vortex circumpolaire et la distribution asymétrique des masses d'air autour du pôle. Il déduit la valeur numérique de la fonction d'excitation à partir des données globales des sondages à haute altitude et en tire un mouvement annuel du pôle

$$x = +0''141 \sin(\odot + 190°6)$$
$$y = +0''158 \sin(\odot + 97°3).$$

Le terme annuel déduit des résultats du SIL étant

$$x_a = +0''095 \sin(\odot + 216°4)$$
$$y_a = +0''079 \sin(\odot + 106°2).$$

Kirita estime qu'une grande part de ce mouvement trouve son origine dans la variation saisonnière du vortex circumpolaire.

Pour ce terme annuel Madame Stoyko trouve

$$x_a = 0''089 \sin(\odot + 201°)$$
$$y_a = 0''087 \sin(\odot + 292°).$$

(b) *Dérive des continents*

L'importance de ce phénomène avait été bien soulignée au Symposium de Stresa. Aussi divers auteurs se sont préoccupés de vérifier la stabilité des stations de mesure de l'heure et de la latitude.

Wm. Markowitz considère que les variations progressives et à longue période de latitude peuvent être représentées par le seul mouvement du pôle moyen et qu'une explication au moyen de rotations et déplacements de la croûte demanderait un ensemble de circonstances très fortuites.

Selon Enslin la latitude moyenne de Hambourg dérive de 0″016/an et cet effet n'est pas dû aux erreurs sur les mouvements propres des étoiles.

Madame A. Stoyko a réduit les données du SIL de 1900 à 1966 dans un système uniforme et en a déduit une dérive du pôle de 0″0032/an sur le méridien de 70° W, une dérive locale de −0″003/an à Mizusawa et +0″003/an à Ukiah.

De leur côté, Yumi et Y. Wako ont calculé la variation séculaire de latitude de ces deux stations:

Mizusawa: $-0''0016$/an
Ukiah: $+0''0011$/an.

Après élimination de ces dérives locales le mouvement séculaire du pôle moyen est selon eux de 0″0022/an sur le méridien 77°7W.

Torao et Okazaki ont discuté la variation de longitude de six observatoires de 1944 à 1965. Ils concluent aux résultats suivants:

G, Pa: − 0·62 ms/an W: + 0·95 ms/an BAg: − 0·37 ms/an To: + 0·04 ms/an.

Sakharov constate que la latitude de Poulkovo est constante depuis 1948.

(c) *Variations de la surface équipotentielle terrestre*

Yumi, Ishii et Sato attribuent à un tel phénomène certaines composantes de la variation non polaire des latitudes (termes en 3λ) tandis que Okuda leur attribue l'existence d'une périodicité de 19 ans dans les mêmes variations locales aux cinq stations internationales.

(d) *Tremblements de Terre et rotation de la Terre*

Un Symposium a été organisé sur ce sujet en 1969 à l'University of Western Ontario (London, Canada).

Selon Mansinha et Smylie la statistique des tremblements de Terre, utilisée pour en estimer l'effet cumulatif, montre qu'il est dans le domaine des possibilités que les tremblements de Terre puissent exciter le mouvement du pôle avec une amplitude conforme à celle qui est observée. D'après ces auteurs, des changements dans la trajectoire du pôle sont concomitants à la plupart des tremblements de Terre de magnitude supérieure à 7·5. Dans certains cas la déformation du mouvement du pôle précède le tremblement de Terre de près de 20 jours.

Cependant plusieurs auteurs sont parvenus à un résultat négatif quant à l'existence réelle de cette corrélation. C'est le cas notamment pour la comparaison des observations combinées aux PZT de Herstmonceux et Ottawa.

De leur côté, Takeuchi, Shimazaki et Hamano traitent le problème de l'excitation du mouvement libre amorti (chandlérien) et constatent que la corrélation avec les grands séismes n'est pas bonne.

Markowitz ne trouve aucune corrélation entre les changements d'accélération ou de vitesse de la rotation de la Terre et les tremblements de Terre.

Haubrich calcule que l'excitation ainsi provoquée est au moins d'un ordre de grandeur inférieur à ce qui est nécessaire pour entretenir le mouvement. Il critique également le niveau de signification de la corrélation trouvée.

8. VARIATIONS DE VITESSE DE LA ROTATION DE LA TERRE

Sidorenkov (Sternberg A. I.) et Korsun (Kiev) ont fait l'analyse spectrale des variations de vitesse.

Enslin déduit des observations du PZT de Hambourg une amplitude de 0^s011 pour la composante semi-annuelle.

Proverbio indique l'existence de périodes voisines de 2, 3·3 et 5·3 ans.

Iijima, Okazaki et Mori constatent une décroissance systématique d'amplitude de la composante annuelle de la variation de vitesse sur l'intervalle 1956·8–1968·2, à part un saut en 1963:

Année:	1957	58	59	60	61	62	63	64	65	66
a(ms):	35·3	31.5	30·3	26·3	21·1	12.3	29·0	20·1	16·0	17·0.

Madame A. Stoyko et M. N. Stoyko ont trouvé que pour la période de 1900 à 1963 l'augmentation de la vitesse de rotation de la Terre (c'est-à-dire l'accélération) correspond à la diminution

(a) de l'activité du Soleil;

(b) de l'énergie des tremblements de Terre, dont la profondeur du foyer dépasse 70 km;

(c) de l'amplitude et de la période de la nutation libre de la Terre;

(d) ainsi que la diminution de la fluctuation de régime de la circulation générale de l'atmosphère terrestre.

Markowitz analyse les observations des PZT de Washington et de Richmond depuis l'introduction du temps atomique (1955·5) et montre qu'on peut représenter les variations par quatre arcs de parabole sans qu'il y ait de brusques changements de la vitesse de rotation. Il y a par contre de brusques changements d'accélération à peu près tous les quatre ans.

9. MESURES DE DISTANCES TERRE-LUNE PAR LASERS

Les mesures de distance Terre-Lune par réflexion de signaux laser sur les rétro-réflecteurs déposés sur la Lune par les missions Apollo pourront atteindre une précision de 15 cm.

Dans ces conditions, la distance de la station terrestre à l'axe instantané de rotation de la Terre devra être corrigée, avec une précision comparable, des oscillations de cet axe par rapport au manteau terrestre. Inversément on pourra déterminer les mouvements de l'axe si trois stations terrestres procèdent simultanément à de telles mesures. La vitesse de rotation de la Terre, les dérives continentales pourront également être contrôlées par ce nouveau procédé. P. L. Bender s'occupe de la mise en œuvre de ces mesures.

Le groupe d'étude de l'Observatoire de Paris étudie aussi les possibilités offertes par cette méthode.

P. MELCHIOR
Président de la Commission

REPORT OF THE INTERNATIONAL POLAR MOTION SERVICE DURING AUGUST 1966–JULY 1969

(prepared by S. Yumi)

Work of the International Polar Motion Service has been carried on smoothly from the beginning of 1962 through this period.

Collaborating stations are: 1. Alger, 2. Belgrade (2 instr.), 3. Besançon, 4. Blagoveshchensk, 5. Borowa Gora, 6. Borowiec (2 instr.), 7. Bratislava, 8. Bucarest, 9. Buenos Aires, 10. Cape,

Table 1. Coordinates of the Pole for Every Month
Unit: 0".001

	1965		1966		1967		1968		1969	
Epoch	*x*	*y*	*x*	*y*	*x*	*y*	*x*	*y*	*x*	*y*
·056	− 75	+ 73	+ 47	+122	+68	+217	− 32	+294	− 80	+301
·139	−161	+133	+ 3	+100	+48	+166	+ 10	+315	−118	+324
·222	−182	+257	− 92	+102	−14	+144	+ 39	+278	+ 6	+391
·306	−204	+344	− 98	+173	+10	+181	+ 68	+257	+ 37	+403
·389	−136	+391	−130	+212	−13	+139	+ 59	+230	+129	+362
·472	− 46	+448	− 98	+257	+19	+163	+ 48	+216	+160	+325
·556	+ 85	+435	− 99	+344	+69	+174	+125	+187	+191	+257
·639	+158	+358	− 24	+339	+18	+202	+ 60	+142	(+176)	(+167)
·722	+212	+283	+ 75	+326	+ 3	+206	− 19	+174		
·806	+225	+215	+126	+306	−30	+228	− 39	+156		
·889	+215	+178	+121	+275	−68	+255	−113	+237		
·972	+ 61	+100	+110	+246	−63	+317	−137	+282		
Annual Mean	+ 13	+268	− 5	+234	+ 4	+199	+ 6	+231		

Table 2. Coordinates of the Pole for Every $0^{y}.05$
Unit: $0''.001$

	x	y		x	y		x	y
1965·00	− 12	+ 49	1966·00	+ 75	+114	1967·00	+98	+234
·05	− 67	+ 69	·05	+ 33	+103	·05	+75	+214
·10	−120	+103	·10	0	+ 98	·10	+53	+193
·15	−160	+153	·15	− 29	+100	·15	+32	+176
·20	−185	+226	·20	− 58	+108	·20	+13	+164
·25	−196	+286	·25	− 86	+124	·25	0	+156
·30	−194	+334	·30	−105	+149	·30	− 6	+153
·35	−174	+374	·35	−116	+181	·35	− 7	+153
·40	−130	+408	·40	−119	+215	·40	− 2	+155
·45	− 72	+434	·45	−115	+255	·45	+11	+159
·50	− 3	+444	·50	−104	+298	·50	+37	+166
·55	+ 71	+433	·55	− 86	+330	·55	+55	+174
·60	+127	+399	·60	− 57	+344	·60	+47	+184
·65	+168	+349	·65	− 10	+345	·65	+26	+195
·70	+201	+303	·70	+ 52	+337	·70	+ 6	+207
·75	+221	+259	·75	+ 96	+324	·75	−13	+221
·80	+227	+221	·80	+117	+308	·80	−31	+237
·85	+220	+186	·85	+125	+291	·85	−49	+253
·90	+194	+156	·90	+123	+273	·90	−62	+273
·95	+138	+131	·95	+115	+253	·95	−64	+292

	x	y		x	y
1968·00	− 56	+302	1969·00	−127	+289
·05	− 37	+308	·05	−120	+310
·10	− 14	+308	·10	−102	+330
·15	+ 8	+302	·15	− 73	+350
·20	+ 31	+290	·20	− 33	+370
·25	+ 51	+276	·25	+ 10	+386
·30	+ 64	+260	·30	+ 52	+392
·35	+ 67	+245	·35	+ 91	+386
·90	+ 64	+231	·40	+125	+367
·45	+ 60	+216	·45	+154	+337
·50	+ 88	+202	·50	+174	+302
·55	+119	+183	·55	(+185)	(+260)
·60	+106	+166	·60	(+187)	(+213)
·65	+ 54	+157			
·70	+ 8	+156			
·75	− 27	+161			
·80	− 56	+172			
·85	− 84	+197			
·90	−109	+233			
·95	−123	+265			

11. Carloforte, 12. Dresden, 13. Engelhardt, 14. Gaithersburg, 15. Gorky, 16. Greenwich, 17. Hamburg, 18. Irkutsk (2 instr.), 19. Kitab (4 instr.), 20. La Plata, 21. Leningrad (VNIIM), 22. Leningrad (Univ.), 23. Mendielievo (2 instr.), 24. Milano (2 instr.), 25. Mizusawa (4 instr.), 26. Mt. Stromlo, 27. Neuchâtel, 28. Nikolaev, 29. Novosibirsk (4 instr.), 30. Ottawa, 31. Paris, 32. Pecný (3 instr.), 33. Poltava (2 instr.), 34. Potsdam (2 instr.), 35. Prague, 36. Pulkovo, 37. Quito, 38. Richmond (2 instr.), 39. Rio de Janeiro, 40. Roma, 41. San Fernando, 42. Santiago de Chile, 43. São Paulo, 44. Shternberg, 45. Tashkent (2 instr.), 46. Tokyo, 47. Turku Tuorla, 48. Uccle, 49. Ukiah, 50. Warsaw and 51. Washington. No data were available from Zi-Ka-Wei.

Preliminary values of the instantaneous pole referred to the Conventional International Origin which was adopted at the last General Assembly in Prague were calculated from the results of the five ILS stations. They were published in the Monthly Notes of the IPMS as a rapid service together with the brief summaries of latitude observations at the collaborating stations.

Full data of the collaborating stations and coordinates of the pole for every month and for every 0ʸ05 were published in the Annual Report of the IPMS, the last issue of which for the year 1967 was published in 1969. No time data but only the latitude data were dealt with in them.

Star pairs for the use of the five ILS stations were renewed at 1967·0 so that they would be observed without any change for at least twelve years. No change was made in the observing programme. Details on them were published in the "Annual Report of the IPMS for the Year 1967" and also in the *Publications of the International Latitude Observatory of Mizusawa*, **5**, No. 2, 1966.

A total of 39161 observations were made at the five ILS stations – Mizusawa, Kitab, Carloforte, Gaithersburg and Ukiah – during the period from 6 August 1966 to 5 August 1969. Coordinates of the instantaneous pole, derived from these observations for every month (common epoch) and for every 0ʸ05 are shown in the Tables 1 and 2 together with the revised values for the year 1965. The values of x and y for the years 1965–1967 are referred to the Annual Report of the IPMS and those for 1968–1969 are to the Monthly Notes of the IPMS.

20. POSITIONS AND MOTIONS OF MINOR PLANETS, COMETS AND SATELLITES (POSITIONS ET MOUVEMENTS DES PETITES PLANÈTES, DES COMÈTES ET DES SATELLITES)

PRESIDENT: G. A. Chebotarev.
VICE-PRESIDENT: F. K. Edmondson.
ORGANIZING COMMITTEE: S. Arend, P. Herget, H. H. Hirose, H. W. Wood.

INTRODUCTION

The Report in the part dealing with the minor planets and satellites has been compiled at the Institute of Theoretical Astronomy in collaboration of S. G. Makover (minor planets) and V. A. Shor (satellites). Contributions to this report have been received from S. Arend, H. Debehogne, J. Dommanget, P. Herget, S. Herrick, G. M. Jannini, B. G. Marsden, V. V. Michkovitch, B. Milet, Z. M. Pereyra, B. Popovič, E. Rabe, E. Roemer, J. Schubart, W. Strobel, C. Torres, H. Wood. The part on comets has been prepared by E. Roemer.

The main feature of minor planet research during this triennium has been the further application of modern electronic computing machines to the practice of computing. The time has gone when the computations of orbits and ephemerides were performed with the aid of logarithmic tables or with desk calculators. Therefore it is no wonder that all computational work in the field of the motion of minor planets is nowadays concentrated in a limited number of institutions provided with powerful up-to-date computing techniques, like the following.

(1) Institute of Theoretical Astronomy, Leningrad has used till recently a computer BESM-2 and now has at its disposal a new more powerful computer BESM-4.

(2) The Observatory of Cincinnati has used till recently a powerful computer NORC. Unfortunately this machine has been nowadays dismantled, but the observatory can perform some kinds of computations with an IBM-360 computer.

(3) Latvian State University, Riga, has made a large amount of computations with the computer BESM-2.

(4) University of California, Los Angeles, makes a large amount of computations with its own computer.

The four institutions mentioned above are computing nowadays more that 99% of orbits of all planets with permanent numbers as well as of many unnumbered planets.

The Minor Planet Center at the Cincinnati Observatory has continued issuing Minor Planet Circulars with data on minor planet observations made throughout the world, results of their orbit improvements as well as information on all new-numbered planets. By the 1st of October 1969, 2970 MPC's have been published. At the same date 1748 minor planets have received their permanent numbers.

The Institute of Theoretical Astronomy has continued issuing the yearbook *Ephemerides of Minor Planets* in which elements and ephemerides of all numbered planets are published. By the 1st of October 1969 the issue for 1970 has appeared and been distributed to all observatories.

OBSERVATIONS

The Institute of Theoretical Astronomy (I.T.A.) has carried out observations of minor planets at the Crimean Astrophysical Observatory with a Zeiss double astrograph 40×160 cm. These observations, made under the guidance of Mrs L. I. Chernyh, have been made at a gradually increasing rate: 850 planetary positions were obtained in 1967 as compared with 2000 in 1969. Each position has been measured at the Observatory and reduced at the ITA. The large field of the telescope (10 × 10 degrees) enables the observer to find many planets not observed for a long time. About 700 unnumbered planets have been discovered. For 30 of these, elliptic orbits have been computed and some of them have already been reobserved in another opposition.

At the Uccle Observatory the observations of minor planets and comets have been performed on a large scale with the Zeiss double astrograph 40 × 200 cm under the guidance of S. Arend and J. Dommanget. Unfortunately the sky transparency has gradually declined owing to the neighbourhood of the large city of Brussels, which has badly influenced the observations.

At the Nice Observatory observations of minor planets continued with the Zeiss double astrograph 40 × 200 cm under the guidance of B. Milet. About 2000 positions have been obtained annually, and have been measured, and reduced with an electronic computer. Orbits have been computed for some newly discovered objects. Investigations of the brightness variation have been made for a number of planets, aiming at a determination of the periods of their rotation.

A large amount of research has been carried out at the Lunar and Planetary Laboratory, University of Arizona, Tucson. T. Gehrels and coworkers have made photometric observations of asteroids that had not previously been observed for light curves. The 36-inch reflector of the Steward Observatory and telescopes of the Kitt Peak Observatory have been used to obtain light curves for about 10 objects. Observations of polarization-phase relations of five bright asteroids have been compared with the results to be published by J. Veverka.

G. van Biesbroeck has continued his astrometric observations of comets, satellites and a few minor planets with the 154 cm reflector of the Catalina Station.

E. Roemer has continued astrometric observations, photographic photometry of comets, satellites and minor planets. Particular attention has been paid to observations of planets under the investigation of B. Marsden (see below).

Observations of planets with uncertain elements or an insufficient number of observations or not observed during the last ten years have been carried out at the observatories at Córdoba and Santiago.

Observations of ten selected minor planets (1, 2, 3, 4, 6, 7, 11, 18, 39, 40) for the determination of equinox and equator corrections of modern fundamental star catalogues have been obtained at different observatories (Bordeaux, Córdoba, Sydney, Uccle a.o.). These observations are for the greater part to be finished in 1969. The results of these observations have systematically been sent to the ITA for further reducing.

The Minor Planet Center at the Cincinnati Observatory has continued as in former years its works on the maintenance of a complete punched card file of all minor planet observations.

ORBITS, EPHEMERIDES

As in former years the Institute of Theoretical Astronomy has regularly published the yearbook *Ephemerides of Minor Planets*. The ephemerides computed either in the ITA or elsewhere have been published in this edition. For instance in the 1970 issue of the yearbook ephemerides of 1356 planets have been published including 389 sent from Cincinnati, 98 from Riga and 7 from other places. Beginning with 1971 issue the contents of the yearbook will be partly changed. At present an ephemeris of each planet consists of 6 lines in steps of 10 days and covering on the whole 50 days. In future each ephemeris is expected to consist of 8 lines covering altogether 70 days. Such a change will be made meeting the request of some observatories with powerful telescopes able to observe faint planets at least during two months around opposition.

At the same time the publication of extended ephemerides of bright planets will be discontinued.

Every year the ITA improves the orbits of about 150 planets (F. B. Hanina, S. G. Makover). V. I. Orelskaya has continued her work on the collection and reduction of observations of ten selected minor planets. By the 1st of October 1969 about 7500 observations have been collected at the following 18 observatories: Bordeaux, Bucharest, Cape, Cordoba, Golosejevo, Copenhagen, Leiden, Moscow, Nice, Nikolajev, Poznan, Pulkovo, Santiago, Sydney, Tashkent, Tsingtao, Uccle, Zo-Se.

The Cincinnati Observatory and the Minor Planet Center (under the guidance of P. Herget) have made the following contribution: All the ephemerides until 2000 AD of the minor planets which were on the selected list have been completed on microfilm and several copies have been deposited at the ITA and some other institutions. The programs for computing preliminary orbits and for computing plate reductions have been rearranged for an IBM-360 computer. The Schubart-Stumpff N-Body program for computing perturbed planet orbits was in use. C. M. Bardwell has continued his work on the improvement of orbits and testing of identifications. E. Rabe began his work on the improvement of orbits of planet 1011 Laodamia and 1362 Griqua aiming at the improvement of the calculated masses of Mars and Jupiter.

S. Herrick and coworkers (University of California) have given much care in producing elements and ephemerides for the minor planets 1566 Icarus, 1580 Betulia, 1620 Geographos and 1685 Toro. In the period 1967–70 about 20 ephemerides have been published. S. Herrick has completed his two-volume book *Astrodynamics* which is expected to appear in 1970. Some of the original features include his work on universal variables, on the updating of the astrodynamic constants, on perturbed variation techniques in the method of variation of parameters and a new universal, singularity-free solution for an orbit through two positions.

B. Popovič (Belgrade) is preparing a monograph on the theory of the motion of planetoids and comets, using only the heliocentric position and velocity as orbital elements.

M. A. Dirikis (Latvian State University) has completed the improvement of orbits of about 120 planets, including all Trojans, 887 Alinda, 944 Hidalgo. Ephemerides of all these planets have been calculated regularly.

E. Rabe (Cincinnati), G. Zech (Heidelberg) and J. Lieske (Yale) have improved the orbit of 433 Eros using all available observations, and have obtained new values of the mass of the Earth-Moon system. Similar work on the planet 1221 Amor has been done by J. Schubart (Heidelberg). In all four cases the values of the Earth-Moon system appeared to be in a good mutual agreement as well as with those obtained by radar. Schubart has computed an ephemeris of 1221 Amor for 1972. The observations of the planet in this year are of great importance, since after 1972 the observing conditions of the planet will not be favorable for a long time.

B. G. Marsden (Smithsonian Astrophysical Observatory) has calculated new orbits for several numbered and unnumbered planets. The planets selected were ones with high orbital eccentricity (greater than 0·35) or with unusually small or large mean distances (less than 2·0 a.u. or greater than 3·5 a.u.). Five of these planets were reobserved by E. Roemer, including two members of Hilda group and the lost Trojan 1949 SB.

Another reason for doing this work was to verify that the program prepared for computing the orbits of periodic comets (where there are some additional difficulties due to large perturbations by Jupiter and to the effects of nongravitational forces) was working correctly. Two minor planets caused some difficulties: 887 Alinda, until it was realized that, with its frequent approaches to the Earth, this planet provides a dynamical test of the correctness of the radar value of the astronomical unit; and 944 Hidalgo, where it seems likely that the slight systematic trends remaining in the residuals are due to nongravitational forces.

S. Arend, H. Debehogne and J. Dommanget (Uccle Observatory) have computed orbits of planets discovered at the Observatory. Moreover they have made a set of investigations connected with the problem of reducing of astrophotographs. Debehogne has studied a question on the influence of the number of reference stars on the accuracy of the result by different methods of reducing. Dommanget has been occupied with the derivation of formulas for reducing the astrophotographs for the cases when the field of the telescope reaches as much as 45° or 90°. Debehogne has studied the influence of second order terms caused by an error of optical center determination. He has also made a program for rapid identification of plate reference stars.

1566 ICARUS

This planet was in 1968 in a close approach to the Earth ($\Delta = 0{\cdot}04$ a.e.). Accurate ephemerides have been prepared by S. Herrick *et al.* (University of California) and I. I. Shapiro (Massachusetts Institute of Technology). The planet was discovered by E. Roemer at the Catalina station on June 11. Near the date of close approach the planet was moving so fast, that usual photographic methods of observations were hardly suitable. Therefore a special photoelectric observational method was used at the Crimean Astrophysical Observatory, the planet image on a background of stars having been taken from a screen of a television device. About 300 observations of the planet were published, made at 20 observatories. P. Tiffany (University of California) has started to improve the planet's orbit using all old observations as well as all 1968 observations. Some observatories have made a set of photoelectric observations for the determination of the planet's brightness, colour and polarization. A rotational period of approximately 1.13 hours has been suggested. T. Gehrels has made an extensive series of photopolarimetric observations of the planet.

The most striking event in the field of research of the minor planet's motion during the triennium considered has been an attempt at making radar observations of Icarus. This attempt turned out to be successful owing to the recent improvement of Icarus's orbit made by Herrick who used old observations and two observations of 1968 June 11. I. Shapiro *et al.* successfully obtained radar observations from June 13 till June 15. R. Goldstein (Jet Propulsion Laboratory, Pasadena) made a series of radar observations of Icarus from June 14 till June 16. J. Lieske and G. Null (JPL) have made a new improvement of Icarus's orbit using 154 astrometric observations of the planet obtained in 1949–68 and 7 radar observations. This improvement has been made to verify the predictions of general relativity and to estimate the dynamical oblateness of the Sun, the mass of Mercury and some other parameters.

In August 1969 the planet 1620 Geographos was in close approach to the Earth. An extensive program of observations of the planet has been made.

GENERAL

Under the guidance of N. A. Bohan (Institute of Theoretical Astronomy) a library of standard subprograms for the computer BESM-4 has been elaborated. Each of these subprograms is intended for the solution of separate tasks connected with the theory of motion of minor planets and comets. The library is to be used for the composition of complex programs, each of them consisting of a short unifying program and a set of standard subprograms.

E. I. Kazimirchak-Polonskaja has worked out a new method for the integration of the equations of planetary motion, which is considered to be a combination of the methods of Numerov and Subbotin and therefore may be named the Numerov-Subbotin method. In this method only two approximations are needed for receiving the accurate solution, as compared with several in the Cowell method. The method has been programmed for the computer BESM-4.

V. F. Mjachin has elaborated a new method for the simultaneous integration of the equations of celestial mechanics based on ideas of Taylor and Steffensen. In this method a planet's coordinate is expanded into a series containing not high differences, but high order derivatives. The method is particularly convenient for integration with a variable step.

N. S. Chernyh has made an attempt for the determination of the mass of Jupiter from a study of the motion of the planet 10 Hygiea. Annual reviews of minor planet problems have been prepared by N. S. Yahontova.

G. A. Chebotarev and M. Shmakova have written a paper on the statistics of minor planet orbits. The existence of three main zones in the minor planets belt has been shown, each of them having its own physical peculiarities. The existence of "flat" and "spherical" components of the belt has been shown. G. A. Chebotarev, N. A. Beljaev *et al.* have investigated the motion of some minor planets (Thule, Hilda group, Trojans, planets with large eccentricities) for a period of 400 years

(1660–2060). It has been shown that the concentration of planet perihelia near the perihelium of Jupiter takes place only in some epochs.

J. Schubert (Heidelberg) has studied long-periodic effects in the motion of Hilda-type planets. He intends to continue his research on minor planets with a commensurable mean motion. He is studying the long period effects of 334 Chicago and of 279 Thule. One of his graduate students is investigating the possibility to determine the mass of Jupiter from selected planets of the Hilda- and Thule-type.

E. Rabe (Cincinnati) has studied the elliptic restricted problem (periodic solutions and their stabilities for motions of the Trojan type, but for appropriate mass ratios different from the one in the actual Jupiter-Sun case).

V. V. Michkovitch (Belgrade) as well as H. Debehogne and J. Dommanget have studied the problem of the determination of the shortest distance between two asteroid orbits with small mutual inclination.

SATELLITES

In connection with the launchings of space vehicles to Mars and Jupiter already started or planned for the near future we should focus our attention on improvement of the orbits of natural satellites of the planets especially close to their primaries. Such improvements should be based on a large series of highly accurate observations. Unfortunately only a small number of observers devote their energies to these fields. In this connection the fruitful activity of the U.S.A. astronomers should be noted. E. Roemer has obtained a good series of plates of the Martian satellites in 1969. D. Pascu (Naval Observatory) has obtained in 1968 a large series of plates of the Galilean satellites of Jupiter with the McCormac reflector. G. van Biesbroeck has continued his astrometric observations of planetary satellites (emphasis made on Neptune II = Nereid and Saturn IX = Phoebe), with the Catalina reflector. E. Roemer with coworkers has regularly observed the satellites Jupiter VIII to Jupiter XII and Saturn IX; it is intended to add Jupiter VI and VII to the program.

A large series of observations of Jupiter V has been performed by P. V. Sudbury (University of Manchester, England) with the 74-inch reflector of the Kottamia Observatory (U.A.R.). In the U.S.S.R. a small series of observations of the Martian satellites have been made in 1967 (Pulkovo, Golosejevo, Kazan a.o.). Observations of Saturn IX are planned at Córdoba.

During the triennium considered theoretical investigations have been successfully developed. G. A. Wilkins has continued his fruitful investigations of the orbits of the Martian satellites with particular reference to the investigations of tidal friction effects. S. N. Vashkovjak (Moscow) has developed a new analytical theory of the Martian satellites, taking into account not only secular but also periodic perturbations due to Mars' oblateness and the Sun. At the ITA work on the refinement of the elements of the orbits of these satellites has been continued (V. A. Shor and N. I. Glebova).

P. V. Sudbury has obtained a new system of elements of Jupiter V based on all old observations made in 1892–1949 and some new ones made in 1967 by himself (see above). V. N. Kiryushenkov (Moscow) elaborated an analytical theory of Jupiter V.

A. Bec (Bureau des Longitudes) has elaborated a numerical theory of the motion of the outer satellites of Jupiter and has determined Jupiter's mass from the motion of Jupiter IX. P. Herget has completed his large work on the orbits of Jupiter's outer satellites VIII to XII. He obtained precise orbits by process of numerical integration, improved the orbital data, based on all observations available and has computed ephemerides until 2000 AD. M. Charnow, P. Musen and J. L. Maury have published a paper on the application of Hansen's method to Jupiter X, containing new elements and a table of the general perturbations of the satellite. S. Herrick has continued his work on Jupiter IX and Jupiter XII. R. R. Allan (London) has studied the evolution of the Mimas-Tethys commensurability in the system of Saturn's satellites. J. R. Gill and B. L. Gault have determined a new orbit of Triton based on observations made in 1877–1958. G. A. Chebotarev has investigated a problem related to the stability of the motion of natural satellites in the solar system.

PROPOSALS

(1) The Institute of Theoretical Astronomy requests the discussion and approval of its suggestions regarding a new form of the *Ephemerides of Minor Planets*, to start with the 1971 issue.

(2) B. G. Marsden has pointed out that the recent practice of including extended ephemerides for Earth-approaching minor planets in the *Ephemerides of Minor Planets* should be strongly encouraged. It is hoped that the number of planets that will be given this special treatment will be increased so that the list includes eventually all those with highly eccentric orbits (greater than 0·35 say) and not necessarily the best observable during the two month interval around opposition.

(3) E. Roemer states that the uncertainty regarding the accuracy of the ephemerides is still the prime deterrent against the inclusion of additional minor planets on her observing list. The very limited fields of large reflectors means that an O–C of 5′ may well make plates unmeasurable. The date of the last observation presently printed with the ephemerides, does not really give the information we need. It would be more practicable to print with the ephemerides a reference to the elements on which the ephemeris is based, perhaps the year of the "Ephemerides" in which the elements were introduced.

(4) T. Gehrels proposes revision to the present photometric system of asteroids. The proposal is that the UBV photometric system must be adopted. The new values are 0.10 magnitudes fainter, nearly the same difference for all asteroids, which is a small change in view of the precision of the asteroid magnitudes. However fundamentally the change is important, because the UBV system now widely used. A new list photoelectric magnitudes of nearly all numbered asteroids will be submitted before Brighton meeting.

(5) The Institute of Theoretical Astronomy states with satisfaction that the number of observation of minor planets published formerly often in an approximate form is now being reduced rapidly. Taking into account that only accurate observations are of value for computers the Institute appeals to all observatories having modern measuring tools and computing facilities to publish in the future only accurate observations of minor planets. This proposal is concerned not only to planets having permanent numbers, but also to unnumbered ones.

(6) The President proposes to create a new Working Group dealing with the natural satellites' motion.

G. CHEBOTAREV
President of the Commission

21. LIGHT OF THE NIGHT-SKY (LUMINESCENCE DU CIEL)

PRESIDENT: M. Huruhata.
VICE-PRESIDENT: H. Elsässer.
ORGANIZING COMMITTEE: J. Dufay (†), M. F. Ingham, A. Lebedinskij (†), F. E. Roach, G. Weill, J. L. Weinberg.

INTRODUCTION

Studies of airglow, zodiacal light, and galactic light in this triennium were almost as active as in the last. Results of observations obtained during and after the IQSY were the major contributions in this term. Night Airglow studies during the IQSY were widely reviewed in the *Annals of the IQSY*, vol. 4, in which the following articles are included:

"Night Airglow Observations during the IQSY" (F. E. Roach, L. L. Smith and J. R. McKennan);
"Airglow Research during and since the IQSY" (G. Weill);
"Hydrogen and Hydroxyl Emissions in the Nightglow" (N. N. Shefov and Yu. L. Truttse).

Proceedings of the NATO Advanced Study Institute, July 29–August 9, 1968, entitled *Atmospheric Emissions*, contains some ten significant papers on airglow both in observations and interpretations. Many of them are referred to in the following sections.

Many stations have been engaged in airglow observations since the beginning of the IASY in 1969, although the number of stations seems to be a little less than during the IGY. Stations near the equator and in the southern hemisphere, however, are more numerous and active than during the IGY in response to the strong wishes of many airglow researchers. These efforts are being made especially by G. Weill's group of France, V. J. Dachs' and Lange-Hesse's groups of Germany.

It is noteworthy that interferometer measurements have been widely introduced in airglow studies and many valuable works are being performed recently. Airglow measurements from space have been fruitful and appear to be promising in many respects. Spectral domains not observable from the ground have provided valuable information about the very high atmosphere. It is also noteworthy that very faint emissions previously undetectable by filter photometry have come to be regularly observable.

The proceedings of the Symposium on the Zodiacal Light and the Interplanetary Medium was published during this past triennium (NASA SP-150, edited by J. L. Weinberg).

Studies of zodiacal light and the interplanetary medium since the colloquium in Honolulu in 1967 were also very active, including observations from spacecraft. Interesting results about the effects of solar activity on the zodiacal light have been presented. Work on interstellar particles has cast light on the importance of the study of galactic radiation in connection with airglow and zodiacal light studies.

Suggestions for future studies and research items were proposed by some researchers. Among them, the importance of studying the brightness and spectral distribution of the integrated starlight and the method of correcting observations for extinction and scattering are emphasized. Observations of the interplanetary medium and zodiacal light from space probes in the future will be promising directions in this field.

Most of the references to Soviet studies were taken from the report prepared by Dr E. V. Fesenkova, and references to studies on the zodiacal light were mainly contributed by Dr J. L. Weinberg.

I. AIRGLOW, GENERAL

An excellent and extensive atlas of the airglow spectrum was published by A. L. Broadfoot and K. R. Kendall of Kitt Peak National Observatory (*J. geophys. Res.*, 73). The atlas covers the wavelength region 3100–10000Å and has two major advantages over earlier atlases: (1) the intensity scale is strictly linear; and (2) the whole region accessible from the ground by photoelectric techniques is shown to a uniform, absolute scale. This atlas is undoubtedly a great help for airglow researchers, in addition to the photographic atlas by Krassovsky *et al.* published in 1962.

The differential brightness of the airglow spectrum between 1·02–1·13μ was measured at a spectral resolution of 27Å by A. W. Harrison (*Planet. Space Sci.*, **17**).

The night airglow data during IQSY were published by World Data Center A (Report UAG-1). This book contains (*a*) hourly zenith intensities of emissions, 5577Å, Na I, 6300Å, OH, 3914Å, and (*b*) hourly values of the ratio of intensity (north/south) near or at zenith distance 75°.

The international calibration of airglow photometers has been an important problem in assessing the general level of maximum reliability of individual airglow measurements and in assessing how far systematic variations from place to place on the globe are significant. In 1968, before the start of the IASY, M. Gadsden proposed a convenient way to circulate a radioactive phosphor source among the main regional stations for the intercalibration of their photometers. The calibrations were carried out in 1969 at several stations in Europe, India, Japan, as well as in U.S.A. The result will be shortly published by Gadsden.

Based on spectral and electrophotometric observations, L. M. Fishkova (*Geomag. Astr.*, vol. 9) showed that the continuous spectrum of the nightglow in 4700–6700Å corresponds to the spectral type G2, but possesses two broad emission bands near 5950Å and 6200–6300Å.

Airglow stations near the equator and in the southern hemisphere are being active, especially in the African zone. The French and German groups are operating a few stations in South West Africa. The station at Tahiti in the South Pacific, operated by the French group, is important as a low latitude station with the station in Hawaii. They are obtaining observations of many airglow emissions using photoelectric photometers.

Another low latitude station at Mt. Abu in India is continuing observations in the visible and near infra-red regions. An Ebert-Fastie type spectrometer is near completion there to be used for near infrared OH studies.

The importance of the study of airglow continuum is emphasized by M. F. Ingham and P. V. Kulkarni. The brightness and spectral distribution of the integrated starlight is still not known with sufficient accuracy and the method of correcting observations of the diffuse light of the night sky for extinction and scattering are very rough. It would be worth paying more attention to these aspects of our work in the future.

II. NIGHTGLOW VARIATIONS

Correlations between the airglow intensity and solar activity were studied by many researchers. N. Rosenberg and S. P. Zimmerman (*Planet. Space Sci.*, **15**) have analysed 5577Å line intensities observed at Sacramento Peak since 1956, and showed statistically significant positive correlations between the two, being influenced both by solar ultra-violet radiation and solar corpuscular radiation. K. Yano (*Planet. Space Sci.*, **15**) and M. N. Gnevyshev (*Planet. Space Sci.*, **16**) demonstrated a similar tendency in the same period. R. Shapiro (*Planet. Space Sci.*, **17**) considers that there is no significant association between solar corpuscular radiation and the intensity of 5577Å.

Variations of other airglow emissions with the solar cycle are shown by G. Weill (*Ann. IQSY*, vol. 4). The intensities of OH and Na D seem to have continuously decreased throughout the recent solar cycle.

From the measurements of 6300Å during the IGY and IQSY, the Soviet researchers have shown that the emission 6300Å is closely connected with the solar activity, namely the solar activity index $F_{10\cdot7}$. The details will be described in Section III. L. L. Smith and W. R. Steiger (*J. geophys. Res.*, **73**) concluded that the 6300Å emission has a seasonal variation, but it appears to be a function of the solar cycle.

Observations of various nightglow emissions at Tsumeb (South West Africa) were made by J. Dachs (*Phys. Atmos.*, vol. 41), who studied nightly and seasonal variations and their relation with the ionosphere. Airglow inside the southern auroral zone was studied by M. A. Gordon

(*Planet. Space Sci.*, **16**) showing a seasonal decrease and pronounced diurnal maximum of 5577 Å and 6300 Å emissions.

Latitudinal variations of 5577 Å were observed during a shipboard expedition by J. A. Greenspan and J. H. Woodman (*J. atmos. terr. Phys.*, **29**). T. P. Markham and R. E. Anctil (*J. atmos. terr. Phys.*, **29**) measured the 5577 Å and 6300 Å from a jet aircraft.

Spatial movement of nightglow was studied by L. H. Wilkins *et al.* (*J. atmos. terr. Phys.*, **29**) who studied E-W motion of mid-latitude aurora at Moscow, Idaho, by auroral radar.

A chart of isophotes of the 5577 Å line was drawn by M. P. Korobeinikova and G. A. Nacirov in Ashkhabad, and the isophotes are characterized by periodic variations of probability of streamlining of isophote directions with maximum amplitude approximately four days after the beginning of a geomagnetic storm.

At Haute Provence, 6300 [OI] and 5200 [NI] emissions in travelling ionospheric disturbances were observed by G. Weill and J. Christophe-Glaume (*C. r. Acad. Sci.*, **264**). The maximum NI emission is delayed by about 12 minutes with respect to the OI emission.

Variations of hydroxyl airglow during the solar cycle, its seasonal variations, and the latitudinal dependence were extensively studied by N. N. Shefov (*Planet. Space Sci.*, **17**). The work is described in Section VII.

The long discussed problem on the correlation between the 5577 Å line and the green continuum, was treated by F. Shanchez Martinez and R. Dumont (*Semin. Astr. Geodes. Madrid*, No. 60), and very close correlation, even quantitative, was obtained. R. Robley (*C. r.*, **265**) also found a positive correlation.

III. AERONOMY, EXCITATION PROCESSES, AND DOPPLER TEMPERATURE

Many geophysical rather than astronomical studies have been done, especially many important interferometric investigations have been made.

Measurements of the width of the 5577 Å line made with a Fabry-Perot interferometer were reported by E. B. Armstrong (*Planet. Space Sci.*, **16**) who gave many important results: the apparent line width was increased by small magnetic disturbances, and the temperature at an altitude of 90–100 km showed a variation coinciding in phase with the seasonal intensity variation. He also suggested that the wind velocities in the 90–100 km region could be measured by Doppler shifts.

Measurements of the speeds of neutral winds in the thermosphere by Doppler shifts in the 6300 Å line were made also by E. B. Armstrong (*Planet. Space Sci.*, **17**) and they were compared with the neutral wind patterns for altitude 250–300 km shown by theoretical models. A very broad agreement with the models is evident, but the local conditions in the thermosphere may have an influence on winds. The changes in height of the region emitting the 6300 Å line may be responsible for some of the observed effects.

The measurements of atomic oxygen emission at 6300 Å during IGY and IQSY render it possible to make some conclusions about the behaviour of the upper atmosphere higher than 200 km at the time of geomagnetic disturbances (V. I. Krassovsky, N. N. Shefov, Yu. L. Truttse; *Planet. Space Sci.*, **16**). It was shown that the emission 6300 Å is closely connected with solar activity, namely that a close connection exists between this intensity and the solar activity index $F_{10.7}$. For sufficient geomagnetic disturbances ($D_{st} > 100\,\gamma$) the log of emission intensity of the line can be lenearly related with the indices D_{st} and $F_{10.7}$.

Regular observations in Zvenigorod have shown that the increase of this emission is connected with some enlargement of its contour giving evidence of some warming up of the upper atmospheric strata. In the time of reddish polar aurora the emission continuum is increased and this speaks for an augmentation in the atmosphere of molecular NO.

The whole planetary energy released in time of greatest geomagnetic disturbances was determined empirically. It surpasses the energy of the particles in the magnetosphere. Evidently this heating of the upper atmosphere is followed by some variations of its density at different heights.

In Abastumani the emission of O I (6300–6364 Å) was investigated in crepuscular conditions and different regularities of its appearences were established on the basis of observations made in years 1962–68. It turned out to be possible that the maximum of this emission in spring and autumn is connected with low latitude polar aurora.

Twilight and nightglow spectral shapes of both 5577 Å and 6300 Å lines were measured by M. A. Biondi and W. A. Feibelman (*Planet. Space Sci.*, **16**), and they found that no evidence for a dissociative origin of the ^{1}D or ^{1}S oxygen was obtained, and the temperature of O(^{1}D) atoms from the thermal Doppler shape was obtained as a function of time after sunset. The same kind of interferometric measurement of the 5577 Å line in aurora was made by R. V. Karandikar (*Planet. Space Sci.*, **16**), and the Doppler temperatures of various types of aurora were given. A few observations were obtained for the sub-visual aurora, and the Doppler temperature showed the bulk of the values in the range 400–700 K with a pronounced peak around 430 K and also a minor peak in the range 250–300 K obtained for the faint diffuse glows. It was suggested that the sub-visual aurora was a distinct type with temperatures higher than other forms and possibly originating at comparatively higher altitudes.

The effective Doppler temperatures of the 6300 Å line are calculated by R. G. Roble *et al.* (*Planet. Space Sci.*, **16**) using the data of Evans through the night during solar cycle minimum.

Forbidden transitions in the upper atmosphere were widely reviewed by F. E. Roach (*Transitions Interdites Spectres Astres*, Liège, 1968).

A theoretical study of 6300 Å emission in connection with Barbier's formula was made by T. F. Tuan (*Ap. J.*, **157**).

The atmospheric movements of atomic oxygen density with altitude was studied by rocket-borne measurements by M. Gadsden and E. Marovich (*J. atmos. terr. Phys.*, **31**), and showed that the 5577 Å emission depends mainly on the occurrence of pressure waves of vertical wavelength down to several tens of kilometers. M. Gadsden (*J. atmos. terr. Phys.*, **30**) has interpreted the observed distribution of atomic sodium in terms of release from dust particles of meteoric origin in the upper atmosphere.

IV. TROPICAL AIRGLOW

Tropical enhancement of the 6300 Å and the second layer of 5577 Å have been two of the major items of interest in tropical airglow since the IGY. A good number of papers has dealt with this problem during the triennium. Some researchers used observations obtained at Haleakala, Hawaii, which is favorably situated for the study of tropical airglow. W. E. Brown and W. R. Steiger (*Nature*, **216**) compared the brightness of the 6300 Å line with the electron content determined by the HF signals from the synchronous satellite, and found that the large increases in the intensity are preceded by nocturnal increases in electron content. W. E. Brown, W. R. Steiger and F. E. Roach (*Nature*, **220**) have come to the surprising conclusion that a large night-time source of ionization is required in the tropics which can occur at any time between midnight and dawn, but the most likely time is 0150 (L.M.T.).

Using photometric observations of 6300 Å at Mt. Abu and Haleakala, S. R. Pal and P. V. Kulkarni (*Ann. Geophys.*, **24**) showed that the post-midnight enhancement and the superposed 6300 Å hump can be reasonably accounted for by considering the dynamics of the night-time *F*-region including vertical drifts during those hours.

Preliminary results obtained in Debre Zeit (Ethiopia) are presented by G. Weill *et al.* (*Ann. Geophys.*, **24**). The green doublet of N I at 5199 Å was regularly recorded, and it is highly correlated to 6300 Å intensities, being consistent with production of N(^{2}D) by electronic dissociative recombination of NO^+.

V. MID-LATITUDE ARCS

Owing to the unexpected relatively moderate activities in the present solar maximum not many observations have hitherto been obtained, but a few studies on the mid-latitude arcs have been published.

Based on the measurements from Haute Provence and Tamanrasset, G. Weill and J. Christophe-Glaume (*Ann. Geophys.*, **23**) made an extensive study and tried to detect systematic variation during sudden commencement of geomagnetic storms. Storm effects on night airglow are characterized as follows. The A Group (5577, O_2 bands and continuum) increases by a few percent during the main phase, and the green line shows a positive post-perturbation with a marked intensity increase 35 hours after SSC; and the B Group (OH, Na) decreases by about 10% during the main phase; the red line of 6300 Å behaves differently at the two stations.

Some M-arcs observed in 1967 at Moscow, Idaho, were reported by T. Ichikawa and J. S. Kim (*J. atmos. terr. Phys.*, **31**). They obtained the southward motion of the arc with an average speed of 32 ms^{-1}. The arc was analysed together with the visible aurora (I) by I. Ichikawa *et al.* (*J. geophys. Res.*, **74**) who found that the southern edge of the visible aurora was very close to the northern edge of the M-arc. The characteristic motions of the visible aurora and the M-arc were also studied.

Morphology and dynamics of the M-arcs observed in 1968 and 1969 at Whiteface Mt., N.Y., were studied by T. Old *et al.* who found that the observed arcs differ in many respects from the arcs observed during the IGY.

The detailed results of M-arc aurorac observed around the last solar maximum at Memembetsu, Hokkaido, were published (Reprint from *Rep. geomag. geoelect. Obs.*, 1967, No. 8.)

VI. EXCITATION OF 6300 Å BY PHOTOELECTRONS

Pre-dawn enhancement of the 6300 Å [O I] emission as the result of the travel of photoelectrons from the magnetically conjugate region has received wide interest by airglow as well as geophysical researchers.

M. L. Duboin, G. Lejeune, M. Petit and G. Weill (*J. atmos. terr. Phys.*, **30**), based on optical observations of the ionosphere above St. Santin de Maurs made concurrently with Thomson scattering soundings, confirmed that the pre-dawn enhancement of the oxygen red lines occurs in synchronism with electronic heating in the F-region at times of sunrise in the magnetically conjugate area. A preliminary calculation indicated that photoelectron impact is directly responsible for the excitation of $O(^1D)$.

H. C. Carlson and G. Weill (*Ann. Geophys.*, **23**) found many important features about the solar cycle control on the timing of 6300 Å pre-dawn enhancement. At Haute Provence the onset occurs 70 min earlier at solar maximum than at solar minimum activity. The observations at Arecibo in 1965–67 confirmed the effect, and a preliminary calculation showed that it is quantitatively interpreted by the combined variation of upper atmospheric neutral temperatures and of the solar ionizing flux through the 11-year cycle.

Photometric observations made in 1965–68 by R. W. Smith (*Planet. Space Sci.*, **17**) revealed that the enhancement is a regular occurrence, at least up to geomagnetic latitude 59·5° N in the manner reported from lower latitudes. The onset time shows a solar cycle, seasonal and irregular variation. Long period variations are suggested to be related to thermopause temperature and the solar EUV intensity. Heights of emission are deduced from the azimuthal measurements which are mostly in the region 250–300 km. There is a trend towards increasing height with increasing thermopause temperature.

Some theoretical considerations were made by J. R. Catchpoole (*J. atmos. terr. Phys.*, **29**). The latitude dependence of the enhancement was studied by R. T. Bennett (*J. geophys. Res.*, **74**) who found the upper limit $L \sim 3{\cdot}2$, while M. R. Torr and D. G. Torr (*J. geophys. Res.*, **74**) claimed that it was observable in winter at the antarctic station where $L = 4{\cdot}0$.

Using the observations at Tohkatta, Japan, M. Okuda and K. Misawa (*Rep. Ionos. Space Res. Japan*, **23**) studied the time variation of the effect vs. the changes of the solar zenith angle at the conjugate point, and the daily dependence of the effect on the geomagnetic activity.

Evidence was presented for the existence of a flux of fast photoelectrons arriving from the conjugate ionosphere by J. V. Evans in his study of midlatitude *F*-region densities (*Planet. Space Sci.*, **15**).

VII. HYDROXYL (OH) NIGHTGLOW

Extensive studies of hydroxyl nightglow have been made by Soviet researchers. The analysis of observations made in the last years in Zvenigorod near Moscow (V. I. Krassovsky, N. N. Shefov, Yu. L. Truttse; *Ann. Geophys.*, **24**, *Planet. Space Sci.*, **17**, *Nature*, **218**) reveal variations of intensity of the OH hydroxyl band with periods of 5·5 years and 26 months during solar cycle activity. Intensity variations of the OH bands from median vibrational level are greater in the winter and summer months and smaller in spring and autumn. The variations of rotation temperature show a marked maximum in winter and minimum in summer and within the limits of errors coincide with variations of atmospheric temperature at a height near 80 km. The vibration temperature corresponding to these observations attains its maximum in winter (nearly 11 000 K) and minimum in summer (nearly 5600 K), so that the intensity variations of OH at upper and lower vibration levels are quite different. The total energy of hydroxyl emission attains its maximum in summer.

Besides these regular variations of OH emissions their disturbed variations have also been detected. After the maximum of main meteor-showers the intensity of OH increases over a period of 3–4 days. A similar phenomenon is observed also with the appearence of noctilucent clouds.

Moreover, there exist variations of more than 20 days' duration after geomagnetic disturbances; the intensity of OH bands at first increases (by $+80\%$ at $K_p = 9$), but rotation and vibration temperatures diminish. The maximum amplitude of perturbed intensity variations is attained several days after the corresponding disturbance. The analysis of observations made at different latitudes shows that the perturbance spreads from the zone of polar aurora with a speed of ~ 10 m s^{-1} in meridional direction. The period of intensity and temperature variation after geomagnetic disturbance is equal to 6 days on the average according to the Zvenigorod data and increases with growth of index K_p. These variations are a consequence of molecular dissociations resulting from penetrations of corpuscles in the zone of polar aurora. This gives supplementary energy approaching at maximum to 10^{23} ergs when $K_p = 7$.

The seasonal variations of intensity and rotation temperature of OH bands in visible and infrared spectral regions were studied also in the Astrophysical Institute, Alma Ata (Z. V. Kariaguina, V. E. Mozaeve). It was shown that in summer the temperature value of OH and its variations are smaller. The intensity and temperature of OH bands are not correlated between themselves but, on the contrary, a sufficiently strong correlation exists between intensities of OH and hydrogen line Hα.

From the spectra of several bands of OH nightglow, rotational temperatures have been derived by G. J. Kvifte (*Planet. Space Sci.*, **15**), giving an average of 210 ± 10 K. No seasonal dependence or latitude effect is found. The spectrum between 1·02 and 1·13 μ has been measured at a spectral resolution of 27 Å by A. W. Harrison (*Planet. Space Sci.*, **17**) and the rotational temperature of 240 ± 20 K for (4–1) band and 220 ± 20 K for (5–2) band were obtained by comparison with synthetic spectra. Absolute brightnesses of the two bands were also given.

At Abastumani Observatory the bands of hydroxyl OH are investigated in twilight conditions. Two principal types of intensity and temperature variations were established: (1) the increase with brighter twilight, i.e. at lesser solar depressions below the horizon, mainly in the morning and in summer, and (2) relatively small and irregular variations at middle twilight most frequently in winter time and in the evening as much as in the dawn. The first type of variations can be observed most frequently at the fourth day after a geomagnetic disturbance. The second type of variations does not depend on geomagnetic activity. These differences between the morning and evening values for the intensity and temperature of OH emission disappear completely with solar depression greater than 11°.

A seasonal course of OH emission was revealed in winter with a maximum at 18°—12° of solar depressions, and in summer at lesser solar depressions. It is possible that these regularities of the crepuscular OH emission depend on dynamical processes in the mesosphere (T. I. Torochelidse, *Bull. Acad. Sci., Georgian SSR.*, **52**).

VIII. HYDROGEN AIRGLOW

Since the last triennium, the Soviet colleagues have been especially active in these studies.

In the years 1967–69 investigations of the red hydrogen line Hα in the night sky were continued with an interferometric system supplied with a Fabry-Pérot etalon (P. V. Shcheglov). Observations were made at the same time by Z. V. Kariaguina and V. E. Mozaeva at the Astrophysical Institute, Alma Ata, and by L. M. Fishkova in Abastumani. It was discovered that the morning hemisphere of the night sky is more luminous in this line emission in comparison with the evening one, that in the antisolar point there is a deep minimum of Hα emission and that in the northern part of the sky this Hα emission sometimes undergoes changes not related to its twilight course. Calculations from these data of the concentration of neutral hydrogen showed some changes with a period of several tens of minutes. A comparison with similar observations from different places leads to a provisional conclusion that the areas of these disturbances of neutral hydrogen are inhomogeneities of a few hundreds of kilometers in dimension.

It was shown on the basis of spectrographic Hα observations made in Abastumani that the distribution of atomic hydrogen corresponds to the presence of a symmetrical geocorona with some diurnal variation of hydrogen content (at night the amount of hydrogen is 5–6 times greater). Seasonal variations of Hα intensity reduced to a certain Earth's shadow height, namely to 1000 km, are nearly the same during the whole year. Their summer minimum does not exceed 20% of the mean annual value. The observed intensity of Hα also enables the deduction of the vertical distribution of the total quantity N(H) of atomic hydrogen in the column of atmosphere from 250 to 2700 km. In general this quantity increases with a decrease of solar activity. In addition, the observations in Abastumani have shown that the intensity of Hα in the twilight is several times greater than in nocturnal conditions. At the same solar depression some rise in Hα emission in the dawn in comparison with the evening was also observed. From such observations it is possible to obtain the hydrogen concentration at heights up to 200 km.

Continuous observational series during 1962–68 reveal the dependence of Hα emission on solar activity, the phase displacement being 1·5 or 2 years (I. A. Chvostikov, T. C. Megrelichvili).

Very interesting measurements of the emission rate of the Hα line were made with a photoelectric spectrometer at Haute Provence by M. F. Ingham (*M.N.R.A.S.*, **140**). Care was taken to make proper allowance for the integrated star light, the zodiacal light, the airglow continuum, and also for contamination by galactic sources. The results show that there is sufficient neutral hydrogen in the exosphere for multiple scattering of solar $L\beta$ radiation to be important. Ingham believes that measurements of Hα have a certain advantage over $L\alpha$ observations.

Observations of Hα emission were made at Camden, Australia, by E. B. Armstrong (*Planet. Space Sci.*, **15**) and the following results were obtained. There is a definite build-up of the hydrogen content during the night, and the maximum content is attained soon after midnight. The temperature at the exosphere base rises significantly before dawn at that altitude. The concentration of hydrogen towards the plane of the ecliptic is possibly less marked than previously reported. The absolute intensity of the Hα emission is higher than would be expected from reported measurements of the solar $L\beta$ intensity.

Observations of Hβ emission were made with a Fabry-Pérot interferometer at Testa Grigia in the Italian Alps by N. K. Reay and J. Ring (*Planet. Space Sci.*, **17**). They obtained an emission rate of 0·1–0·2 rayleigh, probably representing the addition of two components, galactic and non-galactic. The emission at galactic latitudes greater than $\pm 15°$ is interpreted as being non-galactic in origin and is discussed in terms of both geocoronal and interplanetary emitting regions.

IX. OTHER EMISSIONS IN THE NIGHTGLOW

(N_2^+)

There has been a long-standing argument whether or not the N_2^+ emission is present in the nightglow. A. L. Broadfoot and K. R. Kendall reported in "The Airglow Spectrum" (*J. geophys. Res.*,

73) that the absence of the emission is obvious in this spectrum but the 1*R* limit cannot be lowered owing to the blending of what appears to be a strong unidentified band at about 3908 Å.

(N I)

Very faint emissions previously undetectable by filter photometry now seem to be regularly observable. Observations of the 5200 Å [N I] line at Haute Provence were reported by G. Weill and J. Christophe-Glaume (*C. r.*, **264**). Close correlation between the 5200 Å [N I] and 6300 Å [O I] emissions were revealed, the ratio of radiation rates being about three percent. Implications for the *F*-region recombination processes have been studied by G. Weill (*C. r.*, **264**) with the following results. The lifetime of the $N(^2D)$ state at 250 km is of the order of 12 min, the ratio of excitation of $N(^2D)/O(^1D)$ is of the order of 4, and the recombination of nitrogen molecules at this altitude predominates over that of the oxygen molecules.

The green doublet 5200–02 Å of N I was studied by B. P. Saxena (*Ann. Geophys.*, **25**) who found the intensity maximum in the *F*-region, in agreement with Weill *et al.*

X. TWILIGHT AIRGLOW

A study was made of the intensity distribution among the Meinel bands and first negative system of N_2^+ due to resonance scattering of sunlight by A. L. Broadfoot (*Planet. Space Sci.*, **15**). Absolute transition probabilities were used to calculate the relative populations among the ion states under resonance scattering conditions. Photon scattering rates were calculated for most of the ion bands and it was suggested that an appropriate value for the 3914 Å band would be 0·05 photons s^{-1} per ion. Sunlit auroral spectra of N_2^+ were also discussed. He made a study of the tropospheric scattering of the emission, constructing a model for the whole twilight atmosphere (*Planet. Space Sci.*, **16**).

In Abastumani the investigation of twilight emission of sodium was performed taking into account the screening influence of the low atmospheric strata. The effective height of the screening for this emission varies 40 and 10 km for solar depressions from 6° to 13°. This permits the derivation of the vertical distribution of the atmospheric sodium and the elucidation of its new properties (T. I. Torochelidse). In particular, it turned out that in winter the height of the maximum content of sodium is 10–15 km lower than in summer. It depends also on the atomic concentration.

In Abastumani observations of the twilight intensity at symmetrical points of the solar vertical were commenced following a method of V. C. Fesenkov. Their purpose is to detect cosmic matter in the Earth's upper atmosphere.

The O_2 band at 1·58 μ was observed with a scanning Fabry-Perot system by R. L. Gattinger (*Can. J. Phys.*, **46**). The brightness at a solar depression angle of 6° was about 25 kR, but the emission could not be detected in the morning twilight. A theoretical discussion indicated that the ozone dissociation mechanism is probably the correct one.

Measurements of twilight helium 10 830 Å emission arising from resonant scattering of sunlight in the upper atmosphere were made by B. A. Tinsley (*Planet. Space Sci.*, **16**), and intensities from less than 200 R to greater than 1000 R were found for solar depression angles greater than 10°. A marked maximum in the emission rate was found in the winter of 1966–67. The possible latitude variations in atmospheric constituents and the effects of conjugate point photoelectrons are discussed with regard to the north-south asymmetry and winter maximum. A strong northerly emission accompanying a moderate magnetic storm was observed, and the possibility of considerable solar cycle variations, at least at low latitude, was indicated.

Photometric measurements of the helium line with a Fabry-Pérot instrument were performed by Y. P. Neo and H. N. Rundle (*Planet. Space Sci.*, **17**). They recorded large fluctuating enhancements while visible auroras were present. The intensity decay during twilight is consistent with the excitation of triplet helium atoms by collision of photoelectrons with ground state helium atoms, followed by resonance scattering.

Observations of many twilight emissions are being made at Haute Provence by N. Doan. He pointed out that the Li I and Ca II emissions are connected with meteoric swarms, though they are much affected by atmospheric movements. He also found that the He I line at 3889 Å is fairy strong, being 10–20 rayleighs.

Ca II twilight emissions in the southern hemisphere at Adelie Land, Antarctica, were first observed by G. Weill (*Ann. Geophys.*, **22**). The observations confirmed the seasonal nature of the intensity variations and indicated a large maximum in early June. The Ca I resonance line at 4227 Å did not appear on any spectra. Aircraft observations of the Ca II emission were made by T. P. Markham and R. E. Anctil (*Planet. Space Sci.*, **15**), and they identified the K line with an intensity of 30 rayleighs.

G. I. Kvifte and L. Wallace studied a method of determining the atmospheric sodium concentration from twilight observations (*Planet. Space Sci.*, in press). A. L. Broadfoot is currently making observations of sodium D_2/D_1 emission in twilight at Kitt Peak.

XI. DAY AIRGLOW

The daytime airglow might not belong to the field of our commission, but as it has close connection with the twilight or night airglow, studies of day airglow will be briefly introduced in this section. Some studies by rocket borne instruments are included in this section.

Observations of 5577 Å [O I] emission of the dayglow were obtained with a rocket photometer from the equatorial launch site at Natal, Brazil, by B. S. Dandekar (*Planet. Space Sci.*, **17**). He observed that the emission comes from three different ranges of altitudes: a layer around 96 km which contributes about 40% of the total emission, a second layer 110–150 km, and a third around 150 km of about 30% of the total emission. The probable excitation mechanisms of the layers were given.

The daytime 5577 Å emission was also observed by A. H. Jarret and M. J. Hoey (*Planet. Space Sci.*, **15**), who found the maximum varying from an estimated 50 to 100kR, being possibly auroral emission rather than general O I dayglow.

Observations of O_2 bands in dayglow were made by several researchers. L. Wallace and D. M. Hunten (*J. geophys. Res.*, **73**) made rocket observations of the A band over the range 35–128 km, and (0–1) band 59–95 km. The major production processes have been identified. Calculation of the transmission function for a self-absorbed band and production and transmission of the nightglow emission were also discussed.

The O_2 band at 1·58 μ was observed with a ground-based spectrometer by R. L. Gattinger (*Can. J. Phys.*, **47**), and the brightness was found to be approximately 630 kR.

The altitude profile of the infrared oxygen airglow at 1·27 μ was measured with a rocket-borne photometer by W. F. Evans *et al.* (*J. geophys. Res.*, **73**). They found the emission to peak at 49·5 km and the maximum volume emission rate to be 1030 kR km^{-1}. The emission layer agrees with that predicted from an excitation mechanism of the photolysis of ozone in the Hartley continuum. There was some excess emission above 80 km, and possible explanations were considered. Based mainly on this observation, it was pointed out by D. H. Hunten and M. B. McElroy (*J. geophys. Res.*, **73**) that O_2 can be ionized by the 1027–1118 Å band. This source appears to be able to explain the presence of O_2^+ ions in the *D*-region, and quantitative estimates of the production rate were made.

Surface albedo and the filling-in of the Fraunhofer lines in the day sky was studied by D. M. Hunten (*Kitt Peak nat. Obs. Contr.*) who concluded that the principal effect responsible for smearing of the spectrum is probably the so-called Rayleigh wings.

Rocket-borne spectrometers have been used to examine the spectrum of the daytime airglow in the near infrared from 6000 to 11 500 Å, and a number of fainter emissions, N_2^+ emissions at 7875 and 9212 Å, atomic oxygen emissions at 7774 and 8446 Å, and N_2 emission at about 8850 Å, were observed and studied by L. Wallace and A. L. Broadfoot (*Planet. Space Sci.*, **17**). Rocket observations of visible and ultraviolet dayglow were made by T. Nagata *et al.* (*J. Geomag. Geoelect.*, **20**). The emissions observed were O I 6300 Å, N_2 3914 Å, and ultraviolet between 2000 and 3000 Å. Some theoretical interpretations were given on the basis of calculations made for emission rates for various excitation mechanisms. The ultraviolet dayglow in 2400–3200 Å was measured from aircraft and a total emission rate of approximately 300 kR was obtained by L. R. Doherty (*J. geophys. Res.*, **73**).

XII. ROCKET AND SATELLITE OBSERVATIONS OF AIRGLOW

A good many observations of airglow from rockets and satellites were made in this triennium as in the last.

Observations of 5577 Å, 6300 Å, Na D, and the continuum at 5300 Å in the nightglow were made in Woomera by R. G. H. Greer and G. T. Best (*Planet. Space Sci.*, **15**), and emission profiles were obtained. They consider that up to 80% of the 5300 Å emission observed from the earth may be extra-atmospheric in origin.

The profiles of 6300 Å emission were observed by M. Huruhata and T. Nakamura (*Space Res.*, **8**) with two rockets at Kagoshima, and the heights of maximum emission were found at 260 km and 290 km. The height profiles were found to be fairly different for different geographic positions and different solar altitudes. M. Huruhata *et al.* (*Rep. Ionosph. Space Res.*, **21**), made a rocket measurement of 5577 Å emission and the continuum at 5300 Å with both upward and downward directed photometers. A slightly lower height of the maximum 5577 Å emission was obtained than reported in previous studies.

Observations of the OH band, 5577 Å line and the continuum at 5577 Å were made and their altitude profiles were given by D. J. Baker and R. O. Waddoups (*J. geophys. Res.*, **72**).

Two satellite observations of airglow were reported in this triennium, and very interesting results were given. J. G. Sparrow *et al.* (*J. geophys. Res.*, **73**) employed photometers sensitive to three wide spectral bands, carried on the OSO-B2 which was launched in 1965 into a near-circular orbit at an altitude of 600 km, with an inclination of 33° to the equator. The intensity of the continuum was found to vary between wide limits, ranging from 0·1 to 0·9 R $Å^{-1}$ in the visual spectral region and from 0·06 to 0·5 R $Å^{-1}$ in the blue region. Intensity changes of a factor of 2 were often found between observations one orbit or 24° of longitude apart. Features in the longitudinal airglow profile had typical sizes of 1500–2000 km and life times in excess of several days. No latitudinal variation of continuum intensity was observed although a marked variation in intensity occurred during the ten months.

D. D. Elliott and S. R. LaValle (*Space Res.*, **8**) surveyed the equatorial region repeatedly from a polar-orbiting satellite launched in 1966. The measurements were made at 5577 Å and 6300 Å during four months at a local time near midnight. There were red (6300 Å) arcs every night, and green (5577 Å) 9 times out of 13. The red arcs of median brightness of about 150 rayleighs occurred in pairs spaced symmetrically around the magnetic equator. The green arcs frequently occurred in coincidence with a red arc. The red and green arcs typically occurred at ±15° magnetic latitude and were about 10° wide, and the arcs were not dependent on K_p.

Rocket photometry of the far UV twilight airglow was made by H. W. Moos and W. G. Fastie (*J. geophys. Res.*, **72**). $L\alpha$ showed a height profile similar to that of the dayglow with a peak value of 3·6 kR at a height of 152 km. The O I 1304 Å had a maximum value of 110 R toward the horizon, and low intensity emission in 1360–1600 Å was observed.

XIII. PHOTOMETRIC STUDIES OF ZODIACAL LIGHT

Photometric and polarimetric observations of the zodiacal light were reported by many researchers. Observations from space vehicles were performed and many valuable results were reported.

A comprehensive review of the zodiacal light has been given by D. E. Blackwell *et al.* (*Adv. Astr. Astrophys.*, **5**).

Assuming that if the zodiacal light is due to a heliocentric dust cloud, photometric localization of the Sun's equatorial plane is much easier when the Earth is close to that plane, R. Dumont *et al.* (*Ann. Astrophys.*, **31**) made photometric scans in June and December, 1966, and they got an estimate of the inclination to the ecliptic of the heliocentric dust cloud: 1°9 northward in Leo and 1°3 southward in Aqr. They found the shifts to be in very good agreement with the hypothesis of a heliocentric dust cloud flattened along the invariable plane and in poor agreement with a cloud flattened along the solar equator. They concluded that the best interpretation is a photometric mixture of the two clouds, unequally flattened on their symmetric planes, perhaps indicating a geocentric dust component in the zodiacal brightness. R. Robley (*C. r.*, **265**) found a positive correlation between the brightness of zodiacal light and the intensity of the 5577Å airglow line and suggested the existence of a second zodiacal light component. He is continuing the observations at the Pic-du-Midi Observatory in efforts to confirm the existence of this second component.

Polarimetric observations in narrow bands at 5080Å and at 5577Å were made at Haleakala by J. L. Weinberg *et al.* (*Planet. Space Sci.*, **16**), and they found that the polarization of the continuum at 5080Å is caused both by the polarization of zodiacal light and by atmospheric polarization, and that the polarized component of the 5577Å line emission is, for large areas of the sky, brighter than that of the continuum at 5080Å. An important suggestion was made that broad-band measurements can lead to overestimates of the polarization of the zodiacal light, especially for regions near the ecliptic poles. J. L. Weinberg and H. M. Mann (*Ap. J.*, **152**) made observations near 4300Å, a wavelength region which Powell *et al.* had suggested may contain negative polarization in the region of small elongations ($\varepsilon < 90°$). No negative polarization was detected at these elongations at 4300 Å or at other wavelengths in the visible.

In the interest of studying the particles causing the zodiacal light, M. Daehler *et al.* (*Planet. Space Sci.*, **16**) have observed the profile of the Hβ absorption line in the zodiacal light spectrum at Chacaltaya with a Fabry-Pérot spectrometer. The absorption feature was obscured by Hβ emission of uncertain origin, preventing precise Doppler-shift measurements and no wavelength shift could be shown. Observations of the Hβ line profile in the zodiacal light were repeated in order to decide on the origin of the previously observed emission features with a higher resolving power by P. H. Hindle *et al.* (*Planet. Space Sci.*, **16**). All the scans showed a small emission component which had a concentration toward the ecliptic and showed a shift to the blue side. From this observation, N. K. Reay and J. Ring showed separately (*Nature*, **219**) that the observed radial velocity variation for elongations up to 60° agrees well with that expected of dust particles in circular orbits about the Sun. For this result, L. W. Bandermann and R. D. Wolstencroft (*Nature*, **221**) reached the conclusion that the dust orbits are nearly all direct and of low inclination and low eccentricity. Additional studies have been carried out by D. Clarke *et al.* (*J. Phys.*, **28** and *Modern Astrophys.*, Paris, 1967), J. F. James (*M.N.R.A.S.*, **142**), and N. K. Reay (*Nature*, **224**).

M. F. Ingham and R. F. Jameson (*M.N.R.A.S.*, **140**) have obtained polarization observations of the nightglow and have derived a model for zodiacal dust in a direction normal to the ecliptic.

R. D. Wolstencroft and F. E. Roach (*Ap. J.*, **158**) critically discussed the conflicting values of the brightness of the zodiacal light at the north ecliptic pole reported by various authors during the past four years.

Infrared observations of the outer corona at 2·2μ by R. M. MacQueen (*Ap. J.*, **154**) showed that the isophotes of the infrared corona have an interpolated peak coinciding closely with the invariable plane, and this apparent concentration of particles in the invariable plane would rule out a major geocentric contribution to the zodiacal light. Infrared observations of the corona at 2·23μ by A. W. Peterson (*Ap. J.*, **155**) confirm his earlier predictions of thermal emission by zodiacal dust and indicate the existence of a vaporization zone at 3·9$R_{\odot}$.

F. Sanchez Martinez (*Semin. Astr. Geodes. Madrid*, Publ. No. 60) noticed that residual fluctuations of zodiacal light brightness would be governed by solar chromospheric eruptions.

Some very interesting papers about the effect of the solar activity on the brightness and polarization of the zodiacal light were presented by A. S. Asaad (*Nature*, **214**, and *Observatory*, **87**), G. Weill (*C. r.*, **263**) and J. Dufay (*C. r.*, **263**). Asaad plotted the brightnesses of zodiacal light observed since 1950 and found that the brightness increased with decreasing solar activity by a factor of 2, and he concluded that errors of observations and uncertainties in the method of reduction would not explain this relation. As possible explanations of this phenomenon, he pointed out that during the maximum solar activity the solar wind is rather strong and might destroy, evaporate or push away some of the interplanetary material. Another possibility is that when the corpuscular radiation is strong it may cause the very fine particles to spiral away from the sun, whereas during minimum solar activity the spiralling would be toward the Sun.

An analysis of 13 years of nightglow observations towards the zenith and pole from Haute Provence was made by G. Weill and J. Dufay. They found a solar cycle variation of the brightness of the zodiacal light, of total amplitude 25%, almost in opposite phase with solar activity. The maximum zodiacal light precedes the minimum solar activity by two years.

In the Dudley Observatory, J. L. Weinberg is analyzing nightglow data obtained at Haleakala between 1961 and 1969. R. D. Wolstencroft is continuing observations at the Haleakala Observatory on the anticipated annual variation of the relative brightness of zodiacal light at mirror points of high ecliptic latitude in the two ecliptic hemispheres; the position of the axis of zodiacal light; and the brightness and polarization of the integrated galactic light at 5080 Å.

Observations of Zodiacal Light from Space Vehicles

The first report of observations of zodiacal light from a satellite was given by J. G. Sparrow and E. P. Ney (*Ap. J.*, **154**). The satellite OSO-B2 had a 9-month operative lifetime from its launch in February, 1965. The photometers were aligned along the spin axis of the satellite, and the nature of the satellite restricted the observations to a constant elongation of 90°. The brightness and polarization of the zodiacal light were measured in visual and blue wavelength regions. The brightness at the ecliptic pole obtained by these observations was 50 S_{10} (vis), being far smaller than that obtained by previous ground-based and rocket observations. The observed degree of polarization was also smaller than the previous results, and for none of the strongest magnetic events was an increase in the polarized component greater than 10% observed. The upper limits of the interplanetary electron density during a number of solar events was 17 electrons cm^{-3} in a cloud of thickness 1 a.u. Additional measurements are being made from OSO-F (E. P. Ney; Univ. of Minnesota) and from OSO-G (Ruoy, Carroll and Aller; Rutgers).

The importance of future studies of the zodiacal light and interplanetary medium from artificial vehicles has been emphasized by many researchers. H. Elsässer (*Nuovo Cim.*, **5**; *Naturwissenschaften*, **56**) pointed out some interesting problems to be solved by observations from space vehicles and described some measurements to be performed in the near future. C. Leinert (*Ster. Weltraum*, 1969). D. Lemke and W. Gabsdil (*Mitt. astr. Ges.*, Nr. 25) described a project for zodiacal light observations with balloons.

The group of H. Elsässer is now preparing zodiacal light observations by a balloon-borne telescope and by the German-American space probe HELIOS which will approach the Sun to a distance of 0·3 a.u. Rocket and balloon observations are also being carried out by French and Japanese groups. R. H. Giese and C. Dziembowski (*Planet. Space Sci.*, **17**), derived two types of models of interplanetary dust and suggested measurements that should be made from space probes.

Two zodiacal light experiments are being developed at the Dudley Observatory by J. L. Weinberg and his group: (1) Photoelectric measurements from an Earth-orbiting satellite to be launched in spring 1972. Observations of brightness and polarization will be obtained at ten wavelengths between 4000 and 8200 Å. (2) Photoelectric measurements from the Pioneer F/G Asteroid-Jupiter probes: F to be launched in 1972, G to be launched in 1973. Observations of brightness and polarization will be obtained in two broad spectral regions.

XIV. DUST PARTICLES IN INTERPLANETARY SPACE

The basic problems of the distribution and density of dust particles in interplanetary space and the scattering of light by the particles have been studied from many angles both observational and theoretical. Substantial studies were made especially by Soviet researchers in this triennium.

The essential features of the optical method for surveying material from space that has penetrated into the upper layers of the Earth's atmosphere through analysis of twilight phenomena was described by V. G. Fesenkov (*Sov. Astr.–AJ*, **12**). It was readily possible to identify individual irregularities that develop mainly at a height of about 80 km, and to follow their time variation. Numerical results obtained by reducing the observations were reported.

D. E. Blackwell *et al.* (*M.N.R.A.S.*, **136**) presented a model of the distribution of the interplanetary dust based upon photometric measurements of zodiacal light and *F*-corona and on the distribution function of particle radius obtained from space vehicles. They proposed a model where the dust density decreases and the albedo of the particles increases with distance from the Sun. This model could lead to improved agreement between calculation and observation.

The suggestion of Peale that the earth is surrounded by geocentric dust whose particles are diffusely concentrated toward the ecliptic plane was examined by R. D. Wolstencroft (*Planet. Space Sci.*, **15**), who showed that the contribution of such a cloud to the zodiacal light would be less than 20%.

Purely gravitational capture of cosmic dust by the Sun with the assistance of the planets, and by planets with the assistance of their satellites was proposed by V. V. Radzievskiü (*Sov. Astr.–AJ*, **11**) and approximate quantitative computation of the captured mass was made. The earth captures three orders of magnitude more zodiacal matter than is necessary for sustaining the circumterrestrial cloud with the experimentally observed density. The evolution of the circumterrestrial cloud was also discussed.

The influence of solar radiation pressure on the motion of dust particles in a circumterrestrial orbit was studied by N. B. Divari and Yu. A. Klikh (*Sov. Astr.–AJ*, **11**). They showed that the perigee distance for the orbits of micron-size particles experiences oscillations of substantial amplitude, and consequently these dust particles will sink rapidly into the Earth's atmosphere. Another study of the effects of solar radiation pressure on the particles of interplanetary dust as a function of their radius and density was made by L. M. Gindilis *et al.* (*Sov. Astr.–AJ*, **13**). It was concluded that the micron and submicron components of the dust must primarily consist of comparatively dense particles of an asteroidal origin. For particles of the order of several microns or larger, as well as for very small dimensions, interplanetary space must contain less dense particles of cometary origin in addition to the asteroidal particles. Absorbing particles with radii less than 0·1–0·5μ cannot be present in the interplanetary dust.

A meteor model for the zodiacal cloud was presented by N. B. Divari (*Sov. Astr.–AJ*, **11**), who made calculations of the concentration of dust material as a function of the distance from the Sun and the heliocentric latitude on the basis of meteor orbits, and these results were used to compute the brightness and polarization of the zodiacal light. The results were compared with observed values, showing that the meteor model provided a relatively good representation.

Reference should be made to an important paper on the optical scattering from non-spherical randomly aligned, polydisperse particles by R. S. Powell *et al.* (*Planet. Space Sci.*, **15**).

XV. RELATED STUDIES

A very significant study, which has important relationships with airglow and zodiacal light studies, was reported by A. N. Witt (*Ap. J.*, **152**). He carried out new photoelectric measurements of the intensity and distribution of the diffuse radiation in the Galaxy. Contaminations originating in other sources of the night sky light were investigated and taken into account. The residual intensity was interpreted as general galactic starlight scattered by interstellar dust particles, and many properties of the scattering were presented.

C. F. Lillie and A. N. Witt (*Astrophys. Letters*, **3**) detected no diffuse galactic light in the UV (see, also, C. F. Lillie, Dissertation, Univ. Wisconsin, 1968). J. L. Weinberg *et al.* (*Bull. AAS*, in press) have separately observed the diffuse galactic light, and they find a brightness in polarized light of 1 to 2 S_{10} (vis) units at 5300Å.

Observations and interpretation have also been carried out by F. E. Roach (*Modern Astrophys.*, Paris, 1967), C. J. van Houten (*BAN*, **19**), N. C. Wickramasinghe (*Nature*, **218**), and H. C. van de Hulst and T. de Jong (*Physica*, **41**).

F. E. Roach and L. L. Smith (*Geophys. J.R.A.S.*, **15**) have examined nightglow observations in a search for cosmic light, and they have established an upper limit of 5 S_{10}(vis) units at 5300 Å.

The determination of the brightness of any light source outside the atmosphere from ground-based observations was studied by R. D. Wolstencroft and I. G. van Breda (*Ap. J.*, **147**). The results of calculations of the brightness and the degree of polarization of the light scattered in the Earth's atmosphere were presented in the cases of zodiacal light, the integrated starlight, and an isotropic source.

The scattering of light in the troposphere and the polarization of certain components of night-sky radiation were studied by N. B. Divari (*Sov. Astr.–AJ*, **12**) who showed that a quite high degree of polarization must be taken into account when analyzing polarization measurements of the zodiacal light. It was argued that revisions should be made in data on the zodiacal light intensity at the ecliptic pole obtained without allowance for the light from all sources of night-sky emission scattered by the troposphere.

Rocket observations of galactic light outside the airglow layer are being carried out by the group of Tokyo Observatory. H. Tanabe is measuring star fields on the Palomar Sky Atlas in connection with these observations.

The most recent searches for lunar libration clouds have produced negative results and have established upper limits to the brightness that could be associated with such clouds (R. G. Roosen and C. L. Wolff, *Nature*, **224**; J. L. Weinberg *et al.*, *Bull. AAS*, in press).

M. HURUHATA
President of the Commission

22. METEORS AND METEORITES (MÉTÉORES ET MÉTÉORITES)

PRESIDENT: Z. Ceplecha.
VICE-PRESIDENT: R. E. McCrosky.
ORGANIZING COMMITTEE: P. B. Babadžanov, W. G. Elford, C. L. Hemenway, A. A. Javnel', P. M. Millman.

INTRODUCTION

The limited space made available for this report by recent regulations of the Executive Committee of our Union does not allow the traditional list of all published papers. Also, it omits the data on detection of cosmic dust, which are now published each year in surveys by the COSPAR Dust Panel. This report is mainly based on completely fresh information contained in numerous letters sent to me.

The accompanying report of our Meteorite Committee is done in the old style, but only a limited number of references are given. Evidence for a terrestrial origin of tektites is becoming stronger every year, and this subject is omitted in order to shorten the report.

Before starting the report, I mention with great regret the death of Prof. Dr Cuno Hoffmeister. His name remains connected forever with the famous period of naked-eye observations of meteors. Another great loss for meteor observers was the death of Mr Koiro Komaki, the President of the Nippon Meteor Society.

GENERAL

The period of this report began with Symposium no. 33 of the IAU, Physics and Dynamics of Meteors, held in Tatranská Lomnica, Czechoslovakia, September 4–9, 1967. The symposium volume was edited by Kresák and Millman with exceptional dispatch and published by D. Reidel Company, Dordrecht, Holland, in 1968. It contains 48 papers and all the discussions.

During the last years, the greatest improvement in our knowledge of meteors was achieved in the interpretation of photographic records of fireballs and in the production and study of simulated meteors in laboratories and in the atmosphere. Also completed were several evaluations and interpretations of quite a number of meteors recorded by multistation high-power radar systems. Interest in meteor spectroscopy is continuously increasing, and there has been a great improvement of the observational data on fainter meteors over those of several years ago, as well as achievement of dispersions better than 100 Å/mm for fireballs. Work done in the physical theory of meteors is mostly based on attempts to explain the difference between the dynamically and the photometrically computed meteor masses. Generally, three mechanisms, progressive fragmentation, frothing, and heatshock, were proposed, and some models computed. It seems that no one of these explanations alone is able to explain all the observed effects both for faint meteors and for bright fireballs. There is still much more to be done in meteor physics. It appears also that the atmospheric constituents and variations are more important than we thought.

Statistical studies of photographic meteor orbits, together with some physical parameters of the meteor phenomenon, give an indication that a greater part of sporadic meteors than recently assumed could be of asteroidal origin. This problem is far from being finally decided. The traditional idea that faint meteors are only of cometary origin and bright fireballs only of asteroidal origin has become untenable.

FIREBALLS AND METEOR PHOTOGRAPHY

Precise photographic data on several hundred very bright fireballs recently became available. Most important in this respect was the continuation of systematic photographic work in a multiple station project of the Smithsonian Astrophysical Observatory in the mid-western United States

(Prairie Network). McCrosky found a surprising result from his Prairie Network data on fireballs: The bulk densities of the meteoroids computed for these large bodies by equating the dynamic and photometric mass are the same small values as those previously determined for faint Super-Schmidt meteors (0·4 gr/cm³). A similar value was obtained by Ceplecha (Ondřejov Observatory) for the Příbram fireball, but the bulk density of the recovered meteorites is 3·6 gr/cm³. It is as though the meteorites were the strong portion of an otherwise weak, loose material. One must choose from two possibilities: (a) the majority of the fireballs are produced by material that is completely different from that of meteorites, or (b) our understanding of the physical processes encountered by a high-speed ablating body is fundamentally wrong. If bodies producing fireballs are similar to meteorites, almost all the kinetic energy of the body must be converted into visible light. But the average luminous efficiency, relatively well known from several different experiments, is two orders lower.

Another important result obtained from the Prairie Network is that there are considerably more large meteoroids (greater than 1 kg in mass) in space than was earlier estimated on the basis of meteorite falls. The data indicate a population of large (pre-atmospheric) meteoroids that is perhaps as much as one hundred times greater than expected. Two possible falls have been observed; despite extensive searching, neither of these has been found on the ground.* This failure to recover meteoritic material, as well as photographic observations of explosion-like break-up in the atmosphere, also indicates that the large meteoroids are friable.

Another photographic network, using one all-sky camera with rotating shutter at each station for recording fireballs, continues activity in Central Europe. Twenty-six German stations directed by Zähringer (Max-Planck-Institut für Kernphysik, Heidelberg) cooperate with twenty Czechoslovak stations directed by Ceplecha and Porubčan. Ceplecha and Ježková (Ondřejov Observatory) compute fireball trajectories for the whole European Network. Several searches for meteorites were organized for bodies with significant computed terminal masses. The lowest observed height of a luminous trajectory was 20 km, for a very bright fireball (−20th magnitude?) on 1969 April 10, over southeast Germany. This fireball might have dropped several tens of kilograms of meteorites. An extensive, but unsuccessful search close to Otterkirchen (Passau) was organized in the summer and autumn of 1969 by Zähringer. No meteorites have yet been found for any fireball within the European Network.

A network of 12 camera stations is being constructed in western Canada by Halliday (Dominion Observatory, Ottawa) to provide data for the recovery of meteorites and the study of their orbits. Each station employs 5 cameras using 70-mm film and is equipped with a photoelectric meteor detector. Instrumentation of the network should be completed in 1970.

Millman (National Research Council, Ottawa) maintains a catalog of all bright fireballs observed in Canada. Search is initiated whenever there seems to be a chance for the recovery of meteoritic material. He is also making a statistical study of the orbits of meteorite-dropping fireballs.

Olivier (American Meteor Society, Philadelphia) continues to organize the visual observations of fireballs and to compute the height, location, and orbit for many fireballs for which two or more reports are available.

Levin and Simonenko (Schmit's Geophysical Institute, Moscow) determined true radiants of 14 meteorite fireballs. Simonenko also eliminated computational errors in published data for trajectories and orbits of 5 other meteorite fireballs.

Rajchl (Ondřejov Observatory) searched for further possible explanations for the origin of two groups of fireballs. One he recognized to be connected with the "new" comets, on the basis of the inclination distribution, but the other orbital elements do not fit well. He started to compute a non-gravitational jet deceleration acting on part of the parent comet shortly after a split that decreases a and e, with i being constant.

Ceplecha, Ježková, and Novák continue their systematic double-station program for meteor

* McCrosky photographed a meteorite fall from two stations of the Prairie Network on January 4, 1970 at 2h 14m 17s GMT. G. Schwartz found the first meteorite (bronzite chondrite?) of 9.8 kg within the computed area close to Lost City, Oklahoma. Altogether, 3 meteorites of this fall were recovered up to February 1970.

photography at Ondřejov Observatory, using 30 classical small cameras with rotating shutter, placed at two locations 40 km apart and covering about half the visible hemisphere. This program has been going on for 20 yrs without interruption; in more than 6000 h of exposure time, 40000 glass plates (9 × 12 cm) have been exposed and almost 1000 double-station meteors photographed (mostly sporadic). Only a small part of these data have been reduced and published. Most reductions have been on special bright objects (including the Příbram fireball), in connection with high-dispersion spectra of the same objects, or during special periods like IGY and IGC.

Seven years ago Yabu and Takeuchi from the Nippon Meteor Society began multiple observations of meteors, both visual and photographic, using a baseline of 90 km in the middle of Japan Island. About 50 double-station meteors have been photographed and their reduction is now in progress.

Ayers (NASA Langley Research Center, Virginia) developed two different methods of making photometric measurements of photographic artificial meteors; they can be used for the photometry of natural phenomena as well.

Dohnanyi (Bellcom, Inc., Washington) undertook a statistical analysis of the McCrosky and Posen photographic meteors in an effort to define their distribution more precisely.

Ceplecha (Smithsonian Astrophysical Observatory) studied discrete levels of meteor beginning height in detail. He confirmed the existence of two main levels of beginning height for Super-Schmidt meteors, separated by about 9 to 10 km. Among all possible interpretations of these discrete levels, composition and fragmentation seem to be the most important.

Dalton (NASA Marshall Space Flight Center, Alabama) made a complicated statistical analysis of 413 meteors by Jacchia *et al.*, and of the 333 brightest objects selected from McCrosky and Posen Super-Schmidt meteors.

METEOR SPECTRA, PHOTOELECTRIC OBSERVATIONS AND SIMULATED METEORS

A working meeting on meteor spectra was held at Smithsonian Astrophysical Observatory in Cambridge in April 1969. Its main purpose was to find whether or not the spectra contain any useful information on the composition and structure of meteoritic particles. Even if the conclusion was rather pessimistic, the group recommended that research continue in three different ways: (a) to develop new models of semi-equilibrium for larger bodies penetrating very deep into the atmosphere, (b) to obtain spectra of early portions of faint meteor trajectories, where the free molecular flow regime is certain, (c) to examine the Hα profiles. It was recommended to continue sending a list of all meteor spectra photographed at different observatories to Millman, National Research Council, Canada.

Halliday continues the regular program of meteor spectroscopy at the Meanook and Newbrook Meteor Observatories. Several of the spectrographs have been equipped with new gratings to secure spectra of greater dispersion than previously possible. The Super-Schmidt photographs of spectrum 210, an iron meteoroid, have yielded a light curve and improved values of the deceleration and orbit for this particle. Gault (Dominion Obs.) measured the decay rate of the forbidden O I line at 5577 Å as a function of height in the photographic spectra of six meteors showing multiple images of the line when photographed with jumping-film spectrographs.

Photographic meteor spectroscopy is being continued by Millman, and a revised world list of meteor spectra is being prepared. It will include spectra of over 1200 different meteors to the end of 1968. In August of 1969, a cooperative program with the Dudley Observatory, Albany, was commenced. This involves the recording of meteor spectra by image orthicon, and a correlation of the data with records made simultaneously by visual, photographic, and radio techniques.

Ceplecha and Rajchl (Ondřejov Observatory) continued photographing meteor spectra. Several dozen spectra, most with dispersions better than 100 Å/mm, were recorded, of which 5 are of extremely good quality. One of them with about 260 spectral lines in the visible region belongs to a fireball that penetrated deep into the atmosphere. The end height of this meteor and its photographed spectrum was 30·06 km with 6·5 km/s velocity, while the velocity at 60-km was 18·4 km/s. The dispersion of the spectrum is 56 Å/mm in the first order for most of the long trajectory. The end mass of the main body was computed dynamically to be about 600 g. Meteorites should have landed. The separation into about 10 fragments is visible during the last 20 km of the trajectory, and the stronger spectral lines of some of the fragments were recorded. Two types of wake are observable in this spectrum: (a) a gaseous one with much lower excitation than that of the head spectrum, (b) a particulate wake having no substantial difference in excitation from that of the head spectrum. The Si II lines 6347 and 6371 Å are observable down to 45-km height and 12 km/s velocity as extremely faint features in comparison with the rest of the spectrum. This spectrum is now being reduced, with another extremely good spectrum of a bright meteor flare and wake containing more than 600 individual spectral lines and many molecular bands in the visible region.

Most of the meteor spectra at the Ondřejov Observatory are complemented by the direct double-station photographic data, and thus accurate heights, velocities, and distances are available for each trajectory point.

Kresák (Astronomical Institute, Bratislava) continued photographic observations of meteor spectra at the Skalnaté Pleso Observatory.

Hirose photographed an infrared meteor spectrum at Dodaira Station of Tokyo Astronomical Observatory on January 4, 1968. The spectrum covers the range up to 8700 Å, and preliminary investigations reveal the existence of emission lines of O I, N I and Ca II in the infrared region. As the meteor was directly photographed at the other 5 stations, detailed analysis, including computation of its atmospheric path, is now going on. In the spring of 1968, the automatic meteor cameras at Dodaira station were converted into spectral cameras with gratings of 150 lines/mm.

Etheridge and Russell (University of Southern California, Los Angeles) made a preliminary comparison of the spectrum of a 1966 Leonid with the spectrum of a 1962 Leonid. Differences, if present at all, are far smaller than those between Leonid and Perseid spectra. Russell obtained 21 spectra; on one of these showing two Perseid spectra, the forbidden oxygen line at 5577 Å appears faintly in the fainter spectrum but not in the stronger. The H and K lines are absent from the spectrum showing the green line. This practically eliminates the possibility that changes in the atmosphere contribute to differences in the green-line strength. Systematic study of the green-line appearance in the 24 spectra obtained in 1969 is pending. In 1969 Russell photographed four Perseid meteors simultaneously with from two to four cameras with polaroids in front of the lenses.

Rajchl developed a model of the interaction zone ahead of the meteor body in the near-free-molecular and transitional flow. He considered ionization, excitation, and recombination within this zone. Among other advantages, this model helps to explain head echoes and the 5577 Å forbidden oxygen emission. The transition to the shock wave during the slip flow could explain the observed inverse proportionality between the Ca II emissions and the 5577 Å oxygen-line intensities.

Three meteor spectra were photographed at Krym Station of VAGO. Astapovič is preparing a list of all meteor spectra photographed in U.S.S.R. to 1967.

In connection with the SAO meteor project, Harvey (NASA Langley Research Center, Hampton, Virginia) is developing instrumentation for and performing studies of meteor spectra. The purpose is to improve understanding of the physics of meteors and to determine meteoroid composition. The program consists of a Meteor Spectral Patrol in Southwestern United States, laboratory simulation of meteoric radiation from impact flash and d.c.-arc tests, and absolute photometry. The Meteor Spectral Patrol is equipped with 14 spectrographs with a Maksutov optical system of $f/0{\cdot}83$ to $f/1{\cdot}3$. These instruments have photomultiplier-triggered shutters that open when meteoritic events occur in the field of view; the resolution is as high as 1 to 2 Å. In the past year the Meteor Spectral Patrol has detected over 250 meteor spectra ranging from 0 to -5 magnitude, of which about 50 have more than 50 lines. Preliminary conclusions are the following: (a) The luminous efficiencies obtained from one flow regime are not directly applicable to another flow regime, and the masses of bright optical meteors have tended to be overestimated. (b) The physics of meteors is significantly different for faint meteors as opposed to bright meteors, and this fact should be considered in the analysis of integrated-light data, particularly in analysis of flaring and fragmentation. (c) The present spectral classification does not apply to faint optical meteors.

Observations of the same meteor by the Havana radar (Southworth, SAO) and by a sensitive image-orthicon television system (McCrosky, SAO) have been successful for meteors as faint as $m_v = 7$. A limited discussion of the coincident data confirms that the ratio of ionization to luminosity has been overestimated by about an order of magnitude. Coincident radar and image-orthicon spectral observations are also being made with the same system.

Cook (SAO), Hemenway (Dudley Obs.), and Millman (NRC) are studying observations of the same meteors with the radar and spectrographs at Springhill, Ontario, and with the Dudley Observatory's image-orthicon spectroscope. The observations were made during the 1969 maximum of the Perseids and should guide the analysis of the results obtained at Havana.

Vybornyj (Krym Meteor Station of VAGO, U.S.S.R.) finished construction of a photoelectric device for meteor observations. He has begun to register meteors down to 6th magnitude within a 15 degree field of view. Astavin-Razumin (Kučino Station, U.S.S.R.) finished a photoelectric device to record meteors. The instrument was successfully tested in the laboratory and is now routinely operated for meteor observations.

Hoffman and Longmire (NASA Electronics Research Center, Cambridge) proposed that significant contributions to meteor ion spectra may result from a two-step ionization and excitation mechanism. The first step is the fast ionization of atmospheric molecules in collision with re-emitted molecules, while the second step consists of fast near-resonant charge-exchange collisions that excite the evaporated meteor atoms. Using a first-collision model, Hoffman has found qualitative agreement with meteor observations. He has derived an expression for the spectral luminous efficiency defined for particular spectral lines. The result is velocity independent and proportional to the square of the atmospheric density and the square of the meteoroid radius.

Longmire is studying collisions of Fe^+ ions with N_2 and O_2 by passing an ion beam through a gas target. Preliminary results show that the $3d^64p$ configuration of Fe^+ is readily excited by collisions at an ion velocity of 56 km/s. Decay of this configuration may produce several of the Fe II lines observed in meteor spectra. These lines are detected by ground-based observations, but excitation of their upper levels is indicated by emission of ultraviolet radiation for the same levels. An attempt will be made to determine the velocity dependence of the excitation and to measure the emission cross sections of several lines.

Rajchl has found that for electron densities greater than 10^{10} el./cm, dissociative recombination is more efficient for the ionization decay than the diffusion. He also points out a possible connection between the dissociative recombination of NO^+, O_2^+ and the origin of the infrared emissions. Using a gas-dynamic model, he is searching for the mechanism of energy transfer to excitation and ionization. A significant parameter here is the modified Knudsen number.

Flannery and Levy (Smithsonian Astrophysical Observatory) are engaged in theoretical studies of the excitation and ionization produced in collisions between slow, heavy particles and of electron-ion recombination. Particular reference is made to atoms and molecules of meteoric and atmospheric interests. Multistate and semi-empirical Born calculations are in progress.

Using a plasmatron with 10000 K, Bronšten (VAGO, Moscow) has made laboratory experiments on the ablation of meteorites. A periodical spraying of the melted surface by the incoming air was found to be the main ablation process for large meteoric bodies. Periodic pulsations of the melted surface film were also observed. This effect is qualitatively used to explain the observed meteor flares.

Givens and Page (NASA-Ames Research Center, California) have been performing simulated meteor experiments in a ballistic range, using stainless-steel spheres of 0·6-mm diameter. Observations are made with shadowgraphs, filtered photomultipliers, and spectrographs. Velocities have reached 8 km/s at a Knudsen number of 0·002 (continuum-flow regime). They determined the mass-loss rate and the luminous efficiency. Panchromatic luminous efficiency is 0·002 at 6 to 8 km/s, which compares with 0·01 at 11 km/s from the Trailblazer simulation experiments, and 0·005 at 20 to 40 km/s for laboratory micrometeor observations at TRW Systems. The ablation process is complex: melt from the forebody resolidifies as flanges, increasing the body frontal area, and subsequently breaks away erratically, ultimately to vaporize or decelerate in the wake. Maximum luminosity appears in the wake and extends 250–500 body diameters downstream. Atomic Fe and

Cr lines dominate the spectrum from 3000 Å to 4500 Å, while metallic-oxide bands appear above 4500 Å.

Friichtenicht and his colleagues (TRW Systems, Inc.) measured the luminosity, ionization, and spectra of iron particles accelerated to 20–40 km/s and interacting with air, argon, N_2, and some other gases. They found the luminous efficiency not dependent on velocity and roughly comparable to that from other laboratory and meteor estimates.

Ayer, McCrosky, and Shao are analyzing 10 artificial iron and nickel meteors launched into the atmosphere by rockets and by shaped-charge accelerators. Velocities from 8 to 16 km/s were achieved for "meteoroids" of about 1 g. The results for iron suggest little change from the previous value of $\tau = \tau_0 v$ ($\log \tau_0 = -18{\cdot}1$ cgs and 0 magnitude) obtained from an earlier experiment. The luminous efficiency of nickel is less well determined but is certainly lower than that of iron, perhaps by a factor of 3 at 10 km/s. McCrosky examined meteors with spectra consisting of Fe radiation, but without D lines of Na I. He concluded that they are possibly derived from a source other than that which produces iron-nickel meteorites.

RADAR METEORS

An analysis of radar meteors by Southworth (Smithsonian Astrophysical Observatory) shows that attenuation of Fresnel oscillations is often caused by fragmentation. This phenomenon is common among even the faintest radar meteors (10^9 electrons/cm). His study of 10000 radar meteors with from 5×10^9 to 3×10^{11} electrons/cm at the maxima of their ionization curves shows the following: (a) The heights of radar meteors are nearly independent of magnitude and similar to Super-Schmidt heights. (b) Manning's initial radii are much more nearly correct than Öpik's. (c) Rapid recombination of the electrons is very important for meteors below a limiting height. This limit rises with increasing electron density. Fresnel-pattern analyses confirm the recombination rate of N_2 and O_2. Recombination invalidates all previous estimates of the ratio of ionization to luminosity.

The meteoric program of the Upper Atmosphere Research Section, National Research Council of Canada is based in Ottawa, the observing station being the Springhill Meteor Observatory. The meteor patrol radar was operated continuously from October 1957 to January 1968. The reduction of 10 years of data is continuing, and results are appearing in a series of papers by Millman and McIntosh. Special studies using these records have also been made by Štohl and Šimek. Subjects being investigated include the mass-distribution index for meteoroids recorded by radar, the heights of radar echoes, the physics of the ionization of meteor trails, and the structure of specific meteor showers. In particular, a 12-year study of the Leonid shower is being carried out. Final corrected rates of meteor echoes are being determined for each hour of every day in the year. These are being compared with a large number of visual meteor observations made by teams during the period 1957–1968. Advanced planning has commenced to set up identical meteor radar patrols at Ottawa, Ontario, Canada and Newcastle, New South Wales, Australia (Ellyett and Keay). The data readout will be automated, utilizing online computers, and the whole program will be coordinated to provide a comparison of the rates in the northern and southern hemisphere.

Ellyett and Keay prepared a complete report on the radar meteor survey undertaken in New Zealand from 1963 to 1965, which has now been thoroughly examined; the complete rate data and major findings will appear soon in the Memoirs of the Royal Astronomical Society. Ellyett analyzed 2·5 million meteor echoes from New Zealand and another 15 million from Canada. He established conclusively that Lindblad's idea of an inverse relationship with sunspot number is correct.

Brown and Elford (University of Adelaide) have examined, theoretically, the growth and decay of the radio echo from an irregularly ionized trail. This work has shown that the rate of decay is strongly affected by wind shear, and that the shears known to occur at meteor heights acting on irregularly ionized trails are sufficient to explain the scatter in the observed diffusion coefficients at a given height.

Pupišev, Andrianov, Belkovič, Tochtasev, Kurganov (Radio Astronomy Laboratory, Kazan, U.S.S.R.) studied the problems of meteor radar selectivity as well as those of the systems for meteor observations by oblique scattering, taking into account the initial-radius effect. The nature of monthly variations in meteor-distribution parameters is being investigated. Systematic measurements of density distributions of sporadic-meteor incident fluxes over the celestial sphere for a number of years are carried out to obtain seasonal and secular variations.

The Meteor Physics Section of the Institute of Atmospheric Physics of the Italian National Research Council is developing, under the guidance of Verniani, a multistation radar system for studying meteors and the Earth's upper atmosphere. The project has three main purposes: (a) to determine systematically the temperature and density of the atmosphere as a function of height in the 70–100 km region, (b) to study the physical characteristics of meteors in the mass range $10^{-2} - 10^{-4}$ grams and to measure their flux, (c) to study the details of the interaction of meteors with the upper atmosphere.

At the Radioastronomy Laboratory in Kazan, Sidorov, Belkovič, Kostylev, and Michajlov have been investigating the problems of radiowave scattering by meteor trains of various densities as well as amplitude-phase reflection characteristics by mathematical modeling that takes into account ionization variation, the diffusion coefficient, and the initial radius along the meteor trajectory. Belkovič and Kurganov have developed a statistical theory of meteor radio reflections that considers various effects as completely as possible.

In the period 1967–1969 a new radar system for meteor-echo observations was designed and built at the Institute for the Research on Electromagnetic Waves of C.N.R., Florence, Italy (Ronchi-Abbozzo). The receiving system is completely automatized to detect and count echoes.

Plavcová (Ondřejov Observatory) has reduced recent observations of Geminids and derived the distribution of the particles with their mass. With the help of amateur groups, Plavcová, Šimek and Beránek are organizing simultaneous visual and radar observations. The meteor radar of the Ondřejov Observatory was recently expanded to three receiving stations to obtain the orbits of individual meteors. The new equipment is now tested. The first steps toward digital recording of all the radar data have been taken.

Hoppe (Jena, German Democratic Republic) started a program of simultaneous visual and radar observations in Jena. Some factors affecting the duration of radar-meteor echoes, in particular the dependence on the radiant position, were analyzed by Hajduk (Astronomical Institute, Bratislava). Jones (University of Western Ontario) investigated the meteoroid mass distribution both theoretically and experimentally, using forward/and back-scatter radars, and found a slight mass dependence of velocity necessary to reconcile radio and previous optical data. The mass-distribution index is found to be close to 1·7. A device to measure short-scale ionization fluctuations along radio meteor trains has been constructed, and projects for the photoelectric and image-orthicon observation of meteors are under way. Forsyth and Vogan (The University of Western Ontario) have measured absorption in the ionosphere (below the meteor levels), using three frequency measurements of radar meteor echoes.

Maanders (Eindhoven Technical University, The Netherlands) announced that a station has been built in a quiet place near Eindhoven; it operates at a frequency of 39·5 MHz with a pulse transmitter of 25 kw peak power. With this station the entire meteor activity is being observed. Třísková (Geophysical Institute, Prague) has suggested a model of radiant distribution that effectively characterizes the sporadic meteor activity.

PHYSICAL THEORY

When theoretical equations are applied to the meteor phenomena, one assumption is usually made: the air density is a known value given by some standard atmosphere. Recently, it became known that the air density for typical meteor heights can vary as much as 60% from the standard atmosphere in a very short time. This should be kept in mind as a serious source of errors when theory is applied to meteor observations.

The existence of a large difference between the dynamic and the photometric masses of photographed meteors was also recently recognized for fireballs. This has led to many attempts at explanation. However, the sum of information available from meteor photographs, together with present knowledge of the physics of the meteor phenomena, is insufficient to determine uniquely the physical parameters that describe a meteoroid. The range of solutions consistent with the data heretofore available extends from low-density meteoroids of minimal structural integrity to meteoroids whose structure may be similar to that of some chondritic meteorites. The success of all these models rests on the introduction of a new variable in the classical equations. The variables – progressive fragmentation, frothing, or thermal shock – are not in themselves limited by any known parameter of the meteoroid, and thus sufficient leeway is provided to explain almost any of the observed phenomena of fainter meteors. However, the new data available for very bright fireballs seem to be more favorable to the low-density interpretation of a significant part of the faint-meteor data.

A review of the structure and fragmentation of meteoroids was written by Verniani in 1969. He attempts to examine all the present observational, theoretical, and laboratory data on the luminous and ionizing efficiencies of meteors, with the aim of establishing a mass scale. Equating the dynamic and the photometric masses, he finds the average bulk density to be 0·3 g/cm^3. He concludes that the paramount importance of progressive fragmentation, the behavior of abrupt-beginning meteors, and the low density of nearly all meteoroids (even of large ones) support the interpretation of a porous, fragile structure for most of these.

McCrosky and Ceplecha are investigating the effects of thermal and pressure stresses on meteorites in flight. For strong meteoritic stone, bodies with $R > 10$ cm are unlikely to fragment under thermally induced stresses, and those with $R < 50$ cm will not suffer from pressure fragmentation. They conclude that these effects cannot provide a general explanation for the apparent low density of most fireballs observed by the Prairie Network. The negligible terminal mass (ballistic) of most fireballs offers further evidence that these objects are distinct from meteorites.

Allen and Baldwin (NASA Ames Research Center, California) have worked out a frothing ablational model for Super-Schmidt meteors. They do not need the low-density fragile material to explain the faint meteor anomaly. At the same time, the computed luminous-efficiency factor is found to vary substantially with velocity: above 35 km/s, it is close to that recommended by Verniani, but below 25 km/s it increases to about six times greater.

Baldwin has been investigating faint-meteor ablation processes, using the data of Jacchia, Verniani, and Briggs. Factors contributing to apparent low meteoroid densities that have been considered are the following: production and accumulation of low-density froth during atmospheric flight; delay in light production due to the time required to slough off the froth and vaporize debris in the wake; and variations in meteoroid properties accompanied by differences in luminous efficiency. Recently, these meteors have been found to separate into classes when the ratio of photometric mass to ablation mass is plotted versus the heat of ablation. This classification corresponds roughly to that found by Ceplecha using a method based on beginning heights of luminosity and on the ratio of photometric mass to dynamic mass.

Padevět (Ondřejov Observatory) computed several ablation models in an attempt to explain the difference between dynamic mass and photometric mass on the assumption that the main body loses material in the form of small equidimensional particles continuously fragmented from the surface and exposed to the action of undisturbed atmosphere. However, the results were not satisfying, and he is now developing a more complex model that takes into account the interaction of the flow field behind the main body with the fragmented particles.

Simonenko studies theoretically the behavior of small particles separated from a meteor body after the beginning of intensive evaporation. She constructed light curves for different types of fragmentation, using special nomograms. Analyzing then the meteor light curves observed, she concluded that one of the principal mechanisms of fragmentation of meteor bodies is a separation from the parent body of particles of the order of 150μ in size. The particles appear to be solid, and Simonenko suggests they are structural elements of meteor bodies.

Cook (Smithsonian Astrophysical Observatory) is isolating the Super-Schmidt meteors that behave in accordance with classical theory both in their trajectories and in their light curves. As a class, these show a rather small dispersion in the two constants (one in the deceleration equation, the other in the ablation equation) required to fit the theory.

Kostylev, Tochtaseev, and Aminov investigated meteor-particle flight regimes and ablation by mathematical modelling based on assumptions concerning the structure and physical properties of the particles. Various ionization mechanisms for different interaction energies, as well as the effect of meteor velocity on the ionization coefficient, were studied. Hoppe (Jena) successfully applied his fragmentation theory to a meteor of −5 magnitude splitting into 3 pieces observed at 55 km height.

METEOR ORBITS AND SHOWERS

Anders (Enrico Fermi Institute, Chicago) made more Monte Carlo calculations of meteorite orbits. It seems that approaches to Jupiter cause many difficulties, particularly for orbits of the Příbram type. Sometimes such orbits even become retrograde. A solution of this problem is urgently needed.

At the Astronomical Institute of the Slovak Academy of Sciences, Bratislava, problems of the cometary and the asteroidal origin of short-period meteors were investigated. Kresák applied orbital criteria to distinguish between the two types of meteors, and he compared the distribution of meteor orbits with those of the comets and asteroids and established differences in the observed physical characteristics of meteors moving in orbits of different type. Štohl investigated the connection between the distribution of sporadic-meteor orbits and the annual variation of meteor rates, using simplified models of radiant distribution. The differences between the orbits of extremely bright fireballs and those of ordinary meteors were shown by Kresák to be essentially due to selection effects previously unaccounted for. The reality of hyperbolic meteor orbits was critically examined by Štohl, who found that even the perturbation origin within the solar system is unsound.

Andrianov (Kazan) measured radiants and meteor velocities by oblique scattering. Methods of determining meteor-orbit parameters are being worked out and the problems of measurement accuracy and selectivity are being examined. The singularities of element distributions in high-velocity meteor orbits are being studied.

Sekanina (Smithsonian Astrophysical Observatory) has developed a new method for analyzing the dynamical properties of meteor streams. It is based on the Southworth-Hawkins D-criterion of the similarity of two orbits in terms of the differences in their Keplerian elements, but the statistical rather than the determinate concept is accepted for the definition of a stream. The method emphasizes the significance of the character of the D-distribution of meteors with respect to the mean orbit of a stream. The empirical D-distribution for each stream is assumed to consist of an intrinsic Maxwellian D-distribution of the stream and of a contaminating sporadic background. Model calculations suggest that the D-dispersion increases in value secularly as a result of planetary perturbations. A relation between the mean age of a stream and its D-dispersion is likely. Several new meteor showers have been discovered by his search program among the Radio Meteor Project collection of orbits. In particular, meteor streams associated with the minor planets Icarus, Adonis, Hermes, Apollo, and 1968 AA have been detected. In addition, some meteors of the sample have orbits resembling those of comets not previously associated with major streams.

Lindblad (Lund Observatory, Sweden) applied Southworth's D-criterion program to all published photographic orbits of meteors, including those graphically reduced by McCrosky and Posen. He found 43% of all meteors to occur in streams, many of them previously identified. He also found many new streams: 12 correspond to previously identified visual streams of Denning or McIntosh.

The investigation of the internal structure of meteor streams by Porubčan (Astronomical Institute, Bratislava) showed that in permanent showers no condensations up to the dimensions of about 10^3 km and no meteor twins are present. A filamentary structure of the meteor stream formed by Comet Halley was suggested by Hajduk (Astr. Inst., Bratislava) as an explanation of long-term semiregular variations of meteor rates and peculiar changes in the annual curve of activity during the past 50 years. Some features appearing in a stream structure as a consequence of its evolution were pointed out by Kresák. Work on this problem, based on the dispersion of shower radiants, is in progress. New methods to distinguish dispersed showers from the sporadic background were designed both for the photographic and the radio-echo observations by Porubčan and Hajduk.

Guth (Ondřejov Observatory) computed perturbations of Jupiter on a meteoric stream uniformly distributed along an orbit with half the period of revolution of Jupiter, with aphelion beyond Jupiter's orbit, and with an inclination of 19°. This was the case for the meteor stream of Comet Pons-Winnecke 1915 to 1927. Strong perturbations in all elements are the reason for the non-stability of such a stream, which is dispersed during the very short time of two revolutions.

VAGO, U.S.S.R., organized a special expedition to the Arctic region for observation of the possibly abundant return of the Leonids in 1967. The results showed that the Leonids in 1967 did not exceed 8 meteors per hour. Leonids 1967 were also observed by radar in Frunze, U.S.S.R.

Lindblad continued systematic simultaneous observations of Perseids with radar and with a team of visual observers. He now has 17 years of Perseid observations by the same radar and the same visual group.

Alpha Lyrids were observed visually by VAGO members at Semachinsk Astrophysical Observatory, U.S.S.R., in 1969. About 4000 meteors were recorded, and their radiants determined. Telescopic observations of this stream were also made.

Skrivanek (Air Force Research Laboratories) investigated the Perseid meteor stream during August 1968. Four high-altitude particle-collection rockets were launched before, during, and after the visual peak of the shower. Preliminary results do not indicate any difference between data from rockets flown before and those during the peak of the shower.

Znojil (Brno Observatory) studied the structure of meteor streams, mainly Perseids and Cassiopeids, in the interval of 6–8 m, using recent observations of about 7000 telescopic meteors.

VISUAL OBSERVATIONS

Olivier (American Meteor Society, Philadelphia) is preparing a third daily-hourly rate catalog for naked-eye meteors, to supplement the two already published by him in the *Smithsonian Contributions to Astrophysics*. He is also preparing a paper on telescopic meteors reported from 1950 to 1969 to American Meteor Society.

In recent years, many meteor observations have been obtained in Japan. More than one hundred visual observers are active, and nearly 30000 visual meteors are observed every year.

Visual observations were made at Krym Meteor Station of VAGO with use of a method of calibrated counting. During a 3-yr period, they observed about 12000 meteors, which are now being reduced.

Šulc (Brno Observatory) is studying the errors in the treatment of visual meteor observations by the independent counting method. He uses optically simulated meteors, together with an independent counting method modified by Kvíz. He found that the necessity for observer independence in the group is well satisfied and that the source of error is mainly due to the appearance of meteors in different parts of the field of view, combined with the "straight-line shape" of the phenomenon.

Kresáková (Astronomical Institute, Bratislava) has investigated the angular velocities of meteors and combined them with the results of telescopic meteor observations. She points out that the velocity effect on the visibility and apparent magnitude of meteors is weaker than that accepted by the authors using results from experiments with optically simulated meteors, mainly because the latter method does not account for the occurrence of wakes and trains.

Hajduková (Komenský University, Bratislava) used visual observations of meteors through color filters to obtain information on the color indices of ordinary meteors too faint for photographic spectra to be obtained.

ZDENĚK CEPLECHA
President of the Commission

REPORT OF THE COMMITTEE ON METEORITES

Several monographs and collections of papers were published in *Meteoritics* during 1967 to 1969 (Chirvinskij, 1967; Vdovykin, 1967; Wood, 1968; Cikulin, 1969; Anders *et al.*, 1967; Stanjukovič, 1968). Extensive work on meteorites is continuing in the Meteorite Committee of the Academy of Sciences of U.S.S.R. Fesenkov and Krinov continued publishing the journal *Meteoritika*, where most of the papers of scientists of the U.S.S.R. are published. Hoffleit continued publishing *Meteoritics* of the Meteorite Society. Javnel' continued editing a bibliography on meteorites. A bibliography on the isotopic and chemical composition of meteorites was published by Kielbasinski and Wanat. The UNESCO working group on meteorites under the chairmanship of Orcel edited the *Directory of Meteorite Collections and Meteorite Research* in 1968.

The Meteorite Committee of the Academy of Sciences of the U.S.S.R. organized its 13th meteorite conference in 1968. The Meteorite Society had its 30th to 32nd annual meetings during 1967 to 1969. The International Atomic Agency in cooperation with UNESCO and five other international scientific unions and organizations including the IAU represented by Millman, prepared a Symposium on *Meteorite Research* in Vienna, Austria, 7–13 August, 1968, initiated by the UNESCO Working Group on Meteorites. More than 70 papers were presented at this Symposium, showing the interests of various sciences in meteorites. The Symposium volume appeared very soon after the sessions were closed in the series *Astrophysics and Space Science Library*, published by D. Reidel Publishing Company, Dordrecht, Holland, under the title *Meteorite Research* edited by P. M. Millman (1969, 940 pages).

The information on new meteorite falls and finds continued in the *Meteoritical Bulletin* edited by Krinov. In 1968, another source of data on newly fallen meteorites started to be available in an information circular published by the Center for Short-Lived Phenomena at the Smithsonian Astrophysical Observatory, Cambridge, U.S.A. (Citron). During the period from January 1, 1967 to November 1, 1969, 12 meteorite falls were reported, among which two carbonaceous chondrites, Pueblito de Allende in Mexico and Murchison in Australia, which fell during 1969. Short lived isotopes were detected in 5 of these meteorites. The ratio of Ar^{37} to Ar^{39} in the Alandroal iron meteorite was found to be low by Fireman, which could be due to variations of cosmic rays in space.

Many papers dealt with the isotopic composition of primordial gases in meteorites and their different compounds. The maximum concentration of primordial gases (He, Ne, Ar) was found in fine-grain matrix material by Megrue using a laser microprobe and a mass spectrometer. This could be explained by injection of material from the solar wind.

Radiogenic gases and isotopes were measured by many people. The measurements of the time after the nucleosynthesis until gases were maintained resulted in about 100–200 million years for chondrites, using the ratio I^{129}/Xe^{129} (Gerling, Manuel, Alexander). Another method for the determination of this interval is based on the Pu^{244}/Xe^{136} ratio and gives more or less identical results according to several authors (Kuorda, Rao). The Xe^{136} content is higher in chondrites than in achondrites (Reynolds), which could be connected with the spontaneous fission of a super-heavy element with $Z = 112 - 119$ (Dakowski, Anders, Heymann). The Rb^{87}/Sr^{87} ages were measured for several meteorites. The results on the silicate phase of iron meteorites are in the interval from 4·7 to $3{\cdot}8 \times 10^9$ yr which is in agreement with the K^{40}/Ar^{40} ages (Wasserburg, Burnett).

Extensive rare gas isotope, mineralogical and chemical work is being performed at the Max-Planck-Institut in Heidelberg. A large number of stony meteorites has been studied: Primordial rare gases were detected in several chondrites, e.g., Phum Sambo, Barratta and Hedjas (Müller, Zähringer). The isotopic composition of lithium which is significant for the early irradiation history of meteorites, was determined by Krankowsky and Müller in a large number of meteorites. Radioactive isotopes were detected. The light and dark phases of the amphoterite chondrites showed unusual high alkali contents giving a precise Rb-Sr-age of $4{\cdot}70 \times 10^9$ yr. Carbon and rare gases were found to be correlated to the mineralogical classification by Otting and Zähringer. In 45 meteorites, cosmogenic rare gases were analysed by Nord and Zähringer. The K-Ar age of iron meteorites was found to be about 5×10^9 yr for a few specimens using improved techniques.

Extensive work on meteorites is being performed at the Max-Planck-Institut für Chemie in Mainz. It was shown by Begemann *et al.* that considerable fractions of Ar^{39} and P^{32} activities in stone meteorites are produced by neutrons in the 10 MeV region. The deduced neutron fluxes vary considerably, depending more probably on the orbital elements than on the size of the meteoroids (Begmann and Wänke). Other parameters strongly related to meteoroid orbits were also found: (a) Loss of tritium from iron meteorites and the metal phase of chondrites (Hintenberger *et al.*, Schultz). (b) Loss of He^3 from pyroxene and feldspar (Hintenberger). Both effects are caused by solar heating at the perihelion. From Be^{10}, Cl^{36} and other cosmic-ray products, the terrestrial ages of iron meteorites were found to exceed values of 100000 yr in about 70% of all finds. One meteorite – Tamarugal – was found to have fallen 3×10^6 yr ago (Chang and Wänke). A new method for the determination of the cosmic-ray exposure ages has been developed based on the Ar^{36}/Ar^{38}-ratio found in the metal phase of chondrites (Nyquist *et al.*). Using the K^{40}/K^{41}-method, exposure ages were determined for more than 60 iron meteorites by Voshage. Medium and fine octahedrites of the Ga-Ge group III were produced in two events 600 and 700×10^6 yr ago, while fine octahedrites of the group IVa were created 400×10^6 yr ago. From studies of cosmic-ray produced rare gases, it was found that hexahedrites have exposure ages considerably lower than those of the other groups of iron meteorites (Wänke). Trace element studies in chondrites showed that boron belongs to the group of highly fractionated elements (Quijano-Rico and Wänke) and a solar ratio $B/Li = 7$ was derived, instead of $B/Li = 0{\cdot}2$ as assumed previously. In meteorites with high amounts of solar gases, indium was found to be highly enriched in the dark gas-rich portions (Rieder and Wänke), while the boron contents are nearly identical in both the light and dark portions. These findings fit well to the solar wind model of gas-rich meteorites as boron is expected to be very rare in the solar wind (Wänke). A large excess of Xe^{129} was detected in silicate inclusions of Campo del Cielo meteorite (Hintenberger).

The intensity of the galactic component of cosmic rays was found by Lavruchina to be two times higher at 3–5 astronomical units in comparison with those measured within the Earth's orbit, using models of the atomic reactions compared with meteorite measurements.

Extensive work on meteorites has been carried out in various laboratories in France since 1966. Ion-sputtering mass-spectrometry has been used to study the isotopic composition of lithium (Li^7/Li^6) in several meteorites as compared to terrestrial samples. The very large variations found in the Holbrook chondrite by other investigators were due to instrumental effects (Gradsztajn *et al.*). Residual variations of $\pm 10\%$ remained in some chondrites (Mezo-Madaras, Richardton) while variations in terrestrial samples (granite, hornblende) did not exceed $\pm 2\%$ (Bernas *et al.*). Possible solar cosmic-ray effects induced by slow neutrons in different meteorites have been put forward by Yokoyama, indicating a trend for a cometary origin of chondrites. Galactic cosmic-ray induced nuclides and formation rates by several processes have been measured and calculated for the St-Séverin chondrite (Tobailem *et al.*) and the Granés chondrite (Bibron *et al.*, Nabuchi *et al.*). Track method has been used to study the depth variation of track densities in meteoritic detectors from cores taken in the least ablated regions of St-Séverin (Cantelaube *et al.*). The implications of these results for the time-averaged flux and energy spectrum of galactic cosmic-ray iron-group nuclei during the last 12×10^6 yr have been discussed in collaboration with the Tata Institute of Bombay (Lal *et al.*). Evidence has been shown for the presence of an appreciable flux of low energy protons ($10 < E < 300$ MeV) in the least ablated regions of St-Séverin at distances of the order of 2–3 A.U., by means of correlated studies of track densities and induced activities of some cosmo-nuclides (Amin *et al.*). Observations have been reported of a low energy particle irradiation in some crystals from dark portions of gas-rich chondrites and achondrites. This low energy particle flux is thought to have an early solar origin and to have been recorded in pyroxene crystals before their accretion (Pellas *et al.*, Poupeau *et al.*).

Great progress in the study of the chemical composition of meteorites is characteristic for the last several years. Neutron activation analysis was used for trace elements by many investigators. New methods for investigating the major constituents were developed by Javnel', Djakonova,

Charitonova, Moss, Ahrens, Sprenkel-Segel, Herr and others. Wasson proposed a classification of iron meteorites into 9 groups, based on Ga and Ge content, which differ in cooling rate. In Finland, the analyzing of the major constituent elements in meteorites is going on. Wiik prepared for press a summary of this activity containing 112 complete analyses. He concludes that meteorites are not products of differentiation. The fractionation of some abundant siderophile and lithophile elements in chondrites was examined and some rules for their ratios were found by Javnel' and Ahrens.

Many papers were devoted to the composition and structure of different constituents, phases, minerals and chondrules in meteorites, which is the base for considerations about the meteorite origin. The development of sensitive methods yielded almost 20 new minerals in meteorites (Buseck, Fuchs, Olsen, Keil, Ramdohr, Faust *et al.*). Many measurements of organic compounds (Oro, Nooner, Hayes, Vdovykin *et al.*) point more to the non-biogenic origin in meteorites.

An active group is working on meteorites in Austria (Kurat). Investigations of the plessite formation in iron meteorites showed that phosphorus and carbon exert a great influence on the $\gamma \rightarrow \alpha + \gamma$ transformation (Dörfler). Also the nucleation mechanism of iron-nickel phosphides in iron meteorites has been investigated (Dörfler). Bulk and trace element analyses of several meteorites and meteoritic minerals have been performed (Schaudy, Höfler, Kiesel, Seitner, Hecht, Bauer).

The new hypersthene achondrite Peckelsheim (fall of 1953, March 18) was recovered and studied with the electron microprobe and X-ray techniques (Ramdohr, El Goresy). Detailed mineralogical studies were carried out on stony and iron meteorites by Ramdohr in Heidelberg. Zircon was observed in the chondrite Muizenberg. The composition of coexisting sphalerite, daubreelite and troilite in iron meteorites was discussed by El Goresy from Heidelberg.

An active group in Jena, German Democratic Republic (Heide, Brückner, Doenitz, Illner, Schrön) is engaged in studies of the following problems: new meteoritic stone Wolamo, stony meteorite Aukober, structure of rhabdites, magnetic behaviour of cohenite, schreibersite and/nickel-iron, the schreibersite group, meteoritical glass, motion evidence in meteorites.

Much work was done on conditions of the chondrule formation. Chondrules should have been formed as a mechanical mixture with metal and sulfide particles (Kvaša). The appearance of secondary chondrules (Frederiksson) disagrees with the theory of direct formation by condensation. On the other hand the similarity of chondrule structure and shock-melted material of some chondrites are in favour of it (Wlotzka). Another aspect of the origin of chondrules was discussed by Kurat. His petrological investigation of chondrules points to a possible magmatic origin. Crystallization of the chondrule minerals should have taken place from a strongly undercooled melt. Detailed investigations of glass-bearing chondrules showed further that chondrules from "unequilibrated" chondrites are the products of several complex and overlapping processes and that condensation and/or vaporization took place during the final stage of chondrule formation.

Data on metamorphism in chondrites point to changes of their structure and composition at 400°–820°C at static pressures of about 1–2 kbar, probably during the secondary heating. The structure of iron meteorites demonstrates their pre-atmospheric dynamical deformations (Axon, Baldanza). These deformations for the Ga-Ge group III originated from a shock with a pressure exceeding 130 kbar (Jain, Lipschutz). The dynamically deformed structures in a number of meteorites, both irons and stones, led to the recognition that differences exist between deformations due to a relatively "sudden" dynamic event and others which seem more likely due to a prolonged stress action (Baldanza and Pialli). A new concept was put forward by Suess and Wänke (Mainz) about the metamorphosis and equilibration processes in chondrites.

The thermoresistant nature of meteorite magnetization was measured by Gus'kova for different types of meteorites, by heating them to about 800°C in magnetic fields of 0·2 to 0·6 Oe.

Computations of meteorite orbits point out that the source of meteorites cannot be the Moon or comets, and also the asteroids of the Apollo family crossing the Mars orbit are not a very probable source (Anders, Mellick). A more probable source could be the asteroids with aphelions close to Jupiter (Wetherill), though the observational results are not completely in agreement with it (Levin, Simonenko). A majority of scientists is inclined to assume that meteorites were formed from asteroidal bodies with diameters of several hundred kilometers. There are some theories of lunar origin of some classes of meteorites (Urey, Wänke), which were supported by a similarity in composition of achondrites rich in Ca and lunar materials (Vinogradov). Levin reviewed the data on the origin of meteorites and compared them with the data on the planetary cosmogony. He continued to defend the role of diffusion in the solid state in the formation of different types of meteorites.

Experiments on meteorite ablation and fragmentation in the Earth's atmosphere were done (Bronšten, Lebedinec, Portnjagin, Fadeenko). The extent of atmospheric ablation in the St-Séverin meteorite has been measured by the track-method and corresponds to 25–30% of the preatmospheric mass (Cantelaube, Pellas *et al.*).

The fusion crust of meteorites and chromite variations in chondrites were studied by Ramdohr. El Goresy and Fechtig compared the fusion crust of iron meteorites and mesosiderites with cosmic spherules. The fusion crust was studied also by Judin.

About 1000 individual pieces of the Sikhote-Alin meteorite produced by atmospheric fragmentation were found by the expeditions organized during 1967 to 1969 by Krinov.

Theoretical studies of shock waves produced by meteorites during their atmospheric flight were published by Cikulin (1969). Many papers contain problems of meteorite craters, their recognition by mineralogical, structural, and other symptoms, and theoretical studies of huge meteorite impacts with the solid surface of the Earth and other planets (*Symposium on Shock Metamorphism*, Ed. French and Short, 1968; Stanjukovič, 1968).

REFERENCES

Anders, E., Taylor, S. R., Levinson, A. A. (eds.) 1967, Meteorites and Tektites, *Geochim. Cosmochim. Acta*, **31**, no. 10.

Chirvinskij, P. N. 1967, *Pallasites*, Moscow.

Cikulin, M. A. 1969, *Shock Waves Produced by Motion of Large Meteoric Bodies in the Atmosphere*, Moscow.

Fesenkov, V. G., Krinov, E. L. (eds.) 1968–70, *Meteoritika*, nos. 28–30.

French, B. M., Short, N. M. (eds.) 1968, *Shock Metamorphism of Natural Materials*, Baltimore.

Hoffleit, D. (ed.) 1967–70, *Meteoritics*, **3–4**.

Javnel', A. A. (ed.) 1968, *Bibliographic Guide to the Literature on Meteoritics*, nos. 14–16 (1964–1966).

Javnel', A. A. (ed.) 1970, *Bibliographic Guide to the Literature on Meteoritics*, nos. 17–18 (1967–68).

Kielbasinski, J., Wanat, L. 1968, *Isotopic and Elemental Composition of Meteorites. A bibliography: 1947–1967*, Warsaw.

Krinov, E. L. (ed.) 1967–1969, *The Meteoritical Bulletin*, nos. 39–48.

Millman, P. M. (ed.) 1969, *Meteorite Research*, D. Reidel, Dordrecht, Holland.

Stanjukovič, K. P. (ed.) 1968, *Explosive Craters on the Earth and Planets*, Moscow.

UNESCO 1968, *Directory of Meteorite Collections and Meteorite Research*, Paris.

Vdovykin, G. P. 1967, *Carbon Matter of Meteorites*, Moscow.

Wood, J. A. 1968, *Meteorites and the Origin of Planets*, New York.

A. A. JAVNEL'
Chairman of the Committee

METEORITE RESEARCH IN THE U.S.A. AND CANADA

Chondrites

There have been considerable improvements in the classification of chondrites. Van Schmus and Wood have subdivided each of the major classes into 4 petrological groups, differing in degree of recrystallization. This classification seems to be of rather profound significance: though based entirely on textural criteria, it correlates remarkably well with compositional features, such as content of primordial gases, In, and C. Several detailed studies of individual chondrite classes have appeared: enstatite, amphoterite, type III carbonaceous, and unequilibrated ordinary.

Much effort has been spent on major and minor element abundances. Larimer and Anders have tried to explain the observed trends by fractionation processes in the solar nebula. Wood and Dodd, on the other hand, ascribe most of them to metamorphism in the meteorite parent bodies. The reality of metamorphism has been questioned in a spirited debate, but recent work by Hohenberg and Reynolds (Xe^{129} retention in chondrites) and Dundon and Walter ($Fe^{2+} - Mg^{2+}$ disorder in pyroxenes) and others strengthened the case for it. Arguments on the origin of chondrites continue. Blander *et al.* agree with Wood and Larimer and Anders that the chondrites are basically a direct condensate from the solar nebula, but show that supercooling of the gas may have had an important influence on the nature of the condensate. Wood concludes from a study of carbonaceous chondrites that their chondrules formed in a region of the nebula where dust had been concentrated 5000-fold relative to the gas. Larimer has attempted to reconstruct the formation conditions of enstatite chondrites from petrological considerations. A study of metal particles in chondrites shows that they cooled through 500°C at a rate of 0·2 to 10 deg/10^6 yr (Wood). This suggests that the parent bodies were of asteroidal size. Arguments for a lunar origin were presented by Urey.

Irons

Here, too, a new and superior classification system has been developed by Goldstein. It is based on Ga, Ge, Ir and Ni contents (mainly due to Wasson), and on cooling rates (mainly due to Goldstein and Short). Most iron meteorites fall into one of 14 groups. The range of cooling rates is 0·5 to 500 deg/10^6 yr, which corresponds to parent bodies of asteroidal size. Lipschutz *et al.* noted that one of these groups (III) consistently shows rather severe shock effects, which implies high ejection velocities. They suggest that these meteorites originated from an asteroid whose orbit did not cross that of Mars. In that case only fragments ejected at high velocities had a chance of eventually being deflected into an Earth-crossing orbit. The radiation age of these meteorites is about 600×10^6 yr, which may indicate the date of the collision.

Two minerals often thought to imply a high-pressure origin of meteorites have been reinterpreted. Diamond from Canyon Diablo and several other shocked meteorites turned out to contain a new, hexagonal modification (lonsdaleite), stable only above 130 kbar. Hydrostatic pressures of this order can be attained only in bodies well above lunar size, but such bodies would resist disruption and hence cannot serve as the source of meteorites. Graphite particles of cubic habit (= cliftonite), sometimes regarded to be pseudomorphs after diamond, were synthesized by Brett and Higgins at low pressures, using nickel-iron alloys saturated with carbon.

Stony irons

Pallasites and mesosiderites were studied by Buseck *et al.* and Powell. Their cooling rates are lower than those of irons, suggesting an origin in larger bodies.

Achondrites

Duke and Silver reported a comprehensive petrographic study of 21 eucrites and howardites, and 3 mesosiderites. Two other studies of fairly broad scope were published by Mason and Schnetzler and Philpotts.

Meteoritic minerals

A number of new phosphate minerals were discovered by Fuchs and Olsen. Other new minerals (or first meteoritic occurrences) were: potassium feldspar, amphibole, yagiite, niningerite, garnet and brezinaite. Among the well-established minerals, the following were the subject of intensive study: feldspar, chromite, cohenite, pyroxenes, schreibersite and ilmenite. A review of meteoritic minerals was written by Mason.

Ages

This has continued to be the most active field of meteorite research. Much of the effort was devoted to extinct radioactivity, notably 16×10^6 yr I^{129} and 82×10^6 yr Pu^{244}. An early report by Hohenberg *et al.* suggested that all chondrites began to retain Xe^{129} in high-temperature sites within $\pm 2 \times 10^6$ yr. But more recent study by Podosek shows measurable time differences from one chondrite to another. The total range for 10 chondrites and 2 aubrites is 14×10^6 yr. Retention of Xe^{129} in low-temperature sites took considerably longer, about 100×10^6 yr. Thus the cooling curve of meteorites seems to have been fairly steep initially (greater than 100 deg/10^6 yr?), flattening out afterwards to about 1–10 deg/10^6 yr. It is remarkable how well the cooling rates determined by different techniques agree with each other. Fleischer *et al.* have derived a cooling rate of $\left(1{\cdot}1 \begin{array}{l} +0{\cdot}6 \\ -0{\cdot}5 \end{array}\right)$ deg/10^6 yr for the iron meteorite Toluca, using fission tracks of Pu^{244}. This result agrees quite closely with the metallurgically determined value, which in turn agrees with that based on Xe^{129}. Other studies of fission- or cosmic-ray-induced tracks in meteorites have given much new information. The space erosion rate of the Patwar mesosiderite was found to be less than 0·1 mm/10^6 yr, far below previous estimates. The abundance distribution of heavy nuclei in cosmic radiation has not changed much in the last 10^7 yr.

Careful isotopic analyses of xenon have shown that calcium-rich achondrites contain one kind of fission-produced component, probably from Pu^{244}, while chondrites contain another. It has been suggested that the progenitor of the latter type of xenon might be an extinct super-heavy element of $Z = 112$–119.

A controversy has arisen over the usefulness of Pu^{244} for dating purposes. Kuroda *et al.* maintain that meaningful and concordant Pu^{244}–Xe^{136} and I^{129}–Xe^{129} ages can be obtained by using the initial abundances of the two extinct nuclides predicted by a "constant-rate, continuous nucleosynthesis" model. But Reynolds has pointed out various ambiguities in the interpretation of the data, and Fisher has shown that some of the uranium contents assumed by Kuroda were in error. Moreover, the nucleosynthesis model itself is open to question. Dicke and Hohenberg have re-evaluated the available evidence, and conclude that the continuous nucleosynthesis model is an oversimplification. Numerous other studies on meteoritic Xe and Kr have appeared. Marti has shown that radiation ages of stony meteorites can be determined from Kr^{81}/Kr^{83} ratios.

The Rb^{87}-Sr^{87} dating method has yielded several interesting results. By an extraordinary improvement in technique, Papanastassiou and Wasserburg were able to show that 7 eucrites had formed within 4×10^6 yr of each other, $4{\cdot}39 \pm 0{\cdot}26 \times 10^9$ yr ago. In view of the long half-life of Rb^{87} (50×10^9 yr), the resolution of such fine time differences must be regarded as an extraordinary achievement. The brecciated octahedrite Kodaikanal was found to have Rb-Sr age of only $3{\cdot}8 \pm 0{\cdot}1 \times 10^9$ yr. This is the first well-documented case of a meteorite originating after the era of meteorite formation $4{\cdot}5 \times 10^9$ yr ago.

Several more or less complete surveys of individual meteorite classes were published. Most of the L-chondrites seem to have been involved in a collision 520×10^6 yr ago, and must therefore come from a single body.

Primordial gases

This field is in a state of flux. For some years, the prevailing notion was that primordial gas consisted of two components, differing in elemental composition: an unfractionated ("solar") and a fractionated ("planetary") component. Isotopic differences between the two components were first noted for Ne then for He and Xe. Recent studies suggest, however, that Ne is still more complex: as many as 5 different subcomponents may be present (Black, Pepin). At least one of the unfractionated components may be derived from the solar wind, but the trapping mechanism is still under dispute.

Radioactivity

Fireman has continued his methodical study of H^3, Na^{22}, Al^{26}, Ar^{37}, and Ar^{39} in freshly fallen meteorites. He concludes that the 11-yr cosmic-ray variation is smaller at the average solar distance of most meteoroids than at 1 A.U., and that the heliocentric radial gradient of the high-energy component is $20 \pm 8\%$/A.U. of solar maximum and minimum. Kohman and Bender have done extensive calculations on the cosmic-ray production rate of various nuclides as a function of meteoroid size and depth. A review of similar scope was written by Kirsten and Schaeffer. Several experimental studies on Al^{26} and other radionuclides have been reported.

Orbits

(Prairie Network: see the Commission Report). Wetherill redetermined the collision frequency in the asteroidal belt and used Arnold's Monte Carlo model to test various possible parent objects for their ability to produce meteorites with the proper characteristics: radiation ages, morning-afternoon fall frequency, and elongation of radiant. Anders and Mellick used a similar, though less ambitious approach to search for a parent object that would produce Příbram-like orbits with appreciable frequency. Fisher questioned the predominance of afternoon falls, long accepted as a well-observed characteristic.

Organic compounds

Oró *et al.* have made a comprehensive survey of heavy hydrocarbons in most of the known carbonaceous chondrites. A few meteorites were studied in greater detail by Hayes and Biemann and Studier *et al.* The latter authors showed that most compounds reliably identified in meteorites could be synthesized from CO and H_2 in the presence of iron meteoritic powder, which acts as a catalyst. Similar reactions may have taken place in the solar nebula. An extensive review of this field has been given by Hayes. Shorter reviews were written by McCarthy and Calvin and Nagy.

Meteorite craters

Several detailed descriptions of Canadian craters have appeared. Shock features due to meteorite impact have now been identified in 17 Canadian craters (Beals, Halliday). The ages of the Manicouagan and Clearwater craters were determined as $208 \pm 25 \times 10^6$ yr and $100\text{–}300 \times 10^6$ yr by fission track dating (Fleischer *et al.*). The latter figure is consistent with a more-accurately determined K-Ar age of 300×10^6 yr. French has presented evidence for an impact origin of the Sudbury structure. A brief review of terrestrial craters was given by Beals and Halliday.

E. ANDERS
Vice-Chairman of the Committee

23. CARTE DU CIEL

PRÉSIDENT: W. Dieckvoss.
VICE-PRESIDENT: P. Lacroute.
ORGANIZING COMMITTEE: P. Couderc, A. N. Dejč, P. Herget, H. Kox, W. J. Luyten, P. Sémirot, H. W. Wood.

There seems to be a consensus of opinion that the useful life of the commission has come to an end with the achievement of the principal aim of the historical task of the Carte du Ciel, presenting an overall picture of the Heavens as to positions of the fixed stars up to a definite limiting magnitude. The French presidents in succession with great energy, supported by the IAU, succeeded in overcoming all obstacles and in finishing the printed *Astrographic Catalogue*. Remaining tasks of deriving definitive plate constants, eventually deriving reliable proper motions of all stars in the programme evidently rests with individual observatories or at other commissions of the Union within their proper frame of activities.

Following a suggestion presented at the meeting of the Executive Committee in September 1969 a decision was reached to abolish the Commission 23 proper by merging it with Commission 24 (Stellar parallax and proper motion), and a working group in the latter commission would become a suitable means to continue work on data furnished by the *Astrographic Catalogue*.

Several of those observatories still in possession of Carte du Ciel-astrographs are going on with proper motion (p.m.) work in their zones or are preparing to do so. The application of the astrographs to the programme of attachment of stars to galaxies according to A. N. Dejč's method is in progress.

P. Kustaanheimo of *Helsinki* reports the astrograph of the Helsingfors zone of the Carte du Ciel and the Astrographic Catalogue (A.C.) being in excellent condition after some modernization. The collection of plates, the oldest dating back to 1892, is well preserved. After transferring the work on positions of artificial satellites to other institutions of the country the observatory now is willing to continue its proper motion work on faint stars, work that was stopped shortly after the war. There is a threat to close down the observatory by the authorities who would like to use the observatory buildings for offices. As the observatory is well renowned for its high standard in published results – the Helsingfors zone of the A.C. is one of the most accurate in both hemispheres – a closing down would be detrimental to highly useful long range programmes on faint stars, and a resolution of the IAU asking the authorities of Finland to support the observatory in its tasks would be most welcome.

B. J. Harris reports re-installation of the *Perth*-A.C.-astrograph at the new observatory site Bickley. Re-evaluation of plate constants and tests for a total re-observation of the Perth zone are in progress.

H. Wood of *Sydney* announces completion of the Sydney section of the A.C., the last volume, now in print, will contain a history of the work. It will include particulars as to methods applied and as to information and tables to be used with the catalogue.

A. N. Dejč of *Pulkovo* reports as to the programme of attachment of stars to galaxies by means of the A.C.-astrographs:

1. Aux observatoires de Poulkovo, Moscou et Taschkent on a commencé de photographier les aires choisies dans la zone $+90°$ à $-5°$ et plus australes jusqu'à $-25°$ de déclinaison pour la seconde époque ayant un intervalle de 20 ans.

2. Le programme photographique des galaxies de la zone $-5°$ à $-25°$ a été fini à San Fernando pour la première époque et sera bientôt accompli pour la zone $-25°$ à $-90°$ à Santiago et à Perth (Bickley).

Chr. de Vegt of Hamburg used data of the A.C. extensively for p.m. work on open clusters. Original plates lent by O'Connell of Castel Gandolfo were measured with the Mann comparator of Hamburg Observatory. The mean errors of presentation of the AGK 2/3 reference positions were diminished by 25 to 30% as compared with the original coordinates as printed in the A.C. Rome-Vatican. The association Cep. III is being studied in collaboration with A. Blaauw by re-measuring the original Vatican-plates with the automatic machine of K. Strand at the U.S.-Naval Observatory Washington.

Eichhorn and Gatewood in 1967 published new plate constants for the Northern Hyderabad zone of the A.C., $+35°$ to $+40°$.

Work on definitive plate constants in the zones covered by AGK 3 (-2.5 to $+90°$) went according to plan in Hamburg (H. Kox, A. Günther). The results for the zones Greenwich, Rome-Vatican, Catania, Helsingfors are in print (*Astronomy and Astrophysics*, a European Journal). In France the work under supervision by P. Lacroute may proceed when the AGK 3 will become available south of $+36°$ (approximately in February 1970).

North of $+40°$ the A.C. positions of the reference stars from AGK 3 after due correction on account of systematic errors and by applying definitive plate constants are being used for a general catalogue of positions and proper motions (W. Dieckvoss).

TOPICS FOR DISCUSSIONS AT THE XIV GENERAL ASSEMBLY

Historical Survey of the Carte du Ciel (P. Sémirot)

Report on definitive plate constants,
  Bergedorf (H. Kox)
  France (P. Lacroute)

Magnitude equation of faint stars in the A.C.

Remeasurement of preserved old plates by modern, preferably automatic, machines.

Proper motions of faint stars following repetition of old zones, preferably by modern wide angle objectives, which are less prone to optical aberrations than the original A.C. objectives.

Transformation of A.C. data into machine-readable form in an international scale.

Application of A.C. data to p.m. of clusters.

Coordination with p.m. in Kapteyn's selected areas.

Situation in Helsinki, preparation of a resolution.

Merging with Commission 24 in a joint session, appointment of a working group concerning itself with A.C., as proposed by P. Lacroute.

W. DIECKVOSS
President of the Commission

24. STELLAR PARALLAXES AND PROPER MOTIONS (PARALLAXES STELLAIRES ET MOUVEMENTS PROPRES)

PRESIDENT: W. J. Luyten.
VICE-PRESIDENT: S. Vasilevskis.
ORGANIZING COMMITTEE: A. N. Dejč, V. V. Lavdovskij, J. v. B. Lourens, W. W. Morgan, C. A. Murray, K. Aa. Strand, P. van de Kamp.

At the Pulkovo Observatory a parallax programme has been started by Kanayev for 120 stars chosen mainly from Vyssotsky's lists and the Leningrad list of multiple systems. Proper Motions relative to galaxies have been determined by Fatchikhin for 14669 stars in 85 different fields; the magnitude equation is being investigated and secular parallaxes as well as the solar apex and the constants of galactic rotation are being calculated. The general programme on absolute proper motions is being carried out by Dejč at Pulkovo, and at Cerro Calan in Chile, by Panteleyeva at Sternberg, and by Rahimov at Tashkent.

Another major programme is concerned with the proper motions of stars in clusters, and surrounding field. At Pulkovo the original list has been increased from 75 to 340 clusters by Lavdovskij, and data for NGC6866 and 7789 will be published by Koroleva. Motions and UBV magnitudes for 2000 stars near M 92 are being determined by Kadla, with N. Richter at Tautenburg, and by Panova at Pulkovo for M 15 and M 53.

At Sternberg proper motions for stars surrounding the Pleiades and M 92 have been determined by Artiukhina and Kalinina. At Tashkent Latypov, Ishmuhamedov and Kadyrov are determining proper motions of stars in a large number of areas containing open clusters.

Further proper motions for five novae and surrounding stars as well as for a large number of stars in Tau, and Cyg have been determined at Pulkovo by Sokolova, and Bronnikova, while at Sternberg Karimova and Pavlovskaya determined proper motions for 569 Cepheids and Am and Ap stars and are currently deriving similar motions for 200 B stars. At Tashkent Primkulov determined proper motions for 8 U Gem stars and for a large number of surrounding field stars.

In Bergedorf work is in progress to construct a catalogue of positions and proper motions comprising stars of the AGK 2 by combining the AGK 2 (revised, and brought to FK 4) AGK 3, and positions from the *Astrographic Catalogue* after systematic corrections derived by H. Kox and A. Günther while working on definitive plate constants for the *Astrographic Catalogue* north of $+32°$ Decl.

In a circular area of 5° radius around Alpha Persei positions for all stars on the relevant plates of the Astrographic Catalogue, zones Helsingfors and Catania, were determined. The proper motions will be derived by comparison with new epoch plates taken with the Bergedorf Schmidt. The measurement of these plates with the MANN comparator is in progress.

Eichhorn reports that at the U.S. Army Map Service, and in collaboration with Googe, Lukac, and Murphy, positions for 511 stars in the Pleiades have been determined with an average mean error of $0\overset{''}{.}02$, and proper motions have been determined for 101 stars in König's list. At South Florida Sofia, Eichhorn and Gatewood have determined proper motions for 83 stars near Sco X 1 and suggest that the X-ray object belongs to the Scorpio-Centaurus Association.

Wallerstein disagrees with this, and suggests that the X-ray object is more distant.

At South Florida a new 26-inch Schmidt-Cassegrain reflector has been installed and will be used for parallax and proper motion work.

The Perth Observatory has been re-established at its new site, and concomitant instrumental difficulties have now been largely overcome. Harris states that programmes for determination of proper motions in the regions of clusters, for RR Lyrae variables, and absolute proper motions relative to galaxies have been started. A pilot programme in the field of M 7 = NGC 6475 has been published.

From the Lowell Observatory Giclas reports that the proper motion survey of stars between magnitudes 8 and 17 and with motions larger than 0."27 per year has been essentially completed for the northern hemisphere. Duplicate measures from overlapping plates will be combined and a list ordered in Right Ascension, including references to identification charts will be compiled. Additional results of the survey can be found in Lowell Observatory Bulletins Nos 138 to 151. Upon completion of the catalogue in the northern hemisphere, which will contain over 10000 entries, a programme for proper motions in selected regions will be undertaken, and the survey will be continued in the southern hemisphere.

Gliese has completed the calibration of spectral and photometric parallaxes of main-sequence stars later than about F 5 on the basis of reliable trigonometric parallaxes and presented preliminary results in 1968 at the Symposium on Low Luminosity Stars, and at the meeting of the Astronomische Gesellschaft. These investigations will be continued and published in the following years. A new catalogue of Nearby Stars was published as No. 22 of the *Veröffentlichungen des Astronomischen Rechen Institut* at Heidelberg in 1969. The gathering of data on Nearby Stars will be continued.

Lourens reports that at the Cape Observatory parallax plates taken between 1943 and 1951 have been measured and the results will be published soon. In the current observing programme which now includes 30 southern nearby stars 1535 plates have been taken.

The new 60-inch telescope of the Vienna Observatory, as reported by Meurer will be used in the general programme for the determination of absolute proper motions of star clusters, which programme utilizes a variety of plates taken with different telescopes. For the Pleiades this is planned in conjunction with positions derived by Eichhorn at Tampa.

Morgan communicates that work has continued on increasing the precision of the resolution and of the calibration of the MK classification system. A working group, composed of Blaauw, Eggen, Keenan, Morgan, Osawa, and O. C. Wilson is preparing a report on the present status of the method of spectroscopic parallaxes, to be submitted to the 1970 meeting of the IAU.

At the Royal Greenwich Observatory Murray reports the preparation of a series of computer programmes for parallax determination and the completion of parallaxes with an estimated probable error of 0."005 for 25 stars, mostly M dwarfs.

Proper motions have been determined for a hundred semi-regular and RV Tau variables, mainly brighter than 11 pg at maximum, partly from plate to plate comparisons, partly by using published astrographic catalogue coordinates as first epoch positions.

The parallax programme of low luminosity stars with the U.S. Naval Observatory's 61-inch astrometric reflector, located at Flagstaff, Arizona, has been continued, as reported by Strand. Since the start of the program in April 1964 a total of 15250 astrometric plates has been obtained. A publication is in preparation reporting on the astrometric and photometric results for the first 100 stars. The current program consists of 310 stars with emphasis on the study of the fainter degenerate dwarf, and subdwarf stars.

At the Yerkes Observatory Van Altena reports that the 40-inch refractor as well as the plate measuring machine have been modernized and their efficiency increased. The new parallax programme begun in 1966 contains some 60 objects, mainly large proper motion stars and white dwarfs of average magnitude about 15. Generally using 20–25 plates probable errors of ±0."003 are being obtained for the parallaxes. Proper motions of M stars in the Pleiades have been published by Van Altena and accurate relative motions for 100 members of this cluster have been determined by Burton F. Jones, with errors as low as ±0."0001 per year, and it is planned to determine the absolute motion of the cluster relative to galaxies by using cluster members of the same magnitude as these galaxies. New proper motions and probabilities of membership have been determined for faint stars in the region of the Hyades cluster. These proper motions have been combined with those from other surveys to yield the highest accuracy motions available for faint Hyades stars. Membership determinations based on proper motions obtained from 40-inch plates are under way for the clusters NGC 2420, 6530, 6611, 6823, and 7062.

Van de Kamp and Lippincott report that the Sproul 24-inch refractor and its mounting were overhauled and that tests indicate no systematic differences between the positional results before and after renovation. A number of publications have appeared in the *Astronomical Journal*, dealing with positional studies of selected stars as well as one giving a summary of Sproul parallaxes determined from 1938–1968. Determination of parallaxes and mass ratios of several binaries and studies of perturbations of stars heretofore considered single, such as Ci2354, and PGC588 have been made. An alternate dynamical analysis of Barnard's star indicates the presence of two unseen companions in co-revolving co-planar circular orbits with periods of 12 and 26 years and masses close to that of Jupiter.

At the Lick Observatory, Vasilevskis has expanded the parallax programme to 207 stars, mainly nearby dwarfs, U Gem stars, Hyads, and Faint Blue Stars, 20–24 plates being obtained normally, but at least 32 plates for stars with expected small parallaxes. Data are ready for 20 stars, yielding parallaxes with probable errors of 0$''$003. In the proper-motion-relative-to-galaxies programme a pilot study of 106 fields was initiated and approaches completion under the direct supervision of Klemola. Its purpose is mainly to develop the techniques of measurement, to study possible systematic errors, and to derive ultimately corrections to the precession constant, the constants of galactic rotation and to determine further the proper motions of stars of the 12th and 16th magnitude and for 80 RR Lyr variables. The start of the complete coverage of the sky accessible from Lick is planned for 1970.

Wagman reports from the Allegheny Observatory that parallax observations are being continued with the 30-inch refractor on all good nights, and about 1000 plates per year are being obtained. During the last three years, and with National Science Foundation support, 47 stars have been measured. White dwarf suspects to magnitude 15 and some red stars are added to the list as they appear in Luyten's and Giclas' lists.

Working under grants from the National Science Foundation and the Office of Naval Research Luyten has completed and published the motions of 17500 stars brighter than $m = 21$ pg in the North Polar Cap, north of declination +70. Likewise the motions of 1300 stars between +70 and +68·5 are being published. A general list of the motions of 951 Faint Blue Stars was issued, a catalogue of 1220 Double Stars with common proper motion, and of 130 Double Stars containing degenerate components, while a list of the motions of some 3000 White Dwarfs is in press. More than 5000 proper-motion stars have been found near the South Galactic Pole; these data will probably be published within a year or so.

A re-analysis of Van Altena's catalogue of proper motions in the Hyades region has shown that several of his largest motions are spurious and that, in general, the accuracy of his motions is much less than claimed, and is, in fact, less than that of the motions determined with the Palomar Schmidt telescope.

W. J. LUYTEN
President of the Commission

25. STELLAR PHOTOMETRY (PHOTOMÉTRIE STELLAIRE)

PRESIDENT: A. W. J. Cousins.
VICE-PRESIDENT: D. L. Crawford.
ORGANIZING COMMITTEE: A. N. Argue, M. Golay, R. H. Hardie, G. E. Kron, V. B. Nikonov, E. Rybka, J. Stock.

INTRODUCTION

During the past three years there has been increased activity in many fields where photometric methods are employed; more telescopes have been commissioned or adapted for photometric measurements and new magnitude systems have been proposed and tried, but much of the observing is now of a routine nature. Most of this continues to be done in the wide-band UBV system, but more measurements are now being made in the red and infra-red.

More interest is being shown in narrower band-widths that emphasize particular features of a stellar spectrum, because these offer better prospects of separating the effects of temperature, luminosity, interstellar reddening, composition and duplicity. Attention has been given to finding the most useful bands and practical ways of measuring them, both photoelectrically and photographically.

The use of narrow bands, pioneered by B. Strömgren, and applied to other spectral features by the Cambridge group, continues to be exploited, especially for measuring the strength of Balmer lines.

Spectrum scanners are now in use at a number of observatories and, while spectroscopy is not the responsibility of Commission 25, quantitative data of this sort are of vital importance for the proper interpretation of the results of stellar photometry. There is a need for more accurate spectrophotometry in the UV and IR. Conventional photometry, in its turn, is now frequently used to find the stars that promise to be of interest spectroscopically.

New techniques have been developed for measuring the rapid fluctuations of flare stars and the even faster changes of pulsars. Automatic devices are being used to an increasing extent for observing, measuring of plates, and for reducing the measures. Photometry from rockets and satellites is extending our knowledge of the far UV.

CATALOGUES AND FINDING LISTS

An important event has been the publication of the Washington *Photoelectric Catalogue* (**1**) containing the collected results of virtually all photoelectric photometry of non-variable stars in the UBV system published prior to 1967. This constitutes a "general catalogue" as contemplated at the Moscow meeting of Commission 25 (**2**). As it gives the individual data from each source it not only serves as a finding list but enables the user to form some idea of the reliability of the available information. Data are given for 20705 stars taken from 424 different sources, most of them independent.

C. Jaschek reports that the La Plata photometric catalogue is almost ready for printing. It gives complete references to all photoelectric measurements – in any system – of stars in our Galaxy. If a star has been measured in the UBV system, average values of V, $B-V$ and $U-B$ are given. There is also a code telling in what other systems a star has been measured. 22000 stars have been listed from 1760 references up to 1967·0. The catalogue will give the effective wavelengths and the number of stars included in each reference. It is expected that it will be published in 1970.

Jaschek (**3**) has discussed information problems in astrophysics, including photoelectric photometry. He advocates data centres with card catalogues that are kept up-to-date and available for reference, and the production of printed catalogues at relatively frequent intervals. Members may wish to discuss this matter further at Brighton.

When using a "general" catalogue it is important to remember that published material, even photoelectric, is not uniform in quality and that some of it is no more reliable than if it had been obtained photographically. Some care is needed when selecting zero point stars.

The *Cape Catalogue of Bright Stars* (**4**) is being extended to include all HR stars lying between $+10°$ and $-10°$ declination. These stars were observed at the Cape following the Berkeley meetings of the IAU, when attention was drawn to the desirability of obtaining photometric and other data for the equatorial stars (**5**). The Cape measures have been revised, applying the same corrections as for the brighter stars, and combined with other published material as weighted means.

Rybka reports that the reduction of 9110 RHP (H. A. 50) magnitudes to a uniform system has been completed and they are now being transformed to the V system.

With the completion of the *Cape Photographic Catalogue for 1950* there is now two-colour photometry for a large fraction of the stars brighter than $V=10{\cdot}0$ south of $-30°$ declination. The standard errors are approximately $\pm 0\overset{m}{.}05$ for V and $\pm 0\overset{m}{.}06$ for $B-V$ derived from these catalogues. Work is in progress on the remaining (Cape Astrographic) zone, $-40°$ to $-52°$.

PHOTOGRAPHIC PHOTOMETRY

(A. N. Argue)

The level of activity in photographic photometry has been maintained. This is evident, for instance, from Bok's summary of the Basel Symposium, 1969 (**6**). The Swedish Astronomers in particular have continued steadily with their projects. Schmidt telescopes of moderate size are playing a useful part, but the construction of very large instruments with Ritchey-Chrétien optics, combined with fast automatic plate measuring machines currently under development, indicate trends in the direction of fainter limiting magnitudes combined with data acquisition on a considerably increased scale. There seems little likelihood of photography being rendered obsolete as a photometric method just yet, although viable alternatives appear to be not far off.

Harvey (Cape) reports progress with the 40-inch "Elizabeth Telescope" reflector-corrector system. Stopping down the corrector to 31 inches and removing the Cassegrain baffles give a reliable working field of at least 1°40′. Stephenson reports favourably on a new mercury flotation system for radial support of the primary mirror of the Warner and Swasey Schmidt. Difficulties previously encountered due to differential thermal expansion between mirror and cell have now been overcome. The maintenance of a uniform figure over the mirror is essential for the Schmidt system because different parts of the mirror are used in different parts of the field. Chi and Pop (Cluj) report on methods of field correction. Kristenson (**7**) has described a photographic method of squaring on the plateholder in the Brørfelde Schmidt, and van Breda (**8**) the use of the Hartmann test in the adjustment of Schmidts.

Taylor reports on a combined autoguider and focus servo under construction by Boller and Chivens for the Steward Observatory 90-inch. It consists of a reflecting rotating knife edge for guiding, and Foucault test for focus. An autoguider for the Kitt Peak 36-inch using an image dissector has been described by Ball and Hoag (**9**), and the Anglo-Australian Project Group are developing an autoguider for the 150-inch based on beam-splitting prisms and a "Channeltron" electron multiplier used as image dissector.

Progress continues in the development of plate measuring machinery. The Lick automatic machine is in production, the Edinburgh "Galaxy" nearly so. Kibblewhite is working on a fast laser scanning system at Cambridge (U.K.). Numerous hand-operated machines have been rendered semi-automatic by application of servos to discrete operations such as centring the image or adjusting the iris diaphragm. These modifications generally lead to greater speed and consistency, making less demand on the skill of the operator. Output on to punched tape or cards and automatic computer reduction are commonplace. Wayman (Dunsink) reports on the development of reduction programs. Individual observers usually have their own ideas as to how reductions should be carried out and prefer to devise their own programs, but a stock of readymade programs can also be drawn on from the literature.

No difficulties have been reported over the purchase of emulsions, although rationalizations by certain manufacturers have sometimes resulted in supplies having to be obtained by roundabout routes and at increased cost. Cousins reports a batch of photovisual plates for the *Cape Photographic Catalogue* as giving very uneven results even though to outward appearance everything "looks normal". Stephenson complains of a batch failing to reach the expected limiting magnitude by 1 magnitude on the Michigan Schmidt at Cerro Tololo early in 1967, and would like to hear if other observers had similar experiences at the same time. Bok (**10**) reports favourably on tests with a new red-sensitive emulsion, Kodak Special Plate Type 098-01.

Kron (Flagstaff) reports on an investigation into hypersensitization by pre-exposure being carried out by Ables, Hewitt and Janes, who find significant gains in signal-to-noise are obtained by pre-exposure to a background density corresponding in the case of Eastman IIa-0 to the beginning of the density-intensity curve (D = 0·2 to 0·3). Van Altena and Bessell (Yerkes) have been investigating the effects of pre-exposure. They find no significant dependence on the duration of the pre-flash over the range 0·001 to 1 s, and advocate flsahing to the minimum density for which near maximum gain can be achieved. Typical gains are times four at density 0·05 and two at 0·2, but they advocate that each emulsion-developer combination should be investigated separately. There is a considerable literature on these and other technical matters which cannot be discussed here. Useful sources of reference are *Photographic Abstracts of the Royal Photographic Society* (London) and *Abstracts of Photographic Science and Engineering Literature* of the Society of Photographic Scientists and Engineers (Washington). There is also a new journal for astronomers, *Photo-Bulletin* published by the American Astronomical Society.

Although most photographic photometry is being done in UBV, other systems are being developed. Straižys (Vilnius) reports on the practical development of a seven-colour (UPXYZVS) system employing glass filter combinations (See p. 244). Schreur (Steward) reports on photometry with a system similar to Strömgren's uvby in Selected Areas 27, 56 and 108. The limiting magnitude is $m_y = 19$. Steinlin (**11**) has published a thorough revision of the RGU system including new transformation equations from UBV, and Todoran (Babes-Bolyai) reports on an investigation into transformations of photographic systems. These developments are to be encouraged. The impending data explosion referred to earlier will place astronomers under obligation to optimise their photometry.

Further work has been put into intensity calibration. To obtain mean magnitudes and colours with accuracy approaching photoelectric measurements, each plate needs, in round figures, at least 30 calibration stars for setting up an intensity scale and calibrating field corrections. At high galactic latitudes the number of stars per ten square degrees averages about 40 brighter than 11^m. To this limiting brightness, then, photography is unproductive since virtually every star would need to be calibrated photoelectrically. The alternative of transferring a standard sequence photographically is economical but less accurate.

To fainter limiting magnitudes, and especially in crowded star fields, photographic photometry is certainly profitable, but local sequences are needed and at the faint end photoelectric measurements are costly in telescope time. This has led to a revival of interest in objective gratings and similar step-intensity methods of calibration. Provided the characteristics of the system are carefully determined photoelectrically under the identical observing conditions, and the zero point is securely tied in, most of the objections to the earlier use of this method are overcome. Eelsalu reports on work carried out, to be published in Tartu Publications. Reddish (**12**) has proposed an objective mosaic prism, and Brück *et al.* (**13**) have tried a doubly-refracting crystal filter calibrator. Butler and Wayman (**14**) have used a coarse grating with the ADH Baker-Schmidt at Bloemfontein to extend Magellanic Cloud sequences from 15^m to 18^m. They find the "grating constant" to depend on brightness but to be predictable within $\pm 0\overset{m}{.}15$. Hardie (Dyer) reports on tests with a grating to extend magnitude sequences by 4^m. He finds an unaccountable dispersion in "grating constant" ranging up to about $0\overset{m}{.}08$. P. J. Andrews has obtained promising results with the Radcliffe reflector.

The following new photoelectric calibration sequences have been reported. Wayman and Butler: 140 stars in UBV, 10^m–18^m in the Magellanic Clouds. Bok and Bok (**15**): southern Milky Way regions and LMC, limit 14^m to 16^m, UBV. Elvius (Lund) refers to southern Selected Area UBV sequences calibrated by him and his associates and reported to Commission 33, and reports also on three UBV sequences each comprising about 50 stars 7^m–13^m centred on $1^h46, +62\overset{\circ}{.}0$; $1^h52, +64\overset{\circ}{.}5$ and $22^h04, +52\overset{\circ}{.}5$ to be published by Bern and Virdefors, and a fourth at $6^h35, +9\overset{\circ}{.}5$ by Karlsson (**16**). Fr. McCarthy reports on a sequence in IC4665 in *B* and *V* to limit 14^m (**17**), and sequences in Selected Areas 28, 54, 106 and 107. He stresses "the importance of using systematically determined sequence of standards as opposed to a collection of randomly determined measures of individual stars".

Published sequences for clusters can be located from the *Catalogue of Star Clusters and Associations* of G. Alter, J. Ruprecht and V. Vanýsek, but sequences in the general field are sometimes difficult to find. The following list, while not complete, may serve as a starting point.

Table 1. Published photoelectric sequences for photographic calibration in the general field

Regions	Scales	Limit	References
SA 3, 4, 5	UBV	$13\overset{m}{.}5$	(**18**)
SA 51, 54, 57, 71, 82, 94, 107, Aql	UBV	$17^m, 18^m$	(**19**)
SA 141, 193, 158, LMC, S. galactic pole	UBV	$15^m, 16^m$	(**20, 21**)
SA 169, 205, LMC, Car-Cen, Norma	UBV	16^m	(**15**)
−45° E regions	U_cBV*	15^m	(**22**)
−45° E regions	$U-B$	10^m	(**23**)
Sco	UBV	$13\overset{m}{.}3$	(**24**)
S. Milky Way (l^{II} 237° − 7°)	UBV	$13\overset{m}{.}5$	(**25**)
S. Milky Way (l^{II} 310° −340°)	UBV	$16\overset{m}{.}5$	(**26, 27**)
Pup	UBV	$13\overset{m}{.}8$	(**28**)

* *V* needs a correction $+ 0\overset{m}{.}005$ to conform to the revised standard zero point.

WIDE BAND PHOTOMETRY

Many of the observations made in the UBV system have been of variable stars or for population studies of clusters and other groups of stars of more interest to other Commissions. Some of the sequences are of more general interest as noted under "Photographic Photometry", and some work of this nature is included in this section of the Report. Other observations deal with individual stars selected for a variety of different reasons. Most determinations of *R* and *I* magnitudes fall into this category. In some cases stars have been observed specifically as standard or zero point stars.

The following reports of wide band photoelectric photometry have been received:

B. J. Bok (Steward Observatory, Arizona) has published, jointly with P. F. Bok, standard UBV photoelectric sequences for a number of fields (**15**). J. J. Schreur has established UBV sequences in SA 47 (to $V=16$) and in a nearby field centred on BD + 31°588 (to $V=15$).

Feinstein (La Plata) has published UBV observations of 135 stars in the region of Carina-Centaurus (**29**), and reports that UBVRI observations are in progress of 70 Be stars, 110 metallic line stars and 30 bright red M stars.

E. H. Geyer (Observatorium Höher List) used the ESO 1-m photometric telescope in 1968 to obtain UBV observations of 54 stars ranging from 7·8 to 15·7 m_{pg} in SA 157 and the Scutum Cloud. W. Seggewiss observed 162 stars in 7 sequences in the UBV system at the ESO Observatory in 1969, all in the galactic plane, near $l^{II} = 212°$, 223°, 243°, 265°, 279°, 290° and 300°, respectively. The limiting magnitude is $V=15$.

Landolt (Louisiana State University) reports that he is establishing secondary UBV standard star sequences in Selected Areas on the celestial equator in the approximate range $10\overset{m}{.}5 < V < 12\overset{m}{.}5$ over as wide a colour range as possible. Observations have been begun in SA 92, 93, 94, 95, 96,

98, 99, 100, 101, 102, 103, 104, 105, 106, 107, 108, 109, 113, 114 and 115. Eventually 300 to 400 stars will have an average of 10 observations each. Not only will the resulting photoelectric sequences be useful for photometric studies of the Selected Areas, but well-established secondary UBV standards will be available for large telescopes in either hemisphere. The observations are being obtained at the Kitt Peak National Observatory.

Mendoza (Observatorio Nacional, Mexico) has completed several series of UBVRI observations, giving special attention to R and I and the resulting two-colour relationships for different types of objects (**30**, **31**, **32**, **33**). He has also made measures at 5μ (**34**).

Rybka has re-reduced some of the observations made by himself and associates to the UBV system (**35**), and is supervising the photoelectric observations of additional standard stars of magnitudes 7·5, 9·0 and 10·5 and spectral types A, F-G and K in SA 1—43 which have been started at Abastumani Observatory by O. P. Abuladze.

Velghe (Observatoire Royale de Belgique) reports that the UBV photometry of 350 OB-stars obtained at Boyden Observatory (1966) and at La Silla (1968-9) is being discussed by J. Denoyelle. 230 OB-stars in Circinus are to be measured photoelectrically by E. Elst in UBV.

A. A. E. Wallenquist reports on Swedish observatories as follows:

At Lund Observatory A. Ardeberg, K. Särg and S. Wramdemark have investigated the properties of the 61-cm photoelectric telescope in Jävan when used for UBV photometry. A. Ardeberg and K. Särg have measured UBV magnitudes for about 40 luminous B-type stars in Cepheus and 17 stars in the Ursa Major Cluster as well as in a field centred at $\alpha = 17^h21^m$, $\delta = +58°40'$. A. Ardeberg and S. Wramdemark have measured UBV magnitudes for about 40 early type stars in Ophiuchus. G. Lyngå has determined photometric and spectroscopic data for a number of distant O and B stars and studied blue stars in three regions in Carina, at the Uppsala Observatory Southern Station. C. Roslund has determined magnitudes in the UBV system of 24 stars in a region south of NGC 5128 (Centaurus A). U. Lindoff has determined UBV magnitudes for stars in open clusters, 12 southern and 3 northern.

At Stockholm Observatory, Saltsjöbaden: L. O. Lodén has measured UBV magnitudes for a large number of stars in standard sequences in Cru, Vel, Pup, Car and Nor (the latter with B. Nordström). He has also measured UBV magnitudes for about 800 stars situated in "loose clusterings". J. Arkling has measured UBV magnitudes for stars in standard sequences in Milky Way regions and G. Gahm has started a photoelectric investigation of strongly-reddened OB-stars.

At Uppsala Observatory, T. Oja and L. Häggkvist have made photoelectric measurements of V and $B-V$ for 368 stars in three regions, in Cas and Per, around the north galactic pole and in SA 17. In addition, V and $B-V$ have been determined photoelectrically for more than 600 stars, late-type stars brighter than $V = 5{\cdot}05$ located north of declination $-10°$, and stars brighter than $V = 6{\cdot}05$ in the polar cap, $b > 60°$. They have also published accurate photoelectric V and $B-V$ for 34 standard stars distributed over the northern sky. H. Lindén has investigated the atmospheric extinction on the basis of photoelectric measurements made at Uppsala during 1959–1968. B. Westerlund has determined photoelectric UBV magnitudes of OB-stars near the supernova remnants RCW 86 and RCW 103. He also reports UBVRI photoelectric and photographic photometry of open clusters and OB-stars in Cru and Nor.

T. Elvius, at Lund Observatory, reports photoelectric measurements of stars in a number of SA in the southern sky to serve as standard sequences. G. Lyngå and C. Roslund have continued to make photoelectric observations for this programme at Siding Spring Observatory; A. Ardeberg is working at ESO in Chile. Y. Ekedahl has observed sequence stars photoelectrically in northern SA at Uppsala.

At the Cape Royal Observatory the photometry of the HR (= BS) stars south of the equator in the U_cBV (or UBV) system is virtually complete (**36**, **37**). Cousins has published $U-B$ for 526 stars in the E regions (**23**) and has UBV data (to $V \sim 10$) for more than 200 comparison stars in variable star sequences (**38**). Evans and Wild have used UBV photometry to look for sub-dwarf stars (**39**, **40**).

Atmospheric extinction in wide-band systems has been investigated empirically and by numerical integration. Observations have been reported from Cerro Tololo, Torun and Uppsala. Mendoza *et al.* (**41**) have found significant and systematic variations in the ultraviolet extinction during the night. Stock (**42**) has discussed the subject of atmospheric extinction in photoelectric photometry in some detail. The usual procedures are satisfactory for reducing observations to a uniform instrumental system but can lead to significant errors when used to compute extra-atmospheric values. There is no simple way of computing a strictly extra-atmospheric $U-B$ and it may be better not to do so. It would be simpler to make extra-atmospheric observations conform to the present standard system by employing suitable filters.

M. Jerzykiewicz (Wroclaw Observatory) has continued his investigations of extra-atmospheric extinction (**43, 44**). The effect he has found is small but appears to be significant. If the interpretation is correct one would expect to find periodic changes of zero point in stellar magnitude systems, and this would be a serious matter. The difficulty of avoiding all observational errors correlated with the day, the month and the seasons is, however, very great and more convincing evidence is needed before one can be sure that there is a significant amount of extra-atmospheric or interplanetary absorbing material present.

At various times there have been proposals to replace the usual U by a narrower band and Straižys has suggested several filters, or combinations of filters, for doing this. Such a change would seem to be desirable for photoelectric photometry but might cause too much loss of light for faint-star photography.

The choice of standard stars for the UBV system and a possible change in the U band should be discussed at the next meeting of the Commission.

Several groups have used numerical integration methods to study the effects of atmospheric extinction, interstellar absorption, mirror tarnish and other changes on wide-band magnitude systems. The effect on $U-B$ is rarely a simple function of $B-V$ or $U-B$. Wesselink (**73**) has discussed the relationship between surface brightness and M_v and other stellar parameters.

NARROW-BAND AND INTERMEDIATE-BAND PHOTOMETRY

(David L. Crawford)

At Stockholm Observatory, Saltsjöbaden, U. Sinnerstad and J. Arkling have measured intensities of the absorption lines Hβ, Hγ, He I 4471 and 4922 in bright B0-B5 stars, and U. Sinnerstad and I. Furenlid, J. Arkling, S. H. Alm and P. Brattlund have measured line-photometry photoelectrically for early B-type stars.

At Uppsala Observatory L. Häggkvist has determined Hγ absorption for 365 northern stars, including nearly all stars brighter than $V=5{\cdot}1$ and classified A2 or earlier, using interference filters centered at 4059 Å, 4346 Å, and 4508 Å, with half widths of about 40 Å. P. Malmström has measured monochromatic magnitudes photoelectrically for 276 stars using interference filters with maxima at 3937 Å, 4346 Å and 4508 Å (half-widths 23–56 Å).

A. G. Davis Philip (Dudley Observatory) reports that forty field horizontal-branch stars have now been found in fields at high galactic latitudes by means of Strömgren four-colour photometry (*Ap. J.*, **152**, 1107; *Astr. J.*, **74**, 209; *Astr. J.*, **74**, 812). Hβ photometry has been started. Preliminary results show that, between $(b-y)_0=0{\cdot}00$ and $0{\cdot}15$, the β index is $0^m_{\cdot}06$ smaller than the β index of Population I stars of the same colour. Four-colour and Hβ photometry has been started on blue horizontal-branch stars in the globular clusters M 4, NGC 362, NGC 6809. All A5 and earlier type stars in a 30 sq. degree region surrounding the NGP were photometered (*Astr. J.*, **73**, 1000) in the UBV and four-colour systems. The reddening at the pole was found to be $E_{B-V}=0{\cdot}05$. At $V=11{\cdot}5$, the A stars change from mostly Population I to mostly Population II. A similar study is underway at the south galactic pole.

John D. Bahng (Northwestern University, Illinois) reports that intermediate-band photometry of late-type stars in the infrared was carried out. The three bands are centered at 1·21, 1·59 and 2·15 μ with half-widths of approximately 5 % of the central wavelengths. For normal G, K and M stars a continuum emission peak near 1·6 μ was observed. For carbon stars, this peak was not observed, indicating that neither H^- nor H_2O is an important source of continuous opacity. The infrared colours of carbon stars are not correlated with Morgan-Keenan C-classification nor with the vibrational temperatures derived from CN bands.

John Graham reports that he is continuing at Cerro Tololo his program of uvby photometry of faint high-latitude stars and of white dwarfs. He has recently completed observations of Hβ indices for 450 OB-type stars in Carina.

M. Golay (l'Observatoire de Genève, Switzerland) reports that the multicolour photoelectric system described in *Vistas in Astronomy* (Vol. 14, Part II) has been used at Geneva Observatory since 1963. The new photometric catalogue, by F. Rufener (*Publ. Obs. Geneve*, in press), will be published in 1970. This will contain 1400 stars in 7 colours. The accuracy is better than 0·01 magnitude. There are [U] [B] [V] B_1 B_2 V_1 G colours of (a) metallic line stars, (b) peculiar stars, (c) nearby stars (listed by Gliese), (d) stars of associations Cas VII, Per I, II, III, Aur I, Gem I, Mon I, II, Cyg I, II, III, VIII, IX, Cep I, (e) stars of galactic clusters (Pleiades, Coma Ber, Hyades, Praesepe, h and χ Per, NGC 188, 752, 1545, 1647, 2168, 2244, 2264, 2682, 6633, 6871, 7092, 7160, 7822, IC 348, 1805, 4665 and (f) Giant G, K, M.

Recent work on K and M stars seems very promising and the observing program is being considerably extended.

The Geneva group has studied the properties of colour indices and parameters obtained by linear combination of colour indices. Two-dimensional representation gives a large separation of luminosity effects for stars B 3 to F 8, chemical composition effects for stars A 1 to G 8, rotational effects for stars B0–B7 and between A 6 and F 2 (Golay, *Pub. Obs. Genève*, no 75. A Maeder, *Publ. Obs. Genève,* nos. 74, 75, B. Hauck, *Publ. Obs. Genève*, nos. 74, 75). Three-dimensional representation permits the separation of some kinds of peculiar stars, A VI, Si 3955 Cr, 4012, λ Boo, Sr-Cr-Eu. The effect of the several laws of interstellar extinction on the linear combinations has been studied to be able to calculate parameters free of the interstellar extinction effects (Goy, thesis, Genèva, 1969). This last work, applied to young galactic clusters, shows that we can have different interstellar extinction laws in the same cluster. To calibrate the photometric measurements in the Geneva system in terms of certain physical characteristics of the atmospheres considered, Peytremann (thesis, in press, Geneva) prepared a program for computing a grid of model atmospheres, (for different $\log g$, for effective temperature between 10000 K and 5000 K, for different metal-to-hydrogen ratios and opacity due to the metallic lines). To study the photometric effects of rotation, Maeder and Peytremann (thesis, in press, Geneva) compute the flux emitted by rotating stars employing the method proposed by Collins. The blanketing effect of the hydrogen lines is taken into account. The aim is to try to separate the v from $\sin i$ by photometric means and to estimate the influence of the rotation of a star on the position of the turn-off point.

At Kitt Peak and Cerro Tololo, we are continuing photometry with the uvby and Hβ systems. Observations have been completed on the following programs, some published, the rest being prepared for publication:

1. Standard stars for uvby photometry;
2. Colour and M_v calibrations for B stars, and for A and F stars;
3. O-G0 stars, $V \leqslant 5{\cdot}0$, south of $\delta = -10°$;
4. O-B5 stars, $V < 6{\cdot}5$, north and south;
5. B8-A0 stars, $V < 6{\cdot}5$, $\delta > -10°$;
6. Stars in the areal vicinity of globular clusters, to derive reddening;
7. Open clusters and associations: Hyades, Coma, UMa, Praesepe, Pleiades, α Per, NGC752, NGC6231, h and χ Per, 1C4665, I Sco, III Cep.

Activity in the sub-field of narrow- and intermediate-band photometry, during the past two years, has been primarily in the following subjects (some of the investigators are listed, selected from the main journals):

1. Classification of late-type stars:

Eggen, O. J. 1969, *Ap. J.*, **158**, 225.
Jones, D. H. P. 1968, *Mon. Not. R. astr. Soc.*, **139**, 189.
McClure, R. D., van den Bergh, S. 1968, *Astr. J.*, **73**, 313.
Wallerstein, G., Helfer, H. L. 1966, *Astr. J.* **71**, 350.

2. OB Stars:
Graham, J. A. 1968, *Bull. astr. Inst. Neth., Suppl.* **2**, 397.
Graham, J. A. 1968, *Astr. J.*, **73**, 17.
Gutierrez-Moreno, A., Moreno, H. 1968, *Ap. J. Suppl.*, **15**, 459.
Smith, L. F. 1968, *Mon. Not. R. astr. Soc.*, **140**, 409.

3. High-Latitude Blue Stars:
Graham, J. A. 1967, *Bull. astr. Inst. Neth.*, **19**, 161.
Graham, J. A., Doremus, C. 1968, *Astr. J.*, **73**, 226.
Newell, E. B., Rodgers, A. W., Searle, L. 1969, *Ap. J.*, **156**, 597.
Philip, A. G. D. 1968, *Astr. J.*, **73**, 1000.

4. Hα:
Andrews, P. J. 1968, *Mem. R. astr. Soc.*, **72**, 35.
Tebbe, P. L. 1969, *Astr. J.*, **74**, 920.

5. Open Clusters:
Crawford, D. L., Barnes, J. V. 1969, *Astr. J.*, **74**, 407.
Crawford, D. L., Barnes, J. V. 1969, *Astr. J.* **74**, 818.
Crawford, D. L., Barnes, J. V. 1969, *Astr. J.*, **74**, 1008.
Epstein, I. 1968, *Astr. J.*, **73**, 556.
Geneva workers, see Golay report above.
Graham, J. A. 1968, *Astr. J.*, **73**, 152.
Perry, C. L., Hill, G. 1969, *Astr. J.*, **74**, 899.
Sears, R. L., Whitford, A. E. 1969, *Ap. J.*, **155**, 899.

6. Variable and Peculiar Stars:
Breger, M. 1969, *Astr. J.*, **74**, 166.
Breger, M. 1969, *Ap. J. Suppl.*, **19**, 79.
Hayes, J. E., Heiser, A. M. 1968, *P.A.S.P.*, **80**, 57.
Johnson, H. M., Golson, J. C. 1968, *Ap. J. Letters*, **154**, 77.
Johnson, H. M., Golson, J. C. 1969, *Ap. J. Letters*, **155**, 91.
Kizilirmak, A., Wood, H. J. 1967, *Astr. J.* **72**, 727.
Koch, R. H. 1967, *Astr. J.*, **72**, 411.
Philip, A. G. D. 1968, *P.A.S.P.*, **80**, 171.
Philip, A. G. D. 1969, *P.A.S.P.*, **81**, 248.
Williams, J. A. 1966, *Astr. J.*, **71**, 615.

7. New "systems" and miscellaneous:
Breger, M. 1968, *Astr. J.*, **73**, 84.
Johnson, H. L., Mitchell, R. I., Latham, A. S. 1967, Commun. Lunar Planet. Lab., **6**, 85.
Mitchell, R. I., Johnson, H. L. 1969, Commun. Lunar Planet. Lab., **8**, 1.
Neff, J. S., Travis, L. D. 1967, *Astr. J.*, **72**, 48.
Perry, C. L. 1969, *Astr. J.*, **74**, 705.
Wilson, O. C. 1968, *Ap. J.*, **153**, 221.

8. Interpretation:
Baschek, B., Reimers, D. 1969, *Astr. Astrophys.*, **2**, 240.
Collins, G. W. II, Harrington, J. P. 1966, *Ap. J.*, **146**, 152.
Hardorp, J., Strittmatter, P. A. 1968, *Ap. J.*, **151**, 1057.
Kraft, R. P. 1967, *Ap. J.*, **150**, 551.
Kraft, R. P., Kuhi, L. V., Kuhi, P. S. 1968, *Astr. J.* **73**, 221.

POLARIZATION

(V. B. Nikonov and N. M. Šahovskoj)

Instrumentation and methods of observation

Nearly all new polarimeters belong to one of the two following types.

(1) Differential dual-beam polarimeters are used for precise measurements of polarization of the light of bright stars. An instrument of this type has been described by Elvius and Engeberg (**45**). Polarimeters of the same type have been developed at Konkoly Observatory (Hungary) and at some other observatories. At Yerkes and Mt Stromlo Observatories such polarimeters are being used with rotating telescopes. The principal problems to be tackled, using this type of polarimeter, are studies of interstellar polarization in the vicinity of the Sun (**46**), the establishment of precise standards for polarimetric measurements (**47**) and polarization measures of the light of bright variable stars.

(2) Integrating polarimeters are used for the observation of faint objects. The main principle employed in the design of these instruments is the integration of results obtained from many cycles of a fast-rotating polarizing device. Digital or analogue outputs are used for registration. This permits long exposure times and consequently observations of very faint objects. The method under consideration is very insensitive to changes of observing conditions. Polarimeters of this type, designed for stellar measurements, have been described by Appenzeller (**48**), Leinert *et al.* (**49**) and by Efimov and Šahovskoj (**50**). Polarimeters for planetary research have been designed by Ksanfomaliti (**51**) and Bugaenko *et al.* (**52**).

Some theoretical aspects of the polarimetric method have been discussed by Hall (**55**), Wiedling (**53**) and Efimov (**54**).

Among the other types of polarimeters, recently developed, that of Hall (**55**) for faint extended objects (galaxies, nebulae) must be mentioned. Pratt (**56**) has devised a special polaroid filter enabling simultaneous monitoring of observing conditions in the outer parts of the field.

A new method for photographic polarimetric measurements of stellar radiation has been developed by Treanor (**57**).

General problems of stellar polarization measurements

Efimov studied theoretically the dependence of the accuracy of stellar polarimetric measurements on the observing conditions. The case of a single-beam integrating polarimeter, using photon-counting, for observing variable stars with very fast changes of brightness (flare stars) has been considered. A theoretical analysis of the accuracy of polarimetric measurements has also been given by Bugaenko *et al.* (**52**) and by Wiedling (**53**).

The general theory of instrumental polarization has been considered in detail by Jäger and Oetken (**58**). Some particular aspects of this problem have been considered by Efimov *et al.* (**59**) and by Efimov and Šahovskoj (**60**).

Sahovskoj developed a method of analysis of multicolour polarimetric observations when interstellar and stellar components of polarization are both present (**61**).

Observations of stars having little polarization, which can be used as standards for polarimetric investigations, have been published by Walborn (**47**) and Appenzellar (**62**). Early-type stars observed for polarization by Goyne and Gehrels in seven colours (3200–10000 Å) (**63**) may be used as standard stars with a high degree of polarization.

EXTRA-TERRESTRIAL STELLAR PHOTOMETRY

(M. Golay)

The experiments which were performed in the first half of the present decade provide the bulk of the photometric data which are available to date. These experiments showed that an appreciable

Table 2

Experimenter	λ_{eff}(Å)	$\Delta\lambda$(Å)	Field of view	Absolute calibration accuracy	Observations reported	Spectral type	Mag. of faintest object listed	Estimated accuracy of data
[1] Bless *et al.*	2100	300	3° × 3°		45			
	2500	300	3° × 3°	±15%	43	B 0·5-F 8	5·9(B 9)-4·8(A)	±25%
	2800	300	3° × 3°		44			
[2] Smith	1376	200	2°	50%	96	WC-A 2	6·55(B 5)	10% (relative)
[3] Yamashita	1120	120	5°	50%	6	O 9·5-B 7	3·0(B 2)	
	1290	120	5°		6			
	1415	130	5°		6			
[4] Carruthers	1115	130	7°	±20%	8	WC 7-A 1		
	1270	120	7°		8		3·7(O 8)	±5% (relative)
[5] idem					20	O 8-B 5	6	±5% (relative)
[6] Campbell I	2150	250						
II	2550	270			100	O 9·5-A 0	7	±3% (absolute)
III		2750–5000			50			
[7] idem but III		3500–5000			I-II:20 III:50	O 9·5-A 0	7	±30%

Remarks and References:

[1] 6256B EMI photomultipliers with Al-MgF_2 interference filters. Unguided Aerobee launched September 1964. (1968, *Ap. J.*, **153**, 557.)
[2] Bendix resistance-strip magnetic electron multiplier with SrF_2 filter. Magnetically stabilized satellite launched December 1964. (1967, *Ap. J.*, **147**, 158.)
[3] Ionization chamber with short cut-off window. Unstabilized K-10-3 rocket launched March 1967. (1968, *Astr. Space Sci.*, **2**, 4.)
[4] Ionization chambers with short cut-off windows. Stabilized Aerobee launched March 1967. (1968, *Ap. J.*, **151**, 269.)
[5], [6] and [7] 1969, *IAU Symposium* no. 36.

improvement had been achieved in the observational techniques involved and gave a fairly large amount of data suitable for photometric analysis. Whenever a star was observed several times during the same flight, the internal consistency was found to be better than 20%. In the wavelength range longward of 2000 Å, the measures of Bless *et al.* (**64**) and Bogess (**65**) agreed within $\pm 0^m.25$ for the 12 stars they observed in common. But, as is pointed out by Wilson (**66**), it is probably optimistic to expect such consistency for the whole range of observations. It seems very likely that the major uncertainties do not arise as much from instrumental differences and errors as they do from errors involved in the calibration procedure. Bless (**46**) has suggested the use of a synchrotron radiation source as secondary standard for the ultraviolet range, instead of a sodium salicylate-coated photomultiplier assumed to have a constant quantum efficiency over this range.

The photometric data were primarily used to check the validity of atmospheric models. The extended analysis which was made by Bless *et al.* (**67**) and included results obtained by Byram *et al.* (**68**), Chubb and Byram (**69**), Smith (**70**) and Bless *et al.* (**64**) showed that most of the 35 Main-Sequence B and A stars fell within $\pm 0^m.5$ of the theoretical predictions derived from the line-blanketed models of Morton and Adams (**71**). An average deficiency of about $0^m.25$ was however noted and it was assumed, at that time, that such a discrepancy could fall within the calibration uncertainties. A more recent comparison between models and ultra-violet observations (**72**) suggests that B- and A-type stars may emit less energy in the UV than predicted for models which do not include Si, Mg and C as appreciable sources of opacity. These photometric observations were also used to determine the interstellar extinction curve in the UV (**66**) (**67**).

The photometric observations are listed in Table 2.

A large amount of very valuable photometric data will be available in the near future thanks to the successful operation of the Orbiting Astronomical Observatory, OAO-A2, which was launched in December, 1968. Several hundred stars have already been observed by the photometers and spectrophotometers of the Space Astronomy Laboratory of the University of Wisconsin. The approximate effective wavelengths and widths (in parentheses) of the bands measured with the four 8-inch ($f/4$) stellar photometers are 1300 (540), 1400 (660), 1550 (600), 1900, 2350 (960), 2900 (1000), 3300 (1000) and 4200 (1400) Å. The 16-inch ($f/2$) nebular photometer measures four bands at 2100 ($\Delta\lambda = 300$), 2500 (410), 2900 (410) and 3300 (500) Å. Ascop 541 F photomultipliers are used for $\lambda_{eff} < 2500$ Å; E.M.I. 6256B for $\lambda_{eff} > 1700$ Å, with bands in common. The field stops are 10′ and 20′ for the stellar photometers and 30′ and 10′ for the nebular photometer.

The preliminary results (presented at the 1969 IAU Symposium) suggested an upward revision of the stellar temperature scale for stars earlier than B 3 *V*. Interesting results were also given on the interstellar reddening which seems to have a real "bump" at $1/\lambda = 4{\cdot}5$, and a minimum in the vicinity of 5·5 to $6{\cdot}0\,\mu^{-1}$.

During the last three years a series of stellar observing instruments, both photometric and spectrophotometric, developed by the Space Research Group at the Royal Observatory, Edinburgh, have been flown in ESRO roll-rate controlled Skylark rockets. Successful flights produced the observations as given in Table 3.

Table 3

Date of firing	26.5.67	22.5.67	3.12.67		7.10.67	
ESRO rocket number	S05/2	S11/2	S27/1		S47/1	
Experiment number	R62	R65/1	R100		E121A	R120
Instrument description	[1]	[1]	[1]	[2]	[1]	[3]
Number of stars observed	20	20	50	100	50	11

[1] 2-channel photoelectric photometer, collecting area 120 cm² using interference filters with 250 Å pass-bands centred on 2100 and 2500 Å.

[2] Large aperture (340 cm²) single-channel photoelectric photometer recording the range 1300–2000 Å.

[3] Single-channel photoelectric photometer, aperture 32 cm², for recording sky brightness in the regions 1650–3200, 2400–3200 and 2650–3200 Å.

A payload with objective prism and Schmidt camera having a magazine holding 9 photographic plates will be flown early in 1970. In parallel with this instrument is a wide-band photometer to assist in the absolute calibration of the photographic data.

The Edinburgh group's main effort is at present engaged on a collaborative experiment with Institut d'Astrophysique, Liege, in the TD1 ESRO-Satellite. It is expected to measure the brightness of early-type stars down to $m = 8$.

The Laboratoire d'Astronomie Spatiale et Observatoire de Marseille group (Courtès, Cruveillier, Roussin, Valerio, Viton, Sivan) has flown sounding rockets. The preliminary results obtained with a photoelectric photometer having 4 passbands show an UV excess of the central part of And nebula relative to the sky background. The preliminary results from a second flight, to photograph the sky in UV, give the magnitudes of 201 stars at 2600 Å.

The space research group of the Observatory of Geneva has developed a balloon technique. These experiments are to extend into the UV region the ground-based observations of the Geneva multi-colour stellar photometry. Modern balloons are large enough to carry a 200 kg payload to an altitude of 43 km. During the last four years eleven flights have been carried out with French and American balloons. The observations were recorded photographically with a small Maksutov camera and with a 16-cm Schmidt telescope. These results have revealed the feasibility of UV photometric measures using balloon techniques.

PHOTOMETRIC INSTRUMENTATION

(G. E. Kron)

A conventional UBV photometer designed by Kenneth Janes has been built and is in operation on the 20-inch reflector at the Bethany Observing Station of Yale University Observatory (P. Demarque).

At the Okayama Station of Tokyo Astronomical Observatory, the photometric equipment has been modernized by employing transistor amplification, electronic logarithmic conversion, and digital voltage display. Sky brightness in the vicinity of a star is sampled by displacement of the telescope focal plane by means of a tilted quartz plate (K. Osawa).

A two-channel Elliptical Polarization Photometer has been designed and built by Blitzstein and Wolf for the Flower and Cook Observatory. Measurements with this instrument lead to the determination of the normalized Stokes parameters of partially elliptically polarized radiation. The instrument employs intermediate band interference filters, achromatic quarter-wave plates, an analyzing prism, and a depolarizer. The output of each channel is integrated, and the integrals are read out sequentially by an integrating digital voltmeter controlled by a digital clock (W. Blitzstein).

Two new classical photometers have been put into service by the Geneva Observatory for stellar photometry at stations at Haute Provence, Jungfraujoch, and Gornergrat. Research work is being undertaken to develop photon counting techniques. An automatic guider, to permit stellar photometry with smaller focal plane apertures, is being built for the 100-cm telescope in Haute Provence (M. Golay).

At the Royal Greenwich Observatory stellar photometry has been done with an RCA-Carnegie Institution image intensifier on the 91-cm telescope. With IaO plates, exposure times were about 30 min for $m_v = 18$, with standard errors of $0^m.1$. A magnitude scale was derived from densitometric calibration and integration of the star images, a procedure possible with the fairly diffuse images resulting from the large scale of 15 arc sec per mm. (Communicated by the Astronomer Royal.)

At the Dominion Astrophysical Observatory, Victoria, a single channel photometer that features an offset guider, thermoelectrically refrigerated photomultiplier, and a pulse counting electronics system has been designed and built by Walker, Brealey, Brosterhus and Morbey. Walker, with help from Pfannenschmidt and Smyth, installed his four-channel photometer with a Vidar data acquisition system on the Mount Kobau 16-inch telescope. The Evans 16-inch telescope has been rebuilt with an electronic drive by Andrews. This telescope and a Boller and Chivens 16-inch telescope have been installed on Mt Kobau for photometry (K. O. Wright).

Steward Observatory, Tucson, has acquired a dual beam photometer from Astro-Mechanics for use with the new 90-inch reflector for general purpose photometry. The photometer has a dichroic beam splitter enabling simultaneous measures in red and blue light. A Johnson-type JKL photometer has been purchased from the Lunar and Planetary Laboratory for operation this fall. A polarimeter built in cooperation with Gehrels should be in operation by summer, 1970. The 36-inch and 90-inch photometers now transfer photometric data to magnetic tapes, which are processed in Tucson by the CDC 6400 computer (W. Fitch).

Photoelectric stellar and surface photometry is being done at the Flagstaff Station of the U.S. Naval Observatory with the Navy Electronic Camera on either the 61-inch Astrometric Reflector, or the 40-inch RC reflector. The plates, which give a linear response to exposure, are assessed by a scanning and integrating procedure employing a modified Joyce-Loebl-Tech-Ops Isodensitracer. Data are fed into an IBM card punch, and are reduced by computer. The limiting magnitude for measurable images yielding a standard deviation of $0\overset{m}{.}2$ is $B = 23$ to 24 magnitude, depending on the seeing, for a one hour exposure with the 61-inch. The full accuracy associated with photoelectric photometry is obtained up to about $B = 21{\cdot}5$ to 22 magnitude, depending on the seeing (G. Kron).

A UBVRI photometer employing two photomultipliers is being constructed at the University of St. Andrews. Photomultiplier selection is by interposition of a diagonal mirror into the light beam; multiplier and filter can be selected by remote control. Measurement can be by pulse counting or charge integration at the will of the observer. In the integration mode, digitization is effected by sampling the output voltage of the integrator and then discharging the sampling amplifier at a constant rate. The time of discharge is proportional to the output voltage of the integrator (D. W. N. Stibbs).

The 1955 *F*/5 scanner at the University of Michigan, Ann Arbor, is now fitted with a two-channel pulse-counting photometer constructed by Lasker and Williams. The second channel provides compensation for changes in sky transparency. Diaphragms, filters, and wavelength have been encoded and are recorded automatically for each observation. Output is on punched paper tape (J. A. Williams).

At the Royal Observatory, Cape, conventional photometers are used with d.c. recording or with electronic integration. Output is printed out from a digital voltmeter, and then processed with a keyboard electronic computer. Plans have been made for replacing some of the older photometric equipment which has been much used for many years.

At Northwestern University, Dearborn, a photoelectric spectrum scanner by Bahng has been built for use with the 100-cm reflector. Spectra are scanned very rapidly by rotating the grating. Output from the photomultiplier is sent to an on-line PDP 8/S computer where signal counts from each 10 Å interval are stored in a separate location of the core memory, thus producing an accumulative spectrum. The computer also controls the scanning operation, so programming can be very flexible. An Hα scanning photometer by Bahng and Caplan has also been furnished for the 100-cm reflector. This instrument, which has photoelectric output, scans over a wavelength range of several Angstrom units in the vicinity of the Hα line; resolution is about 0·4 Å. Dispersion is by echelle grating for predispersion and a fixed-space Fabry-Pérot etalon. The dispersing units are in separate chambers and are simultaneously pressure-scanned with nitrogen (J. D. Bahng).

The Catania Astrophysical Observatory at Serra la Nave, Sicily, is now in regular operation with telescopes of 91-cm, 41-cm and 31-cm apertures fitted with photoelectric photometers. Amplifiers employing field effect transistors are employed and a photometer for simultaneous stellar photometry in three colours (**74, 75**) has been constructed to facilitate the observation of flare stars (G. Godoli).

Three groups in Romania, at Bucharest, Timsoara University and Cluj University, are using photoelectric photometers (G. Chis).

A. T. Young has continued his studies of sources of error in photoelectric photometry with a discussion of stripchart reading errors (**76**) and the causes and effects of scintillation (**77**).

STELLAR PHOTOMETRY IN THE U.S.S.R.

(V. B. Nikonov)

1. *Instrumentation*

Two 125 cm Ritchey-Chrétien photometric telescopes (Draft Report 1967, p. 472) are under construction.

A three-channel stellar photometer has been constructed and is now under test at the Radioastrophysical Observatory of the Latvian Academy of Sciences (G. Spulgis, G. Zarevsky).

A two-channel photoelectric photometer designed for the photometry of nuclei of galaxies is being put in operation on the 2·6-m telescope of the Crimean Astrophysical Observatory (A. B. Severny, N. A. Dimov, A. Sčerbakov).

At the Engelghardt Astronomical Observatory an iris photometer automatically accounting for the plate background has been constructed (L. A. Urasin, S. S. Tohtas'ev, S. K. Fomin).

2. *Methods of observation and reduction procedures*

The complex of two (twin) 48-cm Cassegrain telescopes of the W. Struve Astrophysical Observatory, Tartu (Estonian S.S.R.) (Draft Report 1967, p. 471) is being used to study the possibilities of compensating for the effects of atmospheric variations (U. K. Weissmann).

At the Crimean Astrophysical Observatory the work on stellar TV photographic photometry is being continued. It is used mostly for the narrow-band multicolour photometry of faint variable stars showing rapid changes in brightness. Further improvements of the method are being made. The photometric field error is studied and taken into account when reducing observations. It was found that by cooling the target of a sensitive image orthicon down to $-30°C$ its sensitivity is increased by 3^m and the accumulation of charges on the target may be for as long as 10 min (N. A. Abramenko and V. V. Prokof'eva).

Investigations in connection with fundamental photoelectric photometry are being carried out at the Crimean Astrophysical Observatory (Precise account of atmospheric extinction, control of the stability of photometric systems, reductions to standard systems) (V. B. Nikonov, V. Burnašov).

3. *Photometric systems and standards*

Tests of the glass version of the Vilnius (Straižys's) multicolour photometric system UPXYZVS, carried out at Vilnius and Crimean observatories, show that it is somewhat better for classification purposes than the interferometric one (UPXYZTS), reported to Commission 25 at the Prague meeting. The maximum separation of luminosity classes is reached for B and K-M spectral types. The accuracy of classification of A-G stars using the Vilnius system is the same as with the Strömgren one. The Vilnius system is being tested photoelectrically as well as photographically. For the calibration of the Q, Q diagrams, used for classification, nearly 300 stars of different spectral types, luminosity classes and chemical content have been observed with the help of the Crimean Astrophysical Observatory. Subdwarfs, and different kinds of peculiar stars have been observed too. The possibility of detecting unresolved binary stars photometrically has been demonstrated.

To study the photographic version of the Vilnius system the necessary observations have been obtained of the clusters NGC4996 and 7654 at the Main Astronomical Observatory of the Ukrainian Academy of Sciences and at the Radioastrophysical Observatory of the Latvian Academy of Sciences; the Cassegrain focus of a 70-cm telescope and a Schmidt telescope have been used. The photoelectric standards were obtained with the 70-cm telescope of the Crimean Observatory.

Two standard sequences in the UPXYZVS system have been established in Cygnus and Aquila each containing 32 stars from 5^m to 7^m of different spectral types and luminosity classes.

4. *Photoelectric sequences*

V. M. Kovalenko (State Sternberg Astronomical Institute) is making UBV photoelectric observations of the $5^m.5 - 6^m.0$ BS catalogue stars having no accurate photometric data.

5. *Photographic sequences*

N. B. Grigor'eva and E. M. Roščina determined *B, V* magnitudes for 830 stars down to 12^m in a region of 30 square degrees centered at $\alpha = 19^h32^m$ $\delta = +22^\circ.5$. T. A. Uranova made similar observations of 85 stars in the region centered at $\alpha = 19^h08^m$, $\delta = +14^\circ$ and 199 stars in three regions with centers at $\alpha = 19^h27^m$, 19^h32^m and 19^h38^m and $\delta = +8^\circ.5$. She is carrying out *B, V* photographic photometry of O-F 5 stars in three other regions 1° in diameter in field N 1 of Parenago's plan and centered at $\alpha = 18^h24^m$, $\delta = +22^\circ.5$; $\alpha = 18^h52^m$, $\delta = +13^\circ$ and $\alpha = 19^h15^m$, $\delta = +9^\circ45'$. A. S. Sharov determined *B, V* magnitudes for 266 stars brighter than $15^m.5$ in the neighbourhood of NGC 7142. N. M. Artjuhina and E. P. Kalinina, when studying proper motions, determined approximate values of *B* magnitudes for 4228 stars down to $13^m.5$ in an area 4·5° in diameter around the Pleiades and for 2964 stars within a 2·5° circle around the cluster M 39 (Sternberg Institute).

UBV magnitudes have been determined for nearly 10000 stars to $B = 15^m$ in the directions of SA 26 and 27 (S. S. Tohtas'ev) and SA 28 (I. A. Dudjago). Similar observations have been made of 4021 stars in the region of NGC 1664 (L. A. Urasin) (Engelghardt Observatory).

6. *Extra-atmospheric stellar photometry*

Photoelectric photometry of the sky background outside of the atmosphere has been carried out in two wavelength regions (at 2700 Å and in the visual, near to *V* system). The field of view of the photometers was 18°. They worked on Cosmos 51 and 213. The minimum value of the brightness, observed near the Galactic Pole, was ~ 30 stars of 10^m per square degree (N. A. Dimov, A. B. Severny, A. Zvereva) (Crimean Astrophysical Observatory).

Two photoelectric photometers for the wavelength bands at 2270 Å (halfwidth 50 Å) and 2750 Å (halfwidth 80 Å) having object glasses of 50 mm diameter and field of view of 1° were used on Cosmos 215. Pulse-counting was employed and stars down to 5^m have been observed (N. A. Dimov, V. K. Prokof'ev) (Crimean Astrophysical Observatory).

7. *Miscellaneous*

A. S. Šarov and N. N. Jakimova (Sternberg Institute) completed a list of photoelectric catalogues and regional photometric standards.

Two papers (Riives, *Tartu Publ.* no. 38; Eelsalu, ibid. no. 39) where the law of darkening of faint photographic images is derived from the concept of a two-stage reaction of the grains to exposure are awaiting publication.

ACKNOWLEDGMENT

In conclusion, the President wishes to express his thanks to all who have contributed to this Report, either individually or by preparing sections. All contributions have had to be edited to conserve space. It cannot be regarded as a comprehensive report. It is worth considering what form future Reports should take and whether some of the contents should not be published elsewhere.

A. W. J. COUSINS
President of the Commission

BIBLIOGRAPHY

(This list is in no way comprehensive. Some other references will be found in the text.)

1. Blanco, V. M., Demers, S., Douglass, G. G., Fitzgerald, M. P. 1968, *Publ. U.S. nav. Obs.*, **21**.
2. *Trans. IAU*, **10**, 386 (1960).
3. Jaschek, C. 1968, *P.A.S.P.*, **80**, 654.
4. Cousins, A. W. J. 1967, *Magnitudes and Colours of Bright Stars South of +10° Declination* (Mimeogram), R. Obs. Cape.
5. *Trans. IAU*, **11B**, 311 (1962).
6. Bok, B. J. 1969, *Summary and Outlook at the Conclusion of the Basel Symposium on the Spiral Structure of Our Galaxy*. Steward Observatory preprint.
7. Kristenson, H. 1969, *Observatory*, **89**, 72.
8. van Breda, I. G. 1969, *Mon. Not. R. astr. Soc.*, **144**, 73.
9. Ball, W. F., Hoag, A. A. 1968, *Sky Telesc.*, **35**, 22.
10. Bok, B. J. 1969, *Photo-Bulletin am. astr. Soc.*, **1**, 8.
11. Steinlin, U. 1968, *Z. Astrophys.*, **69**, 276.
12. Reddish, V. C. 1968, *Observatory*, **88**, 146.
13. Bruck, M. T., Nandy, K., Caprioli, G., Smriglio, F. 1969, *Astrophys. Space Sci.*, **4**, 213 = *Commun. R. Obs. Edinburgh*, no. 69.
14. Butler, C. J., Wayman, P. A. 1969, *Observatory*, **89**, 109.
15. Bok, B. J., Bok, P. F. 1969, *Steward Obs. Preprints*, no. 24; *Astr. J.*, **74** (in press).
16. Karlsson, B. 1966, *Medd. Lunds astr. Obs.*, II, no. 145.
17. McCarthy, M. F., O'Sullivan, S. 1969, *Ric. astr.*, **7**, 483.
18. Johnson, H. L., Mitchell, R. I., Iriarte, B. 1962, *Ap. J.*, **136**, 75.
19. Purgathofer, A. Th. 1969, *Lowell Obs. Bull.*, no. 147.
20. Bok, B. J., Bok, P. F. 1960, *Mon. Not. R. astr. Soc.*, **121**, 531.
21. Bok, B. J., Basinski, J. 1964, *Mem. Mt. Stromlo Obs.*, **4**, no. 16.
22. Cousins, A. W. J., Stoy, R. H. 1962, *R. Obs. Bull.*, no. 49.
23. Cousins, A. W. J., 1967, *Mon. Notes astr. Soc. Sth. Afr.*, **26**, 151.
24. Roslund, C. 1966, *Arkiv astr.*, **4**, no. 20.
25. Lodén, L. A. 1969, ibid., **5**, no. 9.
26. Lyngå, G. 1964, *Medd. Lunds astr. Obs.* **II**, no. 139.
27. Sarg, K., Roslund, C., Engver, N. 1967, *Arkiv astr.*, **4**, 405.
28. Lindoff, U. 1967, ibid., **4**, no. 28.
29. Feinstein, A. 1969, *Mon. Not. R. astr. Soc.*, **143**, 273.
30. Mendoza, E. 1967, *Bol. Obs. Tonantzintla Tacubaya*, **4**, 106, 114, 149.
31. Mendoza, E. 1968, *Ap. J.*, **151**, 977.
32. Mendoza, E. 1969, *Pub. Dep. astr. Univ. Chile*, no. 7, 106.
33. Mendoza, E. 1969, *Bol. Obs. Tonantzintla Tacubaya*, no. 31.
34. Mendoza, E. 1968, *Bul. Soc. Argentina Ast.*, no. 11.
35. Rybka, E. 1969, *Acta Astr.*, **19**, 229.
36. Corben, P. M., Stoy, R. H. 1968, *Mon. Notes astr. Soc. Sth. Afr.*, **27**, 11.
37. Stoy, R. H. 1968, *Mon. Notes astr. Soc. Sth. Afr.*, **27**, 119.
38. Cousins, A. W. J., Lagerweij, H. C., Shillington, F. A. 1969, *Mon. Notes astr. Soc. Sth. Afr.*, **28**, 63.
39. Evans, D. S., Wild, P. A. T. 1969, *Observatory*, **89**, 15.
40. Wild, P. A. T. 1969, *Mon. Notes astr. Soc. Sth. Afr.*, **28**, 48.
41. Mendoza, E., Moreno, H., Stock, J. 1969, *Pub. Dep. astr. Univ. Chile*, no. 5, 67.
42. Stock, J. 1968, *Vistas Astr.*, **11**, 127.
43. Jerzykiewicz, M. 1968, *Acta astro.*, **18**, 327, 437.
44. Jerzykiewicz, M., Opolski, A. 1968, *P.A.S.P.*, **80**, 101.
45. Elvius, A., Engeberg, M. 1967, *Ark. Astr.*, **4**, 387.
46. Appenzeller, I. 1968, *Ap. J.*, **151**, 907.
47. Walborn, N. R. 1968, *P.A.S.P.*, **80**, 162.
48. Appenzeller, I. 1967, *P.A.S.P.*, **79**, 136.
49. Leinert, Ch., Schmidt, Th., Schwarze, B. 1967, *Veröff. Landssternw.* Heidelberg-Konigstuhl, 20.
50. Efimov, J. S., Sahovskoj, N. M. 1970, *Izv. Krym. astrofiz. Obs.* **44** (in press).
51. Ksanfomliti, L. V. 1967, *Pribory i Tehnika Eksperimenta* no. 2, 167 (in Russian).

52. Bugaenko, L. A., Bugaenko, O. I., Kruglov, V. D., Petuskov, V. G. 1968, *Astrometrija i Astrofisika* (Kiev, Naukova Dumka), **1**, 193 (in Russian).
53. Wiedling, T. 1967, *Ark. Astr.*, **4**, 349.
54. Efimov, J. S. 1969, *Izv. Krym. astrofiz. Obs.*, **41–42** (in press).
55. Hall, J. S. 1967, *Astr. J.*, **72**, 801; 1968, *Lowell Obs. Bull.* **7**, no. 6.
56. Pratt, N. M. 1967, *Publ. R. Obs. Edinburgh*, **6**, 39.
57. Treanor, P. I. 1967, *Mon. Not. R. astr. Soc.*, **138**, 325.
58. Jäger, F. W., Oetken, L. 1968, *Publ. astr. Obs. Potsdam*, **31**, E. 1.
59. Efimov, J. S., Polosuhina, N. S., Sahovskoj, N. M. 1969, *Izv. Krym. astrofiz. Obs.*, **39**, 3.
60. Efimov, J. S., Sahovskoj, N. M. 1970, *Izv. Krym. astrofiz. Obs.*, **44** (in press).
61. Sahovskoj, N. M. 1969, *Izv. Krym. astrofiz. Obs.*, **39**, 11.
62. Appenzeller, I. 1966, *Z. Astrophys.*, **64**, 269.
63. Coyne, G. V., Gehrels, T. 1966, *Astr. J.*, **71**, 355; 1967, ibid., **72**, 887.
64. Bless, R. C., Code, A. D., Houck, T. E., McNall, J. F., Taylor, D. J. 1968*a*, *Ap. J.*, **153**, 557.
65. Bogess, A. 1966, *Ap. J.*, **153**, 561.
66. Wilson, R., Boksenberg, A. 1969, *A. Rev. Astr. Astrophys.*, **7**.
67. Bless, R. C., Code, A. D., Houck, T. E. 1968*b*, *Ap. J.*, **153**, 561.
68. Byram, E. T., Chubb, T. A., Werner, M. W. 1965, *Ann. Astrophys.*, **28**, 594.
69. Chubb, T. A., Byram, E. T. 1963, *Ap. J.*, **138**, 617.
70. Smith, A. M. 1967, *Ap. J.*, **147**, 158.
71. Adams, T. F., Morton, D. C. 1968, *Ap. J.*, **152**, 195.
72. Bless, R. C. 1970, *IAU Symposium* no. 36.
73. Wesselink, A. J. 1969, *Mon. Not. R. astr. Soc.*, **144**, 297.
74. Cristaldi, S., Paterno, L. 1968, *Mem. Soc. astr. ital.*, **39**, 583.
75. Cristaldi, S., Paterno, L. 1969, *Non-Periodic Phenomena in Variable Stars*, Ed. L. Detre. Hungarian Academic Press, Budapest.
76. Young, A. T. 1968, *Observatory*, **88**, 151.
77. Young, A. T. 1967, *Astr. J.*, **72**, 747.

26. ÉTOILES DOUBLES (DOUBLE STARS)

PRÉSIDENT: P. Couteau.
VICE-PRÉSIDENT: J. Dommanget.
COMITÉ D'ORGANISATION: O. J. Eggen, P. Kulikovsky, P. Muller, K. Aa. Strand, P. Van de Kamp.

1. INTRODUCTION ET GÉNÉRALITÉS

On constate que dans l'ensemble les observations classiques se font en quantité égale à celle de 1964 à 1967. Les observations photographiques sont plus nombreuses.

De nouvelles méthodes de mesure sont essayées et détectent des couples inaccessibles aux méthodes classiques. Nous citerons par exemple les observations photoélectriques d'occultation de binaires par la Lune (Rakos), les mesures d'intensité interférométrique par comptage de photons (Twiss). Dans les deux cas des séparations très inférieures à 0″01 sont mesurées. Ces travaux semblent ouvrir la voie à une génération nouvelle de moyens d'observations de binaires.

Les problèmes liés aux vitesses radiales n'ont pas, semble-t-il, retenu l'attention de nombreux observateurs, mis à part ceux de l'équipe de l'Observatoire d'Astrophysique de Victoria (Batten, Fletcher et Scarfe).

Les observations photométriques, par contre, croissent nettement grâce aux efforts des astronomes de l'U.S. Naval Observatory (Strand, Worley, Franz), de ceux de Vilnius, U.R.S.S. (V. Straigys, G. Kakaras).

Les circulaires d'information de la Commission 26 (P. Muller) paraissent régulièrement trois fois par an. Depuis la réunion de Prague 74 orbites y ont été annoncés ainsi que 192 couples nouveaux.

Plusieurs catalogues sont parus ou à paraître depuis la réunion de Prague.

"Catalogue de 240 couples visuels d'étoiles naines rouges" (P. Baize, 1966, *J. Observateurs*, p. 1).

"Catalogue des différences de magnitudes des composantes de 2379 étoiles doubles et multiples" (Wierzbinski, 1969, *Contr. Wroclaw astr. Obs*, No. 16).

"Troisième catalogue d'éphémérides d'étoiles doubles" (P. Muller et Cl. Meyer, 1969).

En ce qui concerne le fichier central de mesures de Washington le responsable, C. E. Worley, communique:

"Since transfer of the double star catalogs to the Naval Observatory in 1965, no report of progress has appeared in either the IAU Draft Reports or the Transactions. Below is a summary of progress to October 1, 1969.

Upon receipt of the catalogs in tape form in March 1965, they were transferred to cards. Insofar as possible, the 1200-odd references were checked in the library; errors of various kinds were found in 129 instances. Subsequently, corrected lists of references were supplied to the three depositories (Lick, Greenwich, Meudon).

Four groups of cards have been prepared and sent to the above depositories. These cards number 20346, and represent data from 63 lists of published data. It has been possible to keep the punching of data reasonably current, thus no long time-delay in distribution of the material has occurred.

Errors of many kinds have been detected and corrected in these catalogs. However, this work has received second priority compared to the acquisition and dissemination of the new data.

Upon request, data has been freely supplied from the catalogs. Foreign requests have constituted an appreciable fraction of the total. (In some cases requestors might have received faster service had they consulted the copies deposited abroad; it is possible that existence of these copies has not been sufficiently publicized.)

Data could be made available more rapidly, and the chance of error decreased, if all observers would strive to follow the publication format described in the 1967 IAU Draft Reports."

Un Colloque UAI sur "La Cohésion entre les procédés d'observations des étoiles doubles" a eu lieu à Nice en septembre 1969. Il groupait trente participants représentant dix nations. Six sessions totalisant 34 exposés ont abordé les problèmes de mesures, les problèmes connexes aux parallaxes, mouvements propres et binaires, les liaisons entre la spectroscopie et les binaires, la photométrie et les binaires. Les résultats seront publiés dans les suppléments de la revue européenne *Astronomy and Astrophysics*. Le Colloque était dédié au Professeur van den Bos en reconnaissance du monumental travail que représente sa longue carrière d'observateur.

La résolution finale du Colloque est la suivante:

"Les participants au Colloque UAI No. 5, tenu à Nice en septembre 1969, sur les étoiles doubles visuelles, constatent la nécessité de constituer, dans un avenir rapproché, un document homogène formé de couples visuels pour lesquels à la fois sont connues, toutes les données orbitales, photométriques, spectroscopiques et astrométriques, document destiné à servir de base aux recherches sur la relation masse-luminosité, sur les problèmes liés à l'évolution stellaire et sur tous les autres problèmes connexes."

2. ACTIVITÉS DES OBSERVATOIRES

(a) *U.S. Naval Observatory, Washington*

Le Professeur K. Aa. Strand et C. E. Worley communiquent:

Visual observations:

A total of 4826 micrometer measures were carried out by Worley (3477 measures) and Walker with the 26-inch (66 cm) and 12-inch (30 cm) refractors at Washington, and the 61-inch (155 cm) and 40-inch (102 cm) reflectors at Flagstaff. Measures of 463 double stars were published by Walker (1969) and a list of 5581 measures has been submitted for publication by C. E. Worley.

Photographic observations:

The program on photographic observations of double stars with the 26-inch (66 cm) refractor at Washington has been continued with about 1400 multiple exposure plates on approximately 300 different systems being obtained during the report period. 2600 plates with a total number of 150000 exposures were measured (or remeasured). Measurements of 3286 plates of 610 double stars have been published (Kallarakal *et al.*, 1969).

Photometric observations:

Worley (1969) published Δm's for 94 double stars based upon the Muller method. Strand (1969) published Δm's for 874 double stars determined from visual estimates on several series of multiple exposure photographic plates.

Spectroscopic observations:

MK classification of 142 visual binaries have been carried out by Christie and Walker (1969), based upon spectra taken with the 40-inch (102 cm) reflector.

Divers:

Worley (1969) étudie les caractéristiques de duplicité des étoiles de faible luminosité (Ch. 9 of *Low Luminosity Stars*, Gordon and Beach, London and New York).

(b) *Sproul Observatory Swarthmore*

Le Professeur P. van de Kamp et S. L. Lippincott communiquent:

Peter van de Kamp and Sarah Lee Lippincott of the Sproul Observatory are continuing their studies of mass ratios and perturbations of nearby stars from plates taken with the 24-inch refractor. The results are published mostly in the *Astronomical Journal*.

Sarah Lee Lippincott has made a statistical study of the distribution of errors and perturbations in astrometric positions. At the IAU Colloquium on Visual Double Stars held in Nice in September 1969, she reported on her work and also on the recent study by B. H. Feierman of the blending effect in unresolved astrometric binaries from wide doubles by overlapping images of either component with the first order grating images of the other component.

Van de Kamp made an alternate dynamical analysis for Barnard's star resulting in two unseen companions in co-revolving co-planar circular orbits with periods of 26 and 12 years, and masses close to the mass of Jupiter.

(c) *University of Minnesota*

Le Professeur W. J. Luyten communique:

W. J. Luyten a continué ses recherches sur les étoiles doubles à mouvement propre commun, sur des plaques prises avec le télescope Schmidt de 1·20 m du Mont Palomar. Il a déjà publié une liste de 1250 bînaires de magnitude photographique inférieure à 21 et à mouvement propre supérieur à 0″04 par an. Il estime à 75000 le nombre de ces binaires dans l'ensemble du ciel et plus proches statistiquement que 250 parsecs. Elles sont donc beaucoup plus nombreuses que les autres catégories d'étoiles.

D'autre part W. J. Luyten a publié un catalogue de 125 binaires contenant au moins une étoile dégénérée, parmi lesquelles se trouvent 15 paires de naines blanches, et probablement quatre couples contenant une étoile dégénérée du type M.

(d) *Kitt Peak Observatory and Lunar and Planetary Laboratory, Tucson*

Le Professeur G. van Biesbroeck communique:

G. van Biesbroeck continue son programme de couples serrés en mouvement orbital à l'aide du télescope de 154 cm de Catalina et du télescope de 213 cm de Kitt Peak. Les résultats seront publiés en 1970.

(e) *Lowell Observatory, Flagstaff*

Dr O. G. Franz communique:

At the Lowell Observatory, O. G. Franz is using a photoelectric area scanner of his own design to carry out a long-range program of determining photometric data and relative positions of the components of double stars (Franz, 1969).

Using the same equipment, Franz and Millis (1969) have initiated a program of photoelectric investigation of visual double stars containing at least one variable component.

(f) *Dominion Astrophysical Observatory, Victoria*

Dr A. H. Batten communique:

A number of visual binaries has been observed spectrographically at the highest practicable dispersion, principally by Fletcher, Batten, and Scarfe (University of Victoria).

Fletcher has observed Procyon and Σ 2173 at a dispersion of 2·4 Å mm^{-1}. The velocities obtained for Procyon agree well with the predictions from Strand's visual orbit (1951). The two components of Σ 2173 have been spectrographically resolved. The relative velocity is close to that given in the catalogue by Dommanget and Nys (1967). A joint paper with F. R. West is in preparation.

Batten is observing 70 Oph. He has continued R. M. Petrie's work on HD 100018 (ADS 8189) and derived the relative velocity of the centre of mass of the close pair and the visual secondary component. The results are not in good agreement with the predictions from visual orbits. A paper is in press (Petrie and Batten, 1969).

Scarfe has continued observations of several systems: α U Ma, ζ Her, γ Leo, β L Mi, δ Equ and has reported on many of these at the 1969 Nice Colloquium.

Batten and Fletcher have investigated the errors in measurements of the separation of very close line pairs in stellar spectra, and their implications for the determination of masses from visual binaries. These results have also been reported at Nice.

(g) *Observatorio Nacional, Rio de Janeiro*

Le Dr R. R. de Freitas Mourao communique:

Les observations visuelles et photographiques d'étoiles doubles sont régulièrement faites à la lunette de 45 cm. Le programme photographique comprend des étoiles en mouvement sensible et de distances supérieures à 4″. De nombreuses mesures seront publiées en 1970.

Les calculs d'orbites sont publiés régulièrement et des recherches théoriques sur les méthodes de calculs ont été effectuées à Uccle en collaboration avec J. Dommanget et S. Arend.

Un programme astrophysique lié à l'astrométrie s'organise avec un nouveau photomètre photoélectrique. Il est prévu de publier prochainement un catalogue photométrique d'étoiles doubles visuelles.

(h) *Observatoire de Nice*

Le Dr P. Couteau communique:

La lunette de 76 cm d'ouverture a été remise en service en juin 1969. L'objectif a été retaillé, la monture et la coupole restaurées. Cet instrument est destiné aux mesures d'étoiles doubles et de parallaxes stellaires, son optique permet de travailler à la limite imposée par la diffraction. Avec la lunette de 50 cm installée en 1967, l'Observatoire de Nice possède maintenant deux bons instruments consacrés à l'observation des binaires.

En trois ans, environ 4000 mesures visuelles de binaires ont été faites par P. Couteau et P.-J. Morel, tandis que 500 clichés étaient pris par E. Fossat. L'ensemble de ces travaux est paru au *Journal des Observateurs* (1969) ou bien est sous presse dans les suppléments de la revue européenne *Astronomy and Astrophysics* (1969).

Les recherches de binaires nouvelles sont menées par P. Couteau et P. Muller à la lunette de 50 cm. Depuis l'Assemblée de Prague 300 couples ont été découverts d'une part entre les déclinaisons +17° à +29° (P. Couteau) et d'autre part au-delà de +60° (P. Muller). Ils sont annoncés régulièrement dans les Circulaires d'Information de la Commission 26.

P. J. Morel a étudié du point de vue théorique le problème du photocentre (thèse de 3e Cycle) et l'amélioration des élements d'orbites (convergence) à l'aide d'ordinateur, en tenant compte de très près des signatures des observations et des difficultés qu'elles représentent, ce dont seul un observateur peut juger.

Un programme de mesures de vitesses radiales relatives ou absolues est en voie d'achèvement pour déterminer des parallaxes absolues, des rapports de masses, des positions de noeuds. Ces observations sont prévues à la caméra électronique associées au spectrographe à très grande résolution de l'Observatoire de Haute-Provence.

(i) *Observatoire de Paris-Meudon*

Le Dr P. Baize et le Dr P. Muller communiquent:

1100 Mesures d'étoiles doubles encore inédites ont été faites à la lunette de 38 cm de Paris depuis 1967 (P. Baize). De nombreuses orbites ont été publiées.

Les observations courantes se sont poursuivies avec le 83 cm (P. Muller) sur un programme remanié récemment (650 objets environ), régulièrement augmenté des couples nouveaux intéressants, surtout ceux de P. Couteau. Une seconde série de mesures depuis la remise en service de l'instrument en 1965 est sous presse à *Astronomy and Astrophysics*.

P. Muller a commencé au 50 cm de Nice une recherche systématique de couples nouveaux au Nord de +60°. En une quinzaine de nuits, un peu plus de 2200 étoiles ont été examinée et 36 étoiles doubles trouvées.

(j) *Observatoire d'Uccle, Bruxelles*

Le Dr J. Dommanget et le Professeur S. Arend communiquent:

(i) *Observations*

Le nouvel objectif Zeiss de 45 cm d'ouverture et de 7 m de distance focale ayant été monté sur l'équatorial, les observations régulières d'étoiles doubles visuelles ont repris dès juillet 1967, suivant le programme décrit dans les rapports précédents (voir par exemple: *Trans. IAU*, **11B**, 256).

Les observations sont de plus en plus gênées par les poussières de la ville rendant le ciel d'Uccle particulièrement diffusant en toutes saisons. On dépasse difficilement la magnitude 9·0.

Des observations photographiques sont envisagées pour les couples les plus brillants, mais relativement serrés, à l'aide d'un agrandisseur d'images réalisé avec un Miflex Zeiss.

(ii) *Recherches effectuées depuis octobre 1966*

Nos recherches ont porté principalement sur les sujets suivants:

(*a*) l'orientation des plans orbitaux des binaires visuelles et des vitesses spatiales,

(*b*) la détermination des caractéristiques physiques d'un corps invisible perturbant le mouvement linéaire ou orbital d'une étoile,

(*c*) la distribution des binaires visuelles et des binaires spectroscopiques dans le voisinage du Soleil,

(*d*) les erreurs affectant les périodes, les demi-grands axes et les masses des couples stellaires,

(*e*) plusieurs orbites circulaires, paraboliques ou elliptiques,

(*f*) l'amélioration d'une orbite elliptique (programmée sur ordinateur par O. Nys),

(*g*) le mouvement relatif rectiligne des composantes du couple ADS818 a été calculé (1969) ce qui conduit à supposer qu'une naine blanche perturbe le mouvement relatif de ce couple.

(iii) *Recherches projetées*

(*a*) poursuite des recherches entreprises dans le domaine de l'évolution des binaires par perte de masse,

(*b*) prospection systématique des couples susceptibles de présenter dans leurs mouvements relatifs, une anomalie pouvant être expliquée par la présence d'un corps invisible,

(*c*) recherches sur les mouvements propres des composantes de couples optiques de mouvement relatif bien marqué,

(*d*) calcul d'orbites et de trajectoires rectilignes.

Le Dr E. L. J. van Dessel (Utrecht):

Au cours d'une recherche sur l'évolution d'une binaire serrée (spectroscopique) par échange de masse entre les composantes, suivie d'éjection violente de matièren on a pu expliquer les caractéristiques des étoiles A, en admettant que l'une des composantes du système évolue vers ce type particulier, l'autre se transformant alors probablement en une naine blanche.

Cette même théorie permettrait d'expliquer la formation de couples classiques plus écartés (visuels) (1968).

(k) *Observatoire de Copenhague*

Le Dr N. P. Wieth-Knudsen communique:

Depuis la XIIIème Assemblée Générale, les observations d'étoiles doubles avec un micromètre Muller sont poursuivies. Bientôt un miroir de 30 cm ouvert à *F*/6 sera monté.

(l) *Observatoire de Munich, République Fédérale*

W. D. Heintz a publié plusieurs milliers de mesures faites à la lunette de 27 cm de Munich et de nombreuses orbites.

(m) *Observatoire de Vienne*

Le Professeur J. Hopmann a déposé à l'Académie des Sciences en Autriche un mémoire destiné à être publié "30 Jahre Doppelsternforschung". Cet ouvrage résume l'activité de l'auteur à Vienne, on y trouve un catalogue de 103 systèmes avec parallaxes, masses, magnitudes absolues et d'autres données.

(n) *Institut Astronomique de Wroclaw*

Le Professeur St Wierzbinski communique:

Depuis 1966 J. Bem entreprend des mesures d'étoiles doubles avec la lunette de 20 cm d'ouverture munie d'un nouveau micromètre Zeiss. La première liste contient 130 binaires.

(o) *Observatoire Astronomique de Belgrade*

Le Dr P. M. Djurković communique:

De nombreuses mesures d'étoiles doubles sont faites à la lunette de 65 cm par Djurković, Popović et Zulević. Depuis quatre ans plus de 2000 mesures sont publiées ou sont sous presse.

Des orbites ont été calculées par Zulević et Maly.

Popović commence des recherches de couples nouveaux entre 40° et 50°.

(p) *Observatoires de l'Union des Républiques Socialistes Soviétiques*

Le Professeur Kulikovsky communique:

Photographic observations of double stars of Strand's two lists have been completed at the Pulkovo Observatory with 26″ refractor. Plates have been measured and now are being reduced. Photographic observations of 61 Cyg were continued.

A. N. Deutsch (1969) proceeded both with the study of proper motions of all double stars using photographs taken with a large difference of epochs and with finding of wide pairs by the similarity of proper motions of stars.

Works on introduction of automatisation for measurings of double stars on the plates are carried out at the Pulkovo Observatory.

Observations of double stars with 200 mm refractor supplied with a polarizing micrometer were continued at the Sternberg Astronomical Institute, Moscow. G. A. Starikova determined rectilinear relative motions in the systems of 33 double stars and corrected orbits of 4 binaires ADS 11484, 7307, 9425, 9626 (BC).

V. Telniuk-Adamtchuk (1967) of the Kiev University Observatory computed the orbit ADS 16836 and studied the orientation of planes of double and triple stars.

Observations of multiple stars of Trapeze Orion type and of nearer stars were proceeded in Goloseevo (near Kiev) for discovering invisible satellites.

Anosova applied dynamical and statistical criteria to the triple of ε Lyr type stars from the catalogue IDS with the aim of separation of possible physical systems from optical ones.

B. I. Fesenko (Pskov) (1967) studied effects of the observational selection on statistical correlations of characteristics of visual double stars.

V. Straizys and G. Kakaras (Vilnius) (1969) proposed an original method for discovering unresolved double stars in accordance with their position of two-colour and on *QQ*-diagrams using seven-colour photometric system *UPXYZVS* proposed in Vilnius. The author insists that such double stars be called photometric and that this term should not be used for the nomination of eclipsing binaries.

Kakaras (1969) confirmed possibilities for the discovery of such doubles from combined photometric data for 32 double stars ADS, obtained by him with 200 mm reflector at the Crimean Astrophysical Observatory.

(q) *Bosscha Observatory, Lembang*

Le Professeur G. B. van Albada communique:

Des mesures photographiques de 168 couples faites avec la lunette de 60 cm sont sous presse dans la revue européenne *Astronomy and Astrophysics.*

(r) *Republic Observatory, Johannesburg*

Le Professeur Hers communique:

With the 26½-inch refractor 2183 micrometer measures were made by van den Bos, Knipe and Newburg, and interferometer measures by Finsen (319 measures).

Orbits were calculated by Finsen, Newburg and Knipe.

The transfer of the 26½-inch refractor to a new and better site and the construction of larger new instruments is at present under consideration. Until such times as a decision is made it is proposed to continue double stars work on the same lines as in previous years.

3. THÉORIE

A la demande du Président de la Commission, le vice-président, J. Dommanget, a transmis l'analyse ci-contre:

Analyse succincte des travaux théoriques et statistiques publiés de 1967 à 1969 sur les binaires visuelles

Les recherches théoriques et statistiques sur les binaires visuelles ont porté principalement sur le calcul des orbites, sur la relation masse-luminosité et sur leur évolution.

Le calcul des orbites a continué à faire l'objet de recherches diverses. Non seulement ont été mises au point de nouvelles méthodes, l'une applicable indifféremment au cas de l'ellipse, à celui de la parabole et à celui de l'hyperbole (V. Janova, 1966), l'autre, graphique, destinée au cas parabolique seul (S. Vlaicu, 1967), mais la méthode d'amélioration par variation différentielle des éléments orbitaux a été généralisée, adaptée à plusieurs cas particuliers et traitée par développements en séries de Fourier (Heintz, 1967). La méthode classique de Thiele-Innes, que l'on aurait pu croire entièrement établie, a cependant encore révélé des aspects nouveaux (S. Arend et R. R. de Freitas Mourao, 1967), quant aux conditions d'existence d'une solution pour le système des trois lieux fondamentaux et de la constante des aires.

Comme prolongement des problèmes orbitaux, ont également été traitées les conditions d'éclipse des composantes d'un couple (P. Couteau, 1967), tandis que l'intérêt de l'observation des vitesses radiales des composantes des binaires visuelles a été rappelé (J. Dommanget, 1967) à l'occasion du Symposium No. 30 de l'UAI.

Les raisons de croire à une dichotomie réelle de la relation masse-luminosité, telle que l'a soupçonnée P. van den Kamp et que semble l'avoir précisée O. J. Eggen, paraissent s'affirmer encore (O. J. Eggen, 1967): les binaires de l'amas des Hyades (et des Pleïades?) conduiraient à une relation différente de celle habituellement considérée et dont le Soleil et Sirius sont des représentants typiques.

Mais les problèmes évolutifs et galactiques ayant trait aux couples visuels ont gardé tout leur intérêt auprès des chercheurs bien que ce soit surtout dans le domaines des couples serrés (photométriques et spectroscopiques) que les problèmes qui s'y rattachent sont les plus délicats et que les travaux correspondants sont les plus nombreux. C'est ainsi que la stabilité de la relation période-excentricité a été réaffirmée (J. A. Zaera de Toledo, 1966), que le problème des couples à orbites paraboliques ou hyperboliques (binaires de "rencontre") a continué à retenir l'attention (J. Hopmann, 1967) et que les effets des sélections observationnelles et statistiques ont fait l'objet de nouvelles recherches (B. I. Fesenko, 1967) destinées à en estimer l'importance et à les neutraliser par des corrections adéquates.

L'action du milieu galactique sur la pérennité des couples stellaires et sur leurs éléments orbitaux a été traitée par des tests statistiques permettant de mettre en évidence la probabilité de "dilution" des composantes dans ce milieu à la suite du passage d'un astre dans le voisinage du système (T. A. Agekajan et N. L. Primak, 1968) et l'évolution statistique des éléments orbitaux des couples non dilués (T. A. Agekjan et Z. P. Anasova).

L'évolution par perte de masse soit continue, soit discontinue (nova récurrente) a encore fait l'objet de divers travaux (J. D. Hadjidemetriou, 1967 et 1968).

Toutefois le problème de l'origine des couples stellaires a particulièrement retenu l'attention. On notera la divergence et la diversité maintenue des hypothèses de travail ou des conclusions: origine différente pour les couples serrés d'une part et des couples écartés de l'autre (Su Shu Huang, 1968); formation possible à partir de la dissociation progressive d'un système d'étoiles (T. S. van Albada, 1967); origine commune pour tous les couples, complétés d'une évolution orbitale consécutive (perte de masse). Une discussion synthétique des diverses hypothèses plausibles a été faite par A. H. Batten (1967). On notera également l'ensemble des travaux présentés sur l'évolution des étoiles doubles lors du Colloque tenu à Bruxelles sous les auspices de l'UAI à ce sujet en 1966.

Binaires à éclipses

Theoretical work (with G. Hill) on reflection effect in close binaries, applied to Algol by J. B. Hutchings. A program was developed to compute the heating effect on a cold companion by a hot star and the resultant light curve of the system at any wavelength, for any sizes and orbital configuration. Numerical results were obtained which match the observations of the Algol system at six wavelengths, all phases. Hence numerical data for the component stars were deduced, compatible with other published conclusions.

The work is now being continued in the development of a similar program which takes into account gravitational distortion of the component stars and it is intended to apply this to a suitable stellar system. Observational work over as wide a spectral range as possible is also planned for eclipsing binary systems.

Distribution des étoiles doubles

Jonckheere étudie la distribution des étoiles doubles.

Tous les types spectraux lui ont indiqué une prépondérance d'étoiles doubles serrées dans l'hémisphère Sud, surtout très importante pour le type F. Un des résultats les plus intéressants fut la découverte de régions ou les composantes sont plus serrées que dans le reste du ciel.

4. PROPOSITIONS À DISCUTER

J. Dommanget propose

(1) Chaque calculateur d'orbites devrait signaler dans sa publication:

(*a*) la méthode utilisée pour le calcul de l'orbite,

(*b*) celle utilisée pour le calcul de la parallaxe.

(2) Lors de la réunion de la Commission No. 26 à Prague, il avait été convenu de l'intérêt de l'établissement d'une "liaison" entre la Commission No. 26 et la Commission No. 30.

Il y aurait lieu de faire le point à ce sujet.

Un groupe de travail me paraîtrait bien convenir pour traiter de certains problèmes intéressant les deux commissions. Je note par exemple:

(*a*) la mise sur pied des programmes d'observation de Vr intéressant les doubles visuelles,

(*b*) la recherche des observations individuelles de Vr pour les composantes des couples d'orbite connue et l'interprétation de ces observations quant à leur signification pour la détermination du nœud ascendant des orbites par exemple. Il existe dans ce domaine des problèmes purement spectroscopiques que seul un spécialiste peut traiter. La publication d'une liste de ces Vr, réduites à un même système homogène serait souhaitable (ex. *Publ. Lick Obs.*, **16**).

(3) Il serait intéressant que le U.S. Naval Observatory effectue une statistique des couples du

catalogue de Jeffers-van den Bos, en fonction de la distance de séparation, des magnitudes apparentes et des différences de magnitudes. (Je n'ai pas connaissance d'un tel travail et de la publication afférente.)

J. Hers suggère

"No satisfactory solution has as yet been reached regarding Finsen and Worley's Double Star Orbit Catalogue. This would seem to be a matter which could well be raised again."

PAUL COUTEAU
Président de la Commission

27. VARIABLE STARS (ÉTOILES VARIABLES)

PRESIDENT: L. Detre.
VICE-PRESIDENT: O. J. Eggen.
ORGANIZING COMMITTEE: A. A. Bojarčuk, G. H. Herbig, R. Kippenhahn, K. Kwee, G. Odgers.

1. INTRODUCTION

This Report in arrangement of the material follows closely that written by Dr. Herbig for the last IAU General Assembly. It consists of four parts, each written by a different author. The first deals with variable stars in general, and was prepared by the President of Commission 27, with the exception of a section on the Theory of Variable Stars, very kindly written at his request by Prof. Kippenhahn. Appendix I on the Spectra of Variable Stars was prepared by Dr M. W. Feast, Chairman of that Committee of Commission 27. Appendix II covers the Variable Stars in Clusters, written by Dr Helen Sawyer-Hogg, Chairman of that Committee. Appendix III is a Report of the Working Group on Flare Stars, prepared by Drs Andrews, Chugainov, Gershberg and Oskanjan. No report has been received from the Working Group for Nomenclature for Magellanic Cloud Variables. About supernovae see the report of the Committee of Commission 28 for Research on Supernovae. On pulsars see the Reports of Commissions 25 and 40. Our thanks are due to all members of the Commission and to other astronomers occupied in variable star research, who have assisted in collecting the material for this report. Great help was a general report on flare stars by the competent Working Group.

The drastic reduction of this report, in accordance with a decision of the Executive Committee, to half its extent in the *IAU Transactions* **XIIIA**, made some changes necessary. We tried to make up for lost extent using references condensed in the text in the most abbreviated form. In addition to the usual brief abbreviations the following have been used: *AA* = *Astronomy and Astrophysics*; *AcA* = *Acta Astronomica*; *AZ* = *Astr. J. U.S.S.R.*; *IB* = *Var. Star. Inf. Bull. Budapest*; *MAI* = *Mem. Soc. astr. it.*; *MVS* = *Mitteilungen über Veränderlichen Sterne, Sonneberg*; *ROB* = *Royal Observatory Bulletins*; *PZ* = *Peremennje Zvjozdy*. About references to Proceedings of Symposia or Colloquia see Sections 2 and 3. References without volume or page numbers are in press.

2. MEETINGS AND SYMPOSIA

An IAU Colloquium on Non-Periodic Phenomena in Variable Stars was held in Budapest in September 1968, the fourth in the triennial series of meetings on variable star topics. A Colloquium on Mass Loss from Stars was organized at Trieste in September 1968 by Mrs Hack. Many papers on variable stars were presented at the IAU Symposium no. 36 on Ultraviolet Stellar Spectra and Related Ground-Based Observations held in Lunteren, Netherlands, from 24 through 27 June 1969 (referred in the following as *Lunteren*). Some material on our topic was presented on the Liège Colloquia on Forbidden Transitions in Stellar Spectra in 1968 (ref. *Liège*, 1968), and on Pre-Main Sequence Evolution in 1969 (*Liège*, 1969), both organized by the Institut d'Astrophysique, Cointe Sclessin, further at the ESO Symposium held at ESO headquarters in Santiago, organized by Blaauw, the scientific director of ESO, following the ESO's dedication ceremonies 1969 March (ref. *ESO*). In May 1969 X-ray stars were discussed in Rome at the IAU Symposium no. 37 organized by Gratton and the Dudley Observatory (ref. *Rome*). A Conference on pulsars was held in New York City on May 20–21, 1968 (*Sky Telesc.*, **36**, 4). A three-day international symposium on low-luminosity stars was held in March at the University of Virginia, Charlottesville, where very young stars and flare stars were the subject of several papers. Several papers on supernovae, pulsars and

X-ray objects were read at the Eleventh International Conference on Cosmic Rays in Budapest, August 1969.

3. CATALOGUES, CHARTS, BOOKS AND INFORMATION BULLETIN

The Moscow team headed by B. V. Kukarkin and P. N. Holopov prepared for publication the third Edition of the *General Catalogue of Variable Stars* containing information on 20437 variables discovered and designated up to 1968. The increased number of data necessitated to publish the Catalogue in two volumes. The first volume containing the list of references for the whole catalogue and the variables in the constellations Andromeda to Grus was distributed in December, 1969.

An Identification List of 1648 new variables nominated in 1968 has been published by the Commission for Variable Stars of the Astronomical Council of the Academy of Sciences in the U.S.S.R. in *IB*, 311.

Nikolov constructed a *Catalogue of the Light and Colour Curves of Cepheids in the U, B, V system* containing about 300 stars. The catalogue was published by the Bulgarian Academy of Sciences in 1968. The possibilities for editing a similar catalogue of RR Lyr stars are under consideration between the Sternberg Institute at Moscow University, the Department of Astronomy at Sofia University and Odessa Observatory. Fernie and Hube published a Catalogue of Fundamental Data for 362 Classical Cepheids in the Galaxy (*AJ*, **73**, 492, 1968). Maffei intends in collaboration with Blaghikh to compile a catalogue of irregular variables. A catalogue of suspected symbiotic objects (mainly southern) is being prepared by Miss Nordström (Stockholm).

H. Huth has so far prepared about 10% of the material for the new edition of the *Geschichte und Literatur des Lichtwechsels* (*GuL*), begun by the late Dr Schneller. Printing will be started as soon as the discussions on the intended international co-operation and the kind of the publication (whether volumes, booklets or cards) have been closed.

Bateson, A. F. Jones and Stranson of the VS Section, RAS of New Zealand are editing charts for southern variables. Ultimately, the complete series, totalling ten, will cover all variables other than those of short period classes south of declination S. 30° and brighter than visual magnitude 13 at maximum. They hope to complete the work before the XIVth General Assembly. Cesevič together with his collaborators prepares the *Atlas of Finding-Charts for Variable Stars*. Four volumes containing charts for 5000 stars have been completed.

The proceedings of the Budapest Colloquium were printed at Academic Press Budapest (ref. *Budapest*), edited by Detre and sold by Reidel Publishing Company as co-publisher. The proceedings of the two Trieste Colloquia were edited by Mrs Hack (ref. *Trieste*, 1966 and *Trieste*, 1968, the latter is Vol. 13 of Astrophysics and Space Science Library). The Symposium volume of IAU Symposium no. 30 on *Radial Velocities*, held in Toronto in June, has also been published containing one section on radial velocities of variable stars (ref. *Toronto*). The symposium volume *Low Luminosity Stars* was edited by Kumar (Gordon and Breach, London, 1969).

A series of monographs dedicated to variable stars will soon be published in the U.S.S.R. The first volume of *Pulsating Stars* (Ed. B. V. Kukarkin) and the second volume of *Eruptive Stars* (Ed. A. A. Bojarčuk, R. E. Gershberg, Crimea) will appear in 1970. V. B. Nikonov's *Methods for the Study of Variables and Non-Stationary Objects and Stellar Evolution* are under preparation.

51 articles on pulsars, which appeared in *Nature* till about September 1968 were collected in a volume entitled *Pulsating Stars* and edited by Macmillan with introductions by Professor F. G. Smith and Dr A. Hewish. A very useful popular book on variable stars was written by Glasby. As more than a quarter of all the studies in the field of astronomy is directly or indirectly connected with the problem of stellar variability, many other books contain topics relevant to our field.

The issue of special magazines (cf. *Peremennje Zvjozdy*) and circulars in the domain of variable star investigations was continued. At the end of 1969 the fourth volume of the *Information Bulletin on Variable Stars* of Commission 27 was completed, each volume containing 100 numbers, edited by Detre and Szeidl at the Konkoly Observatory, Budapest. The support of the *Bulletins* has been guaranteed by the Hungarian Academy of Sciences for a further 3-yr period. In order to avoid unnecessary duplication of work, the *Bulletins* should more often bring information on variables which are being observed or which have been put on an observing program, creating in this way a real communication centre. The Sonneberg Observatory provides further for the edition of the *Mitteilungen über Veränderliche Sterne* (*MVS*).

4. GENERAL PROBLEMS, DEVELOPMENTS AND TRENDS IN VARIABLE STAR RESEARCH

The most significant new development in the field of variable stars was the discovery of pulsars early in 1968 by Miss Bell and Dr Hewish, and especially the optical observation of NP0532, the pulsar in the Crab Nebula by Cocke, Disney and Taylor at Steward Observatory on January 16, 1969, and the identification a few days later by Lynds, Maran and Trumbo at Kitt Peak NO as the south preceding star of the pair near the centre of the nebula. There is agreement that pulsars are neutron stars, and in this way Zwicky's 1933 hypothesis of some supernovae being caused through the collapse into neutron stars seems to have found its confirmation. Yet, there is some controversy about whether the observed periods are rotation periods of magnetic neutron stars (Gold's hypothesis), or pulsation periods of neutron stars with small mass ($\sim 0{\cdot}1 \odot$, Saakian, Vartanian, *Astrofizika*).

The optical identification of the Crab Pulsar was made possible by the use of the new techniques of synchronous signal averaging first developed by Clark (W. A. *MIT Res. Lab. El. Qu. Progr. Rep.* 114, 1968) in order to reduce the effects of the sporadic variations in light common to old novae and similar variables. Using these techniques, the search for optical objects associated with other pulsating radio sources has been continued by Kristian at Mt. Wilson to look for pulsations at the known radio frequencies. No pulsing optical objects have been found in the fields of 11 pulsars, with limits ranging from 20th to almost 25th magnitude. The Crab Pulsar seems to be of exceptionally high absolute luminosity compared with other known pulsars.

In the discovery and observation of pulsars high time resolution techniques are of paramount importance. Application of these techniques to other types of variables, to old novae, X-ray stars, symbiotic variables, also led to interesting results. In addition, high time resolution observation may reveal short time variation in properties of stars previously thought to be stable. In April 1967, Westphal and Sandage began an experimental monitoring of the optical flux of Sco X-1 with time resolutions as small as 2 s. At the Kenneth Mees Observatory, Savedoff, in collaboration with Dutton, Sturch, Rosenberg and Hamann, has assembled equipment which permits the rapid digital recording of photoelectric signals. At Victoria, Hutchings has obtained photoelectric observations of line profiles in B and Be Stars with high time resolution (*Budapest*, p. 191).

A program is being carried out by Hesser, Lawrence and Ostriker (*ApJ*, **148**, L161, 1967; **153**, L151, 1968; **155**, 919, 1969) for acquiring long ($\sim$ 7200 s) continuous, digitized records of the apparent luminosity of various stars in the lower lefthand portion of the HRD. These records are then searched for low-amplitude, periodic variability using modern autocorrelation and power-spectrum techniques. Various noise statistics are also developed which help to distinguish objects possessing intrinsic but non-coherent variability from either quiet objects or objects exhibiting periodic behaviour superimposed upon an otherwise quiescent radiation field. About application of power-spectrum techniques to irregular variables see Plageman, *Budapest*, p. 21.

Sometimes the components of close visual binaries are variable stars. Till now, it was impossible to get accurate photoelectric magnitude differences and colours for the components of very close double stars. A new instrument for variable star work is the photometer developed by Rakos and called *Area scanner* (*Appl. Opt.* **4**, no. 11, *Acta phys. Austria*, XXV, no. 4, 1967; Franz, *Lowell Bull.* 134; J. S. Hall, *ibid.*, 143; A. Elvius, *ibid.*, 142; Boyce, *AJ*, **71**, 847). It enables accurate photoelectric photometry of close binaries with a separation as small as one second of arc. The apparatus is very suitable also for polarimetric work. Complete equipments can be ordered at the Austrian company A. Pear KG., Graz.

Bakos, at the University of Wisconsin, is examining the properties of image orthycons and their use in variable star work. He is now processing the photographic material obtained by him by means of an image orthycon system at Dearborn Observatory (see Appendix **II**, 2b). The television equipment employed at the Crimean Astrophysical Observatory was described by Abramenko and Prokofjeva (*Budapest*, p. 57).

Mt. Cuba Observatory recently initiated a program of photoelectric photometry including differential photometry of variable stars and monitoring of flare-stars using a 70 Å-wide filter centred on the Ca II K line. This is very effective in reducing the sky background more than the signal from flare stars which are characteristically emission-line objects. Ultraviolet monitoring can then be carried further into bright moon periods (R. D. Herr).

A photometer for simultaneous stellar photometry in three colours has been constructed in the laboratories of the Catania Observatory (*Budapest*, p. 51), where four instruments are available for variable star work. An apparatus for the automatic converting of analogic data, given by the photometer, into digital data on punched cards is under construction. Similar equipments are already in use or in construction at several observatories.

An increasing convergence is apparent between the fields of stellar physics and solar physics stellar analogues of solar phenomena are becoming the subjects of specific researches (*Budapest*, pp. 10–12). The example of flare stars is well known. The spectral study of chromospheres of dwarf M and K stars has been fruitful, and efforts are under way to detect the stellar analogues of the solar magnetic cycle. It would be extremely valuable if sunlike activity were discovered in other stars, not only because it might explain the variability of the Sun but also because it should help to answer basic questions about the physics of a hot magnetic plasma. Since 1967 a wide program on stellar activity of solar type was undertaken at Catania (see Godoli *Atti dell'XI. Convegno, Padova*, p. 224, 1967; *Catania Pubbl.* 118, *Nobel Symposium* no. 9, 221, 1968), observing photoelectrically α Boo, α Tau, HD 119850, that show variable H_2 and K_2 emission components. Also the magnetic stars 21 Com and 20 χ Ser, and probably the flare star BD + 51° 2402 belong to this class. Photoelectric flux measurements at the center of stellar H and K lines with the coudé scanner of the 100-inch telescope at Mt. Wilson and its associated pulse-counting equipment started in 1967 in a search for stellar analogues of the solar cycle (O. C. Wilson). A number of main-sequence stars are being followed, covering the spectral-type range F 5 to M 2. The fluxes at the center of the H and K lines of Ca II are measured as frequently as possible in order to detect changes in the chromospheric components of the radiation. A preliminary analysis of the observations shows that nearly all stars whose spectra reveal H and K reversals on 10 Å/mm spectrograms have undergone some changes in chromospheric emission. Liller (*ApJ*, **151**, 589, 1968) has found short term K2 intensity changes of about 20% in α Tau. No periods can be determined yet. In 61 Cyg A *and* B the chromospheric fluxes are varying in a roughly cyclical manner, but these cycles may be the rotation periods of the stars. It is hoped to extend this investigation soon to giants and supergiants. At Washburn Observatory Ch. Anderson expects to utilize the échelle spectrometer to monitor the Ca II K emission in several late-type stars in an attempt to detect solar-cycle-like variations either through changes in the integrated emission flux or through variations in the frequency of K-line flare activity. Boesgaard (*PASP*, **81**, 283, 1969) called attention to the presence of fairly strong Ca II emissions at H and K in the spectrum of the M3S star 4 Ori = HD 30959 which were absent some time ago (Warner, *MN*, **129**, 263, 1965). This evolved star thus shows a substantial variation in the intensity of Ca-emission presumably due to chromospheric changes.

A similarity to the 5-min oscillation of the Sun has been noted by H. John Wood (*ApJ*, **152**, 117, 1968). He developed the use of photoelectric narrow-band photometers for measurements of Balmer-line variations (line-curves, *Leander McC Publ.* **15**, Part II, 1965). Photoelectric line curve observations of HR 9080 show time variations of about 10% in the effective equivalent width of Hβ, Hγ, and Hδ. Quasiperiods of about 35 min are found. The oscillations become unrecognizable after several cycles.

Although the K-lines are the principal features in late-type spectra that have been related to stellar chromospheres, the Mg II lines at 2974 Å and the hydrogen $L\alpha$- and $L\beta$-lines have been found to show similar structure. Zirin has observed the He I line at 10830 Å with an infrared image converter at the 200″ coudé spectrograph and found emission in β Dra and in ε Gem; in the latter star emission was not present a few years ago.

Another possibility to discover sunlike activity in stars may be the observation of emission lines of highly ionized atoms in their spectra, that might reveal the existence of stellar coronas. On the average the intensity of these emissions lines should go up and down in unison with the star-spot cycle in a period not unlike that of the solar cycle although the length of the period might be quite different. Longterm variations in the statistical parameters of irregular variables showing flare activity (T Tau stars, flare stars, X-ray stars, old novae), as found for RW Aur stars by Cesevič and Dragomineskaja (*Sky Tel.*, **34**, 366, 1967), might be the manifestations of activity cycles.

The length of the period of some magnetic variable stars seems to change periodically. Assuming that the stellar surface has some kind of bright or dark magnetic spots and the magnetic and light variations are apparently induced by stellar rotation (oblique rotator hypothesis), the change of period can be explained by spots moving from high latitudes towards the equator or from the equator toward high latitudes. A similar effect is caused by the sunspots during the 11 year solar cycle (Rakos, *PASP*, **80**, 563, 1968). To find the first measurable connection between stellar and solar magnetic fields, Albrecht, Maitzer and Rakos (*AA*, **3**, 236, 1969) succeeded to prove that the sun varies its brightness synchronously with its rotation. The question was investigated using the most extended photometric observations of very high precision carried out at Lowell Observatory during the period 1950–1966 (see e.g. *Lowell Bull.*, 137). They also proved the dependence of the period of the light variation upon the solar cycle. As the amplitude of the variations is very small ($<0{\cdot}01$ magnitude), a continuation of the observations is highly desirable.

Be stars and shell stars exhibit cycle of activity, that may be of magnetic origin. Crampin and Hoyle's (*MN*, **120**, 33, 1960) idea, that a weak initial magnetic field should be amplified by the differential rotation to a maximum value, until an explosive dissipation of the material in the extended atmosphere occurs, was recently discussed and compared with observations by Henriksen (*AA*, **1**, 457, 1969).

Of considerable importance is Eggen's effort to construct a classification system for intrinsic variables based on our present knowledge of the evolutionary and dynamical history of galactic stars. He has given an excellent review of the problem (*Sigma Reviews prepr.*). He makes a basic subdivision on the basis of the region occupied by the variables in the luminosity-temperature plane; a second parameter in his system of classification is the star's age. He uses for the recognition of disk populations data on space motion, ultraviolet and infrared radiations, TiO absorption measured by the colour (**65**, **62**) i.e. the magnitude difference between the bandpasses of narrow band filter at 6250 Å and 6500 Å.

F. W. Wright and P. W. Hodge (*PASP*, **81**, 238, 1969) called attention to HV13055 in the LMC which has a very unusual light curve. We do not know of any other variable star with properties like those of HV13055 and do not know to what class of variables it belongs.

In a manuscript by G. A. Richter (*AN*) problems with regard to the determination of the probability of discovery are discussed. Another publication concerns a general error analysis of the method used to estimate the number of variable stars (together with L. Winkler, of Pennsylvania State University, *MVS*, **4**, 143, 1968).

5. THEORY OF VARIABLE STARS

(by R. Kippenhahn)

After Christy's pioneering work on the theory of *non-linear radial pulsation* one is puzzled with the discrepancy between the masses of cepheids derived from normal stellar evolution theory and the masses determined from the "second bumps" in the light curves derived from the nonlinear pulsation theory (Christy, *ESO*). The recent work by Stobie (*MN*, **144**, 485, 511, 1969) using a similar code confirmed Christy's results, suggesting that stars have undergone a considerable mass loss before they became cepheids. Baker and von Sengbusch (*AG. Mitt.*, **27**, 1969) now are investigating nonlinear radial pulsations with a new numerical method.

The cepheids are still the only pulsating variables where a combined approach of stellar evolution theory and pulsation theory has been carried out. The RR Lyrae variables probably would give good tests for the different ideas of the horizontal branch stars. δ Scuti stars and dwarf cepheids probably offer similar opportunities to get information by comparing light curves with nonlinear pulsation theory.

Work is going on at different places on the theory of Mira type variables, but the work suffers from the insufficient knowledge about the interaction between pulsation and convection.

The question whether the vibrational instability of main sequence stars above 60–90 solar masses can cause the *P Cyg phenomenon* has recently been discussed by Appenzeller (*AG Mitt.*, **27**, 1969) by the nonlinear pulsation theory for a star of 130 solar masses driven by the fluctuating nuclear energy output.

The cause of the β CMa pulsation still seems to be unknown. Two possible mechanisms have been discussed recently:

(a) Stothers and Simon (*ApJ*, **157**, 673, 1969) suggest that a helium enriched layer in the outer part of an otherwise normal (hydrogen rich) envelope changes the pulsational properties of the stars in such a way that the nuclear energy generation produces a vibrational instability. In order to obtain stars of this type they have to assume that all β CMa stars are members of close binary systems and that during the mass exchange a considerable amount of helium enriched material is transferred to the companion. Then for the helium enriched companion the mass critical for vibrational instability (caused by fluctuation of the nuclear sources) is reduced from the well-known 60 solar masses for a normal population I mixture down to the mass range of β CMa stars. In contradiction to the last assumption, in all cases of mass exchange in close binary systems computed up to now never a considerable amount of the helium enriched material is transferred, but mass exchange stops as soon as the top of the helium enriched core approaches the surface of the star which looses mass.

(b) Stars above 10 solar masses form a semi-convective region during the exhaustion of their central hydrogen. At the transition zone a region can be formed which is dynamically stable (i.e. the Ledoux criterion is fulfilled) but vibrationally unstable against nonspherical oscillations (i.e. the classical Schwarzschild criterion is violated) (see Gabriel, *AA*, **1**, 321, 1969). In these layers oscillations could occur and – since the mass range and the evolutionary state fit well to the β CMa stars – the phenomenon could be connected with these variables. Unfortunately, it seems that the region which is able to drive the nonspherical pulsation contains only little mass and probably will not be able to overcome the damping in the rest of the star.

For *late type stars* an appreciable part of the star's mass is in the state of transient ionization reducing the $\gamma = C_p/C_v$ below $\frac{4}{3}$ in a large part of the star and making the star eventually dynamically unstable. This mechanism was used independently by Lucy and Paczynski to explain the formation of planetary nebulae. Up to now all efforts to follow up the evolution of stars into this instability failed and it might be that during their red giant and supergiant phases stars in the HRD never penetrate into the region of instability caused by ionization. But in the case of mass exchange in close binary systems the ionization effect becomes important. If a red giant star fills its critical volume and mass exchange sets in, the outburst becomes rather violent. This mechanism recently was used as an explanation of *novae and U Gem variables* by Bath (1969, reprint).

The Loreta-hypothesis that the light variation of *R CrB* stars is due to obscuration by a soot cloud is worth being discussed again since the amount of radiation missing in the optical range if the star is not of normal brightness has been discovered as black body radiation of about 1000° in the infrared by Stein, Gaustadt, Gillett and Kancke (*ApJ*, **155**, L3, 1969). A similar effect has been found for RY Sgr by Lee and Feast (*ApJ*, **157**, L173, 1969).

6. PHOTOGRAPHIC SURVEYS, SPECIAL FIELDS

The systematic photographic search for new variables inaugurated in 1959 by Strohmeier at Remeis Observatory in Bamberg, and extended to the southern heavens in 1964, is continuing highly successfully, recently in cooperation with Professor F. B. Wood. In addition to the set of cameras working at the Boyden Observatory there is now another in New Zealand, at Mount John University Observatory, Lake Tekapo. A new, six-camera unit will soon be operated at La Plata, Argentina. All the cameras are fitted with inexpensive $f/6$ Tessar lenses of 4″ aperture, and cover a 13-by-13 degree field. The number of new Bamberg variable stars (*BV*) now exceeds 1200. For many of *BV* additional observations back to about the year 1900 have been made using the plate collections at Harvard Observatory and at Sonneberg, GDR. The results are announced from time to time in the *IB*. From December, 1969, U.S. participants from the University of Florida, Gatesburg will supervise the program at the New Zealand station and probably help at Bamberg in plate measurements.

Hundreds of plates have been taken in fields in the southern sky with the Metcalf 10″ telescope at Boyden Station by van Hoof. Tens of new variables have already been discovered.

The search in selected fields for variable stars, begun by C. Hoffmeister, is continued by G. A. Richter at Sonneberg, in order to obtain further material for such statistical investigations as derived from previous material and published in *Ver. Sternw. Sonneberg* **7**, no. 3, 1968. Lists of recently discovered variables and also their newly derived elements are published continuously in the *MVS*. Hoffmeister published 223 new variables in 19 fields (*AN*, **290**, 277, 1968). The elements of light variations of variables near the galactic pole have been published by L. Meinunger and Wenzel in *Ver. Sternw. Sonneberg*, **7**, no. 4, 196, those in areas around ν Gem, δ Aql, ϱ Cyg and ε Pav by Gessner (*ibid.* **7**, no. 5, 1970; *MVS*, **5**, 96, 1969), and those around α Pav by I. Meinunger (*MVS*, **5**, 1969). Ahnert using the 20/30/30-cm Schmidt-cameras derives normal maxima for Cepheids in Cas.

The processing and publishing of faint variable stars in the Cygnus Cloud and adjoining areas in Lyr and Vul have been steadily continued at the Fordham University Astronomical Laboratory. Since the 1967 Draft Report ten more brochures have appeared on 88 Cyg Cloud Variables and 12 variables in Lyr and Cyg (VV 151–250, *Ric. astr.*, **7**, 197, 217, 241, 261, 315, 411, 435, 459, 499). At least 150 additional Cyg Cloud variable stars remain to be studied at the FUA Laboratory, and they will be published in future issues of *Ric. astr.* Accurate positions and comparison star sequences have already been measured for virtually all of these variables by Father Miller.

At his request, in July 1969 Father O'Connell, Director of the Vatican Observatory, loaned 438 plates centered on BD $+53°3167$ to Dr Wachmann of Hamburg Observatory.

Dr Wachmann will process for future publication about 150 variable stars discovered in this part of the Cep-Lac-Cyg star clouds, taken most of these by Miller on the Zeiss 16″ quadruplet astrograph at Castel Gandolfo. The variables will be published under joint authorship in *Ric. astr.*

Romano at Treviso has continued the photographic survey of variable stars using Asiago Schmidt plates. In particular, he has examined the fields of ϱ Cyg (*MAI*, **38**, 138, 1967), γ Cyg and δ Com (*ibid.*, **39**, 236, 1968 = *Pubbl. Padova*, no. 149), γ UMa, 89 Psc and γ Com (*ibid.*, **38**, 621, 1967), finding new variable stars and giving improved elements and magnitudes for those already known. Results of the photographic observations of 27 variable stars in two fields around o Per and k And are reported in *MAI*, **40**, 261, 1969. Researches on the following fields: η Cyg, γ Lyr, $19^h04^m +43°$, are in progress.

Maffei at Frascati started a search for infrared variable stars in 6 fields, using the Asiago Schmidt telescopes. Preliminary work shows these variables have in general long periods. The collection of plates will be continued another year at least.

The Palomar-Groningen search for faint variable stars in four selected regions (see *Trans. IAU*, **XIIA**, p. 361) is now approaching its conclusion. Results on Fields 1 and 2 are published in the *BAN Suppl.* **1**, 105, 1966; **2**, 293; **3**, 1, 1968. The data for Field 3 are now ready for discussion and will be published in 1970. Field 4 is in the hands of de Kort at Nijmegen. The study of magnitude and spectral variability of giant M stars in Field 2 and in VSF 193 Sgr by Miss Houk (*Trans. IAU*, **XIIIA**, p. 544) was published as a thesis (1967).

Kooreman discussed 80 variable stars in the region of CPD-31°5547 (*BAN Suppl.*, **3**, 41, 1968). Oosterhoff and Ponsen published an investigation of 96 variables in the galactic window 17^h53^m, $-29°02'$, 1900 in Sgr (*ibid.*, **3**, 79, 1968).

Landolt is studying two regions of the dark nebulosities in Tau (4^h05^m, $+25\overset{\circ}{.}4$; 4^h30^m, $+24\overset{\circ}{.}8$) using photographic plates extending over a 7-yr interval and obtained at Dyer Observatory's 24″ reflector-corrector camera. *UBV* photoelectric sequences reaching to $V = 16{\cdot}8$ were published in *AJ*, **72**, 1012, 1967.

Geyer started a study of variables at $l^{II} = 180°$, $b^{II} = \pm 60°, \pm 30°, \pm 10°$ with the Bruce 40-cm astrograph at Heidelberg. Gieseking at Observatorium Hoher List uses a 30 cm *f*/5 Sonnefeld-Vierlinser for searching variables in $6° \times 6°$ fields around the associations Cep OB2, Cyg OB9, Lac OB1, Per OB2.

At the Harvard Observatory Miss Harwood and her assistant, Mrs Swam, continue to investigate the variable stars in the Scutum Cloud, using plates made with the Bruce 24″ refractor. Estimates of magnitudes have been made of 316 stars. This number includes the remainder of the new variables discovered at the Maria Mitchell Observatory by Heath. The stars having variations of long period are being edited and prepared for publication. 174 stars, which have short period variations, are being measured also on a series of 50 plates made with the 48″ Palomar-Schmidt telescope by Sandage and loaned recently to Harvard. These plates, of 10^m exposures, were taken throughout each of the five nights of June 20–24, 1955. From these results they hope to derive periods of many RR Lyr stars. Their future program will continue the variables in the Scutum Cloud until all variables old and new, for which type and period are not known, have been investigated.

At the Maria Mitchell Observatory Miss Hoffleit and her associates published several notes on new variables in Sgr (*AJ*, **72**, 711, 1967; **74**, 718, 1969; *Irish AJ*, **8**, 149, 1968, *IB*, 254, 277, 312). Identification charts have been prepared for the majority of recently published and unpublished variables in VSF 193. It is expected that they will publish them in 1970. After the preliminary results have been obtained for all of the new variables, they plan to bring the observations of selected interesting variable stars in this field up to date, and eventually do the same kind of work in Cyg and Com.

L. Evans has begun at Pretoria a new photographic search for red variables in three fields near the galactic centre. At the Konkoly Observatory Jankovics has started a search for faint variables in a region of Cep (22^h6^m, $+52\overset{\circ}{.}2$, 1950) using plates obtained at the 60/90/180 cm Schmidt-camera. The magnitude limit is about 19. The Auckland members of the Var. Star Section, RAS of New Zealand, are photographing with a 20″ reflector variable star fields, especially for recurrent novae.

The photographic program of Weber at Mainterne was continued until his death on 27 July 1969. His last list of new variable stars appeared in *IB*, 321, bringing up the total number of Wr variables to 188. He was one of the most successful private astronomers in our field. Beside his discoveries, he has published many interesting notes on variables of special interest.

A recent search for bright variable stars in M 31 has been made by Moffat on 52″ Tautenburgh Schmidt plates obtained by van den Bergh (*AJ*, **72**, 1356, 1967). Five new variables have been found in M 31 and two in the Galaxy in the direction of M 31. Van Agt discovered 92 variables in the UMa dwarf galaxy on plates obtained by Baade 1953–58 (*BAN*, **19**, 275, 1967).

Since 1895 till 1950 the Harvard College Observatory searched variable stars in the Magellanic Clouds, finding more than 3000 variables to which 150 were added during the investigation from 1962 performed by the Gaposhkins. They thoroughly sieved all announced variables by making more than two million photographic and more than 10000 photovisual observations on some 2000 plates. The final results on the distribution of eleven classes of variability in the two nearest galaxies are given in *IB*, 420.

7. VISUAL OBSERVATIONS

The Variable Star Section, RAS of New Zealand, directed by Bateson, has remained very active and continues to observe southern variables, especially of U Gem, T Tau, recurrent novae, R CrB, UV Cet types as well as the Mira and semiregular variables. Several groups within the Section are conducting researches into specialized fields. A list of variables under observation was published in *IB*, 236. Members of the Section now operate several photoelectric photometers thus determining accurate sequence magnitudes for comparison stars for southern variables. A report has appeared in *Southern Sky*, **23**, 47, 1969.

Acting on suggestion made by our Commission at the last IAU General Assembly, Glasby, Director of the Variable Star Section of the British Astronomical Association extended his program to include more U Gem, flare stars and T Tau variables. Mrs Mayall, Director of the American Association of Variable Star Observers, continues her monthly articles in *J. R. astr. Soc. Canada*. Members of the Variable Star Section of the Swedish Astronomical Society have made visual observations of novae, U Gem stars and long period variables. A local amateur group in Malmö, South Sweden, is also active in this field. The reports of the Variable Star Section of the Danish Astronomical Society appear in *Nord. Astr. Tidsskr.*

The late A. W. Roberts made during two decades with the year 1900 as a "mean epoch" about 75000 observations of southern variables. The observations of six stars (R Ara, U, X and 1 Car, S Mus and RS Pup) were placed at Nielsen's disposal by the Council of the Boyden Observatory where the observations are now kept. The final discussion of about 5000 observations is proceeding.

Beyer in Hamburg continues observing 16 interesting objects at 26 cm refractor. He publishes his work from time to time in *AN*. Peltier at Delphos, Ohio, with his assistant Mrs Hurless concentrated their efforts on faint variables of irregular types and on M-type stars in their faint stages, thus about 80 per cent of their total estimates range from 14th to 16th magnitudes.

At the statutory meeting of the new International Union of Amateur Astronomers (IUAA) at Bologna in Italy April 19–21, 1969, it was decided that the IUAA will mainly be concerned with variable stars.

8. RR LYRAE VARIABLES

Considerable work is directed toward the determination of the absolute magnitudes of the RR Lyr variables. It is clear that only a large-scale concerted effort on proper motions, radial velocities and accurate photometry will yield the desired increase in precision.

Clube, Jones and Evans (*Mem. RAS*, **72**, 101, 1969) presented photometric and spectroscopic data for 62 southern RR Lyr variables brighter than $B = 14$ at minimum, observed at the Cape and Radcliffe observatories. The work may be regarded as an extension to Kinman's paper (*ROB*, **37**, 1961). The immediate aims of the investigation have however been to provide sufficient data for each star to assist in population classification and to determine the statistical parallaxes for pulsating stars with periods less than one day broken down by population groups. The absolute proper motions of most of these stars have also been determined and are being published in *ROB* (Clube I *ROB*, **136**; II in press). The photometry was secured at the Cape with the 24-inch Victoria refractor at first, and then from mid 1966, with the 40-inch reflector. The spectra were obtained at the 74-inch Radcliffe reflector. For the most part they have succeeded in deriving period, brightness in three colours, mean radial velocity, spectral type from Ca II and hydrogen lines, producing an estimate of the Preston-index ΔS.

Members of the Greenwich Observatory have obtained radial velocities for a number of RR Lyr stars at Kottamia. Proper motions for over 80 southern RR Lyraes have been determined.

The Leiden Observatory is engaged in determining or improving upon proper motions of RR Lyr variables. Van Herk is entangled in an investigation of 18 RR Lyr variables based on five-colour photoelectric photometry made by van Genderen, Pel, Velthuyse and Wamsteker in 1966–68 at the Leiden Southern Station in S.A. The investigation is aimed to find a correlation between ΔS and the colours of the variables. This work is still in progress.

By summer 1970 Fitch and his collaborators should have finished the observational phase of the *UBV* photometry program on about 170 RR Lyrae field stars. Fitch hopes that new, accurate photometry combined with improved proper motion data could yield a significant improvement in the absolute magnitude calibration.

At Torun, Burnicki (*AcA*, **17**, no. 2, 1967) made a discussion of colour indices of RR Lyr stars with respect to the population type, using photographic observations made at Torun and photoelectric observations of 74 variables compiled from published data. After correcting for interstellar reddening, he found in the colour-period diagram a shift to the red by $0^{m}11$ of variables assigned to pop I, what can be accounted for by differential line-absorption. The same stars are shifted in the light amplitude-period diagram towards lower amplitude values.

At the Uttar Pradesh Observatory *UBV* observations on AC And, RZ Cep, SV Eri, SS Leo, V 986 Oph, KN Per, RU Psc have been completed by Joshi and Mahra. Observations on AE Boo are in progress. At the Skalnate Pleso Observatory, Tremko is observing W CVn, DM Cyg, TT Lyn and VY Ser. Photoelectric observations of RR Lyr variables with the 16-inch telescope of Edinburgh Observatory have been continued by M. J. Smyth and Sherwood, and some results are being made ready for publication. Stepien is going to reduce and interpret the *UBV* observations of about 30 RR Lyr stars. At the Mees Observatory, photometry of bright RR Lyr stars on the Strömgen *u*, *v*, *b*, *y* system is being pursued by Sturch to try to distinguish surface gravity and metal abundance differences. A program of 4-colour narrow-band photometry is in progress at Pretoria. Identifications and spectral types of stars used as comparison for field RR Lyr stars were published by Kim and Sturch (*PASP*, **79**, 72, 1967). Onderlička and Vetesnik of the Astronomical Institute Brno published (*Mem. Czech. AS*, **13**, 1968) 4106 observations on RR Lyr obtained 1961 Aug.–1963 July with a 24-inch telescope at $\lambda_{\text{eff}} = 4420$ Å. Epstein, Columbia University New York, obtained 4-colour (*u*, *v*, *b*, *y*) photometry at the McDonald, Kitt Peak, and Cerro Tololo Observatories, 1966–68, on 35 RR Lyrae stars (*AJ*, **74**, 1131, 1969). Hill (*PASP*, **80**, **309**, **1968**) published *UBV* observations for ST CVn and VV Vir. At the Observatory Cluj (Roumania) neglected, mostly faint RR Lyr variables are being observed photographically. Todoran published his observations of SX Cnc, TX Com, and XY Eri in *Stud. Astr. Bucuresti* and is continuing the observations of the following RR Lyraes: RU, XX, WW, SV, WZ, Boo, UX, UU Cet, RT Com, AC, BE, BG, **BK**, UZ Eri, SW Psc.

RR Lyr stars are intensively observed at Odessa and Rostov, mostly visually. Elements of variability are regularly revised at Odessa. On the basis of these data new ephemerides are calculated and edited at Cracow (*Roczn. astr.*).

In the investigations mentioned, detailed consideration of secondary periods and period changes was mostly avoided, and have been published, what are effectively time averages for each cycle over several years. At the Konkoly Observatory, the processing of about 100000 photographic and 120000 photoelectric observations obtained since 1934 for the study of the Blashko-effect and construction of O-C diagrams should be completed in 1970 for 39 RR Lyrae stars. Szeidl has found the secondary period of 89^{d} for TT Cnc, and 105^{d} for AR Ser. He studied the variations in the Blashko-effect of RR Lyr (*Budapest*, p. 17). Kanyó has finished a photometric study of RV UMa (*Bp. Mitt.*, 63) complementing Preston and Spinrad's (*ApJ*, **147**, 1025, 1967) study of the same star. RW Dra shows 9–10 year cycles in the variation of the amplitude and period of the Blashko effect (Detre, *Wien, Ann.*), which are correlated with changes of the fundamental period. Similar cycles are apparent in RR Lyr and XZ Dra. Also the magnetic behaviour of RR Lyr is subjected to considerable changes, as Preston has not found a measurable magnetic field in 1963 and 1964 (*The Magnetic and Related Stars*, p. 3), whilst Babcock had observed a strong and variable magnetic field about ten years earlier. At the Observatory Cluj, the Blashko-effect in DP Aqr and WY Dra is being studied. Cesevič supposed that the effect appears due to superposition of two independent radial oscillations of the outer zones of the stars (*IB*).

The number of RR Lyr stars with known secondary period is still very small (see Table I prepared by Szeidl). A rapid increase of our knowledge of the Blashko-effect could only be achieved through a concentrated observational effort to determine secondary periods for RR Lyr stars in a few selected globular clusters containing many variables, particularly in ω Cen, M 3, M 5 and M 15. For the exact determination of O-C diagrams the availability of old plate material, especially that in the files of the Harvard and Mt. Wilson Observatories, would be very important. Robinson is now using the Harvard material for the determination of O-C diagrams of field RR Lyr stars.

479 faint RR Lyr variables were found and discussed by Plaut (*BAN Sup.*, **2**, 293, 1968) in Field 2 of the Palomar-Groningen survey at $l^{\text{II}} = 4°$, $b^{\text{II}} = +12°$ with regions of very heavy interstellar absorption. He attempted to use the material for a first determination of the space density

Table 1. RRab-Stars with Secondary Periods

Star	P	P'	ΔS	References
RS Boo	$0^d.377$	537^d	2	Oosterhoff, *BAN*, **10**, 101.
RR Gem	0·397	40:	3	Budapest, unpublished.
SW And	0·442	36·8	0	Balázs, Detre, *Bp. Mitt.*, **36**.
RW Dra	0·443	41·7	3	Balázs, Detre, *Bp. Mitt.*, **27**.
RV Cap	0·448	221·9:	6	Cesevič, *Sternberg Tr.*, **23**.
XZ Cyg	0·466	57	6	Muller, *BAN*, **12**, 11.
RV UMa	0·468	90·1	8	Preston, Spinrad, *ApJ*, **147**, 1025; Kanyó, *Bp. Mitt.*, **63**.
AR Her	0·470	31·6	6	Almár, *Bp. Mitt.* **51**.
XZ Dra	0·476	77	3	Balázs, Detre, *AN*, **271**, 231.
X Ret	0·492	45:	–	Hoffmeister, *Ver. Sonn.*, **5**, 3, 1.
V 674 Cen	0·494	29·5	–	Hoffmeister, *Ver. Sonn.*, **5**, 3, 1.
RZ Lyr	0·511	116·7	9	Romanov, *IB*, 205.
V 434 Her	0·514	26·1	–	Rozovsky, *PZ*, **15**, 211.
Y LMi	0·524	33·4	–	Martynov, *Eng. B.*, **18**.
V 30 in M 53	0·535	37·0	–	Wachmann, *Hamburg Abh.*, **8**, 114.
UV Oct	0·543	80:	–	Hoffmeister, *Ver. Sonn.*, **5**, 3, 1.
RW Cnc	0·547	29·9	–	Balázs, Detre, *Bp. Mitt.*, **23**.
TT Cnc	0·563	89	7	Szeidl, *IB*, 278.
RR Lyr	0·567	40·8	6	Preston, Smak, Paczynski, *ApJ Sup.*, **12**, 99.
AR Ser	0·575	105	8	Szeidl, *IB*, 220.
DL Her	0·592	33·6	–	Szeidl, *IB*, 36.
V 365 Her	0·613	40·6	–	Cesevič, *AZ*, **38**, 293.
AT And	0·617	82·7	3	Tchumak, *PZ*, **15**, 569.
Z CVn	0·654	22·7	8	Kanyó, *IB*, 146.

The stars are arranged according to the length of the fundamental period. In RR Gem the Blashko-effect ceased to exist since about 1940, and the same happened in SW And around 1956. Probably all RRc stars have, at least temporarily, secondary periods, but their length has so far been determined only for three: TV Boo, $P = 0^d.313$, $P' = 31^d.2$ (Detre, *Festschrift f. C. Hoffmeister*), RU Psc, $P = 0^d.390$, $P' = 28^d.8$ (Tremko, *Bp. Mitt.*, **55**), and V 68 in M 3, $P = 0^d.356$, $P' = 10^d.9$: (Szeidl, *Bp. Mitt.*, **58**, p. 74). The secondary periods are still unknown for DP Aqr, WY Dra, SZ Hyd, CZ Lac, BB Pup, V 703 Sco, AT Ser, all showing strong light curve variations.

of these stars adopting a mean absolute photographic magnitude $\langle M_{pg} \rangle = +1{\cdot}0$ and making use of Hidajat's extensive discussion (Cleveland thesis, 1965) of the interstellar absorption in this region.

An RR Lyr variable projected on the Andromeda Nebula was found by Moffat and van den Berg (*Zf. Ap.* **67**, 246, 1967) on plates obtained with the Tautenburg Schmidt telescope. Its distance is ~ 19 kpc. HD 176387 was shown by Przybylsky (*MN*, **136**, 185, 1967) to be a metal poor RRc star with retrograde orbit.

Rosino and Capelli (*Padova Publ.*, no. 144, 1968) have investigated the statistical properties of galactic RR Lyr stars from the point of view of the frequency distribution of periods and amplitudes, comparing them with the RR Lyraes of globular clusters. It has been confirmed that an important group of RR Lyraes of low galactic latitude have periods distributed around 0^d45. No counterpart of this group exists in the globular clusters of the Milky Way.

9. POPULATION I CEPHEIDS

The theoretical prediction of the cepheid instability strip near its low temperature boundary is complicated by the onset of convection, which dominates the heat transport in the surface layers that are responsible for the instability. Sandage and Tammann (*ApJ*, **155**, 531, 1969) approached the problem in an empirical manner. Using the observational data for the cepheids in galactic clusters, in the LMC and SMC, and in M 31 and NGC 6822, they constructed the ridgeline *P-L* relation (see Appendix II). The dispersion of the individual stars about this line is enclosed by parallel envelopes displaced by $\pm 0{\cdot}6$ magnitude. It seems certain that the known cepheids of pop I all lie in the instability strip (see Eggen's discussion in *Intrinsic Variable Stars, Sigma Reviews*), however, the question is outstanding, whether or not all stars in the instability strip are variable. Fernie has completed a program looking for low-amplitude pulsation in stars which have spectral types like those of cepheids. 49 supergiant having MK classification between F51b and G51b and not previously known to be variable have been monitored photoelectrically. Preliminary indications are that no more than two or three of these are variable by more than about 0·02 magnitude.

It is of considerable interest from the standpoint of the *P-L* relation to find classical cepheids of particularly short or long period which might have main-sequence companion associated with them, because there exist almost no direct sources of information on the luminosities of cepheids with periods less than 3^d or longer than 15^d. The distance of SU Cas ($P = 1^d.9$) has now been determined by Racine at Mt. Wilson using nearby field stars that, like the variable, are seen to illuminate reflection nebulae, and are presumably connected with the same dust cloud. Its absolute magnitude was found to be $\langle M_V \rangle - 2{\cdot}45 \pm 0{\cdot}2$ mag, in good agreement with predictions based on modern calibration of the *P-L* relation. Fernie has determined the absolute magnitude of 1 Car (*AJ*, **72**, 1327, 1967; $P = 35^d.6$), and Sandage and Tammann obtained separate light curves in two colours for the companions of CE Cas. 13 cepheids in the period-interval $1^d.95 - 41^d$ are now available, and they have rediscussed the *P-L-C* relation using the new data (see Appendix II).

Several authors have anew derived the *P-L* relation for cepheids in the Magellanic Clouds. In spite of the efforts of the past decade there is still no general agreement on some of the fundamental questions including the form of the *P-L* and *P*-colour relations in the two Clouds, and whether these relations differ from Cloud to Cloud or between the Clouds and the Galaxy. It is clear they cannot be answered without more accurate data for the shorter-period cepheids. To obtain such data, Gascoigne observed at Mt. Stromlo *B*, *V* light curves for 13 SMC and 7 LMC cepheids with periods in the range 1·24 to 6·69 d. (*MN*, **146**, 1, 1969). The following *P-L* and *P-C* relations were found

LMC $\langle V \rangle = 17{\cdot}58 - 3{\cdot}072 \log P$, $\langle B \rangle - \langle V \rangle = 0{\cdot}49 + 0{\cdot}191 \log P$
SMC $\langle V \rangle = 17{\cdot}73 - 2{\cdot}879 \log P$, $\langle B \rangle - \langle V \rangle = 0{\cdot}26 + 0{\cdot}306 \log P$

Wright and Hodge (*ApJ Sup.*, **17**, 467, 1968) used ADH Schmidt and Harvard Bruce plates for a section of the LMC. Work is nearing completion for a similar study, in the same region of 21 more variables, ten of which were recently discovered. Van Genderen (*BAN Sup.*, **3**, 221, 1969) finished his major work on two-colour photographic photometry of 105 cepheids near the centre of the SMC, based on plates taken by Arp. The variables appeared to fit well into the instability-gaps of the young SMC clusters NGC330 and NGC458. In a further comparative study between the Pop I cepheids of both the Magellanic Clouds, the Galaxy and the Andromeda Nebula, van Genderen (*BAN Sup.*, **3**, 299, 1969) found indications that the cepheids of the two Magellanic Clouds are more luminous than the galactic ones and those of M31. In another investigation based on five-colour photoelectric observations of 12 SMC cepheids he (*BAN*, **20**, 317, 1969) found that the intrinsic colours of these cepheids were bluer than those of the Galactic cepheids. He suggested that metal deficiency might be the cause (see also Christy, *ESO*).

Feinstein and Murzio (*AA*, **3**, 388, 1969) discussed the light curve of the brightest cepheid, 1 Car, from 178 photoelectric observations distributed over a 17 year interval. Beside the main period of $35^d.5330$ periodicities of $190^d.1$ and $29^d.92$ were found, both with an amplitude of $\sim 0^m.02$ in *V*. The main period shows no evolutionary change. Light curves in blue have been obtained at Kottamia (*ROB*, 15).

Eggen (*ApJ*, **156**, 617, 1969) obtained *UBV* photoelectric photometry of 18 cepheids in Cyg and 10 in Mon with the 200″, 100″ and 60″ Palomar-Mt. Wilson and 40″ Siding Spring telescopes. Takasa carried out *UBV* photoelectric photometry for 29 cepheids during 1965–68 with the Okayama Observatory's 91-cm reflector (*Tokyo Bull. II. ser.*, 191, 1969). Mavridis and Tsiomis started a new series of photoelectric three-colour observations of the 18 cepheids observed in 1955–58 at Heidelberg by Bahner and Mavridis. The aim of these observations, carried out with the 38-cm reflector of Stefanion Astronomical Station in Greece, is to get complete and accurate new light and colour curves for these stars. Abaffy at Konkoly Observatory observed in 1967 about 40 cepheids in *UBV*. Detre (*Wien Ann.*) used these observations together with those obtained by him in 1953–54 for a study of period changes. No evolutionary changes were found. Butenko investigated period variations of about 100 cepheids. Makarenko studied cepheids with small amplitude, and determined colour excess for 204 cepheids from *UBV* data. Nielsen at Aarhus is processing with the photoelectric *BV* observations of about ten mostly bright cepheids.

On the basis of his catalogue of light and colour curves, Nikolov derived the principal photometric characteristics of the light curves and colours of about 300 cepheids (*PZ*, **16**, 312, 1968). During 1969 the Department of Astronomy at the University of Sofia came to an agreement for cooperation with the Sternberg Institute for investigation of the morphological, spatial and evolutionary characteristics of cepheids. Nikolov and Kunchev discussed the *U-B* versus *B-V* diagrams (*Aph. Sp. Sc.*, **46**, 3; *Budapest*, p. 481, 1969). They attempted to obtain more precisely the reddening lines of cepheids using values of *U-B* and *B-V* for cepheids at maximum (*ApL*, **1**, 151, 1968).

Pel, Velthuyse and Wamsteker made five-colour photoelectric observations of BK Cen and QT CrA during their stay at the Leiden Southern Station in S.A. The reductions are still in progress.

For the determination of the absolute magnitude of Polaris new *B*, *V*, *B-Y*, *C*, *m* and Hβ photometry of αUMiB have been obtained by McNamara (*AJ*, **73**, S 106, 1968), carefully excluding the light of Polaris (αUMiA) in the measurements. If *A* and *B* form a physical system, the absolute magnitude of Polaris is $M_v = -3{\cdot}7$, a value somewhat brighter than predicted by the *P-L* relation. At Georgetown College Observatory, Kessler is starting observations on several cepheids in *B*, *V*, Hα and Hβ with a new photometer.

Wisniewski and Johnson (*Comm. Lunar Planet. Lab.*, **112**, 1968) have made multicolour photoelectric observations on the *UBVRIJKL* system for 20 cepheids. The data are sufficient to define light curves. These exhibit the well-known shift of phase with wavelength. The combination of the data with the known radial velocity curves makes possible the computation of the absolute magnitude for each of these cepheids entirely empirically which does not depend upon stellar model computations. Eggen is making *UBVRI* photometry of cepheids at Mt. Stromlo in both the disk and halo population.

A critical study on Wesselink's method for determining radii of pulsating stars was published by Fernie and Hube (*PASP*, **79**, 95, 1967). On the basis of a *P-R* relation obtained from radii determined for 15 cepheids by this method, it is suggested by Fernie (*ApJ*, **151**, 197, 1968) that η Aql, W Sgr, β Dor and X Cyg are pulsating in the first overtone, and that U Car may be pulsating in the third overtone (*ApJ*, **151**, 197, 1968). Opolski points out that new observations of radial velocities combined with simultaneous *UBV* photometric data would be very valuable. This problem needs international cooperation.

Fitch extended Schwarzschild's standard model computations to the first six radial pulsation modes of a set of polytropes. Comparison of these theoretical calculations with published results on the cepheids showing two excited radial pulsation modes has led to the result that all are pulsating simultaneously in both fundamental and first overtone mode.

Christy has completed a study of pulsation in cepheid type models. The most useful general results of his study are as follows:

(a) The general cepheid type instability strip can be most simply outlined in a log Period log T_e diagram independent of the Mass-Luminosity Relation. The blue boundary, including dependence on the mass fraction of Helium, *Y*, is approximately given by $\log T_e\,(\text{blue}) = 3{\cdot}797 + 0{\cdot}1\,Y - 0{\cdot}057 \log P_{\text{Fund}}$ and an estimate of the red boundary is $\log T_e(\text{red}) = \log T_e(\text{blue}) - 0{\cdot}06$.

(b) The pulsation calculations show that the light curve structure associated with the Hertzsprung progression is caused by a compression wave originating near the surface at the time of minimum radius and travelling to the center of *the star and out again*, arriving again at the surface some $1{\cdot}1 - 1{\cdot}5$ periods later. The travel time of this wave is proportional to the stellar radius and the appearance of the reflected wave can be used to determine the (mean) radius of certain variable stars where it is found (*Qu. J. RAS*, **9**, 13, 1968).

(c) For such stars (of period near 10 days) the radius, so determined, leads also to a mass determination, and, if the mean T_e is available, also to a mean luminosity determination and to a distance determination. This procedure should be quite useful when it has been appropriately calibrated.

Millis (*Lowell Bull.*, **148**, 1969) called attention to a cepheid with a period of 3·335 d and $A_v = 0{\cdot}16$. The abnormal small amplitude is explained by the fact that it has a close companion of similar brightness. Since the first micrometer measurement in 1886, the separation of the two components has diminished from 0ʺ3 to 0ʺ1. This system may offer the opportunity for the direct determination of the mass of a supergiant.

Radial velocities of cepheids have been obtained at Kottamia (*ROB*, 114). Feast determined velocity curves for some distant cepheids near $l^{II} = 290°$ (*MN*, **136**, 141, 1967, *Radcliffe Com.*, 94). Crampton of Victoria has discussed with Fernie a new determination of Oort's constant *A* from cepheids.

At Harvard Observatory, the Gaposhkins have catalogued 1155 cepheids in the SMC and 1111 in the LMC. They have also determined light curves for many of these stars. The number of known cepheids per square kiloparsec is about four times greater in the SMC than in the Large. With this increased material, previously suspected differences between the cepheids in the two Clouds have been confirmed. For instance the SMC contains an unusually high percentage of short-period cepheids. In the *P-A* relation, SMC cepheids with $P < 10^d$ have the largest amplitudes, while those in the LMC have the smallest. At the longest periods, however, cepheids in our own galaxy have the largest amplitudes, those in the SMC the smallest.

10. POPULATION II CEPHEIDS AND RV TAURI VARIABLES

The photometric study of RU Cam by Szeidl during 1967–69 on over 600 nights shows that the same star can in rapid succession change the character of its light variation from W Vir to RV Tau type, even becoming sometimes semiregular or completely irregular. Eggen also points out that the distinction between the long-period cepheids of the halo and old disk population and the RV Tau stars is obscure (*Sigma Reviews*).

Similar to RU Cam is V 725 Sgr. The peculiar behaviour of this object was first announced in 1936 by Swope. From 1889 to 1926 the star showed only slight, very gradual change around magnitude 13 on Harvard plates. But then the star began varying like a cepheid, with a period that lengthened steadily from 14 days in 1928 to 21 days in 1935. Meanwhile, the amplitude of the variable was diminishing while the average magnitude stayed near 13·2. No observations of the star have been published since 1935. Unpublished data indicated little or no periodic variations in 1959 (Plaut) and in 1960 (A. T. Young). In 1968 Demers began photoelectric measurements of the object on Cerro Tololo. The star was variable with a period of $15^d\!.5$ and an amplitude of only about $0^m\!.2$. The light curve changes from cycle to cycle. The average *V* magnitude is 12·55 and the average *B-V* colour index 1·35. An image-tube spectrum at the dispersion 100 Å/mm revealed a G 5 spectrum with giant or supergiant characteristics (*BAAS*, **1**, 185, 1969).

Kwee's work on galactic population II cepheids, based on *UBV* observations made at the McDonald Observatory in 1963, have been finished and published in a series of four papers (*BAN Sup.*, **2**, 77 and 97, 1967; *BAN*, **19**, 260, 1967; 374, 1968). It was shown that pop II cepheids can have changes in the period amounting to the order of 0·4% of the period. The stars with periods between 13 and 20 days can be divided into two groups with different shape of the light curve, and when data on cepheids found in globular clusters are considered, both groups have different ages and metal contents. In order to extend this investigation to more pop II cepheids, during the years 1967 and 1968, Pel, Velthuyse and Wamsteker have made five-colour photoelectric observations of 27 more variables at the Leiden Southern Station. The reduction of these new data is still in progress.

Woolley pointed out that W Vir stars have not got high velocities relative to the sun characteristic of halo objects. But the fact that some variables with periods > 1ᵈ0 occur in globular clusters suggests that there ought to be some stars with periods > 1ᵈ0 occurring free in the galactic system, and having the kinematic properties of the RR Lyr stars. Attention has been given at Greenwich Observatory to searching for such stars by collecting radial velocities of suitable candidates. At Mt. Wilson and Kottamia, Woolley and Palmer obtained velocities for seven W Vir stars. Three stars, FM Del, UY Eri and V 716 Oph are really fast moving objects and in this way kinematically halo objects. All three have periods less than 2ᵈ5 and seem to be an extension of the RR Lyr stars beyond the conventional limit of a period of 1ᵈ0. There are now 13 variables known to have periods around 1 or 2 days in 5 dwarf galaxies (Swope, *AJ*, **73**, S205, 1968). They are however from 0·4 to 0·8 mag brighter than the usual member variables of the RR Lyr type variables found in Leo II and in van Agt's UMi system. In addition to the meridian observations at Cape and Greenwich Observatories astrometric proper motion for 167 semiregular and RV Tau variables are being determined from plates taken with the astrographic telescope. Only stars which are brighter than $m_{pg} = 11$ at maximum are being measured. First epoch plates for 18 stars in the Vatican zone have been lent by O'Connell. Blackwell and Lowne (*ROB*, 142) published the proper motion of 182 semiregular and RV Tau variables on the system of the FK4. At the Leiden Observatory van Herk is engaged in determining or improving upon proper motions of RV Tauri variables.

Serkowski is continuing his polarimetric observations of RV Tau stars. The position angle of the intrinsic polarization for the RV Tau star U Mon remains close to 0° around and before the deep light minima and close to 105° around and before the shallow light minima occurring half-way between the deep minima. This suggests the non-spherical oscillations of the star. The position angle changes abruptly at the time when the doubling of the spectral lines is observed about 14 days after each light minimum. The changes in polarization of another RV Tau star R Sct are less regular than those for U Mon. For both stars the amount of intrinsic polarization is reaching the largest values in the ultraviolet spectral region during the light minima (*ApJ*).

At Sonneberg, Huth (*MVS*, **4**, no. 7, 1968) studied the period changes of the pop II cepheid CC Lyr. In Hodge and Wright's work (*ApJ Sup.*, 153) on LMC variables two W Vir variables are included, each with changing period and each subluminous for its period in comparison with the classical cepheids. At Dushanbe, Erleksova found that the character of period variations of RV Tau stars is similar to that of W Vir stars.

11. LONGPERIOD AND SEMIREGULAR VARIABLES OF BOTH POPULATIONS

About Eggen's efforts for obtaining population discriminants see Section 4. About spectroscopic observations see Appendix I.

At the Warner and Swasey Observatory, a detailed paper on spectral variations of 121 faint long-period variables is in preparation (*AJ*, **73**, 518, 1968). Houk plans to extend her studies on magnitude variability of giant M stars (*AJ*, **73**, S99, 1968) above 3 additional Palomar-Groningen Fields. This will be carried out at the Kapteyn Laboratory, Groningen, between February and September 1970, in collaboration with Plaut. The research on spectral changes in 51 variable stars and correlations with corresponding light changes was reported by Terrill (*AJ*, **74**, 413, 1969). These stars are of Mira type, semiregular and irregular variables. 19 are S stars, 31 are M stars, and R CMi is a weak carbon star.

Nancy Evans (Thesis) observed 48 Mira stars in *UBV* at maximum light at the David Dunlap Observatory. She found both $(B\text{-}V)_0$ and $(U\text{-}B)_0$ bluer at maximum than would be expected for stars of spectral type M. Students working under van den Bergh's direction are continuing *UBV* photometry of Mira stars near maximum light. Barnes is carrying out an extensive program of near infrared (*V*, *R*, *I*) photometry of Mira stars. Electrophotometry of red variables in the region 3800–11000 Å is carried out at Odessa. The program of *UBV* measurements of various long-period variables has been continued by Landolt at Kitt Peak and Cerro Tololo (*PASP*, **78**, 531, 1966; **79**, 336, 1967; **80**, 228, 450 and 680, 1968; **81**, 134, 1969). Smak observed 11 S-type and 11 C-type stars in *UBV* (*AcA*, **18**, 317, 1968). Szeidl observes UZ Aur and VZ Cam in UBV. At Riga, Alksnis and Alksna observe *UBVR* magnitudes of carbon stars, Kiselev and Kiseleva record fast brightness variations in long-period variables at Dushanbe (*AC*, **436**, 1967). Mendoza presented multicolour data in *U*, *B*, *V*, *R*, *H*, *I*, *J*, *K* for a number of long-period variables (*Tonantz. Bol.*, **28**, 1967). These data can be used to separate carbon stars and M and S unreddened stars. At Cape, photographic curves in B are being obtained for W Cam, U CVn, ME Cep, UZ Gem, WX Her, and VX UMa, all having unknown minima. The regular visual observations of long-period and irregular variables has been continued.

Eggen is making *UBV* and narrow band (**102, 65, 62**) photometry of red variables in both the disk and halo populations. Approximately 50% of the HD stars of type M are variable. Most of these stars are old disk objects. Light variations with amplitude greater than $0^{m}.05$ have been found in 78 bright, M-type stars. The boundary of the red instability region in the (M_B, T_e) plane is near the temperature represented by (**102, 65**)$= -0^{m}.6$ for disk population stars (*IB*, 355).

Price (*AJ*, **73**, S32, 1968) reported on a survey of the southern sky in the 2·2μ region. Approximately 50% of the area between $-30°$ and $-52°$ has been covered to a sensitivity of about $1{\cdot}5 \times 10^{-15}$ W cm^{-2} μ^{-1}. 57 late-type variable stars were detected or suspected. There seems to be a trend that for Miras and semiregular variables the longer the period, the greater the $m_v - m_{2.2\,\mu}$ index for a given spectral type. BM Sco is anomalous, the infrared irradiance from it is too large by a factor of 10 for its listed spectral type.

Light variations in the Taurus infrared object were found by Cannon (*Obs.*, **86**, 150, 1966; **87**, 231, 1967) and Wing, Spinrad and Kuhi (*ApJ*, **147**, 117, 1967). Assuming that the Taurus object is also a Mira variable, continued observations at the Greenwich Observatory show that the only regular period which could fit their observations in the *V* region is $390^{d} \pm 5^{d}$.

A large number of Mira and semiregular variables have been found to exhibit time dependent polarization discovered by Serkowski (Shakhovskoy at Crimea, *Krym. Izv.*, **39**, 11, 1968: Vardanian at Byurakan, *Budapest*, p. 33; Zappala, *ApJ*, **148**, L81, *AJ*, **72**, 838, 1967; Kruszewski *et al.*, *AJ*, **73**, 677; Dyck, *AJ*, **73**, 688, 1968). The polarization rises steeply into the ultraviolet, attaining in some cases 8% or more. Both the degree of polarization and the position angle usually show large variations with time, together with the wavelength dependence of the polarization. The absorption by graphite platelets can work in carbon stars but it cannot explain the generally large polarization in M-type stars. Some authors considered silicate grains which have a higher emissivity than graphite in the far infrared (Gilman, *ApJ*, **155**, L185; Knacke, Gaustad, Gillett and Stein, *ibid.*, L189, 1969).

A rediscussion of the absolute magnitudes of Mira variables derived from statistical parallaxes has been completed by Clayton and Feast. The results agree satisfactorily with absolute magnitudes derived from membership of Miras in clusters and double stars. Feast has concluded that the distance of far-off Miras have been underestimated, and that some of those he observed towards the galactic centre actually lie near the galactic nucleus. Further work is planned on stars in this region.

The lack of population discriminants complicates the interpretation of the observed motions, particularly the shapes of velocity ellipsoids and the dependence of the solar velocity on *z* for the Miras. At present, the only parameter that can be used to subdivide the Miras is the period of light variation. Clearly, there can be no definitive discussion of the kinematical properties of the Miras until some reliable photometric or spectroscopic population discriminant is found. Radial velocities of semiregular variables have been obtained at Kottamia (*ROB*, 114). At the Leiden Observatory van Herk is engaged in determining proper motions of long-period variables.

Ahnert discussed the correlations between the irregularities of the period and radial velocity, length of period, amplitude and spectrum of Mira variables (*Budapest*, p. 325), Fischer (*ibid.*, 331, *Wien Ann.*) has collected and studied the whole material on the light-changes of Mira Ceti. Feuchter (*AJ*, **72**, 702, 1967) supplemented Harrington's (*AJ*, **70**, 569, 1965) statistical studies.

12. BETA CANIS MAJORIS STARS AND OTHER EARLY-TYPE VARIABLES

Percy has given (*Can. JRAS*, **61**, 117, 1967) a review of the observational problems of the βCMa stars. A great advance in this area was Hill's extensive photoelectric search for βCMa stars among a sample of 153 early B stars in the nearest associations and galactic clusters which yielded 24 new βCMa stars. These new observations have extended considerably the limits of spectral type, luminosity class, absolute magnitude, period, and rotational velocity for these variables. The limits now become O9·5 – B3 (sp. type), V-I (luminosity class), $3^h - 10^h$ (period) V, 0–300 km(s /rotational velocity) (*ApJ Sup.*, **14**, 263, 1967).

At Steward Observatory, Fitch is beginning feasibility tests for observations of accurate line profiles and velocities of selected βCMa stars, using the new coudé spectrograph and image tube camera on the 36-inch telescope. He hopes to obtain a fairly complete record of variations in profiles and velocities during individual pulsation cycles and through the course of the long period variation, for one or two bright βCMa stars which display pronounced double periodicities. He intends to make simultaneous photometric measures using the 21-inch telescope.

At Uttar Pradesh Observatory, Bondal made (*Obs.*, **87**, 220, 1967) a photoelectric study of 53 Ari. Shapes of the maxima and minima on different nights vary appreciably. Joshi has found a beat period of $17^d\!.1$ for KP Per. Observations on 25 Sex, 16 Lac and BW Vul are being analyzed by Mahra and Joshi. *UBV* photometry of γ Peg was obtained by Jerzykiewicz at the Lowell Observatory (*IB*, 382). At Radcliffe Observatory Breger has re-observed τ Lup and has found evidence that β Cen may belong to the β CMa stars (*MN*, **136**, 51, 1967). Shobbrock and Robertson obtained from radial velocity measures a period of $0^d\!.1348$ (*Proc. AS Austr.*, **1**, 82, 1968).

Leung (*ApJ*, **150**, 223, 1967) using visual absolute magnitudes from spectroscopic H γ equivalent widths of 17 β CMa stars derived a period-luminosity and a period-colour-luminosity relation. The latter leads to the existence of a set of linear equal period-contours across the instability strip in the HRD. The same problem was treated by McNamara and Matthews (*Mod. Astrophys.*, **127**, 1967) based on measurements of the β index and equivalent widths of the Hγ and Hδ lines.

Van Hoof has given a review (*ZfAp*, **68**, 156, 1968) of all the radial velocity and photometric observations of δ Cet since the beginning of the century, and made new photoelectric observations at Boyden. The period $0^d\!.1611366$ remained constant over the whole time interval considered. Briers finds for the beat period in θ Oph $3^d\!.9$ instead of that found by van Hoof (6^d). At the University of Western Ontario, Gray is observing β Cep. Additional light curves have been obtained to study the period. The intensive photoelectric observations of a few southern β CMa stars (σ Sco, θ Oph, α Lup, HD53435 and HD53974) has been continued both at the Boyden and ESO observatories. The periods proposed for the latter two by Hill have been found to be erroneous (van Hoof). An extensive search for new β CMa stars among fainter early B-type stars is planned. Shobbrook *et al.* at the Siding Spring Observatory have shown that the primary of the binary system α Vir is a β CMa variable. The light shows a $4^d\!.17036$ period of amplitude 1·6%, in addition, there is a 3% variation over the $4^d\!.014$ orbital period, which can be accounted for by aspect changes of the tidally distorted primary (*MN*, **145**, 131, 1969).

Fitch, extending his study of σ Sco (*ApJ*, **148**, 481, 1967) reported (*Budapest*, p. 287) 16 Lac to have three excited pulsation modes, the third being excited by the primary mode through a tidal resonance at $\frac{1}{5}$ the orbital period. Fitch believes that the primary mode is the nearly radial fundamental pulsation mode and the secondary mode is the nearly axisymmetric S-mode discussed by Chandrasekhar and Lebovitz. All three modes have velocity-to-light amplitude ratios which differ significantly from each other. Fitch also has performed an analysis of the velocity measures on β Cep, and found the star is apparently a single-line spectroscopic binary with an orbital period near $10^d\!.9$ and orbital eccentricity of about 0·5. The adopted elements show that the primary pulsation ($P = 0^d\!.19049$) undergoes a strong, non-resonant amplitude modulation around the time of periastron passage, with maximum amplitude probably occurring at maximum apparent tidal compression and minimum amplitude at maximum apparent tidal expansion (*ApJ*, **158**, 269, 1969). Theoretical work on non-radial oscillations is being carried out by Smeyers at Leuven.

Slettebak has given a thorough review of the problems of Of and Be stars (*Budapest*, p. 179). Mass loss from stars in general was reviewed by Deutsch (*Trieste*, 1969, p. 1) and for WR stars by Underhill (*ibid*, p. 17). Rocket observations of mass loss from hot stars was discussed by Morton (*ibid*, p. 36).

A study of the shell star 48 Lib in the period 1950–62 was published by Faraggiana (*AA*, **2**, 162, 1969). The observed radial velocity curves suggest that an eruption took place in the star and material from the shell is now moving outwards. The radial velocity fluctuations are probably the result of an eruptive force and the deceleration due to gravity. Radial velocities of o And were studied by Galeotti and Pasinetti (*Contr. Milano-Merate*, 300, 1968) and Pasinetti (*ibid.*, 276), those of ζ Tau by Hack and Steiner (*MAI*, **40**, 87, 1969).

Mayall has published a light-curve for AG Cas, 1940–69, a star with a spectrum similar to P Cyg (*Can JRAS*, **63**, 221, 1964). Brodokoja carried out three colour observations for EW Lac (*PZ*, **16**, 423, 1968), AB Aur, XX Oph, X Per and MWC 374 (*ibid.*, 435). *UBV* observations of P Cyg by Fernie (*PASP*, **81**, 168, 1968) revealed no short-period variability. Feinstein at La Plata obtained from 1967 to 1969 new observations in *UBVRI* of all Be stars published in *ZfA*, **68**, 29, 1968. Wallerstein made a search for rapid light variations in WR stars (*PASP*, **80**, 483, 1968), as these have been suggested as possible X-ray sources (Bless *et al.*, *ApJ*, **151**, 2117; Wallerstein, *ibid.*, 1421, 1968). For no star were any variations noted that were greater than the limit set by the noise. But there is irregular variability in these stars in their emission-line intensities (Kuhi, *PASP*, **79**, 57, 1966). Mendoza confirmed that Of and Be stars exhibit infrared excesses (*Cerro Galan Publ.*, **7**, 106, 1969). The main cause of these excesses could be the presence of stellar shells.

Polarimetric observations were made for 40 Be and shell stars by Serkowski (*ApJ*). About half of these stars show evidence of intrinsic polarization, in most cases changing with time. Kruszewski *et al.* explained the observed wavelength-dependence of the polarization by scattering from electrons in an asymmetrical envelope together with self-absorption in a hydrogen plasma.

13. MAGNETIC VARIABLES

A great activity was displayed in spectral (see Report of Commission 29) and photometric observations and theoretical studies of magnetic variables, but little is being done in measurements of stellar magnetic fields. This important subject was widely reviewed and found ample attention in the books *The Magnetic and Related Stars* (Ed. R. C. Cameron, Mono Book Corp.) and *Magnetism and the Cosmos* (Ed. Hindmarsh *et al.*, Oliver and Boyd, Section II).

On the theoretical side, the oblique rotator hypothesis seems to come to the front with the model developed by Böhm-Vitense (*Mod. Astrophys.*, 97). She found great inclinations between the magnetic and rotation axes. Preston has demonstrated in a statistical study (*ApJ*, **150**, 547, 1967) that approximate orthogonality of the rotation axis and the hypothetical magnetic axis is a general characteristic of the periodic magnetic stars, if they are treated as oblique rotators. It may be necessary to modify the interpretation if the stellar magnetic fields are not dipolar. Anharmonic variation in HD 125248 is according to Preston due to the location of two regions of opposite polarity, both near the rotational equator, but separated by considerably less than 180° of longitude. Neither is the magnetic field of HD215441 dipolar (Preston, *ApJ*, **156**, 967, 1969). Mester (*Magnetic and Related Stars* p. 101) pointed out that a spheroidal star with axis of symmetry inclined to the axis of rotation might result from an internal poloidal magnetic field, that can exhibit a light variation due to the variation of the area of the disk projected on the line of sight. In a discussion of the variability of HD 19216 (*ApJ*, **156**, 1175, 1969) Preston suggests that this may be the origin of at least a component of the light variation. Kodaira (*Tokyo Ann. II, ser. X*, no. 4, 1967) finds in a spectroscopic study of HD 221 568 inconsistency with the oblique rotator hypothesis.

S. C. Wolff measured magnetic field variations in HR 7575 at Lick Observatory. She found that the amplitude of the magnetic variations derived from measurements of lines of Mn, Gd and Ca were decidedly smaller than the magnetic amplitudes derived from lines of Cr, Eu, Ti and Ce. These groups of elements must, therefore, exist in different location on the stellar surface having different magnetic field intensities (*ApJ*, **158**, 1969).

Preston's Zeeman spectrograms of HD 215441 give evidence for a periodic variability of its large magnetic field in concordance with the $9^{d}.5$ photometric period (*ApJ*, **156**, 967, 1969). He is using Zeeman doublets to study the large (~ 15 kpc) magnetic field of HD 126515. Both the mean surface field and the effective field of this star appear to vary smoothly in a period of 130^d.

At Oxford, Hockey finished off his work on HD125148 (*MN*, **142**, 453). The measurements and reductions of the Lick spectra of HD153882 are now completed, and are being used to test magnetic rotator models. Pyper has made study of the spectrum, radial velocity, magnetic and *UBV* variations of α^2 CVn at Lick, and found that the magnetic configuration is represented by a combination of dipole and quadrupole fields. Preston and Stepien discussed the light, magnetic, and radial velocity variations of HD10783 (*ApJ*, **154**, 971, 1968). Preston at Mt. Wilson is obtaining rotational velocities from coudé spectrograms of all known Ap stars brighter than the 9th magnitude and north of $\delta = -40°$.

In virtually every case if enough data are obtained, it is possible to find periods. A period of $3\overset{d}{.}7220$ has been established for 78 Vir by Preston (*Astrophys. J.* **158**, 243, 1969). He, Stepien and Wolff have derived $P = 5\overset{d}{.}08$ for 17 Com A. A provisional $P = 5\overset{d}{.}00$ was obtained for κ Cnc (*ApJ.* **156**, 652, 1969). Preston has found that the radial velocities and line intensities in the spectrum of 21 Per vary periodically in the $2\overset{d}{.}88$ photometric period derived by Stepien (*Astrophys. J.* **158**, 251, 1969). Steinitz and Pyper obtained for 3 Hya $P = 4\overset{d}{.}606$ (*IB*, 413) from Babcock's measurements. Stepien obtained for HD 19216 $P = 7\overset{d}{.}7$, Wolff (*ApJ*, **158**, 1969) for HR 7575 $P = 224\overset{d}{.}5$, showing that the phenomena of magnetic variability is not confined to periods of the order of a few days. Berg improved the period of 73 Dra to $20\overset{d}{.}2755$ (*Leander McC Publ.*, **15**, Part IV, 1967). Gökkaya found $P = 6\overset{d}{.}143$ for HD 151199 (*Astrophys. Space Sci.* **6**, 141, 1969).

Random fluctuations seem to be superimposed on periodic variations that will have to be explained by some irregular distortions of the lines of force possible due to some activity (see Bonsack, *AJ*, **72**, 876). This theme was reviewed by Jarzebowski (*Budapest*, p. 227).

Finzi and Wolf suggested a model (*ApJ*, **153**, 865, 1968) that may help to explain the random fluctuations as well as peculiar chemical composition and slow rotation velocities. According to them the fields are produced in the convective cores of these stars and they suggest a mechanism by which magnetic energy could be carried from the core to the surface.

Abt called attention (*PASP*, **80**, 637, 1968) to the fact that periods are more easily obtained from spectral variations, even with moderate dispersion, than from the small light variations or by the difficult field measurements. In this way, Guderley succeeded in deriving a period of $2\overset{d}{.}156$ for the magnetic variable HR234 (*PASP*, **79**, 588, 1967).

Rakos has obtained extensive series of photoelectric observations of four magnetic variables, HD25823, 32633, 71866 and 65339 at Lowell Observatory during 1968. The length of the period of HD65339 seems to change periodically within a time interval of $1\overset{y}{.}8$. Similar change in the period ($2\overset{y}{.}7$) is indicated for HD215441. Three other stars, HD32633, 224801 and 71866 also show definite changes in the period (*AJ*, **73**, 114, 1968). Cester at the Trieste Observatory initiated a long range program on Ap and Am stars in *UBV*. Catania Observatory is continuing *UBV* observations of magnetic variables (*Budapest*, p. 243, *MAI*, **39**, 579, 1968). A photometric study of magnetic stars has been carried out at Lick Observatory (Stepien) concurrently with spectroscopic observations (Preston). The obtained *V* curves are in phase as well as in antiphase with the magnetic curve. The $B-V$ and $U-B$ curves present a variety of phase dependence relative to *V*-curves (*AJ*, **73**, S36; *ApJ*, **151**, 577, 583, 1968; **153**, 165; **154**, 945 and 971, 1968; **156**, 653; *Budapest*, p. 239). The photometry of several Ap stars is being continued at the University of Western Ontario and at the University of Hawaii. Schöneich, Zehwanova, Hildebrandt and Folge (Potsdam) obtained *UBV* magnitudes of magnetic stars and of Ap stars in clusters. At Leiden, van Genderen has made five-colour observations of 16 magnetic variables. The work is still in progress.

Polosukhina (*Krym Izv.*, **39**, 34, 1969) found irregular changes in the polarization of HD215441 between 0·5–1·7%. The dependence of the polarization on wavelength is changing too. Serkowski and Chojnacki observed (*AA*, **1**, 442, 1969) for polarization some magnetic stars with negative results.

14. DELTA SCUTI VARIABLES AND DWARF CEPHEIDS

Both types of variables have ultrashort-periods, the dwarf cepheids representing the old disk population, the δ Scuti stars the young disk population among these objects (see Eggen, *Sigma Reviews*, in press). The dwarf cepheids have larger amplitude variations and are slightly hotter than the δ Scuti stars, and in most cases the kinematical separation of the two groups is clear. Yet, the two dwarf cepheids with known trigonometric parallaxes differ considerably in space motion and metal abundances (see Bessell, *ApJ Sup.*, **18**, 167, 1969 and Appendix I) AD CMi and VZ Cnc also have space motions characteristic of the young disk. The solar motion relative to δ Sct stars was determined by Millis (*AJ*, **73**, S26, 1968) with $v_0 = 21$ km/s, $l^{II} = 32°$, $b^{II} = +25$. Therefore, it can be concluded that these stars are older pop I objects. A δ Sct type variable was found in the Coma cluster (see Appendix II).

Breger (*ApJ Sup.*, **19**, 79, 1969) has completed an important survey of 213 bright field stars which were tested photoelectrically for short-period variability with special attention being given to the constancy of the non-variables. 17 new δ Sct variables were found. He has listed 39 such stars known at present. The amplitude and incidence of variability depends critically on the colour. The large proportion of variables with small amplitudes close to the observational limit of detection indicates that most stars inside the instability strip probably pulsate. The variables and non-variables in the instability strip show similar galactic *U*, *V*, *W* velocities. Narrow-band photometric and spectroscopic data have been collected for most of the 39 δ Scuti variables published so far.

Breger (*ApJ Sup.*, **19**, 99, 1969) has also studied the lower instability strip which extends from the main sequence upward to the RR Lyrae region. Hot and cool borders were established which intersect the main sequence at A 4 and F 2. In the instability strip more than 20% of the stars show regular variability larger than 0·010 magnitude. These variables have normal masses and show a wide range in metal abundances and rotational velocities. Metal abundance and binary characteristics appear to have no influence on the incidence of variability. Pulsation is found to be strongly related to rotational velocity, $v \sin i$. The variables of small amplitudes share the high rotational velocity of the non-variables, while the incidence of larger amplitudes depends critically on low rotational velocity. The mean rotational velocity of the variables two magnitudes above the main sequence is less than 0·3 that of similar non-variables. This effect decreases as the main sequence is approached.

A period-luminosity-colour relation is established with an average deviation of less than 0·3 magnitude in M_v. The pulsation constant Q is found to vary slightly along the instability strip and a preliminary value of $Q = 0{\cdot}029 \pm 0{\cdot}007$ is derived indicating that the variables pulsate in the fundamental and/or first overtone. It is proposed that the period differences between classical cepheids, δ Scuti stars, RR Lyrae stars and dwarf cepheids are caused not only by luminosity differences, but also by large mass differences.

A search for short-period variability among the brighter giants and subgiants between spectral types A 7 and F 5 has also been made by Millis with the 21-inch reflector at Lowell Observatory. Of the 59 field stars observed, at least 10 are variable. The newly discovered variables cover a considerably larger range in spectral type, colour, period, and rotational velocity than did the original δ Sct group members. In the colour-magnitude diagram the variables which he observed, lie immediately above the main sequence between $B - V = +0{\cdot}20$ and $+0{\cdot}35$. Their periods range from $0^d_{\cdot}037$ to $0^d_{\cdot}194$ with amplitudes in V between 0·015 and 0·3 magnitude. A few non-variable stars were found which definitely fall among the variables in the colour magnitude diagram. Strong evidence of low-amplitude variation has been found in two Hyades stars.

The bright star HR2707 has been found to be a δ Sct variable by Eggen (*IB*, 250).

Fitch has given a summary of presently available information relating to tidal modulations of δ Sct stars CC And, 14 Aur, and δ Del (*ApJ*, **158**, 279, 1969).

Fullerton (*AJ*, **72**, 797, 1967) has analysed the light curves of all the well observed δ Sct stars for multiple periodicities. The secondary frequencies close to the primary might be due to differing polar and equatorial radii just as may be the case for β CMa stars. The close frequency pairs may be due to tidal effect from binary companion.

Johnston and Chambliss obtained observations in *U*, *B*, *V* and Hβ on several dwarf cepheids and δ Scuti stars on Cerro Tololo. At Leiden Station XS Phe was observed by Wamsteker photoelectrically in five colours during 10 cycles in 1968. Reductions are still in progress. SZ Lyn has been observed by van Genderen (*BAN*, **19**, 74, 1967), Joshi and Srivastava (*ZfAp*, **67**, 456, 1967), Binendijk (*AJ*, **73**, 29) and the Wisses (*BAN*, **20**, 333, 1969). Szeidl observes the same star together with several dwarf cepheids in *UBV*. Mahra and Sanyal observed VZ Cnc (*Obs.*, **88**, 59, 1968). The Joshis and Gurthu have found (*Obs.*, **89**, 112, 1969) that γ UMi is a dwarf cepheid.

Hayes and Heiser published photoelectric observations of 4CVn finding $P = 0^{d}.170934$ with a possible beat period of $0^{d}.855$ (*PASP*, **80**, 57, 1968). Five-colour observations have been made of δ Del by van Genderen and Wamsteker in 1966 and 1968. The reduction of these observations are still in progress. Wehlau at the University of Western Ontario is continuing the study of δ Sct stars. Light curves of δ Sct have been combined with those of Fath for a reinvestigation of the light variability. Other δ Sct stars are being studied to compare the light and radial velocity variations together with Leung (*ApJ*, **149**, 39, 1967). Crampton at Victoria has begun observations of the velocity variations of the three δ Sct stars 4CVn, 20CVn, and κ Boo.

Fitch was led by his theoretical results to the conclusion, that the AI Vel stars SX Phe, RV Ari, AI Vel, V 703 Sco, VX Hya, and the δ Sct star 4CVn are all pulsating simultaneously in both fundamental and first overtone mode, while the AI Vel star VZ Cnc is pulsating in the first and second overtone modes with its fundamental mode unexcited.

15. HERBIG-HARO OBJECTS AND T TAURI STARS

Wenzel has given a review of contracting, extremely young variables at the Budapest Colloquium (p. 61) and Kuhi in *Interstellar Ionized Hydrogen* (Ed. Terzian, p. 13, 1968).

Herbig (*Budapest*, p. 75) has made a very important study of the light variations in Herbig-Haro object No. 2, which consists of a number of bright nuclei together with considerable fainter structure enclosed in an elliptical area about $25'' \times 40''$. Three nuclei show strong variability. On the basis of this investigation it can be ascertained that the spectacular brightening of two nuclei in the early 1950's (*Non-Stable Stars*, *IAU Symposium* no. 3, p. 8, 1957) must not be associated with the birth of two new stars. There is still no completely straightforward interpretation of these light variations in H-H objects.

At Sonneberg, the program of photoelectric measurements in three colours on extremely young variables is continued. The discussion of the large material for RW Aur (*MVS*, **4**, 71) gave hints to a very complex behaviour. The investigation of SU Aur (*MVS*, **5**, 53) showed that the appearance of minima in brightness is apparently not correlated with changes of spectral type. These results and especially those on SV Cep (*MVS*, **5**, part 5) suggest that variable circumstellar extinction might be efficient. The measurements of RY and T Tau are finished, those of BH and BO Cep are well advanced (Wenzel, Götz).

At the Boyden Observatory, a program for investigating the photometric behaviour of RW Aur variables has been started in 1962 by Geyer. He and Seggewiss reported on UBV observations of DI, ES Car, SY, SZ, TT Phe, and BS Vel (*Budapest*, p. 85). Seggewiss obtained at the ESO in Chile *UBV* magnitudes for the RW Aur stars S6326 Cha and S6343 Cha. Vardanian (*Byurakan Soob.*, **39**, 1969) published a series of three colour photoelectric observations of RW Aur variables.

Landolt published photoelectric *UBV* sequences in Taurus to be used in multicolour studies of the irregular variable stars in the Taurus-Auriga dark cloud (*AJ*, **72**, 1012, 1967).

Gratton, Maffei and Martino presented an interpretation of the light curves of RW Aur stars as the result from a large number of correlated flares (the occurrence of a flare might enhance or decrease the probability of another flare occurring soon after the first) at the 9th *Nobel Symposium* (p. 231, 1968). A general catalogue of all known variables of the T Tau-type is being prepared at the Astrophysical Laboratory at Frascati.

Mendoza continued the infrared photometry of T Tau stars and related objects (*AJ*, **72**, 816, 1967; *ApJ*, **151**, 977, 1968). The new *UBVRIJHKLM* observations of 33 stars confirm the previous result that T Tau stars have large infrared excess. According to Poveda (*AJ*, **72**, 824, 1967), very young stars may be accompanied by a thick circumstellar cloud which absorbs most of the visual radiation of the star and reradiates it in the infrared. He interprets FU Ori's hundredfold increase in luminosity during 1936 as a consequence of the coalescence of the particles into large planetesimals, or because they were blown away by the stellar wind.

A search for radio emission from infrared and T Tau stars with the Arecibo 1000-ft telescope ended with a negative result (Cornella, *ApJ*, **149**, L91, 1967).

The intrinsic polarization of T Tau stars confirms the rule (Serkowski, *ApJ*, **154**, 125, 1968), that it is always associated, both in the early-type stars as in the late-type stars, with the emission lines in the spectrum. This may be a clue for finding the explanation for this intrinsic polarization. Serkowski found changes in the polarization of the T Tau star AK Sco from 0·3 to 0·8% (*ApJ*, **156**, L55, 1969). Polarization in the blue spectral region in R CrA increased during 3 weeks from 9 to 18% (*ApJ. L.*, 1969). Efimov found a correlation between polarization and $B-V$ variations in RW Aur, where the polarization changed between 0·5 to 1%.

One Orion-type variable, NU Ori, was observed by the Celescope and found fainter in the ultraviolet than expected by factors of 10 at 2700 Å, 5 at 2320 Å, and 40 at 1500 Å (*Sky Tel.*, **37**, 281).

16. FLARE STARS

The general subject of flare stars has been reviewed by Haro (*Stars and Stellar Systems*, vol. VII, p. 141, 1968). Geršberg has given an introductory report of observational and theoretical results bearing on the classical flare stars of UV Cet type (*Budapest*, p. 111). Light variability of dMe and dM stars was discussed by Krzeminski (*Low Luminosity Stars* p. 57). About cooperative programs see Appendix III.

Geršberg and Čugainov considered about a hundred photoelectric light curves of flares and found the following characteristics:

(a) the total energy of a flare in the blue region is $\sim 3 \times 10^{31}$ ergs;

(b) characteristic rates of the increase of flare energy output is $10^{27} - 10^{28}$ erg/s;

(c) time of light growth 10–30 s for half of the flares, 3–100 s for 90% of them. They discussed various criteria of flare activity and found that the recorded flares released $0{\cdot}1 - 1\%$ of the energy radiated by the quiet star.

Oskanian (*Budapest*, p. 131) proposed a classification of flare light-curves based on two parameters of the light-curve: (a) the rate of brightness increase; (b) the character of the decreasing branch of the light-curve. He distinguished four types of light-curves. Individual flares of a given star may have entirely different characteristics.

Kunkel (*IB*, 294) has detected ultraviolet flares in two stars listed in the *Lowell Proper Motion Catalogue*. Photoelectric patrolling of a number of dM and dMe stars is being carried out by Shakhovskaya at the Crimean Astrophysical Observatory. Two bright dMe stars, BD + 13°2618 and 20 C 1250 are found to flare (*IB*, 361). Flare activity of seven dMe stars within ten parsecs of the Sun has been determined by Kunkel (*AJ*, 1969). He has found CoD-32°16135AB to be the most active flare star, and Wolf 359 the least active. Faint flare stars were discovered by Hoffmeister (*IB*, 203 and 251) and Sanduleak (in the Coal Sack, *IB*, 275). At Catania, the flare stars BD + 51°2402 and PZ Mon of spectral types K6Ve and K2e, respectively, are under observation (*Catania Pubbl.*, 114 and 124; *IB*, 252; *Budapest* p. 149).

The variety of spectral types displaying flares continues to be remarkable. The star SS 199 II (Slettebak-Stock II faint-star list of stars near the NGP, *Hamb. Abh.*, **5**, No. 5), a halo Pop II A0 type star, was found to flare by Philip (*PASP*, **80**, 171, 1968). This star has now been placed on the patrol program of the Tonantzintla Observatory, and Philip also plans to monitor it further. A finding chart is given in *ApJ*, **148**, L143. Butler's flare star, HD 6090, has been classified by Thackeray as dKO (*PASP*, **79**, 368, 1967), and certainly deserves further attention by southern observers. A long monitoring run of 114 h on V 1216 Sgr is reported by Andrews (*PASP*, **78**, 542, 1966). MacConnell has discussed several flares observed by him in BD − 8°4352 AB and AD Leo in the ultraviolet (*ApJ*, **153**, 313, 1968). The much neglected flare star discovered 45 yr ago by Hertzsprung, DH Car, has been observed to flare by Tapia (*IB*, 286). Proxima Centauri has been patrolled by Bateson and members of the New Zealand team (*IB*, 353).

A further attempt has been made to detect optical polarization associated with flares for EQ Peg and AD Leo by Cubičela and Arsenjevič. Efimov has investigated theoretically the accuracy of such investigations (*Budapest*, p. 169). He measured the polarization of AD Leo (*Krym. Izv.*, **39**, 42, 1969). Vardanian made polarimetric observations during the light decrease of YZ CMi and UV Cet. No polarization greater than 0·8% was found (*AC*, 508, 1969).

Colour changes during flares were investigated by Andrews (*IB*, 265). A statistically important project has been undertaken by Kunkel at the McDonald Observatory using sequentially-sampled *UBV* measures of flares in several stars (*Diss. Austin*, 1967; *ApJ*, 1969). At Crimea, a number of flares of EV Lac have been observed by Čugainov (*Krym. Izv.*, 1969) with a photoelectric spectro-colorimeter, that gives the possibility to record simultaneously the radiation in three spectral intervals, 3350–3650, 4155–4280 and 5120–5320 Å. Kunkel (*loc. cit.*) obtained time-resolved spectra of flares in EV Lac, AD Leo and YZ CMi, together with $(U-B)$ and $(B-V)$ colours, the Balmer jump, the slope of continuum emission redward to the Balmer jump, and the Balmer decrement. These data enabled him to construct a two-component spectroscopic model of flare in which the dominant emission process is hydrogen recombination with descriptive parameters similar to those of solar flares. The second component consists of impulsive heating of the photosphere beneath the flare. He also obtained evidence that the surface extent of the flares covers only a few percents of the apparent stellar disk. The ratio of radio to optical energy of the flares was found by him to be the same for dMe stars, as for the Sun. Lovell's earlier result (*Obs.*, **84**, 191, 1964) is found to be in error because bolometric corrections were not applied by him. The values of the Balmer jump obtained by Čugainov for flares on EV Lac are smaller than those given by Kunkel, especially near the flare maximum. Čugainov has also observed that the intensity of the Hβ emission line relative to the neighbouring continuum emission decreases near the flare maximum (*Krym. Izv.*, **38**, 200, 1967; 1969 in press). Taking into account both these facts, Čugainov concluded that the hydrogen recombination makes only a small contribution to the total radiation near the maximum of a flare.

Arakelian investigated the luminosity function of normal and flare red stars in the vicinity of the sun (*Budapest*, p. 161). He was able to conclude that the space density of flare stars in the solar neighbourhood is higher than in the general galactic field.

The most distant red dwarf star, the non-thermal radio emission of which has been detected so far, is the flare star V371 Ori at a distance of ~ 15 pc. Slee, Higgins, Roslund and Lyngå recently reported the probable detection of radio emission from a number of flare stars near the Orion Nebula at a distance of ~ 400 pc (*Nature*, **224**, 1087, 1969). Compared with the flare stars in the solar neighbourhood, there is a dilution factor of ~ 10^3 in intensity on account of the greater distances, but this must be largely compensated for by the much greater amounts of energy released during flares on these stars.

Since flare activity seems to be a common property of all low-mass red stars, the existence of normally invisible examples of such objects of very low luminosity may be established by observing them during flares. Van de Kamp reported on the flare-up of a previously unseen companion of the proper motion star Groombridge 1830 (*Low Luminosity Stars*, Ed. Kumar, p. 199).

17. R CORONAE BOREALIS VARIABLES

Stein *et al.* (*ApJ*, **155**, L3, 1969) have reported a large infrared excess for RCrB. They suggest that this excess is due to blackbody radiation from a cloud of particles at a temperature of 940 K. If their model is correct, a cloud with an angular radius of 0″.07 would be required, and the authors propose that such a cloud could have formed by the ejection of particles during the past deep minimum, about 7 years ago. These ideas are consistent with the Loreta-O'Keefe hypothesis of carbon particle formation to explain the deep minima of RCrB stars. RY Sgr also exhibits a large excess of radiation at long wavelengths. (Lee and Feast, *ApJ*, **157**, L173, 1969). The K-M index is larger for this star than that found for most infrared stars, including NML Cyg. During a recent 8-month interval the visual brightness increased, while the radiation in the 2–3·4μ region decreased.

Serkowski and Kruszewski found changes in polarization of RY Sgr (*ApJ*, **155**, L15, 1969).

During the minimum the percentage polarization increased from 0·5 to at least 1·3%, and the position angle changed from about 10° to 150°.

Extensive three colour photometry and spectroscopy of RY Sgr has been carried out at Pretoria and the Cape since the beginning of the deep minimum of 1967 (see Appendix I). At Ottawa a photometry of RCrB itself will continue for another season, the principal aim being to investigate the stability of the star at maximum light. According to Meinunger (*MVS*, **4**, 8, 1968) the light of BH Lac, classified as an RCrB variable in the GCVS, is constant. Herbig has already shown (*ApJ*, **13**, 632) that the spectrum is not that of an RCrB star.

Hill (P. W.) of University Observatory St. Andrews pointed out the similarity in composition between the hot helium stars and the RCrB variables. Yet, he found no large brightness variations in the He stars (*IB*, 357). Landolt also observed some helium stars with the same negative result (*PASP*, **80**, 318, 1968).

An RCrB star with a single observed minimum of 2300 days was detected in the LMC by Hodge and Wright (*ApJ Sup.*, 153, 1969).

18. SYMBIOTIC VARIABLES

The spectral observations of the symbiotic stars are considered in Appendix I. Photoelectric observations are carried out at a number of observatories. Tempesti observes since November 1968 EG And every clear night in V and B light, in a combined program with spectrographic observations at the Asiago Observatory. Beliakina (Crimea) showed that brightness and colour changes of Z And, AG Per and AG Dra confirm the hypothesis on the duplicity of these stars. The light variations in the blue and yellow regions are caused by the effect of heating of the cool component by the hot component radiation. The ultraviolet variation results from the ionization of the outer atmosphere of the cool star by $L\alpha$-radiation of the hot component. Luud (Tartu) carried out photoelectric and spectral studies of CH Cyg. The peculiarities are probably caused by the excitation of the outer atmosphere of the gM 6 star by the variable radiation of the hot component of low luminosity. The gM 6 star is also variable. Extensive photoelectric three-colour and objective-prism observations on CQ Tau by Götz and Wenzel show a lot of remarkable effects, partly not yet observed at any other star. Martel and Févre observed photoelectrically CH Cyg in *UBV* at Haute Provence (*IAU Cir.*, 2027, 1967). The same star is observed by Cester (*Trieste Publ.*, 368). A long term program of *UBV* photometry of symbiotic stars has been initiated for about two years at Ottawa. The variability of AG Peg in its spectroscopic period has been confirmed. Philip presented *UBV* and four-colour photometry for MHα 328–116 (*ApJ*, **151**, L69, 1968; *PASP*, **81**, 248, 1969). Johnson and Golson (*ApJ*, **155**, L91, 1969) observed HD4174, AG Dra, AG Peg, MHα 328–116 and CH Cyg with narrow band filters. Robinson published the photometric history of AG Dra from 1890 to 1966 (*PZ*, **16**, 507, 1969). Hoffleit discovered a new variable with symbiotic spectrum, MHα 208–51 (*IB*, 254).

19. NOVAE DURING OUTBURSTS

Nova (T) Pyx 1890, 1902, 1920, 1944, 1966 was observed by Landolt photoelectrically.

Nova (GK) Per 1901 exhibited a recent outburst in September 1967 (IAU Circ., 2037).

Nova CrA 1967 was discovered in 1969 by Sanduleak on objective prism plates taken on Cerro Tololo. The nova must have been in maximum light about the 8th magnitude at the beginning of 1967 (*IB*, 368).

Nova (HR) Del 1967 was discovered by Alcock on July 8 at ~5th magnitude. It attained its peak brightness of 3·5 magnitude in December. Tempesti observes the star continuously since August 1967. Photoelectric observations were published by Stokes (*IB*, 224), Il-Slong Nha (*IB*, 238) Mollerus (*AA*, **3**, 376, 1969). Narrow band photoelectric observations were carried out at Catania (*Atti XII. Conv. SAI l'Aquila*, 1968).

Nova Her 1968 = *S 10376*, a very fast nova was discovered by Richter (*IB*, 293). The object must be in the intergalactic space with z = 42 kpc.

Nova (RS) Oph, 1898, 1933, 1958 recurred again in October 1967; the flare up was detected first by Beyer.

Nova Oph 1967 was detected by Stephenson on objective prism plates taken by Sanduleak on Cerro Tololo. The maximum must have been near $V = 9{\cdot}5$, presumably during the first half of 1967(*IB*, 323).

Nova Sgr 1969 was found by Bateson on Galactic Centre Patrol Plates in June. The maximum brightness was reached on 11th July with 6·5 magnitude.

Nova Vul 1968 (no. 1) was discovered by Alcock on April 15. Photoelectric observations have been obtained by Quast (*IB*, 306), Fernie and Tempesti.

Nova Vul 1968 (no. 2) was discovered by Kohoutek (*IAU Cir.* 2106). The maximum ($B = 9{\cdot}25$) was reached on July 19. From the light curve the nova can be classified as a normal fast nova (Rosino *et al.*, *Astr. Space Sci.*, **4**, 392, 1969).

Nova Ser 1970 was discovered by Honda on February 13 (*IAU Cir.* 2212 and 2214).

Rosino has continued the systematic search of novae in M 31 and M 33. Many novae have been found during the last years.

Gyldenkerne, Mejdahl and West published photoelectric observations of Nova Her 1960 (*Køb. Publ.*, 201) obtained at Brorfelde by means of standard *B* and *V* filters and 6 narrow interference filters.

Intrinsic polarization in the light of an active nova was first reported by Eggen, Mathewson and Serkowski (*Nature*, **213**, 1261, 1967), who found the recurrent nova T Pyx to be variable in the degree, position angle, and wavelength-dependence of polarization. Zellner observed Nova Del 1967, and found variations of polarization, with both a small-amplitude fluctuation with a time scale of days and a secular variation with a time scale of months. A few observations of Nova Vul 1968 (no. 1) suggest that it also has variable polarization (*Trieste*, 1968, p. 300).

Rose discussed a model for the nova outburst (*ApJ*, **152**, 245, 1968). Sobolev published a theoretical study of galactic novae in *Vistas in Astronomy*, **11**. The dynamics of a nova outburst were studied by Sparkes (*ApJ*, **156**, 569, 1969). A computer program has been prepared by Casinelli and Wegner to trace the physical conditions behind the shock wave during an outburst of RS Oph. The calculations predict detectable X-ray emission for about a month after light maximum. According to Kraft's assumption, explosive novalike phenomena occur in binary systems, when a white dwarf-like object accretes mass from an expanded companion. Secco has calculated (*Att. Cong. SAI, Padova*, 292, 1967) first approximation models of the degenerate star for different values of the mass, and the explosion periods have been evaluated for different mass increase rates. Friedjung summarized reasons (*Trieste*, 1968, p. 303) for believing that ejection of matter by novae continues long after maximum light.

20. OLD NOVAE AND HOT SUBLUMINOUS VARIABLES

An excellent review on the general subject of novae and novalike variables has been published by Mumford (*PASP*, **79**, 283, 1967).

A. *Novae near minimum light*

Mustel and Boyarčuk studied photographs of T Aur, GK Per, V 603 Aql and DQ Her, and found that the envelope of novae consists of an equatorial belt and two polar caps. In 1968 the envelope of DQ Her had the electron density of $2 \times 10^3/\text{cm}^3$ and the mass $\sim 10^{29}$ g. With the purpose of studying the interaction of matter ejected in nova explosions with surrounding interstellar material, a number of old novae are being photographed anew with the 200″ by Münch and compared with earlier images obtained by Baade and others. The envelopes of N Her 1934 and N Per 1901 show indications of deceleration but a few of the filaments in the complex envelope appear to be accelerating.

Nova (HR) Del 1967 may have been variable in brightness before outburst according to Van den Bergh and Racine (*IB*, 212). A pre-outburst spectrum was found on a plate exposed July 16, 1960 at 580Å/mm (*PASP*, **79**, 584, 1967). The prenova was a very blue star of magnitude $12 < V < 13$.

Nova (DQ) Her 1934 was studied by Nather and Warner at McDonald Observatory (*MN*, **143**, 145, 1969) using synchronous signal averaging in order to obtain the properties of the 71ˢ7 pulsations discovered by Walker in 1956 (*ApJ*, **123**, 68). They obtained a light curve which is a pure sinusoid within the accuracy of measurement. The pulsation amplitude is $\geqslant 1$ magnitude, it varies by at least a factor of 5. The binary period is increasing probably as a result of mass loss from the system at a rate $2{\cdot}03 \times 10^{-8}\ M_{\odot}$/yr. Variations in light intensity from DQ Her seem to be a superposition of several components. Beside the regular pulsation of the $71^{s}.7$ period and the eclipse phenomenon there are very short sporadic variations (flickering) and longer-term changes in light level which cause the shape of the eclipse curve to vary from cycle to cycle, as well as affecting the light level outside of eclipse. The flickering appears undiminished even at central eclipse and probably arises throughout the nebula.

Nova V603 Aql 1918 undergoes small light fluctuations, and present data by Landolt (*PASP*, **80**, 481, 1968) indicate the object to be fainter than has been reported in the past.

Nova Vul No. 1. 1968 has been studied by L. Meinunger on old plates. He could not find any obvious variability before the outburst (*IB*, 272, 1968).

Nova (WZ) Sge 1913, 1946 has been observed photoelectrically in July 1969 by Krzeminski and Bretz at Haute Provence Observatory. Krzeminski and Smak proposed a new model of the binary system (*Budapest*, p. 371).

B. *Hot variables, and short-period eclipsing systems with additional activity, X-ray variables*

At the Kenneth Mees Observatory, Savedoff *et al.* has assembled equipment which permits the rapid digital recording of photoelectric signals. Preliminary observations made at the rate of 10 per second of *40 Eri B* indicate that the star has no variation greater than 0·01 magnitude in the period range between 1 and 10 s. *HZ 29*, a peculiar, hydrogen deficient white dwarf with broad, apparently double absorption lines of HeI, was shown by Smak to be a variable star with a period of about 18 min (*IB*, 182, *AcA*, **17**, 255, 1967). The available photometric and spectroscopic data support the hypothesis that HZ 29 might be a close binary system. Stothers suggests (*ApJ*, **154**, L91, 1968) that HZ 29 is a luminous star of high mass and heavy composition and that its variability is due to radial pulsation. According to Bomy and Ledoux (*Ad'Ap*, **28**, 358), stars composed essentially of pure He become unstable against nuclear-energized pulsations above 7–8 $M_{\odot}$. Ostriker and Hesser got for the period from new photometric data and autocorrelation analysis technique the improved value of $17^{m}.52$. In addition null results have been established for a group of white dwarfs for which, in some cases, a limit of $A \leqslant 0{\cdot}003$ magnitude can be set for the maximum Fourier amplitude between 300^{s} and 2 . Analysis of extensive observations of novae and nova-like objects is in progress. Smak presented a new interpretation of *HZ 22*, which was considered till now a normal B-type binary with $P = 3^{d}.58$ (*AcA*, **19**, 165, 1969). According to Smak the star is a peculiar hot subdwarf binary with period shorter than one day ($P = 0^{d}43234$). Landolt derived from *UBV* photoelectric data a $P = 12^{m}.5$ variation in the brightness of a white dwarf-like blue star (*ApJ*, **153**, 151, 1968). The variable may be an eclipsing system since the radial velocity varies with an amplitude ~ 500 km/s. He also observed in *UBV* the nova-like variables AG and HR Car. Hesser *et al.* discovered a new, blue, short-period ($P \sim 10^{m}$) variable star, G-44-32, which is catalogued by Eggen and Greenstein as a white dwarf. They found the central stars of planetary nebulae extremely quiet with regards to both intrinsic flickering and coherent, low amplitude variability.

EM Cyg was observed by Mumford and Krzeminski (*PASP*, **79**, 283, 1967; *ApJ Sup.*, **18**, 429, 1969). The observations suggest that the star is an eclipsing system in which the greatest diminution of light at primary minimum may be attributed to an eclipse of a gaseous disk or shell. The light curves exhibit quite complex behaviour. Many features in the light curves are reminiscent of those found for old novae and novalike variables.

TT Ari was observed by Smak and Stepien at Lick and Haute Provence. Over 3000 measurements were made mostly in the ultraviolet. The light variation can be resolved into three activities. It is suggested that the star is a hot, subluminous close binary system with $P = 0^{d}.2658$ (*Budapest*, p. 355).

UX UMa was observed by M. F. Walker photoelectrically in 1967 and 1968 at Lick. The period, which apparently increased from 1937 to 1953 and then decreased from 1953 to 1962, has continued to decrease since 1962, but at a slower rate. Bretz also observed the star in February and March 1967.

Sco X-1 holds a position in the front of interest since its optical identification in June 1966 Rapid variations in the optical radiation has been reported by many observers (cf. Hiltner and Mook, *ApJ*, **150**, 851, 1967; Osawa and Ichimura, *Tokyo Bul.*, 185). Hiltner organized a world-wide monitoring program of this object for 1970. Its X-flux decreased exponentially over the period 1965–68 with a time constant of about 4ʸ1 (Roco *et al.*, *ApJ*, **157**, L133, 1969). Van Genderen published (*AA*, **2**, 6, 1969) an investigation of the object based on photoelectric five-colour observations made at the Leiden SS in 1966–68. He found a semi-periodic light variation with $P = 0^d5276$. Apart from some irregularities, the light curve and the colours are characteristic for a pulsating star. The object is bluer when brighter (Stepien *ApJ*, **151**, L15, 1968). Narrow-band and *UBV* photometry was carried out by Johnson (H. M.) and Golson (*ApJ*, **153**, 1968). Radio emission of the object was detected by Andrew and Preston (*Nature*, **218**, 855, 1968) with the 46-m radio telescope of the Algonquin Radio Observatory. Ables has shown that the radio emission of Sco X-1 is also variable (*Publ. AS Austr.*, **1**, 237, 1969).

Cen X-2 is a highly variable X-ray star, very much like a nova emitting in the visible spectral range (Roco *et al.*, *ApJ*, **157**, L127, 1969). Its optical identification with WX Cen is yet not certain.

Cyg X-2 may be considered fairly well identified with a bluish optical object. Kristian, Sandage and Westphal's photoelectric measures show in the star variations similar to Sco X-1. The star seems to be a spectroscopic binary.

C. *U Gem and Z Cam stars*

Smak discussed in his introductory report at the Budapest Colloquium some problems connected with the interpretation of U Gem type stars (p. 345).

Z Cam is found to be a double-line spectroscopic binary with an orbital period of $0^d.289834$ and mass ratio near unity (Kraft, Krzeminski and Mumford, *Ap. J.* **158**, 589, 1969). Though eclipses have not been detected, the light curves show a number of features which recur with the same period as the spectroscopic orbit. The light curves reveal major eruptions and rapid, irregular fluctuations.

Z Cha was shown by Mumford to be an eclipsing binary with period of $0^d.0745$ (*IB*, 337). The entire eclipse lasts some 7^m, with minimum being $2^m.5$ to 3^m long.

SS Cyg has according to Walker (*ApJ*, **154**, 157, 1968) an increasing period giving a computed minimum mass loss of $\simeq 1{\cdot}8 \times 10^{-7}\ M_\odot$/yr. The periods, masses, and separation of the components are very similar to those of DQ Her. The greater rate of mass loss in the case of SS Cyg may have some direct influence on the nature and frequency of the nova outburst. Walker and Chincarini observed rapid flares having their maximum amplitude in the far ultraviolet, and decreasing rapidly in intensity with increasing wavelength, becoming invisible around 4100 Å (*ApJ*, **154**, 157, 1969, see Appendix I). Reduction of the observations obtained in September 1966 by Bretz and Mirzoyan at the Haute Provence Observatory are well advanced. The star has again been observed in *UBV* July 1969.

U Gem was discussed by Mayall (*Budapest*, p. 377) using 97 eruptions.

EX Hya, an eclipsing variable with two eclipses and $P = 0^d.068$, was treated by Mumford (*ApJ Sup.*, **15**, 1, 1967).

Mumford obtained fairly extensive data for BV Cen, μ Cen, BP CrA, VW Hyi, HP Nor, HH Tel. There are only in the light curves of BV Cen and VW Hyi phenomena that recall such known eclipsing variables as EM Cyg.

Romano has continued the observations of some U Gem stars. Four new variables of this type have been discovered during his survey = GR 135, 141, 166, 177.

Gorbatzkij and his students considered the influence of gas streams in binaries on the brightness and colour of U Gem type stars. It was found that inhomogeneities in a stream can produce rapid brightness fluctuations. Koroviakovsky studied the exchange of matter in U Gem stars (*Budapest*, p. 391). Osaki proposed a mechanism for the outbursts (*BAAS* **1**, 357, 1969).

L. DETRE
President of the Commission

APPENDIX I

REPORT OF THE COMMITTEE ON THE SPECTRA OF VARIABLE STARS

(prepared by M. W. Feast, Chairman)

The field covered by this report overlaps particularly with the work of Commissions 29 and 45 whose reports should be consulted. I have been greatly helped by reports on Soviet work by Professor Kukarkin and Dr Boyarchuk and a bibliography of European literature by Bloch.

1. *General*

The Sonneberg programme for determining spectral types of variables has been described by Götz and Wenzel (*AN*, **290**, 99) and several lists have been published in *MVS*. Bakos is studying high dispersion spectra of a number of variables that are members of visual binary systems.

2. *Mira, semi-regular and RV Tauri variables*

Stephenson and Terrill (*ApJ*, **147**, 148) and Houk (*IB*, 228) have given spectral types for a considerable number of long period variables. Terrill (*AJ*, **74**, 413) has made an extensive study of the variation of spectrum with phase for 51 red variables. Deutsch, Keenan and Garrison have continued their extensive programme of spectral classification of Mira variables. Keenan and Dessy (*ApJ*, **146**, 583) suggest that the relatively early spectral type of the very long period Miras (R Cen, R Nor etc.) showing double maxima indicates that the fundamental parameter is the half-period (between successive maxima). A similar conclusion (*ApJ*, **93**, 380; *MN*, **125**, 367) was reached earlier by others on the basis of the correlation with period of the difference between absorption and emission velocities. Feast has used K subgiant companion of U Men to derive an absolute magnitude of $-0^{m}.5$ for the Mira star. The radial velocity of the companion agrees with the absorption velocity of the variable near maximum (*Obs.*).

Herbig and Zappala have now discussed in greater detail the identification of certain emission lines in χ Cyg as the inverse predissociation spectrum of AlH (*ZfA*, **68**, 423). J. P. Swings reports that studies of the relative intensities of [FeII] and AlO emissions in Miras support the idea of stratification of the emitting layers. The Swings' have also discussed (*Lunteren*) the possibility of molecular fluorescence being of significance for these stars. Maehara reports the identification of two emission lines (4178, 4138) in *o* Ceti as due to SiH. Fluorescence excitation in Miras has been discussed by Gauzit (*Ann. Astr.*, **30**, 55). Herbig (*Liège*, 1968, p. 353) has discussed the presence of forbidden emission lines in Miras and other cool stars.

The study of red variables, including Miras, in the infrared is being actively pursued in several places and has yielded very important results. This work has recently been reviewed by Spinrad and Wing (*Ann. Rev.* **7**, 249).

Tsuji has made a detailed study of the Se variables R And and R Cyg. Strong circumstellar K I and Rb I have been found. A rapid mass loss (10^{-5} $M_{\odot}$/yr) is deduced. This is further evidence that the S stars are short lived. Hα in Se variables is bifurcated (probably by overlying ZrO) (Bretz, *Trieste*, 1966, p. 166). Although the "Potassium Flares" found in some otherwise normal stars are likely to be spurious (*PASP*, **79**, 351), Miss Bretz points out that the K I emission lines she found in 1966 in χ Cyg occurred on several different nights and are quite narrow, indicating that they are stellar emissions. They did not appear at the (brighter) 1967 maximum. Catchpole and Feast are carrying out a spectroscopic survey of southern C and S stars (*Henize and Wray Catalogues*). A considerable number of variables are included in the programme. Stephenson (*ApJ*, **150**, 543) has discussed the violet opacity in the C-S star V 372 Mon. In the C-S category, Catchpole (Pretoria) is studying UY Cen and Culver (Perkins) CY Cyg. Pik-Sin The (*PASP*, **80**, 104) points out that objective prism spectra show R Vol and HD114586 (*KZP*, 1995) to be C and S type respectively. These classifications are confirmed by unpublished Radcliffe spectrograms. R Vol showed Hα emission in May 1953. Boyarčuk is studying high dispersion spectra of semiregular variables.

VY CMa, an irregular M-type variable with several nebulous companions has been found to be an intense infrared object, and OH and H_2O source and to be strongly polarized. Its spectrum is discussed by Herbig (*Liège*, 1969) and Hyland *et al.* (*ApJ*). It is remarkable for containing ScO, YO and TiO in emission. Herbig associates VY CMa with the H II region around NGC 2362 and proposes a detailed model with the star as a pre-main-sequence object though Hyland *et al.* prefer to regard it as highly evolved. The infrared radiation is believed to originate from dust near VY CMa. This is supported by the observation (Feast, *Obs.*), that in one of the nebulous companions (reflection nebulae) the red TiO bands are stronger than in VY CMa itself where they are filled in by an overlying continuum.

A detailed study of μ Cep (M 2Ia) is being made by Deutsch who has discovered a radial pulsation with a 1100 day period. Bond has confirmed with slit spectrograms the report by himself and Bidelman that the red variable V 422 Cen was like R Aqr in showing strong nebular emission lines. The fact that the M 8·5 III star V 1398 Sgr (in the direction of the galactic centre) has only a small light amplitude has been taken to imply a difference from late type stars near the sun (Spinrad, *et al.*, *AJ*, **74**, 525).

The spectrophotometric gradient of the RV Tau star AC Her has been studied by Alania (*Abastumani*, **36**, 16) and a further spectroscopic study made of U Mon (Aliev, *Sov. AJ*, **10**, 794).

3. *Cepheids*

Schneider has continued his work on the classification of Cepheids in the Barbier, Chalonge, Divan System (*J.d'Obs.*, **50**, 419; *AA*, **2**, 195). This has led to new determinations of reddening for 20 Cepheids. Guseva and Tsarevsky (*PZ*, **16**, 292) have attempted to put published spectral types of Cepheids onto a common system. Opolski uses their results to discuss the spectral type-period relation (*AcA*, **18**, 515). Lloyd-Evans (*MN*, **141**, 109) has made new radial velocity observations of cepheids and finds the frequency of occurrence of cepheid binary systems to be at least 15%. V 636 Sco, S Mus and AW Per are spectroscopic binaries. The latter has a composite spectrum as does KN Cen. This work is being continued. From spectroscopic observations Carevsky and Vitričenko (*Crimea*) suspect the duplicity of the blue companion of δ Cep.

Rodgers and Bell have continued their high dispersion studies of cepheids (*MN*, **135**, 121; **136**, 91; **138**, 23; **139**, 175; **142**, 161). They determine atmospheric parameters and fix the minimum width of the instability zone as $\Delta(B-V) = 0{\cdot}24$ at 10 day period. The effect of line blanketing has been studied and chromospheric effects have been found in Hα for β Dor and (markedly) 1 Car. In the latter star one of the chromospheric Hα components is stationary suggesting a (non-pulsating) circumstellar shell. In the disk cepheid κ Pav the *s*-process elements are deficient relative to *e*-process elements. A similar result is found for RT Aur by Bappu and Raghavan (*MN*, **142**, 295). Caputo (*MAI*, **39**, 401) has made a detailed analysis of ζ Gem whilst Breger (*MN*, **136**, 61) determined radius variations for U Sgr. Wallerstein has begun a programme to study spectra of cepheids in the red region. With Grenfell he has determined line shift and depth effects in SV Vul and plans to work on T Mon. Kipper (*Tartu*, **36**, 234) has made a curve of growth analysis of Polaris.

The carbon variable RU Cam continues to attract attention. Wallerstein and Crampton (*ApJ*, **149**, 225) showed that when its light was nearly constant so was its radial velocity. Detailed spectroscopic analyses have been made by Faraggiana and Hack (*ZfA*, **66**, 343) and Wallerstein (*ApJ*, **151**, 1011). Fe/H is normal and C12/C13 $\sim$ 9. Blue shifted Ca II emission gives evidence of mass loss.

4. *RR Lyrae, Dwarf Cepheids, δ Scuti and β CMa stars*

During the last few years detailed studies of δ Scuti stars from high dispersion spectra and spectral scans have considerably increased understanding of these stars (papers by Kuhi, Danziger, Dickens, Bessell etc., see *ApJ Sup.*, **18**, 167). Due to improvements in the absolute spectrophotometric calibration and in blanketing corrections Bessell's conclusions differ significantly from earlier work. In particular he determines much higher masses (~ 2 $M_\odot$) and concludes that these are old A stars evolving off the main sequence. Unlike most pulsating stars some δ Scuti variables have high rotational velocities (cf. *Astr. Lett.*, **1**, 213). Bessell finds from a curve of growth analysis that δ Sct, δ Del and ϱ Pup are similar to the Am stars in microturbulence and abundances. However Reimers (*AA*, **3**, 94) finds different relative abundances for δ Del using a model atmosphere approach and connects this star with the Ap and λ Boo stars. The Am star 28 And appears to show δ Sct type variations (Nishimura, *Astr. Space Sci.*, **3**, 77). Further radial velocity observations have been obtained of 4 CVn (*PASP*, **79**, 330) and 14 Aur (*Astr. Lett.*, **2**, 175).

Bessell has also studied the AI Vel stars (Dwarf Cepheids). Slight metal under-abundances and low masses (~ 0·5 $M_\odot$) are found and he considers them as posthelium-flash stars. The high lithium abundance of VZ Cnc which presumably belongs to this class is notable (*ApJ*, **147**, 151) (see also *C.R.*, **267B**, 657).

Preston and Spinrad (*ApJ*, **147**, 1025) investigated the 90 day cycle of the RR Lyrae star RV UMa. No stillstand phenomenon was found. In BL Her (1·3 day period) H emission and $(U-B)$ excess correlate as for RR Lyraes but the phase lag of luminosity relative to velocity is different (Preston and Kilston, *ApJ*, **148**, 787). Values of the Δs parameter for RR Lyraes were determined by Alania (*Abastumani*, **36**, 24) and Clube *et al.* (*Mem. RAS*, **32**, 101) who also give radial velocities.

Orlov (*Soviet AJ*, **12**, 89, 450) finds no apparent variation with time of the equivalent widths or central intensities of H and He lines in the β CMa stars, γ Peg and δ Cet. Periodic changes of equivalent width of some lines in δ Cet are however reported by Ciurla and in γ Peg by Contel (*Budapest*, p. 297). The rotational velocity of RS Sex is high (*IB*, 260). Breger (*MN*, **136**, 51) discussed α Lup and β Cen and van Hoof (*Toronto*, p. 237) θ Oph.

5. *R CrB, symbiotic and Nova-like variables*

Observers at the Cape and Radcliffe Observatories have obtained extensive photometric and spectroscopic observations of the R CrB star RY Sgr during and since the 1967 minimum. Early results (*Budapest*, p. 253) show a very rich emission spectrum during decline, a blue continuum (interpreted as the electron attachment spectrum of CN), evidence for rapid mass ejection and for heavy reddening during minimum. The star has a marked infrared excess (Lee, Feast, *ApJ*, **157**, L173) as does R CrB itself (Stein *et al.*, *ApJ*, **155**, L3). A periodicity of about 39 days found in the light and colours is being sought in the radial velocities. Greenstein noted weak chromospheric emission in R CrB during a shallow minimum (*IAU Cir.*, 2158). At the University of Western Ontario, Gray, Moorhead and Marlborough are working on spectroscopic problems of R CrB stars. Abundances in R CrB, XX Cam and RY Sgr have been determined. A detailed spectroscopic description of Herbig's C II emission object V 348 Sgr (which may be related to the R CrB stars) has been given by Houziaux (*BAC*, **19**, 265).

Problems of the symbiotic stars have been reviewed by Boyarčuk (*Budapest*, p. 395) and Sahade (*Liège*, 1968, p. 309). There is considerable interest in these stars at several observatories in the USSR (*Soviet AJ*, **10**, 331; 783; **11**, 8; 453; *Tartu*, **35**, 188, *Izv. Krym*, **38**, 155. *Astrofizika*, **2**, 50, 103). Line intensities and continuum energy distributions have been determined in Z And, AG Peg and AG Dra. Boyarčuk finds that on plates taken by Herbig in 1961 during a flare Z And resembles an A3 P Cygni type. MHα 208–51 is a symbiotic object (*IB*, 254). TX CVn has been briefly discussed (*Padova Comm.*, 46). The general mode of excitation in symbiotic stars is discussed by Ilovaisky and Wallerstein (*PASP*, **80**, 155). Bloch *et al.* have a study of Z And in press and Aller and Wares have observed RY Mus. Friedjung has investigated AG Peg in detail (*AA*) whilst HD 4174 and WY Gem have been described (*MAI*, **39**, 633; **40**, 25).

MHα 328–116 (= V1016 Cyg) continues to be observed. The lines of higher ionization are strengthening (*Liège*, 1968, p. 363, 369; *Ap. Lett.*, **3**, 141, *IB*, 372). The underlying continuum in 1965–66 corresponded to a very hot star (O'Dell, *ApJ*, **149**, 373). An object of possibly similar type is HBV 475 recently discovered by Kohoutek (*IB*, 384, *IAU Circ.*, 2174, 2176, 2182).

In June 1967 Deutsch found that CH Cyg had undergone another outburst and that a hot continuum was present. The spectrum has been extensively studied by Faraggiana and Hack (*Liège*, 1968, p. 317; *Trieste*, 1968, p. 336). There is evidence for mass loss. The system seems to consist of an M giant and an unstable blue subdwarf. Similar conclusions were reached by Luud (*Tartu*). There is also evidence for long period orbital motion (Deutsch, Hack) and rapid fluctuations in the ultraviolet continuum (e.g. Walker *et al.*, *ApJ*, **156**, 117). Swings discussed the ultraviolet line spectrum (*Astr. Lett.*, **1**, 54).

η Car continues to receive much attention. The infrared continuum (apparently thermal) is exceptionally strong (at 10 and 20μ it is the brightest object outside the solar system, Westphal and Neugebauer *ApJ*, **156**, L45). Pagel and Lambert (*Nature*, **221**, 325; **223**, 726) used the [FeII] lines to determine a much higher reddening than previously supposed. They consider that the visual continuum may well be due to two photon emission from hydrogen. Viotti (*Astr. Space Sci.*, **5**, 323) has investigated the absolute magnitude, the continuum and spectral variations whilst Thackeray has discussed the line in the 9000 Å–11 300 Å region (*Liège*, 1968, p. 323; *MNASSA*, **28**, 37; *MAI*, **38**, 451; **39**, 105). Aller has also worked on η Car. Thackeray (*Liège*, 1968, p. 327) has described the progressive ionization (up to [FeVII] in RR Tel and other forbidden lines. Evidence for [NiIV] has been strengthened (Lambert and Thackeray, *Astr. Space Sci.*, **5**, 283) and [FeIV] has been rediscussed on the basis of laboratory work (Edlen, *MN*, **144**, 391).

Herbig and Boyarčuk (*ApJ*, **153**, 397) find the spectrum of the strange variable FG Sge changed from B4 I to A5 Ia between 1955 and 1965. This is interpreted as the ejection of a nebular shell from an (invisible) hot subluminous star. The nebula around the star is probably the remnant of a previous ejection (see also *PASP*, **80**, 712; **81**, 274). AG Carinae is being studied by Viotti and Caputo.

6. *Flare stars, T Tauri stars and related objects*

Objective prism studies of some T Tauri stars have been carried out by Götz and Wenzel (*MVS*, **4**, 5). Hydrogen emission in T Tauri stars has been discussed by Anderson, Kuhi, Dibaj and Esipov whilst Walker has shown that matter falls into the UV excess stars at a variable rate (*Budapest*, p. 93, 103, 107). Infrared CaII emission in MWC 1080 suggests that this may be a T Tauri star (MacConnell, *PASP*, **79**, 66).

Haro and others discuss the spectra of flare stars (*Ton. Bul.*, 31) and suggest that the intensity of the CaII emission lines in G 8–K 0 or later types is a good criterion for predicting flare activity. Kunkel (*ApJ*) has obtained time resolved spectra of dMe flares. He finds that these flares cover only a few percent of the stellar surface despite the large energy output. Kandel (*Trieste*, 1966, p. 146) discusses emission in HD 119 850. Wolf 359 (dM8e) was observed spectroscopically during a flare (Greenstein and Arp, *Astr. Lett.*, **3**, 149) Gershberg (*Izv. Krym.*, **38**, 177) has analysed simultaneous spectroscopic and photometric observations of flare stars. Thackeray has classified Butler's flare star as dKO (see Andrews *PASP*, **79**, 368).

Herbig (*ApJ*, **152**, 439) has drawn attention to the fact that R Mon is a non-stellar object and that there is no convincing spectroscopic evidence for the presence of a G or K type star, the absorption spectrum sometimes seen being that of a transient shell.

7. *Novae*

A large number of observational papers have been published on recent novae especially of the peculiar slow Nova Del 1967. Much more material is unpublished. It remains a problem that so little of the available material is ever taken past the purely descriptive stage (if even as far as that). A group has formed at the Institut d'Astrophysique (Paris) specifically to study nova spectra.

Nova Del 1967. A pre-outburst, objective prism observation shows an early type continuum (Stephenson, *PASP*, **79**, 584). It is not possible to list all the papers and work on this object. There have been extensive observations at Lick (cf. *Sky Tel.*, **35**, 77) Victoria (cf. *Trieste*, 1968, p. 290), Haute Provence (cf. *AA*, **1**, 403; *Trieste*, 1968, p. 281), Asiago and elsewhere.

Nova Vul 1968, has been described by Mammano *et al.* (*Budapest*, 1968, p. 271), Bloch (*Haute Provence*, **10**, no. 13) and Dorschner *et al.* (*AN*, **291**, 217).

Nova Vul 1968 no. 2, has been described by Rosino *et al.* (*Astr. Space Sci.*, **4**, 392).

Nova Her 1963, Doroschenko has made a detailed study of this nova (*Soviet AJ*, **12**, 95; *Astrofizika*, **2**, 240). Tremko (Skalnate) is working up Hamburg spectra and Friedjung (Paris) is studying the Balmer absorption line profiles (see also *Tartu Pub.*, **36**).

Nova Her 1960. Work on this nova has been published by Hidajat (*ZfA*, **68**, 164), Luud and Kuningas (*Tartu Pub.*, **35**, 413) and Savel'eva (*Soviet AJ*, **11**, 576).

General. Antipova (*Astr. Zu.*, **46**, 366) considers the great strength of CN in DQ Her at outburst as an abundance effect in the shell.

Recurrent Novae. The equivalent widths of the coronal lines in T CrB have been measured by Bochonko and Malville (*PASP*, **80**, 177) and the radius and density of the coronal envelope determined. The 1967 outburst of T Pyx was described by Catchpole (*MN*, **142**, 119 and Rosino *et al.* (*Budapest*, p. 261). Tolbert, Pecker and Pottasch have reduced a considerable number of plates of RS Oph at its 1958 outburst and Pottasch has proposed a detailed quantitative model (*BAN*, **19**, 17, 227). Spectra of the star between the 1958 and 1967 outbursts have been described by Wallerstein and Cassinelli (*PASP*, **80**, 589) and Barbon *et al.* (*Budapest*, p. 257) whilst Code (*ApJ*, **151**, L149) described 1967 spectra.

8. *Pulsars, X-ray sources, hot subluminous variables*

NP0532 (the Crab Nebula pulsar) has been investigated by a number of workers (e.g. Lynds, *ApJ*, **157**, L11) who, like Minkowski earlier, find the spectrum to be continuous. Oke (*ApJ*, **156**, L49), from spectral scans finds the energy distribution to be the same in the main and the interpulse and to be well fitted in the visual region by a Planck function at 10000 K. The measurements have been extended into the infrared by Neugebauer *et al.* Westphal *et al.* (*ApJ*, **155**, L109) show that the star near CP1919 is almost certainly not associated with the pulsar.

The most detailed spectroscopic study of Sco X-1 so far is probably that of Westphal *et al.* (*ApJ*, **154**, 139). There is a hot continuum with H, He I and II and other emission lines. The spectrum has a general resemblence to that of old novae and U Gem stars. Line strengths relative to the continuum and radial velocities vary in a complex manner which is not yet properly understood. The object may be a spectroscopic binary but this has not been proved. Spectrophotometry has been published (*ApJ*, **149**, L45; **155**, 1). The spectrum of Cyg X-2 is quite complex (Lynds, Burbidge, Sandage, Kraft etc. *ApJ*, **149**, L41; **150**, L95; L99; L83; **155**, L159). He I and II emissions are present and an absorption component which Kraft and Demoulin identify as probably a G subdwarf. Some of the absorption lines appear to be formed in gas streams. Large velocity variations have been found but it has not been conclusively demonstrated that the star is a binary though spectrophotometric scans suggest the presence of a hot continuum (*ApJ*, **151**, L93). Cen XR-2 seems almost certainly identified with the rapid irregular variable WX Cen whose spectrum resembles Sco X-1 (Eggen *et al.*, *ApJ*, **154**, L27). HD310376 (Schild, *ApJ*, **157**, 709) appears to have many similarities to Sco X-1 but is not known as an X-ray source. The spectrum varies from night to night. In addition to emission lines, H and He absorptions are sometimes seen.

The central star of the nebulosity NGC5189 is a WR star with strong O VI 3811, 3834 (*ApJ*, **152**, L135 and *Radcliffe*, unpublished). Because the star was an early (but now unlikely) identification for Cen XR-2 it is interesting that a spectroscopically similar object has been found close to the position of GX3 + 1 (Freeman *et al.*, *Nature*, **219**, 251; Blanco *et al.*, *ApJ*, **152**, L137). Sanduleak (*PASP*, **80**, 470; *IAU Circ.*, 2150) has found another two similar high excitation WR stars (one in the SMC).

Walker and Chincarini (*ApJ*, **154**, 157) find that at minimum in SS Cyg the irregular light variations result entirely from changes in the intensity of the continuum, the emission line intensities being unchanged. The outbursts are shown to be associated with the hot star around which there is probably a rotating ring of matter. Spectrophotometry of SS Cyg at the time of a flare is described by Chalonge *et al.* (*Haute Provence*, **10**, no. 9). This is the first of a series of such investigations on unstable stars. Bartaya (*Astrofizika*, **2**, 249) also obtained spectrophotometric observations of SS Cyg during flares. Kraft *et al.* (*AJ*, **73**, S21) find Z Cam to be a short period double lined binary (sdBe + G(?)) with both stars of about solar mass. Lortet-Zuckermann (*Budapest*, p. 381) finds broad H absorption during a stillstand of this star which she attributes to Stark broadening in the atmosphere of a dwarf G star.

BL Lac has been studied intensively since Shmitt (*Nature*, **218**, 663) pointed out that it coincided with the peculiar radio source VRO 42·22·01 (*ApJ*, **155**, L133; **156**, L41, L135; *IAU Cir.*, 2170). The spectrum appears to be continuous and may be a continuation of the radio synchrotron radiation. The continuum is highly polarized. As the star brightens the continuum reddens. A possible (unidentified) emission line has been found by Oke *et al.* It is not yet clear whether the object is galactic or extragalactic.

APPENDIX II

REPORT OF THE COMMITTEE ON VARIABLE STARS IN CLUSTERS

(prepared by Mrs. H. B. Sawyer-Hogg, Chairman)

Committee Members: M. W. Feast, G. H. Herbig, L. Rosino

Because of space restrictions, this report on the variable stars in clusters differs in form from those in earlier IAU reports, and is more condensed. Not all of the papers which have appeared since the last IAU General Assembly can be mentioned.

A definite trend is apparent in this field. A few years ago the aim of many programs was to detect new variables or to determine the periods of those known. While this is still being carried on, especially for the clusters in the far south, the emphasis has subtly changed to an avid pursuit of information about individual variables. For many variables, much more is known now than just the periodicity of their magnitude variations in photographic light. The number of observers studying these stars and the number of telescopes used in the program have greatly increased.

1. *Variables in globular clusters*

(a) *Catalogues*

The number of clusters searched for variables has now crossed the hundred mark. Of approximately 130 clusters considered to be globular, in and around our galaxy, 102 have now been examined to some degree. In these, more than 1900 variables have been found, but 10 clusters have no variables. The Atlas y Catalogo de Estrellas Variables en Cumulos Globulares al sur De $-29°$ by Fourcade, Laborde and Albarracin (Cordoba, 1966) appeared about the time of the last IAU and is of inestimable value for these little known southern clusters. The third catalogue of variables for all globular clusters by Sawyer-Hogg with references should appear in the *David Dunlap Publications* at about the same time as this report. A summary was given at the IAU IVth Colloquium on Variable Stars, Budapest, 1968 and appeared in this publication (*Budapest*, p. 475).

(b) *Discovery of new variables*

A cluster near the south celestial pole, IC4499, has come into prominence with the announcement by Fourcade and Laborde (*MAI*) that they have found 128 variables in it, with 42 other stars suspected of variation. This places IC4499 on the list of clusters richest in variables, exceeded only by M 3 and Omega Centauri. Most of the variables are of short period, and are probably RR Lyrae stars. IC4499 is a loose globular cluster, class XI, ang. diam. 6.′2.

In addition to this cluster, Fourcade and Laborde have found a few variables in each one of a long list of little known southern clusters, with the data given in their catalogue. New variables reported by other observers include the following: Omega Centauri (Wesselink, 2), NGC5897 (Sandage and Katem, 3), 5927 (Osborn, 1), 6266 (Gascoigne and Ford, 6), 6304, 3 and 6638, 3 (Terzan), 6656 (Hoffleit, 1), 6864 (Rosino and Coutts, 9), 7492 (Barnes, 3), 7089 (Margoni and Stagni, 4), 7099 (Terzan, 8).

Substantial numbers of variables are being reported and investigated in the surroundings of clusters. Published investigations include those of Terzan (*Publ. Haut-Provence* **9**, no. 24, 1968) in NGC6304 (22 variables) and Palomar 6 (28) and 6638 (16), and Kurochkin (*PZ*, **16**, 460, 1968) in 6779 (20). There is need to formulate a rule, perhaps based on cluster radii, as to when a variable will be considered within the cluster region and when in the field. Rosino reports that with the 67-cm Schmidt telescope he is searching the vicinities of the rich globular clusters NGC5024, 5272, 5904, and 6121 to determine if RR Lyrae variables sometimes escape from clusters.

(c) *Determination of periods*

It is to be expected that most of the newly determined periods are those of RR Lyrae variables. New periods have been published in the following clusters: NGC3201 Kukarkin (14), 6205 Osborn (2), 6402 Sawyer-Hogg and Wehlau (20), 6626 Deery and Hoffleit (1), 7006 Rosino and Ciatti (49), 7089 Kukarkin (4) and 7492 Barnes (3). Of these, the non-RR Lyrae stars are one or perhaps two Cepheids and a long period variable in NGC6402, a red semiregular in 7006, and a red giant in 7492.

Clusters under continuing investigation for determination of new periods are NGC6864, (Rosino and Coutts), 6366 and 6638 (Terzan and Sawyer-Hogg), and a group by Sawyer-Hogg and associates, 6626 (with Moorhead) 5897, 6093, 6402, 6656, and 6934 (with A. Wehlau).

(d) *Determination of period changes*

Extensive work on period changes have been carried out in the following clusters: NGC3201 Kukarkin, 34; 5024, Margoni, 17; 5139 Wilkens, 100; 5904, Coutts and Sawyer-Hogg, 66; 6341, Kheylo, 10 and Bartolini, Battistini and Nasi, 12; 7089 Sawyer-Hogg and Poole, 20. Szeidl has extended his work on the O-C diagrams in M 3 to 15 more RR Lyrae stars, bringing his total to 129 now. Miss Barlai continues her work on M 15, and Lovas is working on M 5. The paper on NGC5904 mentioned above (*David Dunlap Pub.*, **3**, no. 1, 1969) gives a summary of period changes in various clusters. There is no pronounced trend. Some periods remain constant, while others may be increasing or decreasing.

(e) *Color studies*

Substantially more work is being done to determine the position of variables on the color-magnitude diagram of a cluster, and to determine the light curves of variables in two or more colors. Clusters in which work on the color magnitude diagrams contributes significantly to our understanding of the variables are NGC 5024 (Margoni, *Padova Contr.*, 198, 1967), 5897 (Sandage and Katem, *ApJ*, **153**, 569, 1968), 6637 (Hartwick and Sandage, *ApJ*, **153**, 715, 1968) and 7006, (Sandage and Wildey, *ApJ*, **150**, 469, 1967).

Color studies of population II Cepheids have been made in Omega Centauri by Dickens and Carey (*ROB*, 129, 1967), with light curves in two colors for seven cepheids. This yields a period luminosity relation for population II Cepheids as

$$M_B = +\,0\overset{m}{.}2 - 1{\cdot}74\,\log P, \quad \text{and} \quad M_V = -\,0\overset{m}{.}13 - 1{\cdot}90\,\log P.$$

They note that these cepheids in globular clusters appear somewhat redder than the polulation I Cepheids. Omega Centauri is also the basis of a continuing color study by Geyer (*ZfA*, **66**, 16, 1967; *AA* **4**, 40, 1970, with Szeidl) with *U, B, V* observations for 62 RR Lyrae stars and photoelectric for 12, including the cepheids and eclipser.

In M 2, Demers (*AJ*, **74**, 925, 1969) has established a photoelectric sequence and determined light and color curves on the *UBV* system for the three cepheids and the RV Tauri variable, and has obtained similar material for M 13 and M 14.

Clusters in which color studies of the RR Lyrae variables are being made include M 5 (Coutts, Margoni and Stagni) and M 2 (Margoni and Stagni, 1969). Woolley and Stoy (*Qu. J. RAS*, **10**, 114, 1969) report that *UBV* curves for 22 variables in NGC6171 have been obtained. Sturch (*ApJ*, **148**, 477, 1967) has made three color observations of a few variables in M 3, M 5, M 92 and NGC 6171 with the particular aim of determining absorption values for the clusters.

Work is also directed toward the absolute magnitudes and other properties of the RR Lyrae variables. Clube (*ROB*, 136, 1968) has used his comprehensive study of the proper motions of 227 RR Lyrae variables to compare the properties of these variables in globular clusters with those in the general field. Jorgensen and Petersen (*ZfA*, **67**, 377, 1967) have applied their model to the RR Lyrae stars in globular clusters, to show that in M 3 the masses of the *ab* and *c* type variables are equal. Christy has discussed the luminosities of the RR Lyrae stars in the richest clusters in his Darwin lecture (*Qu. J. RAS*, **9**, 13, 1968). Thänert (*MVS*, **4**, 123, 1967) has extended his earlier paper on the correlations of characteristics with age in seven globular clusters. He finds that with increasing age the mean values of amplitudes, colors and luminosities decrease, while mean periods increase.

Kukarkin, Kurochkin and Kukarkina are studying the RR Lyrae stars in several globular clusters as well as examining the variability of horizontal branch stars near the instability band.

(f) *Proof of cluster membership*

A gratifying trend is the increasing emphasis on the determination of cluster membership for variables by the position of the variable on the color magnitude diagram, the radial velocity, or the proper motion. Ideally the proof comes from a combination of all three, as carried out by the Herstmonceux observers for Omega Centauri. In NGC5897, Sandage and Katem note that one of their three new variables is a red star near the top of the giant branch. Combining photometric results with a study of proper motions, Zhukov (*Soviet AJ*, **13**, 3–6, 1969) lists variables in M 3 and M 5 whose membership might be questioned, but the great majority of the RR Lyrae stars in the region belong to the clusters and are genetically related to them.

(g) *Variables of special interest*

In Omega Centauri. Attention has been called to two unusual stars. The first is Var. 1, the brightest variable in the cluster. D. H. P. Jones (*MN*, **140**, 265, 1968) reports that this variable (with period 58^d) is intermediate in characteristics between the population II Cepheids and the RV Tauri stars, though it is hotter than most stars of the latter class. The median spectral type of the metal lines is F 0 Ib, and the mean radial velocity is $+210 \pm 2$ km/s, with an amplitude of 25 km/s. The second is Variable 78, shown to be a cluster member by the Herstmonceux observers. According to Sistero (*IB*, **316**, 1968) it is the brightest known eclipsing binary of extreme population II. Woolley gives its *V* magnitude at maximum as 14·15, which compares with the mean *V* of 45 RR Lyrae variables in the cluster, by Dickens and Saunders, 14·52. The light curve is a typical EA, with a period of $1^d\!.16812879$. From a shape depth relation and a nomographic solution, Sistero has derived photometric orbital elements for the binary.

In NGC6637 the radial velocity observations needed to rule on cluster membership have been carried out for Rosino's variables 4 and 10 (with periods around 200^d) by Catchpole, Feast and Menzies (*Obs.*). Hartwick and Sandage from their color work showed these stars to be probable members. Harding at the Cape is monitoring this (as well as some other clusters) to determine periods. For Var. 10 Radcliffe observations give an emission velocity of $+58$ km/s, and an absorption velocity of $+67$; for Var. 4, the emission is $+30$ and the absorption $+39$. Mayall's velocity for the cluster is $+95$ km/s with a standard error of 30 km/s. The investigators conclude that, despite the spread in the determined velocities, there is no good reason for rejecting either variable as a member, noting that the three Mira stars in 47 Tuc have a spread similar in velocity. No Mira stars with periods greater than 220^d have yet been found as members of globular clusters.

In NGC5024 Wachmann (*Abh. Hamburg* VIII, 114, 1968) has shown that Var. 30 is a star with a Blashko effect, with a short period of $0^d\!.53548466$ which is changing, and a secondary period of

$37^{d}.0317$, as derived from very high maxima. As observations of variables in clusters become more numerous, probably more stars of this type will be found.

2. *Variables in open clusters and associations*

Searches continue to add to the number of variables known in open clusters and associations, but the total rises rather slowly. Because the variables in open clusters are relatively few, investigators tend to study them as a class, grouping them together from various clusters. Cepheids seem to have the major share of interest still, but eclipsing binaries and small amplitude irregular variables are getting more attention. So far no RR Lyrae stars have been found to be members of open clusters. No complete catalogue of open cluster variables exists.

(a) *Types of variables in open clusters and associations*

A summary and discussion of the types of variables in open clusters, with special emphasis on their place in the scheme of stellar evolution has been given by Čugainov for stellar associations (*PZ*, **16**, 107, 1967) and by Efremov for open clusters (*ibid.*, p. 176). Čugainov notes that stellar associations contain variables of the following types: eclipsing, variable supergiants, Be stars, type II supernovae, Beta Cephei stars, classical Cepheids, variable red giants, RW Aurigae stars, and flare stars. Variables of these types have ages from 10^6 to 10^7 yr, and associations contain only those variables whose age does not exceed this. Efremov reviewed the data on the different types of variables in open clusters, RW Aurigae, eclipsing stars, magnetic variables (Ap stars), Cepheids and red supergiants. He explains the occurrence of variability at certain stages in the evolution of a star. Later, Breger and Sanwal announced the discovery of the first Delta Scuti type variable in a cluster (*ApL*, **1**, 103, 1968).

(b) *New variables*

The Delta Scuti type variable just mentioned is star HR4864 = Tr 60, shown by Trumpler to be a member of the Coma cluster. Its variability was found in a systematic search of B, A and F stars. Its period is 80 min, and amplitude 0·02 in visual light. According to Breger and Sanwal, it has the earliest color of any Delta Scuti type known, spectral type A5V from a classification by Morgan, and $M_v = 1{\cdot}89$. Breger suggests that the ultra short period variations found by other observers in 21 Com may indicate pulsations similar to those of Delta Scuti. Although NGC752 was systematically searched by Breger, (*ApL*, **3**, 67, 1969) no other Delta Scuti stars were found, and the only variable noted in it was the eclipsing star DS And.

Hill (*ApJ Supp.*, **14**, 301, 1967) has found 20 new variables in open clusters and associations, including 7 eclipsing stars in the latter, and one eclipsing system each in NGC7160 and 7380. Bakos (*JRAS Can.*, **62**, 193, 1968; *Budapest*, p. 475, 1969) reported four probable new variables, one in NGC129 (the cluster with DL Cas) with a period of 30^d, two in NGC1502 with periods of 90^d and 2^d respectively. A suspected variable in NGC6705 brightened by one magnitude in the course of 30^m. He has announced (*Budapest*, p. 159) an unusual variable, HD 160202 in NGC 6405. Results on NGC1502 have formed a thesis by Kim at Waterloo. Bakos is also investigating NGC663, 1528, 2682, 6871 and 6913, by means of an image orthicon system at the Dearborn Observatory. In NGC6819 Barkhatova and Vasilevsky (*IB*, 374, 1967) report a possible variable. Small fluctuations are also reported by Barblishvili for Star no. 98 in NGC2360, noted earlier by Becker, and by Eggen, and Jackisch observed a possible outburst of HD108486 in the Coma cluster. Thackeray (*Radcliffe Repr.*, 67, 1968) reports that Lloyd Evans has found several variables in IC2581, including one Be star and a suspected eclipsing binary. In NGC6381 he has observed several stars near $V = 18$ with rapid variations and UV excesses. Percy (*Dunlap Obs.*) is testing for light variations those stars in h and χ Persei which seem to be rapidly rotating.

(c) *Cepheids in open clusters and associations*

The number of such objects known is not increasing very rapidly, but the analysis of them is becoming more and more detailed, with photometric measures in several wavelengths, as well as spectroscopic observations. Lindoff (*Arkiv f. Astr.*, **5**, 23, 1968) listed clusters with Cepheids and discussed their absolute magnitudes. Sandage and Tammann (*ApJ*, **157**, 683, 1969) have done extensive work on the double Cepheid CE Cas in NGC7790. The cluster NGC7790 is a remarkable one with three Cepheids (CF Cas, CE Cas *a*, CE Cas *b*) and two eclipsing binaries (QX Cas and V822 Aql). Sandage and Tammann think it likely that the 5^d and 4^d Cepheids which are components of the double star CE Cas form a physical pair. At a distance of 3200 pc their separation is 8000 A.U. As an outgrowth of this work, Sandage and Tammann derive the following *P-L-C* relations from 13 Cepheids in open clusters,

$$M_{\langle V\rangle} = -3{\cdot}425 \log P + 2{\cdot}52\,(\langle B\rangle^{\circ} - \langle V\rangle^{\circ}) - 2{\cdot}459,$$
$$M_{\langle B\rangle} = -3{\cdot}425 \log P + 3{\cdot}52\,(\langle B\rangle^{\circ} - \langle V\rangle^{\circ}) - 2{\cdot}459.$$

This compares with a *P-L-C* relation derived by Fernie (*AJ*, **72**, 1967) from galactic Cepheids, and tested on Cepheids in clusters and binaries, $M_v = -2{\cdot}55 - 2{\cdot}85 \log P + 2{\cdot}73\,(B-V) - 1{\cdot}60\,(B-V)^2$.

Breger (*MN*, **136**, 61, 1967) has discussed the velocity, radius and light variations of U Sgr in M 25, and found a radius of 53·8 solar radii by the Wesselink method. The distance modulus for the star, 9·15, is in good agreement with that of the cluster, 9·08. Latyshev (*AZ*, 473, 1968) derived a radius for $52{\cdot}1 \times 10^6$ km for S Nor.

In the field of NGC6649 Kholopov and Efremov (*AZ*, 437, 5, 1967) have shown that the Cepheid found by Roslund, Cuffey no. 64, has a 6-day period with a Blashko effect. Grubissich (*ZfA*, **68**, 176, 1968) thinks that AG Pup may be a member of the association II Puppis. However, Tammann (*IB*, 351, 1969) questions the reality of the cluster NGC6664 of which the 3-day Cepheid EV Sct is supposed to be a member.

(d) *Eclipsing stars*

The contact binaries in many clusters have been investigated and documented in great detail by Eggen (*Mem. RAS*, **70**, 111, 1967) in his massive paper on these stars. Kraft (1967) in a paper on the structure and evolution of the W Ursae Majoris stars gives the physical properties of TX Cnc, a member of Praesepe, which he finds to be very close to the average for all W U Ma stars. He notes the star is several magnitudes below the cluster break off, suggesting that some W UMa stars are essentially on the zero age main sequence. Kholopov and Sharov (*AZ*, 426, 452, 1967) have given new elements for the W UMa stars ES Cep and FQ Cep in NGC188, earlier thought to be RR Lyrae variables.

D. S. Hall (*PASP*, **79**, 630, 1967) reported observations of V 822 Aql in NGC7790 which confirmed a spectroscopic finding by Fitzgerald that its period is twice the previously assumed 2·6 d. From spectroscopic observations of QX Cas he finds a value of Δm between the two components of $0{\cdot}0 \pm 0{\cdot}3$. He considers this confirms Sandage's assumption that this binary is an actual member of NGC7790. Further observations on QX Cas were given by Sandage and Tammann in their cepheid paper already mentioned. Hall reports recent work on BM Ori in the Trapezium, and on V 729 Cyg in the Cygnus VI association. He has a *UBV* light curve of BS Sct, a star which may be a blue straggler in M 11, and has discovered (Hall and Mallama, *IB*, 383, 1969) that BD $-6°4932$ is variable, with a range of $0{\cdot}1^m$ and a period of 0·1591 d. It is in the region of M 11, but probably not a member.

(e) *Red stars*

Stothers (*ApJ*, **156**, 54, 1969), in a discussion of the pulsation hypothesis for massive red supergiants, lists 13 known variable M supergiants in galactic associations, with spectral types, magnitudes and periods. He thinks, the mechanism of pulsation is suggested by the observed and computed properties of these stars.

(f) *Shell stars*

Pleione has long been a fine example of a shell star, and Henriksen (*AA*, **1**, 457, 1969) has developed a model to explain the instability (1938–1951) of its extended atmosphere. Feast and Lloyd Evans (*Obs.*, **87**, 286, 1967) have found that HD170682 in M 25, discovered as variable by Mitchell, is a Be star that has undergone a shell episode like Pleione. Abt and Morgan (*AJ*, **74**, 813, 1969) consider the spectrum of star no. 41 in NGC2516 to be a superb example of a shell star.

(g) *T Tauri and flare stars*

Rosino and Pigatto have continued the study of T Tau stars embedded in nebulosities and belonging to open clusters. Blue and infrared plates centered on the Trapezium cluster, on ζ Ori, M 8 and M 20, NGC7635, h and χ Per have been obtained mostly with the 67-cm Schmidt telescope at Asiago. The material is under advanced reduction.

MacConnell, in his study of the Cepheus IV Association (*ApJ Sup.*, **16**, 275, 1968), found a new flare star and faint variable Hα-emission objects which indicate that they are probably of T Tau type associated with the dark cloud.

Cesevič has completed a detailed investigation of 10 variables from the T-association in Taurus. Ševčenko prepared a Catalogue of T-associations (*Tashkent Tr series II*, 11, 1967).

Maffei found in the region of *NGC 2264* on Loiano and Asiago plates 40 variables. Estimates of the blue and infrared magnitudes of almost 200 variables are in progress. Bodolian and Erastova accomplished a series of three colour observations of stars in the field of the same association. This work resulted in the discovery of 20 new T Tau variables (*Byurakan Soob.*, **38**, 3, *AC*, 435, 436, 439, 1967). Götz at Sonneberg is also investigating this region.

The compact T association near *NGC 7023* has been studied by Mirzoyan and Parsamian (*Budapest*, p. 165; *Byurakan Soob*, **40**, 31, 1969) on plates obtained with the Byurakan 40″ Schmidt telescope. 15 flare stars have been found.

Up to February 1969, Haro and Chavira detected flare-ups in 275 stars in the *Orion Nebula*. They will publish the finding charts of all these stars. There are two outstanding different kinds of flare stars in the Nebula: the fast and the slow ones. The slow flare stars seem to be confined to the Orion Nebula and possibly NGC2264. (Haro and Parsamian, *Tonantz Bol.*, **31**, 45, 1969). Rosino discovered 60 new flares in Orion during 135 hours of effective observations (*Low Luminosity Stars*, p. 181; *MAI*). Maffei and Blaghikh estimated more than 100 variables in the Nebula on 70 Hamburg plates from 1920–21. The behaviour of 56 of them during this period has been compared with that resulting from observations 1940–60 (*MAI*, **29**, 137, 1968). Roslund of Gothenburgh University has made at the Uppsala Station on Mt. Stromlo a 40 hour photographic search for flare stars in the Nebula and found 15 flares (*Ark Astr.*).

The observatories Alma-Ata, Asiago, Budapest, Byurakan, and Tonantzintla organized in 1968–69 a series of cooperative observations of flare stars in *Pleiades*. The immediate result of this work was an augmentation of the number of known flare stars in this cluster to 146. Two papers (Ambarcumjan, *Stars, Nebulae and Galaxies*, Yerevan 1969, in Russian; Ambarcumjan *et al.*, *Byurakan Prepr.*, 1, 1969) were devoted to a detailed analysis of the obtained data. A statistical study developed by Ambarcumjan shows that the total number of flare stars in Pleiades should be greater than 600. At $V = 13{\cdot}29$, there is a sharp border between the photographically observable flare stars and the non-flaring ones. The cooperative investigation of the cluster will be continued during the coming three years.

According to Haro and Chavira (*Tonantz. Bol.*, **31**, 1969) the relative intensity of the Ca II emission lines in stars as late or later than G 8–K 0 can be used as a good criterion for predicting that the stars can show or not the flare phenomenon.

3. *Variables in clusters of external galaxies*

Arp and Thackeray have published their comprehensive work on NGC1866 and its Cepheids (*ApJ*, **149**, 74, 1967). In this cluster near the Large Magellanic Cloud, seven Cepheids with periods from $2^{d}.6$ to $3^{d}.5$ have similar amplitudes, shapes of light curve, and color indices.

S. Gaposhkin reports that he has found some very small clusters (Mini clusters), one between the galaxy and the Large Magellanic Cloud, and others in the Cloud. The first contains at least three variables, two of which are RR Lyrae stars. The others have variables, which are perhaps not RR Lyrae stars. He thinks there are several hundred such clusters in the LMC.

APPENDIX III

REPORT OF THE WORKING GROUP ON FLARE STARS

(prepared by A. D. Andrews, P. F. Čugainov, R. E. Geršberg and V. Oskanian)

The Working Group was organized in order to promote and develop the cooperative studies of flare stars. Up to the present time our observation programmes were restricted to the UV Ceti-type stars only. Organizational activities followed the aims:

(1) To extend the existing network of photoelectric observers in order to provide 24-hour coverage of a star during selected sessions lasting usually 14 days.

(2) To collect data using standardized techniques in order to study frequency of stellar flares and secondary variations.

(3) To provide optical coverage for simultaneous radio, spectroscopic and polarimetric observations.

Optical observatories participating in the international programmes since 1967 have included: Abastumani, Armagh, Boyden, Budapest, Byurakan, Catania, Cerro Tololo, Crimea, Delaware, Lick, Mt. John (RASNZ), Odessa, Okayama, Smithsonian (satellite-tracking stations), and Steward, with collaboration from a number of amateur organizations. Radio observations were made by the Nuffield Radio Astronomy Laboratories at Jodrell Bank and by the CSIRO Radiophysics Division at Parkes and Culgoora. For recent joint programmes, references in parentheses to results published prior to October 1969 are summarized:

1967 YZ CMi (**1, 2, 43**)		UV Cet (**22–26, 43, 44**)
1968 YZ CMi (**3–9, 27**)	AD Leo (**16**)	UV Cet (**13, 27–33**)
1969 YZ CMi (**10–15, 34, 35, 45**)	AD Leo (**17–21, 45, 46, 48**)	

Up to 40% non-overlapping coverage throughout 14 day sessions has been achieved by combining photoelectric, photographic and visual observations. A full discussion of several hundred recorded flares must await the remaining unpublished material. During these sessions major flares (amplitude greater than 1·5 magnitudes in the blue spectral band, integrated intensity (**37**) of 10 minutes over the duration of the flare) were reported for YZ CMi (**1, 2, 3, 6, 10, 13, 14**), for UV Cet (**13, 22, 23, 25, 28, 31**), and for AD Leo (**21**). A number of flares were observed simultaneously by radio and photoelectric techniques. One particularly interesting result was the remarkable flare of YZ CMi observed by Kunkel, Andrews and Perrott during the night of January 18–19, 1969, which was observed simultaneously at two radio frequencies by Lovell (**35**). The radio emission in this flare has been discussed by Kahn (**36**) in terms of succession of shock waves in the expanding corona. Joint radiooptical observations of flares of UV Cet in September–October 1967 have been discussed by Higgins, Solomon and Bateson (**44**). Simultaneous photoelectric and polarimetric observations of YZ CMi and UV Cet have been obtained by Vardanjan (**27**), and of EV Lac by Efimov and Shakhovskoy (**47**). No measurable polarization of flare radiation has been detected.

The importance of long series of homogeneous observations of stellar flares for statistical studies has led in recent years to a change of emphasis from programmes purely designed to provide optical coverage for simultaneous radio observations to a variety of independent investigations which are, in fact, mutually inhomogeneous. Some of the problems of standardization of photoelectric technique in cooperative projects, and difficulties involved in the selection of standard parameters to describe flares have been discussed amongst observers (**37**). The majority of observers have selected the standard blue spectral band of broadband photometry for continuous monitoring. However, largely on the grounds that flares emit more strongly towards shorter wavelengths, the standard

U-band is favoured by some observers. Arguments against the use of the U-band are the low and erratic signal-to-noise ratio due to the intrinsic faintness of quiescent flare stars and the more serious atmospheric effects in the ultraviolet. Also, there are practical difficulties in defining and reproducing the standard U system. Astrophysical reasons for selecting a certain spectral band may, of course, out-rule this type of objection.

A preliminary survey of the available photometric material on flare stars shows that at least four kinds of light variations deserve further cooperative studies:

(1) The well-known, slow, secondary variations are again commented upon by several authors, for UV Cet (**22, 31**), V 1216 Sgr (**38**), and DO Cep (**39**), although, unfortunately, few accurate details are given. More thorough investigations are necessary. It is desirable to pay attention to a possible relation between these variations and flares.

(2) Possible secular variations of stellar brightness are to be examined. For this purpose standard comparison sequences have been measured for YZ CMi (**40**), EV Lac (**41**), AD Leo (**42**), DH Car (**49**) and V 1216 Sgr (**38**).

(3) Stellar flares. 24-hour patrolling of a flare star gives, in principle, the possibility to investigate the time distribution of flares. The departures from Poisson statistics of flare incidence first suspected by Andrews (**50**) are now confirmed by Čugainov (**51**) and Kunkel (**52**). Unfortunately, the last two investigations are based only on the part of cooperative observations.

(4) The long term variations of flare activity were suspected earlier by several authors. Possibly, we may believe each flare star to have its own cycle of activity analogous to the solar 11-year cycle. The present situation is still unclear owing to inhomogeneity of available material.

There are many other problems of flare stars which could be resolved using cooperative observations. The problem of the relation between flare stars of different kinds, from UV Ceti variables to the Orion population variables, is one of them. We hope to collaborate with persons interested in such problems.

BIBLIOGRAPHY TO APPENDIX III

1. Andrews *IB*, 307.
2. Chuadze, Barlishvili, Kuljanishvili *AC*, 408.
3. Eksteen, Schmidt *IB*, 264, *MNASSA*, **27**, no. 1–2, 7.
4. Čugainov *IB*, 266.
5. Osawa, Ichimura, Noguchi, Watanabe *IB*, 267, *Tokyo Bull.*, 180.
6. Oskanjan *IB*, 268.
7. Stepien *IB*, 305.
8. Bateson *RAS New Zeal. Var. Star Sec. Circ.*, 135.
9. Cristaldi, Rodono *IB*, 274.
10. Andrews, *et al.* *IB*, 325.
11. Osawa, Ichimura, Noguchi, Watanabe *IB*, 331, *Tokyo Bull.*, 192.
12. Cristaldi, Narbone, Rodono *IB*, 332.
13. Čugainov *IB*, 338.
14. Shakhovskaya *IB*, 339.
15. Bateson *IB*, 352.
16. Ichimura *Tokyo Bull.* 187.
17. Cristaldi, Narbone, Rodono *IB*, 333, 367.
18. Ichimura, Noguchi, Watanabe *IB*, 334.
19. Andrews, Perrott *IB*, 336.
20. Čugainov, Shakhovskaya *IB*, 340.
21. Szeidl *IB*, 345.
22. Bateson, Köhler *Southern Stars*, **22**, no. 5, 118.
23. Chuadze, Barblishvili *AC*, 451.
24. Solomon *IB*, 296.
25. Solomon, David *IB*, 297.
26. Kunkel *IB*, 315.
27. Vardanjan *AC*, 508.
28. Osawa, Ichimura, Noguchi, Watanabe *IB*, 310, *Tokyo Bull.*, 188.

29. Cristaldi, Narbone, Rodono *IB*, 317.
30. Čugainov *IB*, 338.
31. Čugainov, Havlen, Westerlund, White *IB*, 343.
32. Jarrett, Eksteen *IB*, 349, *MNASSA*, **27**, nos. 11–12, 145.
33. Bateson *IB*, 354.
34. Kunkel *Nature*, **222**, 1129.
35. Lovell *Nature*, **222**, 1126.
36. Kahn *Nature*, **222**, 1130.
37. Andrews, Čugainov, Geršberg, Oskanjan *IB*, 326.
38. Andrews *IB*, 273.
39. Herr, Broich *IB*, 329.
40. Andrews *IB*, 342.
41. Andrews, Čugainov *IB*, 370.
42. Solomon Private Comm.
43. Čugainov *Krym. Izv.* (in press).
44. Higgins, Solomon, Bateson *Australian J. Phys.*, **21**, 725.
45. Herr *IB*, 364.
46. Jarrett, Eksteen *IB*, 364.
47. Efimov, Shakhovskoy Private Comm.
48. Oskanjan *IB*, 345.
49. Tapia *IB*, 286.
50. Andrews *PASP*, **78**, 324.
51. Čugainov *Budapest*, 127.
52. Kunkel Private Comm.

28. GALAXIES

PRESIDENT: G. C. McVittie.
VICE-PRESIDENT: E. M. Burbidge.
SECRETARY: E. B. Holmberg.
ORGANIZING COMMITTEE: D. S. Evans, D. Heeschen, E. B. Holmberg, B. E. Markarjan, R. L. Minkowski, M. Schmidt.

The volume of work published on extragalactic astronomy in the period August 1966 – July 1969 has been large. Limitations of space have precluded the writing of an all-inclusive report; in all sections some selection has had to be made. References are given, whenever possible, by the reference numbers of the Bulletin Signalétique of the CNRS of France, with the omission of the Section number: "2" for vols. 27 to 29 and "120" for vol. 30. A reference number preceded by J is taken from the *Astronomischer Jahresbericht*. When no reference numbers were available, abbreviated journal references have been given.

QUASI-STELLAR OBJECTS, SEYFERT AND RELATED GALAXIES
(by E. M. Burbidge)

1. *Quasi-stellar objects*

A detailed review was published by Schmidt (*Ann. Rev. Astr. Ap.*, **7**, 527). This listed 296 references and covered the years 1967 and 1968. The first review in that series, by Burbidge (29-6492) covered work done between that discussed in Sandage's section of Report of Commission 28 (*Trans. IAU*, **13**, 579) and the period covered by Schmidt.*

Identification

For the 3C revised catalogue, Schmidt (30-7617) listed 44 identified and confirmed QSOs with relevant references. For the 4C catalogue, identifications have been suggested by Wills (29-6401); Barbon, Braccesi and Fanti (29-7858); Bailey and Pooley (29-13304); Olsen has an unpublished list of 68 on which Schmidt is making spectroscopic observations in the zone $+20°$ to $+40°$, with the proportion of correct identifications running about 60 %. Extensive lists of suggested identifications in the Parkes catalogue have been published, by Bolton and Ekers (28-10961), by Shimmins (29-14754), by Bolton, Shimmins and Merkelijn (29-14730), by Merkelijn, Shimmins and Bolton (30-4370) and by Merkelijn (30-11642; *Austr. J. Phys.*, **22**, 237). Two new areas in the 5C survey have been completed and searched for possible identifications: the 5C 2 by Pooley and Kenderdine (30-4379) and the 5C 3 around M 31 by Pooley (*M.N.R.A.S.*, **144**, 101). Wills (30-14525) suggested some identifications for high-frequency sources. Mills, Shobbrook and Stewart-Richardson (30-4371) suggested several identifications among radio sources near Abell clusters of galaxies, observed at Molonglo and Arecibo. Thompson, Kraus and Andrew (30-7620) suggested identifications for some sources with peaked or flat radio spectra in the Ohio survey. The rapidly variable bright object BL Lac has been identified with the Vermilion River Observatory source VRO 42.22.01 by Macleod and Andrew (30-4259) and Schmitt (29-14751); it may be either a QSO or N-type galaxy on the basis of its lineless optical spectrum. Hazard, Gulkis and Sutton (30-9767) have accurate source positions from lunar occultations and suggested several QSS identifications.

* The book *Quasi-Stellar Objects* by the Burbidges (J67-13445) should also be mentioned. – G. C. McV.

For correct identifications, accurate optical positions which enable comparison with the increasingly accurate radio positions have been measured by Lü and Fredrick (29-13506; 30-4366, -4367), by Véron (30-8567) and by Bolton (30-4226).

Candidates for radio-quiet QSOs have been selected by surveys for blue objects with UV excess with the Tautenburg Schmidt by Richter, Richter and Schnell (*Mitt. Taut.*, No. 38) and by Bronkalla (30-9622). Eggen (30-7518) and Barbieri, Erculiani and Rosino (*Publ. Padova*, No. 143) published lists of faint blue stars. Whereas such lists will contain a percentage of white dwarfs and hot subdwarfs, Braccesi (28-12524) showed that the presence of an infrared excess as well as UV excess makes identification virtually certain. Braccesi, Lynds and Sandage (30-604) and Janes and Lynds (30-13060) gave spectroscopic results on QSOs selected in this way. According to (30-604), the number of QSOs down to $B=19{\cdot}4$ mag. may be 5 per square degree. Sandage and Luyten (30-14303) made a further study of 301 blue objects in high latitude, and concluded there may be 10^7 QSOs down to $B=22$ over the whole sky.

Fanti-Giovanni and Fanti (28-12437) found many QSOs selected optically to be weak radio emitters; Lang and Terzian (*Ap. J.*, **158**, L11) observed 38 QSOs and compact emission-line galaxies and found weak emission at 318 MHz in 22 of them.

A catalogue of QSOs was published by Barbieri, Battistini and Nasi (29-14886).

Redshifts; distribution of redshifts

Redshifts have been determined by Kinman and Burbidge (29-3514); Dibai and Esipov (28-11035; 29-364, -5008; 30-4343); Ford and Rubin (*Carnegie Inst. Yrbk.*, **65**, 67); Schmidt (30-7617); Braccesi, Lynds, and Sandage (30-604); Shimmins, Searle, Andrew, and Brandie (30-4388); Schmidt and Olsen (*A.J.* **73**, S117); Lynds and Wills (30-2627); Bolton, Kinman, and Wall (30-11683); Burbidge (30-11684); Searle and Bolton (30-11707); Stockton (30-14612). Tabulations of redshifts can be found in Burbidge (29-6492), Schmidt (*Ann. Rev. Astr. Ap.*, **7**, 527) and Burbidge and Burbidge (30-1456). The latter included 31 related emission-line galaxies as well as 155 QSS and QSOs. The objects found by Braccesi, Lynds, and Sandage included some objects with very small redshifts and colors like Seyfert or N-type galaxies, and their physical properties may be more akin to these than to the "classical" QSS.

Burbidge (28-12527) noted that there seemed to be too many QSO and QSS with absorption-line redshifts at $z\approx 1{\cdot}95$ for a chance distribution. The distribution of redshifts was analysed by Burbidge (30-11688; 30-14561); evidence for two peaks at $z=0{\cdot}061$ and $z=1{\cdot}95$ was presented, and for possible peaks at integral multiples of 0·061. Cowan (30-7615) suggested an intrinsic pattern in the redshift distribution corresponding to $z=1/6$ and 1/15. Barbieri, Bonometto and Saggion (30-13808) drew attention to two peaks in the redshift distribution for quasars (see also discussion by Tinsley, Patterson and Tinsley, *Ap. Letters*, **4**, 55). The peak at $z=1{\cdot}95$ led Shklovsky (29-4670; 29-13519) and Kardashev (29-13040) to consider model universes of the Lemaître type, further discussed by Petrosian, Salpeter and Szekeres (29-144).

The lack of correlation between magnitude and redshift was further discussed by Longair and Scheuer (*Nature*, **215**, 919). The rather abrupt cutoff in redshift near $z=2$ was displayed in a plot by Burbidge and Burbidge (*Nature*, **224**, 21). Brecher and Silk (*Ap. J.*, **158**, 91) discussed the absence of larger redshifts in terms of a Lemaître model universe, while Rees (*Ap. Letters*, **4**, 61) suggested that the apparent deficiency at $z>2$ might be due to the onset of absorption by intergalactic atomic hydrogen if this were largely neutral at epochs corresponding to $z>2$.

Emission line intensities; Physical conditions in emitting regions

Most of the papers giving determinations of redshifts contain approximate eye estimates of relative intensities of lines. Wampler (29-3525; 30-2639, -11682) measured line intensities in some quasars with a photoelectric scanner and discussed physical conditions. Wampler and Oke (29-5027) studied the spectrum of 3C 273 in detail and detected faint features due to Na I and permitted Fe II lines. Burgess (29-6494) calculated that the size of the emitting region of a cosmologically distant quasar in which [O II] is seen must be 10^2–10^3 pc. Osterbrock (*Liège Coll.*, 1968) discussed the calculated and observed intensities of several emission lines; discrepancies led him to calculate new collision strengths for [C III] λ1909 and other semi-forbidden emission lines.

Sargent and Searle (30-7618) found that the Hβ emission intensity was similar in many quasars and related emission-line galaxies and did not agree with recombination theory. An overlying non-thermal continuum and coupling between thermal and non-thermal radiation could explain the observations. Bahcall and Kozlovsky (30-14555) calculated detailed models for 3C273, giving the distributions of various stages of ionization. The results suggested relative abundances of the elements similar to solar values, except that He may be underabundant by a factor ~ 10, as suggested earlier by Osterbrock and Parker.

Absorption-line spectra

Absorption lines have been found in the spectra of some quasars and have proved more puzzling and intriguing than the emission lines. Bahcall, Sargent and Schmidt (29-6490) made a further analysis of 3C 191 and, from measured equivalent widths of the absorption lines, found a carbon-to-silicon abundance ratio $\sim 2{\cdot}5$, i.e. lower than the solar value, but Bahcall *et al.* (*P.A.S.P.* **79**, 425) found a higher ratio ~ 7. Grewing and Schmidt-Kaler (30-9849) made a similar analysis. Bahcall (29-6489) derived lower limits on electron density or radiation density in 3C 191 from the fact that fine-structure states in Si II are populated.

Most interesting are PKS 0237-23, PHL 938 and Ton 1530, with absorption lines belonging to more than one redshift system, quite different from the emission-line redshift. Burbidge, Lynds and Stockton (30-607) discovered five discrete absorption-line redshifts and two possible ones in PKS 0237-23, the largest at $z = 2{\cdot}202$ and the smallest at $z = 1{\cdot}365$, whereas $z_{em} = 2{\cdot}223$. PHL 938, with $z_{em} = 1{\cdot}955$, had absorptions at 1·906 and 0·613. Ton 1530, with $z_{em} = 2{\cdot}046$, had absorptions at 2·055, 1·980, 1·936. Bahcall (30-4334) derived a formal method of searching for and testing separate redshifts with a computer; using it, Bahcall, Greenstein and Sargent (30-4335) detected five redshifts in PKS 0237-23, and there was a good measure of agreement between the independent studies by Burbidge *et al.* and Bahcall *et al.* Bahcall, Osmer and Schmidt (30-14556) found two more redshifts in Ton 1530, at 1·921 and 1·887.

PHL 5200, earlier found by Lynds to have a supernova-like spectrum with emission lines at $z = 1{\cdot}98$ and broad absorptions extending down to $z = 1{\cdot}88$, was observed one and two years later by Burbidge (30-606, -13051) for possible changes due to expansion of the shell of gas. Structure in the form of absorption dips at $z = 1{\cdot}950$ and 1·891 was found. The latter paper also discussed the radio-quiet QSO B194 which was found to have absorption lines at $z = 1{\cdot}895$ and 1·837.

The above papers discussed the possible origin of the absorption lines, either in expanding shells at very high velocity, in intervening galaxies or intergalactic clouds at lesser redshifts, or in shells at different gravitational potentials. All suggested explanations encountered difficulties. Burbidge and Burbidge (30-14561) thought the spectroscopic arguments and statistics favored expanding shells particularly since rather similar absorptions were found in the Seyfert galaxy NGC4151. Peebles (30-10873) suggested "dead" galaxies in intervening space. Bahcall and Spitzer (*Ap. J.* **156**, L63) suggested extended halos around intervening galaxies. Probabilities for screening in Lemaître universes were calculated by Petrosian *et al.* (29-144) and Roeder and Verreault (30-3522, -6355; 30-13834). Roeder (*Ap. J.*, **157**, L153) found Bahcall and Spitzer's suggestion incompatible with Friedmann universes. Wagoner (29-13428) calculated strengths of UV lines expected to be produced in absorption.

Continuum energy distribution, UBV photometry, IR measures, variability, polarization

The papers by Wampler and Oke already listed also give continuum energy distributions. Grewing and Lamla (39-4352) derived a mean continuum. Schmidt (30-7617) and Lari and Setti (29-7870) derived mean continua from *UBV* measures. Bahcall and Sargent (29-3502) and Grewing (29-7868) calculated colors for $z > 2{\cdot}3$ and 2·2. Moroz and Dibaj (30-2630) observed 3C 273 and some Seyfert galaxies at 1–2·5 μ.

Kinman (30-4853) has produced a review of our knowledge of variable galaxies. Optical monitoring of QSOs was continued by Barbieri and Erculiani (30-8549). Cannon and Penston (28-12541) followed the rapidly variable 3C 446 and Wampler (29-3524) measured its continuum distribution during variations. Kinman *et al.* (30-616) found the variations in luminosity and polarization in 3C 345 to be quasi-periodic, with periods ~ 80 and 321 days. Angione (30-9829) discovered rapid variations in 3C 454·3 on old Harvard plates and Lü and Hunter (30-9856) gave a light curve for it with a suggested period of 340 days. Bolton and Wall (*Ap. Letters*, **3**, 177) found large brightness variations in some QSOs. Kunkel (29-14899) made a harmonic analysis of variations in 3C 273.

Kinman (29-3513) found polarization in 3C 279 and 3C 345, and none in PKS 0405-12 and PKS 0837-12. Appenzeller and Hiltner (29-6488) measured polarization in 14 QSOs and found it generally present, though not very strong; Appenzeller (30-7612) confirmed that 3C 273 has no appreciable polarization. Elvius (30-4347) measured 3C 273, 3C 279, 3C 345 and found variable polarization in 3C 279 and 3C 345. Visvanathan (30-2638) observed 14 QSOs for polarization and found it in PHL 658, PHL 1093, 3C 57, 3C 454·3, PKS 0736 + 01, PKS 2145 + 06, PKS 2349 + 09. Visvanathan (30-1432) found large polarization in BL Lac. Low and Kleinmann (30-11682) found large polarization in 3C 273 in the infrared at 1·5 μ.

Models and theories for QSOs; the nature of the redshifts

No satisfactory theory has yet been found. Although a majority favors the cosmological nature of the redshifts, the factors that led this to be questioned originally are still unexplained, particularly the distribution of redshifts, and the inverse-Compton-synchrotron paradox which is not present for variable Seyfert and N-type galaxies of comparatively low redshift. Bahcall, Schmidt and Gunn (*Ap. J.*, **157**, L77) found B264 to lie in a cluster of galaxies at the same redshift, but this is only $z = 0{\cdot}095$ and some of the galaxies in the cluster are brighter than B264. Oke (*Ap. J.*, **158**, L9) found the colors of B264 to be similar to Seyfert and N-type galaxies rather than to QSOs. To avoid the inverse Compton problem, proton-synchrotron emission in strong magnetic fields was suggested by Jukes (29-7869) and Rees (30-3645). Kinman *et al.* (30-616) proposed a rotating object radiating anisotropically. Relativistically expanding sources were discussed by Petschek (29-3544) and Rees and Simon (30-630). A model using the inverse Compton process itself was proposed for 3C 273 and 3C 286 by Takarada (30-2637, -13073). Piper (30-11704) and Dallaporta (29-131) discussed combined cosmological and gravitational redshifts. Shimoda (30-4389, -4390) proposed a theory for redshifts in terms of non-linear effects in a dense plasma.

Evolutionary properties required to explain the redshift distribution and number counts were treated by Rowan-Robinson (29-14937) and Schmidt (30-7617). Grewing, Pfleiderer and Priester (29-6499, -6505) and Mathez (*Astr. Ap.*, **4**, No. 1) discussed interpretations of the line spectra in terms of cosmological or local theories. Arp (29-3501; 30-598) continued work on pairs of radio sources including QSOs about peculiar galaxies, suggesting non-velocity redshifts. Wagoner (29-490) and van der Laan and Bash (30-633) discussed his statistics. Terrell (30-4395) discussed the proper motion of QSOs in his local Doppler theory.

For the energy source, the following were proposed: lagging cores in the expanding universe by Ne'eman and Tauber (29-13044), matter-antimatter annihilation by Alfvén and Elvius (*Science*, **164**, 911), galactic flares by Sturrock and Feldman (29-14906), the evolution of dense star systems by Gold (29-14921) and Ipser and Thorne (30-9289).

2. *Seyfert and related galaxies*

We include in this category Seyfert, N-type, other radio and compact galaxies, also other galaxies with peculiar nuclei or giving evidence of violent fragmentation. All are objects in which explosive or violent events appear to have occurred.

Seyfert galaxies

The proceedings of a conference on Seyfert galaxies held in Tucson in 1968 were edited by Pacholczyk and Weymann and published (30-11682) with plates and an extensive bibliography. Some of the papers at the Tucson conference were published elsewhere, more fully or *in toto*. Morgan (30-2629) described the optical forms of Seyfert and related galaxies. Sĕrsic and Pastoriza (29-1944) presented statistics on galaxies with peculiar nuclei; the paper at the Tucson conference updated this and more than doubled the number of galaxies studied. De Vaucouleurs and de Vaucouleurs (*Pub. Dept. Astr. Univ. Texas,* Ser. II, Vol. II, No. 7) gave a summary of data on classification, frequency, structure and diameters of Seyfert galaxies, their luminosities, colors, spectra and estimates of distances and absolute dimensions. They also (*A.J.*, **73**, S174) measured effective diameters of the nuclei of eight Seyfert galaxies.

Oke and Sargent (29-14903) gave a detailed spectrophotometric analysis of NGC4151. Andrillat and Souffrin (29-13492) announced their discovery of spectral variations in the nucleus of NGC3516 between 1943 and 1967, with an interpretation in terms of the physical conditions. Fitch, Pacholczyk and Weymann (29-13501) discovered optical variability in the nucleus of NGC4151. Colgate (29-13543) discussed his theory of stellar coalescence and multiple supernova explosions for both QSOs and Seyfert nuclei. Sargent (29-14904) discovered the Zwicky compact galaxy I Zw 0051 + 12 to be highly luminous and to show Fe II emission lines (as does 3C 273). Oke *et al.* (29-13510) found the compact galaxy I Zw 1727 + 50 to be optically variable. Morgan, Smith and Weedman (30-620) described spectral variations in the nucleus of NGC4151 between 1956 and 1968.

Visvanathan and Oke (30-634) gave polarization measures for NGC1068 and used them to separate the non-thermal from the thermal radiation. Arp (30-2619) showed the redshift-apparent magnitude relation for Seyfert and N-galaxies to be steeper than the normal curve and to join the QSO region. Demoulin, Burbidge and Burbidge (30-2621) described the spectrum of NGC7714, which appeared related to Seyfert galaxies. Barnothy and Barnothy (30-4337, -9830) discussed their theory that QSOs are produced by galaxies acting as gravitational lenses and focussing the light from very distant Seyfert galaxies. DuPuy (30-4346) gave the redshifts for four Haro galaxies. Westerlund and Wall (*A.J.*, **74**, 335) presented *UBV* photometry for 14 QSOs, 8 N-type galaxies, 39 radio galaxies and 19 radio-quiet galaxies. Rubin and Ford (30-9864) gave velocity and line-intensity measures in NGC3227; the rotation curve gave a mass of 2.6×10^{10} $M_{\odot}$ and discrete high-velocity clouds were found in the nucleus.

Work apart from that described at the Tucson conference includes the following. Bardin, Chopinet and Duflot-Augarde (29-5004) found from a spectrophotometric study of the nucleus of NGC5548 that T_e appeared to vary with time. Ford, Purgathofer and Rubin (30-2625) described the spectra of NGC4151 and some planetary nebulae near 1 μ, noting the great strength of [S III] $\lambda\lambda$9069, 9532 in NGC4151. Souffrin (30-9869, 30-14610) gave a detailed spectrophotometric study of NGC1068, 3516 and 4151, including study of the physical conditions and mechanism of ionization.

Vorontsov-Velyaminov (29-11796) described the structure of the nuclei of Seyfert and related galaxies. Moffet (30-4371) compared the optical and radio positions for NGC1068, 1275 and M87. Takayanagi (30-4394) measured light variations in 3C 120 between 1928 and 1943. Walker (30-4405) measured the motions of high-velocity cloud complexes in the nucleus of NGC1068. Danielson, Savage and Schwarzschild (30-11685) measured the diameter of the nucleus of NGC4151 with the balloon-borne 36-inch telescope and obtained an upper limit of 0.″18. Wampler (30-11713) found that there was considerable reddening, attributable to dust, in the nuclei of seven Seyfert galaxies, from measures of [S II] lines in the infrared and violet. Peterson (30-14599) found variation in the nucleus of NGC1275. Osterbrock (*Ap. Letters*, **4**, 57) showed that the [Fe X] and [Fe XIV] emission lines observed in NGC4151 can be understood on the basis of the photoionization model of Williams and Weymann. Anderson and Kraft (*Ap. J.*, **158**, 859) discovered sharp absorption lines in the nucleus of NGC4151 occurring at three separate redshifts, similar to the multiple-redshift absorptions in QSOs and implying ejection of shells of gas.

N-type galaxies

Oke (29-13509) discovered 3C 371 to vary optically. Cannon, Penston and Penston (29-7864) discovered 3C 390·3 also to be variable. Sandage (29-13517) confirmed variability of 3C 371 during the course of *UBV* photometry of 13 N-type galaxies. Sandage (29-13518) announced optical variations in 3C 109, 3C 390·3 and I Zw 1727 + 50. Visvanathan (29-13525) found that 3C 371 showed no significant polarization. Usher and Manley (30-4398) found long-term variations in 3C 371 from old Harvard plates.

Demoulin and Burbidge (39-9841) discussed the non-thermal optical flux from 3 Seyfert galaxies, 3 N-type galaxies, M82 and M87. The radiation could be synchrotron in origin; the variations set size limits according to $R < c\tau$ that did not violate the conditions for synchrotron losses to exceed inverse Compton losses, in sharp contrast to the case for the QSOs at cosmological distances. Grewing, Demoulin and Burbidge (30-9846) calculated the non-thermal flux from N-type and Seyfert galaxies for a range of possible physical conditions, and showed that *UBV* photometry alone was insufficient to separate non-thermal from thermal components, particularly when dust was present.

M87 and other radio galaxies.

Felten (29-14890) discussed the radiation and physical properties of the jet in M87. Arp (29-13493) discovered a faint counter-jet in M87. Walker and Hayes (29-13529) measured the motions of high-velocity cloud complexes in the nucleus of M87. Walker (30-4406) discussed the "orthogonal fan jet" in M87. Friedman and Byram (29-6497) confirmed their earlier observation of M87 as an X-ray source; Bradt *et al.* (29-13495) also detected it. Johnson (*Can. J. Phys.*, **46**, S481) further discussed this X-radiation. Haymes *et al.* (29-14896) detected M87 as a probable source of hard X-rays.

Solinger (30-13071) discussed the observations of M82, and suggested that it is a Seyfert galaxy seen edge-on but this is questioned by van den Bergh (30-14617). Sandage and Visvanathan (*Ap. J.*, **157**, 1065) measured the surface brightness, colors and linear polarization of selected regions of the outer filaments of M82; polarization reached 32% in one patch and the data supported the hypothesis of synchrotron emission by the filaments produced by electrons up to 4×10^{12} eV in magnetic fields 10^{-5} to 10^{-6} gauss.

Bahcall and Schmidt (29-7442) showed from measures of [O III] lines in radio galaxies over a wide range of redshift that the fine structure constant had not varied with time. Redshifts of radio galaxies were measured by Wills (29-3526); Burbidge (29-7863); and Sandage (29-13330). Moffet *et al.* (29-3518) identified 3C 17 with a galaxy, from precise coordinates obtained by lunar occultation.

Compact galaxies and galaxies with peculiar nuclei

Zwicky (29-7883) discovered the first two actual clusters of compact galaxies. He also (29-13531) described blue compact galaxies in the Leo A and Leo B clusters, which also contain normal galaxies. Zwicky (29-13530) discussed compact galaxies as being studded with neutron stars, and having the essential properties of QSOs. He discussed the nature of the redshifts.

Pastoriza (29-6504) studied the peculiar nuclei of NGC1672, 1808, 2997, 5236, 7552 and measured the [N II] λ6584/Hα intensity ratio. Itoh and Kogure (29-3510) studied the strengths of a number of emission lines in nuclei of galaxies and concluded they were produced by collisional excitation (including the Balmer lines). Shobbrook and Shaver (29-3522) studied the nucleus of NGC4945 from blue and infrared plates, and showed that the position of the radio source is associated with the nucleus. Walker (30-636) discovered the nucleus of NGC4254 to have varied, from comparison of 1921 plates of Lampland with recent observations. Burbidge and Demoulin (*Ap. Letters*, **4**, 89) discovered that a small gas cloud is being ejected at 700 km s^{-1} from the nucleus of NGC4939.

Miscellaneous work on galaxies, groups, or clusters suggesting explosive activity; cases of peculiar velocities

Kiang (29-13504) searched the Palomar Atlas and collected cases of Haro's violet galaxies showing peculiar jets and filaments. Sĕrsic, Pastoriza and Carranza (30-4387) published a description of a triplet of peculiar galaxies with high excitation and suggested that it resulted from a violent event in the past.

Ceccarelli and Dallaporta (29-7443) proposed a model for the formation of clusters of galaxies by the ejection of fragments from an original larger-mass plasmoid. Sĕrsic (30-630; 30-4386) discussed the possible formation of galaxies by fragmentation. Arp (30-4330) found lines of galaxies around large E-type radio galaxies and suggested they were formed by ejection from the radio galaxy.

Rudnicki (30-9865) suggested the nucleus of M31 may have been active in the past. Bonometto and Lucchin (30-13050) suggested a connection between present shapes of spiral arms and past activity of galactic nuclei, and applied their results to M31. Ford and Rubin (*Carnegie Inst. Yrbk.*, **67**, 286) measured velocities of blue objects in the Virgo Cluster and found a wide dispersion. Burbidge and Demoulin (*Ap. J.*, **157**, L155) found a blue shift in IC 3258, a small galaxy near the center of the area of the Virgo Cluster. Sargent (30-4383) found a remarkable distribution of red-shifts in the chain of galaxies VV172; four objects have velocities around 16000 km s^{-1} and one has 36880 km s^{-1}

GALAXIES IN GENERAL

(by G.C. McVittie)

1. Among studies of individual galaxies, the following may be cited. G. and A. de Vaucouleurs and K. C. Freeman (30-4404) deal with the kinematics and dynamics of the Magellanic-type barred spiral NGC4027. They used longslit spectra obtained with the B spectrograph at the prime focus of the McDonald 82-inch reflector. Tammann and Sandage (29-14907) have identified and classified 59 variables in NGC2403. The physical conditions in M51 and M81 that can be derived from the intensities of emission lines and of the continuum at various wavelengths form the subject of a paper by Peimbert (30-9860). NGC6822, a dwarf of the Local Group, and GR8, which may belong to the Local Group, have been studied by Kayser (29-1939) and Hodge (29-5015), respectively. The velocity, neutral H content, globular clusters or star clouds and reddening and distance of M31 are discussed by Rubin and D'Odorico (*Astr. Astrophys.*, **2**, 484), Brundage and Kraus (28-1875), van den Bergh (28-5477, 30-6360) and Schmidt-Kaler (29-6509). M. F. Walker (29-1946) has studied M33 with the aid of a two-color dye-transfer photograph. The distance of NGC3389 has been doubled – to 16·6 Mpc – by de Vaucouleurs (29-13524). A dwarf galaxy in UMi (van Agt, 29-11794), a peculiar galaxy (Burbidges and Shelton, 29-13496) and the peculiar galaxy NGC6438 (Sěrsic, 28-3054) have all received attention. Analyses of the nuclei of the following galaxies have been carried out: NGC4258 (Chincarini and Walker, 29-13498); NGC4486 (Walker and Hayes, 29-13529) and NGC4151 (Andrillat and Souffrin, 29-1931). Appenzeller (30-597) has determined the linear polarization of four spiral galaxies while Bertola (28-4256, 29-5005) gives the luminosity distribution of NGC4618/25 and of an extragalactic system analogous to Mayall's object. A discussion of the binary galaxy system NGC4631/4656 and the hydrogen "bridge" linking these two galaxies is due to Roberts (30-4380).

2. Discussions of the masses, rotations and luminosity distribution of galaxies are found in the investigations of the Burbidges (30-4341, -11689) on NGC1808 and 1832, and of Demoulin and Chan (*Ap. J.*, **156**, 501) on NGC6574. Masses turn out to be small multiples of $10^{10}\ M_{\odot}$. Chincarini and Walker (28-12529, 28-12540) have analysed galaxies of complex structure and rapidly rotating nuclei, Bertola and Bernaca (29-477) have determined the masses of NGC5005 and 6503, while Brandt and Roosen (*Ap. J.*, **156**, L59) argue that M87 has a mass of $2{\cdot}7 \times 10^{12}\ M_{\odot}$ and is the galaxy of greatest known mass. The de Vaucouleurs and de Cesare (*B.A.A.S.*, **1**, 186) have also derived the rotations and masses of nine galaxies. Takase and Kinoshita (30-631) have calculated the mass, angular momentum and energy for some 20 galaxies and found relations between these quantities. They interpret these relations in terms of the evolution of galaxies from proto-galaxies. Takase (30-592) has worked out the corresponding quantities for M31 and our Galaxy. Line widths have been used by N. Heidmann (*Astrophys. Letters*. **3**, 19) to estimate total masses and angular momenta. The rotation curve of M33 is the subject of a paper by Courtès and his colleagues (28-7064). Brosche (28-12525), through a statistical study of the rotations of 53 galaxies, finds a systematic difference in rotation between spiral and elliptical galaxies. Photometric data on galaxies lead J. Heidman (J66-14528, 29-5014) to conclude that there exists a relation of the form $L \propto A^q$ between absolute luminosity L and linear diameter A. Here $q = 2{\cdot}8$ for spirals and 1·9 for ellipticals. Spectro-photometric gradients of galaxies have been compiled by Vandekerkhove (30-9871).

3. Among analyses of the forms, stellar contents and colors of galaxies may be mentioned the classification of the forms and stellar contents of giant galaxies due to Morgan and Osterbrock (30-14596) who find three principal types: the Orion type with H II regions, giant blue stars and high gas density, the Intermediate, with absorption, spectral type F and high compositeness; and the Amorphous. McClure (30-13064) shows that CN absorption is stronger in the central cores of galaxies than in regions surrounding the nucleus. The stars in the cores of spirals and ellipticals are super rich in metals. *UBV* observations of field galaxies lead McClure and van den Bergh (30-14593) to conclude that the metallicity index has a lower value for non-cluster than for cluster glaxies. There are also investigations of three-color photoelectric surface photometry of six Sb galaxies by Simkin (29-11790); of eight-color photometry of galaxy nuclei by H. L. Johnson (27-10711) and of *UBV* observations and color diagrams by Shobbrook (27-12346) and Chiosi (29-482). An H-R diagram for galaxies is proposed by Firsoff (29-5011). The jets and filaments in Haro's violet galaxies have been studied by Kiang (29-13504).

4. The distribution in space and the clustering of galaxies have been discussed in many papers. Fairall (30-4348) has published a catalogue of compact and bright-nucleus galaxies. An examination of 1589 objects in the Tonanzintla Catalogue on the Palomar Sky Atlas by Prata (27-10717) revealed 16 non-stellar objects, 23 which might be stellar and 6 new blue non-stellar objects. The Lick counts have been analysed by Shane and Wirtanen (29-2025) for clouds of galaxies and the distribution in space, and also by Kalinkov (29-5016) for clusters and groups of clusters. From the same counts, Totsuji and Kihara (*Publ. Astr. Soc. Japan*, **21**, No. 3) have determined the correlation function for the spatial distribution of galaxies.

Catalogues of galaxies or of clusters of galaxies have been analysed for the existence of clusters of clusters and there is no general agreement on the matter. The papers by Kiang and Saslaw (30-14586), Kiang (29-484), Karpowicz (29-3512), Silk (29-14940), Reaves (*P.A.S.P.*, **80**, 564), Reaves and Stern (J67-145182) and de Vaucouleurs and Peters (30-7333) bear on these questions. The stabilization of clusters of galaxies by ionized gas is proposed by Woolf (29-3527). Zonn (30-4409) has evaluated the moments of the galaxies in a cluster. He finds that the mean number of cluster centers is 0·1 Mpc^{-3}. Three percent of all galaxies are members of pairs (30-4410) and, when redshifts are known, it turns out that the relative motion of a pair is hyperbolic. The forms and position angles of 4287 galaxies in the Hydra, Ursa Major, Virgo and Eridanus clusters lead F. G. Brown (29-14888) to the conclusion that, in each cluster, there is a preferred range for position angles.

King and Setteducate (29-11789) have calculated the rectangular coordinates of bright galaxies in de Vaucouleurs' catalogue to permit of identifications on the Palomar Sky Atlas. A catalogue of 139 galaxies south of +20° Dec. from the Palomar Sky Atlas has been published by Glanfield and Cameron (29-9915). Questions regarding the relationship between the axes and the ellipticities of elliptical and irregular galaxies are dealt with by Rood and Sastry (29-1942) and Hodge and Hitchcock (27-10710).

With regard to individual clusters and groupings and their contents the review of present knowledge of the Local Group by van den Bergh (30-9872) may be cited, as well as the investigation of the dwarf galaxies it contains by Hodge and Michie (*A.J.*, **74**, 587). The distribution of intensity in the elliptical galaxies of the Virgo cluster is analyzed by M. H. Liller (28-9591) and the identification of dwarf galaxies in the center of the Coma cluster, by Reaves (28-7078). Descriptions and lists of clusters or small groups of southern galaxies are given by Klemola (*A.J.*, **74**, 804) and by Sěrsic (29-7879).

5. In the study of H II regions and of other gaseous contents of galaxies, the catalogue and atlas of H II regions in 20 nearby galaxies by Hodge (30-13059) and the survey of Hindman (29-1938) may be noted. Other investigations of H II regions are those of Hodge (30-13058; *Ap.J.*, **156**, 847) and of Carranza *et al.* (*Astr. Astrophys.*, **1**, 449, 479; **2**, 1). Sěrsic (*Z. Ap.*, **69**, 242) gives coordinates and diameters of 35 H II regions in NGC1313. Courtès and his colleagues have carried out an inter-

ferometer study of ionized hydrogen in M33 and of the HII regions of the N. E. arm of M31 (29-13497, 30-9840). The highly excited state of the gas in NGC4194 is discussed by Demoulin (30-14566) and the emission of Hβ in galaxies with strong emission lines, by Searle and Sargent (30-7618). The distribution of neutral hydrogen in NGC6946 has been analyzed by M. S. Roberts and his co-workers (30-11694). Maps of the integrated HI distribution in M31 have also been published by Roberts (29-9924). Although there is some HI in the central region of a spiral galaxy, the simplest description of the overall distribution is that of a ring, i.e., the HI is not centrally concentrated. In M31 the optically defined arms lie within this ring, close to the peak HI distribution. In M33, M101 and several other spiral galaxies, including the Milky Way, the spiral features lie interior to the HI ring. The optical spiral arms do not occur at the maximum of the beam-averaged brightness temperature distribution. Roberts has also derived an upper limit of 10^{-3} atoms cm^{-3} for the neutral hydrogen density in the halo regions of two edge-on galaxies, NGC4244 and 7640. The connection between neutral hydrogen and the dynamics of galaxies is dealt with by Guelin and Weleachew (30-9851, -9852) from the radio astronomical point of view.

6. Theoretical investigations of the dynamics of galaxies include the density-wave theory of C.C. Lin and his co-workers (29-2501, 29-7871, 30-14590), Ng's theory of a disk-shaped galaxy (29-13507) and the relation of gas and spiral structure due to Fujimoto (30-609). Orbits of stars in galaxies are discussed by Danby and Zahar (29-9909) and by Barbanis and Woltjer (29-19494). Sharpless and Franz (30-9866) look into the question of the degree of central concentration of mass in spiral galaxies. A review of the problem of the evolution of galaxies is due to Reddish (30-8563); a general theory of the development of a galactic nucleus or a compact globular cluster has been produced by Spitzer and Saslaw (27-12365, 28-12547). Questions concerning the background radiation and the upper limit of the mass density of galaxies are discussed by Peebles and Partridge (29-3519, -5021).

COSMOLOGY

(by W. B. Bonnor)

1. *Observational data*

Sandage (29-14469, 30-91) has derived a new redshift-apparent magnitude diagram for galaxies and radio galaxies. The diagram is restricted to galaxies which are the brightest member of a cluster, with a consequent small dispersion of $\pm 0^{m}.3$ about the linear Hubble relation. No significant deviation from this relation is observed out to the largest redshift involved, which is 0·46. Sandage also rediscussed the value of the Hubble constant H and obtained $75\ ^{+19}_{-15}$ km s^{-1} Mpc^{-1}, although a value as low as 50 km s^{-1} Mpc^{-1} cannot be ruled out.

Measurements of redshifts, and of radial velocities generally, have been made by the de Vaucouleurs (29-6511) for 113 bright galaxies and by Rudnicki and Tarraro (*Act. Astr.*, **19**, 171) for six objects. Southern galaxies have been dealt with by Carranza (28-11034), Shobbrook (27-8219) and by D. S. Evans and collaborators (29-6496; 30-11690). All these observations refer mainly to bright galaxies with velocities of recession usually below 6000 km s^{-1}.

Pooley and Ryle (30-4378) have extended the radio source counts to fainter sources by using the 5C survey. Their new log$N$–log$S$ relation refers to an observing frequency of 408 MHz, the limiting flux density being 0·01 flux units. The observed slope is $-1{\cdot}85$ down to 3 flux units, where the curve starts to flatten off, reaching a slope of $-0{\cdot}8$ at 0·01 flux units. The integrated radiation from all the sources is about half the estimated extragalactic background at 408 MHz. A log$N$–log$S$ relation at a higher frequency (2700 MHz) has been obtained by Shimmins, Bolton and Wall (29-9762), who find a slope of $-1{\cdot}4$. It is not clear whether there is a discrepancy with the Cambridge results (29-13326; 30-4249; *B.A.N.*, **20**, 171).

Boynton, Stokes and Wilkinson (*Phys. Rev. Letters*, **21**, 462) have measured the microwave background at 3·3 mm, and obtained a temperature of $2{\cdot}46\ ^{+0.40}_{-0.44}$K. They regard this as consistent with a black body spectrum, the measurements at longer wavelengths giving an average temperature of $2{\cdot}68\,^{+0.09}_{-0.14}$K. References to earlier measurements are given in this paper. Clauser and Thaddeus have published data on CN rotational temperatures based on the spectra of 8 stars lying in various directions in the Galaxy. All their results are compatible with a unique temperature of 2·7K at 2·6 mm. Bortolot, Clauser and Thaddeus (*Phys. Rev. Letters*, **22**, 307, 806) give upper limits to the temperature of the background at 1·32 mm (CN), 0·559 mm (CH) and 0·359 mm (CH^+). The first two limits are in conflict with the rocket measurements of Shivanandan, Houck and Harwit (30-9290) and Houck and Harwit (*Ap. J.*, **157**, 187) (who obtained a temperature of 8K in the wavelength range 0·04 to 0·13 cm), unless this far infrared radiation is confined to narrow lines.

The angular distribution of the microwave background has been studied by Conklin and Bracewell (29-6382), Partridge (*Am. Sci.*, **57**, 37) Penzias, Schraml and Wilson (*Ap. J.*, **157**, L49) and Conklin (30-13813). None of these workers found intensity variations down to a precision of 0·1 %, except Conklin who claims to have detected a significant 24-h variation, which he interprets in terms of the motion of the earth through the microwave background. The observed motion does not correspond to the rotation of the Galaxy or its motion relative to the Local Group of galaxies, but it does agree with the prediction of de Vaucouleurs and Peters (30-7333), who took into account the possible motion of the Local Group in a supercluster centered on the Virgo cluster of galaxies.

The intergalactic medium is the subject of a review article by Gould (30-8554) and the determination of its temperature is discussed by Gould and Ramsey (28-1878). The search for intergalactic hydrogen in the Virgo cluster has received attention from R. J. Allen (30-4327) and S. J. Goldstein (27-9442). The possibility of detecting a dense intergalactic medium by X-ray observations is considered by Henry *et al.* (30-2626). The X-ray emission observations from the Coma cluster seem to give contradictory results according to H. Friedman *et al.* (28-11036). Felten *et al.* (28-9857) argue that the emission must come from intergalactic matter. The diffuse X-ray background has been discussed by Silk (30-4392, -9868) and the general position is reviewed by Bunner *et al.* (*Nature*, **223**, 1222).

The helium problem has been thrown into a confused state by further evidence that the helium abundance in the Galaxy may vary with position. In particular some old stars appear to have a very low helium abundance. A brief review of this problem is given by Burbidge (*Com. Astr. Sp. Phys.*, **1**, 101), who also refers to some of the complications that can be introduced into the cosmological problem of helium formation in the hot big bang. Lynden-Bell (29-3516) has proposed that giant galaxies are perhaps helium rich.

2. *Properties of world-models*

Tables of values of useful quantities in a large number of zero-pressure world-models are now available through the work of Refsdal, Stabell and de Lange (29-6106). World-models with pressure are analyzed by Stabell (29-14470), Harrison (29-6101) and McIntosh (29-14458, 30-3522) and anisotropic models are investigated by Tomita (30-7331), Saunders (30-8285), Singh and Singh (30-3535) and Jacobs (30-3515).

McVittie (J66-4489) has shown that the special relativity formula connecting redshift and velocity of recession can be recovered in any world-model by appropriate definitions of distance and time-coordinate. Using a maximum redshift criterion, Crilly (30-8276) has reduced the number of observationally permissible models. Dulewicz and Zieba (30-3508, -3509) have proved some geometrical properties of Friedmann models applicable to redshift, angular diameter and photometric distance.

Petrosian and Salpeter (30-7327) studied ghost images in inhomogeneous Friedmann models and showed that simple focussing is impossible, the image being split into many images. Gunn (29-13037) has investigated the statistical effects of local inhomogeneities on the propagation of light, while Roeder and Verreault (30-3532) and Wagoner (29-13528) consider the effects of an intervening galaxy. McVittie with Schusterman (J66-134160) has shown from radio source counts that in uniform models these sources must have been more numerous in the past than now, and with

Ringenberg (*M.N.R.A.S.*, **142**, 1), that this conclusion is not altered by the use of differential counts.

Wilson and Edelen (29-1448) have shown that in inhomogeneous world-models the existence of a potential bound, such as the Schwarzschild limit, implies a positive cosmological constant and a lower bound for it can be calculated.

The theoretical consequences of the background radiation are discussed by Hawking and Ellis (29-14452) and by Narlikar and Wickramasinghe (29-3147, -13043). The primordial radiation in homogeneous anisotropic world-models is studied by Misner (30-7326).

3. *Cosmogony, origin of galaxies, etc.*

A general relativistic treatment analogous to the Newtonian treatment by Bonnor of gravitational instability in the expanding universe has been given by Arons and Silk (30-2353), and Nariai, Tomita and Kato (J67-44121). In an investigation of the origin of galaxies Harrison (30-8279) has argued that inhomogeneities were imprinted on the universe from the earliest moment of its expansion. The evolution and early stages of the Lemaître model form the subject of papers by Tomita (*Progress theor. Phys.*, **42**, 9) and Harrison (30-2361). Rawson-Harris (30-12549) has shown, by using perturbation theory in Newtonian cosmology, that a Lemaître model permits the formation of galaxies from a statistically probable fluctuation of the density.

The thermodynamics of a self-gravitating medium in quasi-static equilibrium, with applications to galaxy and stellar system formation has been studied by Saslaw (30-5911).

Peebles (29-143; 30-13035) has predicted the angular momentum of the Galaxy, using the theory of the formation of galaxies by gravitational instability. Tinsley (30-7367) has demonstrated that 'galaxies' resembling all normal types can be formed in about 12×10^9 years from the time when Population I stars begin to form, the sequence of types being, however, a non-evolutionary one.

The early evolution of the expanding universe, and chemical abundances have been studied by Japanese workers (J 67-44148; *Progress Theor. Phys.*, **41**, 840).

Radial motions of fluid spheres including cases of collapse, expansion and bounce have been investigated by McVittie and Stabell (J66-44207; J67-5596, -44113, *Ann. Inst. H. Poincaré*, **9**, 371) without prior imposition of an equation of state. Thompson and Whitrow (29-1723) have worked out a more general method and have applied it to uniform density cases with an arbitrary equation of state at the center. Similar cases are treated by Bonnor and Faulkes (29-7474) and by Bondi (30-10853). Buchdahl (28-10721) has dealt with static spheres with an equation of state while Silk and Wright (30-12588) discuss the collapse of a rotating sphere.

The appearance of a star collapsing through its Schwarzschild radius has been studied by Ames and Thorne (30-7318). Chandrasekhar (28-10722) has investigated the post-Newtonian effects of general relativity on uniformly rotating Maclaurin spheroids and the incorporation of Mach's principle in general relativity has been re-investigated by Lynden-Bell (29-6103).

Brief mention may be made of theories alternative to general relativity. The relation of quantum electrodynamics to the direct interparticle action theory of Hoyle and Narlikar (30-62, 30-13818, and references there given) is examined and found to support the steady state theory. Time-symmetric electrodynamics in Friedmann universes have been investigated by Roe (*M.N.R.A.S.*, **144**, 219).

A class of scalar gravitational Lorentz-invariant theories with a variable light-velocity has been set up by Page and Turner (29-13045). An article by Alfvén and A. Elvius (*Science*, **164**, 911) explains their views on antimatter, quasi-stellar objects and galaxy evolution. Hunter (29-7449) finds that no non-rotating model of Alfvén-Klein type with equal amounts of matter and anti-matter could have established a black-body radiation field at its minimum radius and then expanded to infinity.

Dicke (29-14447) has suggested that the He content of Population II stars could provide a test for comparing general relativity and his scalar-tensor theory. Nariai (30-8281) demonstrates that the Brans-Dicke theory satisfies Mach's principle in a homogeneous isotropic universe.

EXTRAGALACTIC ASTRONOMY IN THE U.S.S.R.

(by B. Vorontsov-Velyaminov)

In 1968 Vorontsov-Velyaminov and Arhipova (30-6364) finished the publication of the Morphological Catalogue of Galaxies (MCG). It contains 30000 numbered galaxies down to $15^{m}_{.}0$ and to $\delta = -33^{\circ}$ with detailed descriptions and notes. Many new morphological types were discovered, one of which was later denoted by W. M. Morgan as cD. Together with Noskova, Vorontsov-Velyaminov (30-635) compared the descriptions of MCG galaxies with the classification proposed by G. de Vaucouleurs. They also published in the "Nonstable phenomena in galaxies" (*Bjurakan*, 1968) papers on galaxies with peculiar nuclei, Haro galaxies and interacting galaxies. Zasov (30-4408) showed that the "bridges" in the latter can be long-lived at the expense of the accretion from the intergalactic gas. Parsamian, Sahakian, Iskudarian and Shahbazian (*Aca**, **4**, 41) developed the statistical characteristics of the central condensations and super-associations in Sc, Sb, SB and E galaxies. Sharov (30-626, -2636, -8565) studied the space distribution of Novae, globular clusters and gaseous and absorbing matter in M31. The inclination of its central layer was studied by Makarenko (28-12535).

Colorimetric and spectrophotometric investigations were published by I. Pronik and Chuvaiev (29-7877; *Astr. Zu.*, **46**, 755) on NGC628 and on the jet in M87. Kallogljan (28-9585; 29-3511) dealt with Stephan's quintet and Parsamian (29-13511) with NGC3310 and 3351. Ogorodnikoff (29-13508), from dynamical considerations, claims that needle-shaped galaxies may be possible. By averaging the forms of the spiral arms Karachentsev (29-5017) concluded that they follow Archimedes rather than logarithmic, spirals. Zasov studied interacting galaxies statistically. Markarian (28-12536) discovered over 70 faint galaxies with conspicuous UV continua. Some of them happened to be of Seyfert type, as revealed by Khachikian (30-13078) and others. Pronik, Esipov, Dibai and Vorontsov-Velyaminov (29-6495, -7866; *Astr. Zu.*, **46**, 237, 240, 725) studied spectrophotometrically many Seyfert galaxies and proposed physical models for them. Studies of their variability were also carried out by the Crimean observers. Hagen-Thorn and Dombrovski (30-4365, -4351) measured the polarization in some radio sources. In M82 in particular, the polarization is ascribed to the dust rather than to the non-thermal emission. Nezhinsky and Ossipkov (29-7872) re-determined the luminosity function of field galaxies. Vorontsov-Velyaminov (29-13526; 30-6363, 30-14619) showed that, according to radio observations of the rotation of galaxies, the spiral arms as a whole are situated inside the region of rigid body rotation in the overwhelming majority of cases. He showed also that in physical pairs of interacting galaxies the directions of the arms (= of rotation?) are equally frequently in the same and in opposed senses. Pickelner (29-13512) proposed his version of the magneto-hydrodynamical hypothesis for the formation of spiral arms, while Marochnik and Suchkov (*Astr. Zu.*, **46**, 319, 524, 762) amplified the theory of spiral arms as the product of density waves in a two component model with differential rotation. Their theory explains the existence of the trailing and of the unwinding arms. Sizikov (29-3523; *Aca*, **5**, 317) improved the theory for calculating the distribution of density from the velocity curve and the photometry of a galaxy. The Genkins revised the tables of the masses of galaxies and their luminosity functions. With Ozernoy (30-6341), they found that for young galaxies the mass is proportional to the cube of the diameter. Zasov and Ozernoy (29-5028) studied the rotational momentum – total and per unit mass. Towmasian (28-5476, -9599; 29-11792, -13523; 30-4396) studied the galaxies of M82 type and measured in Australia, the radio emission from several hundred different galaxies. He has found a correlation between radio emission and optical peculiarities.

Karatchentsev (28-5473, -12531) confirmed the statistical nature of the phenomenon denoted as "superclusters of galaxies" and estimated the density of H I in clusters. He and Lipovezky (30-8558) estimated statistically the light absorption in clusters. Agekian and Suzmina (29-13491) calculated the distribution of intrinsic oblatenesses of galaxies in clusters. Nezhinsky and Osipkov (*Astr. Zu.*, **46**, 685) showed that the dispersion of velocities in clusters is dependent on the mass distribution.

* *Aca* stands for *Astrophysica*, published in the U.S.S.R.

A large calculated value indicates the presence of expansion. Karachentseva (29-13503) analyzed the distribution of dwarf galaxies in three clusters and Zwiagina (29-11799) did the same for the groups in MCG, using the notes in that catalogue.

Radio observations of galaxies with discussions of the results, were carried out by Kurilchik and Lekht (29-485, -3515; 30-4362) and their colleagues. Esipov and collaborators (29-5008, -5009; 30-4343, -8552), using image converters, studied the spectra of faint QSS, etc., by means of a 125 cm telescope. Komberg (29-6500; 30-4361, -8560) discussed the nature of the broad emissions in the spectra of QSS and suggested that they may be due to binaries. Vorontsov-Velyaminov opines that QSS form groups and clusters of galaxies by repeated fragmentation. Gordon (29-11707) is producing models of the structure and masses of the envelopes of Supernovae, QSS and Seyfert galaxies. Shklovsky (29-13519) considered the nature of the standard absorption spectra of QSS on the basis of special relativity. He explains the variation of the optical spectrum of 3C 345 by the clearing of clouds of Mg II in the frequencies of the resonance lines due to the heating by the stream of particles ejected from the nucleus. Ozernoy and Chertoprud (29-3189; 30-4353) analyzed the light curves of the QSS using a magnetohydrodynamical model. Shklovsky (28-11055) showed the possibility of various causes for the X-ray emission. Ozernoy showed that the plasma ejected by a magnetoid can condense into stars. Jointly with Sazonov, he calculated the spectrum and polarization of a radio source produced by a relativistic fission of its components. Ozernoy and Chernin (29-11490; *Astr. Zu.*, **45**, 1137) formulated the hypothesis of a dynamical character of pregalactic "photon vortices" and on this basis they developed a theory of the origin of galaxies.

Zelmanov showed that the space volume, the quantity of particles and the energy in a cosmological model can be infinite in one system of reference and be finite in another. Grĭšcuk (29-7446) elucidated the difference between the group and differential criteria of a space non-uniformity. He has found all metrics satisfying the second criterion proposed by Zelmanov. Grĭšcuk also showed that for a dusty space the general solution of the equations of gravitation contains an essential non-simultaneous physical singularity. The solution containing a simultaneous singularity is not stable. Grĭšcuk, Doroshkevich and Novikov showed that the solutions could be made isotropic. They gave formulae for the anisotropic surviving emission in the models described by the solutions which were obtained. The team headed by Zeldovich, which includes Novikov, Doroshkevich and Sjunjaev studied a large number of problems of relativistic cosmology: the residual emission at different epochs, the expansion of anisotropic models, gravitational instability and collapse, and the formation of galaxies by different types of perturbations of a Friedman model. At the critical density of 2×10^{-29} g cm^{-3} for the intergalactic gas, the universe is open (infinite).

G. C. MCVITTIE
President of the Commission

WORKING GROUP ON THE MAGELLANIC CLOUDS
(by A. D. Thackeray)

1. *General*

A Symposium* on the Magellanic Clouds was sponsored by E.S.O. in Santiago, Chile, in 1969 March in connection with the dedication of the E.S.O. Observatory at La Silla. Review papers on the Clouds have been written by Westerlund and Bok (*J. R. astr. Soc. Can.*, **63**, 105).

2. *Variables and the P-L relation*

Gascoigne (*Mon. Not. R. astr. Soc.*, **146**, 1) has prosecuted further his *P-L* work on Cloud

* Where no references are given, the reader should consult the Proceedings of this Symposium and also forthcoming issues of the *Mon. Not. astr. Soc. S. Africa*.

cepheids and finds

$$V = 17{\cdot}58 - 3{\cdot}072 \log P \quad (7 \text{ LMC cepheids}),$$
$$V = 17{\cdot}73 - 2{\cdot}879 \log P \quad (13 \text{ SMC cepheids})$$

while 25 SMC, 18 LMC and 13 galactic cepheids obey the Sandage relation

$$M_v = a - 3{\cdot}609 \log P + 2{\cdot}52(\langle B \rangle - \langle V \rangle).$$

This implies moduli of 18·73 (LMC) and 19·36 (SMC), but Gascoigne recommends, in the light of other evidence, adoption of moduli 18·75 (LMC) and 19·2 (SMC). The colours of SMC cepheids are $0^m.1$ bluer than those in the Galaxy and they may be somewhat underluminous for their periods.

Arp's SMC sequence has been further tested at Radcliffe by Andrews and at Boyden by Butler.

Hodge and F. Wright have derived periods for 19 LMC cepheids on a 63-year baseline without finding changes in period. The slope of their *P-L* relation agrees well with that of Gascoigne and of the Herstmonceux group. They have also studied two W Vir, two long-period and two eclipsing variables. A peculiar R CrB variable (max. 16·2 m_{pg}) showed one minimum only over 63 years.

Two-colour photographic photometry of 100 SMC cepheids (central part) and 5-colour photoelectric photometry of 12 SMC cepheids has been completed by van Genderen (*BAN Suppl.*, No. 6). The Gaposchkins have continued their very extensive studies of Cloud variables. S Dor has still not recovered from its deep 1963 minimum. Martini has found that Cordoba spectra show that the variable was in minimum phase in 1945–46. Landi Dessy has found 222 variables in his SMC field '*f*' (away from the bar) of which 111 were discovered on blink-pairs with intervals 0·146–0·27 days. Christy has applied his theory of cepheid instability to Cloud cepheids. In particular his interpretation of secondary bumps in light-curves as internal echo phenomena has led to the conclusion that a single variable with clear bumps and period near 10 days can be used to determine a distance modulus. From one of Gascoigne's variables he derives 18·75 for the LMC modulus. Lloyd Evans is searching the Clouds for red variables. Several found in the LMC with $P \sim 200$ days are presumably genuine Mira variables. Meier (29-9920) reports 12 new Bamberg variables and one nova (1966) in the LMC.

3. *Gaseous components, radio, X-ray, etc.*

Faulkner (29-6460) has set up an upper limit of $5{\cdot}10^5\ M_\odot$ to the mass of 30 Dor from Balmer-line isophotes. Le Marne (30-4256) has studied 30 Dor at 408 MHz. The presence of non-thermal sources on rings surrounding supernova remnants has been pointed out by Matthewson and Westerlund.

OH has not been found in the Clouds with the Parkes reflector (29-1941). Of various attempts to observe X-rays from the Clouds (30-4365, 29-13461) only Mark *et al.* claim success with the LMC at a level of order 4 % the intensity of Tau X-1. H. M. Johnson suggests that localized X-ray sources might have non-thermal continua optically detectable.

Westerlund (30-13077) has summarized known data on the Cloud planetaries. Feast and Webster independently find the SMC to be relatively rich in low excitation planetaries. Miss Webster has measured absolute emission line-fluxes (H, [O III], [O II]) and continua photoelectrically in 30 Cloud planetaries. She has also combined measures of the central stars with H, He line strengths to yield temperatures and luminosities. Some similarities with Seaton's evolutionary track are found.

4. *Clusters*

(a) *Globular*

Gascoigne and Hearnshaw (*Astr. Soc. Aust.*, **1**, 208) are deriving *c-m* arrays of the clusters NGC 2010, 2136 which contain cepheids. P. J. Andrews and Lloyd Evans find in their work on integrated spectra of red globulars that LMC globulars match galactic counterparts with moderately large metal deficiency but, in the SMC, NGC416 and Kron 3 have no galactic counterparts.

(b) *Associations and clusters*

Westerlund and P. J. Andre are independently investigating the wing of the SMC and agree in the dominance of a very young population here. Andrews finds NGC371 (SMC) to resemble *h* and χPer in its *c-m* array: in one of three nearby small clusters (20″ of arc) there is found a giant branch at $V \sim 16{\cdot}5$, $B-V$ from 0·45 to 0·95.

Van den Bergh (39-4399) has observed and discussed integrated *UBV* magnitudes of 39 LMC and 13 SMC clusters. G. Walker and S. Morris have published *UBV* photometry of two LMC associations. F. Wright and Hodge (*A.J.*, 73, S184, S210) have discussed the distribution of ages of LMC clusters through the system.

5. *General field*

(a) *Stellar spectroscopy*

Sanduleak (30-6356) has applied the Michigan Schmidt at Cerro Tololo to list 169 probable SMC members through objective prism classification (580 Å mm^{-1}). Fehrenbach and Thackeray independently confirm the reliability of Sanduleak's list. Sanduleak's new members include only 6 brighter than 12·0 m_{pg}. Sanduleak has a similar list of 1200 LMC members in preparation.

Fehrenbach reports that the ESO group at La Silla have now completely examined 13 fields, covering the bulk of the LMC, that have yielded 469 high-velocity stars, 1765 foreground stars and 200 stars of intermediate or doubtful velocity, etc. Maurice *et al.* have measured and classified 150 LMC stars (72 Å mm^{-1}). Florsch is listing 150 SMC members, while Mme Carozzi, in studying the region between the Clouds and between LMC and Galaxy, has found a surprising number of high-velocity stars apparently belonging to the galactic foreground.

Walraven, Fehrenbach and Thackeray all find the LMC to be richer than the SMC in the brightest super-supergiants of type F-K. Wesselink has calculated radii for such stars by a method which circumvents questions of bolometric corrections, but not of intrinsic colour, and finds $R = 10{\cdot}3$ A.U. for HDE268757. 5-colour photometry of 69 LMC supergiants has been completed by van Genderen. The state of evolution of such stars still poses considerable difficulties as indicated by Kippenhahn, Stothers (29-14531) and others.

Przybylski (30-624) confirms the close similarity of HD33579 (LMC) in abundances of metals to αCyg. L. F. Smith (30-4393) in discussing absolute magnitudes and colours of W stars comments on the similarity of LMC objects to Pop. I in the solar neighbourhood, while the SMC Pop. I is compared with that of the outer parts of the Galaxy. Underhill (30-8489) has also discussed W stars in the LMC.

(b) *Kinematics*

Monnet and Courtès report measures of 300 radial velocities of H II regions over the whole LMC, yielding a rotation curve like that for 21-cm radiation. The Fehrenbach group are supplementing their stellar objective prism work with radial velocities from slit spectra (72 Å mm^{-1}).

Feast (30-2624) has published radial velocities of 25 LMC and 11 SMC planetaries. The LMC objects define a rotation curve very similar to that for young objects (centre displaced from the bar) but with a larger dispersion (22 km s^{-1}). A spread in ages is indicated. Webster's (30-13016, -13017, -13018) velocities suggest a centre nearer to the bar. The distribution of Feast's SMC planetary velocities shows a double maximum like Hindman's for the 21-cm velocities. There is some support for the SMC being regarded as a bar seen end-on with matter flowing out at the ends, just as may be occurring with the LMC bar as suggested by de Vaucouleurs.

(c) *Photometry*

Basinski, Bok and Bok (29-6491) have published photometry of stars down to 16^{m} in the core of the SMC, predominantly unreddened blue stars. The Boks have checked Tifft's LMC sequence from $V = 9{\cdot}4$ to 16·4. Mianes *et al.* have carried out 5-colour photometry of more than 300 LMC members at La Silla. Walraven reported further results of his multicolour photometry of Cloud stars at the ESO Symposium, compared with model computations. Five-colour photometry by Wamsteker of 500 blue stars and two-colour photography by Velthuyse have been carried out in the inter-Cloud region at the Leiden Station.

(d) *Absorption and polarization*

Studies of absorption continue to emphasize local obscuration in many parts of the LMC as compared with the relative transparency over much of the SMC, including the core. Gascoigne has examined main sequence B stars and adopts locally in the SMC a reddening of $0^{m}.2$, most of which may be in the foreground, and $0^{m}.05$ in the LMC. Le Marne (*Astr. Soc. Aust.*, **1**, 97) and Faulkner's work on 30 Dor, which combines radio and optical measures, suggests $A_v/E = 3$ for foreground absorption, but $A_v/E = 7$ within the nebula. However, Gascoigne's photometry of the cepheid HV 2749, only 15′ of arc south of 30 Dor, yields $A_v/E = 2{\cdot}8 \pm 0{\cdot}25$.

6. *Tidal Action*

Tidal action between the Clouds and the Galaxy has been considered by a number of workers (29-5003), but it is still uncertain whether the Clouds are bound gravitationally to the Galaxy. Kerr (*A.J.*, **73**, S102) has pointed out that the bulk of the high-velocity H I clouds, with a total mass comparable to that of the Magellanic system, may be moving in an approximately opposite direction to the Magellanic Clouds.

WORKING GROUP FOR RESEARCH ON SUPERNOVAE

(by F. W. Zwicky)

By October 1966, the number of supernovae in extragalactic systems which had been definitely identified was 184. During the following three years an additional 70 supernovae were discovered, bringing the total number by July 1969 to 254.

The search for and the investigation of supernovae has been engaged in by an ever increasing number of observatories and astronomers, the majority of whom strongly support the continuation of the working group on supernovae and, until further notice, of the writer as its chairman.

Circular letter No. 9 was issued to the members of the working group in June 1968, giving data on the supernovae Nos. 171 to 216. A catalogue-monograph on the supernovae found since 1885 is in preparation and should be ready for presentation at the 1970 IAU Assembly. In the meantime K. Rudnicki and A. Karpowiczowa have announced that the manuscript of their catalogue of supernovae, including all major bibliographic data has been sent to the printer.

Two new approaches to the search for supernovae are in the first stages of realization, namely (*a*) automated searches by Professors Hynek and Colgate, the first having already resulted in 40000 bright galaxy reductions, the second being expected to start operation in the fall of 1969 and (*b*) the endeavours by J. Strong and F. Zwicky to have full-size objective gratings built for large (up to 50″ apertures) Schmidt telescopes. With gratings of this type, supernovae to about $m_p = 16{\cdot}5$ may be discovered on single plates, the spectrum being observed at discovery.

The frequency of supernovae appearing in galaxies within about five absolute magnitudes of the brightest has been redetermined by several investigators using extensive material. The frequency found originally by Zwicky of about one supernova per galaxy per 300 to 400 years has been confirmed. It is estimated that most of the supernovae which have appeared in the central cores of the galaxies observed have been missed. Also, no estimate can as yet be made of the frequency of intrinsically faint supernovae such as Eta Carinae. As predicted already from the statistics made 30 years ago there are some special galaxies which produce supernovae at a far greater rate than normal. For instance four supernovae each have appeared in NGC 5236 and NGC 6946 during the past fifty years.

A most interesting issue arises with respect to the frequency of supernovae in quasars and in the most massive compact galaxies, in some of which according to theory up to several hundred supernovae per year may be expected. The spectrum of the first variable compact galaxy, [R.A. $0^h39^m.5$ Decl. $+40°03'$ (1950)] discovered by Zwicky in February 1965 and showing random variations of two magnitudes since 1936 is therefore now being scanned.

Many light curves of supernovae have been established during the past three years and many spectra have been obtained. From an improved redshift-magnitude relation for supernovae a probable value of 150 km s^{-1} per million parsecs for the Lundmark-Hubble constant is derived.

With the discovery of the pulsar in the Crab Nebula, Zwicky's 1933 hypothesis of some supernovae being caused through the collapse into neutron stars seems to have found its confirmation.

No cosmic rays, radio waves or X-rays have as yet been observed coming from recently observed supernovae, as theory would predict.

No tenable interpretation of the spectra of supernovae of the types I and IV has yet been proposed.

WORKING GROUP ON GALAXY PHOTOMETRY

(by G. de Vaucouleurs)

1. A tabular presentation has been adopted and the report does not cover work in the U.S.S.R. During the report period Circular No. 6 (Report on Prague Meeting) and a Technical Supplement, No. 6a, were distributed. Work on standard luminosity profiles for a few well-observed galaxies (NGC3115, 3379, 4486) progressed more slowly than expected, but provisional results will be available for presentation at the 1970 meeting.

2. On the observational side the major event to report is the publication of an *Atlas de Galaxias Australes* by J. L. Sĕrsic; it gives photographs, descriptions and detailed surface photometry of 50 southern galaxies observed with the 150-cm reflector of Cordoba Observatory. Another important contribution is the dissertation of H. D. Ables on "Optical Study of Nearby Galaxies" of interest to radio astronomy. The isophotometric atlas of 70–80 galaxies and *UBV* catalogue of 500 galaxies observed by G. and A. de Vaucouleurs at Mt. Stromlo and McDonald Observatories and by H. D. Ables at Flagstaff, F. Bertola at Asiago and other collaborators is being prepared for publication in 1970-71.

On the technical side the rapid development of automatic digital methods for data recording and processing should lead to a great acceleration in the production of precise isophote maps, but most commercial instruments are still deficient with respect to mechanical precision, optical design (stray light, beam definition) and electrical stability.

On the theoretical side I. R. King (Berkeley) is continuing his work on the structure and dynamics of elliptical galaxies, and lenticular and spiral galaxies have been analyzed by K. C. Freeman in Australia and by J. and N. Heidmann in France; all these studies demonstrate again the basic importance of precise and detailed surface photometry for an understanding of the structure and evolution of galaxies.

Table. Summary of galaxy photometry 1967-70

Authors	Subject	Type*	References
H. J. Rood *et al.*	Coma cluster (315 gal.) pg, pv	C, pg	29-1943, 30-625, 30-4381
N. B. Richter	730 compact gal. to $m = 18{\cdot}5(V)$	A, pg	*Mitt. Taut.*, No. 37
S. van den Bergh and R. D. McClure	32 field E. gal. (*UBV*)	B, pe	30-14593
S. van den Bergh and R. D. McClure	56 gal., 5 spectral indices (0·35–0·45 μ)	B, pe	30-6302
A. Sandage *et al.*	Center M31 (*UBVRIHKL*), M32, M81, NGC7331 (*UBV*)	A, pe	*Ap. J.*, **157**, 55
G. Paal	> 1000 gal. in rich clusters	D, pg	Unpubl.
P. W. Hodge *et al.*	NGC4449 + others	C, pg	30-612
R. H. Miller *et al.*	NGC3115	C, pg, pe	30-2628
S. D'Odorico	NGC1156	C, pg	Unpubl.
C. W. Fraser	50 gal., Virgo cluster	C, pg	Unpubl.
H. D. Ables	NGC1569, 6946; IC342, 356, 1613; A1009	C, pg, pe	Unpubl.
H. Oleak *et al.*	M87	C, pg, pe	30-2631
H. Oleak *et al.*	Virgo cluster	A, pg, pe	Unpubl.
M. Motylinska	Cluster Abell 1781	A, pg	30-14597
W. G. Tifft	64 gal., 4 colors	A, pe	*A. J.*, **74**, 354
I. R. King	Ellipticals	C, D, pg	Unpubl.
J. L. Sĕrsic and	50 southern gal.	C, pg	*Atlas de Galaxies Australes*, Córdoba.
H. Dottori and	6 peculiar gal.	C, pg	*Bol. Inst. math. astr. y fis.*, **3**, No. 2.
E. Aguero	Southern gal. in field + Virgo cluster	C, pg	Unpubl.
H. M. Johnson	Hβ photom. of 13 near gal.	B, pe	Unpubl.
A. Elvius	M82	E, pe	*Lowell Obs. Bull.*, **7**, 117
G. de Vaucouleurs *et al.*	M87 jet and corona	C, pg, pe	30-4403; *Ap. J. Letters*, **4**, 17
G. de Vaucouleurs *et al.*	Fornax	C, pe	30-4401
G. de Vaucouleurs *et al.*	50 gal., 10 bands (0·34–0·55 μ)	B, pe	Unpubl.
T. Borchkadze	7 gal., in 2 groups, *UBV*	C, pg	Unpubl.
W. Zonn	Out-of-focus photom.	F, pg	29-1948

* *Type.* – A = general broad-band photom., etc., magnitudes and/or colors; B = intermediate band photom., energy distribution, spectral indices, line-strengths; C = detailed surface photom., isophotom., drift curves; D = photometric diameters and other special studies; E = polarization; F = theoretical.

29. STELLAR SPECTRA (SPECTRES STELLAIRES)

PRESIDENT: M. W. Feast.
VICE-PRESIDENT: Y. Fujita.
ORGANIZING COMMITTEE: M. K. V. Bappu, G. Herbig, L. Houziaux, I. M. Kopylov, A. Przybylski, J. Sahade.

Material for this report was collected by the President, Vice-President and Members of the Organizing Committee. The President is, however, responsible for the form in which the report now appears. A number of special abbreviations in the references are explained in the report of Committee 27a. In addition, 3rd Harvard = 3rd Harvard-Smithsonian Conference on Stellar Atmospheres (1968). The field of Commission 29 overlaps particularly with those of 9, 27a, 36, 44 and 45 whose reports should be consulted. Since the last IAU meeting 29 has co-sponsored the following meetings: IAU Colloquium No. 4 on *Stellar Rotation* (Columbus, Ohio, September 1969); IAU Symposium No. 36, *Ultraviolet Stellar Spectra and Related Ground-Based Observations* (Lunteren, June, 1969); Second Trieste Colloquium, *Mass Loss from Stars* (September, 1968). We are also co-sponsoring IAU Symposium No. 42 on *White Dwarfs* to be held in Scotland (August, 1970). The thanks of the commission are due to their representatives on the organizing committees of these meetings. Reports from some working groups are appended. The working group with Commission 44 has not felt it necessary to submit a report (its main activity was the organization of Symposium No. 36). Miss Underhill (Chairman) recommends that the working group on Tracings of High Dispersion Stellar Spectra be dissolved.

GENERAL, INSTRUMENTATION, PRACTICAL MATTERS

Volumes containing relevant papers are: Liège, 1968, *Forbidden Transitions in Stellar Spectra*; Trieste, 1968, *Mass Loss from Stars*; Trieste, 1966, *Late-Type Stars*; Paris, 1967, *The Origin and Distribution of the Elements*; Tucson, 1968, *Beam Foil Spectroscopy*; Lunteren, 1969, *Ultraviolet Stellar Spectra and Related Ground-Based Observations.* Abundance work has been summarized (Cayrel, *Ann. Rev.*, **4**, 1; Kuchowicz, *N.E.I.C. Review Report*, No. 34, Warsaw, 1968. cf. also Pagel, *Q.J.R.A.S.*, **9**, 401). Theoretical work on rotational broadening is reviewed by Strittmatter (*Ann. Rev.*, **7**, 665). Allen (*M.N.R.A.S.*, **133**, 21; **139**, 367) advocates a statistical approach to stellar spectroscopy.

There has been widespread use of image intensification devices (Ford, *Ann. Rev.*, **6**, 1) and photoelectric spectrum scanners (Oke, *P.A.S.P.*, **81**, 11). Important results have been obtained from Fabry-Perot interference spectroscopy (Vaughan, *Ann. Rev.*, **5**, 139) and Fourier transform spectroscopy (Johnson *et al.*, L.P.L. **113**; Gebbie and Twist, *Rep. Prog. Phys.*, **29**, 729. Connes *et al.*, *Publ. Obs. Hte-Provence*, **9**, No. 6). Instrumentation completed includes: a high resolution spectrograph with two echelle gratings in series (Barannes *et al.*, *J. Observateurs*, **50**, 289); a high dispersion objective prism system (Treanor, *Vistas*, **11**, 147); application of pulse counting methods to magnetic field measurements (Severny, Dimov); a combined echelle-interferometer device (Wyller); a method for studying polarization effects in line profiles (Clark, Grainger, *Ann. Ast.*, **29**, 355). Under discussion have been: the conversion of line profiles into congruent radio frequency profiles for high resolution work (Van Bueren, *B.A.N.*, **19**, 97); the use of multislit spectrographs (Serkowski, *Observatory*, **87**, 259). Progress has been made in fully automating equivalent width reduction (e.g. Thompson, *Edin. Publ.*, **5**, 245). The determination of equivalent widths on a common (and if possible absolute) system remains a central problem (see 29b and e.g. Hartley, Powell, *Observatory*, **87**, 75). Uncertainties in oscillator strengths remain a problem (cf. Cowley, Warner, *Observatory*, **87**, 117; Garz, Kock, *Astron. Astrophys.*, **2**, 274).

P CYG STARS, SHELL STARS, WR STARS

De Groot (*B.A.N.*, **20**, 225) made a detailed study of P Cyg including line identifications, atmospheric structure and mass loss (2×10^{-4} $M_{\odot}$/year). Hutchings (*M.N.R.A.S.*, **144**, 235) rediscussed Beal's velocities with similar results. Spectrophotometric gradients, velocities, line profiles and (normal) abundances were derived by Luud *et al.* (*Tartu Publ.*, **35**, 196; **36**, 202, 221; *Budapest*, 197; *Astr. Zu.*, **44**, 269; *Astrophys.* **3**, 391), and abundances in four similar stars. Hutchings (*Observatory*, **87**, 289; *Budapest*, 191; *Columbus*, 1969) found rapid variation in line profile for Be and P Cygni stars. In γ Cas this may be related to stellar rotation (period ~ 0.7 days). Rapid changes were also studied by Delplace *et al.* (*Budapest*, 223) and Popova, Dokuchaeva (*Alma Ata*). Ivanova *et al.* (*Budapest*, 215) studied long term continuum changes in γ Cas. The P Cygni star HD 51 585 shows [O I], [O III], [S II], [Fe II] emission (Andrillat, Houziaux, *Liège*, 1968, 343).

Work on shell stars includes: changes (and periodicities) in ζ Tau (Underhill, v.d. Wel, *Toronto*, 251; Delplace, *Publ. Obs. Ht-Provence*, **9**, No. 31; Hack *et al.*, *Trieste Publ.*, 373, 392); variations in 48 Lib (Faraggiana, *Astr. Astrophys.*, **2**, 162); changes in 4 shell stars (including A-type shell stars) (Cowley, Hiltner, *P.A.S.P.*, **80**, 685); velocities and abundances in o And (Galeotti, Pasinetti IB 301, *Pubbl. Oss. astr. Milano-Merate*, 283, 300); variations in HR 2142 (Peters, *Los Angeles*).

The problem of the WR stars has been comprehensively reviewed by Underhill (*Ann. Rev.*, **6**, 39). A symposium on WR stars has been published (Nat. Bur. Stand. Special Pub., 307) and summarized by Thomas (*Astrophys. Letters*, **2**, 147). Kuhi's scanner observations of V444 Cyg have been published (*Ap. J.*, **152**, 89). He finds great differences in the eclipse curves of different emission lines. In the continuum the depth of the secondary eclipse increases with increasing wavelength. Continuum and line intensities show rapid erratic fluctuations. Underhill (*B.A.N.*, **19**, 173) discussed the spectrum of HD 191 765 in detail. Identifications and equivalent widths are given. It is suggested that electron scattering is not important for the line profiles. Bracher (*P.A.S.P.*, **80**, 165) discussed violet displaced He I 3889 in HD 211 853; the velocity varies in phase with the B component. A revised spectral classification scheme for WR stars is proposed by Lindsey Smith (*M.N.R.A.S.*, **138**, 109). Bertola and Ciatti have observed the brightest WR stars in the 1 μ region at 500–800 Å mm^{-1}. Nugis and Luud (*Tartu*) studied HD 191 765, 192 103, 192 163 for effective temperature, bolometric correction and emission line intensity corrections. They find the stars to lie to the left of the main sequence (near the helium main sequence). Galkina (*Izv. Krym. astrofiz. Obs.*, **39**, 44) studied the close binary HD 190 918 (WN 5·5 + O9). Detailed studies of the contour of He II 4686 emission lead to an orbital period of 105 days.

The report on Be stars is the responsibility of a special working group.

Additional material supplied by Dr. Bappu is as follows. P. Swings and J. P. Swings (*Beam-Foil Spectroscopy*, Ed. S. Bashkin) confirm the assignment of 5806 Å to C IV in WN stars. Aller has in progress an observational programme at Cerro Tololo of a number of Of stars that are either the nuclei of planetary nebulae or are prototype stars. Aller and Lindsey Smith have in press a paper on the classification of emission line spectra of planetary nebulae. Underhill (*Astrophys. Space Sci.*, **3**, 109, 1969) has discussed the mass loss characteristics of WR stars. Deutsch finds that the He I 3888 absorption line which is the strongest absorption feature in the spectra of WN-B stars shows time dependent variability of intensities and velocities. He points out that the displaced absorption line of 3888 Å is found only in those WC stars that are known spectroscopic binaries or that have nebular lines in their spectra. L. V. Kuhi and L. F. Smith are working on an atlas of WR line profiles. Bappu and Scaria are engaged on an atlas of intensity tracings of bright WR spectra obtained at 10 Å mm^{-1} and 20 Å mm^{-1} dispersion. Jones, Evans and Catchpole (*Observatory*, **89**, 18, 1969) find the nucleus of Henize 99 to be a WC 9 star with emission bands at 4651 (C III, C IV 4650 Å), 4267 (C II 4267 Å) 4186, 4121, 4070 (C III 4069 Å), 3920, 3757 Å.

The current survey for luminous stars in the Southern Milky Way by Stephenson and Sanduleak has turned up 3 to 4 new WR stars. One of these appears to be a visual binary companion to a G-type supergiant (*IAU Circ.*, No. 2143, 1969). Sanduleak has published (*IAU Circ.*, No. 2150, 1969) data concerning a new Wolf-Rayet star showing O VI 3840 as strong as the 4670 Å blend. He has also (*P.A.S.P.*, **80**, 470, 1968) called attention to a previously known WR star in the Small Magellanic Cloud which has this property and which may also have a variable spectrum. Only about half a dozen such high excitation WR stars are known and two of them have been inconclusively suggested to be X-ray sources. D. C. Morton has derived the ratio of Lyman-continuum to visual fluxes for six WN stars from the free-free radio emission of surrounding nebulae and the apparent magnitudes of the stars. Comparison with the theoretical ratios obtained from model atmospheres which omit the emission lines, but include ultraviolet line blanketing where appropriate, gives estimates of the effective temperatures. Values around 50000K are found for WN4·5 and WN5 35000° for WN6 and 25000° for WN8. Bappu and Ganesh (*M.N.R.A.S.*, **140**, 71, 1968) have measured line intensities of the hydrogenic transitions of C IV using coude spectra of dispersion 10 Å mm^{-1}. Assuming LTE, excitation temperatures have been derived for WC6, WC7 and WC8 spectral types. Van Blerkom and Castor (*M.N.R.A.S.*, **143**, 461, 1969) question the validity of these values, and claim that they cannot be identified with physically meaningful electron temperatures of WR envelopes. Kuhi is currently observing HD 50896 and HD 191765 with scanner and high dispersion spectrograph to detect possible periodicities. Scanner observations of CV Ser as a function of phase are also being obtained.

M. K. V. Bappu and K. S. Ganesh (*Kodaikanal Obsv. Bull.*, No. 183, Series A, 1967) assign a period of 78·5 days for the binary γ_2 Vel. V. N. de Monteagudo and Sahade using Cordoba spectra confirm this value and improve it to 78·5002 days. They have studied the changes in structure of the He I 3888 line in absorption as well as emission which can be interpreted in terms of emitting material located between the two stars and moving from the WR star towards its companion. D. C. Morton, E. B. Jenkins and N. H. Brooks (*Astrophys. J.*, **155**, 875, 1969) have examined far-ultra-violet spectra of γ_2 Vel. Stepien has carried out *UBV* photometry of the WR eclipsing system CV Ser in 1967 and finds no evidence of eclipses. This interesting observation needs confirmation and extension. T. Kogure is measuring radial velocities of the He II lines in V444 Cyg from spectrograms obtained at Okayama. E. D. Carlson (Doct. thesis, Northwestern University, U.S.A., 1969) has studied the spectra of the two southern Of stars HD 148937 and HNZ 759. HD 148937 is a very high luminosity nucleus of the planetary nebula, NGC 6164–5, with a spectrum that displays the Pickering series in addition to He II 4686 and N III 4634, 4640. D. C. Morton, E. B. Jenkins and N. H. Brooks (*Astrophys. J.*, **155**, 875, 1969) have studied the far-ultraviolet spectrum of ζ Pup. P. S. Conti has in progress a detailed study of O and Of stars using high dispersion spectra. Equivalent widths, line profiles and measures of the line broadening parameter are being measured for different spectral types. E. N. Walker has found that HD 153919, on the outskirts of NGC 6281 is a high velocity O7f star. He has also detected weak Of characteristics in the binary supergiant HD 152667 which is probably a member of NGC 6231. Walker has derived a spectroscopic orbit for this star. Small amplitude light variations with two minima and a period of 7·84 days have been detected at the Cape Observatory for this star. Slettebak (*IAU Colloquium on Variable Stars*, Budapest, 1968) has observations of variability of the spectral features of the Of stars λ Cep, 9 Sag, HD 34656, HD 190429N and 29 CMa.

O AND B TYPE STARS

Hack, Stalio, Aydia (*AI*, **39**, 1; **40**, 7) determine temperatures, gravities and abundances in O stars. Underhill (*B.A.N.*, **19**, 500) compares 10 Lac (O9) with model atmospheres and finds non-LTE effects important and helium abundances suspect. Problems of Of stars were reviewed (Slettebak, *Budapest*; Feast, *Lunteren*). There is evidence for mass loss and rapid spectral variations (Hutchings; Slettebak; Brucato (Northwestern); Battistini, *Pubbl. Oss. astr. Univ. Bologna*, **9**, No. 9).

Spectrophotometric data on B stars is given by Aller, Jugaku (*Mich. Publ.*, **9**, 203) and Buscombe (equivalent widths, central depths, $v \sin i$, Balmer profiles *M.N.R.A.S.*, **144**, 1, 31, etc.).

Messerschmidt *et al.* (*Z. Astrophys.*, **66**, 246) find theoretical profiles for diffuse series He I lines fit τ Sco satisfactorily. Observed and computed equivalent widths of Si II 4128, 4130 indicate that the adopted *gf*-values are 6 times too large (Underhill, *Ap. J.*, **151**, 765). Underhill attributes a discrepancy between observed and computed Balmer line profiles in Ia supergiants to an underabundance (× 1000) of hydrogen. Smit (*B.A.N.*, **20**, 274) and Svolopoulos (*Ann. Astr.*, **29**, 29) studied βOri (B8 Ia) and Kikuchi (*P.A.S.J.*, **20**, 190) its variable (PCygni) Hα. High dispersion spectra and spectral scans of τHer (B5IV) are used in a detailed study by Heintze. He also discusses the A stars Vega and Sirius A (which may be a binary). In addition he deduces effective temperatures for 42B stars using a relation between Balmer discontinuity and U − B (*B.A.N.*, **20**, 1, 154). Other detailed studies of B stars at Utrecht are: ζDra (Visser), εOri (Lamers), o^2CMa (Van Helden), ηCMa, 67Oph (Underhill). Snijders (*Astr. Astrophys.*, **1**, 452) following Warner uses FeIII in ζCas, γPeg (B2) and finds an Fe abundance three times larger than previously supposed. Other work includes: abundances in B supergiants (Luud, Sitska, Nugis, *Tartu*, **36**, 210, 194); photoelectric Hβ profiles and comparison with models (Brucato, *P.A.S.P.*, **80**, 263); Hγ equivalent widths in eight associations (Walker, Hodge, *P.A.S.P.*, **80**, 290); continuum scans of 10 stars (Karetnikov, Medvedev, *Astrophys.*, **4**, 303); Scans and radial velocity changes in B3V stars (Kodaira).

In IC4665 the rotational velocities of B8-A3 stars are comparable with Pleiades stars: Those of B4-B7 stars are much smaller (Abt, Chafee, *Ap. J.*, **148**, 459). In Sco-Cen the B0-B6 stars rotate slower than general field stars; the B7-A0 stars faster. In PerI the normal giants rotate faster than those in the general field (Slettebak, *Ap. J.*, **151**, 1043; **154**, 933). Contrary to earlier work, Hyland finds that there is no dependence of atmospheric helium abundance on rotation (*Nature*, **214**, 899). Friedjung (*Ap. J.*, **151**, 779) calculates the effect of gravity darkening on measured rotational velocities. Departures from LTE are not necessary to explain the helium singlet-triplet anomaly (Norris, *Ap. J.*). Mass loss with outward acceleration is found in some OB supergiants (Hutchings, *M.N.R.A.S.*, **141**, 219, 329, etc.).

Much interest attaches to Bidelman's helium variable star HD125823 (Jascheks, Morgan, Slettebak, Norris, *Ap. J.*, **153**, L87; *P.A.S.P.*, **79**, 70; *Nature*, **219**, 1137, 1342). The spectrum varies regularly ($8^{d}.8$ period). At some phases the star is like the weak He stars of Population I and II. At others it is close to a normal B2V. Similar stars are under study (Jaschek, Arnal, *P.A.S.P.*), HD 191980 (Bp) may be a related object (Keenan *et al.*, *Astrophys. Letters*, **3**, 55). Whilst searching for Ap stars in young clusters (NGC2264, 2362, IC5146) Bernacca, Ciatta found some helium-variable cluster members. Two helium weak (Bp) stars in Orion, HD36916, HD37058 have affinities to the Ap stars (Hack, *Astrophys. Space Sci.*; Sievers, *P.A.S.P.*, **81**, 33, 288). Conti (in press) was unable to confirm the magnetic fields reported in the latter star and the similar HD36629 (Sargent, Strittmatter, *Ap. J.*, **147**, 1185). HD19216 (B9V) may be related to the Ap's (Preston, *Ap. J.*, **156**, 1175). The microturbulence in 3Cen A was revised downwards (< 2 km s^{-1}) and abundances redetermined (Hardorp *et al.*, *Z. Astrophys.*, **69**, 429). The neon abundance is normal (Sargent *et al.*, *Ap. J.*, **157**, 757). Preston is surveying a large number of B0-B5 stars to determine the frequence of 3Cen A-like stars. Jaschek, Aquilar (*P.A.S.P.*, **81**, 170) find PII etc. in HR3817 and compare it to 3Cen A. Thackeray (*Observatory*) notes that the peculiar lines are weak and seem mainly associated with the fainter component of the 2″ double. HD96248 (B1 Iab) and HD14443 (B2 Ib) are probably nitrogen deficient (Jaschek, *Ap. J.*, **150**, 355).

HIGH LATITUDE B STARS, HORIZONTAL BRANCH STARS AND RELATED OBJECTS

Amongst 30 Feige high latitude stars are: 7 normal A dwarfs; 7 normal B dwarfs, probably runaway stars and one normal B giant; 4 sdO's (Balmer lines strong for colours); 4 sdB's (Balmer lines broad for colours, some have HeI weak); 6 Bw's (Balmer lines normal, HeI weak); one Aw (metals weak for colours) (Sargent, Searle, *Ap. J.*, **153**, 443). Walker, Hodge (*P.A.S.P.*, **80**, 238) give Hγ equivalent widths for Feige stars. Blue HB stars in globular clusters have weak HeI (Sargent, *Ap. J.*, **148**, L147). Barnard 29 (3^{m} above H-B in M13) is not helium weak (Stoeckly, Greenstein, *Ap. J.*, **154**, 909). Other elements are underabundant and the mass only 0·3 $M_{\odot}$; it may be a binary. Work on H-B stars in NGC6397 was revised (Newell *et al.*, *Ap. J.*, **156**, 597); the masses are $\sim$0·6 $M_{\odot}$; the helium abundance is unknown. Atmospheric parameters from spectra and scans of blue H-B stars in M67 suggest the masses are about solar (Sargent, *Ap. J.*, **152**, 885). For these stars $\langle v \sin i \rangle \sim 70$ km s^{-1} and one is a single-lined binary (Deutsch). The field H-B B2 star Feige 86 is like 3Cen A (P II etc.) (Sargent, Searle, *Ap. J.*, **150**, L33). The He weakness in these stars seems unrelated to abundances in the prestellar material. The anticorrelation of P II and He I may be due to non-LTE effects (Underhill, *B.A.N.*, **19**, 537).

Baschek and Norris interpret the He singlet-triplet anomaly in the hot subdwarf HD205805 as due to high gravity and low He abundance. The high velocity B star HD214539 has a normal He/H ratio; other elements are deficient 10 to 30 times (Przybylski, *M.N.R.A.S.*). Abundances in the metal deficient H-B A stars HD 161817, HD86986, HD109995 are similar to those in HD122563 (metal poor giant); C and O have high abundances (Kodaira *et al.*, *Ap. J.*, **155**, 525; *P.A.S.J.*, **19**, 550). HD6870, a rapidly rotating metal deficient star, seems to be a member of the σPup group (Rodgers, *Ap. J.*, **152**, 109); it may be a close binary. For $+39°$ 4926 Kodaira *et al.* find $T = 7500°$ $\log g = 1$ $M_v \sim -3$ and 775 day period velocity variations; metals are 0.01 solar but C, N, O are normal. HD125924 (B2) may be one of the brightest halo stars (Hill *Obs.*, **88**, 163). HZ29 seems to be a hot star with a helium spectrum (Wampler, *Ap. J.*, **149**, L101). OB subdwarfs are easily found in the Paris system (Berger, Fringant). Davis Phillips observed 40 H-B stars spectroscopically.

A STARS INCLUDING PECULIAR AND MAGNETIC STARS

θLeo (A2 V) has normal abundances (Kriz, *B.A.C.*, **17**, 175), Taffara (*Asiago Contr.*, 184) obtained new atmospheric parameters and abundances for αCyg (A2 Ia). Andrillat, Baylac (*C. r.*, **268**, 1537) studied the variation of hydrogen P_{12} with type and luminosity. Houziaux (*3rd Harvard*) finds oxygen $\sim\frac{1}{4}$ solar in Vega and Sirius. Stromgren's m_1 depends on abundance rather than microturbulence and Ca abundance varies by a factor 2 for A stars near the sun (Henry, *Ap. J.*, **152**, L87). There is a good agreement in effective temperatures determined from the Balmer jump, the continuum and intensity interferometer measures (Wolff *et al.*, *Ap. J.*, **132**, 871 from scanner work). Rodgers (*Observatory*, **87**, 127) discussed spectrophotometry of γCen (A0). Scholz, Hardorp are studying Vega. Strom *et al.* (*Observatory*, **88**, 160) find that Sirius shows Am type abundance anomalies contrary to Warner's view. Conti, Strom (*Ap. J.*, **152**, 483; **154**, 975; **158**) continued work on abundance anomalies in early A stars and their relation to Am stars. γGem, θVir, HD2421 are like Vega except for Ba, oPeg is like Sirius. The secondary (A4) of the double-lined binary ηVir has Am like deficiencies of Ca, Sc, Ti; the primary does not. Both have enhanced heavy metals. There are four early Am-like stars in the Pleiades; one a spectroscopic binary (*A.J.*, **73**, 348). Olson discusses the detection of metallism in early A stars (*P.A.S.P.*, **81**, 97). The λBoo group as originally defined is rather heterogeneous. A small group (including λBoo) seem to be young main sequence A stars with low turbulence and slight metal deficiency (Oke, Baschek, Searle, Kodaira, *Ap. J.*, **150**, 513; **155**, 537; *P.A.S.J.*, **19**, 556).

The conference on Magnetic and Related Stars (Cameron, Mono Book Corp.) has been published and much other work has appeared on these stars. A detailed study of Ap Stars (atmospheric parameters, abundances) is consistent with them being slow rotators, not pole-on stars (Wolff, *Ap. J.*, *Suppl.*, **15**, 21). Spectral scans show most Ap's have atmospheres like B stars (Jugaku, Sargent, *Ap. J.*, **151**, 259). The λ4200 and (possibly) the Mn Ap Stars are Ne deficient (Sargent *et al.*, *Ap. J.*, **157**, 757). Detailed analysis shows no signs of non-LTE effects in HD204411 (cool Ap) (*Ap. J.*, **157**, 1265). The micro-turbulence in γEqu (Ap) is less than previously supposed (Preston, Cathey, *Ap. J.*, **152**, 1113); correlation of microturbulence with spectral peculiarity is therefore unlikely. BD $+5°$2468 may be a halo population Ap star (Greenstein, Sargent). HD1909 is like 53Tau (Mn); it has Hg II (Buscombe *et al.*, *M.N.R.A.S.*, **140**, 369). Several Ap Stars have Pt II (Dworetsky, *Ap. J.*, **156**, L101). Model atmosphere analyses of Mn Ap stars show atmospheres like 'normal' A's and with abundance differences from star to star (Strom, *Astr. Astrophys.*, **2**, 182). Si II 4128 is not always enhanced in λ4200 stars (Megessier, *Paris*) ιCrB (Ap) was studied by both Taffara and Hack (*Asiago Contr.*, 206, *Trieste*, 367). Other stars studied at Trieste are: HD151199 (Gokkaya, *Astrophys. Space Sci.*); γCap, 53Cam, 78Vir, θAur, ν_1 and ν_2Dra, βCrB (Farragiana, Hack) and stars found photometrically peculiar by Golay. Other work on Ap stars is; spectroscopy

(Bertaud, Floquet, *J. Observateurs*, **50**, 425, *Astr. Astrophys.*); $U-V$ spectral classification (Nariai, *P.A.S.J.*, **19**, 180); model atmosphere study of 111 Her (Mn) (Luud, Kuush); α And, π Boo A (Mn stars) (Khokhlova, Aliev, Rudenko); 38 Dra (Mn) (Sargent, Adelman; ξ UMa, κ Psc (Rudenko, Aliev); β CrB, 21 Per (Polosuchlina); line identifications in HR 710 (Ir I present), HD 216533, 41 Tau (Jaschek, Brandi, Garcia); 73 Dra (Pt, Au, U, Os I and II present with CN and CH) (Jaschek, Malaroda); 73 Dra (Os/Fe ≥ 300 ☉) (Guthrie); spectroscopy of several Ap's (Aller *et al.*); rotational velocities of all Ap stars brighter than 9^m (Preston). Bertaud is preparing a supplement to his Am-Ap catalogue. The peculiar star HD 101065 may be "the coolest Ap" (Przybylski, *Postepyi. Astr.*, **15**, 225).

Work on α^2 CnV includes; Cohen *et al.* (*Ap. J.*, **156**, 629) identifications in the visual (Pb II, P II, Gd III, Pr III, Cl II etc.), a theoretical discussion is in progress; variations in spectrum, radial velocity and magnetic field (Pyper, *Ap. J. Suppl.*, **18**, 347; Kodaira, Unno, *Ap. J.*, **157**, 769; Octken *et al.*, *A.N.*, **292**, 1) these support the oblique-rotator hypothesis; abundances (Chumak, *Alma-Ata*); spectral variations (Kumaigorodskya, Kozlova). Studies of similar stars are; HD 124224 spectrum and velocity changes and model (Khokhlova); 73 Dra (Bonsack, Markowitz, *P.A.S.P.*, **79**, 235, Berg, *McCormick Publ.*, **15**, 9), period constant since 1933, rapid variations in H lines; 56 Ari (Bonsack, Wallace), extensive material shows Si II, He I period constant over 8400 cycles, H, Ca II, Mg II vary but in a different period; detailed spectroscopy of Osawa's Ap star HD 221568 (Kodaira, *Ann. Tokyo Obs.*, **10**, 157; *Ap. J.*, **157**, L59); CU Vir, spectral and radial velocity changes (Aydin, *A.I.*, **39**, 721); variations of HR 234 (Guderley, *P.A.S.P.*, **79**, 589); HD 221006, a λ4200 spectrum variable with a shell (Hyland, *P.A.S.P.*, **80**, 559); γ Boo (A7 III) (Baglin *et al.*, *Budapest*, 300), possibly a spectrum variable.

Praderie (*Ann. Astr.*, **30**, 793; **31**, 15, *3rd Harvard*) studied several Am stars in detail and computed $U-V$ spectra for Am and Ap stars. Nishimura, Watanabe followed light and velocity changes in 28 And (Am) (*Ann. Tokyo Obs.*, **11**, 142; *Astrophys. Space Sci.*, **3**, 77). Osawa's group in Tokyo is also studying RR Lyn, δ Cap, δ Del, HR 4816. Provost, Veer-Menneret (*Astr. Astrophys.*, **2**, 218) analyse 22 Ser (Am) and also 9 Aur (F0V) which has an unusually high microturbulence. The Jascheks (*A.J.*, **72**, 806) studied the relation between blanketing and degree of metallicity for Am's. HD 111844-5 is the only reported case of an Am in a visual binary system with an evolved secondary. Cowley *et al.* (*P.A.S.P.*, **79**, 497) conclude that the primary is not a true Am star though related to this class. Classification of some proposed Am's has been discussed (Walker, Cowley *et al.*, *Observatory*, **87**, 80). In normal main sequence A-F and Am stars, microturbulence up to the sound speed is observed (Baschek, Reimers, *Astr. Astrophys.*, **2**, 240). A number of writers have discussed theories of the origin of the Am and Ap stars (e.g., Van den Heuvel, *B.A.N.*, **19**, 309, etc.; Renson, *Ann. Astr.*, **30**, 697; Guthrie, *Astrophys. Space Sci.*, **3**, 542.)

Intensive and continuing work on stellar magnetic fields has been carried out at Lick and later Palomar by Preston, partly in collaboration with Stepiens and others (*Ap. J.*, **147**, 804; **150**, 547, 871; **151**, 577, 583; **154**, 971; **156**, 653, 967; **157**, 247). He has been particularly concerned with the periodicity of magnetic variations and their correlation with light variations. Periodic changes in light, spectrum and magnetic field are established for 17 Com and of light and field for κ Cnc. In HD 10783 light and magnetic maxima coincide, a small velocity variation lags ~¼ cycle; the star is probably a long period spectroscopic binary. In HD 32633, for which the period and range in light and field have remained constant over 8 years, light maximum coincides with magnetic minimum. In 53 Cam the relative phase of light and field is a function of wavelength. Also in 53 Cam partially resolved Zeeman patterns were detected and it may be a spectroscopic binary (Faraggiana IB 388). 73 Dra is an unusual spectrum variable, a spectroscopic binary and a periodic magnetic variable. The field in HD 215441 is periodic and not dipolar. If most magnetic stars *are* oblique dipolar rotators the magnetic inclination must generally be large. Other stars studied by Preston are 78 Vir, 21 Per, β CrB, HD 126515. In HD 188041 (Wolff, *Ap. J.*, **157**, 253) elements showing the largest spectral changes show the smallest field changes, non-uniform abundances over the star are deduced. In HD 125248 (Hockey, *M.N.R.A.S.*, **142**, 543) the field is periodic (9 days) and the star is a spectroscopic binary (4·4 years). The cross-over effect was investigated. Conti (*Ap. J.*, **156**, 661) finds no evidence of magnetic fields in 16Ori or 15Vul (Am's). This throws doubt on previous reports of magnetic fields in Am's. The possibility that some phenomena of magnetic stars are not periodic is discussed by Jarzebowski (*Budapest*, 227). Zirin (*Ap. J.*, **152**, L177) finds evidence of He^3 (from λ10830) in several magnetic stars including γ^1 Vir, a star with high chromospheric activity and lithium content. The primary of the double-lined binary HD98088 is an Ap magnetic variable. The periods of the binary, of the field and of stellar rotation are the same. The primary mass is 2·2 $M_\odot$ suggesting that the Ap star is on the main sequence for the first time (Abt *et al.*, *Ap. J.*, **153**, 177).

F, G AND K TYPE STARS

General

Pagel (*Liège*, 1968, 189) reviews forbidden absorption lines in late type stars. Scanner observations include: Whiteoak (*Ap. J.*, **150**, 521), G5-K7 dwarfs and subdwarfs in the red; Hagen, v.d. Bergh (*Toronto Publ.*, **2**, 479), F-K stars; Komarov, Pozigoon, *Astr. Zu.*, **45**, 133) αAri, βAri, αCet. Other continuum investigations are: Ardeberg (*Astr. Astrophys.*, **3**, 257), O5-G0, energy distribution as function of type and luminosity; Glebocki (*Acta Astr.* **17**, 381) A-K stars for population differences; Rodriguez (*Kiev*) blanketing effects on colours, F-M stars; Smolinski (*Acta Astr.*, **19**, 103) microturbulence from low dispersion spectra.

F and G types

Cayrel is working on F dwarfs (*Ann. Astr.*, **31**, 43 also 15 Peg) to calibrate Strömgren's m_1, against Fe/H. 16Psc is a double lined binary (*Astr. Letters*, **1**, 173); this, rather than metal deficiency, accounts for the star's m_1. Spite continues work on G stars, with increasing Fe/H, Ca/Fe and Ti/Fe decrease, Mn/Fe increases (*Ann. Astr.*, **30**, 211, 685; **31**, 269; *Astr. Astrophys.*, **1**, 52). The absolute magnitude dispersion of G giants is being studied spectroscopically. μCet (F0 IV) and 36Per (F4 III) are being studied (Wellmann). The Griffins are making an extensive study of Procyon. Seven F stars studied by Islik (*Trieste Publ.*, 366) and the single-lined binary HR5317 (F6 IV) (Kondo, McCluskey, *Ap. J.*, **156**, 1007) are probably metallic line stars. Profiles of strong lines in two Hyades F stars give metal abundances 1·5 times solar (as do curve of growth analyses) (Krishna Swamy, *Ap. J.*, **146**, 731). Cayrel also measured line profiles (3rd Harvard) and used spectroscopy and photometry for temperatures and absolute magnitudes of G-K stars. Kipper (*Tartu*, **36**, 241) studied 41Cyg, νHer (F stars).

Spectroscopic work (Kraft, Kuhi, *A.J.*, **73**, 221) shows that differences in Strömgren's m_1 are caused primarily by abundance effects though one case of turbulent induced m_1 variations was found (see also Barry, *Ap. J.*, **148**, L87, McNamara, *Ap. J.*, **149**, L133). Kraft (*Ap. J.*, **150**, 551) measured the rotation of many F-G stars; chromospheric activity and rotation decay on the same time scale (rotation halved in $\sim 4 \times 10^8$ years). Powell who has a curve of growth programme on F stars shows that Hβ photometry can give temperatures for F dwarfs (*Astr. Letters*, **2**, 11). Alexander (*M.N.R.A.S.*, **137**, 41) redetermined Fe/H for 18 G dwarfs. Samson (*Edin. Publ.*, **6**, 225) used objective prism plates for F-G abundances and luminosities (v.d. Bergh system).

Parsons made a detailed study of αPer (F5 Ib) and βAqr (G0 Ib), both stars have large N/C ratios. Further work will use improved model atmospheres (*Ap. J.*, **150**, 263; *Suppl.*, **18**, 127). Line blanketing in supergiants was determined by Rodgers, Bell (*M.N.R.A.S.*, **142**, 161). Hack is obtaining high dispersion spectra of supergiants to study stratification effects.

The lithium-beryllium problem has been reviewed in detail by Wallerstein and Conti (*Ann. Rev.*, **7**, 99). Only recent work, not dealt with by them, is mentioned here. Danziger (*Astr. Letters*, **3**, 115) studied abundances in the Coma cluster and finds that solar mass, main sequence, stars deplete lithium by a factor e in 7×10^8 years. He also gives data for some field G-M stars. Feast (submitted to *M.N.R.A.S.*) finds Li^7 the predominant isotope in 12 F-G stars. ζDor (F8, slightly above the main sequence), a star with a very active chromosphere, probably has a significant amount of Li^6. Both ωSgr (a highly evolved subgiant) and γPav (a metal deficient subdwarf) have high lithium abundances.

Subdwarfs, subgiants, metal poor stars

The mild subdwarf μCas belongs to the α-rich group (also $Y \sim 0{\cdot}26$) (Catchpole *et al.*, *M.N.R.A.S.*, **36**, 403; *R.O.B.*, 154; Cohen, *Ap. J.*, **154**, 179). The cool subdwarf HD25329 has metals ~1/20 ☉ with Mn lower (Pagel, Powell, *R.O.B.*, 124, curve of growth, Krishna Swamy, *Ap. J.*, **154**, 983, line profiles). In the subdwarfs HD140283, HD19445 carbon shares the large metal deficiency (Cohen, Strom, *Ap. J.*, **151**, 623, model atmosphere work). In σBoo Fe/H ~ ¼ ☉ (Kipper, *Astrophys.*, **4**, 303). The oldest stars seem to contain substantial amounts of helium (Strom, *Ap. J.*, **150**, 501). Microturbulence is *not* a function of metal abundance (*P.A.S.P.*, **80**, 269). C^{13} has not been found in metal deficient stars (Cohen, Grasdalen, *Ap. J.*, **151**, L41) which places restrictions on theories of element production. Jones (*R.O.B.*, 126) includes metal content in a spectral classification scheme. Further work on the very metal deficient giant HD122563 shows even-odd elements deficient with respect to even-even elements (Wolff, Wallerstein, *Ap. J.*, **150**, 257 from ultraviolet work). The similar star HD6268 is being studied by Lloyd Evans. Koelbloed (*Ap. J.*, **149**, 299) analysed the metal deficient red giants HD6755, HD2665. An over-deficiency of C in HD6755 was questioned by Cohen (*Astrophys. Letters*, **2**, 163) who used a more refined analysis. Harmer, Pagel (MS) find N is more deficient than metals in the metal deficient star νInd using CN3883. They suggest that various earlier analyses for N using 4215 CN are in error because of blending problems. The C abundance in HD122563 (cf. Pagel, Ball, *Observatory*, **87**, 19) is now being re-examined (Pagel).

K types

Griffins fine atlas of Arcturus has been published and several studies made of it (*M.N.R.A.S.*, **137**, 253; *Observatory*, **87**, 253; **89**, 9, 62); atmospheric parameters and metal abundances (~ ⅓ ☉) were determined: [O I], [Fe II], [Ca II], Rb I absorptions were found and hyperfine line structure discussed. Wavelengths, identifications and equivalent widths for 1036 stellar lines 11407–25129 Å are given by Montgomery *et al.* for Arcturus (*Ap. J. Suppl.*, **19**, 1, high resolution Fourier transform spectra by Connes). Helfer and Wallerstein (*Ap. J. Suppl.*, **16**, 1) completed a curve of growth analysis of 27 early K giants, some have over-abundances of V, Mn, Co, some have subsolar masses. They find Na anomalies as does Cayrel (*Ann. Astr.*, **29**, 413 details of 7K stars). Griffin analyses two Hyades giants and βHer and αSer (CN strong K2 III, metals ~ 2 ☉) (*M.N.R.A.S.*, **143**, 223, 381). Work on 61 CygB is in progress (Kandal, *Trieste*, 1966, 148; Catchpole). 70Oph (K0 V), εPeg (K2 Ib), 61Cyg A (K5 V), ξ Cyg (K5 Ib) have about solar abundances (Boyarchuk, Voihanskaja).

Spinrad, Taylor (*Ap. J.*, **157**, 1279) use a scanner to measure line and band strengths and also visual and infrared continua (for temperatures). They find evidence for supermetallicity amongst substantial numbers of evolved K stars including stars in M67 and NGC188. Lack of agreement with curve of growth analyses of the cooler K stars is attributed to unsuspected opacities affecting the latter. Further work is clearly required, see for instance Cayrel (*Paris*, 1967, 239) and Rodgers (*M.N.R.A.S.*, **145**, 151) who suggests that the old disk star δPav (G8 V) would qualify as a Spinrad-Taylor supermetal star but has in fact solar abundances and high microturbulence. Pagel is studying spectra of M67 stars for signs of super-metallicity.

Schadee (*Ap. J.*, **151**, 239) computes band intensities in G-K stars for different C, N, O abundances. Enhancement of CN, C_2, CH in two metal strong K giants suggests C/O ~ 1 (Greenstein, Oinas, *Ap. J.*, **153**, L91). C, N, O abundances were determined in 4K giants (Greene, *Ap. J.*, **157**, 737); N is enhanced for $M \geqslant 2M_{\odot}$. There are discrepancies between the observed and predicted red CN band strengths. There has been further work on the auto-ionization lines of Ca I (*B.A.N.*, **18**, 259; *Ap. J.*, **146**, 719; *P.A.S.P.*, **79**, 33). Rousseau discusses line blanketing (*Trieste*, 1966, 174). The Cambridge narrow band photometry has continued (*M.N.R.A.S.*, **131**, 467; **133**, 99, 449; **134**, 135,

171; **140**, 21; *Observatory*, **87**, 18). Lines measured include Fe I, Hα, Na. Analysis (Peat, Pemberton, Price, Scarfe) indicates Ca, Na but not Mg are deficient in high velocity stars. Mg is more abundant in the Hyades than in Coma. Weistrop (*P.A.S.P.*, **79**, 546) studied Hα as a luminosity criterion and found variations in Hα in 22 Vul (G0 Ib) over a 3-year period.

M, N AND S TYPE STARS

There are no systematic differences between high dispersion spectra of low and high (intermediate Population II) velocity K2-M6 giants though some high velocity M stars have strong lines slightly weakened (Deutsch *et al.*, *Ap. J.*, **156**, 107). Spectral classification is discussed by Yamashita, Treanor, McCarthy (*Trieste*, 1966, 94, 109). Boesgaard reports: from MgH (5211 Å) in K-M stars, Mg^{25}, Mg^{26} are present in at least their terrestrial proportions (*Ap. J.*, **154**, 185): Little trace of S characteristics was found in a group of 26 M giants, if M stars evolve into S stars they do so very rapidly (*P.A.S.P.*, **81**, 365); Ti/Zr is 5–10 times greater in K, M stars than in MS, S or Ba II stars: She is working with Deutsch on the circumstellar envelopes of M giants (from profiles of Ca II, Na I, Hα). Abundance work on M stars includes: Orlov, Rodriguez, Shavrina (*Kiev*), Kipper (*Tartu*); Yokoo (*Tokyo*) who is analysing 3 high velocity giants. Boyarchuk (*Astrophys.*, **4**, 289, *Izv. Kzym.*, **39**, 114) studied TiO absorption in M stars. A great deal of infrared work on cool stars has been carried out recently. This is adequately covered by the review of Spinrad and Wing (*Ann. Rev.*, **7**, 249).

In the S star HR 1105 the ZrO bands show that the Zr isotope ratios are different from the terrestrial values and neither O^{17} nor O^{18} are found: there is little evidence for Tc (Schadee, Davis, *Ap. J.*, **152**, 169, L13). Tsuji studied atmospheric structure, composition and mass ejection effects in R And and R Cyg (S types). Locanthi (Davis) is preparing a wavelength list for S stars and V Cnc (intermediate S-Ba II). She and Keenan find NbO in R Cyg.

A major attack on the problems of C stars has continued in Japan (*Proc. Japan Acad.*, **43**, 472. 966; **44**, 495; **45**, 272, 484; *P.A.S.J.*, **19**, 342; **20**, 731; **21**, 41; *Trieste*, 1966, 49, 75; *B.A.C.*, **19**, 274). Results and work in progress include: analysis of high dispersion R star spectra using non-grey model atmospheres (Tsuji); identifications and curve of growth analyses of 22 C stars in the visible (Utsumi); isotope work on Merrill-Sanford bands and analysis of early C stars (including CH stars) (Utsumi, Yamashita); analysis of WZ Cas (Hirai); spectral scanning of CN red bands for spectral classification (Fujita); C^{12}/C^{13} ratios (Fujita *et al*). Fujita's monograph *Interpretation of Spectra and Atmospheric Structure in Cool Stars* (Univ. of Tokyo Press) will be published soon. Other C star work is: C^{12}/C^{13} ratios Climenhaga (*Trieste*, 1966, 54); rotational temperatures (CN red bands) Querci *Ann. Ast.*, **30**, 557); detailed computations of CN rotational profiles and band analysis (Fay); low dispersion spectra of C stars (Catchpole, *P.A.S.P.*, **80**, 744); population differences (Krempec (Torun)); calculations of molecular dissociative equilibrium (Morris, Wyller, *Ap. J.*, **150**, 877). Boesgaard has determined lithium abundances in 20 MS, S and Ba II stars. Some S stars have lithium abundances 2–3 orders of magnitude more than the Sun. Catchpole and Feast have so far found one very lithium rich S star in their spectroscopic survey of southern S and C stars. A very lithium rich S-C star was found earlier (Feast, *Highlights in Astronomy*, 232). Boesgaard finds Li^6 is 12 % ($\pm$6 % p.e.) of the total Li in δ Sgr (M2 Ib-II) (*Ap. J.*). This is less than the normally expected spallation ratio. Lloyd Evans has made a survey of a large number of southern M stars for lithium.

HYDROGEN POOR AND OTHER PECULIAR STARS INCLUDING SYMBIOTIC STARS

Work on helium stars includes: revised abundances in HD 168476, HD 124448, oxygen abundance $\sim \frac{1}{4} \odot$ from O I 7774 (Hill, *Mon. Notes astr. Soc. Sth. Afr.*, **28**, 56, *Observatory*); model atmosphere analysis of BD $+10°$2179, both it and HD 124448 have a high carbon abundance (Hunger, Klinglesmith, *Ap. J.*, **157**, 721); HD 264111 which may be a subdwarf has He/H $\sim$1·0 (by number), it resembles HD 96446 (Stephenson, *Ap. J.*, **149**, 35). For υ Sge: Nariai discusses the mechanism of mass loss (*P.A.S.J.*, **19**, 564); Lee, Nariai find a large infrared excess (probably from a shell) (*Ap. J.*, **149**, L93); Sahade, Albano detect dilution effects. For the hydrogen poor star HD 30353 Nariai (*P.A.S.J.*, **19**, 63) finds a general metal deficiency, Wallerstein *et al.* (*Ap. J.*, **150**, 239, 245) do not, N_C/N_N is 10^{-3}. Warner (*M.N.R.A.S.*, **137**, 119) determined abundances for 5 cool hydrogen deficient, carbon rich stars.

For Ba II stars: Griffin (*M.N.R.A.S.*, **143**, 381) finds Ti, V deficient and *s*-process elements enhanced in HD 101013; Nishimura (*Trieste*, 1966, 124) and Cowley (*Ap. J.*, **153**, 169) obtained abundances in HR 774 which may be slightly Fe/H deficient. Ba II stars may be a high temperature extension of the N stars (Gordon, *Ap. J.*, **153**, 915). Hiltner *et al.* (*Astrophys. Letters*, **2**, 153) describe three peculiar emission objects. The Swings (*Liège*, 1968, 99) discuss possibilities of observing forbidden iron lines ([Fe I] to [Fe VII]).

Further support has been found for the binary theory of symbiotic objects. Boyarchuk (*Sov. A.J.*, **44**, 1016; *Izv. Krym.*, **39**, 124) explains Z And as an M2 III + a hot subdwarf and AX Per, CI Cyg and BF Cyg as M5 III stars with hot subluminous companions. Faraggiana, Hack (Liège, 1968, 317) suggest that CH Cyg is M6 III + a blue unstable subdwarf. WY Vel (M3 Ib) now shows evidence of a B3 star (Cowley). Other work is: spectrophotometry of AG Dra and AG Peg (Chkhikvadze); emission line intensities and abundances (~ solar) for AX Per, CI Cyg, BF Cyg, V1016 Cyg (Boyarchuk); scanner observations (3300–11000 Å) of CH Cyg, AG Peg, Z And, AX Per, BX Mon, T CrB, V1016 Cyg (Lambert); studies of SY Mus, Henize 1213, and CPD −36° 8436 (Carlson, Northwestern Thesis); ultraviolet emission in Boss 1985 (Swings *et al.*, *Ap. J.*, **155**, 515); the existence of two distinct envelopes in AG Peg (Lopez, Sahade, *Bol. Arg. Astr.*, **14**, 68). In V1016 Cyg (MHα 328-116) which may be the early stage of a planetary nebula (FitzGerald, Houk, *Ap. J.*) the continuum may be collisional excited in a nebulosity (Gerola, Caputo, Panagia).

BINARY STARS

VV Cep has multiple Hα absorptions, probably arising in a gaseous stream (Wright, Larson, *Trieste*, 1968, 106). Faraggiana, Hack who plan to study this star spectroscopically over its 20 year cycle, suggest it is a contracting star. Glebocki, Keenan (*Ap. J.*, **150**, 529) find evidence for an oxygen cloud around the companion. Cowley (*P.A.S.P.*, **81**, 297) reviews problems of the VV Cep stars. CD −61° 3575 may belong to this class (Calson). The peculiar binary GG Car was described by Calson and is being studied by Sahade. Wellman finds ε Aur fits Huang's model but Hα emission variations are unexplained. Wright *et al.* (*Publ. Dom. Astrophys. Obs.*, **13**, No. 11, *J.R. astr. Soc. Can.*) find the "clouds" producing the multiple K-lines have lifetimes of days or weeks. Work is also in progress on this star by Hack, Kitamura. 1961/62 eclipse spectra of 31 Cyg show the superexcitation is probably chromospheric (Wellman). Kitamura studied the chromospheric K line at the 1963/64 eclipse of ζ Aur. Spectrophotometry of β Lyr supports Sahade's work on emission line structure (Svolopoulos, *A.N.*, **290**, 155). Structure and variations of He I 10830 Å in this star are reported (Knappenberger, Fredrick, *P.A.S.P.*, **80**, 96; Alduseva, Esipov, *Astr. Zu.*, **46**, 113).

Other spectroscopic work on binaries includes: HZ 22 is a hot subdwarf binary ($P < 1^d$) (Smak); In U Cep, satellite H lines appear at quadratures (Batten, Andrews); Kumsishvili did spectrophotometry of V448 Gem and RY Gem; Kitamura, Sato (*Astrophys. Space Sci.*, **3**, 163; *P.A.S.J.*, **19**, 575) studied R CMa, RZ Scu, Andrews (*Ap. J.*, **147**, 1183) discusses Hα emission in Algol. Mass loss from binaries has been extensively discussed (*Trieste*, 1968). Olson (*Ap. J.*, **153**, 187) compares surface gravities for eclipsing systems from orbital parameters and from line profiles interpreted with broadening theory. 57 Peg is a spectroscopic binary (M4 IIIa + A3 V) (Hackos, Perry, *A.J.*, **73**, 504). HD 118216 is a semi-detached system viewed at low inclination (Conti, *Ap. J.*, **149**, 629). 17 Lep is an M + B binary (Cowley, *Ap. J.*, **147**, 609). Supression of Hε in AX Mon may be due to flare activity in the K star (Cowley *et al.*, *P.A.S.P.*, **79**, 21).

Meisel (*A.J.*, **73**, 350) determined types and rotational velocities in 50 visual binary systems. Murphy (*A.J.*) studied binaries with B type primaries. Several secondaries are probably pre-main sequence stars. In composite systems (typically G giant + A Dwarf) the rotational velocities of the early type components are on the average less than in the general field (Markowitz, (Perkins)).

WHITE DWARFS, NUCLEI OF PLANETARY NEBULAE

Greenstein continues his work on white dwarfs. DB star frequency increases with decreasing space velocities. λ4670 stars seem confined to the high velocity group or binary systems. Work with Trimble on the Einstein effect has been published (*Ap. J.*, **149**, 283). Spectroscopy has been extended to the infrared. Other work is: spectral scans (Oke); Sirius B at 10Å mm^{-1} (Kodaira, *P.A.S.J.*, **19**, 172); a new white dwarf (Hiltner *et al.*, *Astr. Letters*, **2**, 153). Emission nuclei of planetary nebulae are classified by Smith and Aller (*Ap. J.*, **157**, 1245).

STELLAR CHROMOSPHERES

Recent reviews are Praderie (IAU Colloquium No. 42; Feast, *Lunteren*). Work on Ca II emission in late type stars includes; variations in giants (sometimes flare-like). Liller (*Ap. J.*, **151**, 589 Deutsch, *P.A.S.P.*, **79**, 431; *Lunteren*); use of a spectrum scanner, chromospheric activity and age, variations in main sequence stars (e.g. 61 Cyg A period ~ 6½ months, 61 Cyg B period ~ 1 year) (Wilson, *Ap. J.*, **153**, 221, etc.); variations in *o'* Ori (MS type); (Boesgaard, *P.A.S.P.*, **81**, 283); use of Ca II emission to define lower main sequence of Hyades and Pleiades (Kraft, Greenstein, *Low Luminosity Stars*, 1969; also Haro, Chavira, *Ton. Bol.*, **31**); widths and intensities in 200 southern stars, emission in γ Vir N (F0 V) (Warner, *M.N.R.A.S.*, **144**, 333; *Observatory*, **88**, 217); Ca II in γ Boo (Praderie); intensification of emission in binaries (Abt *et al.*, *Ap. J.*, **157**, 717); inverse correlation of emission and intrinsic polarization (Dyck, Johnson, *Ap. J.*, **156**, 389). Wilson finds H and K widths are similar in normal and very metal deficient giants. However Pagel, Tomkin (*Q.J.R.A.S.*, **10**, 194) find that K widths underestimate (~ 1^m) the luminosities of metal deficient giants.

Vaughan, Zirin studied chromospheric He I 10830Å absorption (and sometimes emission) in late type stars (*Ap. J.*, **152**, 123). The line originates in clouds or streams, not a homogeneous layer. Work continues and time variations are found. ε Gem (G8 Ia), θ Her (K1 II) have strong chromospheric activity and strong CN absorption which may thus be a general indicator of such activity (Zirin). Chromospheric He I 5876 absorption is present in ζ Dor (F8), other work is in progress (Feast). Deutsch has studied chromospheric Balmer absorption and Hε emission which are strongly time dependent in some red giants (cf. also Deutsch *et al.*, *Ap. J.*, **156**, 107).

M. W. FEAST
President of the Commission

29a. WORKING GROUP ON ABSOLUTE SPECTROPHOTOMETRY

Two new and independent determinations of the absolute spectral energy distribution in α Lyr have been completed during the last three years. The first of these by Hayes has been published in part (Wolff, Kuhi and Hayes, 1968). It was done at the Lick Observatory using the Crossley reflector and two ribbon-filament standard lamps. His published results are based on the old practical temperature scale which is based on a melting point for gold of 1336·16K. The new scale has been adjusted to agree as well as possible with the thermodynamic temperature scale and is based on 1337·59K for the melting point of gold. Hayes' results have been adjusted to this new scale and are shown in Table 1. The numbers listed are $-2{\cdot}5 \log f_\nu +$ const. when f_ν is the flux from α Lyr in ergs s^{-1} cm^{-2} Hz^{-1}. The results are normalized to 0·000 at λ5556.

The new calibration by Oke and Schild was carried out on Palomar Mountain. A four inch reflecting telescope was built and mounted with the prime focus scanner which was built for the 200-inch telescope. Three light sources were used (1) a ribbon-filament standard lamp calibrated with an accuracy of 2% by the National Bureau of Standards was used from λ3300 to λ8000. (2) Radiation from a sight tube immersed in pure copper in the process of melting or freezing was

Table 1

λ	1/λ	Hayes	Oke Schild	Oke 64
3200	3·125	+1·285		
3250	3·077	+1·246		
3300	3·030	+1·134	+1·082	+1·135
3350	2·985	+1·155		
3400	2·941	+1·127	+1·057	+1·110
3450	2·898	+1·113		+1·095
3500	2·857	+1·095	+1·035	+1·075
3571	2·801	+1·065		+1·050
3600	2·778		+1·014	
3636	2·750	+1·050		+1·030
3680	2·717		+0·998	
3704	2·700	+0·995		+1·005
4036	2·478	−0·303	−0·308	−0·225
4167	2·400	−0·273	−0·288	−0·205
4255	2·350	−0·269	−0·275	−0·190
4460	2·242	−0·221	−0·231	−0·165
4566	2·190	−0·198	−0·210	−0·150
4780	2·092	−0·166	−0·166	−0·115
5000	2·000	−0·099	−0·121	−0·085
5263	1·900	−0·039	−0·064	−0·045
5556	1·800	0·000	0·000	0·000
5840	1·712	+0·081	+0·061	+0·040
6050	1·653	+0·126	+0·102	+0·070
6370	1·570		+0·16	+0·115
6439	1·554	+0·175		
6800	1·471	+0·227	+0·24	+0·175
7100	1·408	+0·289	+0·28	+0·215
7550	1·325	+0·278	+0·35	+0·280
8080	1·238		+0·43	+0·35
8400	1·190		+0·46	+0·38
8804	1·136		+0·44	+0·39
9700	1·031		+0·48	+0·40
9950	1·005		+0·50	+0·44
10250	0·976		+0·53	+0·48
10400	0·962		+0·54	+0·50
10800	0·900		+0·61	+0·56

used from γ6300 to γ10800. The temperature of melting copper on the new scale is 1357·9 K and is very accurately known since it is near the melting point of gold. (3) Radiation from a sight tube immersed in melting or freezing platinum was used from λ3500 to λ10800. The melting point of platinum is 2044 K but is not well known. This source was used as a secondary standard. The final results are also shown in Table 1. For comparison the calibration adopted by Oke in 1964 (Oke, 1964) is also shown.

Several comments can be made. (1) From λ4000 to λ7100 the calibrations of Hayes and Oke-Schild are in excellent agreement. They differ substantially from results by Willstrop (1965) and Bahner (1963) which are very similar to the numbers in the column Oke 64. (2) The Balmer jump found by Hayes is approximately 0·06 mag greater than that obtained by Oke and Schild.

Oke and Schild have determined the absolute flux from αLyr at λ5556 from the copper source and from the standard lamp. The two agree and give a flux of $3{\cdot}50 \times 10^{-20}$ ergs s^{-1} cm^{-2} Hz^{-1} or 3.40×10^{-9} ergs s^{-1} cm^{-2} $Å^{-1}$.

This figure is substantially lower than was adopted by Code (1960) or found by Willstrop (1965).

BIBLIOGRAPHY

Bahner, K. 1963, *Ap. J.*, **138**, 1314.

Code, A. D. 1960, *Stars and Stellar Systems*, vol. VI, *Stellar Atmospheres*, Ed. J. L. Greenstein, University of Chicago Press, Chicago, p. 50.

Oke, J. B. 1964, *Ap. J.*, **140**, 689.

Willstrop, R. V. 1965, *Mem. R. astr. Soc.*, **69**, 83.

Wolff, S. C., Kuhi, L. V., Hayes, D. 1968, *Ap. J.*, **152**, 871.

J. B. OKE
Chairman of the Working Group

29b. WORKING GROUP ON LINE INTENSITY STANDARDS

Systematic equivalent width comparisons as initiated by K. O. Wright during the Seventh General Assembly of the IAU in 1948 and continued by G. Cayrel de Strobel since the Twelfth General Assembly in 1964 are meeting less and less response from the astronomical community. Griffin has stressed systematic errors in spectrographic results. For instance he claims in his paper (*M.N.R.A.S.*, **143**, 319) that owing to the light thrown in the wings of the instrumental profiles, the observed equivalent widths of absorption lines in late type stellar spectra are 5 to 10% less than the true values. Furthermore Griffin pointed out the weakness of purely spectrographic comparisons which are all affected by the same type of errors. What is really needed is to know the true equivalent width of a few lines as they can be obtained from an accurate double pass photoelectric scanner. Pagel suggested that this can be done by comparing spectra of integrated sunlight from sky or minor planets with scans obtained by solar spectrometers. Griffin prefers that a bright star be used for this purpose. A complete change in the activity of this working group could very well be decided if the primary object is to have true standards of equivalent widths and if these cannot be obtained from conventional high dispersion spectrography.

G. CAYREL DE STROBEL
Chairman of the Working Group

29c. GROUPE DE TRAVAIL SUR LES ÈTOILES Be

Spectrophotométrie, variations, vitesses radiales

A. Peton (*Meudon*) à partir de spectres à petite dispersion, a évalué une période de 33 ans pour les variations de l'émission de HD4180, une classe B5 IV et des oscillations assez importantes de l'enveloppe absorbante sur 3–4 ans (non publ.).

F. R. Hickok (*Dunlap*) a étudié les structures fines et les vitesses radiales de ϕPer à grande dispersion de telle sorte que les courbes de vitesse sont plus détaillées, mais, la période ne change pas. Probablement binaire. Des difficultés sont soulignées (non publ.).

Des variations spectrales et de vitesse radiale ont été observées par M. Hack et P. Stenner (*Trieste*, dans l'enveloppe de ζTau (publ.) ainsi que par Van der Wel (*Utrecht*) pour la période 1964–66 (non publ.). A. M. Delplace (*Meudon*) a mis en évidence une relation très étroite entre les profils Balmer H et la période de 7 ans. L'enveloppe, en moyenne, est en récession (publ.).

J. P. Swings (*Liège*), N. L. Burnichon, D. Chalonge et L. Divan (*I.A. Paris*) ont déterminé le spectre continu et la classe spectrale de l'étoile HD45677 (B2 IVe) ainsi que l'identification des raies spectrales dans l'ultraviolet (publ.). Grâce à l'étude d'une longue série de spectres, une étude de l'évolution de cette étoile est en cours par J. P. Swings.

A. Peton a étudié de nombreux clichés à grande dispersion de HD45910 et de HD218393. Les points communs sont, essentiellement, le dédoublement des termes élevés de la série de Balmer et la disparition de l'enveloppe métallique en absorption alors que l'émission des raies métalliques est très forte (non publ.).

Une courte Note sur une nouvelle explosion de l'enveloppe de 48Lib a été soulignée par A. B. Underhill et H. G. Geuverink (*Utrecht*) (publ.) alors que Geuverink a fait une comparaison étendue des spectres des années 1957–58 et 1967–68 (non publ.) et que R. Faraggiana (*Trieste*) a étudié les raies les plus intenses dans le jaune-rouge dans les spectres des années 1950 à 1962, pris à Victoria par A. B. Underhill (publ.).

Mart de Groot (*Utrecht*) a mis en évidence, sur des spectres de PCyg, à grande dispersion, pris de 1942 à 1964, plusieurs couches dans l'enveloppe en expansion. Période de vitesse radiale de la couche externe: 114 *j*. Des pulsations peuvent rendre compte de certains phénomènes (publ.). Dans son étude sur la perte de masse dans les supergéantes jeunes, Hutchings (*Victoria*) a étudié particulièrement PCyg (sous presse).

Spectrophotométrie de certaines raies de βLyr par S. N. Svolopoupos (publ.) L. Houziaux et Y. Andrillat (*Mons* et *Montpellier*) ont trouvé une température de $9750° \pm 300$K pour l'enveloppe de HD217050 par comparaison des séries de Paschen et de Balmer de H (publ.).

M. et C. Jaschek et S. Malaroda (*La Plata*) décrivent les spectres de plusieurs Be et signalent que des composantes d'émission à He I semblent plus fréquentes qu'on ne le pensait précédemment (publ.). C. Jaschek suit un certain nombre de Be Sud régulièrement pour surveiller leurs changements spectraux. W. Buscombe (*Evanston*) a photographié un certain nombre de spectres de Be Sud à grande dispersion, ce qui est rare. Les spectres à petite dispersion seront donnés prochainement. Il a obtenu également des spectres de Be Nord ressemblant aux Be Sud étudiées par lui.

D. Briot (*Paris*) a calculé les décréments Balmer pour 56 étoiles de B0 à B9 pour une corrélation entre type spectral et décrément.

K. Kadaira et M. Scholz (*Mt. Wilson* et *Palomar*) ont observé une diminution de l'émission H dans le spectre de HD58343 en 1968–69 par rapport à 1967–68. Les continus Paschen et Balmer sont toujours proportionnels à ν^{-3}. En janvier 1968, Hα/Hβ était égal à 2·3. Ils discutent les diverses théories concernant les enveloppes Be et concluent en pensant que cette étoile est une "shell" (non publ.).

S. Malaroda et Levato (*La Plata*) mesurent actuellement les vitesses de rotation d'un certain nombre de Be Sud alors que W. Buscombe a fait un travail sur ce même sujet et a montré que 2 ou 3 étoiles sont proches du "break-up", ce qui pourrait obtenir des phénomènes intéressants à observer (publ.).

J. B. Hutchings a pu montrer que les étoiles chaudes (en absorption ou émission) ont des profils de raies extrêmement variables avec le temps. Le groupe de Meudon (Delplace, Duval, Herman, Peton) a trouvé des variations de vitesse radiale dans les enveloppes de ζTau et de HD45910, indiquant l'hétérogénéité de l'enveloppe, en particulier, au moment d'une expansion interne (publ.) et a montré également que le profil de la raie Hβ de HD109387 varie rapidement avec une période probable de 1h 10 min, confirmée par la photoélectricité (publ.).

Photométrie photoélectrique

A. Feinstein (*La Plata*) a continué les observations en *UBVRI* pour les Be Sud (publ.). Ces nouvelles mesures vérifient que presque toutes ces étoiles sont variables. C'est également le résultat du travail de C. Jaschek et Lia Ferrer. A. Feinstein veut mesurer Hα, Hβ et Hγ à l'aide de filtres interférentiels étroits et larges pour chaque raie.

Th. Schmidt-Kaler (*Bochum*) a fait la photométrie *UBV* de 200 naines Be à l'aide du télescope photométrique de 61 cm du Chili (*E.S.O.*).

J. H. Bigay, R. Garnier et G. Adam (*Lyon*) à l'aide d'un dispositif comprenant 3 filtres interférentiels étroits: fil. 4740 (100 Å) + fil. Hβ (35 Å) + fil. 5000 (100 Å), ont pu observer des variations rapides de magnitude de Hβ, concommittantes à celles observées par spectrophotométrie à grande dispersion pour HD 109 387. D'autres mesures faites en *UBV* sur HD 109 387, 45 910 et 23 642 sont en cours.

De façon tout à fait différente N. Morguleff et M. P. Veron (*I.A. Paris*) ont utilisé 8 bandes choisies dans le spectre (spectro. Barbier) et ont pu classer 60 Be Nord. Dans l'ensemble, les résultats sont en accord avec la classification Rojas-Herman.

Théorie des Atmosphères

T. Kogure (*Mito*) a déterminé les largeurs de raies d'émission de HD 202 904 qu'il considère comme une Be normale (publ.). L'étude concernant les profils de raies d'étoiles Be "pole-on" a été publié. Il a ensuite étudié la formation des raies Balmer dans les atmosphères étendues en considérant les comportements différents pour $n \leqslant 10$, $10 < n \leqslant 20$ et $n < 20$ dans la série de Balmer. Il explique ainsi les caractères Be et "shell" (publ.). En collaboration avec Duval et S. Itoh (*Kyoto*), il prépare la publication des diagrammes théoriques (Wα – Hα/Hβ calculés sur la base d'une atmosphère statique et qui permettront d'en déduire les propriétés cinématiques des atmosphères étendues.

D. N. Limber et J. M. Malborough (*Yerkes*) ont réalisé une nouvelle approche en ce qui concerne les forces pouvant entrer en considération pour le support des enveloppes d'étoiles Be (publ.) et Malborough a obtenu des profils de Hα en bon accord avec l'observation en utilisant 5 niveaux liés de H et une méthode d'approximation (publ.). Ces travaux vont continuer et, avec A. P. Cowley, ils vont comparer la théorie à l'observation. Un programme est déja écrit pour calculer les profils de raies dans le cas ou le taux de l'éjection est dépendante du temps.

Catalogues ou cartes

D. H. Menzel (*Harvard*) signale que tout le ciel sera bientôt fait en 3 couleurs (B, V, R) Nord et Sud.

L. R. Wackerling (*Evanston*): la liste de 5287 étoiles des premiers types à émission va être bientôt imprimé.

C. Jaschek (*La Plata*): le supplément au catalogue de Merrill-Burwell est prêt à être imprimé.

R. HERMAN
Président du Groupe de Travail

30. RADIAL VELOCITIES (VITESSES RADIALES)

Président: D. S. Evans.
Vice-Président: H. A. Abt.
Comité d'Organisation: A. H. Batten, A. Blaauw, R. Bouige, A. D. Thackeray.

MARSEILLES AND HAUTE PROVENCE OBSERVATORIES

Les mesures de vitesse radiale (VR) ont été continuées par les trois techniques: prisme objectif, spectrographie à fente, interférométrie, avec des instruments en fonction à l'Observatoire de Haute Provence ainsi qu'à celui de l'ESO.

Mesures au prisme objectif

Grand Nuage de Magellan: Le travail de 13 champs est pratiquement achevé et sera publié dès que les mesures de spectres de fente pris au Chili seront complètement dépouillés.

Un catalogue comprenant 1765 étoiles galactiques et 469 étoiles du Grand Nuage de Magellan est achevé. Un supplément comprend environ 200 étoiles particulières de cette région.

Toutes ces étoiles ont été reportées sur des calques directement applicables sur les cartes P. N. Hodge et F. N. Wright (**1**). Ce catalogue sera imprimé dès cette année (1969). Une étude d'ensemble de ce matériel au point de vue cinématique et dynamique est en cours.

Des mesures *UBV* de plusieurs centaines d'étoiles du Nuage complètent ce matériel (J. P. Brunet, P. Mianes, Mlle M. N. Perrin).

Galaxie: Les mesures dans la Galaxie ont été un peu retardées par le travail sur le Grand Nuage. Néanmoins une liste de 309 VR d'étoiles situées dans Carina ont été publiées par Mlle N. Martin (**2**); de 97 étoiles dans la SA8 par Mme M. Barbier (**3**), et aussi une étude de la rotation galactique qui contient 52 VR des étoiles O, A, B (**4**); dans l'Amas NGC752 par Mme E. Rebeirot (**5**); et par Mme Georgelin dans Cassiopée (164 étoiles) (**6**).

Spectrographe à fente

Un spectrographe pour la mesure des VR a été installé sur le télescope de 152 cm de diamètre du Chili. Sa dispersion de 72 Å mm^{-1} permet la mesure des étoiles brillantes des nuages de Magellan. Plusieurs centaines de spectres ont été obtenus par E. Maurice, L. Prevot, Ch. Fehrenbach et ont été mesurés. Des classifications spectrales très précises de ces étoiles ont été faites; résultats prêts à la publication en 1969.

A l'Observatoire de Haute Provence des mesures de VR seront continuées avec les télescopes de 193 cm et de 120 cm. Des cartes tests d'étoiles standard ont été publiées par Ch. Fehrenbach et Imbert (**7**), (**8**), mais de nombreuses autres mesures n'ont pas encore été publiées. Elles seront activement poussées dès la mise en service du nouveau télescope de 152 cm et de son spectrographe coudé.

Des mesures des raies de Ca II, interstellaires sont en cours à Haute Provence et au Chili (Chu-Kit).

Interférométrie

Le groupe de M. G. Courtès s'est beaucoup développé et de très nombreuses mesures tant galactiques qu'extragalactiques ont été publiées. Un catalogue de VR des régions d'emission a été publié (**9**).

Signalons particulièrement les mesures de P. Cruvellier sur les régions H II de la Galaxie, de G. Courtès, G. Monnet *et al.* sur les nébuleuses extragalactiques M 33, NGC 4449, Andromède etc.

RADCLIFFE OBSERVATORY, PRETORIA

(1) *B stars.* D. Hube (David Dunlap Observatory) used the Radcliffe 2-prism Cassegrain spectrograph to determine radial velocities of southern B8 stars in the Bright Star Catalogue (**10**).

E. N. Walker has measured velocities of 101 stars of type B5 or earlier with $m \leqslant 7{\cdot}5$ and unknown or poorly determined velocities.

63 other such stars have been observed by Thackeray. Taken jointly this will complete the survey of bright early-type stars.

R. Wood has observed B8-B9 stars suspected to be members of the Sco-Cen association, the more northerly being selected from Garrison's list, the more southerly selected partly from proper motions. Wood has also observed B8-B9 stars at distances of order 500 ps in two longitude-groups for investigation of galactic rotation.

Thackeray and Emerson (**11**) have published orbits of two O-type spectroscopic binaries. HD 93403 is a massive system ($m_1 \sin^3 i = 5{\cdot}2 \odot$) with eccentric orbit ($e = 0{\cdot}524$). Trackeray is completing a discussion of the orbits of the O sub-dwarf binary HD49798 and the double-lined B3 binary εLup (known as such for 60 years but hitherto without orbit). Other early-type binaries being studied include HD72754, 101131, 101205, 104631, 161756 (Thackeray), HD152667 (E. N. Walker), 151976 (T. Lloyd Evans), 204370 and 209278 (R. Wood). A few spectra have been obtained of GLCar and HD86118.

(2) *Cepheid variables.* Lloyd Evans (**12**) has reobserved velocity curves of many of Stibbs' southern Cepheids and is continuing this work; he finds that Abt's estimate of the frequency of cepheid binaries must be increased from 2 % to about 15 %.

M. Breger has reobserved υSgr (**13**) and SNor (**14**), for velocity and light variations.

(3) *SR variables.* Feast has completed observations of some 70 SR variables for velocity and hopes to commence reduction shortly.

(4) *Solar and late-type stars.* Feast and Catchpole are carrying out an extensive spectroscopic programme on southern C and S type stars for radial velocity and astrophysical purposes, in part with the new 8-inch camera in the Radcliffe coudé spectrograph.

As a by-product of work on lithium isotope ratios, velocities of about 25 bright southern solar-type stars have been observed with the Radcliffe coudé spectrograph at 13·6Å mm^{-1}.

Further plates of the eclipsing variable SEqu have been taken for M. Plavec.

D. Jones and Catchpole continued (until April 1969) the Cape F. D. S. programme. In May 1969 R. Wallis (Herstmonceux) began an extension of this programme on Gliese stars distant between 20 and 25 parsecs.

(5) *Instrumental.* The velocity system of the two new 8-inch Grubb Parsons cameras in the coudé spectrograph (with field flatteners to cover visual and ultra-violet wavelengths) is being examined. R. Wood has observed the 9th magnitude stars HD205650, 220647 as potential faint standard velocity stars, and is conducting tests on the velocity system of the Cassegrain d-camera.

Plans are going ahead with the Royal Greenwich Observatory for the installation of an image-tube spectrograph at the Cassegrain focus.

(6) Other references to Radcliffe work on radial velocities will be found under the Reports of Commissions 28 (Magellanic Clouds group), 29, 33, and 34.

DOMINION ASTROPHYSICAL OBSERVATORY

Since the last report of this Commission, the Cassegrain spectrograph of the 72-inch telescope has been completely redesigned. We now have all-reflection grating spectrographs of dispersions ranging from 120 to 15Å mm^{-1}. Dispersions of 10, 6, and 2·4Å mm^{-1} are available at the coudé focus of the 48-inch telescope. Much of our effort during the past three years has been directed towards establishing new wavelength systems for these spectrographs. Several changes are clearly necessary from the systems set up by Petrie and his associates for prism spectrographs of similar dispersions. Only for the 10Å mm^{-1} spectrograph have we yet completed wavelength selection over the whole range of spectral types. The accompanying table shows the accuracy that we can achieve at the various dispersions as measured both by internal and external agreement. In due course, the wavelengths chosen will be published. In the meantime, we are ready to provide this information on request.

The speed of the coudé spectrograph has been greatly increased by the use of Richardson image slicers (**15**), new, high-reflectance coated secondary mirrors and the Mount Wilson and Palomar developer formula. In addition a mosaic of four gratings has been constructed for use at the highest dispersion of 2·4 Å mm^{-1}. These improvements together allow us to obtain spectrograms of this dispersion of stars as faint as 6th, or even 7th magnitude. Fletcher has investigated the stability of this mosaic for radial-velocity work. It is necessary to readjust the mosaic for each night's work. Careful attention is being given to all possible causes of instability, and it is hoped to reduce the external mean error quoted in Table 1.

Table 1. Accuracy of new Victoria spectrographs

Spectral type	B0–B9		A0–F2		F5-K4	
Dispersion	int.	ext.	int.	ext.	int.	ext.
2·4	–	–	–	–	0·05	(0·10)
6	–	–	–	–	0·12	0·25
10	–	4·4	0·75	0·92	0·23	0·49
15	–	–	–	–	(0·7)	(1·6)
60	–	–	–	–		

Figures given are the mean errors per plate derived from internal and external agreement and expressed in km s^{-1}. Bracketed figures are preliminary only. Too few lines were measurable in the B-type spectra for a meaningful internal mean error to be derived.

One of the difficulties is the absence of good wavelengths for the iron-argon discharge tube now used as a comparison source. At first, the image slicers introduced a small systematic error into the velocities (about $-0{\cdot}3$ km s^{-1} at 10 Å mm^{-1}). Improved adjustment of them, however, has eliminated this.

Radial velocities have been determined in the course of several investigations of spectroscopic and visual binaries, variable stars, novae, and galactic structure and the results will be communicated to the appropriate Commissions.

Finally we should report the publication and distribution of the *Sixth Catalogue of the Orbital Elements of Spectroscopic Binary Systems* (**16**) since it was undertaken at the request of this Commission.

DAVID DUNLAP OBSERVATORY, TORONTO

The grating spectrograph has been modified to include a 12 Å mm^{-1} system which is more efficient than the previous 10 Å mm^{-1} system. With the 12 Å mm^{-1} system a good spectrogram of a late-type star of magnitude 9·0 ptg can be obtained in two hours. Radial velocities with external mean errors of 1·0 km s^{-1} for a single plate have been obtained on tests with standard-velocity stars. The average systematic deviation of our velocities from the IAU standard velocities is 0·3 km s^{-1}.

Heard and Petrie (**17**), have published the radial velocities and luminosity data for 77 stars in the field of the α Per cluster.

From an analysis of radial velocities of early-type Be stars within 1·5 kpc of the Sun, including new velocities obtained here as well as previously published data, Crampton (**18**), has obtained values of the solar-motion and galactic-rotation parameters, and has shown that they support the hypothesis that the commonly adopted local standard of rest has an outward motion from the galactic centre.

Hube (**10**) has published the radial velocities of 335 late B-type stars obtained here and at the Radcliffe Observatories. Combining these data with other published data, and using the age calibration of Kelsall and Strömgren, he has found that these stars were formed intensively about 190 and 150 million years ago and then more continuously during the past 120 million years. His results do not support the suggestion that stars tend to be formed in the extreme portions of their orbits about the galactic centre.

Gorza and Heard have measured the radial velocities from a number of spectrograms of Nova Del and Nova Vul obtained in 1968.

Heard has re-observed the stars listed by him (**19**) as candidates for a proposed list of 9th magnitude standard-velocity stars. At the time of writing, 14 of the 24 stars have been adequately observed; of these, eight are confirmed as having constant velocity, the remaining six have exhibited variations suggesting that they are long-period small-range spectroscopic binaries.

Heard and collaborators have continued radial-velocity observations of spectroscopic binary stars, particularly eclipsing systems. These latter have been reported to Commission 42. Hickok has completed a new radial-velocity study of the Be spectroscopic binary, ϕ Per.

ROYAL OBSERVATORY, CAPE OF GOOD HOPE

1. Data for a further 220 stars in the Fundamental Data for Southern Stars series is now available and is being prepared for publication.

2. Clube, Evans, and Jones (**20**), have obtained radial velocity and photometric curves for 50 southern RR Lyr stars.

3. Orbital elements for 8 single-lined (HD 2070, 15064, 123515, 155099, 159656, 183007, 194184 and 200334) and the double-lined (HD 188164) spectroscopic binaries have been published by Barker, Evans and Laing (**21**). The orbital elements were determined using a computer program developed by Barker.

4. The triple system p Vel has been rediscussed by Evans (**22**).

5. Nine spectra of the 1966 outburst of Nova T Pyx were discussed by Catchpole (**23**).

6. D. H. P. Jones obtained 18 spectra at 86 Å mm^{-1} of the brightest variable in the globular cluster ω Cen. A discussion of the radial velocity and light curves has been published (**24**).

7. Harding obtained radial velocities for the 126 stars in the south galactic cap which are classified as A0 in the Henry Draper Catalogue but for which velocities were not available. An analysis of the velocities (**25**), confirms the increase in velocity dispersion with height above the galactic plane found by Woolley *et al.* (**26**) in the northern cap.

8. Using observations of standard radial velocity stars Clube and Wallis have carried out an investigation of the instrumental errors of the Radcliffe Cassegrain spectrograph (**27**).

9. The Cape-Radcliffe agreement, under which the Cape Observatory was allocated one-third of the time on the Radcliffe 74-inch was terminated on 1969 March 31. In future all 74-inch time for visitors will be allocated by the Large Telescope Users Panel.

Any time which is allocated to the Royal Greenwich Observatory will be used by Wallis to observe those southern stars within a distance of 25 pc (Gliese Catalogue and extension) with no known radial velocity.

10. Jones, in conjunction with the photometric workers at the Cape has observed radial velocity curves for eight δ Sct stars.

11. Radial velocities and line identifications for 17 planetary nebulae were obtained with the Newtonian spectrograph of the Radcliffe 74-inch by Jones, Evans and Catchpole (**28**).

12. Radial velocities of 8 southern galaxies obtained with the Radcliffe Newtonian spectrograph were published by Catchpole, Evans and Jones (**29**).

13. Evans and Wild (**30**), have conducted a photometric area search for subdwarfs and have new velocities for subdwarfs in the area RA 10^h–12^h, Dec. $-20°$ to $-50°$.

14. Evans (**31**) has discussed the possibility that the radial velocity of the central star of the planetary nebula, NGC 3132 may be variable.

15. Bopp, Evans, and Laing (**32**) have determined orbital elements for the spectroscopic binaries, HD 1273, 148704, 167954, 194215, 202940, and 217792.

The paper includes an appendix on period finding by T. J. Deeming.

16. Evans (**33**) has discussed coudé plates of the multiple system HD 118261 in an effort to develop methods of analysis suitable for stars of higher multiplicity.

The preparation of a supplement to the General Catalogue of Radial Velocities commenced by Evans with assistance from Cape Staff approaches completion at Texas.

ROYAL GREENWICH OBSERVATORY, HERSTMONCEUX

The closer relationship between the Cape and Herstmonceux Observatories and interchange of staff makes it difficult to separate the reports for the two institutions.

Work carried out wholly at Herstmonceux includes Palmer *et al.* (**34**), on the radial velocities, spectral types and projected rotational velocities of 633 bright northern A-stars: radial velocity observations of standard stars with the 30-inch coudé spectrograph (**35**); and the determination of orbital elements of the sharp-lined, metallic-lined A-type spectroscopic binary, HD 184552 by Walker and Jones (**36**).

MOUNT STROMLO OBSERVATORY

Work on the radial velocities of OB stars in Puppis and Norma described in preliminary form in the 1967 Draft report has been published (**37**).

Publications by Buscombe and Kennedy include studies of M 7 (**38**); coudé spectrograph measures of 40 early and 30 late-type stars (**39**), and a further list of 124 southern OB stars and supergiants with apparent magnitudes between 4 and 9 (**40**).

C. R. Chambliss has published a radial velocity curve for the eclipsing binary 9 Cha (BV 430) based on spectra obtained at Mount Stromlo (**41**).

Using spectrograms obtained by Henize, Carlson has obtained radial velocity data for 24 southern peculiar emission line stars. He has also investigated 23 visual binary stars using plates obtained at David Dunlap Observatory (**42**).

ALLEGHENY OBSERVATORY

Beardsley (**43**), has published 2935 individual radial velocities of 129 stars based on plates obtained at the Allegheny Observatory between 1906 and 1917. The results are compared statistically with the Mount Wilson General Catalogue. The contribution includes preliminary orbital elements for ϕ Dra and 6 Lac. Some 1400 plates still remain to be analysed.

With the general introduction of image tubes in spectrographic cameras the question of the accuracy of determination of radial velocities has become of general concern. Beardsley *et al.* (**44**), report favourably on the performance of a fiber optics electrostatic image tube by Westinghouse. Workers elsewhere, using varied equipment, seem less optimistic.

INDIVIDUAL WORKERS

Radial velocity programmes used to be undertaken on an institutional basis and could thus readily be reported observatory by observatory. The work is now devolving more on individuals, with interests in special problems, or in the development of special techniques. The remainder of the report is presented in a form suited to this development. The opinion should, however, be stated that systematic programmes ought to be continued, both because they often produce more accurate results than isolated projects, and because the coverage of even bright stars, especially in the south, is so far inadequate.

Griffin reports that his photoelectric radial-velocity spectrometer has been used from time to time in essentially its original form (**45**). The method requires reference stars of known velocity, and four selected stars have been intercompared using the instrument. The standard deviation of a single comparison is 0·6 km s^{-1}, and the closing error round the sky, 0·58 km s^{-1} (**46**). Velocities have been measured for some 700 stars, mostly unobserved, and mostly in the range from 7^m to 10^m in the $+15°$ selected areas. Late-type HR stars north of declination $-9°$, of unknown velocity are also being observed.

Lutz and Yoss (**47**), have investigated the possibility of determining stellar radial velocities from microphotometer measurements, with promising results.

O. C. Wilson (**48**), has published an important list of more than 300 radial velocities of dwarf K and M stars.

Sandage (**49**), reports radial velocities obtained at Mount Wilson and Mount Palomar for 112 stars of large proper motion as part of a general study of new subdwarfs.

Oetken reports that the Cassegrain spectrograph of the Tautenburg two-meter telescope is being used to obtain radial velocity data for magnetic stars such as α^2 CVn, γEqu, βCrB.

Jaschek, in collaboration with Miss A. Goméz has studied the proportion of spectroscopic binaries in different parts of the H-R diagram, and confirms Petrie's conclusion of a constant percentage among dwarf stars, but only 30 % (in place of Petrie's 50 %), among the giants.

Anne B. Underhill (Utrecht), reports (**50**) observations made at Victoria of ten OB stars in the vicinity of the open cluster NGC7380 of which six are definitely double-lined binaries, five being new discoveries, and the remainder may also be binaries. De Groot is engaged in a similar study for the neighbourhood of IC1805 and has published detailed measures of PCyg (**51**).

Bertiau and Grobben (Vatican Observatory) have a computer program for reduction of spectroscopic binaries which distinguishes between single and double-lined binaries, and in the latter case treats the data differently according to whether the lines are of comparable strength or not (**52**). The capacity of the programme is 220 observations. Bertiau also has some radial velocity measures of 26 late B-type stars in the Scorpio region using Pretoria plates.

Kharadze reports that E. A. Vitrichenko, in collaboration with astronomers of the Tartu Observatory (Eelsalu, Kivila and Joëveer), is using the 122 cm reflector of the Crimean Astrophysical Observatory for radial velocity work. A programme on K0 stars is being started and on the velocity curves of the double stars, γCyg, HD175514, δCep C, HD188439 and HD193793.

Abt and Gregory Smith (**53**) have investigated the accuracy of determination of radial velocities and its dependence on dispersion, spectrum-width and line-width. Accuracy has generally improved over the years, but guiding error remains significant. Broadening the spectra produces little improvement to compensate for the extension of exposure time. Accuracy is almost independent of line-width for rotational velocities above about 100 km s^{-1}. Increasing dispersion is the best way to improve accuracy. Measuring engines of greater precision than the conventional long-screw machine are desirable if full advantage of potential accuracy is to be secured.

In a recent paper Abt and Levy (**54**) have investigated 68 high velocity dwarfs and find that short period binaries are rare.

Abt reports good progress on his general bibliography of radial velocities, in excess of 30000 references. His catalogue of individual plate radial velocities measured by Mount Wilson astronomers is complete to 12^h R.A. and has been submitted for publication (**55**).

G. and A. de Vaucouleurs report that the interference filter scanner (**56**), has been further developed for automatic scanning at pre-programmed wavelengths to record with a 400-channel multiscaler with teletype punched paper tape output. The theory of the system and properties of interference filters used for scanning in convergent light at nonnormal incidence have been studied in detail by W. Tolar (**57**).

About 600 scans were secured between 1966 and 1969 with the IFS scanner on the 36-inch and 82-inch reflectors at McDonald Observatory; 118 objects (73 galaxies, 12 clusters, 28 stars, 5 planetaries) were observed with the λ4006 filter used for the H and K lines; 37 objects (15 galaxies, 5 clusters, 13 stars, 4 planetaries) with the λ4408 filter used for G and Hγ; 60 objects (18 galaxies, 7 planetaries and 35 HII regions in M33 and M101) with the λ5055 filter used for N_1, N_2. The radial velocities derived by computer fitting of Gauss curves to the line profiles (filter half-width $\sim$15Å) have an indicated mean error of ~0·5 Å (emission lines) to ~0·1 Å (absorption lines). A detailed rotation study of M33 and M101 from HII regions is in progress. Red shifts of about 70 galaxies have been obtained and will be prepared for publication in 1970.

Davis Philip, as part of a continuing program of study of horizontal-branch stars (**58**), has determined radial velocities of 12 of these stars in the general field using Mount Wilson and Kitt Peak plates. A *z*-velocity dispersion of 113 km s^{-1} is found for seven of them, confirming membership in Population II (**59**).

Perry has measured radial velocities of 109 early A-dwarf stars in the north galactic cap to determine the galactic force law, K_z. The results support the existence of two groups of A-stars in the *z*-direction (**60**). Velocities for 77 A-type stars in a north galactic pole region have also been determined by Slettebak, Wright and Graham (**60a**).

The following contributions, usually dealing with a single star may be noted: 17 Lep (Cowley) (**63**); RV UMa (Preston and Spinrad) (**64**); HD 2665, 6755 (Koelbloed) (**65**); HD 118216 (Conti) (**66**); RU Cam (Wallerstein and Crampton) (**67**); 73 Dra (Preston) (**68**); FG Sag (Herbig and Boyarchuk) (**69**); SS Cyg (Walker and Chincarini) (**70**); HD 10783 (Preston and Stepién) (**71**); CM Lac, V477 Cyg, TV Cet (Popper) (**72**); HD 215441 (Preston) (**73**); 17 Com, κ Cnc (Preston, Stepién and Wolff) (**74**); HR 7484 (Snowden and Koch) (**75**); 12 Cam (Abt, Dukes and Weaver) (**76**); 78 Vir (Preston) (**77**); 21 Per (Preston) (**78**); HD 222317 (Imbert) (**79**); ER Vul (Northcott and Bakos) (**80**); HD 23631 (Conti) (**81**); HD 107904 (4 CVn), (Crampton and Fernie) (**82**); 14 stars (Roman) (**83**); HD 211853 (Bracher) (**84**); HD 35921 (Mayer) (**85**); TX UMa (Swensen and McNamara) (**86**); IIR 3817 (M. Jaschek and M. L. Aguilar) (**87**).

DAVID S. EVANS
President of the Commission

BIBLIOGRAPHY

1. Hodge, P. N., Wright, F. N. 1967, *The Large Magellanic Cloud*, Smithsonian Publications, 4699.
2. Martin, N. 1967–68, *Publ. Obs. Hte-Provence*, **9**, No. 13.
3. Barbier, M. 1967–68, *ibid.*, **9**, No. 38.
4. Barbier, M. 1967–68, *ibid.*, **9**, No. 34.
5. Rebeirot, E. *ibid.*, in press.
6. Georgelin, Y. *ibid.*, in press.
7. Fehrenbach, Ch. 1967–68, *ibid.*, **9**, No. 16.
8. Imbert, M. 1967–68, *ibid.*, **9**, No. 17.
9. Courtès, G., Cruvellier, P., Georgelin, Y. 1966–67, *ibid.*, **8**, No. 34.
10. Hube, D. P. *Mem. R. astr. Soc.*, in press.
11. Thackeray, A. D., Emerson, B. 1969, *M.N.R.A.S.*, **142**, 429.
12. Lloyd Evans, T. 1968, *M.N.R.A.S.*, **141**, 109.
13. Breger, M. 1967, *M.N.R.A.S.*, **136**, 61.
14. Breger, M. in press.
15. Richardson, E. H. 1968, *J. R. astr. Soc. Can.*, **62**, 313.
16. Batten, A. H. 1968, *Publ. Dom. astrophys. Obs.*, **13**, 119.
17. Heard, J. F., Petrie, R. M. 1969, *Publ. Dom. astrophys. Obs.*, **13**, No. 13.
18. Crampton, D. 1968, *A.J.*, **73**, 338.
19. Heard, J. F. *Trans. IAU*, **13B**, 169.
20. Clube, S. V. M., Evans, D. S., Jones, D. H. P. 1969, *Mem. R. astr. Soc.*, **72**, Pr. 3.
21. Barker, E. S., Evans, D. S., Laing, J. D. 1967, *R.O.B.*, No. 130.
22. Evans, D. S. 1969, *M.N.R.A.S.*, **142**, 523.
23. Catchpole, R. M. 1969, *M.N.R.A.S.*, **142**, 119.
24. Jones, D. H. P. 1968, *M.N.R.A.S.*, **140**, 265.
25. Harding, G. A. in press.
26. Woolley, R. v. d. R., Harding, G. A., Penston, M. J., Ali, K. 1967, *IAU Symposium*, no. 30. p. 247.
27. Clube, S. V. M., Wallis, R. E. 1968, *Mon. Notes astr. Soc. Sth. Afr.*, **27**, 57.
28. Jones, D. H. P., Evans, D. S., Jones, D. H. P. 1969, *Observatory*, **89**, 21.
29. Catchpole, R. M., Evans, D. S., Jones, D. H. P. 1969, *Observatory*, **89**, 15.
30. Evans, D. S., Wild, P. A. T. 1969, *Observatory*, **89**, 15.

31. Evans, D. S. 1968, *Mon. Notes astr. Soc. Sth. Afr.*, **27**, 129.
32. Bopp, B., Evans, D. S., Laing, J. D., with an appendix by T. J. Deeming in press.
33. Evans, D. S. 1968, *Q.J.R.A.S.*, **9**, 388.
34. Palmer, D. R., Walker, E. N., Jones, D. H. P., Wallis, R. E. 1968, *R.O.B.*, No. 135.
35. Harding, G. A., Palmer, D. R., Pope, J. D. 1968, *R.O.B.*, No. 145.
36. Walker, E. N., Jones, D. H. P. 1969, *Observatory*, **89**, 202.
37. Bok, B. J., Gollnow, H., Mowat, M. 1967, *Observatory*, **87**, 250.
38. Buscombe, W., Kennedy, P. M. 1968, *M.N.R.A.S.*, **139**, 215.
39. Buscombe, W., Kennedy, P. M. 1968, *M.N.R.A.S.*, **139**, 341.
40. Buscombe, W., Kennedy, P. M. 1969, *M.N.R.A.S.*, **143**, 1.
41. Chambliss, C. R. 1969, *M.N.R.A.S.*, **142**, 113.
42. Carlson, E. D. in press.
43. Beardsley, W. R. 1969, *Publ. Allegheny Obs.*, **8**, No. 7.
44. Beardsley, W. R., De Jonge, J. K., Haring, D. J., Hansen, J. R. 1969, *Publ. Allegheny Obs.*, **11**, No. 1.
45. Griffin, R. F. 1967, *Ap. J.*, **148**, 465.
46. Griffin, R. F. *M.N.R.A.S.*, in press.
47. Lutz, T. E., Yoss, K. M. 1969, *A.J.*, **74**, 91.
48. Wilson, O. C. 1967, *A.J.*, **72**, 905.
49. Sandage, A. in press.
50. Underhill, A. B. 1969, *Astr. Astrophys.*, **1**, 356.
51. De Groot, M. *B.A.N.*, **20**, in press.
52. Bertiau, F. C., Grobben, J. *Ric. astr.*, **7**, No. 20, in press.
53. Abt, H. A., Smith, G. H. 1969, *P.A.S.P.*, **81**, 332.
54. Abt, H. A., Levy, S. G. 1969, *A.J.*, **74**, 908.
55. Abt, H. A. *Ap. J. Supp.*, in press.
56. De Vaucouleurs, G., De Vaucouleurs, A. 1967, *IAU Symposium*, no. 30, p. 9.
57. Tolar, W. Texas, M. S. Thesis, to be published.
58. Davis Philip, A. G. 1968, *Ap. J.*, **152**, 1107.
59. Davis Philip, A. G. 1969, *Ap. J.*, **158**, L113.
60. Perry, C. L. 1969, *A.J.*, **74**, 139.
60a. Slettebak, A., Wright, R. R., Graham, J. A. 1968, *A.J.*, **73**, 152.
61. Perry, C. L., Bond, H. E. 1969, *P.A.S.P.*, **81**, 629.
62. Tatum, J. B. 1968, *M.N.R.A.S.*, **141**, 43.
63. Cowley, A. P. 1967, *Ap. J.*, **147**, 609.
64. Preston, G. W., Spinrad, H. 1967, *Ap. J.*, **147**, 1025.
65. Koelbloed, D. 1967, *Ap. J.*, **149**, 299.
66. Conti, P. S. 1967, *Ap. J.*, **149**, 629.
67. Wallerstein, G., Crampton, D. 1967, *Ap. J.*, **149**, 227.
68. Preston, G. W. 1967, *Ap. J.*, **150**, 871.
69. Herbig, G. H., Boyarchuk, A. A. 1968, *Ap. J.*, **153**, 397.
70. Walker, M. F., Chincarini, G. 1968, *Ap. J.*, **154**, 157.
71. Preston, G. W., Stępién, K. 1968, *Ap. J.*, **154**, 971.
72. Popper, D. M. 1968, *Ap. J.*, **154**, 191.
73. Preston, G. W. 1969, *Ap. J.*, **156**, 967.
74. Preston, G. W., Stępién, K., Wolff, S. C. 1969, *Ap. J.*, **156**, 653.
75. Snowden, M. S., Koch, R. H. 1969, *Ap. J.*, **156**, 667.
76. Abt, H. A., Dukes, R. J., Weaver, W. B. 1969, *Ap. J.*, **157**, 717.
77. Preston, G. W. 1969, *Ap. J.*, **158**, 243.
78. Preston, G. W. 1969, *Ap. J.*, **158**, 251.
79. Imbert, M. 1969, *Astr. Astrophys.*, **3**, 272.
80. Northcott, Ruth J., Bakos, G. A. 1967, *A.J.*, **72**, 89.
81. Conti, P. S. 1968, *A.J.*, **73**, 348.
82. Crampton, D., Fernie, J. D. 1967, *P.A.S.P.*, **79**, 168.
83. Roman, N. G. 1967, *P.A.S.P.*, **79**, 168.
84. Bracher, K. 1968, *P.A.S.P.*, **80**, 165.
85. Mayer, P. 1968, *P.A.S.P.*, **80**, 81.
86. Swensen, P. R., McNamara, D. H. 1968, *P.A.S.P.*, **80**, 192.
87. Jaschek, M., Aguilar, M. L. 1969, *P.A.S.P.*, **81**, 170.

31. TIME (L'HEURE)

PRESIDENT: F. Zagar.
VICE-PRESIDENT: G. M. R. Winkler.
ORGANIZING COMMITTEE: D. J. Belocerkovskij, J. Bonanomi, H. Enslin, W. Markowitz, H. M. Smith, M. Torao.

ASTRONOMICAL TIME

A. *Universal time*

The astronomical observatories which actively cooperate in the network of the system of the Bureau International de l'Heure, are, at present, about 70. Most of the astronomical time determinations are carried out by means of Danjon Astrolabes or transit instruments; 10 observatories dispose of a PZT and 5 others are installing or planning to have it. The results of their observations, φ (observed latitude) and ΔT (difference between local U.T.0 and U.T.C.), are collected at the BIH and – contrary to the procedure used until 1967 – used to simultaneously obtain the instantaneous coordinates of the pole (x, y), referred to conventional international origin (CIO), and longitudes origin on the pole equator. As a consequence of the adoption of the above-mentioned conventional pole, the BIH had to change all conventional longitudes in order to keep the origin formerly defined (1967); they are now reported in BIH *Annual Reports*. A further change was operated on the constant of aberration which, at the beginning of 1968, from 20″47 was changed to 20″496. The BIH had therefore to operate a step of $+0^{s}.0020$ to U.T.1, at the beginning of January 1968, and so did the Greenwich Observatory, with a step of $+0.0069$. Global corrections, due to these variations and to others occurred from 1955 to 1968 (in 1958, variation of the polar reference, from Cecchini's to the mean pole of the date; in 1961, adoption of FK4 and of new longitudes), to the data of the past years (until 1968), are reported on Table 8 of the BIH Annual Report, 1968. The formula supplying the correction U.T.2 – U.T.1 remained unchanged, except for the variable t, which is expressed (since 1967) as a fraction of the Besselian year. Finally, stars coordinates in the catalogs used for time service were revised on the basis of observations made during the latest ten years (Neuchâtel, Greenwich and Mizusawa) and a catalog has been prepared, which includes 807 right ascensions for 1960, drawn from 20 catalogs, with more than 185000 observations of the U.S.S.R. (Pulkova) time service, which is also preparing a new catalog of proper motions. The unification of star positions and proper motions is recommended for the PZT.

The BIH considers the accuracy of the results such as to induce the study of the lunar effects on the vertical, and the diurnal nutation effects as well. Particular researches on observation errors in the determinations of time and longitude have been performed at the Observatories of Milan and Potsdam, while the Dresden Observatory is carrying out practical researches on a photographic method and theoretical researches on a photoelectric procedure for eliminating the trembling of the images due to the atmosphere.

B. *Ephemeris time*

While observers of stellar occultations by the Moon are rather numerous and give some thousand observations per year (400 only from Tokyo Observatory – Hydrographic Dept.), observatories which determine E.T. are rather few, not more than half a dozen. Most of them use the occultations method; Besançon, Washington, Uppsala and others use the photographic method to determine the position of the Moon on the stars background, and only Tokyo and Washington carry out meridian observations of the Moon. A. and N. Stoyko have been carrying out researches on E.T., starting from the data of 1900, and researches on lunar positions since 1955. According to A. Stoyko, the difference ΔT.0 = E.T. – U.T.2 varied from $+31^{s}.575$ in 1955·5 to $+36^{s}.713$ in 1966·5 and, according to Tokyo observations from $+32^s.90$ in 1962·5 to $+35^s.88$ in 1966·5. From the introduction of the new system of astronomical constants in 1964 and correction of some terms of Brown's lunar theory, E.T.0 time (before 1964) is distinguished from E.T.1 (after 1964): they are connected by the relation

$$\text{E.T.1} - \text{E.T.0} = +0^s.3426 \sin \Omega - 0^s.1084 \sin(2\Pi - \Omega) - 0^s.306 \sin D,$$

where Ω is the longitude of the lunar ascending node, Π the longitude of the lunar perigee and D the difference between lunar and solar longitude. The difference ΔT.1 = E.T.1 − U.T.2, similarly to ΔT.0 varies, according to A. Stoyko, from +30.981 (1955·5) to $+36^s.748$ (1966·5) and, according to the Tokyo Observatory, Hydrographic Dept., from $+29^s.72$ (1956·5) to $+34^s.52$ (1965·5) to $+37^s.13$ (1968·5). E.T.1 eliminates the difference between the results obtained for E.T.0 from occultations at the lunar dark edge and those at the bright edge. The comparison with atomic time proves that E.T.1 (for the short period 1955–66, so far considered), on the contrary of E.T.0, is rigorously uniform; therefore it can be stated that, according to observations of occultations, the Moon represents the most uniform ephemeris time.

ATOMIC TIME

About 20 observatories and laboratories care for time measured by atomic clocks, with approximately 60 standards. Most are caesium clocks but also hydrogen and rubidium clocks are used. Most of these standards are utilized for radiofrequency transmissions emitted by about 40 stations, of which only half are standard frequencies (coming from 16 observatories or laboratories). Differences between relative frequencies $\Delta f/f$ of these standards are of the order of a few units on 10^{12}; hydrogen standards are more stable, considering that two clocks controlled by maser standards do not differ more than 3 ns/day stability and 3 microseconds per year (3×10^{-14}) (NASA Center); for short time intervals also quartz crystal oscillators reach very high accuracies, for intervals from 1 to 10 s they are as accurate as atomic oscillators (NBS Halford).

Some of these Institutes have established an atomic time scale, for instance: A.3 of Paris BiH, A.T. (F) of Paris CNH, A.T. of Boulder NBS, A.T.1 of Braunschweig PTB, G.A.2 of Greenwich and A.1 of Washington NO, the origin of which differs by fractions of a second, while their times agree within a few μs. The first of these scales has been redefined by the BIH on January 1st, 1969, as the mean of time scales of more than 20 commercial standards (10–16 from the NO, 5 from the French group, 3 from the PTB and 3 from the RGO), continuously compared by synchronized transmissions of Loran-C chains; this is the former mean standard A3 which, until 1968, was based mainly on VLF and LF (frequency and phase difference) transmissions of many laboratories, and is now called international atomic time scale and indicated by A.T. From October 1969, laboratories which contribute to the formation of the A.T. became 7. If all precautions taken to avoid disturbances in the frequency were valid, this standard should secure an accuracy of some parts in 10^{13} over a period of some months. According to what has been said, the two time scales E.T.1 and A.T. are uniform and differ only in the initial epochs: E.T.1 − A.T. = $+31^s.868$ (A. Stoyko).

TIME UNIT

In October 1967, the 13th Conférence Générale de Poids et Mesures, adopted the atomic time second to be the fundamental unit of time in the international system, defining it as the duration of 9 192 631 770 periods of the radiation corresponding to the transition between two hyperfine levels of the ground state of the caesium atom 133, thus replacing the definition adopted in 1956 of the ephemeris second, and meanwhile inviting organisations and laboratories concerned to continue their researches on the argument.

This decision did not consider a recommendation of Commissions 4 and 31 of IAU on August 1967 in Prague, which requested CGPM to include in the new definition of the international second the phrase: "The ephemeris second, which is part of the IAU system of astronomical constants" (*Trans. IAU*, **13B**, 41), thus causing the protests of competent astronomers (Clemence, Fricke,

Sadler, Wilkins) and also an intervention of the IAU President. Fricke suggests that during the next General Assembly, a recommendation is made to use always the form "ephemeris second" when this is necessary to avoid misunderstandings.

COORDINATED TIME

The universal coordinated time scale (U.T.C.) continued to be based on the international atomic time scale of Paris (A.3), according to the definition: U.T.C. $-$ A.T. $= S(t-t_0)+b$, $|$U.T.2 $-$ U.T.C$| < 0^s.1$, using frequency offsets and time adjustments indicated by the BIH. But preoccupations raised with the introduction of U.T.C., not even mitigated with stepped atomic time (S.A.T.) which completely excludes the frequency offset, have increased and caused deep discussions on the argument, promoted mainly by the CCIR at the Boulder (August 1968) and Geneva (October 1969) meetings, and by the CIPM at the Paris meetings (May and October 1968). The CCIR appointed expressly the working group VII-1 (chairman H. M. Smith) while the CIPM formed the preparatory commission for the international coordination of time scales (chairman L. E. Howlett).

After a first examination the CCIR has proposed in the document VII-70-E some conditions for a solution which included abolition of offset, maximum difference of $0^s.5$ between U.T.C. and U.T.2, time adjustments of exactly 1^s (S.A.T. system) and a simple code incorporated in radio emissions to determine this difference within 10 ms. The similar document CIPM/68-5, after confirming the need to improve conventions concerning U.T.C. and supporting CCIR conclusions, has also proposed that BIH be equipped with all necessary means for the coordination of all the matter relative to U.T.C. This document also invites the CIPM to take the initiatives for reaching an international agreement on U.T.C. As this Committee took care of time unit, it is also expected to take care both of an international time scale and of an atomic time scale based on the new definition of the second, as an international scale should require a governmental agreement.

Only the first item of the two documents (abolition of offset) found nearly unanimous consent while for the others discussions are still open. According to the authoritative intervention of D. H. Sadler and G. M. Winkler, who examined thoroughly the serious dissensions between those concerned with standard frequencies and those who request a time system connected with the Earth's rotation (astronomers, geodesists, navigators etc.) and in particular the world-wide collision avoidance system for aircraft (CAS), which cannot admit stepping time adjustments, serious objections can be made as regards the proposals of the two documents; also because said Commissions ignore the needs of a great minority who cannot prescind from the definition of above mentioned U.T.C. and in particular cannot accept differences greater than $\pm 0^s.1$. It is stated as a necessity that activities concerning time service should be completely independent of the activities of frequency standards laboratories.

Discarding the suggestion (for practical reasons and to avoid confusions) of two time scales, one approaching U.T. (the present U.T.C.) and the other without offsets and adjustments, only three alternatives remain: (*a*) step adjustment of $0^s.1$ or $0^s.2$ to maintain the U.T.C. sufficiently near to U.T.2 to permit to ignore the difference in most of the applications; (*b*) complete disuse of U.T.C. system replacing it with a coordinated uniform time scale without offsets and steps and therefore not approaching U.T.; (*c*) step adjustment of 1^s exactly; anyway only further discussions will solve the problem.

The PTB has expressed indeed the urgent need that a Consultative Committee for Time Scales be created by the CIPM, being charged with the preparation of the time scale definition work and giving outlines to a laboratory working under these auspices, at best the BIH. This proposal, supported by the members of the PTB, with a long report by G. Becker, at the meetings held in autumn 1969 by the URSI (Ottawa), CCIR (Geneva) and CIPM (Paris) has been recommended in a resolution of the URSI, which also definitively adopted the S.A.T. system, to be commenced before 1st January 1971, with the explicit statement that the definition of astronomical time scale does pertain to the IAU. But only one month after, in October 1969 at the Geneva meeting, the working party VII-1 in its documents 1004 and 1008 rediscussed the question and decided to submit to the next

Plenary Assembly of CCIR (New Delhi, January 1970) both the offset and the S.A.T. system (the latter to be commenced in January 1972), for a final decision, asking in the meanwhile also the opinion of the Scientific Unions concerned, among which the IAU. The variation of the BIH statute, ventilated during the discussions, did not obtain the approval of the astronomers and the only result obtained was the appointment of a BIPM member to the BIH directing Board.

TIME SIGNALS AND SYNCHRONIZATION

An intense activity is devoted to the problems of synchronization of the different time scales, for the near, the far and the very far ones. Most of the emitting stations, i.e. about 40, even when based on atomic standards, transmit the coordinated time scale (U.T.C.) by radiofrequencies; only a few stations emit the A.T., while some emission is made simultaneously in A.T. for carrier frequency and U.T. for time signals. Comparisons among different standards are still carried out in most cases by very low frequencies (VLF); about 10 institutes use the Loran-C chains system, and ten others the television system. Undoubtedly the two latter methods are destined to be largely developed due to their higher accuracy (100 ns for time and 10^{-12} for frequency) and some pairs of atomic standards are now periodically synchronized utilizing television programs, as for instance, since 1968 the standards of the Paris Observatory and of the Prague Electronic Radiotechnical Institute, and since 1969 those of the PTB (Braunschweig) and DHI (Hamburg). But the limited range (maximum distance 1500 km) and difficulties of connection require further researches on equipment and wave propagation in these methods. On the other hand the PTB pointed out that the VLF transmissions, besides being the only method for intercontinental continuous synchronizations, can give results not at all inferior to the above-mentioned, such that for certain experiments they have been used together with the flying clock method, as equivalent procedures.

Some institutes have carried out experiments of synchronization based on the reflection of the waves by artificial satellites, by meteor ionized trails and recently by the Moon. After the experiments with Telstar I (1962) and Relay II (1965), synchronizations have been performed in 1968 between California and Hawaii with the satellite ATS-1. For comparisons similarly based on meteor trails (produced by collisions of meteors with air particles), discontinuity of the phenomenon apart, analogous equipments are available which give similar results, but for distances not exceeding 2000 km (due to the limited altitude of the trail); on the contrary the experiment with the Moon has been performed between Canberra and Goldstone stations. These methods are advantageous because they use the VHF system for the transmissions, requiring a very simple and not expensive equipment securing on the other hand an accuracy of the microsecond.

Finally direct comparisons by flying clocks have proved satisfactory, especially over great distances. Clock transportations organized by Hewlett Packard all over the world with stops and comparisons up to 53 stations, covered about 100000 km and at the end, relative differences amounted to a few microseconds, corresponding to an offset of 10^{-12} in the standard frequencies.

More recently the USNO organized travels with a Caesium standard, which supplied the difference between the U.T.C. (USNO) and the U.T.C. (local) of the laboratories visited; for Paris, for instance, the difference was, in 1968, April 18: $+130\,\mu$s, in 1969 February 24: $+113\,\mu$s, i.e. a variation of $-17\,\mu$s over 1 year. A similar travel was realized by the PTB to Paris, in 1968, which made possible a continuous comparison, with an uncertainty lower than $1\,\mu$s. Even more efficient seem to be the experiments of flying over, where the time scale of the clock on board is sent to the ground station from the aeroplane, possibly on the vertical of the station at a constant altitude, avoiding stops, unloading and transshipments of materials: the application of the SINFRAL (Synchron. Franco-Allemande) since 1969 between the BIH in Paris and the PTB in Braunschweig, foresees accuracies of the order of $0{\cdot}1\,\mu$s.

RELATIVITY

It is known that relativistic phenomena take place in clocks with relative motion or in different gravitational fields and their entities become measurable only when remarkable velocities and sensible potential differences are involved. Therefore only effects measurable by space flights (artificial satellites and space probes) can be expected. As regards the difference between atomic time considered as proper time, in a system connected with the Earth, and ephemeris time, considered as coordinated time in an inertial system with respect to which the Earth moves, after the researches carried out by Aoki and Clemence, who calculated an annual variation of semiamplitude of $0^s.0017$, there are only theoretical researches (Hudson, Becker). Similar researches concern the motion of clocks around the Earth (Polishchuk, Takagi and Hara). On the contrary as regards the only mass effect on the frequencies, there are some experiments (Mungel, Sadeh and collaborators) carried out by means of comparisons between VLF signals and the results of comparisons with HP flying clocks which, anyway, have not yet given reliable results. A satisfactory answer on these problems will come from numerous and long series of ET observations and from rigorous measures of the differences of the above-mentioned times.

REPORTS OF OBSERVATORIES AND LABORATORIES

Australia. Canberra

PZT observations for U.T.0 have been continued, improved star places and proper motions being adopted from 1st July, 1968.

The clock system includes a commercial caesium beam frequency standard and two crystals which have kept the time without loss of continuity since 23rd September 1967. The scale for U.T.C. was compared with overseas scales by means of portable clocks, and the rate is recorded continuously against VLF transmissions from overseas stations and NWC in Western Australia. Time comparisons are also made continuously by courtesy of the deep space stations at Goldstone in California and Tidbinbilla near Canberra, these being linked by signals that are reflected from the Moon. The standard error of the comparisons is about 5 μs, Preliminary time comparisons by means of television transmissions are being carried out against an atomic standard at the National Standards Laboratory, Sydney.

Belgium. Uccle

L'Observatoire Royal a fait l'acquisition de deux standards de fréquence; un standard au rubidium, Rohde & Schwarz et un standard au césium, Hewlett Packard, respectivement mis en service en avril et en décembre 1968.

En outre deux oscillateurs à quartz, 1 Mc, sont asservis par des régulateurs de fréquence depuis 1969; la précision à long terme est comparable à celle d'un rubidium. Un comparateur de fréquence a été mis en service en avril 1968. Une seconde chaîne de réceptions horaires comportant un récepteur Racal, un oscilloscope et un compteur d'intervalle de temps a été installé en janvier 1968. Un second Astrolabe de Danjon sera installé dans le courant de l'année 1970.

Observations effectuées à l'Astrolabe; 1966: 90 séances, 1800 étoiles; 1967: 148 séances, 2777 étoiles; 1968: 123 séances, 2458 étoiles. Réceptions horaires: 20 par jour, soit 7200 par an. Les résultats des observations astronomiques sont publiés dans le BULLETIN HORAIRE de l'Observatoire et dans le *Annual Report* du BIH.

Brasil. Rio de Janeiro

The astronomical time determinations are being performed by two transit instruments, the installation of a PZT and the possibility of an equatorial astrolabe station at Bélem, 2 °S of the Equator, is being studied.

The time service has continued the cooperation with the BIH, sending time signals through the PPE station and checking the quartz clocks by HF and VLF signals from the network of time stations. A caesium beam standard, to be installed in 1970, has been ordered.

Canada. Ottawa

Three commercially built caesium atomic clocks are now in use at Ottawa, two in the time laboratory, and one to replace the rubidium standard formerly used as time and frequency source at CHU. Travelling clocks under the auspices of Hewlett Packard and then of U.S. Naval Observatory have visited Ottawa; and in 1967 the Swiss atomic clock visited Ottawa en route to Expo 67 Montreal. A Lorchron receiver, acquired in 1969, permits a continuous comparison with Loran C. Time service bulletins giving the results of PZT observations and time signal comparisons continue to be published quarterly.

Chile. Santiago

According to an agreement with the ESO, a Danjon Astrolabe is in operation at the National Astronomical Observatory on Cerro Calán (Lat. $-33\overset{\circ}{.}4$). The equipment comprises a Rohde & Schwarz, three quartz clock battery, an Ebauches integrating chronograph and a Rohde & Schwarz VLF receiver, which makes phase comparisons between the local oscillators and WWVL or NBA.

Results are sent weekly to BIH, and monthly to IPMS. Time and latitude bulletins were published quarterly up to 1968. Since 1969, they are published monthly. The yearly results in time and latitude are being published in Publicaciones del Departamento de Astronomía (PDA), Santiago de Chile.

Czechoslovakia. Prague

The new television method for microsecond clock comparison (*Trans. IAU*, **13**A, 663, 1967) was regularly used. During 1968 and 1969 a series of experiments were carried out using TV programs for the time comparison between Prague and Paris. Some of these experiments involved several institutions in Switzerland, German Federal Republic and Italy. The timing resolution of the order of 10 ns was shown. Weekly comparisons are made between Prague and Potsdam. A PZT is to be installed by the Ondřejov Observatory. The OMA signals give Coordinated Time since 1964.

France. Besançon

Le programme d'observations à l'Astrolabe Danjon OPL No. 31 a été poursuivi sans modification.

Le temps des éphémérides est determiné par photographies de la Lune sur le fond des étoiles. La photographie est faite par un dispositif spécial à l'astrographe de Secrétan-Couder de l'Observatoire. La mesure des clichés au mesureur Zeiss et leur exploitation sont effectuées à l'Observatoire et au Laboratoire de calculs de la Faculté des Sciences.

Jusqu'à la fin de l'année 1966, le programme quotidien des réceptions comportait la mesure de 30 signaux sur oscilloscope. A partir de 1967, sur la suggestion de Monsieur le Directeur du BIH le programme a été réduit et limité, après essais, aux 3 signaux HBG et MSF (5 MHz et 10 MHz); en outre, la publication du bulletin horaire mensuel a été suspendue.

Depuis le 1er Mars 1969, nous recevons par coaxial la fréquence d'un rubidium C.S.F. appartenant à l'École Nationale Supérieure de Chronométrie et de Micromécanique de Besançon.

Contrôle en état:

le 6 Octobre 1967 avec une "horloge volante" Hewlett Packard venue de Genève;

le 25 Février 1969 avec une horloge au rubidium de l'Observatoire de Paris.

France, Paris. Bureau International de l'Heure (BIH)

As the work of BIH for 1967 and 1968 was presented in the *Annual Reports*, we shall give here only general information and some complements.

Universal Time and Coordinates of the Pole

The computations are based on 58 series of U.T.0 measures and 44 series of latitude measures (October 1969). Observations of each star group (or sometimes each night) are dealt with individually in order to give a rough solution for every consecutive 5-day interval. Definitive values of U.T.1 − U.T.C., x, y, for the month m are published at the beginning of the month $m+2$ months in Circular D.

We computed U.T.1 − U.T.C., x and y, for 1964·0–1968·0 period, referred to the new system of the BIH and to the new aberration constant: these results were published, for every 1/20 of a year and without smoothing, in the *Annual Report* for 1967.

These new methods of computation will be extended back to the period 1955–64, so that we shall obtain an homogeneous series of values of U.T.1, x, y, since the atomic time exists. This work is in process. The 1962·0–1964·0 period will be available in 1970.

On the request of some laboratories, we supplied short-term extrapolations of x and y and also provisional values of U.T.1. We studied the possibility to establish an "express-service", giving U.T.1 to about 3 ms with a delay reduced to a few days.

Atomic time

The mean atomic time of the BIH, denoted here A.T., is linked to the so-called coordinated time U.T.C. by a rigorous formula; therefore we shall consider them as equivalent in this section of our report. Until the end of 1968, A.T. was computed by using the VLF transmissions as explained in the Annual Reports. From the 1st January 1969, we used the only time standards (or groups of standards) which are linked at intervals shorter than a few days, with an approximation of a few 0·1 μs. This is presently obtained by the reception of the Loran-C pulses by ground wave and of the television pulses. In October 1969, 7 laboratories contribute to form A.T.

Coordinated time

This time gives an approximation of U.T.2 to ±100 ms. The laboratories and observatories i keep generally a coordinated clock which gives an approximation U.T.C. (i) of U.T.C.; Circular D of the BIH gives the differences U.T.C. − U.T.C. (i) when they are known to an accuracy of 1 μs or better.

France. Paris. Laboratoire de l'Heure

La Commission Nationale de l'Heure, en France, a fédéré les étalons atomiques possédés par l'Observatoire de Paris, le Centre National d'Études des Télécommunications, le Centre National de la Recherche Scientifique, le Laboratoire de l'Horloge atomique à Besançon. Les comparaisons faites par l'intermédiaire des impulsions de télévision permettent de construire une échelle moyenne atomique T.A.(F), indépendante. Cette échelle participe aux travaux du BIH.

A l'Office National d'Études et de Recherches Aérospatiales (ONERA), Besson et Cumer (1969) ont développé une méthode de synchronisation précise du temps de bases simplement survolées.

France. Sèvres

Une action du Comité International des Poids et Mesures concernant l'échelle de temps physique peut être justifiée par les arguments suivants:

(1) une échelle internationale nécessite un accord entre gouvernements;

(2) le Bureau International de l'Heure s'occupe déjà des échelles atomiques, mais il lui manque l'appui d'un organisme scientifique international qui donnerait un patronage plus officiel à son activité;

(3) les effets relativistes ne seront plus négligeables dans un proche avenir. Il faudra donc, par un accord international, définir un temps coordonné conventionnel pour les besoins des mesures de temps à la surface du globe ou en proximité.

France. Paris. Stoyko

Longitudes et leur variation. – Mme A. Stoyko a fait la discussion des résultats de la détermination des longitudes obtenus pendant l'AGI et la CGI (Troisième Opération des Longitudes Mondiales). Cette discussion a fait l'objet du mémoire (369 pages), publié dans les Annales de l'AGI, contenant les résultats de la détermination de longitudes de 54 stations avec 97 instruments pour la détermination de l'heure.

D'après les résultats des trois Opérations des longitudes mondiales Mme Stoyko a trouvé qu'il existe un rapprochement entre l'Europe et l'Amérique du Nord de l'ordre de $0\overset{s}{.}0005$ par an. M. N. Stoyko a trouvé aussi l'existence de ce rapprochement en étudiant la période de 1871 à 1965. Ce résultat a été à nouveau confirmé par Mme Stoyko et M. Stoyko d'après l'étude des services horaires permanents de 1925 à 1967.

Germany, Braunschweig. Physikalisch-Technische Bundesanstalt (PTB)

A caesium beam standard (CS1), newly developed at the PTB, was put into operation early in 1969. Its principles of construction are: Hexapole magnets, magnetic field homogeneous to within 4×10^{-4} and in parallel to the beam, a new resonator, interaction length 79 cm, the high frequency magnetic field vector in the beam direction.

In June 1966, the second hydrogen maser of the PTB was put into operation.

A contribution to an international comparison of hydrogen masers by means of a flying clock experiment has been published in 1968. Atomic clock research has been done at the PTB since 1967 in a well equipped "Atomuhrenhaus".

Early in 1967, the PTB, taking into account the theory of general relativity, found necessary to define, for different purposes, two separate coordinate system time scales, "Temps terrestre" and "Temps céleste".

G. Becker made an inquiry in the Federal Republic of Germany about the usefulness of the proposed 1-s-S.A.T. system. The opinions received were of the kind: The 1-s-S.A.T. system can be used, but being a "compromise time system" it has some disadvantages for both types of users (A.T. and U.T.). It was considered to be preferable to distribute independently A.T. and U.T.

The daily time of emission of the standard time and frequency transmission DCF 77 (77·5 kHz) has been extended to 16 hours a day since 1st December 1969.

A 24h daily emission time is expected to start 1st April 1970. 1-s time signals of the PTB will then be emitted following the time system S.A.T.

A monthly DCF 77 bulletin distributed to users and published in the *Nachr.-Techn. Zeitschrift* (NTZ) communicates the daily time differences and the phase time differences in 0·1 μs between DCF 77 and A1 (PTB) and as well the fractional daily frequency difference in 10^{-12}. DCF 77 is synchronized to WWVB (NBS) through the BIH. Publications in *PTB-Mitteilungen* 77–79 (1967–69).

Germany, Hamburg. Deutsches Hydrographisches Institut

Determination of U.T. with PZT were continued. The times of reception, versus U.T.2, of time signals were published quarterly.

A Hewlett Packard caesium beam clock was installed at the DHI in August 1968. It has been compared with the atomic time scale of the PTB, Braunschweig, using the Loran-C and television methods.

The radio time signals of the DHI have been synchronized with U.T.C. (BIH) to 1/10th of a millisecond. From October 1967 onward, a time signal of the English type has been transmitted via DAO, 2775 kHz, following the "Onogo" time signals at 0^h and 12^h U.T.

The DHI took part in three flying clock experiments of the U.S. Naval Observatory. Loran-C measurements have been used to keep available to a few microseconds, at the DHI, the U.T.C. of the U.S. Naval Observatory.

Germany. Potsdam. Zentralinstitut Physik der Erde

Yearly, about 250 astronomical time observations have been made with a Danjon prismatic astrolabe and with 2 transit instruments. Their results related to the time systems U.T.C.-Potsdam and U.T.2-Potsdam were published every two months in bulletins together with the results of oscillographic measurements of time signals.

The visual observations of star occultations were continued.

The coordinated time signal DIZ (4525 kHz) is controlled by the time service. By means of television method of clock comparison and using the results of the flying clock comparisons the time system U.T.C.-Potsdam of this time signal could be brought in coincidence within 0·1 ms with the time system T.U.C. of the BIH.

Italy. Milan. Astronomical Observatory

Determination of the ephemeris time by visual occultations has been carried out by means of the Zeiss refractor. Astronomical time determinations have been secured with the transit instrument AP100 and with the Danjon astrolabe; the clock equipment includes a rubidium standard and three quartz oscillators. Comparisons of time and frequencies, and synchronization by HP flying clock in 1967 and by USNO flying clock in 1969, have been realized.

All the results are published in the Circulars of the Observatory. In May 1968 the Observatory has organized a colloquium on time problems, the results of which are published in Proceedings of the Colloquium on the Problems of Time Determination, Keeping and Synchronization (350 pp., 1968).

Italy. Rome. Astronomical Observatory

The present observation program is the same in use at Brera (Milan) Observatory; a new program based on FK4 star pairs symmetrical with respect to the zenith is under development and test. The reference clock is the rubidium standard of the Istituto Superiore Poste e Telecomunicazioni (Rome) which controls the U.T.C. transmitter IAM.

Italy. Turin. Istituto Elettrotecnico Nazionale (IEN)

The present definition of frequency and time at the IEN, is obtained by means of two caesium atomic resonators and eight piezooscillators; frequency and time unit result known with an uncertainty of $\pm 2 \times 10^{-11}$.

The IEN time scale resulting is based on the mean of five oscillator-clock systems completely independent the one from the other.

Check cycles executed by Hewlett Packard with a flying standard in 1966 and 1967, and by U.S. Naval Observatory, showed differences lower than 100 μs over a period of about 18 months. After a cycle of measurements proposed and initiated by Czechoslovakia, the IEN and the ISPT (Rome) time scales are compared weekly by means of TV signals, which secure a resolution better than 0·1 μs.

The examination of the effects of the intrinsic noise of frequency standards is now being studied, with particular reference to those which are reflected on time scale preservation. In order to obtain a better continuity in the synchronization, a new automatic system for the regulation of a piezoelectric standard on an automatic resonator, is being studied.

Japan. Mizusawa

We pursued the astronomical time observations by both PZT and Danjon astrolabe throughout this period. We installed the Caesium Atomic Clock made at the Hewlett Packard Co., in November 1969. Since then, the frequency comparison among the VLF time signal radio waves (WWVL, NSS, GBR) are pursued regularly every day and these results are sent to the BIH.

Japan. Tokyo Astronomical Observatory (TAO)

The astronomical observations for time and latitude were continued with the PZT, using the star system α 59/δ 66.

Two caesium beam tube oscillators made by the Hewlett Packard Co. (C1 and C2) were installed at the end of 1967 and 1968 respectively. The U.T.C. (TAO) has been kept with the master clock (C11) driven by the oscillator C1. A step adjustment of $-100\mu s$ was applied to the master clock at 00^h U.T. on July 1st, 1969, in order to bring the U.T.C. (TAO) in a closer agreement with the U.T.C. of the BIH.

International clock comparisons among U.T.C.'s of the TAO, the RRL, and the USNO were made with flying clocks by the USNO party 5 times in 1968–69.

Reception of Loran-C signal emitted from Iwôjima, master station of the north-west Pacific chain, was begun tentatively in April 1969 in order to connect the U.T.C. (TAO) internationally with high precision.

Radio Research Laboratories (RRL). – Three caesium beam tube oscillators made by the Hewlett Packard Co. were installed in January 1965, May 1967 and September 1968 respectively. The U.T.C. (RRL) has been kept by the master clock driven by the oscillator 2. Loran-C signal emitted from the Iwôjima master station has been received from the beginning of 1969 for connecting the U.T.C. (RRL) internationally.

Rumania. Bucarest. Observatoire Astronomique

On a effectué des séries fixes d'observations de passage, des réceptions des signaux horaires par la méthode oscillographique avec une installation de trois oscillateurs Rohde & Schwarz, des synchronisations avec le Hewlett Packard flying clock en décembre 1967, et comparaison de phase avec fréquence de 17·8 kH de la station NAA. On est en train de terminer l'installation pour la comparaison entre Bucarest et Prague avec la méthode télévision. L'Observatoire rédige un bulletin T.U.C. pour le BIH et pour Mizusawa.

South Africa. Johannesburg

In October 1966 a Hewlett Packard caesium beam frequency standard was installed. Being adjusted to furnish atomic time rather than universal time, it was initially used as a reference standard only, for checking the frequencies from the older quartz crystal oscillators. As from April 1967, however, all ZUO carrier frequencies have been derived directly from the caesium standard, i.e. without offset, while the ZUO time signals and audio frequencies have continued to be derived from an offset frequency, so as to define U.T.C.

Daily VLF phase comparisons have been continued throughout the period, using the signals from GBR, NBA and, since January 1967, Omega Trinidad.

Flying clock comparisons, combined with the results obtained from daily VLF phase comparisons, have made it possible to keep the ZUO signals within about 1/10 ms of U.T.C. for most of the time. As far as the radio signals are concerned, there is at present no particular need for a higher accuracy, and no attempt has therefore been made to maintain the difference to a lower value.

Some tentative studies have been made in connection with the possible installation of a photographic zenith tube in South Africa.

Spain. San Fernando

Time observations are made since April 1968, with the Danjon astrolabe. The study of the 11 fundamental groups of stars used in time and latitude programs is expected to be complete by 1970. Time bulletins (now time and latitude bulletins) are published yearly.

The number of observed occultations of stars by the moon has been increased in the past three years. More than 250 times of occultations were sent to RGO for analysis in the interval September 1967 to September 1969.

Crystal clocks are maintained in synchronization within $\pm 0\overset{s}{.}0001$ with the coordinated scale of time for the needs of astronomical observations. During 1968 propagation delays of time signals used for the above purpose were confirmed by the visit of a BIH portable rubidium clock.

Switzerland. Neuchâtel

A la fin de l'année 1969, l'Observatoire de Neuchâtel possède 3 horloges atomiques dont 2 au césium. Cette batterie d'horloges permet de maintenir un temps atomique avec une uniformité de $\pm 2 \times 10^{-13}$. Le temps est coordonné avec le BIH. Les moyens de comparaison sont: (*a*) le transport d'une des horloges par voiture entre Neuchâtel et les instituts intéressés; (*b*) la mesure de phase des transmissions à ondes très longues (NSS, NAA, NBA, Oméga Trinidad); (*c*) le système Loran-C (chaînes méditérranéenne et norvégienne).

L'exploitation de l'émetteur HBG sur 75 kHz, débutée le 1er Janvier 1966, s'est poursuivie. Les signaux horaires suivent le temps coordonné avec un écart inférieur à 100 μs, la fréquence porteuse est maintenue à sa valeur nominale avec un écart inférieur à 2×10^{-12}.

Les mesures ininterrompues depuis 1956 au moyen de la lunette zénithale photographique Grubb Parsons se sont poursuivies. Les résultats sont communiqués mensuellement au BIH et à l'IPMS et publiés dans un bulletin.

U. K. Royal Greenwich Observatory

Time determinations have continued with the PZT. The observations made in the years 1958–67 were analysed, and from 1968, January, the results were based on revised star co-ordinates. Although the total effect on U.T. throughout the year of using these revised values is zero, there was a discontinuity on January 1st, such that U.T. (Herstmonceux) became earlier 4·5 ms.

The Cape astrolabe ceased operation on April 30, 1969, and it will shortly be returned to Herstmonceux, where it is adopted to convert it to a reflecting astrolabe incorporating a CER-VIT prism.

It was decided to adopt an atomic time scale, G.A.2, with a constant difference from G.A. such that G.A. – G.A.2 = $0\overset{s}{.}957700$, which is in close accord with A.T., the atomic time scale of the BIH. G.A.2 is determined from the mean of the caesium beam frequency standards at the Royal Greenwich Observatory.

The differences between the Earth's rotational time system U.T.2 and the atomic time scale G.A.2 have been published quarterly in the Greenwich Time Reports (formerly Royal Observatory Bulletins).

A new series of monthly circulars commenced in July 1967, containing the daily readings of the relative phase difference between U.T.C. (Herstmonceux) or G.A.2, as appropriate, and the received carriers of selected LF and VLF radio emissions. From April 1969, measurements have been made of Loran-C from Sylt and the results published in the monthly circulars. Co-ordination with the U.S. Naval Observatory has been maintained to a microsecond by means of travelling clocks.

U.K. National Physical Laboratory, Teddington

The long caesium beam continues in operation as the national standard, together with two H.P. caesium beam equipments as working standards. The long standard has been improved by the incorporation of an oven and detector at each end enabling the direction of the beam to be reversed and any error due to phase difference to be eliminated (Essen and Sutcliffe, 1969).

U.S. Naval Observatory

The determination of Universal, Ephemeris and Atomic Times was continued at Washington,

D.C. and Richmond, Florida, utilizing the PZT, Danjon Astrolabe, Dual-Rate Moon-Position Camera, and atomic clocks.

A PZT is being constructed in the Observatory shop to be used in Washington for experimental purposes. A stepping motor will be used to drive the plate carriage and will eliminate the use of contacts for the timing. This PZT will be used in Washington concurrently with the existing PZT No. 3.

A new hydrogen maser has been installed at the Naval Observatory to improve the short and intermediate term uniformity of the Observatory clock time scales. The long term stability of the time scales will still be determined by the means of about 16 caesium beam frequency standards.

The Loran-C chains of the U.S. Coast Guard are controlled in time to 5 micro-seconds and frequency to 1×10^{12}. The carrier frequency of Omega VLF stations are controlled to 1×10^{-11}.

The precise Time Synchronization Service of the U.S. Naval Observatory carries time to one microsecond world-wide and frequency to 1×10^{-12} world-wide. Caesium-beam portable clocks are used. The results of clock comparisons with observatories and standards laboratories are published weekly as part of the *U.S. Naval Observatory Time Service Bulletin*, Series 4.

The difference A.1 − U.T.C. (USNO) is listed together with other time information (U.T.2 − U.T.C., etc.) on *Time Service Bulletin*, Series 7. The formula which is currently (1969) in use is: A.1 − U.T.C. (USNO) $= 7{\cdot}0859497 + (\mathrm{MJD} - 40221) \times 2592\ \mu$s.

U.S.A. Milwaukee. Physics Department

Monitoring of Loran-C, VLF and Omega was continued. The daily probable error in phase of the antipodal VLF station, NWC, Australia, was 2 microseconds. W. Markowitz found, in a differential Loran-C experiment, that the received frequency of radio transmission is independent of distance.

U.S.S.R.

12 Observatoires et laboratoires de l'U.S.S.R. ont effectué les déterminations astronomiques du temps, en employant surtout les instruments photo-électriques de passage et les astrolabes à prisme Danjon.

Les calculs des moments des émissions de signaux horaires dans le système T.U.1 se sont fait d'après le traitement mathématique des données des observations de tous les services horaires de l'U.S.S.R., de deux Services horaires de la République socialiste de Tchècoslovaquie, du Service horaire de la RDA, de la Pologne, de la Bulgarie, de la République démocratique populaire du Corée, qui ont participé au travail du Service horaire de l'U.S.S.R. à titre de libre coopération. Les résultats définitifs ont été publiés dans les bulletins mensuels "L'heure étalon aux moments moyens des émissions des radiosignaux".

Les stations radio de l'U.S.S.R. ont effectué l'émission des signaux horaires dans le système du temps atomique uniforme (T.A.C). On a déterminé les valeurs des décalages de fréquence pour l'approximation du T.U.2 selon les études des mesures de rotation de la Terre, qui se faisaient régulièrement à VNIIFTRI, et compte tenu des récommendations du BIH. Pour les années 1968 et 1969 la valeur du décalage de fréquence adopté est $+300 \times 10^{-10}$.

Pour ne pas avoir la différence excédant 50 ms entre T.A.C et T.U.2 on a effectué le 1er avril 1969 la correction des moments des émissions pour + 90 ms.

Deux instruments photoélectriques de passage de construction originale, munis de l'isolement thermique renforcé et d'autres modifications, sont actuellement en cours de fabrication dans les ateliers de l'Observatoire de Poulkovo. Quelques services horaires travaillent sur le perfectionnement des appareils pour l'enregistrement photoélectrique de passage des étoiles.

Les listes des travaux, concernant à la détermination du temps universel, les mouvements du pôle et la rotation de la Terre, sont publiées régulièrement dans les numéros de décembre du bulletin "L'heure étalon" à partir de 1967. La bibliographie des travaux nationaux accomplis dans ce domaine avant cette date est en cours de préparation pour l'édition.

Yugoslavia, Belgrade

Le Service de l'Heure de l'Observatoire de Belgrade disposait en 1967, 1968 et 1969 d'une batterie de trois horloges à quartz Rohde & Schwarz et se servait d'un oscilloscope cathodique pour la réception de signaux horaires.

On recevait en moyenne 26 émissions de signaux horaires, on déterminait l'heure demi-définitive et on l'envoyait au BIH.

Dès le 1er Janvier 1969, on déterminait les mêmes données pour les besoins de l'Observatoire même et on envoyait chaque semaine T.U.0 – T.U.C. au BIH. Tous les résultats sont publiés périodiquement dans le bulletin de l'Observatoire.

F. ZAGAR
President of the Commission

33. STRUCTURE AND DYNAMICS OF THE GALACTIC SYSTEM (STRUCTURE ET DYNAMIQUE DU SYSTÈME GALACTIQUE)

PRESIDENT: G. Contopoulos.
VICE-PRESIDENT: S. W. McCuskey.
ORGANIZING COMMITTEE: A. Blaauw, B. J. Bok, E. K. Haradze, F. J. Kerr, G. G. Kuzmin, C. C. Lin, M. Schmidt.

I. INTRODUCTION

The present report is drastically shorter and, therefore, different in form from previous reports. While galactic research has increased considerably in recent years, the financial situation in the IAU made it necessary to reduce the size of the reports to half their previous size. This obliged us to adopt an almost telegraphic style in our report. However, an extended version of the report, including also the necessary references, will be published by the University of Thessaloniki and distributed to members of Commission 33. Any other interested astronomer may write to ask for a copy.

The report has been prepared by G. Contopoulos (Sections I, V, VI) and S. McCuskey (Sections II, III, IV). In our work we have been helped by Drs Kerr (radio astronomy) and Haradze (Russian contributions). Dr Elvius has written the Report of the Committee "Selected Areas". We could cover the literature up to the fall of 1969, plus work in progress, reported by members of our commission.

The reports of the following Commissions are also of interest to our Commission: 7 (Dynamics), 24 and 30 (Distances and Motions), 25, 29 and 45 (Photometry and Spectra), 27 (Variable Stars), 28 (Galaxies), 34 (Interstellar Matter), 37 (Clusters and Association), 40 (Radio Astronomy), 43 (Astrophysical Plasmas) and 44 (X-, γ- and cosmic rays, etc.).

Among the main developments in galactic research during the last three-year period we should mention: (1) The extensive application of newly discovered radio lines in the study of the structure and kinematics of our Galaxy. (2) The development of X- and γ-ray astronomy, as well as infrared astronomy. (3) A better understanding of the spiral structure by optical and radio observations. (4) The development of the gravitational theory of spiral structure. Recent developments include suggestions for the origin of spiral structure, non-linear effects, resonances, etc. Many applications of the theory to specific galactic problems have been also made. (5) The development of new methods in stellar dynamics, similar to those of plasma physics.

The most important event in galactic research during this period was probably the IAU Symposium no. 38 on *The Spiral Structure of our Galaxy*. The Proceedings, containing about 85 papers, are edited by W. Becker and G. Contopoulos.

Other meetings connected with galactic research are the IAU Symposia no. 37 (*X- and γ-Ray Astronomy*), and 39 (*Gas Dynamics of the Interstellar Medium*), the IAU Colloquia on the "Gravitational *n*-Body Problem" and on "Problems of Interstellar Dust", the International Conference on Plasma Instabilities in Astrophysics, Pacific Grove, California, the Fourth Texas Symposium on Relativistic Astrophysics, Dallas, Texas, and the Summer Institutes at the State University at Stony Brook, N. York (1968) and Lagonissi, Greece (1969); the Proceedings of the last Institute are edited by L. N. Mavridis.

Among the books that have appeared recently I would like to mention: (1) *Galactic Astronomy*, D. Mihalas, P. M. Routly. W. H. Freeman and Co., San Francisco and London, 1968. (2) *Interstellar Ionized Hydrogen*, Ed. Y. Terzian, W. A. Benjamin, Inc., New York, 1968. (3) *Nebulae and Interstellar Matter*, Vol. 7 of "Stars and Stellar Systems". Univ. Chicago Press, Chicago, 1968. (4) *Annual Reviews of Astronomy and Astrophysics*, Vols. 4, 5 and 6. Ann. Reviews, Inc., Palo Alto, 1967, 1968. (5) *Advances in Astronomy and Astrophysics*, Vol. 6, Academic Press, New York and London, 1968. (6) *Relativity Theory and Astrophysics. 2: Galactic Structure* (Vol. 9,

Lectures in Applied Mathematics), Ed. J. Ehlers, Amer. Math. Soc., Providence, R. Island, 1967. (7) *Determination of Radial Velocities and their Applications*, Proceedings, IAU Symposium no. 30, Eds. A. H. Batten, J. F. Heard, Academic Press, London and New York, 1967. (8) *An Introduction to Stellar Statistics*, R. Kurth, Pergamon Press, Inc., New York, 1967. (9) *Stellar Kinematics*, W. M. Smart, Longmans, Green and Co., London, 1968. (10) *Proceedings of the Tenth International Conference on Cosmic Rays*, Ed. M. D. Wilson, *Canadian J. Phys.*, **46**, no. 10, Parts 2–4, 1968. (11) *Astronomy 1966, Kinematics and Dynamics of the Stellar Systems*, Moscow, 1968 (Review Articles in Russian). (12) *High Energy Astrophysics*, Eds. C. de Witt, E. Schatzman, P. Véron, Gordon and Breach, New York, London, Paris, 1967.

A number of new observational techniques have introduced recently new elements into galactic research. Such are the ultraviolet, X-ray, γ-ray, cosmic-ray, and infrared astronomy, the observation of many new molecular lines, radio or infrared, and the detection of pulsars and gravitational waves.

Cosmic rays are mentioned in part VI (Dynamics). Here we include some information about the distribution of X-ray, γ-ray, infrared, and molecular line sources, as well as about gravitational waves from our Galaxy.

Pulsars show a distribution near the galactic plane. Mills found a concentration of pulsars in the local and Sagittarius arms. Their mean distance from the galactic plane is consistent with that of known supernova remnants. Indications for the concentration of pulsars in spiral arms were also provided by Davidson and Terzian (1969).

X-ray and γ-ray sources

Review papers on X-ray sources were written by Friedman (*Stars and Stellar Systems*, vol. **7**, 685, 1968), and Giacconi, Gursky and van Speybroeck (*Ann. Rev. astr. Astrophys.*, **6**, 373, 1968). A Joint Discussion of Commissions 28, 33, 34 and 44 on X-ray astronomy took place during the Prague Meeting of the IAU. More recent information is contained in the articles by Maran and Cameron (*Earth and extraterr. Sci.*, **1**, 27, 1969; report about the fourth Texas Symposium on relativistic astrophysics) and Mumford (*Sky Telesc.*, **38**, 96, 1969; report from the IAU Symposium no. 37, in Rome).

Evidence is accumulating that the discrete X-ray sources are located in spiral arms and their distribution is like that of novae. O'Dell found a correlation between the positions of O-associations and X-ray sources. Dupree found no correlation of X-ray sources with cepheids or W-stars, but a good correlation with novae and planetary nebulae. Fujimoto *et al.* found a good correlation with the distribution of interstellar gas.

The region of the galactic center emits low-energy X-rays as well as hard X-rays. The diffuse X-ray radiation has probably a galactic component but part of it may be extragalactic.

General articles on the possible sources and mechanisms of γ-radiation were written by Fazio (*Ann. Rev. astr. Astrophys.*, **5**, 481, 1967) and Fichtel (*Sky Telesc.*, **35**, 85, 1968).

γ-radiation from the galactic plane was found by Clark, Garmire and Kraushaar, using a detector flown on the third Orbiting Solar Observatory. The intensity of the radiation has a broad maximum in the direction of the galactic center. Similar results from balloons were found by Fichtel *et al.*, and by Sood. The γ-ray flux observed is more than an order of magnitude larger than expected from the interaction of cosmic rays with interstellar matter, magnetic fields, and photons. Cowsik and Pal and Shen explain this flux as due to an inverse Compton scattering of the recently observed far-infrared radiation by cosmic ray electrons. However O'Connell and Verma point out that this mechanism predicts a very anisotropic X-ray flux, which is not observed. Ögelman considers the γ-radiation to be due to unresolved discrete sources.

The possibility of observing galactic quarks was discussed by Pacini and Salpeter. Observational upper limits for quarks were set by Ashton *et al.*, Fukushima *et al.*, and by Libby and Thomas. McCusker and Cairns found the first evidence for a quark from cosmic rays.

Observations of intense *Ly* α galactic background radiation were made by Venus 2 and 3 (Kurt and Syunyaev), Venus 4 (Kurt and Dostovalov), and Mariner 5 (Barth).

Infrared radiation

A general article on infrared astronomy was written by Neugebauer and Leighton (*Sci. Amer.*, **219**, no. 2, 51). A large number of infrared sources have been found in recent years. Hughs gives the distribution of 5614 infrared stars. The observed concentration towards the galactic plane indicates a mixed population of giants and supergiants.

A discussion on infrared astronomy, with many participants, was organized by Massey and Ring (*Phil. Trans. R. Soc. Lond.*, A **264**, 107–304). Much work on infrared stars has been done by groups at Caltec, Berkeley, Tucson, Houston, etc.

Besides infrared stars, infrared nebulae were also found. The infrared radiation from the galactic nucleus was studied extensively by Becklin and Neugebauer and by Low, Forbes and Aumann, and Hoffmann and Frederick. A strong source of dimensions of the order of 1pc appears in this direction.

There is an intense far infrared radiation which is probably galactic. The observations now cover all the range up to the radio region. Observations of the galactic center in the short microwave region were discussed by Foster. Some problems of submillimeter astronomy were reviewed by Salomonovic. Interstellar dust may radiate significantly in millimeter wavelengths (Litvak).

Interstellar molecules

A general review article on interstellar molecules was written by McNally (*Adv. Astr. Astrophys.*, **6**, 173, 1968). The distribution of OH is similar to that of HI (Goss). The association of OH emission sources with various galactic objects (clusters, nebulae, or special types of stars) was considered by Menon and by Turner. Results from the distribution and kinematics of OH sources begin to add valuable information to our understanding of the Galaxy.

Indications for the existence of H_2 are provided by the observed decrease of H in dark clouds, where presumably the total gas density increases. The formation of H_2 molecules in grains was discussed by Wentzel and Nishimura. Stecher and Williams calculated the amount of H_2 ejected from M giants.

Recent observations of water vapor in several regions of the Galaxy were reported by Cheung *et al.*, Knowles *et al.* and Meeks *et al.* Ammonia was also observed near the galactic center (Cheung *et al.*). The possibility of observing many other molecules has been discussed by various authors, and preliminary results have been reported.

Of special interest are the recent observations of formaldehyde (Snyder *et al.*, Buhl *et al.*, Palmer *et al.*, Zuckerman *et al.*, Whiteoak and Gardner), because of its wide distribution in the Galaxy.

Gravitational waves

The discovery by Weber of a rather large flux of gravitational waves reaching the earth (10^4 erg cm^{-2}s^{-1}) gave rise to speculations about their origin. There is a possibility that they come from the center of our Galaxy, which may contain a Schwarzschild "hole". At the same time considerable theoretical advance has been made in understanding by behavior of masses emitting gravitational radiation (Chandrasekhar). Lynden-Bell has suggested that the galactic nucleus may be a collapsed old quasar.

II. BASIC DATA AND CALIBRATION PROBLEMS

(S. McCuskey)

A. *Basic data*

No attempt will be made here to summarize completely the data on spectral types, colors, proper motions radial velocities, H I and H II observations, polarization, etc. which enter into the varied problems of galactic structure. But some of the investigations completed, or now under way, should be mentioned.

Apriamaschvili *et al.* at Abastumani have published catalogues of stellar magnitudes, color indices, spectral and luminosity classes for several of the areas embraced in the Parenago Plan for the study of galactic structure.

Particular interest centers around the high luminosity objects (OB stars, WR stars, etc.) at low galactic latitudes in the southern hemisphere. Lodén reports that the Stockholm Observatory Survey of the Southern Milky Way continues and a catalogue of very early and very late type stars brighter than magnitude 13 is in preparation. Stephenson and Sanduleak at the Warner and Swasey Observatory are completing a survey of OB stars along the galactic belt $b^{II} = \pm 6°$ in the southern Milky Way. This survey is an extension of the earlier work carried out in collaboration with the Hamburg Observatory and published in *Luminous Stars in the Northern Milky Way I–VI.*

A revised spectral classification system for the Wolf-Rayet stars and a new catalogue of these stars have been published by L. F. Smith. New criteria for classifying WN and WC stars, have been applied to southern WR stars. The catalogue containing 107 stars, of which 13 are new, gives intrinsic colors and color excesses in the *UBV* system as well as the revised classification for these objects.

While basic data for studies of variable stars is properly the concern of Commission 27 the work being done at the Royal Greenwich Observatory on statistical parallaxes for RR Lyrae stars should be mentioned here.

Two catalogues of considerable significance for investigations of the interstellar reddening in connection with problems of galactic structure have appeared during the triennium under review. The first contains photoelectric color excess data in the *UBV* system for 4697 stars and clusters by Th. Neckel. The second, by Blanco *et al.* contains *UBV* or U_cBV photoelectric magnitudes and colors for some 20000 stars, together with references, spectral classes where known, and coordinates.

In addition to the above a new photometric catalogue containing photoelectric data for some 25000 stars with 1770 bibliographical references to all kinds of photoelectric measures is under preparation at Observatorio Astronomico, La Plata, Argentina.

In connection with the studies of proper motion the listing of MK spectral and luminosity classes, *UBV* photometry and other data for 1849 stars with $\mu > 0''.5$ per year by Cowley *et al.* should be most useful. This catalogue contains data for all such stars north of $\delta = -23°$ and brighter than 10th magnitude.

Eggen has provided a catalogue of luminosities, colors and motions for 1008 giant K and M stars, with $V < 5{\cdot}5$ which should be valuable for studies of the region within 1 kpc of the sun.

Many topics concerning the role played by radial velocity measures in problems of galactic structure are described in the report of IAU Symposium no. 30, *Determination of Radial Velocities and Their Applications*, 1967.

B. *Intrinsic colors and interstellar reddening*

New determinations of intrinsic colors for various spectral groups have appeared since the last Commission 33 Report. Among them are determinations by Herr of the MK spectral classes corresponding to the OB^+, OB°, OB^- "natural groups" and hence the values of $(B-V)_0$ for these. The groups have been separated into "non-emission" and "emission" objects. Intrinsic colors do not differ significantly between these groups. By statistical methods the intrinsic colors for BOVe–B2Ve stars have been found by Rehpenning to be $(B-V)_0 = -0{\cdot}23 \pm 0{\cdot}09$ (s.d.). This value agrees well with that found by Herr for the OB-stars.

The catalogues of *UBV* photoelectric data by Neckel and by Blanco *et al.* have been used by Neckel and by FitzGerald to provide rather detailed exhibits of the run of color excess E_{B-V} with distance out to 4 or 4·5 kpc from the sun. This material is of considerable value in detailed planning of observational programs for research on galactic structure.

The known data on interstellar absorption along the northern part of the galactic equator $b^{II} = 0$ to $\pm 20°$ is being studied by E. P. Polishchuk with a view to determining the distribution of absorbing clouds in and between the spiral features of the galaxy within 3 kpc of the sun.

In concluding this section we mention the new intermediate band photometric system introduced by McClure and van den Bergh, which yields equivalent spectral type and luminosity data, as well as the interstellar reddening. The new filters supplement the standard *UBV* system. It is a five-color system. The authors present data for stars, clusters and galaxies to illustrate the advantages of the system. And the multicolor photometry (*uvby*) coupled with the Hβ index described by Crawford and now extensively practiced, should contribute substantially to the determination of intrinsic colors and luminosities so necessary for progress in unravelling particularly the spiral galactic structure.

C. *Absolute magnitudes*

Basic to the calibration of absolute magnitudes is the distance modulus of the Hyades cluster. Following the suggestion by Hodge and Wallerstein in 1966 that the distance modulus of the Hyades should be corrected by $+0{\cdot}39$ magnitude critical discussions of the problem by many investigators indicated that a correction, if any, should not exceed $+0{\cdot}2$ magnitude. Most recently an extensive investigation by van Altena using new proper motions for 700 stars in the Hyades region together with photoelectric data (*UBV*) yielded a mean correction of $+0{\cdot}20 \pm 0{\cdot}05$ mag (p.e.) to the distance modulus of the cluster. Van Altena derives a modulus, $m - M = +3{\cdot}23 \pm 0{\cdot}05$ magnitude for the cluster.

Petrie and Lee have published a list of 571 spectroscopically determined absolute magnitudes for O9 to B9 stars. The Hγ equivalent width method at the Dominion Astrophysical Observatory was used. Mean M_V and dispersion σ_v as a function of spectral type are given.

Herr, from his correlation of OB star groups with MK spectral classes, derives mean absolute visual magnitudes for non-emission OB^+, OB° and OB^- stars of $-5{\cdot}9$, $-4{\cdot}8$ and $-3{\cdot}6$ respectively. For the emission stars of the same groups Herr finds $M_v = -5{\cdot}1$, $-5{\cdot}0$ and $-4{\cdot}4$ respectively. The last value does not differ significantly from the value $M_v = -4{\cdot}1$ for the BOVe – B2Ve found by Rehpenning.

A new calibration of M_v as a function of spectral class has been derived from 74 clusters by FitzGerald for luminosity classes, III, IV and V. In addition he lists M_v for the entire range of spectrum-luminosity groups.

A determination of the absolute magnitudes of A stars from data on trigonometric parallaxes and proper motions is underway at LaPlata by C.O.R. and M. Jaschek, A. and C. Cowley. At the same time Jaschek reports that the Am stars are under investigation by him and S. Rocha. This is a catalogue of about 1700 B9 to A9 stars north of $\delta = -21°$ which are listed in the original *Bright Star Catalogue* by Schlesinger. MK spectral types and *UBV* photometry are given.

Absolute magnitudes, as well as intrinsic colors, for the Wolf-Rayet stars have been derived by Smith, through five color narrow band photometry of 43 WR stars in the Magellanic Clouds and 77 WR stars in the Galaxy. She finds for WN3 to WN8, M_v ranges from $-4{\cdot}5$ to $-6{\cdot}2$; for WC5 to WC9, M_v ranges between $-4{\cdot}4$ and $-6{\cdot}2$.

Mention should be made of the work by Gordon on the absolute magnitudes of carbon stars.

Jõeveer *et al.* and Jõeveer have studied the kinematic properties of the K0 – K5 stars with $V < 6{\cdot}5$ magnitude in a galactic belt $b^{II} = 0° \pm 6°$, and have derived luminosity distributions for three magnitude groups, $V_1 \leqslant 5{\cdot}5$, $V_2 = 5{\cdot}5$ to $6{\cdot}0$ and $V_3 = 6{\cdot}0$ to $6{\cdot}5$.

Calibration of the 4-color (UVBY) photoelectric photometry and Hβ photometry in terms of intrinsic colors and absolute magnitudes has made rapid progress at the Kitt Peak National Observatory. In particular calibrations for the B, A and F type stars are nearly finished. D. L. Crawford reports that a relation between the β index and absolute magnitudes M_v seems to be solidly established. Observational errors in the β index of $\pm 0{\cdot}01$ magnitude would yield an error in M_v from this source alone ranging from $0{\cdot}2$ magnitude for late B stars to $0{\cdot}7$ magnitude for early B supergiants.

D. *The luminosity function*

New investigations of the distribution of absolute magnitudes for the general stellar population near the sun continue to appear. Perhaps of greatest importance is that by Luyten who has determined from his large-scale proper motion survey a new general luminosity function. He finds a maximum at $M_{pg} = +15{\cdot}7$ for this new function, a value 1·2 magnitude larger than that derived in 1938. For $+3 < M_v < +12$ the 1968 luminosity function does not differ significantly from the earlier one. Luyten finds 0·064 solar masses per pc^3 for the space density of stars for stars nearer than 10 pcs.

Differential methods for determining $\log \phi(M)$ by summing the luminosity functions for spectral groups have been applied by Drilling for a region in Norma ($l^{II} = 330°$, $b^{II} = -2°$) and by McCuskey for a region at the galactic anticenter ($l^{II} = 186°$, $b^{II} = +1°$). In both instances the resulting $\log \phi(M)$ for $-1 < M_v < +3$, the range over which the data is most complete, does not differ greatly from the similar functions for the northern sky previously published.

An extensive study of the proper motions for stars brighter than $m = 15$ in a 1 sq deg area near the cluster M67 by Murray has yielded a luminosity function for $+3 < M_v < +10$. The run of $\log \phi(M)$ agrees reasonably well with the van Rhijn standard function for the region near the sun.

The methods used for determining the general luminosity function are being examined critically at LaPlata by C. Jaschek in collaboration with J. C. Muzzio. Their findings thus far indicate that the maximum in $\log \phi(M)$ depends very critically on the limits in m and μ imposed by the data.

III. LOCAL GALACTIC STRUCTURE
(S. McCuskey)

A. *General surveys*

A summary of photometric techniques employed at Basel for investigating the spatial distribution of stars in the environs of the sun has been published by W. Becker. This sets forth references to work done by Becker and his colleagues for regions in the galactic plane and in the galactic halo.

Currently, W. Becker reports that measurements in the *RGU* system and partially in the *UBV* system are being continued. New plates were taken at Mt. Palomar with the 48-inch Schmidt-telescope and at Asiago for other fields in the galactic disk and halo. Density gradients now exist for six fields in high or intermediate galactic latitude and for six fields in the galactic disk.

With the exception of S.A. 94 and 107, where faint photoelectric measures are still missing, the halo program with the 48-inch Schmidt is on its way to being finished and extended by use of plates taken with the 200-inch Palomar telescope. Space density gradients for S.A. 51 and 57 to distances of 15 and 12 kpc respectively will ensue.

Several investigations concerning the overall distribution of OB stars along the galactic equatorial belt have appeared. Sim has summarized by charts and histograms the distribution of OB stars from the catalogues *Luminous Stars in the Northern Milky Way, I–VI*. She finds an abrupt discontinuity in the frequency of OB stars at $l^{II} = 140°$. A comparison of the OB star distribution with that of the Wolf-Rayet stars shows the same gap in the galactic distribution of the latter group. This dearth of WR stars in the range $140° < l^{II} < 220°$ has been known for some time.

The space distribution of the OB stars in the northern Milky Way (data from the *Hamburg-Warner and Swasey Observatory Catalogues*) and in the southern Milky Way (data from the Heidelberg survey) have been exhibited by Klare and Neckel. The OB^+ and the OB° stars appear to be concentrated markedly in parts of three spiral arms. But the OB^- stars do not exhibit such a general trend. The half-thickness of the stellar system defined by the OB stars was found to be 50 pc, while that of the general interstellar absorbing layer was found to be 30 pc. The OB star complex as presently observed extends to about 4 kpc from the sun.

Distances for individual Wolf-Rayet stars have been determined by Smith and have permitted her to outline their role in the general picture of galactic structure around the sun to a distance of some 6–8 kpc. While there is some evidence for concentrations of these objects in the recognized spiral features in Cygnus, in Sagittarius and in Carina, there is also a considerable scatter of WR stars between the sun and the inner parts of the Galaxy. All WC9 stars are within 7 kpc of the galactic center; all WN6 and WC7 stars are within 9 kpc of the galactic center. Other types are more or less scattered over the galactic plane. Smith concludes that, in general, the WR stars are concentrated along the edges of the spiral structure, as outlined by H I and OB stars, toward the center of the Galaxy. There are no WR stars between $l^{II} = 140°$ and $l^{II} = 220°$ at distances from the galactic center, $R > 10$ kpc.

Hidajat and Blanco have investigated the space density of giant M stars detected by infrared techniques in Groningen-Palomar variable star fields 1, 2, and 4.

Classification of M stars with a view to obtaining the run of space density with distance in three southern Milky Way fields is underway by Pik-Sin The, in a collaborative program between the Warner and Swasey Observatory and the Bosscha Observatory at Lembang, Indonesia. These fields are LF 13, 14 and 15 (*Trans. IAU*, **XIIIA**, p. 687, 1967) at $l^{II} = 281°$, 298° and 330° respectively. The numbers of M-type stars in these $3° \times 3°$ fields are 301, 459, 1007; the increase in number of stars toward the galactic center is obvious. The magnitude limit of the survey is about $I = 13{\cdot}5$.

Mavridis has discussed in considerable detail the distribution of the M stars, the S stars and the carbon stars in the Galaxy. He has now completed a study of the space distribution of these types in an area centered on the Galactic cluster NGC752, ($l^{II} = 137°$, $b^{II} = -23.°4$). Similar studies are underway in two areas centered on the clusters NGC7789 ($l^{II} = 116°$; $b^{II} = -5.°4$) and NGC129 ($l^{II} = 120°$; $b^{II} = -2.°5$) by A. Tsioumis and C. Poulakos respectively.

The detection, cataloging and charting of carbon stars continues. B. Westerlund reports that positions have been measured and infrared and visual magnitudes have been estimated for about 1200 carbon stars within $\pm 5°$ from the galactic equator along the southern Milky Way. The limiting infrared magnitude is about $I = 12{\cdot}5$. For the northern sky C. B. Stephenson at the Warner and Swasey Observatory has compiled a catalogue of carbon stars, classified into R and N groups. The limiting B magnitude varies between 12 and 14 depending on the type of star. This is being extended to the southern Milky Way as a corollary of the OB star survey reported above.

Stars and other objects showing Hα in emission have been the subject of search, cataloguing and statistical study by Wray. The survey covers a band in galactic latitude 20° wide along the galactic equator from $l^{II} = 240°$ through 0° to 10°; V magnitudes (approximate) are given. This catalogue combined with all previous listings of these objects has now been incorporated in an extensive work by L. Wackerling of Northwestern University containing data for some 5000 Hα emission stars and other objects.

Additions to the listing of Hα objects, principally faint stars and planetary nebulae, are being made by The.

In the process of completing Parenago's complex plan for investigation of the Milky Way, the distribution of hot stars and absorbing clouds in Taurus is being studied in detail by N. B. Kalandadze.

Gould's belt continues to be of interest. Clube has shown that a group of OB stars 500 pc long and 200 pc wide with its axis toward $l^{II} = 160°$ can be considered a satisfactory model for this stellar aggregate. He concludes that the group moves as a kinematic unit around the galactic center, its internal intrinsic velocity dispersion being about 3 kms^{-1} and as a unit it is identifiable with the Pleiades group.

The distribution of stars and absorbing matter in the local region around the sun is being studied by A. D. Chuadze at Abastumani on the basis of colorimetry and two-dimensional spectral classification for some 5000 stars in Selected Areas 25, 48, 50, 73, 75, 97, 99 and 122.

Among the many contributions made by Eggen we mention here two. Luminosities, colors, motions, and the distribution of some 500 faint blue stars and 150 possible white dwarfs of late-type have been studied by Eggen. He finds the space density near the sun to be 0·01–0·02 stars per pc^3. Eggen also has observed many of the M-dwarf stars detected years ago by Vyssotsky. A catalogue of 472 objects, observed in the *UBV* and *RI* systems, is given. This contains colors, proper motions, radial velocities and space velocity data. The space motions seem to indicate that the stars are about evenly divided between young and old disk population and that the list probably includes at most a dozen members of the halo population.

Klare and Schaifers have identified from objective prism spectra in the near infrared (λ6800–8800 Å) some 1571 late-type stars in a field at $l^{II} = 175°$, $b^{II} = +41°$ for which proper motion plates were also available. By blink methods the stars of measurable proper motion among those detected spectroscopically were sorted out as being probable main-sequence stars. These have a space density distribution given by $D(z) = 46 \exp. [-z/185]$ stars per 10^6 pc^3, where z is the distance in parsecs perpendicular to the galactic plane.

The study of population indices at the Astronomical Observatory of N. Copernicus University at Torun by W. Iwanowska, C. Iwaniszewska and others continues.

In a study of a limited region of the sky, which may or may not be typical, Heiles has shown that most of the 21-cm emission comes from sheets of hydrogen, and only a small proportion from discrete clouds. Several groups have concluded from hydrogen absorption observations that the gas is partly in cool dense clouds and partly in a hot tenuous medium. The two forms seem to be intermingled, however, so that the assumption of a uniform mean temperature is still a reasonable approximation for large-scale structure work.

The most prominent local features in the radio continuum distribution are the galactic spurs. Supernova explanations for the spurs now seem very unlikely (Seaquist), and several workers have discussed a relationship with the local magnetic field configuration. Bingham has estimated, from comparisons with optical polarization data, that the effective distance of the north polar spur is about 100 pc.

Ellis and Hamilton, and Bridle have studied the ionized hydrogen in the solar vicinity through studies of low-frequency absorption. The column density of ionized hydrogen in various directions has been measured by several groups through dispersion observations on pulsars, the pulsating radio sources which are believed to be neutron stars.

B. *Regional surveys – low galactic latitudes*

1. *The galactic center* ($350° < l^{II} < 10°$)

An extensive study by Ardeberg of an area in Scorpius ($l^{II} = 353°$, $b^{II} = +3°$) is underway at the European Southern Observatory. Photoelectric *UBV* photometry in a six sq. deg. area for all early-type stars to a limiting magnitude $B = 12{\cdot}8$ is aimed at an evaluation of the space distribution of these stars and the question of their connection with possible spiral structure.

In one rather homogeneous part of the field the space distribution of all stars by means of spectral types determined from objective prism spectra, and *UBV* photometry is contemplated. A study of the radial velocity distribution of the A stars and the B stars is planned.

Cailliatte has made an extensive study of the region around the globular cluster OHP 1 ($l^{II} = 357°$, $b^{II} = +2{.}^{\circ}1$) in star cloud B in Sagittarius. In the distance interval 1–3 kpc from the sun there is an infrared absorption of 2·2–2·3 magnitude (blue 5·5–5·8) with several absorbing clouds at large distances, each absorbing about 1 magnitude. Red giant stars of class M have a peak frequency at $m_{ir} = 14{\cdot}5$–15. From the study Cailliatte deduces a distance of 9 to 10 kpc to the center of the galaxy.

A re-examination of the photometric-spectral data in a field at $l^{II} = 0°$, $b^{II} = 0°$, at the Warner and Swasey Observatory, previously discussed (McCuskey), has strengthened the space density analysis in this region toward the galactic center. The original photometry and spectral classifications were made by K. Purbosisswojo. New photoelectric photometry for the area was undertaken by P. A. Wehinger and B. Hidajat to strengthen the magnitude sequences used in the original study. For nearly all groups of stars, the space density decreases in the first 1·5 kpc from the sun. For the G8–K3 giant stars, an increase in density appears to set in beyond that point. The interstellar absorption is very spotty and heavy in this region.

2. *$10° < l^{II} < 60°$ (Sct, Aql)*

Two studies by Kuznetsov of the galactic structure in Aquila have appeared. In the first, an area at $l^{II} = 52°$, $b^{II} = 0°$, spectral types, photographic magnitudes and color excesses for five sub-areas have revealed that the B8–A5 stars appear to be concentrated at distances of 100–300 pc from the sun, and that no O–B2 stars are present. The interstellar absorption over the area is quite variable; in the darkest clouds $A_{pg} \sim 2$ magnitude/kpc.

The second paper by Kuznetsov displays the interstellar absorption characteristics, A_{pg}, over the galactic plane for $0° < l^{II} < 54°$ to a distance of 5 kpc. A series of histograms depicting the space densities of the B8–A0 and the A1–A5 stars as functions of galactic longitude indicate that fluctuations in these densities can be ascribed almost completely to variations in A_{pg}. Kuznetsov concludes that there is a concentration of dust between the Sagittarius and Carina-Cygnus spiral arms of the Galaxy.

Ardeberg reports that, in collaboration with S. Wramdemark, he has a region at $l^{II} = 36°$, $b^{II} = +8°$ under extensive study.

The spectrophotometric survey of stars along the Milky Way, initiated by Malmquist, Schalén, and Westerlund at Uppsala, continues. The objective is to include all stars brighter than $B = 10$ in a galactic belt $b^{II} = 0°$ to $\pm 3°$. The survey will exhibit the overall stellar distribution and that of interstellar absorbing material, and will serve for the selection of areas for future studies of the distribution of faint stars. The results for $43° < l^{II} < 73°$ are now almost complete, according to T. Oja, and plates for $133° < l^{II} < 153°$ have been taken.

Fedorchenko has made a comprehensive study of the galactic region $44° < l^{II} < 54°$. Early B stars appear to be scarce for $r < 1{\cdot}7$ kpc. For A stars the space density is rather uniform for $r < 1$ kpc but there is a general decrease at greater distances. The concentrations of A stars and of late type giant stars appear to coincide in distance with the concentrations of interstellar dust but there is no correlation with the concentrations of H I. Fedorchenko puts $l^{II} = 37°$ as a northerly limit for the Sagittarius spiral arm, whereas others indicate $l^{II} = 22°$ for this limit.

Two investigations closely allied with that by Fedorchenko are those by Voroshilov and by Voroshilov and Polishchuk.

3. *$60° < l^{II} < 150°$ (Cyg, Cep, Cas, Per)*

The spectrophotometric survey carried on at Uppsala, reported by T. Oja, and indicated in the previous section of this report, extends into the galactic longitude range embraced here. The investigation for $133° < l^{II} < 153°$ will be particularly of interest because of the spiral galactic structure (local and Perseus arms) and of the peculiar discontinuity in the populations of OB stars, and WR stars which occurs at $l^{II} = 140°$.

Observational and statistical investigations of galactic structure underway in Russia are mainly concerned with the spiral nature of the Galaxy within distances of 5 to 8 kpc of the sun, as well as with the relationship between diffuse matter and stars of various spectral types. Dr Haradze reports that these studies are based upon photometric work in the surroundings of seven galactic clusters (the Main Astronomical Observatory of the Ukrainian Academy of Science in Gollosseyevo) and upon spectral classification in these and other areas (Abastumani Observatory). T. A. Uranova and G. A. Starikova have made an extensive study of color excesses and interstellar absorption in the Rift region of the Milky Way in Aquila and Cygnus. At 50–250 pc from the sun there appears to be a dense front of interstellar material where the rate of absorption is 7 magnitude/kpc. The dust appears to border the spiral arm in Cygnus on the side toward the galactic center. A statistical analysis indicates that the fluctuations in surface brightness of the Milky Way in this region can be explained generally by interstellar dust clouds within 400–500 pc of the sun.

Among the other current activities of the Russian astronomers reported by E. K. Haradze are the investigations by L. N. Kolesnik and G. L. Fedorchenko of the distributions of stars and interstellar matter out to distances of 3·5 to 5 kpc towards NGC 6913 ($l^{II} = 77°$), NGC 6823 ($l^{II} = 59°$), NGC 7086 ($l^{II} = 94°$).

At the Warner and Swasey Observatory, an intensive study of the Cygnus-Cepheus region $70° < l^{II} < 100°$ is being undertaken by M. R. Chartrand. The general stellar population and its relationship to the spiral structure is under study by means of spectral type distributions and *UBV* photometry.

A region at $l^{II} = 106°$, $b^{II} = +0\overset{\circ}{.}1$ in Cepheus has been studied by Masnauskas. Evaluation of data on colors and spectral types yields values of interstellar absorption in 8 subregions. The absorption at 1 kpc varies from 0·5 magnitude to 3 magnitude. This is relatively near a region at $l^{II} = 103°$, $b^{II} = +1°$ now being investigated for stellar space density at the Warner and Swasey Observatory. And it is close to the field at $l^{II} = 104°$, $b^{II} = +4°$ in which A. Ardeberg (ESO) reports that he, in collaboration with K. Särg, has observed photoelectrically in the *UBV* system some 40 stars of OB type. These authors also plan to undertake spectrophotometric observations of the same stars. It is a complex region bordering the association Cep OB2.

Several investigations of galactic structure in Cassiopeia may be mentioned. Oja has summarized 7 papers concerning the stellar and interstellar distributions at $l^{II} = 128°$, $b^{II} = -2°$. This is a region rich in galactic clusters and somewhat heterogeneous as regards the surface density of stars. The interstellar matter appears to be concentrated in clouds of the order of 40 pc in diameter. Most are within 1 kpc of the sun, but a few are at distances as great as 3 kpc. The implication is that these dark clouds are connected with the local and the Perseus spiral arms. Some concentration of the more luminous stars in the Perseus spiral arm is evident, but no spiral structure beyond this is apparent. This conforms to the results of a recent study of the OB stars brighter than B = 11·5 in LF5 ($l^{II} = 129°$, $b^{II} = -2°$) from MK spectral classes and photoelectric photometry (*UBV*) by G. A. Snyder at the Warner and Swasey and the Kitt Peak National Observatories. No evidence of any concentration in space of the OB stars beyond the Perseus spiral arm was found.

The space density functions for B0 – B2 stars and A0 stars in a 45 sq. deg. region at $117° < l^{II} < 124°$ have been derived by Grigoreva.

Georgelin has determined objective prism radial velocities and *UBV* photoelectric magnitude and color data for some 150 stars in Cassiopeia at $l^{II} = 112°$, $b^{II} = +0\overset{\circ}{.}3$. These stars range in brightness from $m_{pg} = 7$ to 12, and in MK spectral type from early B to late K, and are of all luminosity classes. Distances were determined photometrically. A frequency plot of radial velocity for the BAF stars as a group shows considerable asymmetry, in the sense of a predominance of velocities between –40 and –70 km s^{-1} over that predicted by a gaussian function which best fits the maximum and the positive side of the frequency distribution. This may imply that some of the stars are distant enough to show the effects of galactic rotation appropriate to the Perseus spiral arm. The distribution of Ia, Ib, and II luminosity class stars in distance shows three maxima which coincide well with the concentrations of H I in this region. These are the local spiral arm, the Perseus arm complex from 2 to 4 kpc and another maximum at 7–8 kpc. The latter coincides with the "outer spiral arm" suggested by the radio astronomers. A very definite gap exists between 5 and 7 kpc both in the stellar and in the H I distributions. A second feature indicated by the radial velocity data, corrected for solar motion, is the considerably less rapid decrease in radial velocity with distance than is predicted by the circular rotation model.

Kharadze reports a study by V. I. Voroshilov of the region in the direction of NGC 7654 ($l^{II} = 113°$, $b^{II} = +0\overset{\circ}{.}5$) where, at distances between 8 and 9 kpc, the presence of the stellar population in an outer spiral arm of the Galaxy has been detected. Whether this is truly an outer arm, or has some connection with the Perseus spiral arm complex is being investigated.

A photometric investigation (*UBV*) by the Becker method of a region at ($l^{II} = 143°$, $b^{II} = +3°$) has been published by Karimie.

This entire region near $l^{II} = 140°$ is of considerable interest because of the discontinuities there in the galactic longitude distribution of several elements of galactic structure. Brück *et al.* and Reddish have discussed this feature.

An extensive survey by Houk and McCuskey covering the entire galactic belt $b^{II} = 0°$ to $\pm 3°$, and some regions at higher galactic latitude, for detecting B8–A3 stars to $V = 13$ magnitude is underway at the Warner and Swasey Observatory. Thus far, the ranges $55° < l^{II} < 75°$, an interarm region, and $95° < l^{II} < 150°$, a spiral arm section of the Milky Way, have been finished. Counts of A stars indicate sizable concentrations at $l^{II} = 130°$ to 140° above the galactic plane and at $l^{II} = 98°$ to 108° below the galactic plane which cannot be accounted for by fluctuations in the interstellar absorption. Most of the variations in surface density of these stars in the interarm region seem to be due to fluctuations in the absorption there. Studies of space density are underway.

4. *The anticenter region* ($150° < l^{II} < 210°$)

Karlsson has been investigating the stellar space distribution for a region in Monoceros at $l^{II} = 203°$, $b^{II} = +2°$. The Becker *RGU* photometric method of analysis has been applied by Yilmaz to a field of 0·31 sq. deg. at $l^{II} = 179°$, $b^{II} = +1{:}2$.

Two investigations of a field at $l^{II} = 186°$, $b^{II} = +1°$ near the galactic anticenter have been completed at the Warner and Swasey Observatory. The first (McCuskey) gives spectral data, *V* magnitude and color data (*UBV*) for 3621 stars to a limiting $V = 12{\cdot}5$ for an area of 18·5 sq. deg. Interstellar absorption in the area increases linearly to $A_V = 1{\cdot}8$ magnitude at 2 kpc and much more slowly beyond that distance. The space density of OB stars decreases monotonically outward from the region of the sun. There is no trace of a continuation of the Perseus spiral arm into this anticenter region. The second paper, also by McCuskey, evaluates the distribution of gM stars in an 8 sq. deg. part of the region discussed above.

The analysis of the anticenter region at $l^{II} = 186°$, $b^{II} = +1°$, together with similar data for the galactic center, permits the determination of a cross-sectional view of the run of space density *from* 2 kpc in the anticenter direction, through the sun, *to* 2 kpc in the galactic center direction. The results will be published in the report of IAU Symposium no. 38, Basel, 1970.

5. $210° < l^{II} < 300°$ *(Pup, Vel, Car, Cru)*

Extensive and intensive studies of the southern Milky Way are underway at present. The development of first class observational facilities in Australia, in Chile, and in other southern hemisphere locations promises to remove the longstanding slow rate of growth of our knowledge of this part of the Galaxy.

An outstanding example of the way in which optical data on the distributions of OB stars, H II regions, dust clouds, young clusters, long-period cepheids, WR stars etc. can be combined with radio data to provide a comprehensive view of galactic structure has been given by Bok *et al.* This is a progress report prepared for the Basel Symposium (IAU no. 38) on *Spiral Structure of Our Galaxy*.

Velghe and Denoyelle are studying this heavily obscured region in the range $260° < l^{II} < 275°$. The absence of OB giants and supergiants between 265° and 275° seems quite definitely established. The work is being extended to $l^{II} = 284°$.

On the inner edge of the Carina complex, $290° < l^{II} < 311°$, Lyngå has observed some 284 faint OB stars and has classified 19 of them on the MK system. *UBV* photometry in $298° < l^{II} < 306°$ reveals a lane of low interstellar absorption. Here $A_V \sim 2$ magnitude at 3 kpc and there appears to be no appreciable absorption as far as 6 kpc (Lyngå).

Investigations of many Milky Way fields in the southern sky are underway at the Stockholm Observatory. A descriptive summary of the type of the present spectrophotometric work is given by Lodén. A catalogue of very early and very late type stars brighter than $V = 13$ is in preparation by Lodén and Miss Sundman. An investigation of the surroundings of η Carinae is being undertaken by Miss B. Nordström.

One of the regions being studied in depth at the Warner and Swasey Observatory is LF 13 at $l^{II} = 281°$, $b^{II} = +3{:}9$. A good photoelectric sequence (*UBV*) for stars to $V = 13{\cdot}6$ magnitude has been established in the area. Spectral classification is now being done on plates taken at the Bosscha Observatory, Lembang, Indonesia and at Cerro Tololo Interamerican Observatory in Chile. The observations and analysis are being done by W. H. Wooden II.

6. *300° < l^{II} < 350° (Cen, Cir, Nor, Sco)*

As a part of the investigations of galactic structure at the Warner and Swasey Observatory the region of LF 15 (l^{II} = 330°, b^{II} = −2°) has been studied by Drilling.

Two studies by Westerlund of the OB stars near the supernova remnants RCW 86 and RCW 103 indicate the coexistence in space of these emission nebulae and the hot stars. Westerlund's observations indicate distances for the OB associations in this region consistent with two spiral arms of the Galaxy. From I Ara at l^{II} = 337° to OB I Nor at l^{II} = 328° the Sagittarius arm becomes evident, the distance from the sun increasing from 1·4 to 2·5 kpc. The Norma-Scutum arm is identified by the Ara grouping at 3·5 kpc, l^{II} = 337° and the RCW 103 association and Norma grouping at 3·9 kpc, l^{II} = 332°. This arm contains most of the thermal radio sources and is rich in dust. A strongly absorbing dust cloud seems to be present well within 1 kpc of the sun on the north side of the galactic plane.

C. *High galactic latitude studies*

Considerable activity centers around the local galactic structure in the intermediate and higher latitudes near the sun. In particular the problem of disentangling the disk and the halo populations is in the forefront. While the researches at Uppsala and at Lund by T. Elvius and others should properly be recorded in the report of the *Sub-commission on Selected Areas* we shall also mention them here. The program involves spectrophotometric data, colors, and magnitudes for stars brighter than B = 14 in 33 Selected Areas in the northern sky and 19 in the southern. Data for 19 areas have been published. Most of the plates have been obtained for the southern areas and photoelectric sequences are being established by observations at the Siding Spring Station of the Mt. Stromlo Observatory. The purpose is to investigate the stellar space density as a function of distance from the galactic plane. Kirsten Lodén at Stockholm is associated with Elvius in this work.

An investigation of the south galactic polar region (including S.A. 141) is being made by Eriksson from plates taken with the Uppsala Schmidt telescope at Mt. Stromlo. In an area of 40 sq. deg. BV photometry has been finished for all stars brighter than V = 15; in 30 sq. deg. the photometry extends to V = 16 and is completed by spectrophotometric data for stars brighter than B = 14. All plates have been taken and reductions are almost complete.

An analysis of the region of 120 sq. deg. around the north galactic pole has been continued by Malmquist. In a preliminary analysis of the data he has found absorbing clouds at 200 pc and 400 pc from the sun. The determination of the stellar distribution is complicated by the irregular distribution of absorbing dust clouds, as found by Ljunggren from a study of accurate color excesses of selected stars in the same region. The stellar distribution of late-type giant stars perpendicular to the galactic plane is being studied intensively by Häagkvist and Oja at Uppsala.

An extensive program involving star counts in Scorpius, Ophiuchus, Corona Australis and Sagittarius, begun many years ago, has been completed by van Hoof.

The important research by means of 3-color photometry by W. Becker and his colleagues at Basel for intermediate and high galactic latitudes has been summarized by Fenkart. Space density gradients for several Selected Areas in the galactic halo are compared with models of the halo. Details for three such investigations are given by Becker and Fenkart.

A major attack on problems of galactic structure at high latitudes is underway at the Dudley Observatory under the supervision of A. G. D. Philip. Forty-two areas are under observation; spectral types, colors and magnitudes required for studies of the space density fluctuation perpendicular to the galactic plane are being determined. The magnitude limit for these surveys is $B \simeq 13$.

Philip reports that photoelectric photometry is being done in the UBV the Strömgren 4-color, and the Hβ systems of all stars of early spectral type found in these high latitude fields. In one paper it was shown that the A stars at the NGP fainter than V=11·5 are predominantly of Population II, while the brighter ones are of Population I. The reddening at the NGP was found to be E_{B-V}= 0·05 magnitude. A similar study for an area at the south galactic pole indicates that the reddening there is at least twice as great as at the NGP and the change from predominantly Pop. I to Pop. II is not as abrupt. This is consistent with the idea that the sun is somewhat north of the galactic plane.

J. S. Drilling is collaborating with Philip in the analysis of the areas at $l^{\text{II}} = 0°$, $b^{\text{II}} = -45°$ and $l^{\text{II}} = 180°$, $b^{\text{II}} = -45°$. Analysis of these areas, together with an analysis of the SGP area, will help in determining the direction of surfaces of equal star density near the sun. J. Stock is collaborating in the study of a series of fields along a strip from the SGP to the galactic equator to determine the change in distribution of the early-type stars with galactic latitude.

A study of stellar space densities at intermediate galactic latitudes is reported by A. R. Upgren. The objective is to determine the inclination to the galactic plane of layers of equal star density.

In addition to the research mentioned here several other studies of the regions around the galactic poles have been made. Slettebak *et al.* have obtained slit spectrograms and *uvby* color data for 77 A stars with magnitudes between 7 and 9·5 at the NGP. Spectral types were estimated and rotational and radial velocities were measured from the spectrograms. An investigation of an 840 sq. deg. region around the SGP has also been reported by Slettebak. An objective prism survey with the Michigan Schmidt telescope at the Cerro Tololo Interamerican Observatory in Chile has yielded a list of some 900 stars, F0 and earlier, brighter than magnitude 14. In the course of this study seven carbon stars, four of which are new, were detected. With the exception of the bright variable R Scl, all are early R-type stars at distances between 2 and 5 kpc from the galactic plane (Slettebak *et al.*).

Eggen has continued his observational investigations of stellar population samples at the galactic poles. For 130 stars at the south galactic pole listed in the Bruce proper motion survey he has given *UBV* photometry. The majority of the stars from the Bruce Proper Motion survey have elliptical orbits with $e < 0{\cdot}5$ and appear to belong to an old disk population.

The detection and study of faint blue objects in high galactic latitudes is a fertile field of galactic research at present. Barbieri *et al.* have used the Schmidt telescope at Asiago to search for stars with negative $U-B$ color index in a field at $l^{\text{II}} = 243°$, $b^{\text{II}} = +67°$. Here 129 objects in an area of 25 sq. deg. were found. In connection with this type of search Luyten reports on a technique developed by him and Sandage, whereby beginning with three-image plates taken in ultraviolet, blue and yellow, and following with determinations of photoelectric colors and proper motions, it has become possible to identify correctly the quasars, sub-dwarfs, white dwarfs and other objects from among the faint blue stars in high galactic latitudes.

IV. OVERALL GALACTIC STRUCTURE

(S. McCuskey)

The large-scale structure of our Galaxy depends to a large extent upon observational techniques sufficiently powerful to penetrate the interstellar absorption in our neighborhood. These include predominantly optical photometric techniques in those sections of the spectrum (the infrared for example) where the dust in space has the least effect, the radio techniques for studying the distribution of H I and H II and other gases, and polarization techniques, both optical and radio.

In the period under review, three large 21-cm line surveys of the Milky Way strip have been published by Westerhout, by Kerr, and by Weaver. There have also been extensive surveys of recombination line sources, OH emission and absorption, and of the radio continuum at various wavelengths. Studies of several other lines are still in their early stages. The Commission 40 report contains details of these surveys. The available information has been very considerably enlarged. In what follows, an attempt will be made to assess the current state of affairs concerning the large-scale view of the Galaxy.

A. *Spiral structure*

By far the most significant discussion and evaluation of spiral features in the Galaxy took place in IAU Symposium no. 38 in Basel in August 1969. In due time, a volume detailing the symposium and containing an excellent summary by B. J. Bok will appear. Here only a few essential features concerning the spiral structure will be noted.

(1) It is clear from the researches on stellar distribution to distances of 4 to 6 kpc from the sun that OB^+ and OB° stars, WR stars, early Be stars, O associations, some S and N (carbon) stars show concentrations into spiral-like features resembling similar features in external galaxies.

(2) The classical cepheids of $P > 10$ days ($M_v < -4{\cdot}3$), possibly some of the early giant M stars, possibly the upper main sequence (B5–A2) stars, may be loosely identified with local spiral structure but the evidence is not strong.

(3) Other star groups in the solar vicinity to distances of 2 to 3 kpc show no correlation with groupings under items (1) and (2).

(4) Extended studies by W. Becker and his colleagues confirm the association of young clusters with the other optical spiral features. The Perseus, Orion, Sagittarius and Norma-Scutum spiral-like structures have very interesting counterparts in the outer regions of NGC1232. And there is a gap in the local longitude system of NGC1232 corresponding to the spiral-structure gap at $140^\circ < l^{II} < 180^\circ$ in our Galaxy where only one spiral arm appears (the local arm). It was also pointed out that a gap of 1·5 kpc extent exists between the sun and the Carina spiral feature. The large pitch angle of these optical spiral features compared to the small angle shown by the HI spiral structure persists.

(5) Optical HII regions studied by Courtés and colleagues outline spiral-arm structures with pitch angles about 20° in the following regions:

Perseus	$105^\circ < l^{II} < 190^\circ$	3 kpc at 120° minimum distance
Orion	$59^\circ < l^{II} < 254^\circ$	0·5 kpc at 180° minimum distance
Sagittarius	$274^\circ < l^{II} < 32^\circ$	1·5 kpc at 330° minimum distance
Norma-Scutum	$308^\circ < l^{II} < 333^\circ$	3·5 kpc at 330° minimum distance

The inclination of these is similar to that outlined by the young clusters (item 4). According to the French astronomers the OB stars, long period cepheids and HII regions all have the same velocity characteristics.

(6) A study of the dust lanes in 17 Sc-type galaxies by B. Lynds indicates rather conclusively that these appear first close to the galactic nucleus and gradually assume a position bordering the luminous arms on the inside edges. The dust lanes define clearly a spiral pattern in which the brightest HII regions are generally found at the edges of the dark material. Presumably a similar state of affairs would prevail in our Galaxy.

It is clear that more detailed studies of regions showing apparent spiral structure are now required. Two such investigations were reported at IAU Symposium no. 38. The first by H. Dickel and colleagues is a study of the Cygnus region $70^\circ < l^{II} < 90^\circ$, $-8^\circ < b^{II} < +8^\circ$. Here 193 optical nebulae have been catalogued and their shapes and orientations determined. Distances have been obtained for about one half of these in the Cygnus X complex by comparison of optical and radio data. The model developed by Mrs. Dickel and colleagues indicates that (1) the pitch angle of this spiral feature is about 13°; (2) the angle of tilt relative to the galactic plane is about 1°; (3) the sun is approximately 0·8 of the distance from the central axis toward the edge of the spiral structure; (4) the semi-major axis of the arm cross-section is 0·2 kpc; (5) the ratio of minor to major axes of the cylindrical cross-section is more than 0·5 and is close to 1.

A second important study is that by B. J. Bok *et al.* of the spiral-like feature in Carina, $270^\circ < l^{II} < 300^\circ$. Optical data, plus radio data from Australia and Argentina on the HI distribution and the HII distribution indicate a sharply bounded, elongated concentration in Carina extending from 2 to 8 kpc from the sun. Along $l^{II} = 283^\circ - 285^\circ$ the OB star and supergiant star distribution is sharply bounded. The other boundary of the feature is less sharply delineated at $l^{II} = 295^\circ - 300^\circ$. This inside edge of the spiral structure is not very rich in dust. At 5 kpc from the sun, the major elements of the spiral feature are between 100 and 300 pc below the galactic plane.

Clearly it is imperative to have more studies similar to the two mentioned above in order to gain any adequate picture of the large-scale spiral structure in the neighborhood of the sun.

In addition to the spiral-arm tracers mentioned above the use of stars embedded in reflection nebulae for this purpose appears feasible. Racine and van den Bergh have introduced these

"R-associations". Racine has shown by spectrographic and *UBV* photometric observation of 188 stars associated with reflection nebulae that groupings (R-associations) exist whose galactic positions and distribution correlate well with the OB associations which appear to trace out the local spiral arm. The inner part of this structure is about 350 pc from the sun toward the Perseus arm. All of the R-associations are within 1 kpc of the sun. These data imply that the local spiral arm is tilted some 10° with respect to the galactic plane. Its center is some 250 pc toward $l^{II} = 160°$; for $l^{II} < 160°$, $z > 0$ and for $l^{II} > 160°$, $z < 0$ for objects delineating the R-associations.

Dixon has made a comprehensive study of the 3-dimensional spiral structure from distributions of O and B stars, H I and H II gas. He considers both non-gravitational and gravitational forces, the former causing stars to fall away from the interstellar clouds from which they were formed. The local spiral feature, according to Dixon, has a ribbon-like structure about 200 pc thick, and 1500 pc long; it is tilted with respect to the galactic plane, rising to a $z = +100$ pc at 800 pc toward $l^{II} = 70°$ and falling to $z = -150$ pc at 700 pc toward $l^{II} = 250°$. The motions of the early and the late B stars within 1 kpc of the sun indicate that the older stars (age 3×10^7 yr) drift away from the younger (age $< 10^7$ yr).

Following his study of the Orion complex and Gould's Belt, Clube called attention to the modification in H I concentrations which would be necessary if the velocity field of Gould's Belt were taken into account.

Fernie has made a new study of the relationship of classical cepheid variables to the galactic spiral structure. No *well-defined* spiral structure is evident in the distribution of the cepheids. If one sees in this distribution some semblance of spiral structure the sun would appear to be toward the *outer* edge of the concentration.

The large-scale distribution of hydrogen in the Galaxy has been reviewed by Kerr. This paper includes a hydrogen spiral diagram, in which the arms have a small pitch angle. Weaver has derived a more open type of spiral pattern from his own observations. The various observational surveys give quite similar results; the differences are in the interpretation, mainly in the manner of joining together the observed features into large-scale patterns. In the solar vicinity, the main difference between the two patterns is that Weaver joins the Sagittarius and Carina arms, while Kerr has a Carina-Cygnus arm passing through the Sun, and a separate Sagittarius arm.

Burton and Shane have carried out a detailed model-fitting study of their observations in a restricted range of longitude $l^{II} = 22° - 56°$. They have taken the Lin density-wave theory into account, but find it necessary to introduce additional motions in order to improve the fit. They find two arms in the inner region, with a low pitch angle.

The difficulties in the interpretation of the 21-cm data arise from the restriction to kinematic distances, which are based on a limited understanding of the large-scale velocity field, and the existence of a large amount of fine structure superposed on the large-scale pattern.

The hydrogen spiral arms are found by all workers to be patchy on a scale of 500–1000 pc, and also they show a considerable amount of splitting and cross-linkage. An individual spiral arm is often centered a few tens of parsecs above or below the galactic plane, maintaining approximately the same displacement over a substantial distance (Henderson).

Recombination-line surveys can also be used to investigate large-scale structure through the distribution of H II regions (Burke, Mezger and Palmer, McGee and Gardner, Dieter, Wilson). It is found, however, that the H II observations cannot lead to a spiral structure map on their own, primarily because the velocity dispersion of the individual sources tends to smear out the spiral pattern. The H I and H II results are generally compatible, but there is an important difference between the radial distributions. The "giant" H II regions which are responsible for the recombination-line and OH emission are almost wholly confined to a ring between 4 and 7 kpc from the galactic center, whereas the maximum density of H I is found outside this ring.

Absorption studies show that hydroxyl and formaldehyde molecules are widely distributed through the Galaxy (Goss, Snyder *et al.*). The close agreement between the velocities measured in the OH, H_2CO and H I lines indicates that the molecules must have the same general distribution as the H I, and must therefore be concentrated to the spiral arms.

A number of continuum surveys have been carried out in various parts of the Galaxy. The most detailed are the 11-cm surveys by Altenhoff, and by the Australian group, which show the concentration of thermal sources to the spiral arms.

B. *The galactic disk*

The structure of the galactic disk as regards the stellar population is discussed in Section III. Here we indicate those radio astronomical studies of this part of the Galaxy which have been made during the triennium under review.

Henderson has carried out a new study of the shape of the neutral hydrogen layer on the northern side. He has delineated the small variations from a plane in the inner parts, and the twist of the outer parts. The possibility that the outer twisting was produced by the Magellanic Clouds has been rediscussed by Avner and King, and Hunter and Toomre.

Altenhoff and Andrew have shown from radio continuum studies that the ionized hydrogen is concentrated in discrete regions, with little or no evidence of any general disk-like distribution.

There have been several new studies of the spectrum of the galactic continuum emission at high latitudes (Bridle, Yates and Wielebinski, Getmantsev *et al.*).

C. *The galactic nucleus*

Optical observations of the galactic nucleus are impossible in the usual wavelength regions because of the interstellar absorption. Much effort recently, however, has been devoted to observing this galactic center region with far infrared techniques.

Becklin and Neugebauer have surveyed with relatively high resolution at $\lambda = 1{\cdot}65$, $2{\cdot}2$ and $3{\cdot}4\,\mu$, the galactic center region. They find a dominant source of infrared radiation 5′ arc in diameter, together with a point source centered on the dominant source. There is an extended background of radiation at these wavelengths, and additional extended hot-spots. The dominant infrared source coincides in position with Sagittarius A as determined by radio observations. Its observed energy distribution is similar to that of a black body at 900 K. The small point source has a diameter of 5″ arc, or less. Across the region surveyed the energy distribution compares favorably with that across the nuclear region of M 31.

In a second contribution Becklin and Neugebauer have extended their observations to include wavelengths from $1{\cdot}65\,\mu$ to $19{\cdot}5\,\mu$. An extremely strong source 20″ arc in diameter was observed at $\lambda = 10$ and $20\,\mu$. In summary, there appear to be three sources of infrared radiation in this region: (1) a general background source which is some 150–200 pc in diameter in which the intensity decreases with distance from its center; (2) a core source about 1 pc in diameter which stands out at $\lambda = 5$ and $10\,\mu$; (3) a pointlike source as mentioned in the paragraph above.

An intense source of far-infrared radiation, $\lambda = 100\,\mu$, extending for $6\overset{\circ}{.}5$, at least, along the galactic equator in the galactic center region has been observed by Hoffmann and Frederick. It is indicated that, if this is assumed to be thermal emission from interstellar dust grains, the observed intensity is 30 times greater than might be anticipated by such a mechanism.

It is clear from the paucity of present optical data concerning the nuclear region of our Galaxy that much new information should result from these important infrared techniques in the immediate future.

Radio investigations of the nuclear region have been reviewed by Kerr and Lequeux. More recent work has largely concentrated on studies of line emission and absorption in the vicinity of the galactic center. A number of OH clouds can be seen in the region. One of these, which may be very close to the center, appears to be a discrete rotating cloud of 10 pc diameter (Kerr and Sandqvist), which also contains ammonia (Cheung *et al.*) and formaldehyde (Snyder *et al.*).

Sgr B2, at $l^{\mathrm{II}} = 0\overset{\circ}{.}7$, $b^{\mathrm{II}} = 0\overset{\circ}{.}0$, is a complex object, which shows strong OH and H_2O emission, weak NH_3 and recombination line emission, and H I, OH and H_2CO absorption effects. The object shows typical characteristics of an H II region and also those of a molecular cloud, approximately coincident in position.

From lunar occultation observations, Maxwell and Taylor have derived an improved position for the source Sagittarius A, which is believed to be at the galactic center.

D. *The galactic halo*

Section IIIC has dealt with the optical stellar distributions in the high latitude regions and into the galactic halo. A brief summary of the radio astronomical investigations during the past three years follows here.

Burke and others have thrown doubt on the existence of the synchrotron-emitting halo which has been discussed for the last decade. It seems likely that, if a halo exists at all, it must be smaller in size and have a lower volume emissivity than was originally proposed by Baldwin and Mills.

Miss Keppner, Weaver and others have found neutral hydrogen in many parts of the Galaxy which appears to be at large z-distances from the plane (1 kpc or more), but is still related to the spiral structure in the plane. In these cases, the main spiral-arm concentration of gas is close to the plane, as known before, but there are lower-density extensions above and below the plane to a considerable distance.

V. KINEMATICS

A. *Stars*

Galactic rotation

Extensive work in deriving Oort's constants from proper motions of stars has been done by Fricke. From FK4 proper motions he found: $A = 12{\cdot}3 \pm 2{\cdot}4$, $B = -7{\cdot}6 \pm 2{\cdot}4$, while from distant stars (> 100 pc) he found: $A = 14{\cdot}2 \pm 1{\cdot}9$, $B = -11{\cdot}8 \pm 1{\cdot}9$ (FK4 system), and $A = 17{\cdot}1 \pm 1{\cdot}9$, $B = -10{\cdot}0 \pm 1{\cdot}9$ (N30 system). Dieckvoss found $A = 14{\cdot}8 \pm 0{\cdot}7$, $B = -11{\cdot}3 \pm 0{\cdot}6$ from AGK3 proper motions.

A quite different value of $B = -24 \pm 2$ has been suggested by Aoki. He assumed that the unexplained part in the secular motion of the obliquity is due to a motion of the equatorial plane caused by friction between mantle and core of the earth. An examination of Aoki's basic assumption has been made by Lieske and Fricke, showing that there is no significant motion of the equatorial plane of unknown origin.

Crampton calculated the solar motion and galactic constants for Be stars, cepheids, young clusters and H II regions. He found the average values $U = -5{\cdot}5$, $V = +14{\cdot}0$, $A = 13{\cdot}5 \pm 1{\cdot}5$ and $K = -2{\cdot}0$. Crampton suggests that the commonly adopted local standard of rest moves outwards at 5 km s^{-1}. This value is smaller than the uncertainty in the local standard of rest, pointed out by Blaauw. In another paper Crampton and Fernie found the values $U = -8{\cdot}0$, $V = 14{\cdot}5$, $A = 12{\cdot}5$ from available data on classical cepheids. Takase found the values $U = -8{\cdot}1$, $V = 13{\cdot}9$, $W = 7{\cdot}3$, $A = 14{\cdot}3$, $B = -6{\cdot}1$, $K = -2{\cdot}6$ from 146 cepheids. Preston found values of A from 6 to 25, from the motions of Mira variables. Georgelin found a value $A = 6$ from a kinematic and photometric study of a region in Cassiopeia. The solar motion with respect to stars with enhanced metallic lines was found by Komarov. Barbier found that the velocities of O, B, A stars in a region in Cassiopeia are larger than in Schmidt's model. He derived also a value $A = 16{\cdot}0 \pm 2{\cdot}2$. The galactic motions of early type stars within 3 kpc from the sun were found by Yokoo through a statistical analysis of their radial velocities. Feast studied the radial velocities of distant cepheids. A value of $R_0 = 9{\cdot}8 \pm 1{\cdot}4$ for the distance of the galactic center was derived. Fokker derived the values $R_0 = 9{\cdot}7$ and $A = 14{\cdot}3$, using the radial velocities of distant stars. Petrie and Petrie found $A = 15{\cdot}1 \pm 0{\cdot}4$, $K = -0{\cdot}6 \pm 0{\cdot}3$ from 688 O9–B8 stars.

The solar motion with respect to the local group and exterior galaxies was found by Byrnes (341 ± 47 km s^{-1} towards $l^{\text{II}} = 83^\circ\!.8 \pm 9^\circ\!.3$, $b^{\text{II}} = -12^\circ\!.4 \pm 7^\circ\!.7$). Conklin found that the velocity of the earth with respect to the cosmic microwave radiation is the same as with respect to the local supercluster of galaxies. de Vaucouleurs and Peters found the velocity of the sun with respect to galaxies of various distances. While the velocity with respect to nearby galaxies is 315 ± 15 km s^{-1} towards a direction $l^{II} = 95° \pm 6°$, $b^{II} = -8° \pm 3°$, giving essentially the rotation around the center of the Galaxy, the velocities with respect to more distant galaxies are much larger, reaching about 4000 km s^{-1}; these velocities reflect mainly the motion in the "local supercluster".

The connection between galactic kinematics and an "absolute" system of proper motions was discussed by Clube. Further astrometric problems connected with galactic research were discussed by Murray and Blaauw during the IAU Meeting in Prague. The needs of radial velocities for deriving the local and large-scale galactic structure were pointed out by Blaauw and Thackeray. The accuracy of the calculation of the galactic constants has been considered by Mennessier and Crézé.

Velocity distribution

Martinet studied the kinematic properties of some groups of dwarfs, selected from physical criteria. He found similarities with the kinematics of A stars (vertex deviation, unusual ratio of axes of the velocity ellipsoid, etc.). Martinet and Mayor found the kinematic properties of 1000 field giants near the sun, in relation to their location in the H–R diagram.

The velocity ellipsoids of stars of different spectral classes and luminosities were found by C. Jaschek and Miss A. Gomez. The motions of bright F stars were considered by Eggen. He found that most stars belong to a few groups. Rydgren studied the distribution and kinematics of local early F stars. He found a concentration of F stars near the sun, which, however, must be a transient phenomenon. Jõeveer and Jóeveer *et al.* studied the kinematics of K stars. Philip found the radial velocities of 12 field horizontal branch stars. He derived a *z*-velocity dispersion larger than 100 km s^{-1}.

Smak studied the kinematics of Mira-type variables. Iwanowska considered the kinematics of late-type stars. Pavlovskaya and Karimova are determining the proper motions of many long-period cepheids, and Ap, Am stars. Crampton and Hill are determining distances and velocities of B stars, central stars of H II regions, and stars in the direction of high-velocity clouds. Froeschlé studied the kinematics of OB stars belonging to the local group. Shatsova studied the kinematics of O-B5 stars up to a distance of 3 kpc from the sun. She found that the dispersion of velocities increases towards the galactic center while it decreases slightly, or remains constant towards the anticenter. The dispersion of velocities in the region of the local spiral arm has a minimum at a distance less than 1 kpc.

The kinematics of planetary nebulae were considered by Perek, Feast, and by Webster.

Some problems of the distribution of velocities of nearby stars were discussed by Delhaye. He found that Am stars and many Ap stars have an elongated distribution in velocity space, near an axis with a vertex deviation 25°; Eggen's groups lie also along this axis.

A number of stars with velocities larger than 100 km s^{-1} perpendicular to the galactic plane were found by Eggen. Eggen found several groups of stars in the old disk population.

Ogorodnikov and Latyshev studied the kinematics of local star streams. Many stars in the solar neighborhood belong to the Hyades stream, and the Ursa Major stream. Ogorodnikov, Latyshev and Antonov plan to study the neighborhoods of known clusters to find further streams.

Groups of stars of the same spectral type were found by Meurers to have the same expansions as OB-associations. Groups of young stars were found by Crampton to have a common motion, differing by 5–10 km s^{-1} from the average motion of all the stars. Vandervoort found many moving pairs among the A-type stars within 20 pc. He suggested that they have common origin, being extreme cases of moving groups.

The kinematics of Gould's belt were considered by Clube and by Lesh. Clube found that the B stars of Gould's belt have a group motion corresponding to the Pleiades group, plus an expansion, giving an expansion age of $3{\cdot}7 \times 10^7$ yr. The A stars have the same kinematics as the B stars. Lesh proposed two models: One with a mixed population and expansion age $9{\cdot}0 \times 10^7$ yr and another one, with an expanding subset, which has an expansion age $4{\cdot}5 \times 10^7$ yr.

Crampton and Byl found evidence that some stellar rings are not physical groups.

An ellipsoidal law for the distribution of stellar velocities, which is similar to a Planck law, was proposed by Shatsova. Rudnicki has pointed out that the dependence of the velocity distribution of stars on space location is not known even in rough outline.

Bouvier found the influence of a finite velocity of escape on the ellipsoidal distribution of velocities. Massonie considered the effects of observational selection on the maximum of the velocity distribution.

B. *Interstellar matter*

Galactic rotation – non circular motions

The 21-cm rotation curve is consistently defined by radio astronomers, with only small differences, that can be attributed to differences in beamwidth and bandwidth. The "bumpy" form of the rotation curve, the north-south asymmetry and considerable fine structure are established. These effects are small with respect to the total velocity, but they are important in estimating distances. A review of these problems is given by Kerr (*A. Rev. Astr. Astrophys.*, 7, 39, 1969).

The "bumpy" character of the rotation curve is due mainly to the gravitational forces of the spiral arms (Barbanis and Woltjer; Lin, Yuan and Shu; Yuan; Burton and Shane). The north-south asymmetry may be due to an oval distortion of the Galaxy (Hunter). Deviations from axial symmetry are observed in many external galaxies (M. S. Roberts, Robinson).

Bash found the galactic rotation curve, without assuming low optical depths, which give lower than usual velocities by 10 km s^{-1}. Petrovskaya studied the rotation of neutral hydrogen in the inner regions of the Galaxy for $z \neq 0$, assuming that the surfaces of equal angular velocity are spheroids.

The rotation curve derived from recombination line observations (Mezger *et al.*) agrees well with the H I curve. The kinematics of ionized and neutral hydrogen thus appears to be identical (Dieter, Mezger and Höglund, Kerr *et al.*, Mezger, Wilson).

Optical observations of the kinematics of H II regions were made by J. S. Miller and by Courtès and his associates. The kinematic behavior of H II regions is the same as that of other Population I components, like Cepheids and neutral hydrogen.

Velocities derived from H I, OH and H_2CO absorption lines agree with H I emission lines. The extension and intercomparison of observations of this kind will add greatly to our knowledge of gas kinematics.

Large non-circular motions have been observed in particular regions. Lindblad found several extended features in the anticenter region with radial motions up to at least 30 km s^{-1} in the Perseus arm. Explanations of such phenomena involve gravitational effects, due to the spiral arms, or big explosions. Systematic radial motions of about 10 km s^{-1} are predicted by Lin's theory of spiral waves (Yuan). Henderson has found an inward motion of the Perseus arm; however no such motion has been found in the Sagittarius arm (Kerr).

Blaauw has stressed the uncertainty of the local standard of rest, which leaves room for a (possible) lag of H I and the associated youngest stars by up to 8 km s^{-1}. An expansion of $+7$ km s^{-1} of matter near the sun, as proposed by Kerr, is also possible.

Many sections of spiral arms show a rolling motion which seems to be related to helical magnetic fields. Westerhout found a helical streaming of gas with velocity of the order of 10 km s^{-1} and a skew angle $\sim 30°$ in the direction of galactic rotation.

Dixon presented evidence that the gas moves at about 15 km s^{-1} less than the circular velocity, because of non-gravitational forces. However Penston *et al.* found that non-gravitational forces produce a lag of the gas behind the stars by only about 3 km s^{-1}. There is much evidence that gas and the common stars move together (Takakubo, Courtès *et al.*, Kerr). Even deviations from circular motion seem to be the same for gas and stars (J. S. Miller).

This conclusion is supported by calculations of the solar motion with respect to gas and stars. Recent values of the solar motion with respect to neutral hydrogen have been provided by Takakubo ($U = -10{\cdot}4$, $V = 15{\cdot}2$, $W = 10{\cdot}9$), Venugopal and Shuter ($U = -12{\cdot}5$, $V = 14{\cdot}5$, $W = 8{\cdot}8$), Mast

($U = -11{\cdot}2$, $V = 15{\cdot}4$, $W = 8{\cdot}5$), and Nesterczuk ($U = -10{\cdot}2$, $V = 15{\cdot}4$, $W = 10{\cdot}9$); all are close to the standard solar motion.

The solar motion with respect to H II regions and Oort's constant A were found by J. S. Miller ($U = -6{\cdot}4 \pm 2{\cdot}1$, $V = 13{\cdot}3 \pm 1{\cdot}7$, $A = 15{\cdot}3 \pm 1{\cdot}4$) and by Georgelin ($U = -7{\cdot}5 \pm 1$, $V = 13{\cdot}1 \pm 1{\cdot}5$, $A = 14{\cdot}3 \pm 0{\cdot}5$).

Venugopal and Shuter, and Venugopal studied the velocity ellipsoid of the local hydrogen. The vertex direction was found aligned with the local magnetic field. The anisotropy of peculiar motions of the gas has been considered by Takakubo. Goldstein and MacDonald have studied local gas motions in the directions of early type stars with $|b^{II}| > 20°$.

A number of dust clouds have been shown to be sources of OH line emission (Heiles, Cudaback and Heiles), H_2CO absorption (Zuckerman *et al.*), and H I absorption (Riegel and Jennings). The association is so well established that the velocities of gas can be regarded as referring to the dust clouds themselves; thus the kinematics of the dust can be studied for the first time.

Central region

Besides the outward motion of the bulk of the observed gas along the galactic plane evidence for outward motions at high angles has been provided by Kerr and Sinclair and by van der Kruit. The detailed motions near the center are studied by the newly discovered radio lines, particularly in the vicinity of Sgr A and Sgr B2.

Large velocities were found in the direction of the center by Mezger and Höglund through observations of the 109α recombination line. Data for ammonia have been published by Cheung *et al.*, and for formaldehyde by Snyder *et al.*

High velocity clouds

Dieter has recently (*Publ. astr. Soc. Pacific*, **81**, 186, 1969) reviewed our knowledge about the high- and intermediate-velocity interstellar gas. Detailed studies of the neutral hydrogen at intermediate latitudes have been made by Takakubo, and at high latitudes by Blaauw and his associates and by Hulsbosch. There is a correlation between deficiency in low-velocity gas and excess in intermediate-velocity gas. This may indicate a streaming in the nearby gas. Some of the intermediate-velocity clouds appear to be connected to the local disk of hydrogen (Verschuur) and it seems probable that these clouds are nearby. There are also indications of an interaction between gaseous masses of high and intermediate velocities (Dieter).

Heiles gave a dynamical model of the intermediate-velocity clouds, considering them as contracting in the halo and accelerating in the galactic gravitational field.

The nature and origin of the high-velocity clouds are still in doubt. Oort's theory of falling intergalactic matter, that sweeps the gas of the halo, is well known. Puppi, Setti and Woltjer studied theoretically the deceleration of intergalactic clouds. The collisions of high-velocity clouds with galactic gas were also studied by Savedoff *et al.*

On the other hand Kerr considers these clouds as satellites of our Galaxy, like the Magellanic Clouds, while Verschuur considers them to be extragalactic. According to Toomre the high-velocity clouds may have been broken away from the Magellanic Clouds during an earlier close passage. Rickard suggested that these clouds were produced by a large explosion, postulated to explain the non-circular velocities in the Perseus arm.

Many ambiguities concerning the high-velocity clouds will be solved if we find their distances. Kepner has published a list of 70 stars in the directions of the clouds, in which it may be possible to observe interstellar absorption lines.

VI. DYNAMICS

Stellar orbits – third integral

The general properties of orbits in two-dimensional potentials have been studied extensively by

Contopoulos and his associates (orbits in galaxies), Deprit and his associates (restricted three-body problem), Danby and Zahar, and Freeman and de Vaucouleurs (barred spirals). Particular emphasis is put on resonance phenomena (resonant periodic orbits and "tube" orbits). In the case of small perturbations the properties of orbits can be predicted by means of a "third" integral, which has special forms when conditions of near resonance are satisfied. Theoretical and experimental results connected with the "third" integral were discussed by Hénon, Saaf, Hori, Goudas, and Barbanis.

One of the most important papers in this area is Moser's construction of convergent series expansions for quasi-periodic motions in systems of 2 or more degrees of freedom.

The properties of orbits in spiral galaxies have been studied by Contopoulos and by Barbanis. Most orbits are of the general "ring-type" while resonant "tube" orbits are important in particular cases.

Indications for the existence of a "third" integral in the finite amplitude oscillations of Maclaurin spheroids were given by Rossner, and in the orbits of artificial satellites by Danby. Further applications of a "third" integral are given in the non-linear theory of spiral galaxies (Contopoulos).

Strongly perturbed dynamical systems display, to a certain degree, an ergodic behavior. Contopoulos has studied the properties of families of periodic orbits and their stability in such systems. He found many "regular" families of periodic orbits, that can be followed continuously from the unperturbed system, and many "irregular" families, independent of the above. It seems that the breakdown of the invariant surfaces is related to the appearance of "irregular" families of periodic orbits.

The breakdown of the "third" integral for large perturbations is understood as an overlapping of resonances (Contopoulos, Chirikov *et al.*, Ford *et al.*). Similar work has been done in plasma physics, in connection with the breakdown of adiabatic invariants or the destruction of magnetic surfaces (Rosenbluth *et al.*, Filonenko *et al.*). An interesting theoretical approach to these problems is contained in the book of Arnold and Avez, *Ergodic Problems of Classical Mechanics*, Benjamin, New York, 1968.

Work on systems of $N(>2)$ degrees of freedom is still at its first stages. In such cases the invariant surfaces are of N dimensions and do not separate the phase space. Thus a small dissolution of the invariant surfaces may secure ergodicity (Arnold diffusion). However recent work on three dimensional systems by Hénon and Froeschlé casts doubt on the effectiveness of such a diffusion.

Orbits in particular galactic models were calculated by Kuzmin and Malasidze, Innanen, Innanen and Southern, Ampel and Lebno, Ollongren, and Andrle.

Hori gave useful formulas for plane galactic orbits. Yuan calculated the orbits of some stars in Lin's spiral model of the Galaxy; he found that all these stars were formed in spiral arms.

Woolley and Candy found the perturbations of orbits due to irregularities in the galactic field.

Extensive numerical work has been done on the gravitational N-body problem. Many papers were published in the Proceedings of the Paris IAU Colloquium (*Bull. astr.*, 1968) by Aarseth, Cohen and Lecar, Feix and Hohl, Hayli, Hénon, Hockney, Hohl, Lecar, Peters, Standish, Szebehely, Thüring, Ulam, von Hoerner, and Wielen. This work deals with systems consisting of a few stars up to a few thousand stars. On the other hand calculations with hundreds of thousands of stars were made by Miller and Prendergast and by Hohl and Hockney.

Further work of the above type was done by Aarseth, Hockney, Hohl and Campbell, Goldstein *et al.*, Cuperman *et al.*, Hayli, and Wielen. Monte Carlo type N-body experiments were made by Hénon (many bodies), and by Agekian and Anosova, and Anosova (triple systems). Worall and van Albada studied systems of a few stars and the formation of binaries. Three special meetings, dealing mainly with plasma numerical experiments, took place in 1967 (Williamsburg, Virginia; NASA SP-153), 1968 (Los Alamos, New Mexico; LA-3990) and 1969 (Culham, England).

A comparison of various runs of the same 25-body problem calculated by different authors was made by Lecar at the Paris Colloquium. Good agreement was found only for very short times. It seems that only a few average quantities may be expected to be reliably computed by numerical experiments. A new IAU Colloquium (no. 10) on the gravitational N-body problem, will take place in Cambridge, England, during August 1970.

Models of the Galaxy

Self consistent models of galaxies have been developed by Ng, Perek, Shu, and by Vandervoort. Another model is considered by Kuzmin and Kutusov. A series of papers on galactic mass models have been published by Ejnasto, Ejnasto and Kutusov, and Ejnasto and Rümmel. Galactic models have been used by Innanen and Fox, and by Innanen and Kellett to derive the z-velocity dispersions. Relations among several dynamical quantities in galaxies were found by Takase and Kinoshita, and by Ejnasto.

Agekian and his associates constructed models of stellar systems by numerical integration of the N-body problem.

The force perpendicular to the galactic plane has been studied by Woolley and by Perry. The density near the sun was found by Woolley and Stewart to be 0·11 $\odot/pc^3$. If we compare this value with the value of 0·08 $\odot/pc^3$ of known objects, we see that the problem of missing matter is not so acute.

Spiral models of the Galaxy were used by Barbanis and Woltjer, and by Yuan.

Collisionless dynamics – stability problems

One of the most important branches of Stellar Dynamics is Collisionless Dynamics. This is applicable to systems with large relaxation times. Whenever the effects of collisions (or encounters) can be ignored we deal with cooperative phenomena. The basic equations used in such cases are the collisionless Boltzmann equation (known as Vlasov's equation by plasma physicists) and Poisson's equation.

In many cases we assume that we have an unperturbed self-consistent system undergoing small perturbations. E.g. we have an axisymmetric self consistent model of a galaxy and we study its perturbations, omitting all terms higher than the first in the collisionless Boltzmann equation. This is the linearized problem, which can be subdivided into the problem of modes and the initial value problem. The problem of modes (and, in particular, the stability problem) has attracted most of the interest.

In the case of gas we have the corresponding linearized hydrodynamical problem. In some papers the coupling between stars, gas and magnetic fields is considered.

A useful introduction to the theory of cooperative phenomena can be found in an article by Lynden-Bell (in *Relativity Theory and Astrophysics, 2, Galactic Structure*, Amer. Math. Soc., p. 131).

Hunter made an extensive study of the oscillations of disk galaxies. He studied first cold disk galaxies and he found growing leading spiral modes. He found also the appearance of continuous spectra of modes in some cases. Then he found large scale oscillations of galaxies that may explain the observed asymmetry between the northern and southern rotation curve of our Galaxy.

Local stability problems were considered by R. Graham, and by Niimi. Yabushita found that a central condensation can stabilize a differentially rotating disk. Julian investigated the possibility of the appearance of overstabilities. Lee, and Marochnik and Ptitsina studied oscillations of cylindrical systems. Hockney and Hohl found, by numerical experiments, the velocity dispersion of a disk of stars necessary to secure stability; their results agree with Toomre's theory.

Wu studied the stability of density waves in a self-gravitating stellar system with uniform rotation. Genkin found a critical wavelength for homogeneous rotating systems. The influence of a halo on the stability of ellipsoidal systems was considered by Miyamoto. Yabushita studied the instability of finite isothermal gas spheres. Maksumov studied collective effects in bounded systems.

Lynden-Bell and Ostriker gave a powerful variational principle for the stability of differentially rotating fluid bodies. Other criteria for stability were discussed by Lynden-Bell and Sanitt, Lynden-Bell, Milder, Lebovitz, Chandrasekhar and Lebovitz, Kovetz, Hohl and Feix, Robe and Brandt, Fridman, etc.

The oscillations and stability of particular types of fluid bodies were studied by Chandrasekhar and his associates, and others. A book on this subject has been prepared by Chandrasekhar.

Relativistic effects were introduced in the stability theory of spherical stellar systems by Ipser and Thorne.

Many papers on fragmentation and collapse have appeared recently. However, these problems are more related to cosmogonic problems, or problems of formation of interstellar clouds and stars.

Some particular solutions of the collisionless Boltzmann equation and Poisson's equation were given by Ehlers and Rienstra, Trümper, and Shimizu and Kitamura.

The bending modes of the Galaxy have been studied by Hunter and Toomre. The actual bending of our Galaxy is ascribed to a close passage of the Large Magellanic Cloud 5×10^8 yr ago. The influence of the Magellanic Clouds on our Galaxy was also considered by Avner and King.

An extreme case of collisionless stellar dynamics is when we have a violent mixing of a stellar system. Such extreme cases were considered by Lynden-Bell, who found a new statistical distribution, similar to the Fermi-Dirac. A number of numerical experiments were made by Hénon, Lecar and his associates, Feix and Hohl, and Hohl and Campbell, to check the applicability of this new statistics. It was found that in many cases where a violent mixing is produced this statistics represents well the observed energy distribution. In cases of less violent mixing, however, there are large discrepancies.

Dynamics of spiral structure – central region

Much work has been done on the gravitational theory of spiral structure in recent years. As indicated in the subsection on Magnetic Fields the strength of the magnetic field of our Galaxy is too small to account for the major spiral features. Thus the emphasis is placed on gravitational effects.

A review of the gravitational theories of spiral structure was given during the Basel Symposium by Contopoulos. Bok gave a very good summary of the whole Symposium.

A problem that attracted the interest of theoreticians was the "antispiral theorem" of Lynden-Bell and Ostriker. The theorem states that no stable spiral modes of oscillation exist, in general, in a gaseous galaxy. Similar theorems for a stellar galaxy were given by Yabushita, and by Shu. Kalnajs derived a similar theorem from general considerations about the symmetry of the equations of motion with respect to the plane of the galaxy and to time. However such theorems do not apply to slightly unstable systems. Thus a slightly growing wave does not contradict the antispiral theorem.

Lin and his associates have given a number of extensions and applications of the Lindblad-Lin theory of spiral structure. Shu has given the basic integral equation from which one can derive the spiral modes. A similar equation was given previously by Kalnajs. Yubashita gave a matrix equation for the eigenvalues and eigenfunctions which he solved numerically.

Shu studied density waves in the post-epicyclic approximation. He found that the action density of the waves is transported inwards with Toomre's group velocity. The oscillation modes appear to be real in this approximation also, thus no indications of instability (growing waves) or damping appear.

Lin, Yuan and Shu have pointed out a number of applications of the theory. A value of 13·5 km s^{-1} kpc^{-1} is adopted for the rotational velocity of the spiral pattern, taking into account the effects of thickness. Yuan has shown that the hypothesis of missing hydrogen between spiral arms cannot explain satisfactorily the observed irregularities of the rotation curve of our Galaxy. On the other hand a good explanation can be given by means of a spiral field amounting to about 5%, in force, of the axisymmetric field. An explanation of the bumps of the rotation curve in terms of a spiral field was given also by Barbanis and Woltjer. Recently Burton and Shane made an analysis of the observations of neutral hydrogen in the first quadrant of galactic longitude and found agreement with Lin's theory.

Yuan calculated the 21 cm profiles that should be observed in the adopted spiral field and found good agreement with observations. The theory predicts also a radial inward component in the spiral arms of 10 km s^{-1}. Such motions have been observed in the Perseus arm but not in the Sagittarius arm.

The interstellar gas moves through the spiral arms forming a shock wave on the inner edge of the arm. This phenomenon was studied by Fujimoto, and by W. W. Roberts. The shock produces a compression of interstellar matter and enhances star formation. An exposition of the applications of the gravitational theory of spiral structure was provided by Lin at the Basel Symposium.

The linearized problem for a galaxy with gaseous spiral arms was considered by Fujimoto. He used the linearized hydrodynamical equations and searched for self-consistent solutions. He found that the angular velocity of the spiral pattern is of the order of 17 km s^{-1} kpc^{-1}, and that the proportion of gas is larger when the spiral arms are more open.

Marochnik and Suchkov considered spiral waves in a galaxy composed of two Populations. Population I rotating in the usual way, and Population II with no average rotation. Then a Landau-type instability occurs, which generates a spiral pattern rotating with angular velocity smaller than that of Population I. Trailing and leading spirals have the same growth rate.

Kalnajs has developed a theory of spiral waves similar to, but independently from Lin. He attacked the linearized problem numerically, without making Lin's asymptotic assumption of tightly wound spirals, and found open spiral waves, rotating faster than the stars, which are slightly growing. He found also some global conservation theorems for spiral patterns.

Spiral modes were found by Yabushita, by integrating numerically the basic integral equation, written in a matrix form. Numerical investigations of spiral modes were made also by Miyamoto. Finite thickness effects in a spiral galaxy were considered by Shu, and by Vandervoort.

Mayor studied the local irregularities of the velocity field near the sun, due to spiral waves. One can explain, in this way, the vertex deviation of the velocity ellipsoid. On the other hand Woolley found that the vertex deviation is confined to young stars only, therefore it is probably connected with their origin and not with the spiral field of the Galaxy.

The problem of preference of trailing spiral waves attracted also the attention of theoreticians. It is known that local theories give always trailing waves. However, global theories favor equally well leading and trailing waves if boundary conditions are not taken into account. In our Galaxy a discrimination between trailing and leading spirals can be provided by Lindblad's inner resonance. Kalnajs presented some arguments that an imposed growing wave, trailing or leading, produces trailing waves near the inner Lindblad resonance. Contopoulos found a strong preference of trailing waves, by numerical calculations, using Lin's asymptotic assumption of tightly wound spirals. An imposed leading field has a trailing response near the inner Lindblad resonance.

The initial value problem can be solved relatively easily, once the problem of modes is solved, by a method similar to Landau's method in plasma physics. Contopoulos formulated this problem in Lin's terminology. In a certain sense the problem of origin of spiral waves is trivial. Any initial perturbation excites Lin's modes. These modes, however, do not have, in general, the same angular velocity. Thus there is a differential rotation of the spiral "patterns", although not equally marked as the differential rotation of the stars.

Toomre has considered the problem of the evolution of a system of waves satisfying Lin's dispersion relation. He found a "group velocity" of the waves which is directed inwards and amounts to ~ 10 km s^{-1}. In a numerical example Toomre found further that the amplitude of the waves is decreasing as the groups of waves move inwards from the co-rotation distance and vanishes near the inner Lindblad resonance.

Near the inner Lindblad resonance the linear theory is not applicable except for short times. As the wave grows from zero non-linear effects become soon very important and mask the linear effects (Contopoulos). The stars are arranged near two periodic orbits similar to two perpendicular ellipses and the system has a rough quadruple symmetry. Lin suggests that such non-linear effects transfer the energy of the inward moving groups of waves into an open spiral mode which extends outwards up to the co-rotation distance. Thus a steady state circulation may be formed.

Spiral waves were found experimentally by Miller, Prendergast and Quirk by *N*-body problem calculations. The waves seem to extend between the inner Lindblad resonance and the co-rotation distance.

The mechanism of origin of spiral waves has been considered by Toomre, who studied mainly the effects of an approach of the Magellanic Clouds. On the other hand Lin ascribed the spiral patterns to large scale turbulence near the co-rotation distance (in our Galaxy at about 15 kpc). At the co-rotation distance another resonance effect is important. Matter is escaping away from arms and concentrates near the Lagrangian points L_4, L_5, and the spiral arms are broken (Barbanis).

Another mechanism for the origin of spiral waves may be provided by the differential rotation (Lynden-Bell). This mechanism was considered by Kalnajs as a two-ring instability.

Non-linear spiral problems were also considered by Marochnik and Suchkov, Berry and Vandervoort, and by Niimi.

A number of authors have pointed out the importance of a bar near the center for the origin of spiral patterns. Some work on the theory of barred spirals has been done by Prendergast, Pikel'ner, and by Freeman. Freeman and Harrington, Freeman, and de Vaucouleurs and Freeman calculated orbits in barred and irregular spirals in order to explain the appearence of rings and some other characteristics of such spirals. However, no wave theory of barred spirals has yet been developed. Such a theory would be desirable in view of the continuity of form between normal and barred spirals.

Magnetic theories of spiral patterns were provided by Greyber, Piddington, and by Pişmiş. Pikel'ner pointed out that a magnetic field is essential for the existence of spiral arms, because it prevents their collapse. Arp suggested that the spiral arms are the tracks of material ejected from the nuclei of galaxies.

The dynamics of the central region of our Galaxy have been studied by Moore and Spiegel, and by Spiegel. A steady circulation of matter was found that may account for the expanding 4 kpc arm. A flow of matter from the nucleus, like the solar wind, reaches the 4 kpc arm, while its velocity drops abruptly from supersonic to subsonic. Then matter is moving outside the plane and returns at an angle to the center.

Collisional dynamics – evolution

Collisional Dynamics deals with stellar systems in which encounters are important. A particular problem of special importance is the long term evolution of stellar systems. Little progress has been made in these problems, because we are faced with many of the difficulties of the general *N*-body problem.

Prigogine and Severne have attacked this problem using their methods of non-equilibrium statistical mechanics. They have stressed the non-markovian character of stellar evolution, the memory effects and the fact that the entropy does not necessarily increase.

Some new approaches have been considered by Saslaw (gravithermodynamics), Gilbert ("dressed" particles) and Miller (polarization). The time of relaxation has been calculated more carefully by Ostriker and Davidsen and by Miller. Genkin found that resonance effects (motions "in phase") reduce considerably the relaxation time in rotating systems.

The effect of density waves on the relaxation of stellar systems was considered by Julian, Thorne, and by Marochnik and Suchkov. Julian found important increases of stellar velocity dispersions due to encounters of stars with density waves.

Thorne finds that cooperative phenomena at most double the dynamical friction. Ogorodnikov, Latyshev and Antonov are studying the effects of vast galactic stellar streams on stellar relaxation. Marochnik found that relaxation is faster if stars move in a field of condensing interstellar clouds. A similar mechanism for the increase of the dispersion of velocities of older stars was proposed earlier by Barbanis and Woltjer. A local model for the increase of dispersion of velocities through interactions with interstellar clouds was proposed by Woolley and Candy.

Lee considered random motions of stars in a stellar system and derived a transition probability. He concluded that the Fokker-Planck equation is not strictly valid. Shimizu studied the effects of a random force on the stars. Petrovskaya and Kaliberda considered the motion of a star in a field of irregular forces.

Relativistic effects in stellar dynamics were considered by Lee, Ipser and Thorne, Ipser, and by Fackerell.

Many hopes for an understanding of the collisional processes and of the evolution of stellar systems are based on *N*-body numerical experiments. In all numerical experiments up to now a dense nucleus and an extended halo are formed. The central density increases while stars escape continuously. In a few examples studied by Hénon a spherical cluster (or galaxy) tends to a homologous system, which has a (mathematically) infinite central density. According to von Hoerner infinite central density is reached in 20 initial relaxation times. The properties of homologous systems were studied also by Agekyan.

A sensational description of the "gravothermal catastrophe" of a stellar system by continuous increase of the central density was provided by Lynden-Bell. However no complete theory exists as yet for the collisional evolution of stellar systems.

The evolution of a stellar system under certain assumptions was studied by Chandrasekhar. A number of authors used, or extended, the virial theorem. A tensor virial theorem for stellar systems was proved by Chandrasekhar and Lee. Bouvier extended the Lagrange-Jacobi identity to a system of stars and diffuse matter. Bouvier and de la Reza considered the applicability of the virial theorem to an expanding system. Tadokoro used the virial theorem for the study of the local group.

Bouvier considered the distribution of residual velocities when the escape velocity is finite. The evolution of the velocity spheroid due to star cloud encounters was studied by Kitamura. Kato studied the velocity ellipsoid by using hydrodynamic equations for the stellar "gas". Genkin used the Chapman-Enskop theory to derive the Schwarzschild velocity distribution.

Magnetic fields – cosmic rays

The first positive observations of 21-cm Zeeman splitting were made by Verschuur. He found fields of 10–20 μG in the Perseus arm (Cassiopeia A). However he found no such fields in four local features. Thus Verschuur argues that fields of the order of 10 μG may be quite local. Davies *et al.* found the same fields in the Perseus arm and fields of 2·5 μG in the Orion arm.

Polarization measurements in Cassiopeia were made by T. Schmidt and a field of 12 μG was found. Polarimetric measurements of nearby stars show that there are systematic differences in the galactic field, even within 100 pc from the sun (Appenzeller). Jokipii *et al.* also stressed the turbulent structure of the Galaxy, as derived from polarimetric observations. Klare and Neckel made polarimetric observations of 1421 southern O-B stars.

There are many observations that give a field along the direction of the Orion arm. This field may be helical. In fact a number of authors explain the optical and radio polarization measurements by means of a helical field near the sun (Hornby, Mathewson, Mathewson and Nicholls). However Appenzeller could not give a satisfactory explanation of his observations by means of such a field. Bingham and Shakeshaft studied the polarization and distribution of non-thermal radio emission within 1·3 kpc from the sun and found that the magnetic field is related to the local spiral arm, but not to Gould's belt. Clube on the other hand considers that the field is related to Gould's belt.

The north galactic spur is possibly related to the radio tracers of the helical structure of the magnetic field (Mathewson). Pronik found that the local magnetic field is quite irregular. Appenzeller found that near the axis of the local arm the field is parallel to the galactic plane, while near the sun it is inclined by 27°. Davies gave two models for the local magnetic field; in both of them the field has opposite directions above and below the galactic plane.

Some of the most recent measurements of galactic magnetic field were provided by observations of pulsars. There is evidence that pulsars are concentrated in spiral arms. Observations of Faraday rotation and dispersion measures in pulsars give the field strength and the electron density. Ekers *et al.* found a field of $0{\cdot}73 \pm 0{\cdot}13\,\mu$G along the inner face of the Orion arm. Similar results were found by Radhakrishnan *et al.* Staclin and Reifenstein found fields between 0·9 μG and 2·8 μG. F. G. Smith found fields between 0·2 μG and 3·5 μG in the directions of some pulsars. Jokipii and Lerche considered the effect of the turbulent structure of our Galaxy on the Faraday rotation and the dispersion in pulsar signals. They concluded that the average value of the magnetic field is 3 μG with a disperson $\pm 1{\cdot}5\,\mu$G.

Reviews about the galactic magnetic fields were provided by van de Hulst, Davis and Berge, and more recently by G. Burbidge, and by Woltjer. A field of about 3 μG is generally acceptable to-day.

Such a small field is not sufficient to counteract the effects of differential rotation, therefore it cannot be the main cause of formation and persistence of the spiral arms (Wentzel, Woltjer). However there are some local effects where magnetic effects play an important role. Fujimoto and Miyamoto developed magnetohydrodynamical models of helical fields in spiral arms. The importance of magnetic fields in the interstellar medium has been stressed by many authors. Review articles on the dynamics of the interstellar medium were written by Spitzer, Pikel'ner, and by Field. According to Pikel'ner the spiral arms are in magnetohydrodynamic equilibrium, and the arm thickness is determined by the pressure of the magnetic field and cosmic rays. This equilibrium is unstable, thus the gas breaks into clouds.

There are two basic models of interstellar clouds. In one (Field *et al.*) the major role is played by the gas pressure. The gas is heated by subsonic cosmic rays (Pikel'ner, and others) and cooled by collisional excitation and ionization of hydrogen and helium.

In the second model (Parker and Lerche) the magnetic pressure is considered larger than the pressure of the gas. The system gas-magnetic field is subject to a universal Rayleigh-Taylor instability. We can consider a hydromagnetic self-attraction, which is 10 times larger than the gravitational self-attraction and produces the clouds.

As regards the origin of the magnetic fields of our Galaxy, Parker ascribed it to interstellar turbulence and differential rotation. He proved that any inital weak field reaches at least 1 μG in 10^9 yr. The opposite view, of a primordial magnetic field, has been supported by Thorne.

The effects of non-gravitational forces (magnetic fields, turbulence and cosmic ray pressure) on the motion of interstellar gas around the center of the Galaxy have been estimated by Penston *et al.* They find that the gas lags behind the stars by a velocity of the order of 3 km s^{-1}.

The motion of cosmic rays in a random magnetic field was studied by Jokipii, and by Jokipii and Parker. Lerche and Parker studied also the bulk motion of the cosmic ray gas in our Galaxy, by taking moments of the collisionless Boltzmann equation. Hydromagnetic waves excited by cosmic rays were studied by Lerche and by Wentzel. Evidence for the existence of such waves comes from the scintillations of pulsars.

Weltzel found how the magnetic field of the clouds is separated from the general magnetic field. The influence of a magnetic field on the stability of the galactic disk was considered by Tassoul, and Aggarwal and Talwar. A magnetic field has a stabilizing effect.

The dynamics of cosmic rays were reviewed by Parker, and by Meyer.

The scattering of cosmic rays by waves is more important than the scattering by magnetic inhomogeneities (Fermi mechanism). The interactions between cosmic rays and Alfvén waves were investigated by Kulsrud and Pearce.

According to Ginzburg and Syrovatskii cosmic rays are mainly galactic, while according to the Burbidges and Hoyle they are universal, and according to Sciama they extend throughout the local group of galaxies. We have evidence that the electron component of cosmic rays is probably galactic, as well as protons and heavier nuclei with energies up to about 10^{15} eV, while the highest energy cosmic rays are probably extragalactic.

We have not discussed abundances, composition, and other properties of cosmic rays. A large number of papers on cosmic rays can be found in the Proceedings of the Tenth International Conference on Cosmic Rays (*Canadian J. Phys.*, **46**, no. 10), and the Fourth Texas Symposium on Relativistic Astrophysics.

G. CONTOPOULOS
President of the Commission

REPORT OF THE COMMITTEE OF "SELECTED AREAS"

Progress of research

Introductory notes

At the IAU Prague meeting it was considered desirable to continue to report on work performed in the Kapteyn Selected Areas (below referred to as SA) and other similar fields, as those belonging to the Parenago, Shajn and Cleveland plans, cf. also the main report of Commission 33.

A more detailed version of the present Report, including references to published work, will be distributed to those specially interested.

Photometry

At the Abastumani Observatory, Chuadze has determined photographic *B* and *V* magnitudes for 4000 stars in SA 25, 48, 50, 73, 75, 97, 99, and 122. Kalandadze has continued her work in Fields III and IV of the Parenago Plan and will publish a catalogue for about 8000 stars in Taurus; Metreveli will publish similar data (photographic *UBV* and spectral types) for 3000 stars. In co-operation with the Golosseievo Observatory a development of the Parenago Plan with photometric studies of regions centered at open clusters has been initiated; a catalogue of *B* and *V* magnitudes for 12000 stars by Voroshilov *et al.* has been published.

Papers dealing with *RGU* photometry have appeared from Basel for SA 51 (W. Becker) and SA 54 and 57 (Fenkart); catalogues will appear later. Landolt (Baton Rouge) is establishing *UBV*-sequences in the *V*-range 10·5 to 12·5 for the 0°-zone of SA; observations have begun at the Kitt Peak Observatory in SA 92–96, 98–109, and 113–115. King (Berkely) has completed his *BVR* photometry in SA 51, 57, and 68. Schmidt-Kaler (Bochum) has at ESO in Chile initiated photoelectric observations of 20 stars down to magn. 9 in each of 107 extended southern SA for calibration of quantitative spectral observations. From Bonn is reported that Geyer had made *UBV*-measurements for 18 standard stars in SA 157 in the *B*-range from 7·8 to 15·7, and Seggewiss for 22 stars brighter than $V = 14{\cdot}5$ in SA 87 (to be used as standards for Schmidt plates).

Rybka (Cracow) has reduced anew the previously reported Crimean photometry for SA 1-139. For Rybka's scheme of pairs of stars near *V*-magnitude 7·5, 9·0 and 10·5 in SA 1–43 Abuladze has started observations at the Abastumani Observatory. Pronik and Sharapova (Crimea) have published pg-magnitudes and colour indices for 1500 stars in a Shajn field at $18^{h}54^{m}$, $+5°0''$. Urassin (Engelhardt, Kazan) reports on work by him, Tochtasjev and Dubjago; photographic *UBV* data are ready for about 10000 stars in SA 26–28.

T. Elvius (now in Lund) continues the measurements of plates for southern SA. For this programme Ardeberg observes photoelectric sequences at ESO in Chile, beginning with SA 92, 165, 188, and 200. Lyngå made in Australia photoelectric observations for about 70 stars in SA 201, 202, 204, and 205. For SA 195 Sarg, Roslund, and Engver have made photographic *UBV* photometry. Bigay (Lyon) reports that the photoelectric *UBV* determinations for O, B, and A0 stars have been continued; for SA 40 and 49 results will soon be published, for SA 64, 74, 87, and 110 the measurements are complete. Bigay adds that a similar programme has been commenced in four Cleveland fields.

Priser (Naval Obs., Flagstaff) has started a new SA-programme for photoelectric *UBV*. Sequences of 5 to 6 stars from the central star (8–10 magnitudes) down to 16 or 17 magnitude will be measured in SA 44, 46, 48, 50, 52, 54, 56, 58, 60, 62, 64, 66, 92, 94, 96, 98, 100, 102, 104, 106, 108, 110, and 114. Grigorieva and Rostshina (Sternberg, Moscow) have determined photographic *B* and *V* for 830 stars in a Shajn field at $19^{h}32^{m}$, $+22°5$, and Uranova (also Sternberg) similar data in Parenago field I. From Steward Obs., Tucson, is reported that B. J. and P. F. Bok have determined *UBV* sequences for SA 169 and 205, and that Schreur makes four-colour photometry (Strömgren type) photoelectrically and photographically in Sa 27, 47, 56, and 108.

From Uppsala is reported by Oja that photoelectric and photographic observations have been commenced in SA 32, 55–58, 80, and 81 with special emphasis on late-type giants; Häggkvist and Oja have given photoelectric results for 38 stars in SA 17, and 18 stars in Sa 8. From the Vatican Observatory (Castel Gandolfo) is reported that all plates have now been assembled and reductions initiated for SA 28, 54, 106, and 107. Roman (Washington) reports that she has completed photoelectric *UBV* photometry for 764 stars in the previously listed SA (*Trans. IAU*, **10**, 504, 1960); similar work has been performed also for SA 161, 164, 167, and 185. Zonn (Warsaw) has in collaboration with Kraskiewicz and Serkowski started photographic observations with the Schmidt telescopes at Torun and Mount Stromlo for a catalogue of galaxies in SA; as a by-product magnitudes of faint stars will be obtained.

Proper motions

Murray (Greenwich Obs., Herstmonceux) reports that work on improving the proper motions in the northern SA is in progress. Modern plates to match all the Radcliffe plates are being taken with the 26-inch refractor at Herstmonceux. When the programme is complete, there will be a total of eight plates covering an interval of 50 to 60 yr in most of the 115 areas. In addition a series of 170 plates on many northern SA have in recent years been taken with the 26-inch refractor to match plates taken with the same telescope, when at Greenwich, for photometric purposes during 1914 to 1920.

Radial velocities and spectral types

At Abastumani, Haradze and Bartaya are performing two-dimensional classification to magnitude 11·5 in SA 1–45. To magnitude 13 similar work is done by Chuadze for SA 25, 48, 50, 73, 75, 97, 99, and 122, and by Metravili in the Parenago region III. Such work is also planned for the open cluster fields, mentioned in the section on photometry.

The Haute-Provence and Marseille programmes have been continued; for SA 8 an investigation by M. Barbier is ready. T. L. Evans (Radcliffe, Pretoria) determines radial velocities and spectral types for 30 late-type stars in SA 141.

For SA and similar fields spectral determinations are also performed at the Bochum, Lund, Sternberg, and Uppsala observatories, and by Miss Roman, in connection with the photometric programmes mentioned above. Upgren (van Vleck, Middletown) has just compiled a finding list for stars of types B8–F0 in SA 28, 54, 106, and 107.

Radio astronomy

Kerr (College Park, Maryland) reports on 21-cm hydrogen observations made at Green Bank for SA 1–163, and at Pereyra, Argentina, for SA 164–206 and an overlapping strip. The central points, and four points stepped off one beamwidth, were observed for each SA. When fully reduced, these observations will provide data at a grid of points over the whole sky for systematic studies of the distribution and flow pattern of the nearby hydrogen, and for detailed comparisons with the wealth of optical data available in the SA.

T. ELVIUS
Chairman of the Committee

34. INTERSTELLAR MATTER AND PLANETARY NEBULAE (MATIÈRE INTERSTELLAIRE ET NÉBULEUSES PLANÉTAIRES)

PRESIDENT: D. E. Osterbrock.
VICE-PRESIDENT: F. D. Kahn.
ORGANIZING COMMITTEE: L. H. Aller, R. D. Davies, J. M. Greenberg, S. B. Pikel'ner, L. Rosino.

PLANETARY NEBULAE

Although additional surveys with large Schmidt telescopes continue to yield small catches of additional planetaries (Kohoutek) one cannot escape the conclusion that most planetaries available to existing equipment have been detected. Much remains to be done with powerful Schmidt equipment in the southern hemisphere, especially in the Magellanic Clouds. A vast body of photometric and spectroscopic observation needs to be garnered for the numerous faint nebulae so far discovered.

Spectroscopic and spectrophotometric studies have been carried out for most bright planetaries ($\lambda < 5800$) but much remains to be done in the red and near infrared. Important advances have been made in the far infrared ($\sim 10\mu$) (Gillett, Low, Stein, Woolf) where a number of planetaries seem to show abnormally strong continua. This abnormally intense radiation has been attributed to non-thermal emission, effects of many faint lines, and to thermal emission by dust grains (Krishna Swamy, O'Dell) with perhaps the bulk of the evidence favoring the last-mentioned hypothesis. An increasing number of radio observations from 9·5 mm to 73 cm (Thompson, Colvin, Stanley, LeMarne, Kaftan-Kassim, Babieri and Ficarra, Terzian, L. Aller and Milne, Hughes) all indicate that planetaries are thermal sources.

Morphological forms and models (Khromov and Kohoutek) permit broad, general descriptions of planetaries but do not do justice to the importance of condensations and filaments evident in best photographs (Minkowski, Kron and Walker) or in spectroscopic observations (Walker, Aller, Czyzak, Kaler, *et al.*), although the reason for the formation of such condensations cannot be understood in terms of thermal instability (Field, Harrington, Mathis). Studies of line profiles in bright planetaries suggest that nebular shells may actually be rather thin, that turbulent velocities are small, that nebulae have shapes of prolate spheroids, and that there is evidence for a stellar wind (Weedman). Attempts to calculate dynamical models (Mathews, Sofia and Hunter, Tamura) show that some of the features of planetaries can be interpreted in terms of fairly simple dynamical models but fairly grave quantitative difficulties remain.

Attempts to handle stratification problems and variation of ionization, temperature and radiation field with depth in the nebula have been made (Goodson, Williams, Flower, Harrington).

Theoretical studies of the hydrogen spectra of nebulae (Cox, Mathews, Walmsley, D. George, Van Bloerkom) and of the helium spectrum (Robbins, Capriotti) have not resolved all difficulties as is evident from the mutually contradictory electron temperatures derived from features of the helium and hydrogen spectra (Lee, Kaler).

Transition probabilities for forbidden lines seem to be in a satisfactory state, while the situation for collision strengths is now much improved (Seaton *et al.*, Czyzak and Krueger, Smith). From these data additional forbidden line ratios can now be interpreted in terms of ionic densities and temperatures.

The problem of the distance scale remains a most vexing one (Gordon, Seaton, Webster). The best clues to this difficulty would appear to be offered by the Magellanic Clouds. The masses of the planetaries most likely lie between 0·09 and 0·26 solar masses (Webster). Doubts on the quantitative correctness of the Harman-Seaton evolutionary track have been raised by Abell and by Webster who suggest that the H-R diagrams observed for central stars represent not the evolutionary track of a nucleus of single mass and chemical composition but rather a combination of an evolutionary track and a mass-dependent sequence.

Fortunately, new observational data are being obtained for central stars of planetaries, magnitudes and colors (Liller) spectra (Lindsey-Smith, Aller, Heap) while new model atmosphere calculations incorporate increasingly sophisticated refinements (K.-H. Böhm, Hummer and Mihalas). Unfortunately the model atmospheres do not predict the observed absorption line shapes and it would appear that conditions of radiation or hydrostatic equilibrium (or both) will have to be relaxed.

The possibility of an association of X-ray sources and planetary-nebulae nuclei remains a challenge (Blanco *et al.*) but data at present are not convincing.

Arkhipova, Dockuchaeva and Kostjakova (Sternberg Astronomical Institute) have carried out absolute photometry of more than 100 planetary nebulae, and the work on the spectrophotometry of planetaries in the central region of the Galaxy is still going on. Observational study of the ultraviolet spectra of these objects in the region $\lambda < 3650$ Å has begun, and the study of their infrared spectra in the region $\lambda < 1{\cdot}1\,\mu$ is planned. Systematic *UBV* electrophotometric observations of the nuclei of planetary nebulae are in progress, particularly a search for variability. Khromov and Moros (Sternberg Astronomical Institute) have begun a study of continuous emission from planetary nebulae in the spectral region 1–2·5 μ to improve estimations of their electron temperatures and the temperatures of their central stars. Kazarian (Burakhan Observatory) produced a colorimetric study of central stars of 20 planetary nebulae and 10 of them were investigated spectroscopically.

Khromov in cooperation with Kohoutek (Astronomical Institute of Czechoslovakia, Prague) completed a morphological study of planetary nebulae. Khromov has begun work on the determination of their structure from intensities of forbidden lines originating in different luminosity zones. Feklistova (Institute of Physics and Astronomy, Tartu) computed the expected spectrum of a planetary nebula in the far ultraviolet region. About 100 transition probabilities were computed for different transitions in ions of C, N, O, Si and Ne.

Perinotto has started a systematic survey of high and low galactic latitude fields for the identification on objective prism plates of new planetary nebulae. He will also continue with Wurm the spectroscopic and photometric researches on the Orion Nebula. Studies in the relatively far infrared of the spectrum of gaseous and planetary nebulae have been carried out at Asiago with an RCA-Carnegie infrared image tube. Perinotto will measure the ratio Hα/[N II] in faint planetary nebulae. Wurm and Rosino will prepare a new and more extended edition of the Monochromatic Atlas of the Orion Nebula.

Y. Andrillat has made a series of researches in the near infrared ($\lambda\lambda$ 5800–8750) with the 120-cm Newtonian telescope in Haute-Provence, using a grating spectrograph with a mean dispersion of 230 Å/mm. She has determined the intensity of the hydrogen Paschen lines and the following other lines: He I 6678, 7065, 7281; [A III] 7330, 7319; [S II] 6731, 6717; [O I] 6363, 6300. Some nebulae (NGC 1976, NGC 6572, IC 418, I I 4997) have been found to present a strong O I 8446 emission. The problem of the excitation of this line through a fluorescence process starting from $L\beta$ has been considered. With measures of intensity of other oxygen lines, such as [O II] 7330, 7319, [O I] 6363, 6300 it is possible to derive the density of the $L\beta$ radiation.

Future projects consider the continuation of the spectrophotometric researches on the planetary nebulae using image tubes or electronic cameras extending the infrared observations up to 1·3 μ. The Paschen decrement will be determined by infrared observations and compared with the values given by the theory of hydrogen recombination, in order to see whether there is a disagreement as is found in the highest terms of the Balmer series. The study of the line He I 10830 will bring new information to the problem of helium abundance. Finally, a further study of O I lines at 9625 and 11830, is devised with the purpose of a better understanding of the $L\beta$ fluorescence mechanism.

Enumerated here are certain of the outstanding remaining problems in connection with planetary nebulae.

Observational

A more intensive study of planetaries in the Magellanic Clouds, which will require the very largest telescopes (apertures of about 150-inches), is urgently needed. At the same time observers will want to fill in the photometric and spectroscopic gaps for known planetaries.

High-resolution photographs such as probably can be obtained only with space telescopes are most urgently needed. A more intensive exploration of the infrared without any interference from the earth's atmosphere is needed. Certainly an exploration of the ultraviolet spectra of a number of bright planetaries is required.

Structure and kinematics

More emphasis must be placed on fine structure of nebulae, both observationally and theoretically. We do not yet understand how condensations can form. Kinematical and stratification effects need to be considered together from a unified point of view. It should be possible to solve for radiation field throughout interiors of model nebulae designed to fit real nebulae as closely as possible. A knowledge of radiation fields should make it possible to solve for chemical abundances.

Theoretical studies

Studies of the hydrogen and helium spectra do not yet explain the observational data quantitatively. There remain discordances in the relative intensities of the helium lines, the Balmer decrement, and the ratio of the Balmer lines to continua.

The problem of distances and masses

This problem still poses formidable questions, which underlie any attempt at a physical theory of stellar evolution of the central stars. The evolutionary sequences of the central stars cannot be regarded as established since we appear to have a mixture of stars of different ages, masses and luminosities.

As for many *central stars* themselves, it cannot be overemphasized that there remains a serious discordance between temperatures estimated from the appearance of the spectra and those obtained from the Zanstra method. The model atmospheres do not represent the observed spectrum lines; this may be attributed perhaps to failure of the assumption of hydrostatic equilibrium (stellar winds), non-LTE effects, etc.

H II REGIONS

Riegel has searched for 21-cm emission of H I associated with galactic H II regions, and found that in some cases no H I can be detected, while in other cases it is present, but there is no case in which the velocity dispersion is large enough to demonstrate conclusively the presence of a surrounding expanding H I region.

Observational studies by Wurm and by O'Dell and his co-workers have demonstrated the presence of dust in many H II regions. Likewise infrared observations of several H II regions by Neugebauer, Low and others have shown that the continua are much brighter than can be explained by the free-free and free-bound emission of the gas, and must be due to thermal emission by the dust.

Theoretical studies by Mathews, Code, O'Dell, and Krishna Swamy have demonstrated that radiation pressure on the dust grains can have important effects on the internal dynamics of the H II regions, and may be responsible for the central holes in some, such as NGC2244.

Radio free-free continuum and recombination-line observations of many H II regions have been carried out by Mezger, Hoglund, Schraml, Henderson, Kieter, and others. Approximate distances are derived from the observed radial velocities, and temperatures from the ratio of line to continuum strength. However theoretical work by Goldberg, Dyson, and Hjellming, and observations of several different lines by Zuckermann, Palmer, Penfield and Lilley show that the simple LTE assumptions are not correct, and the temperatures are therefore not well determined by this method. Observations of several lines may be used in principle to solve for both electron temperature and density, using the theory to find the dependence of b_n on these quantities. However spatial density and temperature variations complicate this method in practice. These observations show that most H II regions have turbulent (rms) velocities of order 10 to 25 km/s. In addition helium recombination lines have yielded the relative He/H abundance in nebulae.

In some H II regions several small, high-density condensations have been revealed by the radio observations. They appear to be ionized shells (cocoons) around recently formed O stars.

Dyson has computed the b_n factors for hydrogen in gaseous nebulae taking into account $n \pm 1$, 2, 3, 4 collisional transitions. The computed b_n factors have been used to compare observations of line temperatures with theoretical predictions of line enhancement. Ratios of the peak temperatures of lines of the same frequency originating from different upper quantum levels indicate line enhancement, but the interpretation depends critically on the estimate of Stark broadening.

Line profiles of the 94 α line in the Orion Nebula measured by Gordon and Meeks show good agreement with low-resolution optical measurements. Dickel, Gebel, Schmitter, Ishida and Kawajiri mapped the distribution of interstellar extinction within and in front of several H II regions by comparison of radio free-free and optical Hα or Hβ surface brightness measurements. Fluctuations of extinction of several magnitudes are found in several of these nebulae. The classical extinction law was confirmed in the region of galactic longitude 5°–130°, while possible variations were found in the region 130°–210°. Lee has studied the extinction in the Orion region by multicolor photometry of the involved stars. He confirms the abnormal extinction laws in the center of the nebula and near ζ Ori. In all cases the stars with abnormal extinction are very early-type stars closely associated with bright nebulosity but not all such stars have abnormal extinction. Similar comparisons of radio free-free and infrared P_{12} surface brightnesses in the Orion Nebula have been made by Werner, Pipher, Terzian and Houck.

In connection with the work by Courtès' group in Marseille a cooperative program of study of H II regions was begun in early 1968 by astronomers in Institut d'Astrophysique de Paris, Observatoire de Meudon, Observatoire de Lyon. The program includes identification, photoelectric and spectroscopic observations of the exciting stars ($m < 12$) (Fringant, Garnier Lortet); radio observations in the continuum at 11 cm and in the H and He recombination lines (Gribenski, Lauque, Kazes). The first regions investigated were small H II regions (diameter less than 15′). Measurements of Hα, Hβ, Hγ and forbidden lines intensities are planned, in different points of some extended (bright) regions, in order to deduce detailed physical conditions in these nebulae.

Meaburn has begun a program measuring shapes, widths and intensity ratios of [S II], [N II], [O III], [O II] and H emission lines in H II regions, supernova remnants, and planetary nebulae using scanning Fabry-Perot interferometers on the telescopes at Pic-du-Midi. Similar measurements have also been made at Kitt Peak by Weedman and Smith.

Considerable theoretical work has been done by Vandervoort, Mathews, Lasker, Hjellming and others on the expansion of H II regions, driven ultimately by the ionizing radiation from the stars. A shock runs ahead of the ionization front, compressing the surrounding neutral gas, while inside the ionized region radiative cooling keeps the temperature more or less isothermal, and the expansion velocities are generally subsonic. Some of the detailed effects of density fluctuations, self-absorption of infrared cooling lines, etc. on the temperature structure of H II regions have been worked out by Rubin.

Dyson has calculated the properties of spherically symmetric self-gravitating condensations of neutral hydrogen immersed in an H II region. A condensation is represented by an isothermal gas sphere with a D-critical ionization front incident on the surface. Typical parameters of such a condensation compatible with the estimated ultraviolet radiation field in the central region of the Orion Nebula are: mass $\approx 1 M_{\odot}$; radius $\approx 10^{16}$ cm; mean density $\approx 10^{-15}$ gm cm^{-3}. The majority of condensations are stable configurations and act as sources of ionized gas which flows into the surrounding nebula.

Dyson has also studied the interaction of the supersonic gas streams from two adjacent condensations. A system of stationary shock waves is set up. Theoretical line profiles through the system give good agreement with those observed in the bright central curves of the Orion Nebula.

Goldsworthy is working on the conditions under which shocks will form within H II regions and on possible reasons for filamentary structure. His research students are working on the problem of ionization front structure – a condition has been obtained which allows ionization fronts to be treated as discontinuities and, at the same time, gives the unique solution. It is the analogue of the Chapman-Jouguet condition in conventional combustion theory. Numerical computations on finding the flow are in progress, similar to Mathews's calculations, but careful watch is being kept upon problems of stability and uniqueness. Another research student is looking at the interaction of ionization fronts with globules – in the first instance improving Dyson's treatment.

Rasiwala is studying the evolution of H II regions, with the exciting star in motion. The motion of the star produces a spherical H II region. The evolution of these regions after the birth of the star in a hydrogen cloud has been studied in two steps:

1. Finding the shape of the H II region.
2. Solving the flow problem within the region.

The hydrodynamic equations have been solved using a semi-analytical method called the "method of integral relations".

Optical spectrophotometric observations by Aller, Czyzak and Walker of NGC604, the brightest H II region in M 33 show that it has approximately the same composition (abundances of He, O, Ne and S relative to H) as the H II regions in our Galaxy.

The problem of the discrepancy between the relatively low values of the electron temperatures in H II regions indicated by low radio-frequency continuum measurements (either emission or absorption) and the higher values indicated by optical emission-line intensity ratios still remains. Temperature fluctuations along the line of sight, as suggested by Peimbert, can partly resolve the discrepancy; however they cannot explain temperatures as low as the value $T_e = 3500°$ reported for IC1805 by Rogers or 3000° reported for NGC1976 by Terzian, Mezger and Schraml. Peimbert and also Rubin have discussed how these temperature variations must be taken into account in deriving abundances from spectroscopic observations of H II regions.

On the other hand Field determined $T_e = 6100°$ for IC1805 from 38 MHz measurements, so no doubt there are still substantial observational problems connected with finite beam size, correction for foreground radiation, and absolute calibration.

Akabane, Morimoto and their colleagues are making continuum measurement of several H II regions at 4170 MHz with 30-m dish. The results obtained were compared with optical data. Yamashita and Watanabe also observed W 49 at 9·4 GHz with 10-m dish. Kawajiri and Akabane studied a statistical relation between the size of an H II region and its electron density. Sato, Akabane and Kerr estimated the distances of the W 49 components using Parkes 210-ft data.

Losinskaya and Doroshenko (Sternberg Astronomical Institute) have been investigating the velocity field in the nebulae IC443, S-22, NGC6888, and the Cygnus Loop by observations of Hα and [N II] lines using a Fabry-Pérot etalon coupled to an image-converter. The results support the idea that IC443 and S-22 are supernova remnants, and that NGC6888 originated from the ejection of the matter from the WR star HD192163 into the interstellar medium. It was shown that the filaments in Cygnus Loop nebula are one-dimensional structures. Dibay (Sternberg Astronomical Institute) has completed an observational study of some cometary nebulae, which allowed him to determine their masses, structure, dynamics and inner physical conditions. The evolution of cometary nebulae and their associated stars is being studied theoretically on the basis of the theory of the gravitational condensation and magnetohydrodynamics. Parsamjan (Burakhan Observatory) obtained observational data for a spectroscopic study of cometary nebulae.

Sorochenko and others (Lebedev Physical Institute) observed radiofrequency recombination lines of H in several bright diffuse nebulae in the millimeter and centimeter wavelength range for determining T_e distribution, inner dynamics, hydrogen level population and its deviations from thermodynamic equilibrium. Preliminary results show that the theory of the Stark-effect must be revised. It is intended to receive radiofrequency images of nebulae at $\lambda = 8$ mm with an angular resolution $2' \times 2'$; the search for new recombination radio lines will be continued.

Pikel'ner in cooperation with Sheglov has considered the influence of the stellar wind from O stars upon surrounding interstellar clouds, which possibly drives fast motions of the matter. Similar computations are in progress for the envelopes of WR stars. Dibay (Sternberg Astronomical Institute) attempted to interpret some morphological peculiarities of diffuse nebulae as "elephant trunks", globules and biconical nebulosities. All these peculiarities appear to be readily explained in the framework of the gravitational condensation of the gas possessing magnetic field during the formation of T-Tauri stars.

Courtès, Louise and Monnet have measured the widths of Hα and [N II] 6583 emission lines in several H II regions, using a photographic Fabry-Pérot interferometer. From these line widths they find effective temperatures of order 5000–11000 K in various nebulae, and turbulent velocities of 6 to 12 km/s. They have also measured the Hα/[N II] λ 6583 intensity ratios in many of the same H II regions, and have deduced lower limits to the relative abundance N/H, which in some cases are considerably larger than the solar abundance ratio.

By means of spectrophotometry the thermal and ionization structure of H II regions has been studied theoretically by Gürtler. He finds that the "on-the-spot" approximation fails in the main parts of the region and yields electron temperatures which are too high by up to 2000 K in comparison with values computed with the exact diffuse radiation field. Density gradients and fluctuations have large effects on temperature, ionization structure and radii of H II regions.

Gerola and Panagia have indicated new possibilities of interpretation of the spectra of gaseous nebulae. They have considered the continuum emission in a gaseous nebula of pure hydrogen, in which the excitation occurs by radiative and collisional processes. In the case of collisional excitation a much greater efficiency for two-photon emission is found, which may drastically change the optical continuum.

Viotti has studied the excitation of Fe II permitted and forbidden lines in some peculiar objects and in the Orion Nebula. He found that the continuum of Eta Carinae is mainly nebular in origin, with a strong contribution of two-photon emission.

HI REGIONS

A picture of the standard interstellar cloud has been current in the astronomical literature for some time. It is taken to have a diameter of 10 pc, a density of 2 hydrogen atoms per cm^3, a mass of 20 $M_\odot$ and an internal velocity dispersion of 2 km/s. Recently Heiles has made a survey with a resolution of 10′ arc in angle and 1 km/s in velocity of a region 160 sq. deg. in area centered at $l^{II} = 120°$ $b^{II} = +15°$ which shows that the standard model is a gross oversimplification. He finds that concentrations similar to the standard cloud are rare and identifies three main groups of features. The first are sheet-like structures elongated parallel to the galactic plane with a width perpendicular to the plane of 20 pc or less and length of 100 pc or more. Their density is ~ 2 cm^{-3} and their velocity dispersion is 1·4 km/s. The second group comprises the "concentrations" which typically have a radius of 20 pc, a mass of 3×10^3 $M_\odot$ and a neutral hydrogen density of 5 cm^{-3}. Then there are the cloudlets identifiable with a surface density of 5 per sq. deg. in the region surveyed. They have a radius of 3 pc, a mass of 10 $M_\odot$, a neutral hydrogen density of 2 cm^{-3} and a velocity dispersion of 1·2 km/s which implies a gas temperature of 100 K or less. Each of these groups seems to contain about the same amount of neutral hydrogen in total.

Heiles finds no significant excess of neutral hydrogen emission in the dust clouds. This is consistent with the conclusions of some of the earlier work on this problem and suggests that the hydrogen in these regions is bound up in the form of molecules. Subsequent observations have shown 18 cm OH emission from a number of dense dust clouds. Numerous OH absorption clouds have also been found in regions along the galactic plane; however it is not possible to identify these objects with particular dust clouds because of the large optical absorption in these regions although most of these features have associated neutral hydrogen absorption indicating densities of 10–

100 atoms cm^{-3}. Formaldehyde (H_2CO) has been found in absorption both in known dense dust clouds and in the denser H and OH absorption clouds indicating clearly the presence of large concentrations of molecules in the densest clouds. Regions of even smaller diameter ($\sim 3 \times 10^{16}$ cm) have been found to emit anomalous (masered) OH radiation near many of the brighter H II regions. Densities of ~ 1 OH molecules cm^{-3} have been inferred for these regions, which may then have hydrogen densities of $10^{5\pm1}$ cm^{-3} assuming the normal abundance ratio of H/OH. Such regions may well be collapsing protostars. Anomalous H_2O emission has been found in many of these regions and shows in some cases variability time scales of the order of weeks implying size scales of 3×10^{16} cm or less in agreement with the OH results.

There has been renewed observational and theoretical interest in the temperature of the interstellar gas. A number of heating mechanisms have been investigated, including low-energy cosmic rays, supernova explosions and X-rays. These mechanisms as well as photoelectric heating tend to make the low density clouds and intercloud regions a factor of an order of magnitude or two hotter than the dense clouds with hydrogen densities of 10 cm^{-3} or more. The neutral hydrogen data now appear to be showing evidence that an appreciable number of clouds have temperatures 10^3 K or more; cloud temperatures of a few tens of K have been shown for some time. More neutral hydrogen absorption observations to give temperatures and pulsar dispersion observations to give electron densities will help decide the validity of the proposed heating mechanisms.

The heating of H I regions by low energy cosmic rays has been reconsidered by Spitzer and Tomasko and by Spitzer and Scott with detailed computations of the efficiency with which the energy of secondary electrons goes into heating the gas. The assumption that low energy particles are produced by supernovae gives an upper limit on the heating rate; the corresponding temperatures exceed 10^4 degrees if n_H is less than 0·1 cm^{-3}, and the resultant thermal instability may account for the existence of interstellar clouds.

Measurements of the equivalent widths of interstellar $L\alpha$ absorption in the rocket UV spectra of stars in Orion by Jenkins and Morton, ζ Pup by Morton, Jenkins and Brooks, and γ Cas by Morton, Jenkins and Bohlin all indicate an average line-of-sight H I density of about 0·1 atom cm^{-3}. On the other hand, observations of stars in Scorpius by Jenkins, Morton and Matilsky show a closer agreement with a value of 1 atom cm^{-3} from 21-cm emission surveys. A review of the $L\alpha$ data has been presented by Jenkins.

Nishimura calculated the structure of the radiative cooling layer behind the shock front due to a cloud collision at a velocity corresponding to Mach number 20, with and without a magnetic field. He studied the abundance of H_2, which depends greatly on the photo-destruction rate. He and K. Takayanagi discussed the destruction rate and showed that the life time of H_2 may be substantially prolonged at the center of a molecule-rich cloud.

Takayanagi and Itikawa studied the rotational excitation of CN molecules by thermal electron impact, applied the result to the interstellar CN molecules and concluded that the observed population in the rotationally excited level cannot be explained by collisional excitation unless the molecules are in a high-density H II region.

Syunjaev (Institute of Applied Mathematics, Moscow) has shown that the heating of the gas in H I regions must be caused predominantly by the metagalactic X-ray emission with energy about 270 eV.

Schmidt has emphasized that no exact value of the density ratio H/dust is available because of the lack of knowledge of the spin temperature of interstellar H and H_2. In high-latitude H I clouds the radiation pressure from stars of the disc population ejects dust. This produces a decrease of the interstellar heavy-element abundance which may be compensated by supernovae events (van den Bergh and Schmidt). Zimmermann has shown that in young H I clouds, i.e. new ones formed after cloud collisions, the heavier particles are concentrated in a thin sheet in the inner parts, while the lighter particles are distributed more or less uniformly through clouds. Schmidt has shown that reflection nebulae are probably in a cosmogonic relation with the illuminating stars, and not distributed by chance.

LARGE SCALE DISTRIBUTION, DYNAMICS AND CONDENSATIONS

Gossachinsky (Pulkovo Observatory) is studying the brightness distribution of the galactic radiation in the 21-cm line. Pariisky and others (Pulkovo Observatory) are engaged in observations of the thermal galactic radio-emission for studying the continuous radio background, the fine structure of H II regions and magnetic fields.

Ariskin (Lebedev Physical Institute) has studied the connection between spatial distribution of the neutral and ionized hydrogen between the longitudes $20^\circ.8 < l^{II} < 32^\circ.8$. Similar researches are underway at Lebedev Physical Institute for other galactic regions as well.

Herbig has analyzed in detail the rich interstellar line spectrum of ζ Oph, and found the total numbers of Na I, Ca II, K I, CH and CN absorbers/unit area, as well as upper limits to several other atoms, ions, and molecules. These data were then used together with a model of the interstellar matter distribution to find the concentrations/unit volume. There is evidence for a thin ($\approx 0{\cdot}15$ pc) dense ($n \approx 10^3$ cm^{-3}) H I layer somewhere along the line of sight to the star. Hobbs has made high-resolution ($\approx 0{\cdot}01$Å$=0{\cdot}5$ km/s) photoelectric scans of the Na I interstellar absorption lines in 77 stars using a pressure-scanned Fabry-Pérot interferometer. These observed profiles generally show multiple overlapping lines. The profiles of single components are Gaussian in some cases, but exponential in others.

Interstellar lines in coude spectra of southern early-type stars (6·8 Å/mm) taken by Thackeray are being studied by Picton (University of London Observatory) who is planning to extend the survey.

Takakubo compared the interstellar absorption lines Ca II K observed by Adams with the 21-cm H I emission line at intermediate galactic latitudes and found that the general natures of optical and radio profiles are quite similar. The correlation between the radial velocity of strongest K component and that of 21-cm is very high, while it is low for high velocity components. He also treated 21-cm G components to obtain kinematical properties of the interstellar gas.

Crampton at the DAO has obtained 60 Å mm^{-1} spectra at the 72-inch telescope and also 2·4 Å mm^{-1} and 6 Å mm^{-1} spectra at the 48-inch telescope to study interstellar K-lines in order to obtain distances of H II regions. Better results should be obtained when the re-imaging camera for the image tube, which is now being assembled, is commissioned.

Hutchings' work on Nova Del included measurement of interstellar Ca II H and K lines in more than 100 spectra. High dispersion plates showed H and K to be double and also revealed the CN lines at 3874·0, 3874·6Å.

From two-color diagrams at high galactic latitudes which are derived by using the kinematical properties of stars of different uv excesses Pfau found that in the North Galactic Polar Cap the reddening is smaller than $E_{B-V} = 0{\cdot}02$, as determined from stars 200 to 500 pc above the galactic plane. Interstellar grains which left clouds and are moving in the intercloud medium under the influence of the pressure of the interstellar radiation field and of the gravitational K_z-force must constitute a dust layer at a distance of about 150–200 pc from the galactic plane (Schmidt). The motion of dust grains during collisions of interstellar clouds, the growth and the destructive processes and the influence of the dust particles on star formation after cloud collision have been studied by Zimmermann.

Observation by means of the 21-cm line continues to provide the fullest set of data concerning the distribution of interstellar atomic hydrogen with respect to galactic longitude and latitude and line-of-sight velocity. The recently published Maryland-Green Bank Survey covers about 340 sq. deg. of sky close to the galactic plane, between $l = 11^\circ$ and 235° in the new coordinate system. The results are presented in the form of contour maps of temperature, taken at constant declination δ, plotted with respect to line-of-sight velocity $v_{\parallel}$ and right ascension α. On a typical map the contours are spaced 1 K apart; a line drawn at constant α would intersect some 60 or 80 contours, typically; a line drawn at constant $v_{\parallel}$ would intersect some 20 contours. The collection of maps therefore contains well over a million significant data points.

Kerr's review article surveys recent developments in hydrogen line studies, and also discusses their interpretation in terms of a galactic rotation curve, a map of the galactic rotation curve, a map of the galactic hydrogen arms and an estimate for the total mass of hydrogen. The 21-cm measurements only lead to a mass per unit area in a column along the line-of-sight, but for any given feature of the interstellar distribution the area within the antenna beam depends on the square of the distance assigned to that feature. In any model of the Galaxy the estimated mass therefore depends quadratically on the value adopted for R_0, the distance from the Sun to the galactic center, since this value sets the linear scale of our model. The correct measure of R_0 is still uncertain by about 20%, and its exact determination cannot be achieved by studies of interstellar matter, but only by means of astrophysical arguments.

Another simple uncertainty arises from the calibration of the antenna temperature. Thus Westerhout states that the observations of the Maryland survey have to have their temperature scales multiplied by a factor 1·18 to bring them into accord with the Australian data, in those regions where the two surveys overlap. Combining the various uncertainties it seems that there may be an error of as much as a factor 1·7 in the estimate for the mass of atomic hydrogen in the Galaxy.

The more detailed interpretations of the hydrogen distribution begin with the construction of a rotation curve and the consequent transformation of the α, δ, $v_{\|}$ map into a three-dimensional map in position space. As has always been done, the rotation curve and all subsequent work are based on the assumption that the systematic motion of the interstellar gas is everywhere in the azimuthal direction, and that it has axial symmetry with respect to the galactic center. This procedure eventually yields a rotation curve whose form depends on whether the observational results used come from the Northern or Southern terrestrial hemisphere. Clearly there is here an indication that some drastic oversimplification has occurred. If one cannot safely assume that the calculated rotation curve has much significance, then one cannot rely upon the location assigned to particular features of the interstellar hydrogen distribution. No doubt the map of the spiral arms is at least topologically correct, but it might be a rather distorted representation of the true structure. This must be borne in mind in discussions on the location of various features of the galactic disk, for example the twist at the galactocentric longitudes $\lambda = 90°$ and 270°.

It has of course long been realized that no model galaxy can continue to support a spiral structure if its interstellar gas is in a state of differential rotation with axial symmetry. To explain the spiral pattern one has to adopt a theory such as that of Lin and Shu who proposed that the spiral pattern is due to a density wave which moves around the Galaxy with a uniform angular velocity Ω_P (equal to 11 km s^{-1}/kpc in their model). There are two spiral arms, each wound about $1\frac{3}{4}$ times around the galactic center, with inner and outer ends at 4 and 14 kpc galactocentric distance, respectively. The arms trail, and Ω_P is everywhere less than Ω, the angular velocity of the interstellar gas in orbit about the galactic center. For example $\Omega = 25$ km s^{-1}/kpc at the Sun's position. Consequently both population I stars and the interstellar gas move so as to overtake continually the spiral pattern. The gravitational field due to the spiral arms perturbs stellar orbits by a displacement with a $\sin [2\lambda - \phi(r)]$ dependence. Here λ is the galactocentric longitude, and $\phi(r)$ a phase term dependent on the distance from the galactic center. The diagram shows how this leads to a perturbed density distribution among the stars and thereby to a spiral component of the gravitational field.

But the gas is affected differently. As the diagram shows a shock is expected to form on the trailing side of each arm. Behind the shock the gas density is much enhanced, and the internal energy is large enough so that the gas motion is now subsonic relative to the spiral pattern. The gas then accelerates forward, away from the shock, passes through a sonic point and runs supersonically into the next shock (Fujimoto).

If there are such large velocity perturbations then there must also be considerable changes made in the interpretation of the 21-cm profiles. The theoretical approach to the problem of interstellar gas dynamics must also be changed. The gas flow proposed in the shock model becomes possible only if the gas which has passed through the shock can retain its internal energy long enough so as to accelerate the flow through the next sonic point downstream, and into the next shock.

The increased internal energy in the gas behind the shock might be ascribed to increased speeds

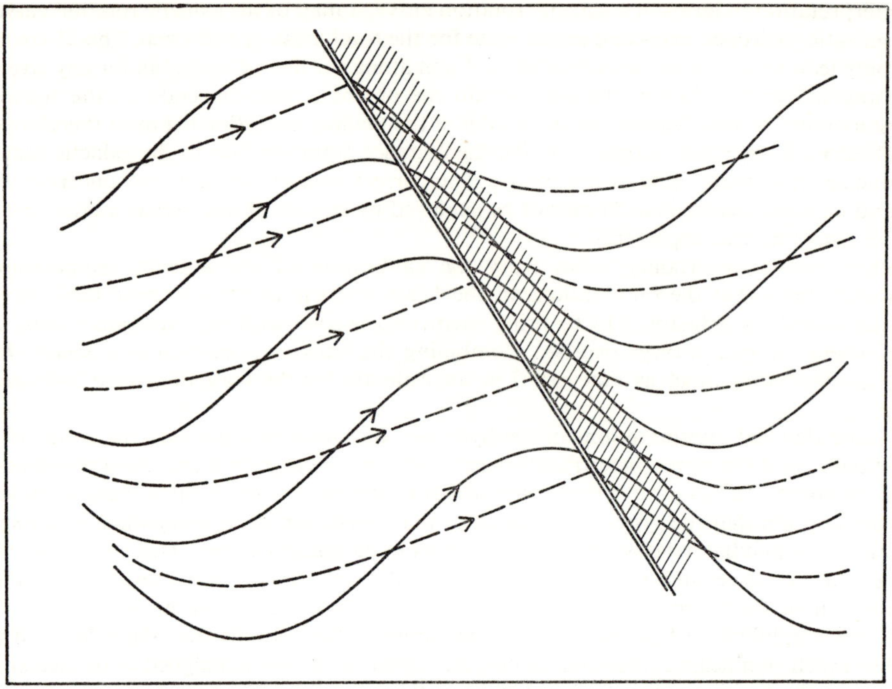

Fig. 1. To illustrate schematically how the gravitational field due to a spiral arm imposes a sinusoidal pattern on the motion of stars (continuous curves), and affects the motion of fluid elements in the interstellar gas (broken line). The shock is indicated by the double line. The crowding of the stellar orbits produces a spiral arm in the shaded region. The stars and the gas move relative to the spiral pattern in the direction as given by the arrows. The horizontal axis represents half of a circle around the Galaxy, straightened out as in Mercator's projection.

in the random motions of the gas clouds, to increased temperature or to an enhanced cosmic ray content associated with the gas. Energy of random motions would decay too fast, thermal energy would radiate too rapidly: the enhanced energy content of the cosmic ray gas thus seems the most probable. Perhaps a rather large rate of production of cosmic rays and suprathermal particles takes place behind the shock, on the side of the spiral arm where the rate of star formation is known to be above average.

Recent work by Spitzer and Tomasko and by Field, Goldsmith and Habing has shown that cosmic rays and/or suprathermal particles are most probably the main sources of heat for the H I regions in interstellar space. One might therefore expect systematic differences in the thermal state of the gas in regions behind and ahead of the spiral shocks. It would also become interesting to study, as Skilling has done, how the cosmic ray particles diffuse relative to the thermal gas. Skilling derives a maximum diffusion rate in his paper; his argument relies on the conclusion that a hydromagnetic instability occurs when the "diffusion speed" of the cosmic ray gas exceeds the local Alfvén speed. The diffusion speed is defined as the ratio of the magnitude of the net energy flux vector to the energy per unit volume for the cosmic rays, measured relative to the ambient interstellar gas. There are two further interesting consequences: following a large injection of cosmic ray energy into interstellar space, the gas and cosmic rays are effectively locked together. This might occur in regions surrounding recent supernova explosions (and probably pulsars), where the cosmic rays would tend to drive the gas. Katgert's observation of a remnant of a large scale explosive event may well be explained in some such manner.

Further, in regions where the instability occurs it will lead to fluctuations of the interstellar electron density. Rickett has found evidence for such fluctuations by his interpretation of pulsar scintillations. The typical length scale of the fluctuations is so short that they could not be produced by any hydrodynamic effect, but only by a plasma interaction.

The cosmic rays will also considerably influence the thermal state of globules and other interstellar condensations. These objects have been discussed from various points of view, either as sources of ionized gas in HII regions (Dyson) as potential sites for the formation of stars or associations of stars (McNally, Disney, Wright, Penston).

Penston has also claimed to find a good correspondence between model globules and observed features of low optical depth superposed on the Orion Nebula, but this deduction rests on a very low value for the assumed continuum opacity of a gas and dust mixture at optical wavelengths. In all these calculations some very simple law is adopted for the thermal state of the gas in the globule.

But Kahn has shown that the gas temperature is the most sensitive parameter in any calculation of the structure of a globule. The flow of heat from the exterior into the globule depends on the properties of the cosmic rays and suprathermal particles, and the rate of loss of heat by the globule should really be computed with allowance being made for the possible self-absorption of the infrared radiation which carries away the heat. This is likely to be a very delicate calculation. It seems therefore that no definite statement can yet be made about the temperature inside a globule. In particular it is not at all clear that an interstellar grain can become cool enough for solid hydrogen to condense on it. Various far-reaching hypotheses have been based on the possibility of the existence of solid hydrogen grains (Hoyle, Wickramasinghe, Reddish), but calculations by Field and by Greenberg and de Jong have shown that conditions in the Galaxy at large are never suitable for the freezing of hydrogen. It would be interesting to know how dense a globule would have to be before the grains in its deep interior become cool enough to allow the hydrogen to condense, and whether such globules are ever likely to be found in nature.

Pikel'ner (Sternberg Astronomical Institute) considered conditions of the formation of stellar groups of the condensation from massive interstellar clouds. If the number of hydrogen atoms in the column above a unit area in the cloud exceeds 10^{21}, then soft metagalactic X-rays are absorbed, the temperature in the center of the cloud decreases, and as a consequence gravitational condensation tends to begin. This process results in the formation of an association of cluster of young stars.

THE GALACTIC MAGNETIC FIELD

Significant progress has been made during the last three years in measurements of the galactic magnetic field using both optical and radio techniques. New evidence has been obtained which provides a better understanding of its strength and configuration.

A major study of the optical polarization of southern stars by Mathewson when combined with available northern data permits a realistic model to be made of the galactic magnetic field. The bulk of the stars studied lie within 500 pc of the Sun and a fraction extend out to 2 kpc so the field orientation can be mapped in depth as well as in galactic coordinates. Mathewson's field model is essentially based on the optical polarization data and at the same time makes a good fit to the distribution of polarization of the galactic non-thermal radio emission and to the Faraday rotation data of extragalactic radio sources. The model has magnetic lines of force forming tightly wound right-hand helices of pitch angle 7°. These helices which lie on the surface of tubes with elliptical cross-sections have been sheared through 40° on the galactic plane in the anticlockwise sense when viewed from the North Galactic Pole. The Sun is displaced towards the galactic center and below the magnetic axis. Such a configuration readily explains the reversal in sign of the Faraday rotation of extragalactic sources on crossing the galactic plane and gives a reasonable fit to the large amount of optical polarization data. The regions of high galactic background polarization lie along the band in the sky which is at right angles to the magnetic field – the direction in which there will be minimum galactic Faraday depolarization. This is also the band where the surveys of extragalactic radio sources show minimum Faraday rotation. Other investigations suggest that it may be possible to describe the Faraday rotation data using closed loops of field rather than tightly wound helices which pose problems for their formation.

An important characteristic of this helical or closed field is its limited extent. Neither the optical polarization nor the Faraday rotation data appear to be compatible with such a field extending right around the Galaxy; it is very local, extending perhaps ± 500 pc in the direction of galactic rotation and has a diameter of 300 pc in the galactic center direction. This magnetic field configuration would appear to be coextensive with the Local (Gould Belt) System which consists of neutral hydrogen and other population I material. However as yet it is not possible to make any detailed correlations between such parameters of the two systems as their major axis directions or precise dimensions.

In addition to the looped field both the radio galactic Faraday rotation and the optical polarization data suggest the existence of a more extended field directed approximately along the local spiral arm in the direction of galactic rotation. Mathewson claims that this field, which is probably the general galactic disc field, lies outside the looped field. The observations made so far indicate therefore that the magnetic field in the vicinity of the Sun is not typical of the spiral arms in general.

Two noteworthy observational developments have been made which allow realistic measurements to be made for the first time of the magnetic field strength in certain regions of the interstellar medium. The first is the successful detection at Green Bank and Jodrell Bank of the Zeeman splitting of the 21-cm line in interstellar neutral hydrogen clouds. A clearcut Zeeman effect is found in the Perseus arm features of the absorption spectrum of Cassiopeia A which indicate line-of-sight components of 10 to 20 μG in various absorption features. The field direction is the same as that derived from Faraday rotation measurements of extragalactic sources and pulsars in the same region of the galactic plane. The field components in the two main Orion arm absorption features are a tenth or less of the Perseus arm strengths. Magnetic fields covering a similar range of magnitudes have been detected in the neutral hydrogen absorption spectra of a number of strong radio sources. These measurements are an indication of the mean field in the denser interstellar clouds. Field strengths more typical of the general interstellar medium can now be derived from observations of the dispersion measure of the recently discovered pulsars. Values have now been obtained for the rotation measure of a few pulsars which, along with their dispersion measures, give directly the line-of-sight component of the magnetic field. Alternatively the dispersion data from all the pulsars can be used to derive a model electron density distribution which can be compared with the Faraday rotation measure distribution to provide an estimate of the magnitude of the magnetic field. A mean field strength near the Sun of 2 to 3 μG is indicated for both the looped field and the general disc field. This *mean* value appears to be a factor of 5–10 lower than the field *magnitude* necessary to account for the galactic radio synchrotron emission assuming that the cosmic ray electron density measured near the Sun applies throughout the galactic disc. Several authors have concluded that this implies that the general field is tangled and that the Faraday rotation data measure the mean component only.

NON-THERMAL EMISSION

A number of surveys of galactic radio emission have been published recently which cover large areas of the sky and have an angular resolution of 1° or better. In addition to giving the overall distribution of radio emission, they also show a great deal of structure even in regions well away from the galactic plane where the emission is synchrotron emission from relativistic electrons. Some of the most interesting structure here is in the spurs and ridges which extend from the galactic plane and in some cases show sharp edges rising to half intensity in $\leqslant 20'$ arc and extending for 10° or more. Only a few of these spurs are correlated with optical emission and their origin is still a matter of debate. A new association has been noted between the steep gradient on the outer edge of these spurs and areas of neutral hydrogen emission showing a velocity dispersion about twice that in neighboring regions. Furthermore most of the spurs are polarized and indicate magnetic fields directed along the arc of the spur. These properties are in line with expectation for an expanding front which sweeps up local magnetic field and compresses neutral hydrogen in front of the expanding shock. The event responsible is probably more energetic than a normal supernova explosion.

Observations of non-thermal emission closer to the galactic plane have now revealed about 90 objects which are believed to be supernova remnants. Nearly 20 of these have been investigated with enough angular resolution to show that they are shell sources. Polarization measurements of several show a field parallel to the edge of remnant as expected for a compression of the interstellar field. However Cassiopeia A is thought to have a radial field; perhaps this is the intrinsic field associated with the supernova explosion.

DUST

The temperature of various grain models have been computed by Krishna Swamy and Wickramasinghe using the interstellar radiation field of the solar neighborhood and the best available optical data for graphite and ice. The expected intensity of the infrared radiation from interstellar grain models of dirty ice, graphite and graphite core-dirty ice mantle has been calculated, and it has been shown that the calculated intensities from various grain models at 100μ agree quite well with the recent observations of Hoffman and Frederick. Krishna Swamy, Wickramasinghe and Hunter have calculated optical properties of interstellar silicate grains and graphite core-ice mantle particles (down to 1000Å).

The "dirty ice" model as interstellar grains has been shown to be unsatisfactory on various grounds. On the basis of the detection of complex molecules in interstellar space it is plausible to assume that these molecules are also present in the grains. The presence of these molecules gives rise to an increase in the absorption coefficient. Therefore, Krishna Swamy, Jackson and Donn have done extinction calculations for $n=1{\cdot}3$ to $2{\cdot}0$ and for $k=0{\cdot}0$ to $0{\cdot}5$ for the refractive index $m=n-ik$. Calculations show that $n=1{\cdot}3$ and $k=0{\cdot}2$, $0{\cdot}5$ gives a good fit with the observed reddening curve. The proposed absorbing icy grain model for $k=0{\cdot}2$ also satisfies the albedo requirements. Calculations of the strength of the $3{\cdot}1\,\mu$ band for the model shows that the band is not detectable even for H_2O content of about 20%.

Jones and Spitzer have made a detailed analysis of magnetic dissipation in various substances, which indicates that grains containing appreciable iron impurities, either in superparamagnetic clumps, or in one or more ferromagnetic grains, can probably be oriented by interstellar magnetic fields of 10^{-6} G or less.

Low energy cosmic rays will disrupt very small grains, and if the interstellar gas is heated by such radiation the grains can probably not have formed from condensation nuclei.

Walker is continuing his analysis of the polarization of the λ 4430 interstellar line, and is studying the interstellar reddening curve between 4000Å and 5000Å, with the low-resolution 30Å mm^{-1} scanner attached to the 72-inch telescope, by comparing the energy curves of reddened and unreddened stars in Cygnus; with the help of Morris and Younger, the tracings have been digitized and processed in the available computers and seem to show very interesting discontinuities. It has not yet been possible to extend this work to other regions of the Milky Way. The variation of interstellar extinction with polarization is also being examined.

Hutching's spectrograms of stars in the cluster NGC6231 indicated a possible variation in time in the λ 4430 diffuse band in three stars. These bands were measured for depth, equivalent width, and profile on plates taken over a few years.

Maron at the Torun Observatory has studied the possibility that carbon stars, which are known to be associated with dust clouds, might have acquired the excessive carbon in their atmospheres through the process of accretion of interstellar graphite. Taking into account in his calculations the effects of radiation pressure and of mixing by convection he arrives at the conclusion that graphite grains of size about 10^{-3} cm might be accreted by cold giants and supergiants in sufficient amount to produce a carbon star spectrum.

Sharov (Sternberg Astronomical Institute) has investigated the distribution of the obscuring matter in the Andromeda galaxy and has compared the results with those for our Galaxy. Together with Pavlovskaya he is studying the influence of interstellar extinction upon the visible structure of the Milky Way to evaluate the possibilities of investigating the spiral structure from the surface photometry of the Milky Way.

Rozhkowsky and Kurchakov (Astrophysical Institute, Alma-Ata) have compiled a catalog of 118 reflection nebulae, which serves as the basis for the determination of their physical characteristics. Glooshkov, Kjachushev, Kurchakov and Sabitov performed spectrophotometric, colorimetric and polarizational studies of NGC7023, IC2118, the Pleiades, reflection nebula, and the Orion nebula. Some other reflection and diffuse nebulae and their illuminating stars are under investigation now. Rozhkowsky and others are studying the diffuse galactic radiation to determine the albedo of interstellar grains. According to the results already obtained, the grains in the region of the bifurcation of the Milky Way have a rather low albedo (0·3–0·4).

Minin (University of Leningrad) has computed the brightness, polarization and colors of reflection nebulae. A work on the computation of the light scattering by grains of refractive index $m = 1{\cdot}50$ with several different types of the size distribution function is being completed.

Kalandadze (Abastumani Astrophysical Observatory) has investigated interstellar extinction and the distribution of dust and stars of different spectral classes in Taurus over an area of approximately 180 sq. deg.

Working in cooperation with Kolesnik (Goloseyevskaya Astronomical Observatory), Kalandadze has studied light absorption in the direction of NGC6913. The light absorption value is 1·4–2·2 in stellar magnitudes for the first kiloparsec. The main mass of absorbing matter is situated at a distance of 400–1600 parsec.

Chuadze (Abastumani Observatory) is engaged in the study of light absorption and distribution of stars in the local system.

Schmidt has shown that most O stars have circumstellar dust envelopes. If the interstellar dust is produced in circumstellar regions with dimensions of 10^{17} cm, the interstellar Fe, Ti, Ca must condense on or in dust grains. With the assumption that the graphic particles originate in the outer layers of carbon stars several frequency distributions of the particle radii can be determined. The observed extinction curve can then be explained supposing that the graphite particles contribute only a small fraction to the interstellar dust (Friedemann, Schmidt). Apart from graphite other chemical compounds can condense in the atmospheres of cool stars – especially SiC which probably is another component of interstellar dust grains (Dorschner, Friedemann). The dust particles in the outer layers of cool stars give a possible explanation of the observed time dependent polarization of some red long period variables. The essential observational facts concerning the intrinsic polarization of these stars are explained with this hypothesis which further represents a helpful tool calculating the properties of circumstellar dust from observed variations of the intrinsic polarization (Friedemann). The silicate grains as constituent of interstellar dust are studied with regard of their origin from circumstellar phenomena, especially from planetary systems. In this connection crushing processes are treated in a quantitative way. Further the spectra of silicate grains in the middle infrared are studied (Dorschner).

Barbieri examined the problem of the illumination of reflection nebulae for several values of the electron density and dust scattering coefficient, both for optically thin and optically thick nebulae.

Sancisi with the 25-m radiotelescope of Dwingeloo has observed the region of the reflection nebulae NGC2068 finding a maximum of peak brightness temperature and a minimum of velocity width near a group of T Tauri stars. He has also reported preliminary results on the H I distribution in the Taurus region.

Future work

(1) Radiation transfer in circumstellar dust clouds and in reflection nebulae.

(2) Studies of high latitude clouds (polarization, ratio between polarization and reddening). Photoelectric *UBV* measurements in or behind high latitude clouds.

(3) Lyman α line radiation transport in H II regions with "dust globules". Explanation of the observed anomalous Balmer decrement.

(4) Attempt to explain interstellar bands by transitions in silicate grains.
(5) Star formation in connection with cloud collisions.
(6) Molecule formation, the dust: gas ratio in H I clouds, star formation and early stages of stellar evolution.
(7) Molecules and interstellar chemical abundance (time-dependence).

Dust (continued)

This is a report of activities, many still in progress, in the subject of interstellar grains and related topics. Where conclusions are not yet substantial the various theories and observations are presented in order to serve as possible guidelines for the ongoing work.

Although no bibliography is appended at the end of this report there will be available, upon request to J. M. Greenberg, a topical list of references.

Extinction

Rocket and satellite measurements indicate a general but varying degree of continued rise in the extinction out to $\lambda^{-1} \approx 9\mu^{-1}$. The exception to this is θ Orionis which shows perhaps less extinction in the far UV than at $\lambda^{-1} = 3\mu^{-1}$. This result is a confirmation of the anomaly in the shape of the extinction curve which had been earlier recognized in the $1\mu^{-1} \leqslant \lambda^{-1} \leqslant 3\mu^{-1}$ region and which has been further confirmed for other Orion stars.

The diffuse extinction hump at $\lambda^{-1} \approx 4{\cdot}8\mu^{-1}$ has been confirmed in all the stars observed in this region of the UV.

A general value of the ratio of total to selective extinction appears to be $A/E = 3{\cdot}2$ with usually small deviations. Significant deviations with much higher values are apparently of a very local nature.

Detailed investigations of the extinction curve in the $1\mu^{-1} \leqslant \lambda^{-1} \leqslant 3\mu^{-1}$ range indicate the possible existence of a number of discontinuities in the slope which seem to be associated with diffuse absorption features.

Polarization

Considerable evidence is accumulating for sizeable variations in the wavelength dependence of interstellar polarization. Some degree of correlation with variability in the shape of the extinction curve indicates differences in grain properties. However, a substantial fraction of the stars observed exhibit a maximum of polarization in the $1{\cdot}5\mu^{-1}$ to $2\mu^{-1}$ range.

A suggestion has been made that temporal variability of polarization could be due to microstructure in the interstellar clouds. It has been generally assumed that time variations can occur only in circumstellar regions. The idea of small scale structure in the interstellar medium is undergoing investigation also in connection with two phase theories of the H I medium.

The confirmation of circumstellar polarization has been obtained.

Scattering

Additional detailed observations of the colors and brightness of reflection nebulae have been made. A number of detailed calculations of scattering by clouds of spherical, cylindrical and spheroidal (small) particles of varying optical properties have been made. Effects of double and – by implication – multiple scattering have been calculated in some special cases. Present comparisons with observations lean toward interpretation in terms of particles of a dielectric nature and not necessarily ices.

Preliminary interpretations of the brightness of reflection nebulae in the rocket and satellite UV are in progress.

New measurements of the diffuse galactic light have received similar interpretations by several independent workers, leading to the exclusion of simple graphite or metallic grains as the dominant interstellar scattering constituents. The tendency is a bit closer to graphite core plus ice mantle than to ices alone.

Chemical composition

The search for the contributions of a $3{\cdot}1\,\mu$ ice absorption associated with extinction has led to conflicting interpretations and the evidence for and against a significant amount of ice in or on the grains does not yet appear conclusive. The interpretation of absorption and emission features in the region between 8μ and 12μ has implied a silicate component of dust in interstellar and circumstellar regions. The diffuse absorption feature at $\lambda^{-1} \approx 4{\cdot}8\,\mu^{-1}$ has been attributed to a characteristic graphite absorption. Although interpretations of infrared spectral characteristics are not subject to detailed physical grain properties the intrinsic absorption features in the ultraviolet may be subject to strong modifications produced by grain surface properties because the wavelength is comparable to the size of the dust particles. Effects of absorption edges of ices and silicates are similarly modified. Laboratory work using the microwave analog method is to be applied to this consideration.

Grain models

High speed digital computers have been exploited to obtain exact (and in some cases approximate) calculations of extinction and polarization (also emission and absorption, and scattering) for (a) ice spheres and cylinders – varyingly oriented (b) silicate spheres and cylinders (c) spherical and spheroidal graphite (d) graphite with ice mantles (e) mixtures of graphite spheres and silicate cylinders (f) graphite cores with solid hydrogen mantles. Laboratory studies have been made of extinction by iron, carbon, silicon carbide and silica and comparison has been made with theories for "equivalent" spheres and cylinders. Agreement seems possible only for silica – which is spherical. Extinction measurements for laboratory mixtures of polycyclic aromatic molecules have given a rough approximation to the interstellar extinction in the visible. For every interstellar dust model which has been considered, counter evidence of varying degrees of strength has been adduced by different workers. On the basis of *optical* properties alone some of the arguments raised *counter* to each model – based on calculations for *smooth* particles – are respectively: (a) no positive spectroscopic evidence of ice, absorption edge of ices leads to lack of continued rise of extinction beyond $\lambda^{-1} \approx 5\,\mu^{-1}$, incorrect shape of absorption hump at $\lambda^{-1} \approx 4{\cdot}8\,\mu^{-1}$; (b) incorrect shape of absorption hump at $\lambda^{-1} \approx 4{\cdot}8\,\mu^{-1}$; (c) poor representation of reflection nebulae and diffuse galactic light, lack of continued rise in extinction etc., simultaneous reproduction of λ dependence of polarization and extinction inconsistent with maximum observed polarization; (d) lack of continued rise in extinction etc., simultaneous reproduction of wavelength dependence of polarization etc.; (e) simultaneous reproduction of λ dependence etc. It appears at present that no simple material model based on calculations for *smooth* particles can simultaneously reproduce the totality of observations of extinction and polarization and scattering. Heuristic approaches to resolving the dilemmas are not sufficiently satisfying and substantial additional theoretical and laboratory work as well as more observations appear to be required. Also other factors than purely optical ones need be considered.

Physical interactions

The formation, growth and destruction of grains have been reconsidered by many authors for a variety of conditions. The formation possibilities appear to be: (1) the "classical" theory of accretion of condensable atoms on nuclei in the interstellar medium – but in a modern context of the physical and dynamical conditions associated with spiral structure. This leads to dirty ices unless modification of chemical structure to larger molecules is achieved by way of radiation effects; (2) growth of graphite and/or silicates in cool stars. These can be either dominant components in the interstellar extinction or the nuclei for process "1"; (3) grinding down and subsequent ejection of "interplanetary material"; (4) condensation during "solar nebula" phase of star evolution. Destruction by cosmic rays and soft X-rays have been considered.

The interaction of the interstellar radiative field with dust grains in normal regions and in dense clouds have been used to calculate grain temperatures for a variety of models of grains. The most complete results appear to preclude the possibility of solid hydrogen condensing on dust. However the lack of observable neutral hydrogen in dense clouds strongly suggests the formation of molecular hydrogen on the low temperature grains. Calculations of molecular hydrogen formations have been carried out.

More elaborate calculations of grain orientation by magnetic fields and by mechanisms bringing about streaming of dust through gas have been made. Cosmic rays have been included in the orientation phenomena. Magnetic fields of the order of 2–5 μG have been shown to orient the ices and silicates significantly but not graphite. The question of achieving the maximum observed ratio of polarization to extinction by means of magnetic and streaming orientation processes has been examined quantitatively for elongated dielectric particles and for graphite cores with either ice or solid hydrogen mantles. In the light of dynamical questions for interstellar matter serious questions have been raised concerning the relaxation times for the various orientation mechanisms.

Diffuse bands

The structure of the unidentified (about 25) diffuse bands has been subjected to further detailed study. Contrary to older ideas it is now believed that the λ4430 band is highly asymmetric with an emission wing at shorter wavelength comparable with the absorption at 4430Å. Calculations of extinction and polarization through the λ4430 band produced by impurities have been performed for grains with background absorptivities ranging from very small (dielectric) to medium and large (metallic). The asymmetry of the polarization parallels that for extinction in the band and its existence would be a relatively unambiguous means of discriminating between dielectric and metallic (or graphite) grains. It would also be used to assign relative contributions by postulated grain constituents.

Laboratory studies of absorption by Ca impurities in matrices simulating possible interstellar grain materials have indicated the possibilities of a group of bands being associated with a single type of impurity.

The distribution and correlation of λ 4430 band with extinction has been further studied.

D. E. OSTERBROCK
President of the Commission

BIBLIOGRAPHY

In accordance with the request of the IAU Executive Committee, bibliographic references have been held to a minimum. References are given only for books, conferences and symposia, and review articles. Each of these references contain very many lists of specific papers.

Books

Middlehurst, B. M., Aller, L. H. (eds.) 1968, *Nebulae and Interstellar Matter*. Univ. of Chicago Press, Chicago.
Perek, L., Kohoutek, L. 1967, *Catalogue of Galactic Planetary Nebulae*. Academia, Prague.
Spitzer, L. 1968, *Diffuse Matter in Space*, Interscience Publishers, New York.
Wickramasinghe, N. C. 1967, *Interstellar Grains*. Chapman and Hall, Ltd., London.

Conferences, symposia, etc.

Planetary Nebulae:
Osterbrock, D. E., O'Dell, C. R. 1968, *Planetary Nebulae* IAU Symposium 34, D. Reidel, Dordrecht-Holland.

H II Regions:
Terzian, Y. 1968, *Interstellar Ionized Hydrogen*. W. A. Benjamin, Inc., New York.

Dust:
Greenberg, J. M., Roark, T. P. 1967, *Interstellar Grains*. NASA, Washington.

Review articles

H II Regions:
Mathews, W. G., O'Dell, C. R. 1969, *A. Rev. Astr. Astrophys.*, **7**, 67.
Osterbrock, D. E. 1967, *P.A.S.P.*, **79**, 523.

H I Regions:
Robinson, B. J., McGee, R. X. 1967, *A. Rev. Astr. Astrophys.*, **5**, 183.

Large Scale Distribution, Dynamics and Condensations:
Dieter, N. H. 1969, *P.A.S.P.*, **81**, 186.
Kerr, F. J. 1969, *A. Rev. Astr. Astrophys.*, **7**, 39.
Pikel'ner, S. B. 1968, *A. Rev. Astr. Astrophys.*, **6**, 195.

Magnetic Fields, Non-Thermal Emission:
van de Hulst, H. C. 1967, *A. Rev. Astr. Astrophys.*, **5**, 167.

Dust:
Lynds, B. T., Wickramasinghe, N. C. 1968, *A. Rev. Astr. Astrophys.*, **6**, 215.

35. STELLAR CONSTITUTION (CONSTITUTION DES ÉTOILES)

PRESIDENT: A. G. Masevič.
VICE-PRESIDENT: E. Schatzman.
No Organizing Committee.

During the period under review, theoretical work in the field of stellar constitution and evolution of stars has been effectively continued by several teams permanently working in Belgium, Denmark, France, German Democratic Republic, German Federal Republic, Italy, Japan, Poland, the United Kingdom, the U.S.A., the U.S.S.R. as well as by individual scientists in Argentina, Bulgaria, Czechoslovakia, India, Sweden and others. Most attention has been given during the past three years to analysis of late stages of stellar evolution, rotation, evolution of binaries, loss of mass on various stages of evolution, stellar stability, neutrino astrophysics, physical processes in the pre-supernovae stage. The discovery of pulsars has raised extremely the general interest for white dwarfs and neutron stars, and consequently the number of papers with new hypotheses concerning quasi-stellar sources has substantially dropped.

Many new evolutionary sequences of stellar models for various masses and different chemical compositions have been computed using improved numerical techniques and large computing programs. It is regretted that many of these results are difficult to compare mainly because of the differences in the input physics and the way the numerical results are being published. This is a field in which a closer international co-operation is highly desirable, especially on the following points: exchange of ideas on the programs presently under way and the planned programs in order to avoid unnecessary duplications: agreement on the numerical results which should be published; exchange of technical informations: (input physics, computing methods, detailed results of computations, opacity tables, etc.). Several attempts in this direction have been undertaken during the period under review (for example, Commission 35 Circular Letters NN 1–7 containing information on current work, publication by the U.S.S.R. Academy of Sciences of the Cox and Stewart's Opacity Tables for 23 mixtures, several joint international projects for using the same programs, etc.), but the problem still remains to be solved.

A large discussion has followed the new results of the solar neutrino experiment resulting mainly in a reexamination of the nuclear reactions rates and the theoretical analysis of the solar interiors. It is somehow disappointing that several investigators (luckily only a few) have taken this opportunity to shed doubt on the theory of nucleosynthesis in stars in general.

It should be mentioned that less attention has been devoted recently to comparisons between theoretical results and observations. This is due to the fact that such a comparison requires uncertain theoretical transformation of observational data – data that are for most cases incomplete. This is especially true for the effective temperature scale of Red giant stars on very late stages of evolution on which theoretical work has been mostly concentrated. On the other side those comparisons, that actually have been made, are carried out on a noticeably higher level than it has been done before. It should be emphasized that at least for Main Sequence Stars the present theory allows to some extent to improve observations, for example to determine masses and the helium abundance of individual, binary and cluster stars with a very satisfactory degree of accuracy. Evolutionary tracks for late stages of evolution allow to predict observable phenomena concerning ratios of some isotopes in stellar envelopes, etc.

In the past period several international symposia and colloquia have been held, handling problems of stellar constitution, and the respective proceedings have been, or are to be published. As the length of the IAU commission reports has been recently considerably restricted, this Report contains only the following sections: hydrodynamical problems of stellar interiors, stellar instability, evolution of binaries, nuclear reactions in stars, neutrino astrophysics, opacities and numerical techniques relevant to stellar constitution. In choosing the above-mentioned problems it has been taken into account that the problems of mass-loss, pre-main sequence evolution and stellar rotation have been covered recently by the Proceedings of the Colloquium *Mass Loss from Stars*, Trieste 1968, Astrophysics and Space Science Library, Vol. 13; the 1969 Elsinore Colloquium "On Mass Loss and Evolution in Close Binaries"; the 1969 Liège Astrophysical Symposium "Pre main Sequence-Evolution" and the Colloquium on Stellar Rotation held at the Ohio State University, Columbus 1969. (See also a review paper of P. A. Strittmatter "Stellar rotation", *A. Rev. Astr. Astrophys.*, 1969, **7**, 665). Quasi stellar objects have been reviewed thoroughly in *A. Rev. Astr. Astrophys.*, **5**, 399, 1967, "Quasi Stellar Objects", by E. Margaret Burbidge; *ibid.*, **6**, 351, 1968, "Variable Radio Sources", by K. I. Kellermann and I. I. K. Pauling-Toth; and *ibid.*, **7**, 527, 1969, "Quasi Stellar Objects", by Maarten Schmidt.

Proceedings of several national colloquia on pulsars are being published (see also an extended review by Maran and Cameron on the Fourth Texas Symposium in 1968). Besides, special meetings on Pulsars and on white dwarfs will take place before the coming IAU Meeting in London.

Also according to the new rules, only selected bibliographical references are given in the Report. A complete bibliography on stellar constitution will be circulated to all the members of Commission 35 before the General Assembly.

I. HYDRODYNAMICAL PROBLEMS OF STELLAR INTERIORS

A. *Radiative zones*

Hydrodynamics enters the theory of convectively stable (sub-adiabatic) stellar zones through the possible presence of non-uniform rotation and through the deviations from a spherically symmetric thermal-gravitational field due to centrifugal and magnetic force. These problems have acquired a renewed interest following the observations by Dicke and Goldenberg (**1**) of a solar oblateness of $5\text{–}10^{-5}$, about five times that induced by the observed surface rotation. Dicke's interpretation in terms of a solar quadrupole moment due to a rapidly rotating radiative core has raised queries about the stability of this model. Howard *et al.* (**2**) and Spiegel (**3**) have suggested that a process would occur analogous to the Ekman pumping, so that a dynamically-driven circulation is set up, with viscous force balancing Coriolis force in the thin Ekman boundary layer. Continuity forces the flow to extend through the bulk, yielding the very short spin-down time.

The treatment of this problem contrasts with that customary for a non-barotropic stellar gas, obeying the law $P \sim \rho T$. A non-conservative field of centrifugal (or magnetic) force can be balanced by suitable variations of ρ and T over isobaric surfaces. The consequent breakdown in radiative equilibrium yields buoyancy forces that drive a generalized Eddington-Sweet circulation. This already sets a possible difficulty for the Dicke model, for in the absence of an inward gradient of mean molecular weight μ, capable of suppressing the circulation as was found by Mestel and Kippenhahn, the postulated large gradient of Ω near the base of the convective zone will yield a substantial modification of the Ω-distribution within a solar lifetime. The question is whether an analogue of the Ekman suction will yield a much shorter spin-down time. The thermally-driven Eddington-Sweet circulation is replaced by a dynamically-driven flow if the scale of variation of Ω is less than $d_c \simeq r \times [(\Omega^2 r/g)(\lambda/r)]^{\frac{1}{2}}$, where λ is the local scale-height. One is led at first to a model in which a rapidly rotating core and a more slowly rotating envelope can co-exist but with the transition region between them always comparable with *dc*. The braking by the magnetically-controlled solar wind would be limited effectively to the convection zone.

The subsequent evolution of the Ω-field (in the absence of magnetic coupling between the two zones) would be given by the Eddington-Sweet theory, with the fairly sharp Ω-gradient and any μ-variations playing an important role. However, a much shorter spin-down time will result if the transition layer were to become unstable. One would then arrive at a picture in which the slow but persistent braking of the star would drive a weak turbulence in the sub-photospheric radiation zone.

According to Howard *et al.* (**2**) one may suggest that the Li depletion could be qualitatively explained by the presence of turbulence below the convection zone. A very important observational correlation between the Li content and the rotational velocities of solar type stars has been found by Conti (**4**). He finds that stars that still have appreciable rotation have the greater Li content. This means that they had less Li destruction. Conti's work has to be considered as the first quantitative result in favour of the above-mentioned phenomenon (see also the review paper by Wallerstein and Conti (**5**)).

The importance of mass motions in radiative zones arises in studies of gravitational separation of the elements. Schatzman (**6**) considers main-sequence stars of masses ranging from 1 $M_\odot$ to 1·6 $M_\odot$, with sub-photospheric convection zones that decrease greatly in depth and mass along the sequence. The violent turbulence in the convective zone is expected to keep the composition uniform, but gravitational separation of the elements can occur in the radiative zone, at a rate proportional to the microscopic diffusion coefficient λ_D. The separation of the elements would be negligible in the sun, but would be effectively complete in stars of mass greater than 1·25 $M_\odot$. However, mass motions in the radiative zone – e.g., those associated with modified Ekman-type spin-down or meridional shear flow yield a macroscopic diffusion coefficient λ which can be much larger than λ_D. The fall of the heavier elements in an initially uniform star is then negligible. Schatzman argues that the fact that we observe e.g., metals or lithium at the surfaces of F stars can be explained only by a sufficiently fast mixing below the convective zone.

Gravitational setting has been appealed to also to explain the under-abundances of He, Ne and O in those Ap stars with appropriate values of surface temperature and gravity, with selective radiation pressure explaining simultaneously the over-abundances of Mn, Sr, Y, Zr and the rare earths (Michaud (**7**)). Greenstein *et al.* (**8**) have similarly tried to explain the weakness of the He lines in old horizontal branch B stars with presumed slow rotations. Such mechanisms require the atmospheres to be non-turbulent. The surface layers of an Ap star will probably satisfy local radiative equilibrium, without any circulation of gas so that Michaud's explanation remains plausible. The suggestion of Greenstein *et al.* will also break down if circulation persists (Smith (**9**)). Again it is possible that a photospheric magnetic field, too weak to be optically significant, is present and allows the star to achieve local radiative equilibrium without circulation.

Large-scale meridian circulation over the bulk of a rotating magnetic star can influence the structure of the superficial magnetic field. In a rapidly rotating star the magnetic flux may ultimately be trapped beneath the surface (Mestel (**10**); Wright (**11**)). Magnetic coupling with a stellar wind may explain the slow rotations of the Ap stars (Mestel (**10**)), and will also tend to change the angle between the rotation and magnetic axes (Mestel (**12**)); but whether towards the large values required by the oblique rotator model, seems to depend critically on the details of the superficial flux distribution (Mestel and Selley (**13**); Selley (**14**)), and so ultimately on internal stellar magnetohydrodynamics.

B. *Convective zones*

The hydrodynamical problems of super-adiabatic zones are those of laboratory turbulent convection, greatly increased by the enormous density changes over at least sub-photospheric zones. The linear stability analysis for an incompressible (Boussinesq) fluid can be plausibly applied to a star if a density scale-height is used for the depth: the critical Rayleigh number then predicts that a very small degree of superadiabaticity is sufficient to ensure instability.

A start has been made on non-linear (developed) convection in Boussinesq fluids (Gough *et al.* (**15**)), with some success in predicting the Rayleigh-number dependence of the Nusselt number (the ratio of total heat transport to that by conduction alone), but we are a long way from an extension to compressible fluids. The conventional phenomenological model adopted for stellar structure work is the mixing-length theory with the size of the convective elements put equal to the density (or pressure) scale-height. The super-adiabaticity required to transport the stellar luminosity is very small except in the ionization zones near stellar surfaces, or in very rapid late phases of stellar evolution, so that it is only in these cases that lack of an adequate theory of convection is serious for studies of stellar structure and evolution.

Recently there has been some discussion as to whether or not an inversion of density occurs in the convective zones of stars with $T_{eff} \leqslant 10^4$ K. According to Chitre and Shaviv (**16**) density inversion occurs always if the pressure scale-height is applied. Krishna Swamy (**17**) obtains similar star models in which, depending on the assumed efficiency of convection, the inversion of density may or may not appear. A. Ergma (**18**) showed that density inversion for evolved stars of large masses ($M > 10 \mathscr{M}_{\odot}$) cannot be avoided and that the structure of the convective zone for these masses is independent of the efficiency of convection. For small masses the density inversion is the smaller the more efficient the convection and may disappear completely if a critical (for the given mass of a star) value of the efficiency is reached (for $1 \mathscr{M}_{\odot}$ the critical value of $\alpha = 1{\cdot}5$ where α is the coefficient in the mixing length theory).

The granulation pattern observed at the solar photosphere can plausibly be described in terms of irregular Bénard-type cellular convection, typical velocities being $\simeq$ 2–3 km/s, and a characteristic cell-size $\simeq$ 1000 km, as compared with a scale height of $\simeq$ 400 km. Superposed on this is the supergranulation with horizontal velocities $\simeq 0{\cdot}4$ km/s and a characteristic scale of $\simeq$ 30000 km. To account for this second preferred scale, Simon and Weiss (**19**) have suggested a modification to mixing length theory, by which convective motions can extend over several scale-heights. They show that in a polytropic atmosphere the variation of the scale-height yields maximum efficiency when the vertical dimension of a cell is about three density scale-heights.

The question has been asked whether a *primeval* general magnetic field could interfere seriously with convective heat transport over the whole sphere, especially in the pre-main sequence phase; this with a view to explaining the infrared objects in the forbidden region to the right of the Hayashi tracks. Moss (**20**) has discussed in detail how the requirement of a fair degree of superadiabaticity (due to interference with or suppression of convective transport) can restore the vertical Hayashi tracks to the roughly horizontal Henyey-type tracks. However, Gough and Tayler (**21**) and Moss and Tayler (**22**) have applied the hydromagnetic energy principle to show that only very strong magnetic fields can suppress convection: in regions where the field has a large vertical component the field energy needs to be comparable with the kinetic energy of the turbulence that would be driven by the assumed superadiabatic gradient. Where the field is nearly horizontal, stabilization becomes much more difficult, as rolling motions about the field-lines are hardly impeded. The general conclusion is that only magnetic fields strong enough to be hydrostatically important (at least in the fraction of a star with temperatures below 10^5 K) could seriously reduce the Hayashi superluminosity, and there is no evidence that even the stars with the strongest surface magnetic fields contain that amount of magnetic flux (Wright (**11**)).

C. *Interaction between rotation and convection*

By considering local disturbances of sufficiently low symmetry, Cowling showed that although rotation has a stabilizing effect for axi-symmetric modes, there always exist non-axisymmetric modes without detailed angular momentum conservation and which are hardly affected by Coriolis force. Thus even in a rapidly rotating star we may expect just some reduction in convective efficiency rather that complete suppression. The Coriolis interference with general motions can be thought of as due to the fluid's attempting to satisfy the Taylor-Proudman theorem for slow inviscid motions, a constraint that is a maximum at the poles and decreases towards the equator. This shows up in Busse's study (**23**) in the Boussinesq approximation of the growth of large-scale modes in a spherical shell. He finds that the Rayleigh number of his modes is unaltered to the first order in rotation, but the diminishing inhibiting effect of rotation with approach to the equator is shown up by the first unstable mode's having a pronounced maximum at the equator. Busse also shows that the dependence on latitude of the disturbance velocities necessarily yields in the second order a non-uniform rotation similar to that in the sun, both in sign and order of magnitude.

This work is in contrast with the earlier model of Biermann and Kippenhahn, which employs a phenomenological anisotropic viscosity tensor to describe the dynamical effect of the thermal turbulence. The consequent radially-dependent rotation low is inconsistent with hydrostatic support in a nearly adiabatic convective zone. The non-conservative centrifugal force field therefore drives a meridian circulation, which in turn sets up a latitude-dependent rotation field, with equatorial acceleration at the surface if the circulation is from pole to equator at the surface.

The most radical hypothesis on the interaction between rotation and convection is that of Gough and Lynden-Bell (**24**) following an earlier suggestion by Scorer. Noting the analogy between vorticity and magnetic field, and studies on expulsion of magnetic flux by systematic fluid motions (Parker, Weiss) they suggest that a convection zone attempts to achieve a state with zero large-scale vorticity, so that $\Omega \varpi^2 \simeq$ constant, ϖ being the axial distance. Because it would imply a singularity on the axis, this asympotic state is never reached; instead, the star continually spews out angular momentum a rival picture to the more popular magnetic braking. Some experimental support for the picture is claimed, though this has recently been questioned (Strittmatter (**25**)).

D. *Semi-convection*

A "semi-convective" region is one in which the stability criterion is initially violated, but which becomes stable if mixing of material reduces the opacity, e.g., by mixing core helium with envelope hydrogen. There is a continuing debate as to what criterion to use in stellar model computations.

Franzman, and Varshavsky (**26**) studied how different assumptions about mixing in the semi-convective layer influence the results of computations for massive stars. It appears that the structure and the evolution of the star are not affected substantially, but the time of evolution increases when partial mixing is taken into account (for 60 $\mathscr{M}_\odot$ Δt–0·4 × 10^6 y).

Tayler (**27**) has published a model of a slightly evolved massive star without a semi-convection zone. He has suggested that this model might be unstable and that this might lead to the setting up of a semi-convection zone.

Tutukov and Varshavsky (**28**) have investigated the behaviour of semi-convection during the hydrogen exhaustion phase with special attention to the turning points of the track. An intermediate fully convective zone appears near the hydrogen burning shell for the last stages of hydrogen exhaustion.

Gabriel (**29**) studies the overstability problem in details, finding that the stable layers above the semi-convective zone prevent overstability from arising. Gabriel concludes that if overstability is the only means of mixing, the Sakashita-Hayashi models are more likely to be accurate.

E. *Time-dependent convection*

The difficult question of time dependent convection and its effect on pulsations and stability has been considered by Kamijo (**30**) using Unno's method (**31**), by Gough (**32**) and by Castor (**33**) while the question of the boundary conditions was discussed by Takeuti (**34**) and further work on the one-zone model is due to Okamoto and Unno (**35**).

Edwards (**36**) has studied the influence of the failure of the quasi-static approximation on the helium flash. He finds the dynamical terms can be much more important than appears from a quasi-static calculation and the core of the star may explode; whether or not this happens depends on the behaviour of time dependent convection. Fraley (**37**) used Cowling's method for the derivation of the convective model (time dependent convection) for the investigation of supernovae explosions.

BIBLIOGRAPHY

1. Dicke, R. H., Goldenberg, H. Mark 1967*a*, *Phys. Rev. Lett.*, **18**, 313.
2. Howard, L. N., Moore, D. W., Spiegel, E. A. 1967, *Nature*, **214**, 5095.
3. Spiegel, E. A. 1968, *Highlights of Astronomy*, Reidel, Dordrecht-Holland.
4. Conti, P. S. 1968, *Astrophys. J.*, **152**, 2.

5. Wallerstein, G., Conti, P. S. 1969, *A. Rev. Astr. Astrophys.* **7.**
6. Schatzman, E. 1969, *Astr. Astrophys.*, **3**, 331.
7. Michaud, G. 1969, *Astrophys. J.* (in press).
8. Greenstein, G. S., Truran, J. W., Cameron, A. G. W. 1967, *Nature*, **213**, 871.
9. Smith, R. C. 1969, *Observatory*, **89**, 208.
10. Mestel, L. 1967, *Plasma Astrophysics*, Ed. P. A. Sturrock. Academic Press, London, p. 185.
11. Wright, G. A. E. 1969, *Mon. Not. R. astr. Soc.* (in press).
12. Mestel, L. 1968, *Mon. Not. R. astr. Soc.*, **140**, 177.
13. Mestel, L., Selley, C. S. 1970, *Mon. Not. R. astr. Soc.* (in press).
14. Selley, C. S. 1970, *Mon. Not. R. astr. Soc.* (in press).
15. Gough, D. O., Spiegel, E. A., Toomre, J. 1969 (in preparation).
16. Chitre, S. M., Shaviv, G. 1967, *Proc. nat. Acad. Sci.*, **57**, 573.
17. Krishna Swamy, K. S. 1969, *Observatory*, **89**, 968.
18. Ergma, A. 1970, *Nauchnye Informatsii, Moscow*, **16** (in press).
19. Simon, G. W., Weiss, N. O. 1968, *Z. Astrophys.*, **69**, 435.
20. Moss, D. L. 1968, *Mon. Not. R. astr. Soc.*, **141**, 165.
21. Gough, D. O., Tayler, R. J. 1966, *Mon. Not. R. astr. Soc.*, **133**, 85.
22. Moss, D. L., Tayler, R. J. 1969, *Mon. Not. R. astr. Soc.*, **145**, 217.
23. Busse, F. H. 1969. Preprint.
24. Gough, D. O., Lynden-Bell, D. 1968, *J. Fluid Mech.*, **32**, 437.
25. Strittmatter, P. A. 1969. Private Communication.
26. Franzman, Yu. L., Varshavsky, V. I. 1969, *Nauchnye Informatsii, Moscow*, **11**, 51.
27. Tayler, R. J. 1969, *Mon. Not. R. astr. Soc.*, **144**, 231.
28. Tutukov, A. V., Varshavsky, V. I. 1970, *Nauchnye Informatsii, Moscow*, **16** (in press).
29. Gabriel, M. 1969, *Astr. Astrophys.*, **1**, 321.
30. Kamijo, F. 1967, *Z. Astrophys.*, **67**, 227.
31. Unno, W. 1967, *Publ. astr. Soc. Japan*, **19**, 140.
32. Gough, D. O. 1967, *Astr. J.*, **72**, 799.
33. Castor, J. 1968, *Astr. J.*, **73**, S169; *Astrophys. J.* (in press).
34. Takeuti, M. 1969, *Astrophys. Space Sci.*, **3**, 219.
35. Okamoto, I., Unno, W. 1967, *Publ. astr. Soc. Japan*, **19**, 154.
36. Edwards, A. C. 1970, *Mon. Not. R. astr. Soc.* (in press).
37. Fraley, G. S. 1968, *Astrophys. Space Sci.*, **2**, 96.

II. STELLAR STABILITY AND STELLAR OSCILLATIONS

A. *General*

A review paper by Lebovitz in *A. Rev. Astr. Astrophys.*, **5** is mainly concerned with the effects of rotation on equilibrium configurations and their stability. Chapter 27 of the second volume of J. P. Cox and R. T. Giuli's, *Principles of Stellar Structure* 1968, New York presents a fairly complete survey of the theory of stellar radial pulsations and vibrational stability. A set of lectures given by P. Ledoux at the "11e Cours de perfectionnement de l'Association Vaudoise des Chercheurs en Physique" Saas-Fee, Suisse 1969 (to be published) covers many of the results up to the beginning of 1969 including the case of non-radial perturbations but neglecting mostly the effects of rotation, magnetic fields and external gravitational fields. An attempt is also made there to present the general time-dependent problem as it should apply for instance to rapid phases of stellar evolution.

Successful methods to find directly the complex solutions of the complete fourth order problem of linear non-adiabatic pulsations have been or are being developed in various groups at Colorado, Liège, Paris.

This approach seems to have been mainly directed at the problem of vibrational stability but, in principle, it covers also the problem of secular (or thermal) stability. It might also prove useful to learn something on the difficult case where two or more of the typical time-scales become comparable. This might have interesting applications, for instance, in stars of very large mass or very high density on the verge of dynamical instability.

Glansdorff and Prigogine's generalized principle of minimum entropy production has guided

Unno (1) in a reformulation of the general problem of stellar stability. Perdang (2) has tackled the general problem of the effects of the kinetics of nuclear reactions on the stability problem with a discussion of the new associated time-scales and of the stability of the reactions themselves.

B. *Dynamical stability and adiabatic oscillations*

1. *Radial modes*

Dynamical instability towards radial perturbations occurs only if some appropriate average of Γ_1 is smaller than $\frac{4}{3}$. This can happen in the very early prestellar stages of contraction as shown again by Upton, Little and Dworetsky (3) using the energy principle. Their model has also been studied by Van der Borght using the small perturbation method which, in particular, shows that, for unstable modes contrarily to the cases of stable ones, $\delta r/r$ may decrease strongly towards the surface. Dynamical instability could also occur for the same reasons in a strongly evolved star when it reaches the red giant phase at very high luminosity and could be responsible for the formation of planetary nebulae or for the excitation of the oscillations of Long-Period Variables. Langer (University of Colorado) has computed a line in the H-R diagram, above and to the right of which models become dynamically unstable.

Very advanced stages of stellar evolution with nuclear equilibria getting established at very high temperatures and densities in a central core provide another range of conditions favourable to dynamical instability.

The problem has been discussed anew by Rakavy *et al.* for carbon and oxygen stars and by Itoh for high temperature non-degenerate stars taking into account most of the possible nuclear equilibria as well as the general relativity corrections. The results show that the instabilities due to the creation of electron pairs and more effectively, to the decomposition of Fe^{56} and He^4 extend outside the relativistic instability region. It is however no longer clear whether this dynamical instability plays a dominant role in the origin of the supernova phenomenon, other types of instabilities, more akin to secular instability albeit with a very short time-scale (neutrino emission as in White and Colgate's theory or explosive nuclear reaction as in Imshenik and Nadezhin's (4) and in Arnett's models (5) being perhaps more significant).

Another possibility of dynamical instability is related to relativistic degeneracy ($\Gamma_\mu \rightarrow \frac{4}{3}$) combined with extrafactors like nuclear equilibrium between β-decay and electron capture and (or) general relativity corrections which reduce further the effective value of Γ_2. A. Baglin found that, under the effects of these factors, a white dwarf would become dynamically unstable for $\rho_c \simeq 8 \times 10^8$ g/cm^3 while Wheeler, Hansen and Cox, taking into account the finite relaxation time of the equilibrium, push back this critical density to $\rho_c \simeq 10^{10}$ g/cm^3.

Another destabilizing mechanism (secular rather than dynamical instability) which might operate at $\rho_c \simeq 2 \times 10^9$ g/cm^3 has been further studied by Wheeler *et al.* (6) who find severe difficulties with this mechanism.

There are still problems with the equation of state for matter at very high densities but it seems established that no stable configurations can exist in the intermediate range between the most condensed stable white dwarfs ($\rho \simeq 10^9$–10^{10} g/cm^3) and neutron stars ($\rho \simeq 10^{13}$–10^{16} g/cm^3). A summary of work done in this field may be found in *Relativistic Astrophysics* by Zeldovic and Novikov, Moscow 1968.

The discovery of pulsars also renewed interest in the periods of radial oscillations of very condensed stars. (See Proc. 4th Texas Symp. 1969).

2. *Non-radial modes*

Smeyers results on the complete 4th order problem of the non-radial oscillations of main sequence stars of fairly high masses have been published (7).

Robe (8) has cleared up completely the question of the classification of the $f-p$ and g-modes, all modes subsisting but the lower ones acquiring extra-nodes, due to the apparent singularities, as the central condensation increases.

The asymptotic behaviour of the non-radial modes has been studied by Vandakurov, Iwens and Smeyers using Langer's method while Tassoul has improved the results by extending Olver's method to this case.

A. Baglin and J. Heyvaerts have appealed to *g*-modes of non-radial oscillations of neutron stars to explain the periods of pulsars. However Thorne and his coworkers (**9**) in an exhaustive analysis of these non-radial modes for fully relativistic models have shown that gravitational radiation is emitted by such pulsations and damps them rapidly, typically in a time of the order of 10^3 to 10^4 times the period for a neutron star.

Perdang has applied a group-theoretical treatment to the case of non-radial oscillations in presence of various perturbing factors.

A general variational principle for the oscillations of a differentially rotating configuration as developed by Lynden-Bell and Ostriker (**10**) was applied by Clement (**11**) to the problem of the β Canis Majoris stars. The same approach is used by Lebovitz (**12**) to show that rotation has practically always a stabilizing influence.

The same general problem can also be tackled by the virial tensor approach as shown by Tassoul and Ostriker (**13, 14**) and Simon has discussed it for an arbitrary low of rotation up to terms in Ω^2 by a generalized perturbation method (**15**).

For uniformly rotating polytropes, Tassoul and Ostriker (**16**) find that secular and dynamical instabilities occur respectively for ratios of the kinetic to the gravitational energy $T/|W| \simeq 0{\cdot}14$ and $T/|W| \simeq 0{\cdot}26$ whatever the value of the polytropic index. Influence of other factors such as a tidal force has also been taken into account by Tassoul (**17**) and by Zahn (**18**) who discusses especially the excitation of fairly high *g*-modes in a component of a binary system.

Kato (**19**) has also considered the problem of the excitation of non-spherical waves in a differentially rotating stellar convective envelope with possible applications to the Sun.

C. *Vibrational stability and variable stars*

1. *Radial modes*

A number of investigations concerning especially large mass stars (around or above the critical mass M_c) are due to Simon and Stothers with application to the blue supergiants (**20**) and Wolf-Rayet stars (**21**) some of which, in their views, might derive from masses originally much larger than M_c. They found a correlation between vibrational instability and the eigenfrequency of the fundamental mode (**22**). They also point out a new mechanism for enhancing vibrational instability (**23**) namely, an increase of the mean molecular weight $\bar{\mu}$ in the external layers as might result from accretion of evolved material in close binary systems. They suggest that this may be responsible for the observed oscillations of the β Cephei stars. In case of subsequent mixing, instability may still arise due to the uniform increase of $\bar{\mu}$ in the star which may account for the variations of some Wolf-Rayet stars. Boury and Ledoux have also discussed the interpretation of Wolf-Rayet stars in terms of vibrational instability and mass-loss but from homogeneously evolving stars of large masses.

Other applications to a helium rich variable star (**24**) and to M supergiants (**25**) have also been considered by Stothers. Kamijo (**26**), on the other hand, has studied the vibrational stability of the primary in a close binary on its way to the white dwarf stage following mass loss and determined the phase of maximum instability.

The direct linear approach (complex solutions) was applied at the university of Colorado by Langer (**27**) to models of Long Period Variables with some success and by Ziebarth for a general stability survey.

Non-linear effects also have been receiving more attention either following the more or less classical approach by series developments or analytical approximation not only for the adiabatic case but also to ascertain the development of the instability due to nuclear reactions in large mass stars. This important problem has now been tackled directly by numerical methods by various people: J. P. Cox at Los Alamos, Ziebarth at Colorado, Talbot at the Massachusetts Institute of Technology, who find that it implies some loss of mass.

Of course the problem of finite pulsations in classical variables like the Cepheids or RR Lyrae star is still receiving a good deal of attention and general accounts of the present status of the problem have been published. A long-range program in collaboration between J. P. Cox (Colorado), D. S. King (Univ. New-Mexico) and A. N. Cox (Los Alamos) is being continued. Christy has obtained some general results (unpublished) for Cepheids concerning the value of $Q\ (=P\sqrt{(\bar{\rho}/\bar{\rho}_{\odot})})$, the instability region and the theoretical Period-Luminosity relation. He confirms also his interpretation of secondary humps.

Oscillation of a red supergiant with $M=M_{\odot}$ has also been studied (**28**), the non adiabatic convection just outside the H-ionization zone playing a prominent part in the excitation of the pulsation. Let us note also in this respect the work of Graham (**29**) who finds instability and finite somewhat erratic pulsations for a star which is just reaching the Hayashi track after gravitational contraction.

The problem of vibrational stability towards non-radial perturbations is also receiving attention and Zahn (**30**) has studied Baker's one-zone model in this respect showing that instability can occur in the region of the H-R diagram where the δ Scuti-variables occur. Somewhat similar techniques were also used by Ishizuka to derive a criterion for the excitation of non-radial pulsations.

The behaviour of the amplitudes of g^{+} modes in small mass stars with a large convective envelope (relatively small amplitudes) and a small radiative core (relatively large amplitudes) also points to a possibility of vibrational instability which is studied by Robe (Liège).

D. *Secular or thermal instability*

Gabriel and Noels (**31**) have found secular instability in carbon stars at Deinzer and Salpeter's minimum mass.

Rakavy and Shaviv (**32**) have discussed the stability of nuclear burning using an "effective specific heat" which is probably related directly to the integral criterion of secular stability as treated by Ledoux (cf. for instance "Stars and Stellar Systems", Vol. VIII). Unno in collaboration with Kato, has also discussed the thermal instability of a shell-source model (**33**).

Rose and Giannone and Weigert have been able to follow the non-linear development into successive pulses of instabilities of this type. Rose (**34**) has shown that the star can become vibrationally unstable during such a pulse.

J. P. Cox, Hansen and Miss Hertz at the University of Colorado have started a general investigation and numerical program allowing for both real and complex solutions.

Secular stability towards non-spherical perturbations of a chemically inhomogeneous zone has been investigated by Kippenhahn. Although possibilities of instability are present, P. Biermann has found none up to now in applications to transition zones from hydrogen rich envelopes to helium rich interiors (unpublished).

These non-radial perturbations are especially significant for the secular stability of stars in presence of rotation, magnetic fields or tidal field. In the case of rotation, Goldreich and Shubert (**35**) and Fricke (**36**) have obtained the same conditions for secular stability of differentially rotating stars. According to Kippenhahn (**37**) the time-scale of these instabilities is however rather long, at least of the order of the Kelvin-Helmholtz time-scale. Extension to the case when magnetic fields with various geometries are present have already been made.

REFERENCES

1. Unno, W. 1968, *Publ. astr. Soc. Japan*, **20**, 356.
2. Perdang, J. Ph.D. Thesis (Liège); to be published.
3. Upton, E. R. L., Little, S. J., Dworetsky, M. N. 1968, *Ap. J.*, **154**, 597.
4. Imshennik, V. S., Nadezhin, D. K. 1967. Preprint; 1969, *Nauchnye Informatsii, Moscow*, **13**.
5. Arnett, W. D. (Orange preprint, Caltech.); *Nature*, **219**, 1344, 1968; *Astrophys. J.*, **157**, 1369, 1969.
6. Wheeler, J. C. 1969, Ph.D. Thesis, University of Colorado.

7. Smeyers, P. 1967, *Bull. Soc. r. Sci., Liège*, **36**, 357, 1967.
8. Robe, H., Ph.D. Thesis, Université de Liège, 1968; *Ann. Astrophys.*, **31**, 475, 1968.
9. Thorne, K. S. 1969, *Astrophys. J.*, **158**, 1.
10. Lynden-Bell, D., Ostriker, J. P. 1967, *Mon. Not. R. astr. Soc.*, **136**, 293.
11. Clement, M. 1967, *Astrophys. J.*, **150**, 589.
12. Lebovitz, N. R. 1969, *The Effect of an Arbitrary Law of Small Rotation on the Oscillations and the Stability of Gaseous Masses* (to be published).
13. Tassoul, J. L., Ostriker, J. P. 1968, *Astrophys. J.*, **154**, 613.
14. Ostriker, J. P., Tassoul, J. L. 1969, *Astrophys. J.*, **155**, 987.
15. Simon, R. 1969, *Astr. Astrophys.*, **2**, 390.
16. Tassoul, J. L., Ostriker, J. P. 1970, *Astr. Astrophys.* (in press).
17. Tassoul, M., Tassoul, J. L. 1967, *Astrophys. J.*, **150**, 213.
18. Zahn, J.-P. 1970, *Astr. Astrophys.* (in press).
19. Kato, S. 1969, *Astrophys. J.*, **157**, 287.
20. Stothers, R., Simon, N. R. 1968, *Astrophys. J.*, **152**, 233.
21. Simon, N. R., Stothers, R. 1969, *Astrophys. J.*, **155**, 247.
22. Simon, N. R., Stothers, R. 1969, *Astrophys. J.*, **156**, 377.
23. Stothers, R., Simon, N. R. 1969, *Astrophys. J.*, **157**, 673.
24. Stothers, R. 1968, *Astrophys. J.* **154**, L91.
25. Stothers, R. 1969, *Astrophys. J.*, **156**, 541.
26. Kamijo, F. 1968, *Z. Astrophys.*, **68**, 21.
27. Langer, G. E. Ph.D. Thesis, University of Colorado (to be published).
28. Keeley, D. A. Red or Long Period Variables, Ph.D. Thesis.
29. Graham, G. XVIe Coll. d'Astrophysique de Liège (in press).
30. Zahn, J. P. 1968, *Astr. Lett.*, **1**, 209.
31. Gabriel, M., Noels-Grotsch, A. 1968, *Ann. Astrophys.*, **31**, 167.
32. Rakavy, G., Shaviv, G. 1968, *Astrophys. Space Sci.*, **1**, 347.
33. Unno, W., Kato, S. 1969, *Publ. astr. Soc. Japan* (in press).
34. Rose, W. K. 1967, *Astrophys. J.*, **150**, 193; 1968, **152**, 245.
35. Goldreich, P., Schubert, G. 1967, *Astrophys. J.*, **150**, 571.
36. Fricke, K. 1968, *Z. Astrophys.*, **68**, 317.
37. Kippenhahn, R. 1969, *Astr. Astrophys.*, **2**, 309.

III. EVOLUTION OF CLOSE BINARIES

Most of the work in the last three years in this field dealt with the mass exchange during post main sequence stages of close binary components. According to the evolutionary stage of the primary at the onset of mass exchange three types of mass exchange have been distinguished, the socalled cases A, B, C:

Case A (onset of mass exchange while the primary is burning hydrogen in its center): Several case A systems are investigated now by Kippenhahn and Weigert (**7**), Paczynski (**13**), Plavec (**14**), Smith (**18**), Ziolkowski (**19**) and Snezhko (**20**) and therefore quite a number of theoretical post mass exchange systems are available in order to compare with observed semi detached systems. Comparisons (**5**, **18**) give good agreement for systems with total masses higher than 3·0 solar masses.

Case B (onset of mass exchange after the primary has finished its central hydrogen burning, but before the ignition of helium): After mass exchange of this type the star which has lost mass ends up as a helium burning star or as a white dwarf, both having only a thin hydrogen envelope (**6**, **7**, **8**, **10**, **14**, **19**). In some cases the mass exchange becomes rather slow (**16**) and the systems acquire the properties of semi detached systems for a while. There are good reasons to believe, that the observed Algol systems with a total mass below 3 solar masses are case B systems in their final phase of mass exchange. There are reasons to believe that there exist many systems which via a case B mass exchange will form a white dwarf in the future (**7**, **8**). WR stars may be end products of case B mass exchange (**1**, **13**, **18**). Although the computed models resemble many properties of WR binary systems the mechanism of mass ejection in these systems is still not known. There is also indication, that mass has been lost from the system during mass exchange.

Gorbatzky and Koroviakovsky (**21**) have studied the behaviour of the gas stream in semi-detached binaries including the effects of gas pressure in the flow.

Case C (onset of mass exchange after the primary has finished its central helium burning, but before the ignition of carbon): Only one system has been investigated (**11**). Similar to certain case B systems (**6, 16**) a quick and effective phase of mass exchange is followed by a second phase of slow mass transfer. The original primary ends up as a white dwarf of one solar mass. The possibility has been discussed (**11**) that the white dwarfs in systems like Sirius and Procyon might have been produced by case C mass exchange.

Besides the mass exchange calculations based on the normal simplifying assumptions (spherical models, conservation of total mass and orbital angular momentum of the system, circular orbits) work is in progress or should be started in the following topics:

1. Details of the flow of matter from one star to the other during mass exchange. Especially the problem, whether the system loses mass and angular momentum. Already a two-dimensional solution would be worthwhile (**9, 21**). Some Algol systems seem to have undergone mass loss during their case A mass exchange phase (**13**). U Gem binaries give similar suggestions. The influence of mass exchange on the eccentricity of the system is unknown. In principle mass exchange could feed enough mechanical energy into the system, even that the system can be disrupted.

2. Work has been done on the Novae and U Gem problem along two lines:

(a) assuming that the contact component of a semi detached close binary system becomes unstable (**2, 12**)

(b) assuming that hydrogen is accreted on top of a white dwarf component of a close binary system (**4, 17**).

There is observational evidence that the cool giant component in U Gem systems is responsible for the outburst but in T Cr B the white dwarf seems to cause the outburst (**15**).

3. No system up to now has been computed which undergoes mass exchange forth and back. In principle one should be able to derive a crude picture of this type of event, for instance from the scheme of the generalized main sequence (**7, 5**) introduced to determine properties of post mass exchange systems.

4. Although the post main sequence evolution of binary systems has been computed for many cases no efforts have been made recently to understand how the close binaries have been formed during the pre main sequence evolution. The fission theory might become attractive again since considerable progress in stellar rotation theory has been made during the last years (**3**).

5. Contact systems seem to be still difficult to explain. The theory still fails to give contact systems of the right mass ratio.

BIBLIOGRAPHY

1. Barbaro, G., Giannone, P., Giannuzi, M. A., Summa, C. 1969, *Mass Loss from Stars*, Trieste 1968, Astrophys. and Space Sci. Library, Vol. 13, p. 217.

2. Bath, G. T. 1969, Dynamical Instabilities in semi-detached Close Binary Systems with possible applications to Novae and Nova-like Variables. University of Sussex. Preprint 1969.

3. Bodenheimer, P. 1969, 16th Liège Symposium.

4. Giannone, P., Weigert, A. 1967, *Z. Astrophys.*, **67**, 41.

5. Giannone, P., Kohl, K., Weigert, A. 1968, *Z. Astrophys.*, **68**, 107.

6. Giannone, P., Refsdal, S., Weigert, A. 1969, Elsinore Colloquium *On Mass Loss and Evolution in Close Binaries.*

7. Kippenhahn, R., Weigert, A. 1967, *Z. Astrophys.*, **65**, 251; Colloquium on "Low Luminosity Stars", Charlottesville.

8. Kippenhahn, R., Kohl, K., Weigert, A. 1967, *Z. Astrophys.*, **66**, 58.

9. Kitamura, M. 1969, Elsinore Colloquium *On Mass Loss and Evolution in Close Binaries.*

10. Kříž, S. 1969, *Mass Loss from Stars*, Trieste 1968, Astrophys. and Space Sci. Library, Vol. 13, p. 257.

11. Lauterborn, D. 1969, *Mass Loss from Stars*, Trieste 1968, Astrophys. Space Sci. Library, Vol. 13, p. 242; Elsinore Colloquium *On Mass Loss and Evolution in Close Binaries.*

12. Osaki, Y. 1969, *A Mechanism for the Outburst of U Geminorum Stars*, Department of Astronomy, Columbia University.
13. Paczyński, B. 1966, *Acta astr.*, **16**, 231; 1967, **17**, 1, 7, 193, 355; 1967, *Comm. Obs. r. Uccle*, **B 17**, 111.
14. Plavec, M. 1968, *Astrophys. Space Sci.*, **1**, no. 2; 1968, *Adv. Astr. Astrophys.*, **6**, 202; 1969, *Mass Loss from Stars*, Trieste 1968, Astrophys. and Space Sci. Library, Vol. 13, p. 242; 1968, *B.A.C.*, **19**, 24; 1969, *B.A.C.*, **20**, 41.
15. Popper, D. M. 1969, Elsinore Colloquium *On Mass Loss and Evolution in Close Binaries.*
16. Refsdal, S., Weigert, A. 1969, *Mass Loss from Stars*, Trieste 1968, Astrophys. and Space Sci. Library, Vol. 13, 253; 1969, *Astr. Astrophys.*, **1**, 167.
17. Secco, L. 1968, *Publ. Oss. astr. Padova*, no. 145.
18. Smith, L. F. 1968, in *Wolf-Rayet Stars*. NBS Special Publication no. 307.
19. Ziólkowski, J. 1969, *Mass Loss from Stars*, Trieste 1968, Astrophys. and Space Sci. Library, Vol. 13, p. 231; 1969, *Astrophys. Space Sci.*, **3**, 14.
20. Snezhko, L. I. 1967, *Collection of Papers on Astronomy*. Ural State University, Sverdlovsk.
21. Gorbatzky, B. G., Koroviakovsky 1969, Proc. Symp. *Variable Stars*, Kishinev.

IV. PROGRESS IN THE STUDY OF NUCLEAR REACTIONS IN STARS

In his talk "The Empirical Foundations of Nucleosynthesis" (**1**) at the Paris Symposium, W. Fowler presented a comprehensive review of the evidence as of early 1967 from nuclear physics experiments and from astrophysical observations that forms much of the empirical foundations of the theory of nucleosynthesis. Ordinary stars, supermassive stars, and the early high-temperature stage of the expanding universe were included in his analysis.

Considerable progress has been made in the study of massive objects, of supermassive stars, and of the synthesis of elements at very high temperatures (**2, 3, 4**). The study of explosive nucleosynthesis has been particularly active (**5, 6, 7, 8**). Wagoner (**9**) has reviewed the thermonuclear reaction rates of interactions that are important in the synthesis of elements within objects exploding from very high temperatures. He has used these rates to analyze the relative abundances of the elements produced.

Fowler *et al.* (**10**) have reviewed the available experimental data on cross sections for the nuclear interactions of neutrons, protons, deuterons, and α particles with light and medium-mass nuclei that are important in nucleosynthesis. They compiled tables of resulting reaction rates, nuclear lifetimes, and energy-generation rates. Bahcall and Fowler (**11**) have examined the effects of excited nuclear states on stellar reaction rates and have used available experimental data to determine the reaction rates for several endoergic (p, n) and (α, n) reactions and their reaction rates. In their work on these reaction rates Fowler and his collaborators have been materially aided by the compilations of nuclear data in *Nucl. Phys.*, (**78**, 1) 1966–68 by Lauritsen and Ajzenberg-Selove, Endt and van der Leun (*ibid.*, **A105**, 1) and Ajzenberg-Selove and Lauritsen (*ibid.*, **A114**, 1). In addition they have used data from private communications or published papers for the reactions $^{14}N(p, \gamma)^{15}O$; $^{13}C(\alpha, n)^{16}O$; $^{9}Be(\alpha, n)^{12}O$; $^{3}He(^{3}He, 2p)^{4}He$; $^{27}Al(p, \gamma)^{28}Si$; $^{24}Mg(\alpha, \gamma)^{28}Si$ and $^{3}He(\alpha, \gamma)^{7}Be$.

The recent study of the reaction $^{12}C + ^{12}C$ at low energies by Patterson *et al.* (**12**) has brought the data for that reaction much closer to energies of astrophysical interest and has shown that the cross section is lower at these energies than previously believed. Arnett and Truran (**13**) have used these new data to reexamine the problem of carbon-burning nucleosynthesis.

Theoretical calculations by Bahcall and May (**14**) have made possible improved reaction rates for proton capture by protons, a reaction for which there are no experimental data.

Audouze and Reeves (1969, Preprint) have analyzed the experimental data available on reactions that destroy ^{6}Li and ^{7}Li by proton capture. Audouze *et al.* (**15**) have studied spallation reactions induced in CNO targets by proton capture.

Bodansky *et al.* (**16**) have examined the silicon-burning reactions and their effects on abundances of nuclei between $A = 28$ and $A = 62$. They found good agreement with solar-system abundances and that the reactions are strongly exoergic. In a further study of these reactions Clayton and Woosley (**17**) have examined the long-standing problem of the production of ^{58}Ni.

At the Paris Symposium, Clayton discussed three mechanisms: the *s*-process, the decay of short-lived *r*-process transbismuth progenitors, and the decay of long-lived *r*-process transuranic progenitors and their effects on the abundances of the isotopes of lead.

Imshennik *et al.* (**18**) have pointed out the important role of the β-processes on early stages of gravitational collapse of a stellar core.

In their study of the spectrum of the hydrogen-poor star HD 30353, Wallerstein *et al.* (**19**) found high nitrogen abundance suggestive of equilibrium abundances in material that has been processed through the CNO bi-cycle.

REFERENCES

1. Fowler, W. A. 1968, *Origin and Distribution of the Elements*, Ed. L. H. Ahrens. New York, Pergamon Press, p. 3.
2. Wagoner, R. V., Fowler, W. A., Hoyle, F. 1967, *Ap. J.*, **148**, 3.
3. Fowler, W. A. 1967, *High Energy Physics and Nuclear Structure*. North-Holland Publishing Co., p. 203.
4. Wagoner, R. V. 1967, *Science*, **155**, 1369; *A. Rev. Astr. Astrophys.*, **7**, 553.
5. Arnett, W. D., Cameron, A. G. W. 1967, *Can. J. Phys.*, **45**, 2953.
6. Truran, J. W., Arnett, W. D., Tsuruta, S., Cameron, A. G. W. 1968, *Astrophys. Space Sci.*, **1**, 129.
7. Arnett, W. D. 1969, *Ap. J.*, **157**, 1369.
8. Truran, J. W., Arnett, W. D. 1969. Preprint.
9. Wagoner, R. V. 1969, *Ap. J. Suppl.*, **18**, no. 162, 247.
10. Fowler, W. A., Caughlan, G. R., Zimmerman, B. A. 1967, *A. Rev. Astr. Astrophys.*, **5**, 525.
11. Bahcall, J. N., Fowler, W. A. 1969, *Ap. J.*, **157**, 645.
12. Patterson, J. R., Winkler, M., Zaidins, C. S. 1969, *Ap. J.*, **157**, 367.
13. Arnett, W. D., Truran, J. W. 1969, *Ap. J.*, **157**, 339.
14. Bahcall, J. N., May, R. M. 1969, *Ap. J.*, **155**, 501.
15. Audouze, J., Epherre, M., Reeves, M. 1967, *Nucl. Phys.*, **A97**, 144; 1968, *High Energy Nuclear Reactions in Astrophysics*, Ed. B. S. P. Shen. Benjamin, New York.
16. Bodansky, D., Clayton, D. D., Fowler, W. A. 1968, *Ap. J. Suppl.*, **16**, no. 148, 299.
17. Clayton, D. D., Woosley, S. E. 1969, *Ap. J.*, **157**, 1381.
18. Imshennik, V. S., Ivanova, L. N., Nadezhin, D. K. 1967, Preprint; 1969, *Nauchnye Informatsii, Moscow*, **13**.
19. Wallerstein, G., Green, T. F., Tomley, L. J. 1967, *Ap. J.*, **150**, 245.

V. NEUTRINO ASTROPHYSICS

During the last few years the rates of many different neutrino-antineutrino pair producing reactions have been calculated, based upon the assumption that there is a universal Fermi interaction which provides a direct electron-neutrino coupling to lowest order in the coupling constant. From these investigations it appears probable that the principal contributions to energy loss in stellar interiors are from the following three processes:

(a) electron-positron pair annihilation; (b) photoneutrino process (an electron scattering process in which a photon is converted into a neutrino-antineutrino pair); (c) plasma neutrino process (transformation of longitudinal and transverse plasmons into neutrino-antineutrino pairs). An approximate analytical formula by Beaudet, Petrosian and Salpeter has been derived, (**1**) which reproduces the energy loss rate to 15% or better in the range 10^8–10^{10} K and with less precision outside this temperature range. Included in these results are additional calculations of the rate of the photoneutrino process by the same authors (**2**) and for plasma effects on this process (**1**).

The same authors have also reviewed the rates for several other processes: the recombination process, the bremsstrahlung process, electron-electron scattering, photon-photon scattering and the URCA process. The recombination process may give a significant contribution to the energy loss in models near the helium flash. The URCA process (alternate electron captures and beta decays) may be important in certain circumstances.

Radiative corrections to the neutrino-electron interaction have been calculated by Hebb (**3**) and have been shown to alter energy loss rates by only a few per cent.

Neutrino-antineutrino emission from stellar interiors with intense magnetic fields, due to a synchrotron process, has been calculated by Landstreet (**4**) and has been shown to be unimportant.

Electron-electron neutrino bremsstrahlung rates have been calculated by Cazzola and Saggion (**5, 6**), who have found that it contributes not more than one percent of the energy loss rates due to other processes under conditions of astrophysical interest.

Neutrino-pair bremsstrahlung, in which a hot degenerate electron gas Coulomb scatters on imbedded nuclei, has been calculated by Festa and Ruderman (**7**). They find that this process will dominate at high degeneracy, since it is not quenched by degeneracy as is the plasma process. The process may also dominate at low temperatures (below 10^8 K).

Bandyopadhyay (**8**) has suggested that there may exist a photon-neutrino coupling process which could give neutrino-antineutrino emission in all of the usual ways and also by a photon-photon scattering process. The rates are smaller than for the usually-considered current-current coupling in the common process. Chaudhuri (**9**) has argued that such a coupling may give a very important synchrotron process. However, the photon-photon process may be much more important than in either of these schemes if it is mediated by an intermediate vector boson (**10**).

The URCA process has also received additional attention (**11, 12, 13**). Tsuruta and Cameron (**13**) have calculated URCA rates for shells in white dwarf stars in which the electron Fermi level is in the vicinity of an electron capture threshold, and have shown that the process can dominate the cooling.

The solar neutrino experiments of Davis, Harmer and Hoffman (**14**) gave results approximately half as large as were expected on the basis of theories of the solar interior. Extensive reviews concerning the physical and the astrophysical aspects of this problem have been published by Bahcall *et al.* (**15**), Iben (**16**), and Kotcharov (**17**). Iben (**18**) has shown that the upper limit on the solar neutrino flux places an upper limit on the solar helium abundance that is low compared to a mean *Y* found for other objects in the galactic disk and in the halo. The discrepancy persists even if the pertinent nuclear cross section factors are varied as far as possible within experimentally estimated limits. The possibility that the presence of a solar convective core will reduce the discrepancy (**16**) suggests that every effort should be made to improve solar opacities.

BIBLIOGRAPHY

1. Beaudet, G., Petrosian, V., Salpeter, E. E. 1969, *Ap. J.*, **150**, 979.
2. Petrosian, V., Beaudet, G., Salpeter, E. E. 1969, *Phys. Rev.*, **154**, 1445.
3. Hebb, J. 1967, *Ap. J.*, **147**, 359.
4. Landstreet, J. D. 1967, *Phys. Rev.*, **153**, 1372.
5. Cazzola, P., Saggion, A. 1969, *Nuovo Cim.*, **63**, 354.
6. Cazzola, P., Saggion, A. 1969, *Nuovo Cim.*, **63**, 367.
7. Festa, G. G., Ruderman, M. A. 1969, *Phys. Rev.*, **180**, 1227.
8. Bandyopadhyay, P. 1968, *Phys. Rev.*, **173**, 1481.
9. Chaudhuri, P. R. 1969, preprint.
10. Campbell, J. A. 1969, *Ap. J.*, **157**, L177.
11. Imshennik, V. S., Nadezhin, D. K., Pinaev, V. C. 1967, *Astr. Zu.* (*U.S.S.R.*), **44**, 768; 1968, *Soviet Astr.*, **11**, 617.
12. Nadezhin, D. K., Chechetkin, V. M. 1969, *Soviet Astr.*, **13**, 213.
13. Tsuruta, S., Cameron, A. G. W. 1969, preprint.
14. Davis, Harmer and Hoffman 1968, *Phys. Rev. Lett.*, **20**, 1205.
15. Bahcall, Bahcall and Ulrich 1969, *Ap. J.*, **156**, 559.
16. Iben, I. Jr. 1969, *Ap. J.*, **155**, L101.
17. Kotcharov, G. E. 1969, Proceedings of VI Winter School on Cosmophysics, Apatite, p. 137.
18. Iben, I., Jr. 1969, *Ann. Phys.* **54**, no. 1, 164.

VI. CALCULATIONS FOR STELLAR MATERIAL PROPERTIES INCLUDING OPACITIES, CONDUCTIVITIES AND EQUATION-OF-STATE

Since 1965 when opacities were given for twelve new astrophysical mixtures there have been only a few additional opacity tables published. However, with computing programs, which were being frequently changed, an additional series of opacity tables was computed at Los Alamos. The mixtures of general interest are given in Table 1.

Table 1. Opacity Mixtures

Name	X	Y	Z	
Demarque I	0·97	0·00	0·03	
Demarque II	0·00	0·97	0·03	
Demarque III	0·999	0·001	0·001	
Demarque IV	0·000	0·999	0·001	
Demarque V	0·99	0·00	0·01	
Demarque VI	0·00	0·99	0·01	
Baerentzen	0·700	0·297	0·003	
Forbes I*	0·913	0·080	0·007	
Forbes II*	0·8750	0·1249	0·0001	
Forbes III	0·75	0·25	0·00	
Forbes IV*	0·600	0·396	0·004	
Forbes V*	0·5000	0·4997	0·0003	
Forbes VI	0·0	1·0	0·0	
Hydrogen	1·0	0·0	0·0	
Cameron III*	0·500	0·479	0·021	
Cameron IV*	0·250	0·729	0·021	
Cameron V*	0·100	0·879	0·021	
Cameron VI*	0·050	0·929	0·021	
Torres I	0·00	0·94	0·06	
Torres II	0·7	0·2	0·1	
Torres III	0·0	0·9	0·1	
Torres IV	0·3	0·6	0·1	
Demarque VII	0·94	0·00	0·06	
Schatzman I	0·70	0·27	0·03	
Schatzman II	0·60	0·37	0·03	
Schatzman III	0·700	0·252	0·048	
Schatzman IV	0·9000000	0·0983364	0·0016635	
Lambert Mix*	0·7877	0·1974	0·0149	
Lambert I*	0·49255	0·49255	0·01490	
Iben I (Demarque III)	0·999	0·000	0·001	
Iben II	0·800	0·199	0·001	
Iben III	0·500	0·499	0·001	
Iben IV	0·200	0·799	0·001	
Iben V (Demarque IV)	0·000	0·999	0·001	
Iben VI	0·9999	0·0000	0·0001	
Iben VII	0·8000	0·1999	0·0001	
Iben VIII	0·5000	0·4999	0·0001	
Iben IX	0·2000	0·7999	0·0001	
Iben X	0·000	0·9999	0·0001	
Iben XVIII	0·00000	0·49995	0·0001	Z(C) = 0·49995
Iben XIX	0·0000	0·0000	0·0001	Z(C) = 0·9999
Iben XX	0·0000	0·0000	0·0001	Z(C) = 0·49995 Z(O) = 0·49995
Iben XXI	0·0000	0·0000	0·0001	Z(O) = 0·9999
Kutter I	Z(O) = 0·13; Z(Ne) = 0·32; Z(Na) = 0·07; Z(Mg^{24}) = 0·043; Z(Mg^{25}) = 0·05			
Vila I	Z(C) = 0·2	Z(O) = 0·8		
Vila II	Z(C) = 0·05	Z(O) = 0·90	Z(Ne) = 0·05	
Vila III	Z(O) = 0·8	Z(Ne) = 0·1	Z(Mg) = 0·1	

* Mixtures do not have the same inter-metal ($Z > 2$) ratios as in the Aller (1961, *The Abundance of the Elements*. Interscience Publishers, Inc. New York) mixture.

It is to be emphasized that these tables are not necessarily consistent with each other. They are, however, from the same version of the programs when they all have the same name except for the number (for example, Demarque I–VI), 23 mixtures have been published in the publication of the U.S.S.R. Academy of Sciences (**1**) and 10 more mixtures are to be published in the Astrophysical Journal with all 33 from the same (most modern) generation of the opacity programs.

Auman (2) has given monochromatic absorption data for water vapor which can be added to the other sources of opacity and then integrated to give Rosseland mean opacities. The water vapor can appreciably affect the opacity of stellar material below about 3500 K. No tables of opacity exist, which include the water vapor absorption; but water vapor has been considered in construction of stellar atmospheres, and has been introduced into computations for late stages of evolution by the Warsaw and the Moscow groups. Tsuji (**3**) has also discussed molecules in a more general but less complete manner, and some general graphical results have been given. The effects of CO and OH, as well as other molecules, might be important to stellar opacities. The addition of molecular absorption to usable opacity tables is a current high priority job.

The hydrogenic approximation has been criticized by Carson and Hollingsworth (**4**) and a small table of opacities is given. Opacities depend strongly on the wave functions of the bound and free electrons, and large differences from the less elaborate hydrogen-like results have been found.

To overcome the assumption of the hydrogen-like approximation, Carson, Mayers and Stibbs have gone to the Thomas-Fermi atomic model and have shown that the total effect of this model on bound-bound, bound-free, and free-free absorptions can reach to almost a factor of three at temperatures like a million degrees and near unity density. Again not enough data are given to evaluate corrections to the Los Alamos tables, and indeed the corrections must be somewhat composition dependent.

Another important factor in increasing opacities near 10^6 K is the absorption by autoionizing lines as discussed by Watson (**5**). A paper by Watson (**6**) gives ten small tables which can be compared with the new Los Alamos results for the same mixtures. Large factors like 3 or 4 can again be found at 2×10^5 K and $\rho = 10^5$ gm/cm^3. The addition of the Thomas-Fermi equation-of-state or even more elaborate schemes as discussed by Rouse (*Ap. J.* in press) and the inclusion of autoionization lines to opacities are two more high priority opacity table requirements. Watson (*Ap. J.* in press) has discussed the collective effects of Thomson scattering when densities become high enough to cause some ordering of the free electrons. This effect is also discussed by Diesendorf and Ninham in an Imperial College of Science and Technology preprint. Decreases up to perhaps 20% or more may occur in the opacity of stellar matter in the high density hydrogen burning phases of stellar evolution.

To the radiative opacity one can add the effects of electron conduction. In the Los Alamos tables this has been done using the Mestel (**7**) results. Discussions by several authors such as Hubbard (**8**), Iben (**9**), Lampe (**10**), and most recently Hubbard and Lampe (Orange Aid Preprint) have resulted in tables (neglecting relativistic effects at $\rho > 10^6$ gm/cm^3) which can be used to produce effective stellar opacities.

There are six desirable additions to the production of stellar opacity tables.

1. Addition of H_2O absorption.
2. Addition of CO, OH and possibly other molecular absorptions.
3. Use of Thomas-Fermi model for bound-bound, bound-free and free-free absorptions.
4. Allowance for the autoionization lines.
5. Corrections for Thomson scattering for collective effects.
6. Addition of recent electron conduction theories and extension to higher densities.

Some of these improvements are now being pursued at Los Alamos.

BIBLIOGRAPHY

1. Cox, A. N., Stewart, J. N. 1969, *Nauchnye Informatsii, Moscow*, **15**, 1–103.
2. Auman, J. Jr. 1967, *Ap. J. Suppl.*, **14**, 171.

3. Tsuji, T. 1966, *P.A.S.J.* **18**, 127.
4. Carson, T. R., Hollingworth, H. M. 1968, *Mon. Not. R. astr. Soc.*, **141**, 77.
5. Watson, W. D. 1969, *Ap. J.*, **157**, 375.
6. Watson, W. D. 1969, *Ap. J. Suppl. Ser.*, **174**.
7. Mestel, L. 1950, *Proc. Camb. Phil. Soc.*, **46**, 331.
8. Hubbard, W. B. 1966, *Ap. J.*, **146**, 858.
9. Iben, J., Jr. 1968, *Ap. J.*, **154**, 557.
10. Lampe, M. 1968, *Phys. Rev.*, **170**, 306; *ibid.*, **174**, 276.

VII. NUMERICAL TECHNIQUES RELEVANT TO STELLAR CONSTITUTION

A. *General theory of quasi static calculations*

Over the past three years, at least two extensive reviews of the relaxation techniques for solving the stellar structure equations have been prepared (**1**, **2**). Valuable discussions have appeared regarding (a) the choice of variables appropriate for specific situations, (b) alternate schemes for treating sets of variables explicitly or implicitly, and (c) alternate choices for space centering variables (**3**, **4**, **5**).

Special schemes have been demonstrated to be advantageous for several specific situations. For example, numerical instabilities that may occur in calculating the helium flash may be avoided by an appropriate choice of the luminosity variables or by an appropriate space centering of the flux variable (**6**). The calculation of shell burning phases may be speeded up by taking advantage of the near invariance of the shell wave form. For some investigations it is adequate to replace the detailed variation of shell quantities by analytic boundary conditions on either side of the shell. This approximation has been extended by Eggleton to shells of finite thickness (**7**). Uus (**8**) and Paczynsky (in press) have improved this method for computation of late stages of evolution of Red giants and Paczynski has calculated evolutionary tracks up to the carbon ignition. Modifications necessary when surface mass loss is important have been discussed (see References Section III).

Methods other than that of relaxation have received attention. In particular, the possible advantage of employing standard integration techniques, but replacing a single interior fitting point by multiple fitting points has been demonstrated (**7**, **9**). Ruben, Lomnev and Vedjashkina have showed that the "particle in cell" technique holds promise for the calculation of evolution when the symmetry is other than spherical (**10**).

B. *Instabilities*

The very slow growth rate of pulsation amplitude relative to pulsation period has hampered attempts to determine whether or not the pulsation amplitude of massive main sequence stars will grow without limit. Growth rate may be artificially speeded up by multiplying $\varepsilon - \partial L/\partial m$ by an arbitrarily large factor. It has been shown (**11**) that the pulsation amplitude usually reaches a finite limiting value. However, it is not entirely clear that the artificial damping introduced may not be responsible for this. The nature of the thermal instability that arises in thin helium burning shells has been further clarified (**12**, **13**). The beginning of the transition from an isothermal core under non-degenerate conditions to a degenerate isothermal core and giant envelope has been shown by a linear stability analysis by Gabriel and Ledoux to correspond to a neutral point. Further insight into the nature of the Schönberg-Chandrasekhar limit has been given by the discussion of an analogous phenomenon in the structure of stellar systems (See References to Section II).

The machinery for detecting the onset of "one-way" instabilities according to Rakavy and Shaviv and for calculating their growth has been further developed by May and White (**14**).

C. *Dependence of realistic models on input physics*

There have been some suggestions in the literature that evolutionary tracks and the corresponding internal development are influenced by the mode of calculation: standard integration as opposed to relaxation of difference equations, one mode of space and time centering as opposed to another, etc. While it is true that numerical differences must occur simply because any computational scheme is an approximation, it is also true that in most instances such differences, are, from a practical standpoint, extremely minor. The major differences among computed tracks are due to differences in the input physics. This is particularly true if a modicum of common sense is used in applying criteria for adequate approximation.

Opacity and convection theory differences account for many of the variations in track details as do differences in nuclear cross section factors, particularly during helium burning phases (see discussion and references in the review paper by Iben (**15**) also 16–20). Mass loss may be important even for stars not in binaries and can significantly affect internal and track behavior (**21**, **22**). The manner in which the semiconvective zone is handled may strongly affect whether or not massive stars can burn helium in the core as red super-giants.

A discussion on the influence of input physics on the calculation of evolutionary tracks for stars of large masses can be found in the paper by Massevitch *et al.* (**25**).

REFERENCES

1. Kippenhahn, R., Weigert, A., Hofmeister, E. 1967, Chapter in *Methods in Computational Physics*, Ed. B. Adler, S. Fernbach, M. Rotenberg. Academic Press, New York.
2. Ruben, G. 1969, *Nauchnye Informatsii, Moscow*, **14**, 3.
3. Nurai, T., Sugimoto, D., Hōshi, R., Hayashi, C. 1968, *Progr. Theor. Phys. (Kyoto)*, **39**, 619.
4. Sugimoto, D. 1969, *Ap. J.* (in press).
5. Varshavsky, V. 1969, *Nautschnye Informatsii, Moscow*, **13**.
6. Heney, L., Forbes, J. E., Simpson, E., Ulrich, R. K. 1969 (preprint).
7. Eggleton, P. P. 1967, *Mon. Not. R. astr. Soc.*, **135**, 243; 1968, **140**, 387.
8. Uus, U. 1969, *Nauchnye Informatsii, Moscow*, **13**, 126.
9. Franzman, Yu. L. 1968, *Nauchnye Informatsii, Moscow*, **7**, 9.
10. Ruben, G., Lomnev, S. P., Vedjashkina, K. A. 1969, *Nauchnye Informatsii, Moscow*, **11**, 1.
11. Talbot, R. 1969, Thesis, M.I.T.
12. Hōshi, R. 1968, *Prog. Theor. Phys.*, **39**, 957.
13. Rose, W. K. 1969, *Mass Loss from Stars*, Ed. M. Hack. Reidel, Dordrecht-Holland.
14. May, M. M., White, R. H. 1967, Chapter in *Methods in Computational Physics*, Ed. B. Adler, S. Fernbach, M. Rotenberg. Academic Press, New York.
15. Iben, I., Jr. 1967, *A. Rev. Astr. Astrophys.*, **5**, 581.
16. Demarque, P., Hartwick, F. D. A., Naylor, M. D. T. 1968, *Ap. J.*, **154**, 1143.
17. Demarque, P., Schlesinger, B. 1969, *Ap. J.*, **155**, 965.
18. Meyer-Hofmeister, E. 1969, *Astr. Astrophys.*, **2**, 143.
19. Paczynski, B. 1970, in preparation.
20. Stothers, R., Chin, C. 1968, *Ap. J.*, **152**, 225.
21. Forbes, J. E. 1968, *Ap. J.*, **153**, 495.
22. Bisnovaty-Kogan, G. S., Nadezhin, D. K. 1969, *Nauchnye Informatsii, Moscow*, **11**, 27.
23. Varshavsky, V. I., Franzman, Yu. L. 1969, *Nauchnye Informatsii, Moscow*, **11**, 51.
24. Edwards, A. C. 1968, Thesis, Univ. of Wisconsin.
25. Massevitch, A. G., Tutukov, A. V., Franzman, Yu. L., Uus, U. K., Ergma, A. 1970, *Nauchnye Informatsii, Moscow*, **17**.

Below are listed books and review papers dealing with the above mentioned problems in general.

1. Clayton, D. D. 1968, *Principles of Stellar Evolution and Nucleo-Synthesis.* McGraw-Hill.
2. Chiu, H.-I. 1968, *Stellar Physics.* Blaisdell, Waltham, Mass.
3. Cox, J. P., Giuli, R. T. 1968, *Principles of Stellar Structure.* Gordon & Breach, New York.
4. Minory, Nishida 1968, *Stellar Evolution.* Blaisdell, Waltham, Mass.
5. Reeves, H. 1968, *Stellar Evolution and Nucleo-Synthesis.* Gordon & Breach, New York.
6. Tayler, R. J. 1968, Stellar Evolution, *Rep. Prog. Phys.*, **31**, 167.

In closing this report, it is a pleasure to thank all members of Commission 35 who have communicated information on their work (I am sorry to have to mention that only about 60% of the Commission members have done so, notwithstanding several appeals).

I am especially grateful to Prof. Mestel, Prof. Ledoux, Prof. Kippenhahn, Dr Georgeanne Couhlin, Prof. Cameron, Dr A. Cox and Dr Icko Iben, Jr. who have kindly prepared very complete reviews for the corresponding sections of this Report and have in that way substantially simplified my task in writing it. The full texts of these reviews will be published in the circular letters of Commission 35.

A. G. MASSEVITCH

President of the Commission

36. THEORY OF STELLAR ATMOSPHERES
(THÉORIE DES ATMOSPHÈRES STELLAIRES)

PRESIDENT: A. B. Underhill.
VICE-PRESIDENT: R. Cayrel.
ORGANIZING COMMITTEE: K. H. Böhm, V. V. Ivanov, J. T. Jefferies, R. N. Thomas, S. Ueno.

INTRODUCTION

This report has been prepared by the President with the assistance of the Vice-President, the Members of the Organizing Committee, V. V. Ivanov, P. Conti and of D. G. Hummer. The work of Commission 36 comes in contact with that of many commissions. Some subjects may be dealt with more fully in those reports; some subjects may have been "unlawfully" included here. Considerable choice has been exercised by the President about what to mention and in which section of the report to place it. The final product is only indicative of the large amount of work going on. An extensive bibliography has been compiled and will be distributed to commission members.

Commissions 12, 27, 29, 44 and 45 deal with observational material which our theories attempt to interpret, thus it is essential to study the reports of these commissions to appreciate the diversity of material for which theories are needed and against which theories may be checked. The theoretical treatment of the problems of planetary atmospheres, Commission 16, is not entirely divorced from that of stellar atmospheres. To set realistic inner boundary conditions on our atmosphere models and on our thinking, we must consider the material of Commission 35 while to set the outer boundary conditions we may consider the physical state of planetary nebulae which is one subject of Commission 34. When we wish to apply ideas about the interactions between radiation, atoms, ions and molecules we use material which lies in the province of Commission 14.

In the report period two conferences directly concerned with model atmospheres and the prediction of stellar spectra were held, namely the Third Harvard-Smithsonian Stellar Atmospheres Conference in 1968 on the subject of model atmospheres and predicted spectra based on the hypothesis of LTE, and IAU Colloquium no. 2 in Munich in 1969, on the subject Spectrum Formation in Stars with Steady-State Extended Atmospheres. The proceedings of the former conference contain a homogeneous grid of models with effective temperatures between 4000 and 50000 K and values of log g between 2 and 5. Information about opacities and Hγ profiles are also given. The calculations were done by Carbon, Gingerich and Kurucz. This material has been edited by Gingerich and was published by the M.I.T. Press. The proceedings of the Munich Conference have been edited by Wellmann and Groth and have been submitted to *Astronomy and Astrophysics* as a supplement. Papers on the theory of stellar atmospheres can also be found in colloquia and conferences sponsored by the Commissions mentioned above.

A. THE STUDY OF STELLAR ATMOSPHERES

1. *Model atmospheres and synthetic spectra*

a. *Based on the hypothesis of LTE*

In addition to the work reported at the Third Harvard-Smithsonian Symposium, systematic studies of model atmospheres and synthetic spectra for B stars and for the sun have been conducted by Aller, Ross, Leckrone and others, and for B 6 to O 9 stars by Underhill. Morton and Adams, Bradley, and Hickok have produced line-blanketed models of types B 4 to O 5 from which they have derived a scale of effective temperature and bolometric correction for early-type main-sequence stars. The question of whether the absorption-line spectrum which could be predicted matches the spectra observed for the above line-blanketed models has not been studied. Strom and Strom have continued to study the effects of opacity from silicon and the light elements on model atmospheres and have predicted the continuous spectra of late B and A stars. Böhm-Vitense as well as Hunger, Klinglesmith and others have made models for helium-rich stars. Models for white dwarfs have been computed by Matsushima and Terashita, by Kolesov in Leningrad and by Bues in Kiel. Models for A, F and G stars are being made by Peytremann *et al.* at Geneva. A grid of models for yellow supergiants has been prepared by Parsons while Krishna Swamy as well as F. Praderie have published models and synthetic spectra for F and G stars of various sorts. Cayrel-de Strobel has made sets of scaled solar models to investigate abundance differences in G and K stars. Matsushima has worked on solar type models. A considerable range of spectral type has been surveyed roughly by De Jager and Neven who published models based on an adopted temperature law and calculated synthetic spectra using simplified concepts. Problems connected with late type model atmospheres have been studied by Auman, Bodenheimer, Kandel, Kumar *et al.*, Tsuji, and by Vardya. The analytical representation of models and of synthetic spectra is being studied by Sapar, Kuusik and Viik at Tartu Observatory.

b. *When the hypothesis of LTE is not made*

In this case the level populations of the atoms and ions causing the opacity (continuous and line) must be found from the equations of statistical equilibrium and the source function must be modified in an appropriate manner. Mihalas, Stone and Auer have studied early type models when the continuous spectrum due to hydrogen and to helium is not in LTE and when radiative transmission in Lyman, Balmer, Paschen and Brackett α is explicitly considered. It is found that the temperature distribution ceases to decrease rapidly outwards, as in LTE line-blanketed models, and that a small temperature rise may occur. Feautrier has also shown that such a temperature rise may result as τ becomes very small. Strom, Strom and Kalkofen, Mihalas, and Lambert and Pagel have discussed the observable effects of non-LTE on the shape of the Balmer and Paschen continua of models. It appears that the change in colour and the size of discontinuities are at the limit of detection by present methods of observation. Pagel has considered the effect of atomic collisions with reference to establishing LTE in stellar atmospheres.

2. *Problems in line formation; astrophysically oriented spectroscopic information*

See also Section C. Athay and Skumanich have considered the theoretical interpretation of the Wilson-Bappu effect as well as the effects of non-coherent scattering on the shape of the curve of growth. The question of how to represent the source function in multiplets has been studied by Avrett. The model atmosphere studies often include predictions of the Hγ line and sometimes a few other lines. Warner has considered the effects of pressure broadening on solar and stellar curves of growth while Gussmann has indicated how the weight-function method may be cast in a more useful manner. Study of the regimes of quasistatic and of impact Stark broadening of hydrogen has been continued by N. Feautrier, F. Praderie and van Regemorter while Kuusik has discussed the significance of scattering and absorption in the hydrogen lines of stellar spectra. Possible sources of opacity in late type atmospheres and their dependence on temperature and pressure have been investigated by Linsky, Morris and Wyller, and by Vardya, while Fay has studied the CN molecule in detail.

3. *The spectra of O and B stars; the helium problem*

Temperature scales for A and B stars have been established by Heintze and by Wolff, Kuhi and Hayes as well as by Morton, by comparing measured energy distributions in the continuous spectrum with those predicted by means of model atmospheres. Strom and Peterson have made surface-gravity determinations for main-sequence B stars by fitting observed and predicted Hγ profiles. Hobbs and C. R. Cowley have used LTE theories to obtain the magnesium abundance in B stars; Snijders has estimated the iron abundance in γ Peg and ζ Cas. The meaning of O type spectra was investigated by Underhill who found that no consistent interpretation could be obtained using LTE theories. Observations of the spectra of halo and globular cluster B stars have been presented by Greenstein, Münch, Sargent and Searle and they have attempted interpretation in terms of T, P_e and helium abundance using simple theory. The helium lines are weak relative to other lines in many of these stars although the colours of these stars are like those of B stars. Theoretical studies by H. R. Johnson and Poland, by Hearn and by Shipman and Strom to estimate the importance of non-LTE effects on the He I line strengths in normal main-sequence stars have yielded contradictory results. Extreme helium poverty and extreme helium enrichment appear to be evident in stellar spectra but spectroscopic analysis is not yet so secure that the helium abundance can surely be determined within a factor two. Underhill has pointed out contradictions in the spectra of B Ia supergiants which may indicate a modest helium enrichment. Hardorp and Scholz have derived the helium abundance in τ Sco by fitting the wings of profiles and using LTE theory. Baschek and Norris are studying the He I line strengths in B stars.

4. *The spectra of A stars; abundance anomalies*

Many quantitative analyses of the spectra of Ap and Am stars have been made as well as of normal A stars and attempts have been made to determine abundances either by differential curve-of-growth techniques or by LTE model-atmosphere analysis. There is the problem of first finding the correct model ($T_{\rm eff}$, $\log g$ and composition) as well as adopting an appropriate theory of line formation and adequate line broadening mechanisms. Since LTE theories leave no free parameter other than abundance, differing line strengths lead to differing abundances, other factors being equal. Guthrie has summarized much of this work. The variety of results obtained is bewildering and suggests that the methods of analysis used may not be so reliable as the various nvestigators indicate. Unsöld has reviewed the question of the abundance and origin of the elements and concludes that the elements must have been created "en gros". The techniques of abundance determination require a thorough examination with careful consideration of the spectroscopic description of each line, where on the curve of growth it may lie, and whether its intensity is likely to be sensitive to the mode of line formation adopted. Preston has shown that the microturbulence in the Ap stars β CrB and γ Equ is small.

5. *The F, G and K stars; the Li/Be problem*

During the report period a considerable number of curve-of-growth, differential curve-of-growth and fine analyses of normal F, G, K stars (main-sequence and supergiants), subdwarfs, cepheids and high-velocity stars has been made and the effects of the lines on the measured colours have been studied. The correction for line blanketing plays an important role in determining where in the HR diagram subdwarfs fall. The helium content of subdwarfs and of population II stars has been deduced by Cayrel and by Strom and Strom from the position of these stars in the HR diagram. Rouse has indicated that the helium content of the solar photosphere may be found from the shape of the wings of strong lines. The abundance analyses for F, G and K stars lead to a variety of results the true significance of which is not entirely clear although it is evident that the value of the not fully understood parameter microturbulent velocity plays an important role. It is generally recognized that the presence of emission components in the Ca II resonance lines indicates the presence of a hot chromosphere (see report of IAU Colloquium no. 2). Conti, Danziger and Wallerstein have studied the question of the abundances of lithium and beryllium in stars. Nearly all the observations are consistent with the following statements: (a) a solar-type destruction mechanism (as yet unspecified) operates in G stars and in later types. Its *e*-folding time is a few 10^8 years for G stars and it is more efficient in later type stars. (b) A dilution destruction mechanism operates in giant stars. The observations are qualitatively in agreement with predictions by Iben. (c) Pre-main-sequence destruction operates in K and M stars. The observations are qualitatively in agreement with predictions by Bodenheimer. (d) The initial Li^7 content is roughly constant along the main sequence, at least from 0·5 to 3 solar masses. (e) The initial Be and Li^6 content is lower in F stars than in G stars, or else an additional destruction mechanism operates. This could be an atmospheric destruction by suprathermal protons (Epherre and Reeves). (f) The origin of Li-Be must be spallogenic since stars without any destruction mechanism operating have spallation ratios of Li^6, Li^7 and Be^9. (g) It is not yet known whether the Li-Be content observed in stars is galactogenic or autogenic in origin.

6. *Stars with molecular bands in their spectra*

Analysis of spectra has been attempted in a few cases and models have been made by Vardya and colleagues, Tsuji, and Myerscough. Attention has been directed chiefly at finding the abundance of lithium and of the light elements C, O and N in M stars as well as in the hydrogen-deficient carbon stars. One section of the book *Low Luminosity Stars* (Ed. S. S. Kumar) is devoted to the problems of the atmospheres of cool dwarf stars. Polarization and obscuration by particles in the extended atmospheres of late type stars is receiving attention from Dyck and H. R. Johnson and from Krishna Swamy and Wickramasinghe.

7. *Microturbulence and rotation*

The question of microturbulence *vs.* abundance for causing the strengthening of lines in stellar spectra particularly in stars of types A to F has been studied by Baschek and Reimers, Barry, Bonsack and Culver, Hardorp and Scholz, McNamara, and Preston and Cathey. Definite evidence from line profiles has been obtained by Hardorp, Bidelman and Prölss that the microturbulent velocity is less than 2 km s^{-1} in 3 Cen A and by Preston that it is less than 3 km s^{-1} in β CrB and γ Equ. It is doubtful that spectral peculiarity is connected with microturbulence or that microturbulence is the major factor causing stars of the same F, G type or $(b-v)$ colour, to have differing values of the m_1 index. The effects of rapid rotation on continuous and line radiation from stars has been investigated by Collins and his colleagues and by Hardorp and Strittmatter. Underhill has done some numerical experiments to demonstrate how the line shapes are deformed when the rotational broadening is of the order of the natural halfwidth of the line. Rapid rotation of a star should give the star a nonspherical form equivalent to a somewhat extended atmosphere in the equatorial plane. This lack of spherical symmetry has been confirmed by the discovery of the intrinsic polarization of Be and shell stars. The geometric effects of the rotation on the line profile produces broad, shallow lines while the extended atmosphere will produce deeper cores. The true interpretation of the observed lines, which are intrinsically strong lines will be reached when theories that take account of the Doppler effect as well as give a physically true representation of line formation in an extended atmosphere are available. The theoretical predictions to date fall short of this goal.

8. *Wolf-Rayet, Be and variable stars*

The problems of Wolf-Rayet stars are reviewed and a bibliography is given in the *Proceedings of the Symposium on Wolf-Rayet Stars* (Eds. Gebbie and Thomas). The observational material has increased very much over the last years and theoretical models have been stimulated with, as yet, no final choice between them possible on most of the conceptual features. There is a general agreement that the excitation state originates in a mechanical energy source of some kind and that the thermal structure consists of at least an initial outward rise of T_e into line-producing regions from the continuum-producing regions. The atmosphere is, thus, chromosphere-like. Castor, Hummer, and van Blerkom are investigating the influence of optical-depth effects on line profiles. An atlas of WR line profiles is in preparation (L. Smith, Kuhi) to be published by JILA-NBS in 1970. Questions of the influence of structural problems of the WR phenomenon center around the suggestions of possible binary character and of mass interchange. The aerodynamical, non-LTE, radiation-transfer coupling remains paramount in atmospheric considerations.

Keller and Mutschlecner at Indiana are initiating studies of atmospheres of cepheids from the standpoint of radiative transfer in moving atmospheres for comparison with the flux from static models. This problem has also been studied by Hillendahl. The physical problem is to determine how well static models, based on the flux coming from the pulsating model at each phase, can represent the atmosphere. Hutchings has continued to study mainly geometric models of the expanding envelopes around OB supergiants while Kogure has attempted models for the shells of Be stars as have Limber and Marlborough.

B. THE SOLAR SPECTRUM

1. *The continuum spectrum and the inferred solar model*

The temperature distribution in the solar atmosphere has been a topic of active research during the last three years, supported by new observations of the solar continuum from space in the UV and in the IR. At the Bilderberg Conference, the Utrecht Reference Photosphere was superseded by a new model referred to as the Bilderberg Continuum Atmosphere (BCA). This model is largely based on the inversion of limb-darkening laws in various spectral regions. The BCA is characterized by a very flat temperature minimum at about 4600 K running from $\tau = 10^{-2}$ to 10^{-4} and is somewhat cooler than the Utrecht Reference Photosphere in deep layers. The BCA at the time of the Conference conflicted only with non-LTE computations of the K line of Ca II. Later on, new observations in the UV by Parkinson and Reeves and in the IR by Eddy, Léna and McQueen have shown the necessity of having a temperature minimum of about 4300 K, considerably higher ($\simeq$ 500 km over the limb) than supposed before. The BCA was revised accordingly by Gingerich and co-authors, and will be published as the "Harvard-Smithsonian Reference Atmosphere" in *Solar Physics*. It may be concluded that a very substantial advance has been made during the past three years in knowledge of the temperature distribution in the solar photosphere and low chromosphere, as well as in opacity sources in the continuum. Comparison between these improved semi-empirical models and purely theoretical models in radiative equilibrium, corrected for convective transport in deep layers, and taking into account the blanketing effect are mostly still under way (see Subsection 8).

2. *Abundance of the elements in the solar photosphere*

A very important result is that the disagreement between the iron abundance in the photosphere and in the corona has been shown to be spurious and due to an erroneous absolute scale of Fe I oscillator strengths. New absolute measurements made by Huber and Tobey, by Martinez, King and Whaling, and by Grasdalen, Huber and Parkinson have shown that oscillator strengths of Fe I lines were about 10 times weaker than believed before. As a result of this, the abundance of iron in the photosphere is raised by a factor of ten and brought to the value log(Fe) = 7·6 in the usual scale in which log(H) = 12·00. The high resolution and low noise observations of the solar spectrum made by Delbouille and Neven at the International Station of the Jungfraujoch have made possible the identification and the measurement of very weak solar lines. A number of forbidden lines of Fe II have been identified by J. P. Swings. Grevesse and Swings have shown that these lines lead to an abundance of iron log(Fe) = 7·5 in good agreement with the coronal abundance and now also with the new photospheric abundance derived from permitted lines. The abundance of lead has been revised by Grevesse (blue and UV lines), and Peach. Both authors agree on an abundance close to log(Pb) = 2·0. Ross, Aller and Mohler in a separate study have found about the same result. The abundance of Li was found to be lower than believed before by Lynds and by Müller. A value log(Li) = 1·0 now seems plausible if blends are taken into account. The abundance of beryllium has been investigated by Envold and Hauge, and by Warner. Warner and various co-authors have made an extensive study of the abundance of the elements in the solar photosphere and have published this material in the *Monthly Notices of the Royal Astronomical Society* (seven papers). Grevesse, Lambert and Mallia have contributed widely to the same task (see mainly *Solar Physics*). Baschek and Holweger have obtained an abundance log(C) = 8·5 for carbon from analysis of center-limb variation of atomic C I lines. They found an excitation temperature of 5200 K from the CH($A^2\Delta - X^2\Pi$) band. Müller, Baschek and Holweger have derived an abundance for O of log(O) = 8·8 from the center-limb observation of O I lines. Ross and Aller have shown how the method of synthetic spectra can be used to determine the chromium abundance and Aller in *Proceedings of the Astronomical Society of Australia* has summarized all recent results on solar abundances.

3. *Non-LTE computations*

The solar atmosphere has been quite naturally chosen as a suitable medium for non-LTE line and continuum computations. Computations of Ca II and Mg II spectra by S. Dumont have led to the conclusion that the theoretical K line profile could be brought into rough agreement with observed profiles only if the temperature was continuously decreasing outwards to 4200 K at 450 km above the limb. A rather similar conclusion was also reached by Linsky and by Athay and Skumanich. It must be recognized that, at the Bilderberg Conference in 1967, non-LTE computations of Ca II and Mg II were the only evidence in favour of a low temperature minimum in conflict with evidence coming from the level of the UV continuum. New UV and IR measurements strongly favour a value near 4300 K and it may be the first time, in the tormented history of non-LTE work, that a non-LTE analysis has preceded any other source of information for finding the actual temperature somewhere in the solar atmosphere. This non-LTE result does rely on the fact that the source function of the K line can be proved to be in LTE, even at 400 km above the limb. Avrett has shown that difficulties arise in the non-LTE analysis of the D line profiles if the chromospheric rise is placed very much above the limb. Cuny has computed the full spectrum of hydrogen even allowing for non-LTE in the continuum. She has also computed the effect of the far wing of $L\alpha$ on the UV solar continuum (Third Harvard-Smithsonian Conference). Dalgarno has criticized the validity of the formula used by Cuny at very large distances from the core of the line. Feautrier has computed departures from LTE for H in a theoretical solar photosphere without lines. Jordan as well as Thomas have thoroughly discussed this problem.

4. *Velocity field*

The statistical description of the velocity field in the photosphere is not much clearer than before. There is a trend to admit a lower microturbulence ($\approx$ 1·0 km/s) and accordingly a somewhat higher macroturbulence. Parnell and Beckers have found that the one-dimensional power spectrum of the velocity does not fit Kolmogoroff's law.

5. *Inhomogeneities*

Margrave Jr. and Swihart have shown on a two-stream model that if the assumption of pressure equality on horizontal planes is dropped the effect on the photospheric structure is quite significant. Léna has discussed the effect of inhomogeneities on the interpretation of the infrared radiation of the photosphere. Wilson has studied the stratification of temperature fluctuations in the photosphere.

6. *Polarization*

S. Dumont has shown that the polarization at the edge of the disk comes mainly from neutral hydrogen and not from free electrons of the chromosphere.

7. *Miscellaneous*

Lambert and Mallia have shown that the absolute wavelengths of Fraunhofer lines are accounted for by convective motions and the gravitational redshift of general relativity.

8. *Work under way*

Most of the work under way has been reported at the Pasadena meeting of the A.A.S. on solar astronomy (1969).

a. *The continuum spectrum and the solar model*

Mutschlecner and Keller are comparing the new empirical solar models with theoretical models in radiative equilibrium using Strom's treatment of the blanketing. Further work is under way by Gay, Lequeux, Verdet and co-authors in France, and by Mankin and Strong in the U.S.A. to squeeze the infrared data towards a better knowledge of the solar model. The Harvard-Smithsonian group is preparing a new solar model in which non-LTE effects are allowed for in the interpretation of the observational data. There is still a problem in the interpretation of the solar continuum around 1700–1800 Å. Peytremann is computing a theoretical model of the solar atmosphere.

b. *Abundance of the elements*

Baschek, Garz and co-authors are ready to publish the iron abundance in the photosphere derived from ionized iron lines. The revision of Fe I oscillator strengths also affects the absolute scale of Fe II oscillator strengths and the abundance of iron derived is log(Fe) = 7·6 in agreement with results derived from Fe I or from [Fe II].

c. *Non-LTE work*

H. R. Johnson is working on the spectrum of O I. The blanketing effect in non-LTE is at present under study at Boulder and at Meudon.

9. *The solar corona and chromosphere*

Abundance estimates from permitted UV and from forbidden coronal lines in the visible have yielded values, especially for the transition group elements, which are substantially higher than those from standard LTE photospheric analysis and considerable dispute about this question has taken place (Athay; Dupree and Goldberg; Jordan and Pottasch). No arguments have been advanced to contradict the results of the coronal forbidden line analyses which support the higher abundance values. See, for instance, analysis of [Fe XIII] 10747 eclipse data by Eddy and Malville and analyses of coronal lines observed in the 1965 and 1966 eclipses (Jefferies, Orrall and Zirker). Delache has suggested that the corona is enriched in heavy metals. Since the suggested mechanism would not apply to the chromosphere, chromospheric abundances have assumed added importance. From analysis of some chromospheric emission-line data Athay has concluded that the chromospheric abundances are in keeping with the GMA photospheric values. However, Pecker and Pottasch have questioned Athay's analysis and they conclude that the data of Dunn *et al.* indicate chromosphere abundances close to the coronal values.

To use measurements of coronal forbidden line intensities for diagnostic purposes an understanding of the dependence of the excitation of highly ionized species on electron density and temperature is required. Bély and Blaha have studied Fe XV while Rozelot has studied Fe XII. Zirker has studied various coronal ions in several stages of ionization while Chevalier and Lambert have considered Fe XIII and Ca XV. These studies indicate that certain line ratios are especially valuable for inference of electron densities.

Noyes and Kalkofen have used OSO IV observations of the Lyman continuum to construct models of the quiet chromosphere. A two-component model is required to reconcile UV data, brightness temperatures at mm wavelengths and chromospheric electron densities obtained from eclipse observations. Withbroe has analysed OSO IV limb darkening in spectral lines in the 300–1400 Å region obtaining results consistent with the model of the transition region due to Dupree and Goldberg.

Photoelectric observations of the limb darkening in the visible continuous spectrum obtained at the 1966 eclipse by Weart and Faller have been studied in part by Weart. Further analysis is being done by K. Gebbie and R. N. Thomas. These studies and those by Henze of 1962 eclipse data on the hydrogen lines indicate that the electron density and probably the temperature are higher than given by the BCA model. Thomas has pointed out that this is probably due to omission of the fact that the hydrogen ionization will not be in LTE. Studies of the NaD lines (Athay and Canfield; Finn, Mugglestone and Young) lead to similar conclusions.

Methods for inferring data directly from line spectra and the shortcomings of these methods have been considered by Jefferies and White, de Jager and Neven, and by Wilson and Worrall. Applications to the solar D-lines have been made by Curtis and Jefferies while de Jager and Neven applied their method to the O I triplet at 7700 Å. Jefferies and White showed that data on a single line do not allow a generally unique interpretation; Wilson and Worrall concluded that the assumption of frequency independence is at fault. Hearn has pointed out the non-uniqueness of interpretation of the solar $L\alpha$ line profile and he has shown that the He I lines 584 and 537 can only be uniquely interpreted if supplementary data on the Balmer α line is available.

C. THE THEORY OF RADIATIVE TRANSFER

1. *Model atmospheres*

Clearly the outstanding technical advance in the present report period has been the computation of fully non-LTE model stellar atmospheres in radiative equilibrium by Auer and Mihalas using the technique of linearization about trial solutions, which they first applied to the problem of temperature correction in LTE radiative models. The full development of this idea appears to make possible the elimination of many problems which have restricted the application of non-LTE theory to highly-artificial models. A brief exposition of this technique has been given by Skumanich. A considerable number of model stellar atmospheres have been computed in which the assumption of LTE is replaced by one of detailed balancing in the resonance lines. Such calculations have been reported by Mihalas; Mihalas and Stone; Feautrier; Strom, Strom and Kalkofen; and Peterson, as well as by others. Further advances have been reported on the problem of computing a radiative equilibrium model atmosphere in LTE by Feautrier, Henyey, Böhm-Vitense, Baerentzen, and Wickramasinghe and Strittmatter.

2. *Non-LTE line formation*

Considerable progress has been made in the theory of non-LTE line formation. Efficient and flexible new computational procedures have been developed by Athay and Skumanich, Rybicki and Hummer, Grant, Grant and Hunt, Kalkofen, and Kurucz. Among the many papers discussing multi-level atoms are those of Avrett; Avrett and Kalkofen; Athay; Avrett, Beebe, Johnson and Cuny; Finn; Finn and Jefferies; Kalkofen; Peterson and Strom; Peterson and Kalkofen; and Beebe, Johnson and Poland. The last two examine the range of validity of the detailed balance assumption for resonance lines entering statistical equilibrium model calculations. All of the above work is based on the assumption of complete redistribution. Line formation by general non-coherent scattering has been studied by Mathis; Hummer; and Weymann and Williams, who have obtained results substantially different from those based on complete redistribution. The effects of non-coherent electron scattering on the line profiles of early type stars have been discussed by Hummer and Mihalas, and by Auer and Mihalas. An extremely valuable discussion of line formation problems in expanding atmospheres was given by Rybicki at the Munich Colloquium of Commission 36. Finally, substantial advances have been made by Wilson; Jones and Skumanich; Krook and Rybicki and by Cannon in our ability to solve numerically transfer problems in media with two and three spatial dimensions. Ivanov has considered the problem of the time variation of a Doppler-broadened resonance line. The extensive use of the Monte-Carlo technique has developed for the solution of various transfer problems. Auer, and Avery and House have investigated transfer with general redistribution. Avery, House and Skumanich have examined the transfer of resonance radiation in finite cylinders and House and Cohen have treated polarized radiation in magnetic fields. Magnan has studied line formation in expanding and rotating atmospheres. House and Avery give a review of applications of the Monte-Carlo technique to transfer problems.

3. *Curve-of-growth theory*

Further refinements of the curve-of-growth theory have been reported by McClatchey and by Pecker. Rosendhal has examined the effects of electron scattering and Edmonds has considered the relative effects of micro-turbulence on various forms of curves of growth. Athay and Skumanich have published Milne-Eddington curves of growth for non-coherent scattering. Useful tables of equivalent widths have been given by Golden and a useful approximation by Goldman. Band models have been developed by Yamamoto and Masaru and by Zachor. Golden and Kyle have independently worked out the Doppler analog to the Elsasser band model and Golden has found an expression for the Voigt analog. Van Blerkom and Hummer have solved the non-LTE overlapping line problem using the Golden-Kyle formula.

4. *Miscellaneous*

The vexing problem of a transfer equation for dispersive inhomogeneous media has been discussed by Zheleznyakov and Pomraning. Generalized Eddington approximations have been discussed by Sherman, by Huang and by Pomraning. Pomraning has also given a variational form of the transfer equation. The computation of special functions has received much attention. The efficient computation of exponential integral functions is discussed by Armstrong, and by Cody and Thatcher. The Voigt function is discussed by Armstrong, by Reichel, and by Chiarella and Reichel. Formal radiative transfer and some related computational procedures have also received much attention. However, limitations of space prohibit a discussion of the field. Books of interest that have been published recently are *Spectral Line Formation* by Jefferies, *Linear Transport Theory* by Case and Zweifel, *Theory of Stellar Spectra* by Sobolev and the collection *Resonance Lines in Astrophysics*, edited by Athay and Mathis.

5. *Work in the Soviet Union*

Research on the theory of stellar atmospheres in the U.S.S.R. centered around radiative transfer problems. Two groups have studied transfer of radiation in line frequencies. These are the Leningrad University group (V. V. Sobolev, D. I. Nagirner, V. V. Ivanov) and the group from the Institute of Atomic Energy, Moscow (Yu. Yu. Abramov, A. M. Dickhne, A. P. Napartovitch). The problem is studied under the usual assumptions of two-level atoms and complete frequency redistribution. The approach is mainly analytical. Both stationary and non-stationary problems have been considered. Attention is paid primarily to the study of the asymptotic properties of the solutions of transfer equations for various standard problems. The usual place for the publication of the results of these groups is the journal *Astrophizika*. A comprehensive review of the analytical results will be found in the book *Radiative Transfer and Spectra of Celestial Bodies* by V. V. Ivanov, which is now in press (Nauka Publ. House, Moscow, approx. 400 pp.). Extensive tables of the generalized H- and related functions for the case of the Voigt profile with the continuous opacity taken into account are being prepared for publication by D. I. Nagirner. The tabulation of the generalized X- and Y-functions with Doppler profile is now under way in Leningrad. Substantial progress in another branch of radiative transfer theory, namely, anisotropic scattering, has been achieved by V. V. Sobolev. In particular, he has found the complete solution of the so-called "pseudo-problems" of anisotropic scattering as formulated by Chandrasekhar some 20 years ago. Quite general results are also obtained in the study of the asymptotic properties of the solutions of transfer equation for anisotropic scattering. Some of these asymptotic results were found independently by H. C. van de Hulst in Leiden. The results of V. V. Sobolev are being published in *Astronomiceskij Zurnal* and in *Astrophizika*. Using the ideas of V. A. Ambartzumian, Byurakan astronomers have attacked analytically several model non-linear radiative transfer problems (V. Y. Terebizh *et al.*).

6. *Problems related to planetary atmospheres*

During the report period much work on multiple light scattering in plane atmospheres has been done by H. C. van de Hulst and his colleagues K. Grossman, R. E. Danielson, D. R. Moore, J W. Hovenier and J. E. Hansen. This work will be summarized in book form by Van de Hulst.

7. *Radiative transfer in extended atmospheres*

Radiative transfer in spherical geometry has been studied by Bellman, Kagiwada, Kalaba and Ueno using the method of invariant imbedding in particle-counting technique. The diffuse transmission of light from a central point source through an inhomogeneous spherical shell as well as diffuse reflection of solar rays by a spherical shell has been studied. Ueno, Kagiwada and Kalaba have derived the functional equations for the scattering and transmission functions and intensity in an inhomogeneous, anisotropically scattering spherical shell atmosphere with radial symmetry allowing for an internal radiation source, an absorbing core and a reflecting boundary. Leong and Sen have obtained some of the same analytical results using the probabilistic method. Uesugi and Tsujita have applied Chandrasekhar's invariance principle as well as the principle of invariant imbedding to some aspects of the problem in cylindrical and spherical media and Matsumoto has looked at time-dependent problems. Although the functional equations for the scattering and transmission functions of the search-light problem in three dimensions have been obtained, it has not yet been possible to solve the complicated integro-differential equations numerically. Radiative transfer in an inhomogeneous infinite cylinder has been studied by Heaslet and Warming and by Buell, Kalaba, Ruspini and Ueno. Multiple Compton scattering in a spherical atmosphere was studied by Code. Aspects of radiative transfer in moving atmospheres have been considered by Kulander, by Rybicki and Hummer, by Magnan and by Bellman, Kalaba and Ueno. The integral-equation method has been applied by Kalkofen to this problem. Krook and Rybicki have studied the source function and emergent intensity in a dynamically unstable, three-dimensional plane-parallel atmosphere. A similar problem has been studied by Jones and Skumanich. An extension of Sobolev's dynamical treatment of the radiation field in extended atmospheres has been examined by Hirata and Uesugi. The problem of line formation in extended atmospheres including non-coherent scattering and recognizing that LTE is not valid is under study in Japan and at the Munich Observatory as well as elsewhere.

D. HYDRODYNAMICS AND HYDROMAGNETICS IN STELLAR ATMOSPHERES

1. *Convection and turbulence in stellar atmospheres*

Progress in the hydrodynamic theory of convection has been slow. In most applications we still have to use the mixing-length theory. However, a few different approaches have been tried: R. Ulrich has applied the concept of a "thermal" (widely used in meteorology) to the convective bubble in the outer convection zone. K. H. Böhm calculated the approximate amplitudes of the different convective modes, assuming that the effects of turbulence can be described by a turbulent "viscosity" and a "conductivity". The basic question of the instability criterion has been re-examined by N. R. Lebovitz and A. Kovetz. P. Souffrin and E. A. Spiegel made the interesting discovery that a system consisting of a convectively unstable and an overlying stable zone can be unstable with respect to the generation of gravity waves. Yamaguchi studied growth rates for an atmosphere with a varying superadiabatic gradient. Ponomarenko reinvestigated the development of turbulence in a conducting fluid. Very useful empirical information about convective modes (in the solar outer convection zone) has been presented by E. Frazier. The mixing-length theory (partly with some minor modifications) has been used in the following studies: K. H. Böhm pointed out the importance of outer convection zones in cool white dwarfs and calculated such zones; K. S. Krishna Swamy calculated the onset and the efficiency of convection in normal and in hydrogen deficient stars. He also compared the onset of convection in ε Eri (normal metal abundances) and in GMB 1830 (metal deficient). S. Mizuno and M. Nishida showed that a change of the assumed values of parameters in the mixing-length theory can lead to large differences in the structure of the convection zone. K. H. Böhm and E. Stückl studied thermal boundary layer effects in a convection zone. W. Unno has tried to improve the mixing-length theory by taking into account (a) the non-locality of convective flow and (b) non-grey radiative transfer in a convective element.

2. *Shock waves and gas dynamic phenomena including mass loss*

Shock wave phenomena have been studied in order to understand the heating of chromospheres and coronas and atmospheric phenomena in pulsating stars. A recent book useful for astronomers working in this field has been published by Zel'dovich and Raiker (*Physics of Shock Waves and High Temperature Hydrodynamic Phenomena*, Academic Press, 1967). An interesting survey of the Chisnell-Whitham method for shock wave calculation and its application to stellar atmospheres and envelopes is contained in R. P. Stefanik's thesis. The Brinkley-Kirkwood theory has been applied by P. L. Sachdev. Radiation phenomena in shock waves have been investigated by C. J. Bartlett, A. J. Skalafuris and V. V. Zaitsev. Line profiles emitted by acoustic waves have been computed by G. Eriksen and P. Maltby. Considerable effort has been devoted to the study of the hydrodynamic heating of stellar chromospheres and coronas (P. Ulmschneider, R. F. Stein, R. A. Kopp, M. Kuperus, R. Mäckle). Ono, Ischizuke and Taira generalized the Chisnell method so that the propagation of any shock can be treated in arbitrary layers. The problem of conductive heating of the solar chromosphere-corona transition region in a magnetic field and its dynamical consequences has been studied by M. Kuperus and R. G. Athay and by R. A. Kopp and M. Kuperus. The problem of mass loss from stars continues to be of great interest. Using energy considerations and estimating efficiences for different mass-loss processes, I. P. Williams has derived upper limits for the mass loss. The mass loss from OB supergiants has been determined by D. C. Morton. The theory of this effect was discussed by Lucy and Solomon. T. Schmid-Burgk made a detailed numerical study of the mass loss of central stars of planetary nebulae (due to radiation pressure effects). G. S. Bisnovatyi-Kogan and Ya. B. Zel'dovich studied mass loss induced by high atmospheric opacity. The connection between mass loss from stellar coronae and stellar rotation has been investigated by K. Nariai.

3. *Turbulence and convection in hydromagnetics*

The fundamental and difficult problem of the dynamics of a magnetic field in a turbulent medium has been reinvestigated by S. I. Vainshtein and E. N. Parker. A stellar atmosphere problem related to this question is the accumulation of magnetic field lines at the boundaries of convection cells (A. Getling and R. W. Milkey). Overstability and oscillatory convection in the presence of a strong magnetic field were studied by S. I. Syrovatskii and Yu. D. Zhugzhda and by M. Saito and S. Kato. The first two authors discovered an instability with respect to a magnetosonic type of oscillation.

4. *Hydrodynamic and hydromagnetic noise in stellar atmospheres; hydromagnetic waves*

Most of the work in this field is stimulated by the hope to understand the heating of chromospheres and coronas and the observed oscillations in the solar photosphere. Very useful new empirical information about the latter problem has been obtained by E. Frazier. Studies of the generation of waves (acoustic, gravity or magnetohydrodynamic) have been made by S. Kato, B. D. Savage, M. S. Howe, J. E. Pfowes Williams. Problems of wave propagation in an atmosphere have been treated by R. F. Stein, A. Getling, M. A. Gintsburg. In the last paper large amplitude waves in a two-fluid model are considered. The response of an atmosphere to gravitational-hydrodynamic waves has been studied by Y. Uchida. P. Ulmschneider suggested that the 300 seconds oscillation may be understood as standing gravity waves (Väisälä mechanism). Uchida has explained the Moreton wave emitted by flares as the MHD fast mode propagated through the corona and refracted back to the chromosphere. S. P. Tarafdar developed a magneto-gas-dynamic model of a stellar corona. Acceleration of electrons in a collisionless shock was studied by C. Lacombe and A. Mangeney.

5. *Hydrodynamics in atmospheres of rotating stars*

It was stated by E. Spiegel during the Nice Symposium that Rossby waves might be of importance in the sun and in other stars. This subject has been studied by S. Kato and at present is being investigated by R. Sakurai and P. A. Gilman.

6. *Miscellaneous problems in hydromagnetics*

H. S. Yun has generalized the magnetohydrostatic sunspot models by Deinzer by including the pressure due to the curvature of the magnetic field as part of the surface boundary condition. In another study an azimuthal component of the magnetic field was taken into account. A. Unsöld and G. Godoli have discussed magnetohydrodynamic problems arising in connection with the study of stellar and solar activity.

A. B. UNDERHILL
President of the Commission

37. STAR CLUSTERS AND ASSOCIATIONS
(AMAS STELLAIRES ET ASSOCIATIONS)

PRESIDENT: M. Golay.
VICE-PRESIDENT: G. Larsson-Leander.
ORGANIZING COMMITTEE: Mrs. K. A. Barhatova, S. B. C. Gascoigne, I. R. King, L. Rosino, J. Ruprecht, Miss H. H. Swope, M. F. Walker.

INTRODUCTION

The data contained in this report have been taken from two sources: (1) Information received from astronomers active in the field of Commission 37 in response to a circular letter mailed July 1969; (2) Surveys of special fields, prepared by W. Becker on "Open star clusters and spiral structure", by G. Larsson-Leander on "Clusters and stellar evolution", by M. Walker on "Young clusters", and by P.-B. Bouvier on "Dynamical models and numerical computations". It is a pleasure to thank them and all of those who have contributed to the preparation of this report.

CATALOGUES

G. Alter reports that after a long delay the second edition of the *Catalogue of Star Clusters and Associations.* (G. Alter, J. Ruprecht, V. Vanýsek), which was discussed at a meeting of our commission at the Congress in Hamburg (1964) (*Trans. IAU*, **12B**, 1966, 336), will now be published by Publishing House of the Hungarian Academy of Sciences. It is considerably enlarged, since it includes the contents of the annual Supplements published in *B.A.C.* between 1959 and 1967.

S. van den Bergh and G. L. Hagen initiate a project aimed at the production of an atlas of cluster colour-magnitude diagrams on the *UBV* system. For each cluster the area of the sky for which data are plotted is chosen in such a way as to optimize the contrast between the color-magnitude diagram of the cluster and that of the background stars. Furthermore, available radial-velocity and proper-motion data are used to eliminate as many field stars as possible. All diagrams will be printed on an uniform scale. To facilitate intercomparison of clusters each color-magnitude diagram will be printed on a separate card of size 21 × 23 cm.

R. E. White at Arizona has prepared a bibliography of all existing color-magnitude diagrams for globular clusters, including reproductions of the diagrams.

ASSOCIATIONS

Within the framework of a project for detection of intermediate-type stellar associations, a number of suspected objects (320) in the Southern Milky Way have recently been observed by L. O. Lodèn. There is a strong indication that a considerable number of these objects ("quasi-associations") do constitute physically real associations or clusters, which are too dispersed to be detected only by inspection of direct-photographic plates. They are disclosed at inspection of objective-prism plates, if two or more spectra of the same type are situated so close on the plate that they form a conspicuous configuration.

According to Pik-Sin The, the OB-associations Vul OB1 and Vul OB4 are located in about the same direction. Their estimated distances are 2 and 1 kpc, respectively. In the direction of these associations we also have the open star clusters NGC6823 and NGC6830. In collaboration with van Paradys, Pik-Sin The has begun the study of these associations in order to determine:

Table 1. Associations

Name	Observer and references	Photometry, method and limiting magnitude		Other data obtained, remarks
Vul OB1	Pik-Sin The, van Paradys	*UBV*	10·5	s.t. planned mag. 15
Vul OB1	Walker, Hodge (1968)	Hγ		s.t.
Vul OB4	Pik-Sin The, van Paradys	*UBV*	10·5	s.t. planned mag. 15
Cyg OB3	Rufener *et al.*	$UBVB_1B_2V_1G$	pe	
Cyg OB1	Rufener *et al.*	$UBVB_1B_2V_1G$	pe	
Cyg OB1	Walker, Hodge (1968)	Hγ		s.t.
Cyg OB1	Schewick			p.m. P Cygni
Cyg OB9	Walker, Hodge (1968)	Hγ		s.t.
Cyg OB2	Rufener *et al.*	$UBVB_1B_2V_1G$	pe	
Cyg OB2	Walker, Hodge (1968)	Hγ	pe	s.t.
Lac OB1	Valtz, Dluzhnevskaya			HR diagram
Cep OB1	Rufener *et al.*	$UBVB_1B_2V_1G$	pe	
Cep OB1	Walker, Hodge (1968)	Hγ	pe	s.t.
Cep OB3	Valtz, Dluzhnevskaya			HR diagram
Cas OB6	Walker, Hodge (1968)	Hγ		s.t.
Cas OB7	Rufener *et al.*	$UBVB_1B_2V_1G$	pe	
Per OB1	Rufener *et al.*	$UBVB_1B_2V_1G$	pe	
Per OB3	Rufener *et al.*	$UBVB_1B_2V_1G$	pe	
Per OB2	Rufener *et al.*	$UBVB_1B_2V_1G$	pe	
Per OB2	Valtz, Dluzhnevskaya			HR diagram
Aur OB1	Rufener *et al.*	$UBVB_1B_2V_1G$	pe	
Aur OB2	Walker, Hodge (1968)	Hγ	pe	
Gem OB1	Rufener *et al.*	$UBVB_1B_2V_1G$	pe	
Gem OB1	Valtz, Dluzhnevskaya			HR diagram
Mon OB1	Rufener *et al.*	$UBVB_1B_2V_1G$	pe	
Mon OB2	Rufener *et al.*	$UBVB_1B_2V_1G$	pe	
Ara OB1	Valtz, Dluzhnevskaya			HR diagram
Sco OB2	Valtz, Dluzhnevskaya			HR diagram

Abbreviations to Table 1. s.t. = spectral type; p.m. = proper motion; pe = photoelectric.

(1) Which of the OB-stars found in the direction of the associations belong to them.
(2) More accurately the distances of these associations.
(3) Whether there are any relations between the open clusters and the associations.

Van Schewick determined the absolute proper motion of the variable star P Cyg and compared it with the proper motions of the open cluster IC 4996 and the association Cyg OB1. The star is a probable member of the OB-association and its absolute magnitude is $M_v = -8{\cdot}1$.

Martinet (1969) has suggested that the high-velocity A-star 7 Sextantis could have been born in the upper-Centaurus Lupus association, $1{\cdot}5 \times 10^6$ years ago, and ejected from it as "run away" star.

G. A. H. Walker and S. M. Hodge (1968) carried out a study of the associations Aur OB2, Vul OB1, Cyg OB9, Cyg OB1, Cep OB1, Cas OB6, Cyg OB2. Spectral types and Hγ equivalent widths have been derived from low dispersion spectrograms of the brightest stars. The associations appear to be divided roughly into two groups. Associations in group 1 contain very bright H II regions and the spectra of their early B stars show stronger Hγ absorption than those of the corresponding stars in group 2 associations. Group 1 is probably younger than group 2, and radio-continuum and 21-cm surveys appear to support this. Trapezium-type systems occur in both groups; they contain at least one O star, and the B stars of latest spectral type do not appear to have contracted to the main sequence.

Table 2. Open clusters

GC	Observer and references	Photometry, method and limiting magnitude			Other data obtained, remarks
29	Tsioumis, Poulakos				space distribution
38	Cannon (in press)				p.m.
	Sharov, Salukvadze				
	Rufener *et al.* (1970)	$UBVB_1B_2V_1G$	pe		
7	Panorin, Starikova				
	Rufener *et al.*	$UBVB_1B_2V_1G$	pe		in progress
69	Lindoff (1968)	*UBV*	pg + pe	14·5	
09	Burkhead	*UBV*	pg		photoelectric calibration
166	Burkhead	*UBV*	pg		photoelectric calibration
2	Cannon (in press)				p.m.
	Mavridis				space distribution
	Vasilyevsky				spectra
	Rufener *et al.* (1970)	$UBVB_1B_2V_1G$	pe		
9-884 (h + χ Per)	Vasilyevsky				spectra
	Rufener *et al.* (1970)	$UBVB_1B_2V_1G$	pe		
1805	Rufener *et al.* (1970)	$UBVB_1B_2V_1G$	pe		
7 (King 4)	Karimie (1968)			15·7	c-m
	Larsson-Leander				in progress
45	Zakharova				
348	Rufener *et al.* (1970)	$UBVB_1B_2V_1G$	pe		
eiades (M45)	McCarthy, Treanor (1968)				mem, p.m., var, Hα, em
	Rufener *et al.* (1970)	$UBVB_1B_2V_1G$	pe		
	McCarthy				s.t.
02	Rufener *et al.*	$UBVB_1B_2V_1G$			in progress
13	B.K.S.Z.				photometry
28	Panorin, Starikova				
	B.K.S.Z.				photometry
	Larsson-Leander				in progress
45	B.K.S.Z.				photometry
	Rufener *et al.* (1970)	$UBVB_1B_2V_1G$	pe		
yades	Rufener *et al.* (1970)	$UBVB_1B_2V_1G$	pe		
	Nalimov				
	Vasilyevsky				spectra
	Kholopov				spatial density
	Straizys, Kavaliauskaite				evolution
47	Rufener *et al.* (1970)	$UBVB_1B_2V_1G$	pe		
62	Vasilyevsky				spectra
	Rufener *et al.* (1970)	$UBVB_1B_2V_1G$	pe		in progress
46	Vasilyevsky				spectra
07	Panorin, Starikova				
99	Panorin, Starikova				
29	V.K.P.F.K. *et al.*				absorption of light
68	Rufener *et al.* (1970)	$UBVB_1B_2V_1G$	pe		
69	Rufener *et al.*	$UBVB_1B_2V_1G$	pe		in progress
44	Rufener *et al.* (1970)	$UBVB_1B_2V_1G$	pe		
5	Burkhead	*UBV*	pg		photoelectric calibration
64	Rufener *et al.* (1970)	$UBVB_1B_2V_1G$	pe		
81	Vasilyevsky				spectra
01	van Schewick (1967)				p.m.
r 121	Feinstein (1967)	*UBV*	pe	10	
62	Lloyd Evans		pe	12	

Table 2 (continued)

NGC	Observer and references	Photometry, method and limiting magnitude			Other data obtained, remarks
2420	Cannon (unpublished)				p.m.
	Zakharova				
2483	Lindoff (1968)	*UBV*	pg + pe	15	
2489	Lindoff (1968)	*UBV*	pg + pe	15	
2506	Burkhead	*UBV*	pg		photoelectric calibration
2516	Feinstein, Mirabel	*UBVRI*	pe	11	materiel collected
2533	Lindoff (1968)	*UBV*	pg + pe	14·5	
2546	Lindoff (1968)	*UBV*	pg + pe	14·5	
Pismis 1	Lindoff (1968)	*UBV*	pg + pe	15·5	
2567	Lindoff (1968)	*UBV*	pg + pe	14·5	
2571	Lindoff (1968)	*UBV*	pg + pe	14·5	
2579	Lindoff (1968)	*UBV*	pg + pe	15	
Cr 185	Lindoff (1968)	*UBV*	pg + pe	15	
Praesepe (2632)	Tucker, Blackwell				p.m.
	Straizys, Kavaliauskaite				evolution
	Rufener *et al.* (1970)	$UBVB_1B_2V_1G$	pe		
	Serkowski		pe		pol. 4 stars
2682 (M 67)	Murray (1968)				p.m.
	van Schewick (in press)				p.m.
	Rufener *et al.* (1970)	$UBVB_1B_2V_1G$	pe		
	Serkowski		pe		pol. 6 stars
2818	Andrews		pe + pg	17	in progress, spectra of brightest stars
Tr 12	Lindoff (1968)	*UBV*	pg + pe	15	s.t.
3292	Andrews		pg + pe	16	in progress
Tr 14	Pik-Sin The	*UBV*	pg	13	in progress
Tr 15	Pik-Sin The	*UBV*	pg	13	in progress
Tr 16	Feinstein (1969)	*UBV*	pe	12	
	Pik-Sin The	*UBV*	pg	13	in progress
3532	Landolt				s.t., r.v.
	Serkowski		pe		pol. 4 stars
3680	Serkowski		pe		pol. 3 stars
3766	Landolt				s.t., r.v.
	Serkowski				pol. 3 stars
	Lloyd Evans		pg + pe	15·5	
3960	Lindoff	*UBV*	pg + pe	15	in progress
	Ardeberg	*UBV*	pe		32 stars within 3′ from centre
Coma Berenices	Straizys, Kavaliauskaite				evolution
	Rufener *et al.* (1970)	$UBVB_1B_2V_1G$	pe		
4609 (Coal Sack)	Feinstein, Marraco	*UBV*	pe	13	material collected
4755	Panorin, Starikova				
	Serkowski		pe		pol. 30 stars
4756	Vasilyevsky				spectra
Stock 16	Lyngå (in press)	*UBV*	pe	12·5	s.t., r.v., OB-stars
5138	Lindoff	*UBV*	pg + pe	15	in progress
5460	Lindoff	*UBV*	pg + pe	15	in progress
Lyngå 2	Lindoff (1968)	*UBV*	pg + pe	14	
5617	Lindoff (1968)	*UBV*	pg + pe	14	
Hogg 17	Lindoff (1968)	*UBV*	pg + pe	14	
Tr 22	Lindoff (1968)	*UBV*	pg + pe	14	
Pismis 20	Lyngå (1968)	*UBV*	pe	13	s.t. OB-stars
6025	Feinstein	*UBVRI*	pe	12	material collected
	Lindoff	*UBV*	pg + pe	15	in progress

Table 2 (continued)

NGC	Observer and references	Photometry, method and limiting magnitude			Other data obtained, remarks
6031	Lindoff (1968)		pg	16	
6129	Vasilyevsky				spectra
6134	Lindoff	*UBV*	pg + pe	15	in progress
6192	Lindoff	*UBV*	pg + pe	15	in progress
	Andrews		pg + pe	17	under observation
6208	Lindoff	*UBV*	pg + pe	15	in progress
6231	Andrews		pe + pg	17	in progress
	Feinstein, Ferrer (1968)	*UBV*	pe	10	
	Seggewiss (1968)	*UBV*	pe + pg	15·2	295 stars, nucl. of Sco OB1
6242	Seggewiss (1968)	*UBV*	pe + pg	13·5	128 stars
	Andrews		pe + pg	17	in progress, spectra of brightest stars
Tr 24	Seggewiss (1968)	*UBV*	pe + pg	13·5	northern section of Sco OB1
6268	Seggewiss (1968)	*UBV*	pe + pg	14·0	75 stars
	Andrews		pe + pg	17	in progress, spectra of brightest stars
6281	Lindoff	*UBV*	pe + pg	15	in progress
	Andrews		pe + pg	15	in progress, spectra of brightest stars
IC4651	Lindoff	*UBV*	pe + pg	15	in progress
6383	Lloyd Evans				var
TR 27	Pik-Sin The (in press)				distance
TR 28	Pik-Sin The (in press)				distance
6400	Lindoff	*UBV*	pe + pg	15	in progress
6405 (M 6)	Serkowski				pol. 5 stars
6416	Pik-Sin The (in press)				distance
6425	Pik-Sin The (in press)				distance
IC4665	McCarthy, O'Sullivan (1969)		pg		c-m
	Rufener *et al.* (1970)	$UBVB_1B_2V_1G$	pe		
6494	Panorin, Starikova				
6530	Panorin, Starikova				
6611	Panorin, Starikova				
6613	Lindoff	*UBV*	pe + pg	15	in progress
	Vasilyevsky				spectra
	Rufener *et al.* (1970)	$UBVB_1B_2V_1G$	pe		
6645	Lindoff	*UBV*	pe + pg	15	
IC4756	Seggewiss (1968)	*UBV*	pe + pg	14·2	471 stars
	van Schewick (in press)				p.m.
6705	van Schewick (in press)				p.m.
6716	Lindoff	*UBV*	pe + pg	15	in progress
IC4996	van Schewick (1967)				p.m.
	Rufener *et al.*	$UBVB_1B_2V_1G$			in progress
6716	Lindoff	*UBV*	pe + pg	15	in progress
6802	V.K.P.F.K. *et al.*				absorption of light
6811	Vasilyevsky				spectra
	Lindoff	*UBV*	pe + pg	14·5	in progress
6819	B.K.S.Z.				photometry
	Vasilyevsky				Spectra
	Lindoff	*UBV*	pe + pg	15	in progress
	King	*UBV*	pe + pg	16	
	Burkhead	*UBV*	pg		photoelectric calibration
6823	V.K.P.F.K. *et al.*				absorption of light
6834	V.K.P.F.K. *et al.*				absorption of light

Table 2 (continued)

NGC	Observer and references	Photometry, method and limiting magnitude		Other data obtained, remarks
Mel 227	Ardeberg	*UBV*	pe	51 stars within 20′ from centre, s.t
6871	van Schewick (in press)			p.m.
	Rufener *et al.* (1970)	$UBVB_1B_2V_1G$	pe	
6866	B.K.S.Z.			photometry
	Koroleva			p.m.
6913	V.K.P.F.K. *et al.*			absorption of light
6939	Cannon (in press)			p.m.
	Burkhead	*UBV*	pg	photoelectric calibration
7063	Sharov			
7086	V.K.P.F.K. *et al.*			absorption of light
7092 = M 39	Artyukhina, Kholopov			
	Straizys, Kavaliauskaite			evolution
	Rufener *et al.* (1970)	$UBVB_1B_2V_1G$	pe	
7142	van den Bergh	*UBV*	pe	
	Sharov			
7160	Rufener *et al.* (1970)	$UBVB_1B_2V_1G$	pe	
IC1434	Larsson-Leander			in progress
7654	V.K.P.F.K. *et al.*			absorption of light
King 11	Burkhead	*UBV*	pg	photoelectric calibration
7789	Cannon (1968)			p.m.
	B.K.S.Z.			photometry
	Vasilyevsky			spectra
	Tsioumis, Poulakos			space distribution
	Zakharova			
	Koroleva			p.m.
7822	Rufener *et al.* (1970)	$UBVB_1B_2V_1G$	pe	
Ru 58	Lindoff (1968)			

Abbreviations to Table 2. V.K.P.F.K. *et al.* = Voroshylov, Kolesnik, Polishtchuk, Fedorchenko, Kalandadze *et al.* B.K.S.Z. = Barkhatova, Kuz'mina, Shashkina, Zakharova.
c-m = colour magnitude diagram; em = emission; pe = photoelectric; pg = photographic; p.m. = proper motion; pol. = polarization; r.v. = radial velocity; s.t. = spectral type; var = variable; mem = membership.

OPEN CLUSTERS

Proper motions

The relative and absolute proper motion of the young cluster IC4996 has been determined by van Schewick (1967). The material consists of three pairs of plates taken with the 30-cm refractor ($f = 5{\cdot}1$ m) of Bonn Observatory with epoch differences of 40 to 43 years. The cluster IC4996 contains no yellow giants. Van Schewick (1967) has also determined the proper motion of NGC2307 from measurements at Groningen and Bonn. The cluster contains no yellow giants. New fundamental proper motions on the FK4 system, with standard errors between $\pm 0\farcs002$ and $\pm 0\farcs004$ per annum, have been determined by Tucker and Blackwell for 19 members of Praesepe, from data in the G.C. supplemented by modern observations. The mean proper motion of the cluster is:

$$15\ \mu_\alpha \cos\delta = -0\farcs0355$$
$$\mu_\delta = -0\farcs0164 \pm 0{\cdot}0007 \qquad \text{(s.e.)}.$$

Murray (1968) has completed his proper motion study of the region of M67 (NGC2682). Outlying red-giant members have been found up to about 0°.5 from the cluster centre. The main sequence of the outlying cluster members shows the marked gap below the turn-off point, which is well-known for the stars in the central region.

Cannon (1968 and in press) has carried out proper motion and photometric studies in NGC188, 752, 6939, 7789, 2420. All five clusters have a clump of red giant stars, which appear to be equivalent to the horizontal branches of globular clusters; comparison with theoretical models by Faulkner and Iben supports this interpretation.

L. S. Koroleva has been putting finishing touches to a proper motion catalogue, containing some 4000 stars in the open clusters NGC6866 and NGC7789.

Miscellaneous studies of open clusters

At the Vatican Observatory two-colour photometry on IC4665 has been published by McCarthy and O'Sullivan (1969) who used 6 plates exposed in the blue and 6 in yellow wavelengths. Standard plate-filter combinations were employed and the photoelectric standards of Hogg and Kron and those of Johnson plus an extension to fainter magnitudes by the photoelectric photometry of McCarthy and of Coyne provided the basis for photographic interpolation. All stars selected for proper-motion studies by Vasilyevskis were observed and colour-magnitude diagrams constructed for the "probable" and "possible" members as defined by the proper-motion measures. Both diagrams indicate the presence of late type stars in the subgiant domain and a gap of 0·3 magnitude near $B-V=+1{\cdot}0$. Abt and Snowden have concluded that there is no extension of the main sequence of this cluster beyond $V=12{\cdot}0$.

In continuation of their work on the Pleiades cluster McCarthy and Treanor (1968) have published a Finding List of 144 possible candidates as late-type low-luminosity members of this cluster. This listing has been made from all available material and considers flare activity, photometric properties, proper-motion data, luminosity criteria and Hα-emission features, which have been observed for each star. Finding charts are provided for 120 stars which could be securely identified. McCarthy has completed a study of 22 of the late-type stars detected in the original Vatican objective-prism study. All 22 stars proved to be late-type dwarfs.

Pik-Sin The is studying Tr 27, Tr 28, NGC6416, NGC6425 in collaboration with N. Stokes. The distances obtained so far for these clusters are 1050 pc, 1470 pc, 800 pc, 840 pc. The data indicate that Tr 27 and Tr 28 are located in the Sagittarius spiral arm (as delineated by optical spiral tracers), while NGC6416 and NGC6425 are interarm objects.

T. Lloyd Evans obtained photometric and spectroscopic data in IC2581 (in press). He has found a distance of 2500 pc and age $\sim 10^7$ years. The cluster contains 2 super-giants, an eclipsing binary and a variable Be star. The evolutionary significance of HD90772 (A7 Ia) in the cluster is discussed. In NGC6383 (age 5×10^6 years) he found a number of rapid variables with *UV* excess (possibly T Tauri stars).

In a paper (1968), Lloyd Evans has concluded that the M super-giants in $h+\chi$Per are unlikely to be pre-main sequence objects.

Burkhead at Indiana is determining color-magnitude arrays in NGC609, 2506, 6819, 6939, IC166, Trumpler 5, and King 11; the magnitudes are photographic *UBV* with photoelectric calibration.

Polarimetric observations have been made by K. Serkowski in M67, Praesepe, NGC3680, NGC4755, NGC3532, NGC3766 and M6. The K0 II star BM Scorpii in M6 has a large intrinsic polarization, changing with time.

V. V. Lovdovski is preparing an extensive observational program comprising over 300 open clusters.

The three-colour photometric determination of the distances of galactic clusters was continued at the Astronomical Institute of Basle using mainly plates of the Palomar and Asiago Schmidt Telescopes. The list of clusters treated with methods of three-colour photometry at different observatories comprises now about 230 objects. The spiral structure is well defined by the distribution of young clusters for distances up to 3 to 4 kpcs.

At the Main Astronomical Observatory of the Ukrainian Academy of Science in Golosseyevo, with the Abastumani Observatory also taking part, absorption of light in space in the directions

Table 3. Globular clusters

NGC	Observer and references	Photometry, method and limiting data			Other data obtained, remarks
104 = 47 Tuc	Royal Greenwich Obs.				c-m, p.m., var
	Serkowski				pol.
288	Philip	*UBV*	pe		sequence in area of NGC288
	Menzies		pe + pg		
362	Philip	4 colours			in progress
	Serkowski				pol.
1261	Alcaino	*UBV*	pe	16·5	
1851	Alcaino (1969)	*UBV*	pe	16·5	
2808	Philip	*UBV*	pe		in area of NGC2808
	Ford, Gascogne			18	c-m 1000 stars
	Alcaino (1969)	*UBV*	pe	16·5	
3201	Philip	4 colours			in progress
4372	Ford, Gascogne			18	c-m 1000 stars
	Alcaino	*UBV*	pe	16·5	
	Andrews		pe + pg	17·5	
4590	Andrews		pe + pg	17	var
4833	Alcaino	*UBV*	pe	16·5	
	Menzies		pe + pg		
5024 = M 53	Panova				
5139 = ωCen	Dickens, Woolley (1967)	*B, V*	pg	19·5	HR, var
	Serkowski				pol.
5272 = M 3	Serkowski				pol.
5904 = M 5	Simoda, Tamikawa				HR
	Philip	4 colours			in progress
	Serkowski				pol.
	Zhukov				
5927	Menzies			pg + pe	
6121 = M 4	Philip	4 colours			in progress
	Landolt	*UBV*	pe		
6139	Andrews		pe + pg	19	
6171	Dickens	*UBV*	pg		var
6205 = M 13	Simoda, Kimura (1968)				*L*(*M*)
	Simoda, Tamikawa				*L*(*M*)
6266	Ford, Gascogne (1967)	*B, V*			var 50 stars
6341 = M 92	Philip	4 colours			in progress
	Kadla	*UBV*			p.m.
6352	Alcaino	*UBV*	pe	16·5	
	Menzies		pg + pe		
6362	Alcaino (1970)	*UBV*	pe	16·5	
6397	Cannon	*B, V*	pg	19	p.m. 450 stars
6541	Menzies		pe + pg		
6752	Alcaino (1970)	*UBV*	pe	16·5	
6809 = M 55	Philip	4 colours			in progress
	Menzies		pe + pg		
6981	Dickens	*UBV*			c-m, var
7078 = M 15	Panova				
7099	Dickens	*UBV*			c-m
	Andrews		pe + pg	19	
Pal 12	Andrews		pe + pg	18	

Abbreviations to Table 3. c-m = colour-magnitude diagram; p.m. = proper motion; var = variable; pol. = polarization HR = Hertzsprung-Russell diagram; *L*(*M*) = luminosity function; pg = photographic; pe = photoelectric.

of NGC2129, 6802, 6823, 6834, 6913, 7086, 7654, is estimated by V. I. Voroshylov, L. N. Kolesnik, E. P. Polishtchuk, G. L. Fedorchenko, N. B. Kalandadse *et al.*

Goy studied IC1805 with the Geneva photometric system $UBVB_1B_2V_1G$. He has found that the interstellar extinction law is not the same for all O stars of the cluster. He has shown that the wavelength of the knee of the extinction law at 4300 Å is not the same for all the stars.

The 200-inch telescope has been used by S. van den Bergh to obtain the faint photometric *UBV* sequence in NGC7142.

Photoelectric *UBV* photometry of the high-latitude galactic cluster Mel 227 has been obtained by A. Ardeberg. Within a distance of 20′ from the centre 51 stars have been measured. Objective-prism plates with a dispersion of 110 Å mm^{-1} are available.

An extended source detected (Price, 1968) recently at 2·2 μ is in excellent positional agreement with a heavily reddened cluster previously found by Westerlund (1961) in Ara. The observed brightness at 2·2 μ may be fully explained by the radiation from the cluster members.

L. Mavridis completed a study of the space distribution of the M, S, C type stars in an area centered at the open cluster NGC752. Similar studies are currently carried out in two areas centered at the open clusters NGC7789 and NGC129 by A. Tsioumis and C. Poulakos, respectively.

GLOBULAR CLUSTERS

Luminosity function

The luminosity function for M13 has been obtained by M. Simoda and H. Kimura (1968) down to the visual magnitude 19, by means of star counts on plates taken with the 74-inch reflector of the Okayama Astrophysical Observatory. Particular attention has been paid to the counting method. A significant deviation has been found between their luminosity function and the M3 luminosity function determined by Sandage. The results for M13 give a generally steeper gradient as compared with Sandage's results for M3.

By use of the new luminosity function and the Population II stellar models given by I. Iben, the life-time and effective helium-core mass have been obtained for each evolutionary stage up to the giant tip. Agreement between the present results and the theoretical ones is good, both for low ($Y_e = 0{\cdot}10$) and high ($Y_e = 0{\cdot}35$) helium cases, contrary to the results obtained by Sandage for M3. The life-time of the horizontal-branch stars have been found to be 9×10^7 years for $X_e = 0{\cdot}9$ and 5×10^7 years for $X_e = 0{\cdot}65$. Some evidence which seems to favor the high helium content is discussed.

Simoda and Tamikawa work also on the luminosity function for M5 and M13, using Kitt Peak and Palomar 200-inch plates for the former and Kitt Peak plates for the latter. Results hitherto obtained suggest a similarity between M5 and M13. The result for M13 seems to be essentially the same as the previous one. Thus, the luminosity function for M5 and M13 seems to agree with that expected from the theory. For M5, a preliminary count down to the fainter magnitudes suggests the mass segregation effect to be consistent with the theory and giving evidence for decreasing original luminosity function with increasing magnitude. The brighter part of the HR diagram for M5 determined by Arp has also been revised, using Arp's standards and plates taken by 84-inch telescope at Kitt Peak Observatory. Results strongly suggest that intrinsic widths of giant, asymptotic, and yellow horizontal branches are all very small.

Wilson and King, at Berkeley, are studying luminosity functions of globular clusters, using dynamical models to extend limited counts to a whole cluster. When treated in this way, M13 and M3 have very similar luminosity functions.

Hartwick at Victoria has determined the luminosity function of M92.

Miscellaneous studies of globular clusters

R. J. Dickens and R. Woolley published (1967) two papers on the structure and dynamics of the cluster NGC5139 (ωCen) and on the Cepheid variables. Photographic B, V photometry down to $V=19{\cdot}5$ has been obtained. The colour-magnitude diagram of members of NGC104 (47Tuc), selected from a proper-motion study based on plates taken with 24-inch refractor, shows a well defined sequence of stars brighter than, and asymptotic to the giant branch. The RR Lyrae variable HV814 is almost certainly not a member. Dickens has obtained light curves in U, B, V for 22 variables in the metal-rich cluster NGC6171 from plates taken at Mount Wilson. Very good correlations between period, amplitude and colour have been obtained. Analyses have shown a complete separation in colour between a- and c-type variables in marked contrast to the variables in metal-poor clusters such as ωCen. This cluster contains large-amplitude variables with periods $\sim 0{\cdot}4$ days, which probably represent the cluster counterparts of short-period large-amplitude a-type variables, previously only known to occur in the general field. Dickens has also obtained colour-magnitude and two-colour diagrams for the cluster NGC6981. The horizontal branch occurs at $V=16{\cdot}9$ with approximately equal populations on each side of the variable gap. Provisional values of the reddening and ultra-violet excess are $E_{B-V}=0{\cdot}07$ and $\delta(U-B)=0{\cdot}20$ respectively. Dickens has obtained colour-magnitude and two-colour diagrams for the cluster NGC7099, which appears to be very metal-poor. The horizontal branch occurs at $V=15{\cdot}5$ and is heavily populated on the blue side of the variable gap. Provisional values of the reddening and ultra-violet excess are $E_{B-V}=0{\cdot}10$, $\delta(U-B)=0{\cdot}28$, respectively. Cannon has measured proper motions for about 450 stars with $B \leqslant 15^m$ in the field of NGC6397. Photographic B, V photometry down to $V=19{\cdot}0$ has also been obtained.

J. W. Menzies is studying globular clusters (the suspected metal-rich clusters, NGC5927, 6352 and the early-spectral-type clusters NGC288, 4833, 6541, 6809) down to below the horizontal branches, in order to study chemical-composition effects on the colour-magnitude diagram. The colour-magnitude diagrams show considerable differences, even among clusters suspected of having the same metal abundance.

A program of four-colour measures of blue horizontal-branch stars in globular clusters has been started by A. G. Davis Philip. Measures of blue horizontal-branch stars in M4, NGC362, and NGC6809 have been already made, and measures are planned in M5, M92, NGC3201. It is hoped that these measures will be of interest for the helium-abundance problem and in defining the relationship between field and globular horizontal-branch stars.

V. Ford and S. C. B. Gascoigne (1967) obtained two-colour light-curves (B and V) for about 50 short-period variables in NGC6266. This cluster shows strong differential absorption, reaching about one magnitude in B. Comparison of colours and magnitudes of variables in the reddened part with those in the unreddened part of the cluster gives a value for A_v/E of $2{\cdot}9$.

K. Serkowski has measured the integrated polarization of globular clusters M3, M5, NGC362, 47Tuc, ωCen. Line-strength estimates in the integrated spectra of 43 galactic globular clusters have been made by S. van den Bergh, using image-tube spectra obtained at the Cassegrain focus of the 200-inch. The purpose of this program was to compare the globular clusters in our own galaxy with those in the Andromeda Nebula. Such a comparison shows that the average metallic-line strength of globular clusters in the Andromeda Nebula is significantly greater than it is in the Galaxy.

Star counts in 54 globular clusters made by J. King, have been published (1968). The results are now being analysed, along with all other data available in the literature, to determine radii of as many clusters as possible. The tidal limiting radii can be determined in more than 40 clusters. The results are also being analyzed for correlations between the various quantities.

At the Geneva Observatory L. Martinet is undertaking a compilation of various data about the known globular clusters (HR diagram, integrated photometry, variables, radial velocity, position, etc.) comparing them with those for other old objects (RR Lyrae, subdwarf, etc.) in order to specify more precisely some physical and dynamical features related to the initial evolution of the Galaxy.

NEW OPEN CLUSTERS AND EXTRAGALACTIC CLUSTERS

Sanduleak and Philip (1968) published spectra and photometry of a new open cluster in the line of sight of the Large Magellanic Cloud. The cluster was confirmed by Murray, Dickens, Walker (1969). The colour-magnitude diagram resembles that of the Hyades. Further astrometric and photometric work on some 600 stars in this region is in progress. The distance modulus calculated for the cluster is $7^{m}_{.}8$, which when corrected for $0^{m}_{.}2$ of absorption yields a distance of 330 pc to the group. It is 185 pc below the galactic plane.

M. Walker used the McGee Spectracon to obtain direct electronographic images in *B* and *V* of globular clusters in the Magellanic Clouds. The intent is to use the existing photometry down to about 19th magnitude for zero point and colour equation, and to continue the c-m diagrams to the limit of the Spectracon. Stars as faint as $V = 22{\cdot}6$ and $B = 23{\cdot}2$ have been measured. The following clusters were observed at Tololo with Spectracon, in SMC: Kron 3, NGC121, 339, 361, 419, 458; in LMC: NGC1783, 1866, 2004, 2164, 2209, 2231, 2257, Hodge 11.

Photoelectric *UBV* observations were made by G. A. H. Walker (1968) for stars in the associations Anon b4 and NGC2081 in the Large Magellanic Cloud.

OPEN STAR CLUSTERS AND SPIRAL STRUCTURES

(Prepared by W. Becker)

Since 1967 the number of galactic open star clusters observed in three colours has increased to about 216 objects. Ninety of them are young enough to demonstrate the spiral structure of our Galaxy. They have confirmed and improved the existing picture of the space distribution. But in certain directions gaps exist, which require new observations. This is particularly the case between the galactic longitudes $l^{\mathrm{II}} = 290°$ and 330° and between 35° and 65°. These observations should help to decide upon two questions of interest. First, one can expect from such observations the answer to the question whether a continuous spiral arm exists, going from Carina over Sagittarius to Scutum, or whether there is a gap near $l^{\mathrm{II}} = 320°$, indicating that the Carina arm extends in the direction of the Sun and continues in the direction of Cygnus. The second question refers to the new results presented by Weaver at the Basel symposium, according to which the so-called local spiral arm is in fact not a real spiral arm but only a spiral-arm-like feature, emerging in the direction $l^{\mathrm{II}} = 50°$ from a Carina-Sagittarius-Scutum arm and fading out in the direction of Puppis. In this way the large pitch angle of 21° of this local feature becomes understandable.

In the future the photometric observations will be extended to fainter objects, in distances up to 5 kpc, in which the photometric method will give results reliable enough for detecting spiral structure.

New spiral arms may be detected in this way, spiral arms which are indicated up to now only by a few objects (clusters and H II regions).

A detailed study of the space distribution of clusters of different ages has not been made up to now. It may well be that the number of clusters in the critical age of 50 to 100×10^{6} years is yet too small.

CLUSTERS AND STELLAR EVOLUTION

(Prepared by G. Larsson-Leander)

The star clusters are important tools for testing theoretical computations of stellar evolution. Detailed comparisons between theory and observations were made by Iben (1967). Considering the young clusters, Iben suggested that the red super-giants of h and χ Per might be stars in pre-main-sequence contraction. Schild (1967) segregated various groups of stars formed at different times, and according to his interpretation the red super-giants are highly evolved objects. Walker (1969) has recently investigated the stars in the vicinity of the Orion Nebula. No evidence for subgroups of different ages were found. The gravitationally contracting stars tend to lie closer to the main sequence than predicted by theory.

Statistical studies of the distribution of red and yellow giants in the colour-magnitude diagrams for groups of several clusters of medium age, and comparisons with theoretical evolutionary tracks were made by Lindoff (1968) and by Barbaro, Dallaporta and Fabris (1969). Lindoff found the giants situated too low, as compared with the theoretical tracks, which would indicate substantial mass loss, while Barbaro *et al.* found smaller discrepancies. As shown by Barbaro *et al.* the dissimilar results are partly due to the different bolometric and colour corrections used in the two studies.

Meyer-Hofmeister (1969) computed a theoretical HR diagram to fit the Large Magellanic Cloud cluster NGC1866. Similar computations by Schlesinger (1969) has produced theoretical colour-magnitude diagrams for clusters of ages between $22{\cdot}5 \times 10^6$ years and 4×10^9 years. Comparisons were made with several clusters, including M11 and M41. The agreement between theory and observations was on the whole found satisfactory. It is stressed by Schlesinger that to make the comparisons really convincing, spectral types should be determined for the brighter cluster stars and careful membership studies should be made for clusters with published colour-magnitude diagrams.

The colour-magnitude diagrams of older Population I clusters, like M67 and NGC188 exhibit gaps just off the main sequence. The gaps are caused by the rapid evolution of stars with convective cores during the phase of central hydrogen exhaustion. The size and position of the gaps may be used for determining chemical composition and age, but the results are highly dependent upon the opacity law assumed. This is shown by Demarque and Schlesinger (1969) and by Aizenman, Demarque and Miller (1969) discussing the gaps in NGC188 and in M67 and NGC188, respectively. The most consistent results for M67 are obtained for the range of $Y = 0{\cdot}38$–$0{\cdot}32$, corresponding to $Z = 0{\cdot}03$–$0{\cdot}04$, which would lead to the surprisingly low age of about 3×10^9 years.

The colour-magnitude diagrams of the globular clusters were interpreted by Iben and Faulkner (1968) in terms of metal-poor stars with high helium abundance ($Y \approx 0{\cdot}33$). On this interpretation the giants and the horizontal-branch stars, evolving towards higher effective temperature, have essentially the same mass (about $0{\cdot}8\ M_\odot$) and the cluster ages are about 9×10^9 years. The relevant arguments were elaborated upon by Rood and Iben (1968). They also showed that neutrino losses would reduce the helium abundance to $Y \approx 0{\cdot}30$ and the cluster ages to about 8×10^9 years.

On the other hand, theoretical studies of the RR Lyrae pulsations by Christy (1968), demand masses of about $0{\cdot}5\ M_\odot$. Spectrographic and narrow-band photoelectric observations by Newell, Rodgers, and Searle (1969) of horizontal-branch stars in NGC6397 corroborate this, in yielding a mean mass of $0{\cdot}55\ M_\odot$. According to Christy's models there is a correlation between the blue-boundary colour of the RR Lyrae instability strip and the helium abundance. This was used by Sandage (1969) in a discussion of new photometry of M3, M13, M15, and M92. A mean helium abundance of $Y = 0{\cdot}32$ was found, and Sandage states that within observational errors Y is the same for M3, M15 and M92, despite large differences in metal abundances.

Considerable evidence has thus accumulated to the effect that the helium abundance of old Population II stars is in agreement with the theory of the primordial fire-ball. If the masses of the horizontal-branch stars are in the interval $0{\cdot}5$–$0{\cdot}6\ M_\odot$ it seems necessary to postulate substantial mass loss subsequent to the giant phase. Otherwise the ages of globular clusters will considerably surpass the Hubble age of the universe.

YOUNG CLUSTERS

(Prepared by M. Walker)

These objects overlap somewhat with associations; in this report, I shall discuss only investigations of clusters since the associations will undoubtedly be discussed elsewhere. In conformity with the guide-lines for the preparation of the current Draft Report, I list here only those programs or results that appear to me to be of particular significance.

I believe that the most important observational discoveries are those of Becklin and Neugebauer (1967) of an infrared star in the Orion nebula, having a black body temperature of 700K and perhaps representing a very early stage of stellar evolution, and by Kleinman and Low (1967) of an infrared nebula, 20″ from the above-mentioned infrared stellar object, having a diameter of $> 30''$ which they interpret as an interstellar cloud of mass $\sim 10^2$–$10^3\ M_\odot$ at a temperature of 70K which is just beginning to contract into stars. That star formation does occur in condensations having masses of this size has been discussed by Aveni and Hunter who cite as examples the BM And, IC 348,

NGC7129, and NGC7023 groups, each of which have masses $\leqslant 200 M_\odot$. A related discovery of importance is the finding by Raimond and Eliasson (1967) of an OH-emission source in the Orion nebula that apparently coincides in position with the infrared stellar object found by Becklin and Neugebauer. These observations emphasize the fact that we now have, with the development of radio and far infra-red detectors, the observational means to observe objects at the very beginning of their evolutionary history when their temperatures are still so low that they have been undetectable by methods heretofore available. Further observations in these wavelength regions are clearly of prime importance.

Other information on the earliest stages of stellar evolution may be provided by the work of Penston (1969) who has studied the distribution of matter in the Bok globules in the Orion nebula. From the variation of optical depth with projected radius, the masses of the globules are found to be of the order of 1 $M_\odot$ and their temperatures about 10K.

Reddish (1967) finds that the increase in interstellar reddening with increasing luminosity first described by Blanco and Williams (1959) occurs in all clusters and associations having stars brighter than $M_v = -9$ (ages $< 10^5$ years). He concludes that very young, early-type stars are surrounded by obscuring clouds having diameters 1 pc, and masses $\sim 30\ M_\odot$.

It is becoming increasingly clear that in order to understand the early stages of stellar evolution that we observe in the young clusters, observations of the c-m diagrams of the clusters are not enough. Detailed spectroscopic observations of the individual stars are also needed. Thus, the photometric study of stars in the Orion nebula cluster by Walker (1969) is of interest since it provides *UBV* data for stars down to $V = 18$, from which objects for spectroscopic or further photometric study can be selected. This study indicates that stars redder than $(B-V)_0 = -0{\cdot}09$ are still in the gravitational contraction stage, corresponding to an age of 3×10^6 years, and that there is, from the *UBV* data, no evidence for an abnormal reddening law in the Orion nebula.

Williams and Cremin (1969), using theoretical evolutionary tracks, have extended the work of Iben and Talbot (1966) and have concluded that there is strong evidence for a definite spread in the ages of the stars in NGC2264, NGC6530, IC2602, and IC5146. This interpretation depends, however, on the assumption that T_e can be obtained directly from the observed $(B-V)$, and thus hinges on the correct explanation of "Walker's anomaly" – the fact that many of the intrinsically faint stars lie close to or even below the main-sequence. As indicated above, spectroscopic observations will be required to investigate this question. Evidence for a spread in the ages of the stars in IC2602 has also been found by Hill and Perry (1969) from the discordance between the nuclear age of the brightest star and the contraction age.

The gravitationally contracting stars in IC1805 have probably been detected by Ishida (1969) who has obtained *UBV* observations of 54 stars of $V \leqslant 14$ that have probabilities of cluster membership 0·5 according to the proper-motion study by Vasilevskis, Sanders, and van Altena (1965). Ishida finds that the stars redder than $(B-V)_0 = 0{\cdot}2$ lie above the main-sequence, corresponding to an age of 1×10^6 years. The cluster is irregularly reddened, and spectroscopic observations are needed to establish membership and reddening of the individual stars.

Schild (1967) has studied the ages of stars in h and χ Per, finding differences in the ages and distances of the two clusters, while Slettebak (1968) has studied the spectral types and rotational velocities of 83 stars in h and χ Per, and finds that the rotational velocities of the giants in the clusters are greater than for field stars of the same types.

DYNAMICAL MODELS AND NUMERICAL COMPUTATIONS

(Prepared by P.-B. Bouvier)

King is preparing for publication dynamical models of globular clusters, based on a realistic luminosity function. These will include models made specifically to fit the best observed clusters.

Prata has extended King's models to a calculation of dynamical evolution, including the effects of the galactic tidal force, the escape of stars, the loss of mass through stellar evolution as well as the energy input and star-loss induced by gravitational shocks due to passing interstellar gas clouds.

Results include a specific study of M 67, showing that the original mass function must have differed from the Salpeter birth function. A more general study shows how initial factors affect dynamical evolution and which clusters are most able to survive. As a by-product, it appears that anisotropy is removed from open clusters by the combined effect of galactic tidal forces and internal relaxation.

Hénon (1970) is studing the dynamical evolution of star clusters by a Monte Carlo technique, which is much faster than the exact *N*-body integration. A number of cases with 1000 bodies have been computed. Among other results, it is found that a cluster develops a very dense core and an extended halo; ultimately, almost all the energy is in the core, while almost all the mass is in the halo. Hénon has generalized a previous theoretical computation of the escape rate from an isolated cluster to the case of an arbitrary mass distribution (1969), the rate of escape of energy has also been computed. These escape rates have been confirmed by the numerical integrations performed by Wielen for 11 cluster models each containing 100 stars; some of the models take into account the galactic field, different stellar masses and a mass loss of evolving stars, in order to allow a direct comparison with open clusters.

These models indicate that the observed age distribution of open clusters can be explained as caused by a dynamical dissolution of open clusters due to star escape. Work on computer simulation of star clusters and clusters of galaxies was continued by Aarseth. Since the Paris colloquium (1967), three more cases of the 250-body problem, including different mass spectra, have been integrated for longer times. Special attention was drawn towards the influence of binary stars on the escape rate and the disruption of the cluster. A description of a perturbation method as applied to the *N*-body problem will be given in a forthcoming paper.

Hayli investigated the influence of the tidal galactic field on the evaporation of stars from an open cluster.

Bouvier and Janin, who had also taken part in the collective attack on the 25-body problem at the 1967 Paris meeting, examined later the disruptive influence of interstellar gas clouds passing near a star cluster; this led to an extension of Janin's numerical program for the *N*-body problem. Investigating also the early evolution of self gravitating systems, Bouvier and Janin dealt with the problem of the waterbag model for a spherical system. The waterbag model for a system of parallel plane sheets is also being reexamined by Janin with a more efficient numerical method, in order to test some former results obtained by Lecar and by Hohl and Feix.

U. Veltmann has discussed the system of globular clusters in the Galaxy, evaluating the possibilities of determining from observations its spatio-kinematical properties, in order to see to what extent they agree with the theory of stellar hydrodynamics.

Spitzer has shown that in a mixture of two stellar masses equipartition becomes impossible if the total mass of the heavier stars exceeds a critical value.

M. GOLAY
President of the Commission

38. EXCHANGE OF ASTRONOMERS (ÉCHANGE DES ASTRONOMES)

(Committee of the Executive Committee)

PRESIDENT: M. G. J. Minnaert.
VICE-PRESIDENT: A. Reiz.
ORGANIZING COMMITTEE: D. Ja. Martynov, P. M. Routly, J. Sahade, J. P. Wild.

Commission 38 gratefully acknowledges the perfect and efficient cooperation with the General Secretary.

During the three-year period 1-1-1967 to 1-1-1970 the following travel subsidies have been granted:

177.	Dr M. Khairy Aly	Helwan to Herstmonceux (R)
178.	Dr K. C. Freeman	Austin to London
179.	Mrs Dr Paris Pishmish	Mexico to Heidelberg
180.	Dr F. Roddier	Nice to Tucson (R)
181.	Dr M. Hockey	Oxford to Santa Cruz (R)
182.	Dr P. Raghuveer	Hyderabad to Middletown (R)
183.	Dr J. Monaghan	Cambridge to Santa Cruz (R)
184.	Dr J. Hadjidemetriou and	
184a.	Dr G. Bozis	Thessaloniki to Manchester (R)
185.	Dr K. Ishida	Tokyo to Austin (R)
186.	Dr A. Tlamicha	Ondřejov to Charlottesville (NRAO) (R)
187.	Mrs Dr A. Ringuelet-Kaswalder	La Plata to Liège (R)
188.	Dr I. Kawaguchi	Pic-du-Midi to Kyoto
189.	Dr M. J. Rees	Cambridge to Pasadena (R)
190.	Dr A. Uesugi	Kyoto to Amherst (R)
191.	Dr J. Kleczek	Ondřejov to Utrecht (R)
192.	Miss Dr S. Itoh	Kyoto to Meudon
193.	Dr J. Kubota	Kyoto to Pic-du-Midi
194.	Dr H. Haupt	Vienna to Haute-Provence (R)
195.	Dr K. A. Hämeen-Anttila	Oulu to Meudon (R)
196.	Dr F. Bertola	Padova to Pasadena
197.	Dr Y. Sobouti	Shiraz to Philadelphia (R)
198.	Dr B. Valníček	Ondřejov to Malta (R)
199.	Dr M. Werner	New York to Cambridge (R)
200.	Dr E. P. J. van den Heuvel	Utrecht to Santa Cruz (R)
201.	Dr E. L. J. van Dessel	Utrecht to Nice (R)
202.	Dr K. K. Sen	Singapore to Nice (R)
203.	Dr R. C. Canfield	Denver to Utrecht (R)
204.	Dr D. P. Gilra	Hyderabad to Madison
205.	Dr P. Koubský	Ondřejov to Haute Provence (R)
206.	Eng. C. Slottje	Utrecht to Nançay (R)
207.	Dr N. A. Doughty	New Zealand to Hyderabad (R)
208.	Miss Dr E. Dekker	Utrecht to Maryland (R)
209.	Dr F. Pacini	Rome to Ithaca (R)
210.	Miss Dr Beverley J. Harris	Canberra to Columbus (Ohio)
211.	Dr P. Koubský	Ondřejov to Utrecht (R)
212.	Dr J. E. Felten	Cambridge to Bombay (R)

213.	Miss Dr H. Demoulin	Paris to Austin (R)
214.	Dr J. Kubota	Pic-du-Midi to Kyoto
215.	Mrs M. Querci	Paris to Berkeley (R)
216.	Mrs E. Ribes	Paris to Sydney
217.	Dr I. Fejes	Budapest to Groningen (R)

In agreement with a decision, taken at the Prague meeting of 1967, the names of the recipients have been mentioned, since the allocation of such a subsidy is considered to be a honour and a distinction. All cases in which roundtrip was paid have been marked by (R). Where this indication is not found, it means that the stay abroad extended over more than one year, or that the return fare was paid from other sources.

The number of subsidies in this period reached 44 (including the subcommission). Since the existence of the commission the numbers appear to increase steadily, though there are statistical fluctuations (see figure 1). In 11 cases the requests had to be refused. In 5 cases the promise of a

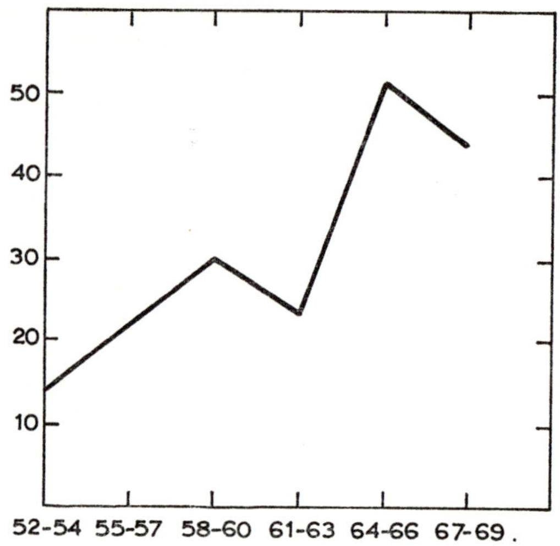

Fig. 1.

subsidy enabled the recipient to accept a fellowship abroad, but afterwards it proved possible to obtain money from other sources and the money was refunded to the IAU.

In the preceding report, we noticed already the increasing percentage of intercontinental travels, which then reached 67 %. In this last period this trend persisted; the percentage now amounts to 76 %. As a consequence the average amount of a subsidy now amounted to $562.–.

Apparently there are nowadays many sources from which travel grants for short trips can be obtained, while the IAU remains one of the few bodies making long-distance journeys possible. The percentage of journeys to the U.S.A. reached 45%.

From the Executive Council a sum of $22000.– was gratefully received for the whole of the three years period 1967–70. Our Commission also highly appreciates a generous contribution of $4000.– by the ICSU; thanks to that additional subsidy it will probably be possible to satisfy all reasonable applications. During the three years 1966–69 the grants totalled $24741.–, including those of the Sub-commission. For the next period a new IAU subsidy of $22000.– will be necessary, assuming that the ICSU will be prepared to grant again a subsidy, comparable to that of the last period.

It should be stressed that the commission has been very economic with the means available. The fares paid correspond to economy class, the peak period was avoided as much as possible and preference was given to the cheapest airlines. Notwithstanding the recommendation of the members at the Prague meeting, we have been obliged to keep strictly to our rules and to refuse any subsidy in view of symposia or costs of living. It has not been possible to accumulate a modest reserve fund.

The reports, received from the host observatories and from those who received a grant, show clearly that the visits to foreign observatories have been without exception useful and profitable. Our travel grants have been particularly beneficial during 1969, when unexpected economic difficulties in some countries made the obtention of official subsidies for scientific travels almost impossible.

It may be useful to remind the members of the IAU that the grants of Commission 38 are available not only for young people, but for astronomers of any age (having reached at least the master's degree). We expect, however, that first of all subsidies from national funds will be applied for.

The Commission tries to work with maximum speed and a minimum of formalities. Nevertheless it appears all the time that applicants are not aware of the information which they are expected to give:

(1) a short indication about their scientific degree and specialization;
(2) the plan for the journey, date of departure, duration of stay abroad, planned investigations;
(3) a letter of the director of the home observatory, expressing his agreement with the travel;
(4) a letter of the director of the host observatory, expressing his agreement with the visit.
(5) When should the money be received and to what address (personal, institute, bank account)?

At the Prague meeting the members of the commission expressed the wish that competent astronomers be sent for some time to countries, not fully participating in the modern development of astronomy, in order to teach there and to direct research. A first attempt in this direction has been made. A circular was sent to 11 countries, of which 3 showed interest in the project and expressed the wish to participate. The Executive Committee transmitted this project to Commission 46, but our commission may have to take care of the travelling expenses.

Information about the activities of Commission 38 was transmitted at their request to: The Dominion Bureau of Statistics (Ottawa), the Secretary of ICSU (Rome), the Grants Register (London), the Consejo Nacional (Buenos Aires).

M. MINNAERT
President of the Commission

REPORT OF SUB-COMMISSION 38

The sub-commission takes care of applications for travel grants from graduate students, having not yet reached the master's degree.

Five requests have been submitted to sub-commission 38 and two grants conceded, for a total of $1107.90.

An amount of $707.90 gave the possibility to Mr. B. De (India) to pay his trip Calcutta-Detroit (U.S.A.) in order to enter, in 1968, the graduate school of the University of Michigan. According to Dr. O. Mohler, Chairman of the Astronomy Department, Mr. De has done very well indeed, his marks being without exception in the highest possible category.

Mr. J. D. Finkel (Canada) has received $400.–, in 1969, for his roundtrip Montreal-Stockholm in order to work with Professor H. Alfvén.

M. MIGEOTTE
President of the Sub-Commission

40. RADIO ASTRONOMY (RADIO ASTRONOMIE)

PRESIDENT: J. P. Wild.
VICE-PRESIDENT: D. S. Heeschen.
ORGANIZING COMMITTEE: N. Kardašev, M. R. Kundu, J. Lequeux, M. Ryle, S. F. Smerd, G. Westerhout.

A. INTRODUCTION

Radio astronomy continues to develop at explosive pace. In addition to containing many notable pieces of observational work carried out with improved sensitivity, angular resolution and spectral detail – and aided by data-handling techniques of increasing sophistication – the last three years have seen a flow of new and sometimes startling results. These are exemplified by the discovery of organic matter in the galaxy, and by the first observations pertaining to what are almost certainly neutron stars. The latter, of course, came with the dramatic discovery of pulsars, which are the subject of an Invited Discourse at this General Assembly.

Following the custom of this Commission, I have asked different members to review the work done in each of the main fields. On this occasion a special section has been added on pulsars; since the announcement of their discovery in February 1968, the rate of publication within so narrow a field could scarcely have been equalled in the history of science. As in the past, radio studies of the planets are incorporated in the Report to Commission 16.

Over recent years this Commission has evolved a format for its Draft Report which is highly concentrated in information content. Few words have been wasted. Regrettably the advantages of this policy have proved illusory: for now that it has been necessary to reduce the size of the Reports, the reduction ratio has been applied on the basis of the size of the 1967 Reports (whether succinct or verbose) rather than on the magnitude of the work covered by each Commission. Each of the contributors to this report has nevertheless preferred to retain, within their limitation of space, the general spirit of previous Reports. Their task has been made more difficult by their having to decide what work to omit (much of it of considerable significance). I thank them sincerely for their efforts and co-operation.

J. P. WILD
President of the Commission

B. SOLAR RADIO EMISSION

(prepared by S. F. Smerd)

1. *Introduction*

The literature for the period late 1966 to late 1969 has been very extensive. Since space does not permit a complete bibliography, some contributions cannot be referred to at all, while some others are referred to by general reference to the proceedings of meetings (**1**, **2**) in which they, or their abstracts (**3**, **4**) have been published; in these cases only will authors be named to identify the paper. Though it has not been possible to list the many instrumental developments, particular mention may be made of the Culgoora 80 MHz radioheliograph (**5**) which came into operation in September 1967.

An international co-operative study culminated in the publication of the Proton Flare Project (the July 1966 Event) (**6**).

High-resolution observations have revealed multiple-source structure from cm (**7**) to m-wavelengths (Clavelier **1**; **8**, **9**), polarization structure (**7**, **10**) and, at times, unsuspected source complexity (**11**). Interaction between widely separated centres on the Sun has been recognized (**12**) in the form of separated burst sources which appear (i) simultaneously, (ii) ~ 10s apart, and (iii) several minutes apart; the implications on magnetic structure are new and important.

The suppression, by the ambient medium, of low-frequency synchrotron radiation has been discovered in type-IV spectra (Boischot and Clavelier **1**). The theory of the medium effect and its application to solar bursts have been extended to lower electron energies (**13**) and to emission at the first few harmonics (**14**).

In two theories of the band-splitting in type-II bursts (**15**, McLean **2**) the emissions are not generated at the same point; published (**11**) and unpublished observations support the theory of separate sources.

Observations of solar emissions at 9·6 and 21 cm showed marked effects (**16**) due to the passage near the Sun of the comet Ikeya-Seki. A strong association has been reported (**17, 18**) between microwave bursts and interplanetary electrons of high (MeV) and moderate (> 40 keV) energies; the latter also show strong correlation with type-III bursts (**18**).

2. *Reviews*

An introductory book on solar radio emission has been published (**19**). Reviews that have appeared include the theory of solar bursts (**20**), and emission mechanisms (**21**), radio evidence of coronal instabilities (**22**), connected activity (**12**), a summary of the Proton Flare Project (**1**), solar radar (**23**), occultation and scintillation studies (**24**) and, in part, radio investigations of the corona (**25**).

3. *The quiet Sun*

Brightness distributions have been derived from eclipse observations, e.g. on 1963 July 20 at 4·6 cm (Higgs and Broten (see **26**)), on 1966 May 20 at 1·2 mm (**27**) showing limb-brightening and on 1966 November 12 at 7 GHz (**28**) showing opposite polarizations from the two hemispheres.

Limb brightening was also detected at 1·2 mm (**29**), at 3·3 mm (Shimabukuro **3**), at 8·6 mm (**30**), asymmetrically at 9·1 cm (**31**) and at 49 cm (**32**), but not at 4·6 cm (see above), nor at 74 cm (Little **2**).

The spectrum has been studied at mm (**33**) and also cm-wavelengths (**26**); cyclic changes have been reported at 169 MHz (**34**).

Theoretical work includes interpretations (**26, 35**) of a possible minimum at 6 mm, and a determination of a corona temperature of 10^6 K from dm and m-wavelengths observations (Sinha and Swarup **2**).

4. *Active centres and the slowly-varying component*

Numerous eclipse observations include 1963 July 20 (**36**) in the range 1 to 33 cm, 1966 May 20 in the range 3·2 cm to 2·7 m (**37**, where several other observations are quoted) and 1966 November 12 (**38**) with polarization structure at 4·28 cm.

Studies of the slowly-varying component include those at 3·4 mm (Mayfield *et al.* **3**), at 8 mm (Efanov *et al.* **1**), at 2 cm (Nagnibeda **1**), in the range 3·2 to 10 cm (**39, 40**), at 9·1 cm (**41**), at 1424, 696 and 408 MHz (Cole *et al.* **2**), at 408 MHz (Little **2**) and at 160 MHz (**42**). A statistical study has been made (Kleczek *et al.* **1**) and spectra have been derived from cooperative observations (**43**).

Several theoretical models have been discussed (e.g. **44, 45**).

5. *Microwave bursts*

Bursts have been observed at wavelengths down to 1·2 mm (**46**). Multi-source and polarization structure has been detected at 3·75 and 9·4 GHz (**7**). Among statistical results are spectra (**47, 48**, Castelli and Aarons **3**), source heights (**49**), and associations with Hα-features (Koeckelenbergh **1**).

Interpretations of microwave bursts have been given in terms of radiation from strong shock waves (**50**), coherent plasma radiation from very strong-field regions (**51**), coherent gyroradiation (Yip **14**) and synchrotron radiation (**13, 52**). The last two references arrive at different conclusions regarding the possible common source of microwave and X-ray bursts. Contracting and expanding source models have been investigated (Gopasyuk *et al.* **1**, Barney **3**); in another model (Kai **2**) upward movement in the sources of microwave type-IV bursts has been adduced.

6. *Storms and type I bursts*

New fine-structure features ('split pairs' and others) have been discovered (**53**) in the range 25–100 MHz. Linear polarization has been detected (**54**) at 74 MHz in type-I and stationary type-IV bursts. Multiple-source structure of storms has been reported at 408 and 169 MHz (**8**) and 80 MHz (**9**). Two apparently correlated storm sources, one compact, the other spread, have been observed (**55**).

Theories of type-I radiation include a 'Laser' theory (**56**) and amplified low-harmonic gyro-radiation (Fung and Yip **14**). The latter theory has also been applied to split pairs (**57**).

7. *Type-II bursts*

The position, large size and fluctuating brightness structure of type-II sources at 80 MHz has been explored (**58, 12**) by radioheliograph; apparent outward movement has been observed at 80 MHz (**58**) and $\lesssim$ 40 MHz (Warwick **3, 59**). In some connected activity the interaction between distant centres occurs at type-II source speeds (**22**).

The enhanced density in a type-II source, assumed to be a shock wave, has been invoked (**60**) to explain the different polarizations of types-II and III bursts and to account for II-derived densities to be twice those of the Newkirk model (Meyer **4**). As noted in Section 1, two band-splitting theories predict separate sources: one (**15**) invokes a Doppler shift between emissions from front and back of the shock wave, the other (McLean **2**) leads to bright emission from much more widely separated regions of greatest contiguity between shock front and plasma level; well separated sources have been observed (**11**).

Coronal magnetic fields have been derived from type-II observations (Meyer **4, 61**). Homologous type-II bursts have been found (**62**) up to 2 days apart.

8. *Type-III and type-V bursts*

A theory of type-III bursts in terms of fundamental dipole and harmonic quadrupole radiation has been tested (Krämer **4**). Fine-structure spectra have shown some association between type-III bursts and the new split-pair chains (**53**). Combined satellite and ground-based observations (**63, 64**, Haddock and Graedel **3**, Stone *et al.* **1**, Malitson *et al.* **3**) have yielded spectra of type-III bursts from 0·7 to 600 MHz; the lowest observing frequency was 0·3 MHz. Densities and temperatures of the outer corona have been estimated from these observations; earlier conclusions that the type-III sources show deceleration at great heights have now been questioned. The average starting frequencies of type-III bursts reached a minimum near the minimum of the solar cycle (**65**). Radioheliograph observations have shown (**9**) a marked displacement between the positions of type-III and type-V sources, and the occurrence of correlated bursts some 10s and $\sim 1\ R_0$ apart (Kai (see **12**)).

Various type-III theories have been reviewed (**66**) and the non-linear treatment of stream stabilization (**67**) developed further. In one interpretation (**68**) type-V bursts have been interpreted as type-III bursts from superheated source regions, in another (**69**) as second-harmonic radiation. The degree, longitude dependence, and linear content of type-III polarization have been investigated (**70**).

9. *Type-IV bursts*

The sources of stationary type-IV bursts have frequently shown double structure at 169 and 408 MHz (**8**). Large-scale motion, first outward and with a later return, has been reported (**59**) in the range 10–40 MHz. Radioheliograph observations have shown (**71, 12**) moving type-IV bursts associated with triggered prominence eruptions and one such source near the top of an expanding magnetic arch (**10**); strong circular polarization in a moving type-IV burst was of opposite polarity to that of the concurrent, more intense stationary type-IV (**72**). The low-frequency cut-off in the spectrum of a moving type-IV burst has been attributed to the Razin effect (Boischot and Clavelier **1**). A model for the source of this burst has been developed (**73, 74**) in which electrons are accelerated to 'synchrotron' energies inside an advancing shock front. In another interpretation (**75**) the hot electrons are injected into the shock front and the Razin effect becomes effective only at much lower frequencies.

Other theoretical discussions of type-IV bursts refer to the ejection of a magnetic cloud (**76**), the conditions under which an ejected plasma cloud emits type-II and type-IV bursts (**77**), interpretations in terms of amplified cyclotron radiation at the long wavelengths (Fung **14**) and at the micro-wavelengths (Yip **14**) and the use of anisotropic pitch-angle distributions to explain a high-frequency cut-off (**78**).

It has been suggested that the overall radio spectrum of proton flares is U-shaped with a minimum at the dm-wavelengths (Castelli and Aarons **3**, Howard **3**, Urbarz **4**); such events have been found (**79**) to occur non-randomly in solar longitude. Different phases of m and Dkm type-IV bursts have been related to different stages of spot-field development (Böhme **1**).

10. *Radar observations*

The 1959 Stanford observations and those taken at 38·25 MHz at El Campo since 1961 have been reviewed (**23**) and interpreted in terms of backscattering from magneto-acoustic waves (**80**).

BIBLIOGRAPHY

1. IAU Symposium no. 35, *Structure and Development of Solar Active Regions*, Ed. K. O. Kiepenheuer, Part VII, 'Proton Flare Project (PFP)', pp. 513–550; Part VIII, 'Radio Structure of an Active Region', pp. 553–601; D. Reidel Publishing Co., Dordrecht-Holland, 1968.

2. *Proc. astr. Soc. Austr.* **1**, (2), 38–68 (1967), 'Solar Research and Observational Techniques'.

3. *Astr. J.*, **73** (5, Part II), S53-85 (1968), A.A.S. Special Meeting on 'Solar Astronomy', Feb. 1968.

4. *Mitt. astr. Ges.*, no. 25, 202–209 (1968), Astr. Ges. meeting, 24–27 September 1968.

5. 'The Culgoora Radioheliograph', Ed. J. P. Wild, *Proc. Inst. Radio Elect. Eng. Aust.*, **28** (9), 277–384 (1967).

6. *Ann. IQSY* Vol. 3, 'The Proton Flare Project (The July 1966 Event)', Ed. A. C. Strickland. The M.I.T. Press, Cambridge, Mass., 1969.

7. Énomé, S., Kakinuma, T., Tanaka, H. 1969, *Solar Phys.*, **6**, 428.

8. Clavelier, B., Jarry, M. F., Pick, M. 1968, *Ann. Astrophys.*, **31**, 523.

9. Kai, K., McLean, D. J., Payten, W. J., Sheridan, K. V., Wild, J. P. 1968, *Proc. astr. Soc. Austr.*, **1**, 79.

10. Wild, J. P. 1969, *Solar Phys.*, **9**, 260.

11. Labrum, N. R. 1969, *Proc. astr. Soc. Austr.*, **1**, 191.

12. Wild, J. P. 1969, *Proc. astr. Soc. Austr.*, **1**, 181.

13. Holt, S. S., Ramaty, R. 1969, *Solar Phys.*, **8**, 119.

14. Fung, P. C. W., Yip, W. K. 1966, *Austr. J. Phys.*, **19**, 759; Yip, W. K. 1967, *Austr. J. Phys.*, **20**, 421; Fung, P. C. W. 1969, *Can. J. Phys.*, **47**, 179.

15. Fomichev, V. V., Chertok, I. M. 1967, *Astr. Zu.*, **44**, 495. – *Soviet Astr.*, **11**, 396.

16. Ariskin, V. I., Berulis, I. I., Levchenko, M. T. 1966, *Astr. Tsirk.* no. 380, 1.

17. Cline, T. L., McDonald, F. B. 1968, *Solar Phys.*, **5**, 507.

18. Lin, R. P. 1968, *Can. J. Phys.*, **46**, S757.

19. Smith, A. G. 1967, *Radio Exploration of the Sun*, D. van Nostrand Co. Inc., Princeton N. J.

20. Takakura, T. 1967, *Solar Phys.*, **1**, 304.

21. Smerd, S. F. 1968, *Proc. astr. Soc. Austr.*, **1**, 124.

22. Wild, J. P. *Radio Evidence of Instabilities and Shock Waves in the Solar Corona*, Proceedings of Conference on 'Plasma Instabilities in Astrophysics', Asilomar, California, Oct. 1968, Ed. D. A. Tidman and D. C. Wentzel, p. 119.

23. James, J. C. 1968, 'Radar Studies of the Sun', ch. 7, *Radar Astronomy*, Ed. J. V. Evans and T. Hagfors McGraw-Hill Book Co., N.Y.
24. Lotova, N. A. 1968, *Usp. fiz. Nauk.*, **95**, 293 – 1968, *Soviet Phys. Usp.*, **11**, 424.
25. Newkirk, G., Jr. 1967, *A. Rev. Astr. Astrophys.*, **5**, 213.
26. Shimabukuro, F. I., Stacey, J. M. 1968, *Astrophys. J.*, **152**, 777.
27. Newstead, R. A. 1969, *Solar Phys.*, **6**, 56.
28. Kaufman, P., Matsuura, O. T., Marques dos Santos, P. 1969, *Astrophys. J.*, **156**, 43.
29. Noyes, R. W., Beckers, J. M., Low, F. J., Davidson, A. W. 1966, *Astr. J.*, **71**, 866.
30. Kalaghan, P. M. 1966, *NEREM Record 1966*, 140 (Nov.).
31. Roosen, J., Goh, T. 1967, *Solar Phys.*, **1**, 242.
32. Swarup, G., Kapahi, V. K., Isloor, J. D., Sinha, R. P. 1966, *Nature*, **212**, 910.
33. Fedaseev, L. I., Lubyako, L. V., Kukin L. M. 1969, *Astr. Z.* **44**, 1191; 1968, *Soviet Astr.* **11**, 953.
34. Leblanc, Y., Le Squeren, A. M. 1969, *Astr. Astrophys.*, **1**, 239.
35. Kaplan, S. A., Tsytovich, V. N. 1967, *Astr. Zu.*, **44**, 1036. – 1968, *Soviet Astr.*, **11**, 834.
36. Castelli, J. P., Clemens, J. 1966, *Astr. J.*, **71**, 652.
37. Fürstenberg, F., Keiser, J., Priese, J. 1967, *Astr. Nachr.* **290**, 183.
38. Kaufmann, P. 1968, *Solar Phys.*, **4**, 58.
39. Ikhsanova, V. N., 1967, *Izv. glav. astron. Obs. Pulkovo*, **24**, 229.
40. Korzhavin, A. N., Peterova, N. G. 1968, *Astr. Zu.*, **45**, 36; 1968, *Soviet Astr.*, **12**, 28.
41. Gopala Rao, U. V. 1968, *Solar Phys.*, **4**, 428.
42. Leblanc, Y. 1969, *Astr. Astrophys.*, **1**, 467.
43. Tlamicha, A. 1968, *Solar Phys.*, **5**, 377.
44. Zlotnik, E. Ya. 1968, *Astr. Z.*, **45**, 310, 585; 1968, *Soviet Astr.*, **12**, 245, 464.
45. Lantos, P. 1968, *Ann. Astrophys*, **31**, 105.
46. Clark, C. D., Park, W. M. 1968, *Nature*, **219**, 922.
47. Tlamicha, A., Olmr, J., Fürstenberg, F., Krüger, A. 1967, *Bull. Astr. Insts. Csl.*, **18**, 70.
48. Kakinuma, T., Yamashita, T., Énomé, S. 1969, *Res. Inst. Atm. Nagoya Univ.*, **16**, 127.
49. Eryushev, N. N. 1967, *Izv. Krym. astrofiz. Obs.*, **36**, 121.
50. Zaitsev, V. V. 1966, *Astr. Zu.*, **43**, 1148; 1967, *Soviet Astr.*, **10**, 919.
51. Tandon, J. N., Deshpande, S. D., Bhatia, V. B. 1968, *Nature*, **220**, 1213.
52. Takakura, T. 1969, *Solar Phys.*, **6**, 133.
53. Ellis, G. R. A. 1969, *Austr. J. Phys.*, **22**, 177.
54. McNarry, L. R. 1966, *Bull. Radio elect. Eng. Div. NRC Can.*, **16**, 6.
55. McLean, D. J. 1969, *Proc. astr. Soc. Austr.*, **1**, 188.
56. Mogilevskii, E. I. 1966, *Geomagn. Aeronom.*, **6**, 809; 1966, *Geomagn. Aeronom.*, **6**, 625.
57. Elgaroy, Ö. 1969, *Astrophys. Lett.*, **3**, 39.
58. Kai, K., McLean, D. J. 1968, *Proc. astr. Soc. Austr.*, **1**, 141.
59. Warwick, J. W. 1968, *Solar Phys.*, **4**, 446.
60. Fomichev, V. V., Chertok, I. M. 1968, *Astr. Zu.*, **45**, 28; 1968, *Soviet Astr.*, **12**, 21.
61. Zaitsev, V. V. 1968, *Astr. Zu.*, **45**, 766; 1969, *Soviet Astr.*, **12**, 610.
62. Stewart, R. T., Hardwick, B. 1969, *Proc. astr. Soc. Austr.*, **1**, 185.
63. Slysh, V. I. 1967, *Astr. Zu.*, **44**, 94, 487; 1967, *Soviet Astr.*, **11**, 72, 389.
64. Hartz, T. R. 1969, *Planet. Space Sci.*, **17**, 267.
65. Malville, J. M. 1967, *Solar Phys.*, **2**, 484.
66. Smith, D. F. 1969, SUIPR Report no. 309, Stanford Univ., California (May).
67. Kaplan, S. A., Tsytovich, V. N. 1967, *Astr. Zu.*, **44**, 1194; 1968, *Soviet Astr.*, **11**, 956.
68. Warwick, J. W. 1967, *Astrophys. J.*, **150**, 1081.
69. Zheleznyakov, V. V., Zaitsev, V. V. 1968, *Astr. Zu.*, **45**, 19; 1968, *Soviet Astr.*, **12**, 14.
70. Fomichev, V. V., Chertok, I. M. 1968, *Astr. Zu.*, **45**, 601, 773; 1968–9, *Soviet Astr.*, **12**, 477, 615.
71. Wild, J. P., Sheridan, K. V., Kai, K. 1968, *Nature*, **218**, 536.
72. Kai, K. 1969, *Proc. astr. Soc. Austr.*, **1**, 189.
73. Boischot, A., Clavelier, B., Boischot, A., Daigne, C. 1968, *Ann. Astrophys.*, **31**, 445, 531.
74. Lacombe, C., Mangeney, A. 1969, *Astr. Astrophys.*, **1**, 325.
75. Warwick, J. W. 1968, *Solar Phys.*, **5**, 111.
76. Kawabata, K. 1966, *Rep. Ionosph. Space Res. Japan*, **20**, 107.
77. Fomichev, V. V., Chertok, I. M. 1967, *Geomagn. Aeronom.*, **7**, 3; *Geomagn. Aeronom.*, **7**, 1.
78. Ramaty, R. 1969, *Astrophys. Lett.*, **4**, 43.

79. Haurwitz, M. 1968, *Astrophys. J.*, **151**, 351.
80. Gordon, I. M. 1968, *Astrophys. Lett.*, **2**, 49; 1968, *Astr. Zu.*, **45**, 1002; 1969, *Soviet Astr.*, **12**, 796.

C. CONTINUUM RADIATION FROM THE GALAXY

(prepared by G. Westerhout)

Several hundred papers in this area have been published since the last reporting period. We can only quote here typical papers and papers of particular importance.

During the reporting period, tremendous progress was made with the study of H II regions. The most recent overall review, discussing many of the points mentioned below, appeared in *Interstellar Ionized Hydrogen* (**2**); the long-awaited volume on *Nebulae and Interstellar Matter* (**4**) finally appeared. Other reviews dealing with H II regions are given in (**6**), (**8**) and (**9**). Electron temperatures appear to vary widely; values range from 3400 K to 14000 K for different objects (**32, 36, 40, 66, 67, 79, 82, 94**). Many compact H II regions or parts of H II regions have been discovered, with typical diameters of 0·1 to 0·5 pc, emission measures of 10^7 to 10^8 cm^{-6} pc, electron densities of 10^4 cm^{-3} and masses of 1 to 10 $M_{\odot}$ (**65, 66, 94**). Often these compact regions are closely associated with OH emission sources. A number of surveys concentrated on H II regions (**12, 30, 37, 50, 58, 74**); the Cygnus-X complex was studied in great detail (**34, 87, 90**), and attempts were made, still rather unsuccessfully, to find distance criteria (**45, 90**). Radio emission by planetary nebulae is extensively reported in (**5**). Many observations of the flux densities of individual nebulae have been reported in the last few years.

Several new catalogues of galactic sources have appeared, as well as a number of surveys of supernova remnants (**11, 48, 51, 69, 77**). A survey on the subject of supernovae was published in (**3**). A review of non-thermal source spectra appeared in (**7**). The occurrence of supernovae in the Galaxy is now generally believed to be approximately once every 50 yr (e.g. **52**). Important progress was made on the study of the intensity and the polarization distribution in several larger remnants. Two different types of magnetic field structure are present: either a radial field or a circular field in the shell (**24, 54, 55, 56, 64, 68, 91**). A correlation is suggested between the surface brightness and radius of supernova remnants, providing a distance criterion (**77**).

The Crab nebula has continued to be the most intensively studied source. Earlier measurements indicating a steep increase of the flux density with decreasing wavelength below 5 mm were confirmed (**19, 53**) although one group reports no effect of this kind (**73**). Accurate flux measurements in the visible and infrared, confirming a steeper decrease in flux with decreasing wavelength than in the radio region are reported in (**70**). Overall polarization measurements at cm-wavelengths have been reported by many authors. Only the detailed measurements of polarization distribution are reported here (**35, 64**). The small source reported at long wavelengths has been studied further (**20, 42, 62**) and is coincident with the central star and the Pulsar.

The most recent review of our knowledge about the galactic centre source is given by Lequeux (**59**). Since then, a number of occultations of the centre by the moon have been observed, showing the presence of a nonthermal source with dimensions of about 2×3 min of arc and a temperature spectral index of 2·25 (**63, 84**). This source coincides with the infrared source (**18**); the infrared point source has not (yet) been detected at radio frequencies.

The most recent review of galactic nonthermal emission is given by Baldwin in (**1**). Surveys of large portions of the sky were made at 4170 MHz (**47**), 3240 MHz (**49**), 2700 MHz (**14, 16, 17**), 1410 MHz (**16, 46**), 610 MHz (**88, 89**), 178 MHz (**29**), 150 MHz (**92**), 85 MHz (**92, 95, 97**), 13 MHz (**15**), and 10 MHz (**43**). High-resolution studies show that the ionized hydrogen is mainly concentrated in discrete sources (**14, 15**). It is not clear whether this is also the case with the nonthermal emission, but studies at higher galactic latitudes show that the non-thermal "halo" largely consists of local features or spurs; perhaps, a low-emissivity spherical halo could still exist (**28, 93, 97**).

Further study of the polarization of the background radiation and Faraday rotation in extragalactic sources is reported in numerous papers of which we can mention only a few (**22, 31, 33, 81**). A model emerges which describes the field as helical locally, possibly connected with Gould's belt, and more or less longitudinal further out, connected with the general disk. This model is based on an overall interpretation of interstellar optical polarization, radio background polarization and Faraday rotation (**60, 61**). The North polar spur appears to be due to this general field and not a supernova remnant (**23, 80**). Estimates of the field strength range from 3×10^{-6} G in between the spiral arms and in the halo to 5×10^{-5} G in dense clouds in the arms (**72,** Woltjer in **10**). Local nonthermal emissivity could be determined by observing the nonthermal emission between the observer and opaque H II regions at low frequencies (**26, 79**). Background polarization and Faraday depolarization show the presence of irregularities with several scale sizes (100 pc, 10 pc, $<0{\cdot}5$ pc) (**33**).

The study of the nonthermal galactic spectrum has progressed considerably. The temperature spectral index between 4000 and 400 MHz is 2·9 (**76**), decreasing to 2·6 between 400 and 200 MHz (**57, 96**) and about 2·4 between 400 and 10 MHz (**25, 38. 39, 75**). Variations of the spectral index with direction are reported in (**25, 57, 85**). The spectrum is now extended by means of satellite observations down to 0·4 MHz with good agreement between observers and with ground-based higher-frequency data (**13, 44, 86**). All of these low-frequency observations clearly show the effect of absorption by galactic ionized hydrogen. Interpretation of these data, now in progress, promises new information on the density and temperature structure of the ionized hydrogen and the distribution of galactic cosmic ray electrons and magnetic fields, especially in the solar neighbourhood. "High" resolution low-frequency observations of parts of the sky are reported at 2·1 MHz (**78**), at 6 and 9 MHz (**21**), at 10 MHz (**43**) and at 13 MHz (**15**). Low-frequency observations, and the agreement of the radio spectrum with the energy spectrum of the primary cosmic-ray electrons indicate the possibility of ionization of part of the cold hydrogen gas by low-energy cosmic ray electrons (**27, 41, 71, 83**).

BIBLIOGRAPHY

Reviews and Symposia

1. *Radio Astronomy and the Galactic System*, IAU Symposium no. 31, Ed. H. van Woerden, Academic Press, London, 1967.
2. *Interstellar Ionized Hydrogen*, Ed. Y. Terzian. Benjamin, New York, 1968.
3. *Supernovae*, I. S. Shklovskii, Wiley and Sons, New York, 1968.
4. *Nebulae and Interstellar Matter*, Vol. VII. *Stars and Stellar Systems*, Ed. G. P. Kuiper, B. M. Middlehurst. University of Chicago Press, 1968.
5. *Planetary Nebulae*, IAU Symposium no. 34, Ed. D. E. Osterbrock and C. R. O'Dell. Reidel, Dordrecht, 1968.
6. 'Structure and Dynamics of the Interstellar Medium' S. B. Pikel'ner, *A. Rev. Astr. Astrophys.*, **6**, 165.
7. Radio Spectra, P. A. G. Scheuer and P. J. S. Williams, *A. Rev. Astr. Astrophys.*, **6**, 321, 1968.
8. 'The Large-Scale Distribution of Hydrogen in the Galaxy', F. J. Kerr, *A. Rev. Astr. Astrophys.*, **7**, 39, 1969.
9. 'Evolution of Diffuse Material', W. G. Mathews and C. R. O'Dell, *A. Rev. Astr. Astrophys.*, **7**, 67, 1969.
10. *The Spiral Structure of our Galaxy*, IAU Symposium no. 38, Ed. G. Contopoulos, W. Becker. Reidel, Dordrecht, 1970.

Papers

11. Akabane. K., Morimoto, M., Miyazawa, K., Suguri, Y., Ishizawa, Y., Nakajima, T., Kawajiri, N. 1967, *Publ. astr. Soc. Japan*, **19**, 88.
12. Akabane, K., Mizazawa, K., Taksahashi, T., Ojima, T., Nakajima, T., Tshirzawa, Y. 1968, *Tokyo astr. Bull.*, **186**, 2193.
13. Alexander, J. K., Brown, L. W., Clark, T. A., Stone, R. G., Weber, R. R. 1969, *Ap. J. Lett.*, **157**, L163.
14. Altenhoff, W. 1970, in press; also p. 519 in (**2**).
15. Andrew, B. H. 1969, *Mon. Not. R. astr. Soc.*, **143**, 17.
16. Beard, M., Kerr, F. J. 1969, *Austr. J. Phys.*, **22**, 121.
17. Beard, M., Day, G. A., Goss, W. M., Manchester, B. A., Thomas, B. M. 1969, *Austr. J. Phys.*, *Astrophys. Suppl.*, **11**.
18. Becklin, E. E., Neugebauer, G. 1968, *Ap. J.*, **151**, 145.

19. Beckman, J. E., Bastin, J. A., Clegg, P. E. 1969, *Nature*, **221**, 944.
20. Bell, S. J., Hewish, A. 1967, *Nature*, **213**, 1214.
21. Benediktov, E. A., Efimova, T. V., Skrebkova, L. A. 1969, *Astr. Zu.*, **46**, 286.
22. Bingham, R. G., Shakeshaft, J. R. 1967, *Mon. Not. R. astr. Soc.*, **136**, 347.
23. Bingham, R. G. 1967, *Mon. Not. R. astr. Soc.*, **137**, 157.
24. Braun, L. D., Yen, J. L. 1968, *Ap. J. Lett.*, **153**, L123.
25. Bridle, A. H. 1967, *Mon. Not. R. astr. Soc.*, **136**, 219.
26. Bridle, A. H. 1968, *Mon. Not. R. astr. Soc.*, **138**, 251.
27. Bridle, A. H. 1969, *Nature*, **221**, 648.
28. Burke, B. F. 1967, p. 361 in (**1**).
29. Caswell, J. L., Crowther, J. H., Holden, D. J. 1967, *Mem. R. astr. Soc.*, **72**, 1.
30. Caswell, J. L. 1967, *Mon. Not. R. astr. Soc.*, **137**, 141.
31. Clube, S. V. M. 1968, *Observatory*, **88**, 243.
32. Cunningham, A. A. 1968, *Ap. J.* **151** 945.
33. Davies, R. D. 1968, *Nature*, **218**, 435.
34. Dickel, H. R., Wendker, H., Bieritz, J. H. 1969, *Astr. Astrophys.*, **1**, 270.
35. Downs, G. S., Thompson, A. R. 1968, *Ap. J. Lett.*, **152**, L65.
36. Field, C. 1967, *Mon. Not. R. astr. Soc.*, **137**, 419.
37. Gardner, F. F., Morimoto, M. 1968, *Austr. J. Phys.*, **21**, 881.
38. Getmantzev, G. G., Karanov, V. S., Sazonov, Yu. A., Tarasov, A. F. 1968, *Izv. Vyssih Učeb. Zaved. Radiofiz.*, **11**, 191.
39. Getmantsev, G. G., Karanov, V. S., Korobkov, Y. S., Tarasov, A. F. 1968, *Astr. Zu.*, **45**, 936.
40. Goss, W. M., Dieter, N. H. 1967, *Publ. astr. Soc. Pacific*, **79**, 575.
41. Gould, R. J. 1969, *Austr. J. Phys.*, **22**, 189.
42. Gower, J. F. R. 1967, *Nature*, **213**, 1213.
43. Hamilton, P. A., Haynes, R. F. 1968, *Austr. J. Phys.*, **21**, 895.
44. Hartz, T. R. 1969, *P. I.E.E.E.*, **57**, 1054.
45. Higgs, L. A., Ramana, K. V. V. 1968, *Ap. J.*, **154**, 73.
46. Hill, E. R. 1968, *Austr. J. Phys.*, **21**, 735.
47. Hirabayashi, H., Ojima, T., Morimoto, M. 1969, *Nature*, **223**, 49.
48. Holden, D. J., Caswell, J. L. 1969, *Mon. Not. R. astr. Soc.*, **143**, 407.
49. Hughes, V. A., Routledge, D. 1969, *Astr. J.*, **74**, 604.
50. Kawajiri, N., Akabane, K. 1967, *Publ. astr. Soc. Japan*, **19**, 476.
51. Kesteven, M. J. L. 1968, *Austr. J. Phys.*, **21**, 369.
52. Kesteven, M. J. L. 1968, *Austr. J. Phys.*, **21**, 739.
53. Kislyakov, A. G., Na'umov, A. I. 1967, *Astr. Zu.*, **44**, 1324.
54. Kundu, M. R., Velusamy, T. 1967, *Ann. Astrophys.*, **30**, 723.
55. Kundu, M. R., Velusamy, T. 1969, *Ap. J.*, **155**, 807.
56. Kundu, M. R. 1969, *Ap. J. Lett.*, **158**, L103.
57. Landecker, T. L., Wielebinski, R. 1969, *Proc. astr. Soc. Austr.*, **1**, 210.
58. LeMarne, A. E. 1968, *Observatory*, **88**, 248.
59. Lequeux, J. 1967, p. 393 in (**1**).
60. Mathewson, D. S. 1968, *Ap. J. Lett.*, **153**, L47.
61. Mathewson, D. S., Nichols, D. C. 1968, *Ap. J. Lett.*, **154**, L11.
62. Matveenko, L. I. 1968, *Astr. Zu.*, **45**, 695.
63. Maxwell, A., Taylor, J. H. 1968, *Astrophys. Lett.*, **2**, 191.
64. Mayer, C. H., Hollinger, J. P. 1968, *Ap. J.*, **151**, 53.
65. Mezger, P. G., Schraml, J., Terzian, Y. 1967, *Ap. J.*, **150**, 807.
66. Mezger, P. G., Altenhoff, W., Schraml, J. 1967, *Ap. J. Lett.*, **150**, L157.
67. Mills, B. Y., Shaver, P. A. 1968 *Austr. J. Phys.*, **21**, 95.
68. Milne, D. K. 1968, *Austr. J. Phys.*, **21**, 201.
69. Milne, D. K., Hill, E. R. 1969, *Austr. J. Phys.*, **22**, 211.
70. Ney, E. P., Stein, W. A. 1968, *Ap. J. Lett.*, **152**, L21.
71. Okuda, H. 1968, *Bull. astr. Inst. Netherl.*, **19**, 352.
72. Okuda, H., Tanaka, Y. 1968, *Bull. astr. Inst. Netherl.*, **20**, 129.
73. Oliver, J. P., Epstein, E. E., Schorn, R. A., Soter, S. L. 1967, *Astr. J.*, **72**, 314.
74. Pariisky, Yu. N., Gol'nev, V. Ya., Lipkova, N. M. 1967, *Izv. glav. astr. Obs. Pulkove*, **24**, 169.
75. Parthasarathy, R. 1968, *Proc. astr. Soc. Austr.*, **1**, 94.

76. Penzias, A. A., Wilson, R. W. 1966, *Ap. J.*, **146**, 666.
77. Poveda, A., Woltjer, L. 1968, *Astr. J.*, **73**, 65.
78. Reber, G. 1968, *J. Franklin Inst.*, **285**, 1.
79. Roger, R. S. 1969, *Ap. J.*, **155**, 831.
80. Seaquist, E. R. 1968, *Observatory*, **88**, 269.
81. Seymour, P. A. H. 1968, *Nature*, **217**, 147.
82. Shaver, P. A. 1969, *Mon. Not. R. astr. Soc.*, **142**, 273.
83. Sironi, G. 1969, *Planet. Space Sci.*, **17**, 1.
84. Thompson, A. R., Riddle, A. C., Lang, K. R. 1969, *Astrophys. Lett.*, **3**, 49.
85. Tokarev, Yu. V. 1969, *Izv. Vyssih Učeb. Zaved. Radiofiz.*, **12**, 161.
86. Weber, R. R., Stone, R. G., Somerlock, C. R. 1969, *Astr. Astrophys.*, **1**, 44.
87. Wendker, H. 1967, *Z. Astrophys.*, **66**, 379.
88. Wendker, H., Yang, K. S. 1968, *Astr. J.*, **73**, 61.
89. Wendker, H. 1968, *Astr. J.*, **73**, 644.
90. Wendker, H., Dickel, H. R. 1968, *Mitt. astr. Ges.*, **25**, 148.
91. Whiteoak, J. B., Gardner, F. F. 1968, *Ap. J.*, **154**, 807.
92. Wielebinski, R., Smith, D. H., Cardenas, X. G. 1968, *Austr. J. Phys.*, **21**, 185.
93. Wielebinski, R., Peterson, C. E. 1968, *Observatory*, **88**, 219.
94. Wynn-Williams, C. G. 1969, *Mon. Not. R. astr. Soc.*, **142**, 453.
95. Yates, K. W., Wielebinski, R., Landecker, T. L. 1967, *Austr. J. Phys.*, **20**, 595.
96. Yates, K. W., Wielebinski, R. 1967, *Ap. J.*, **149**, 439.
97. Yates, K. W. 1968, *Austr. J. Phys.*, **21**, 167.

D. LINE RADIATION IN THE GALAXY

(prepared by R. X. McGee)

The investigation of radio line radiation has reached the status of radio-astronomical spectroscopy in the period covered here – September 1966 to November 1969. The recent discovery of three molecular gases emitting or absorbing a number of lines each, the additional recombination lines of three elements and continuing discoveries in the hydroxyl gas spectrum have completely re-oriented the importance of the subject. With such rapid growth no suitable overall review has been made at this time. The search continues for such lines as those of sulphur hydride, carbon hydride, helium and molecular hydrogen.

1. *The new molecular lines of ammonia, water and formaldehyde*

Two lines corresponding to inversion transitions of rotational levels in the vibrational ground state of the NH_3 molecule ($\lambda \sim 1{\cdot}25$ cm) were detected in late 1968 (**1**) in the source Sagittarius A. Within two months the same authors (**2**) announced the observation of emission from the $6_{16} \rightarrow 5_{23}$ rotational transition of H_2O (λ 1·35 cm) in the directions of Sagittarius B2, the Orion Nebula and W49. In the following month interstellar formaldehyde (H_2CO) was detected in the directions of 15 radio sources (**3**). It was seen in absorption of the $1_{11} \rightarrow 1_{10}$ ground state rotational transition at λ 6·2 cm. So far the formaldehyde has been found in the same sources as the water and the hydroxyl molecules; ammonia and formaldehyde do not appear to occur in the same regions. The H_2CO is of great interest to biologists because it is known to be one of the by-products in the breakdown of amino acids. Again it supplies indirect evidence for the existence of methane, thought to be required in early forms of living matter.

The existence of these molecules is of extreme astrophysical importance, occurring as they do in the cool regions of gas and dust; their investigation must lead to a better understanding of star formation; the very presence of complex molecules indicates that the intensity of ultra-violet radiation is not as great as previously imagined.

Four inversion lines of NH_3 have been observed (**4**) since the early discoveries. The spectra, variability, size and polarization of the H_2O emission sources were reported (**5**). Formaldehyde absorption was noted in the directions of four dark nebulae (**6**) and examined in source W75 (**7**).

H_2CO was observed in the directions of 31 sources (**8**) in southern skies. A high abundance of the isotope ^{13}C near the galactic centre has been indicated from observations of $H_2\,^{13}CO$ (**9**).

2. *The hydroxyl lines*

Two quotations from a review article (**10**) in 1967 still seem to be most appropriate for the present state of the study of the hydroxyl gas. "Many observations have revealed totally unexpected anomalies in the radio-spectral properties of interstellar OH. As a result no reliable astrophysical information has been derived from the OH observations, but a large body of information is waiting to be unravelled." And "When OH emission was discovered it was given the Latin name 'mysterium'; a more appropriate name would be difficult to find". Other reviews (**11–16**) extend only to 1968.

a. *General OH source surveys*

Various aspects of absorption, emission, line widths, circular and linear polarization, satellite lines, association with dust and catalogues of OH sources appear in a number of surveys (**17–25**).

b. *Infrared stars and protostars – possible association with OH emission*

A relation between an OH source and an infrared object in the Orion Nebula has been suggested (**26**). Strong emission at the 1612 MHz line was discovered from the infrared star NML Cygni (**27**). The matter is further discussed in (**28**).

c. *Interferometry of OH sources* (**4**)

A review of very long baseline interferometry (v.l.b.i.) is given in (**13**). The use of v.l.b.i. has established that the sizes of many OH emission sources approach stellar dimensions. Intercontinental v.l.b.i. is discussed in (**29**).

d. *Occultations*

The lunar occultations of Sagittarius A and B2 were observed (**30, 31**). Interplanetary scintillations of the 1665 MHz emission from Sagittarius B2 are reported (**32**).

e. *Discussions of OH-line mechanisms*

Explanations of OH-line radiation have been offered in more than 20 papers. Recent treatments and references may be found in (**33**) and (**34**).

f. *Detection of other OH transitions*

The $^2\pi_{1/2}\ J=\frac{1}{2}$ transition at λ 6·3 cm (**35**), the $^2\pi_{3/2}\ J=\frac{5}{2}$ at λ 5 cm (**36**) and the $^2\pi_{1/2}\ J=\frac{5}{2}$ at λ 3·8 cm (**37**) have been detected.

g. ^{18}OH

The most recent published information is contained in (**38**) where ^{18}OH has been observed in both the 1637 and 1639 MHz absorption lines in Sgr B 2 as well as Sgr A. Abundance ratios ($^{18}OH/^{16}OH$) above and below the terrestrial value were found in separate regions.

3. *The recombination lines*

The investigation of recombination lines (from transitions between levels with large principal quantum numbers) expanded greatly up to 1968. Two important reviews of experimental (**39**) and theoretical (**40**) work are given in (**42**). The application of recombination-line radial velocities to galactic structure and kinematics is explained in (**41**).

a. *Hydrogen α-lines* $(n+1\rightarrow n)$

Hα lines have been reported over the range $n=253$ at 404 MHz (**43**) to $n=56$ at 36, 466 MHz (**44**).

The various observations have established that recombination lines are a property of H II regions and that they afford an excellent tool in the investigation of the physics of the nebulae. Close attention has been given to the calculation of electron temperatures, departures from local thermodynamic equilibrium, the population of atomic levels by dielectronic recombination and Stark broadening. However, as can be seen, for example, in the two reviews mentioned above much more caution is needed in the interpretation of results than has so far been shown.

Earlier surveys of 16 nebulae in H109α (**45**), of 39 nebulae in H158α (**46**), and 34 sources in H126α and H127α (**47**) have given way to large surveys (**48, 49**) in which 75 and 131 sources respectively have been located in H109α.

Useful tables have been published recently which greatly aid the study of recombination lines: tables of oscillator strengths (**50, 51**) and tables of frequencies of the α, β, γ, δ and ε H lines and α and β He lines for quantum numbers $n=40$ to $n=889$ (**52**).

b. *Discoveries of other recombination lines*

The first two-level transition lines, H158β and H159β were reported early in 1967 (**53**). The H148δ line in Orion Nebula (**54**) and the H197β and H225γ lines in six intense H II regions (**55**) followed.

An unidentified line near He 109α was detected in NGC2024 and IC1793 (**56**). It has been suggested that the line is Carbon 109α (**57**).

c. *Helium abundance*

It was found that the mean abundance ratio of helium to hydrogen for 5 nebulae is $0{\cdot}084 \pm 0{\cdot}003$ (**58**). Other authors (**59**) find a ratio of 0·12 for 9 nebulae but with much poorer accuracy.

4. *Neutral hydrogen*

a. *The two components of interstellar HI*

The most important contribution was the firm establishment of a model of the two components of H I – dense cool absorbing clouds are immersed in a tenuous hot medium (**60–62**). It was found that the temperature of this medium in at least one direction was >1000 K; in the cool clouds with an average density of 10 H atoms cm^{-3} the temperature is about 50 K. Supporting observations are reported (**63–65**). Using a model in which heating is effected by low-energy cosmic rays a similar conclusion is reached (**66**). Earlier work along these lines is reviewed in (**67**). Pulsar data have been applied to support the above temperature values (**68**). Finally, it has been pointed out that low spin temperatures in clouds may be the reason for the apparent discrepancies between the densities of H atoms observed at 21 cm and in Lα absorption experiments (**69, 70**).

b. *The Zeeman effect*

Considerable activity has again arisen in experiments of H I Zeeman effect. The recent achievements are reviewed and a value of 2 ± 1 μG placed on the mean interstellar field (**71**).

c. *HI surveys and galactic structure*

A succession of general H I surveys have continued to appear and are reviewed in (**72**). Further references and discussions may be found in the Proceedings of IAU Symposium no. 38. Of great interest was a presentation (**73**) of a recent derivation of H I spiral structure differing entirely from the traditional, albeit recently updated, Leiden-Sydney model.

d. *Miscellaneous*

A number of other H I topics have been discussed or reviewed. Included are hydrogen at intermediate latitudes (**74**), high-velocity clouds (**75, 76**), "intermediate cloudlets" (**77**), the lack of apparent association of H I gas and dust (**78**), the proposal that hydrogen becomes molecular in dust clouds (**79**), the use of H I observations in pulsar work (**80**), and determinations of the Astronomical Unit (**81, 82**).

BIBLIOGRAPHY

1. Cheung, A. C., Rank, D. M., Townes, C. H., Thornton, D. D., Welch, W. J. 1968, *Phys. Rev. Lett.*, **21**, 1701.

2. Cheung, A. C., Rank, D. M., Townes, C. H., Thornton, D. D., Welch, W. J. 1969, *Nature*, **221**, 626.

3. Snyder, L. E., Buhl, D., Zuckerman, B., Palmer, P. 1969, *Phys. Rev. Lett.*, **22**, 679.

4. Cheung, A. C., Rank, D. M., Townes, C. H., Welch, W. J. 1969, *Nature*, **221**, 917.

5. Knowles, S. H., Mayer, C. H., Cheung, A. C., Rank, D. M., Townes, C. H. 1969, *Science*, **163**, 1055.

6. Palmer, P., Zuckerman, B., Buhl, D., Snyder, L. E. 1969, *Astrophys. J.*, **156**, L147.

7. Rydbeck, D. E. H., Ellder, J., Kollberg, E. 1969, *Astrophys. J.*, **156**, L141.

8. Whiteoak, J. B., Gardner, F. F. 1970, *Astr. Lett.*, **4**, 5.

9. Zuckerman, B., Palmer, P., Snyder, L. E., Buhl, D. 1969, *Astrophys. J.*, **157**, L167.

10. Barrett, A. H. 1967, *Science*, **157**, 881.

11. Robinson, B. J., McGee, R. X. 1967, *A. Rev. Astr. Astrophys.*, **5**, 183.

12. Weaver, H. 1968, *Interstellar Ionized Hydrogen*, Ed. Y. Terzian. W. A. Benjamin, Inc., N.Y., p. 645.

13. Moran, J. M., Jr. 1968, *Interstellar Ionized Hydrogen*, Ed. Y. Terzian. W. A. Benjamin, Inc., N.Y., p. 681.

14. Rogers, A. E. E. 1968, *Interstellar Ionized Hydrogen*, Ed. Y. Terzian. W. A. Benjamin, Inc., N.Y., p. 695.

15. Litvak, M. M. 1968, *Interstellar Ionized Hydrogen*, Ed. Y. Terzian. W. A. Benjamin, Inc., N.Y., p. 713.

16. Gold, T. 1968, *Interstellar Ionized Hydrogen*, Ed. Y. Terzian. W. A. Benjamin, Inc., N.Y., p. 747.

17. Goss, W. M. 1968, *Astrophys. J. Suppl. Ser.*, **15**, 131.

18. Weaver, H., Dieter, N. H., Williams, D. R. W. 1968, *Astrophys. J. Suppl. Ser.*, **16**, 219.

19. Palmer, P., Zuckerman, B. 1967, *Astrophys. J.*, **148**, 727.

20. Ball, J. A., Meeks, M. L. 1968, *Astrophys. J.*, **153**, 577.

21. McGee, R. X., Gardner, F. F., Robinson, B. J. 1967, *Austr. J. Phys.*, **20**, 407.

22. Ball, J. A., Staelin, D. H. 1968, *Astrophys. J.*, **153**, L41.

23. Robinson, B. J., Goss, W. M., Manchester, R. N. 1970, '18-cm observations of galactic OH-Longitudes 350° to 50°'. Submitted to *Austr. J. Phys.*

24. Turner, B. E. 1969, *Astrophys. J.*, **157**, 103.

25. Cudaback, D. D., Heiles, C. 1969, *Astrophys. J.*, **155**, L21.

26. Raimond, E., Eliasson, B. 1967, *Astrophys. J.*, **150**, L171.

27. Wilson, W. J., Barrett, A. H. 1968, *Science*, **161**, 778.

28. Mezger, P. G., Robinson, B. J. 1968, *Nature*, **220**, 1107.

29. Cohen, M. H., Jauncey, D. L., Kellermann, K. I., Clark, B. G. 1968, *Science*, **162**, 88.

30. Kerr, F. J., Sandqvist, A. 1968, *Astrophys. Lett.*, **2**, 195.

31. Manchester, R. N., Goss, W. M., Robinson, B. J. 1969, *Astrophys. J.*, **4**, 93.

32. Robinson, B. J., Goss, W. M. 1968, *Astrophys. Lett.*, **2**, 5.

33. Litvak, M. M. 1967, *Astrophys. J.*, **156**, 471.

34. Litvak, M. M., Zuckerman, B., Dickinson, D. E. 1969, *Astrophys. J.*, **156**, 875.

35. Zuckerman, B., Palmer, P., Penfield, H., Lilley, A. E. 1968, *Astrophys. J.*, **153**, L69.

36. Yen, J. L., Zuckerman, B., Palmer, P., Penfield, H. 1969, *Astrophys. J.*, **156**, L27.

37. Schwartz, P. R., Barrett, A. H. 1969, *Astrophys. J.*, **157**, L109.

38. Gardner, F. F., McGee, R. X., Sinclair, M. W. 1970, *Astrophys. Lett.*, **5**, 67.

39. Palmer, P. 1968, *Interstellar Ionized Hydrogen*, Ed. Y. Terzian. W. A. Benjamin, Inc., N.Y., p. 335.

40. Goldberg, L. 1968, *Interstellar Ionized Hydrogen*, Ed. Y. Terzian. W. A. Benjamin, Inc., N. Y., p. 373.

41. Burke, B. F. 1968, *Interstellar Ionized Hydrogen*, Ed. Y. Terzian. W. A. Benjamin, Inc., N. Y., p. 541.

42. Terzian, Y. (ed.) 1968, *Interstellar Ionized Hydrogen*. W. A. Benjamin, Inc., N.Y.

43. Penfield, H., Palmer, P., Zuckerman, B. 1967, *Astrophys. J.*, **148**, L25.

44. Sorochenko, R. L., Puzanov, V. A., Salomonovich, A. E., Shteinshleger, V. B. 1969, *Astrophys. Lett.*, **3**, 7.

45. Mezger, P. G., Hoglund, B. 1967, *Astrophys. J.*, **147**, 490.

46. Dieter, N. H. 1967, *Astrophys. J.*, **150**, 435.

47. McGee, R. X., Gardner, F. F. 1968, *Austr. J. Phys.*, **21**, 149.
48. Reifenstein, E. C. 1968, Thesis, Mass. Inst. Tech.
49. Wilson, T. C. 1969, Thesis, Mass. Inst. Tech.
50. Goldwire, H. C. 1968, *Astrophys. J. Suppl. Ser.*, **17**, 445.
51. Menzel, D. H. 1969, *Astrophys. J. Suppl. Ser.*, **18**, 221.
52. Lilley, A. E., Palmer, P. 1968, *Astrophys. J. Suppl. Ser.*, **16**, 143.
53. Gardner, F. F., McGee, R. X. 1967, *Nature*. **213**, 480.
54. Gordon, M. A., Meeks, M. L. 1967, *Astrophys. J.*, **149**, L21.
55. Williams, D. R. W. 1967, *Astrophys. Lett.*, **1**, 59.
56. Palmer, P., Zuckerman, B., Penfield, H., Lilley, A. E., Mezger, P. G. 1967, *Nature*, **215**, 40.
57. Dupree, A. K., Goldberg, L. 1969, *Astrophys. J.*, **158**, L49.
58. Palmer, P., Zuckerman, B., Penfield, H., Lilley, A. E., Mezger, P. G. 1969, *Astrophys. J.*, **156**, 887.
59. McGee, R. X., Batchelor, R. A., Brooks, J. W., Sinclair, M. W. 1969, *Austr. J. Phys.*, **22**, 631.
60. Grahl, B. H., Hachenberg, O., Mebold, U. 1968, *Beitr. Radioastron.*, **1**, 1.
61. Mebold, U. 1969, *Beitr. Radioastron.*, **1**, 97.
62. Radhakrishnan, V., Murray, J. D. 1969, *Proc. astr. Soc. Austr.*, **1**, 215.
63. Riegel, K. W., Jennings, M. C. 1969, *Astrophys. J.*, **157**, 563.
64. D'Odorico, S., Sancisi, R., Simonson, S. C. 1969, *Astr. Astrophys.*, **1**, 131.
65. Gosachinskii, I. V., Bystrova, N. V. 1968, *Astr. Zu.*, **45**, 691. English transl. in *Soviet Astr.*, **12**, 548.
66. Field, G. B., Goldsmith, D. W., Habing, H. J. 1969, *Astrophys. J.*, **155**, L149.
67. Pikelner, S. B. 1967, *Astrophys. Lett.*, **1**, 43.
68. Hjellming, R. M., Gordon, C. P., Gordon, K. J. 1969, *Astr. Astrophys.*, **2**, 202.
69. Carruthers, G. R. 1969, *Astrophys. J.*, **156**, L97.
70. Jenkins, E. B., Morton, D. C. 1967, *Nature*, **215**, 1257.
71. Verschuur, G. L. 1969, Observational aspects of galactic magnetic fields. Paper at *IAU Symposium* no. 38, Basel.
72. Kerr, F. J. 1969, *A. Rev. Astr. Astrophys.*, **7**, 39.
73. Weaver, H. 1969, *Proc. IAU Symposium* no. 38.
74. Takakubo, K. 1968 *Bull. astr. Insts. Netherl.*, **20**, 107.
75. Hulsboch, A. N. M. 1968, *Bull. astr. Insts. Netherl.*, **20**, 33.
76. Dieter, N. H. 1969, *Publ. astr. Soc. Pac.*, **81**, 186.
77. Heiles, C. 1968, *Astrophys. Lett.*, **2**, 31.
78. Heiles, C. 1969, *Astrophys. J.*, **156**, 493.
79. Varsavsky, C. M. 1968, *Astrophys. J.*, **153**, 627.
80. Radhakrishnan, V. 1969, *Proc. astr. Soc. Austr.*, **1**, 254.
81. Williams, D., Davies, R. D. 1968, *Mon. Not. R. astr. Soc.*, **140**, 537.
82. Knowles, S. H. 1969, *Astr. J.*, **74**, 291.

E. PULSARS

(prepared by S. P. Maran)

As of 1 Sept. 1969, about 330 papers on pulsars had appeared since their discovery (**1**). Only typical references can be listed here; a complete bibliography is available from the reporter. General reviews are given in (**2**) and (**3**); a review of theories also exists (**4**). Several pulsar conferences have been held; the proceedings of one are published as a tape recording (**5**). An anthology of early pulsar articles has appeared (**6**), as has a catalogue of the first 37 pulsars (**7**).

Except for highly variable sources or those with atypical spectra, it is likely that most of the pulsars that are detectable with present equipment have been found; the "dispersion remover" technique offers good prospects for future discoveries (**8**). Pulsar search techniques are reviewed in (**9**).

PSR 0833-45 is probably associated with the supernova remnant Vela X (**10**). NP 0532 (**11**) is associated with the Crab Nebula (**12**), is the source of optical (**13**) and X-ray (**14**) pulses, and is identified with the central star of the nebula (**15**). No other pulsars are associated with known supernova remnants and no pulsar except NP 0532 is optically identified (**16**). Deliberate searches of other supernova remnants by radio (**17**) and optical (**18**) methods failed to reveal pulsars.

Pulsar periods range from 0·033 to 3·7 s and are lengthening in all cases of observed secular trends (**19**), (**20**), (**21**). Pulse widths are typically 5% of pulse periods. Structure as fine as 100 μs has been found (**22**), (**23**) in pulses. The mean pulse profile, averaged over $\sim 10^3$ pulses, is a reproducible "signature" for each pulsar (**24**). At least 7 pulsars exhibit periodic modulation of pulse intensity at rates incommensurable with the pulse frequencies (**25**), (**26**), (**27**), (**28**). Wavelength-dependent intensity variations on time scales up to months in length have been reported (**1**), (**24**), (**29**), as has been time-variable spectral fine structure (**30**), (**31**), (**32**). Some variations may be intrinsic to the pulsars, but the scintillation data yield valuable models for the interstellar medium (**33**), (**34**), (**35**). A discontinuity was observed in the period and period slow-down rate of PSR 0833-45 (**36**), (**37**), as was a second time derivative for the period of NP 0532 (**38**).

Distances are available for the two pulsars associated with supernova remnants. Distance estimates based on galactic rotation models have been made for several sources in which 21-cm absorption is found (**39**), (**40**). Distances based on interpretations of dispersion measures, models for the interstellar medium, *etc.* have been widely discussed (**41**), (**42**), (**43**), (**44**).

Radio spectra of pulsars usually show high-frequency cut-offs (**45**); the emission implies very high brightness temperatures, $\sim 10^{21}$ K (**46**). Some low frequency cut-offs are also found (**47**). The infrared measurements of NP 0532 (**48**), combined with the optical (**49**), (**50**), (**51**) and X-ray (**52**), (**53**) fluxes, clearly delineate a spectral component that is distinct from the radio spectrum. However, the radio, optical and X-ray pulses occur simultaneously, to the limit of measuring accuracy (**54**), (**55**). There are important spectral differences between sub-pulses (**56**). The linear polarization of radio pulses (**57**) is accompanied by strong circular polarization in only one case (**58**). A smooth variation of position angle during the pulse occurs in the radio emission of PSR 0833-45 (**59**), (**60**) and the optical emission of NP 0532 (**61**), (**62**). The highly variable polarization of other pulsars shows complex but reproducible characteristics in the synchronously integrated parameters (**58**). Time-resolved image tube spectra of NP 0532 show no absorption or emission lines (**63**). Recent searches for Cerenkov radiation due to pulsar gamma rays are negative (**64**), (**65**) as are searches for nanosecond optical pulses from NP 0532 (**66**), (**67**). Measurements of interstellar Faraday rotation in pulsar radiation are combined with dispersion measures to evaluate the galactic magnetic field (**68**), (**69**). A review of the observations (**70**) emphasizes that pulsars are concentrated to the galactic disk (**71**).

The "old" idea of a rotating, highly magnetized neutron star as the present energy source of the Crab Nebula (**72**) provides a natural clock mechanism for pulsars (**73**). In fact, the observed slow-down of NP 0532 (**74**) corresponds to a loss of rotational energy sufficient to account for the nebular emission (**75**), (**76**). The pulsar emission may originate at distances from the star of order $r = c/\Omega$, where the radiating particles have been accelerated by magnetic dipole radiation (**77**), (**78**) or by electrodynamic processes in a magnetosphere (**79**). It is shown that these two phenomena are complementary, so that the total loss to the star is independent of the obliquity of the magnetic axis (**2**). Alternatively, the emission may arise on the stellar surface near the magnetic poles (**80**) or in plasma above but near the poles (**81**). Although all supernovae may generate cosmic rays, pulsars are cosmic ray producers (**82**), (**83**) with distances and ages (**84**), (**85**) appropriate to the sources of the galactic particles actually observed at the Earth today (**86**). Oscillations of neutron stars cannot account for the secondary modulation of pulsars (**87**), but structural effects might cause period changes like that found in PSR 0833-45 (**88**), (**89**), (**90**). A number of pulsar emission mechanisms have been discussed (**91**), (**92**), (**93**), (**94**). Nearly all theories require very high magnetic fields, which may be due to compression or Landau orbital ferromagnetism (**95**). The dragging of inertial frames (**95**) requires that careful attention be paid to general relativistic effects in rotating neutron stars (**97**), (**98**). Theoretical explanations for the paucity of long-period pulsars have been given (**99**), (**100**).

BIBLIOGRAPHY

1. Hewish, A., Bell, S. J., Pilkington, J. D. H., Scott, P. F., Collins, R. A. 1968, *Nature*, **217**, 709.
2. Radhakrishnan, V. 1969, *Proc. astr. Soc. Austr.*, **1**, 254.

3. Maran, S. P., Cameron, A. G. W. 1969, *Earth and Extraterrestrial Sciences*, **1**, 3.
4. Chiu, H. Y. *Publ. astr. Soc. Pacific*, in press.
5. Anonymous 1968, *Amer. Assoc. Advancement Sci. Recording*, No. 20/68.
6. Anonymous 1968, *Pulsating Stars, A Nature Reprint*. Macmillan and Co., London.
7. Taylor, J. H. 1969, *Astrophys. Lett.*, **3**, 205.
8. Large, M. I., Vaughan, A. E., Wielebinski, R. 1969, *Astrophys. Lett.*, **3**, 123.
9. Burns, W. R., Clark, B. G. 1969, *Astr. Astrophys.*, **2**, 280.
10. Large, M. I., Vaughan, A. E., Mills, B. Y. 1968, *Nature*, **220**, 340.
11. Staelin, D. H., Reifenstein, III, E. C. 1968, *Science*, **162**, 1481.
12. Comella, J. M., Craft, Jr., H. D., Lovelace, R. V. E., Sutton, J. M., Tyler, G. L. 1969, *Nature*, **221**, 453.
13. Cocke, W. J., Disney, M. J., Taylor, D. J. 1969, *Nature*, **221**, 525.
14. Fritz, G., Henry, R. C., Meekins, J. F., Chubb, T. A., Friedman, H. 1969, *Science*, **164**, 709.
15. Lynds, R., Maran, S. P., Trumbo, D. E. 1969, *Astrophys. J.*, **155**, L121.
16. Kristian, J. *Publ. astr. Soc. Pacific*, in press.
17. Reifenstein, III, E. C., Brundage, W. D., Staelin, D. H. 1969, *Astrophys. J.*, **156**, L125.
18. Carleton, N. P. 1969, private communication.
19. Cole, T. W. 1969, *Nature*, **221**, 29.
20. Davies, J. G., Hunt, G. C., Smith, F. G. 1969, *Nature*, **221**, 27.
21. Reichley, P. E., Downs, G. S., Morris, G. A. 1969, Meeting U.S. Natl. Comm., Internat. Sci. Radio Union, Washington, 21-24 Apr.
22. Craft, Jr., H. D., Comella, J. M., Drake, F. D. 1968, *Nature*, **218**, 1122.
23. Ekers, R. D., Moffet, A. T. 1968, *Nature*, **220**, 756.
24. Lyne, A. G., Rickett, B. J. 1968, *Nature*, **218**, 326.
25. Conklin, E. K., Howard, H. T., Craft, Jr., H. D., Comella, J. M. 1968, *Nature*, **219**, 1239.
26. Drake, F. D., Craft, Jr., H. D. 1968, *Nature*, **220**, 231.
27. Lovelace, R. V. E., Craft, Jr., H. D. 1968, *Nature*, **220**, 875.
28. Taylor, J. H., Jura, M., Huguenin, G. R. 1969, *Nature*, **223**, 797.
29. Downs, G. S., Morris, Jr., G. A., Reichley, P. E. 1969, *Nature*, **222**, 1257.
30. Komesaroff, M. M., McCulloch, P. M., Hamilton, P. A., Cooke, D. J. 1968, *Nature*, **220**, 358.
31. Huguenin, G. R., Taylor, J. H. 1969, *Astrophys. Lett.*, **3**, 107.
32. Huguenin, G. R., Taylor, J. H., Jura, M. 1969, *Astrophys. Lett.*, in press.
33. Scheuer, P. A. G. 1968, *Nature*, **218**, 920.
34. Salpeter, E. E. 1969, *Nature*, **221**, 31.
35. Rickett, B. J. 1969, *Nature*, **221**, 158.
36. Reichley, P. E., Downs, G. S. 1969, *Nature*, **222**, 229.
37. Radhakrishnan, V., Manchester, R. N. 1969, *Nature*, **222**, 228.
38. Richards, D. W., Rankin, J. M., Counselman, III, C. C. 1969, *IAU Circ.*, no. 2164.
39. Guelin, M., Guibert, J., Huchtmeier, W., Weliachew, L. 1969, *Nature*, **221**, 249.
40. Gordon, C. P., Gordon, K. J., Shalloway, A. M. 1969, *Nature*, **222**, 129.
41. Davidson, K., Terzian, Y. 1969, *Astr. J.*, **74**, 849.
42. Davies, R. D. 1969, *Nature*, **223**, 355.
43. Grewing, M., Mebold, U., Rohlfs, K. 1969, *Nature*, **221**, 751.
44. Shuter, W. L. H., Venugopal, V. R., Mahoney, M. J. 1968, *Nature*, **220**, 356.
45. Daintree, E. J., Davies, J. G., Horton, P. W., Walsh, D. 1968, *Nature*, **218**, 1035.
46. Ryle, M., Bailey, J. A. 1968, *Nature*, **217**, 907.
47. Lyne, A. G., Rickett, B. J. 1968, *Nature*, **219**, 1339.
48. Neugebauer, G., Becklin, E. E., Kristian, J., Leighton, R. B., Snellen, G., Westphal, J. A. 1969, *Astrophys. J.*, **156**, L115.
49. Oke, J. B. 1969, *Astrophys. J.*, **156**, L49.
50. Nather, R. E., Warner, B., Macfarlane, M. 1969, *Nature*, **221**, 527.
51. Duthie, J. G., Sturch, C., Richer, H. B., Rodney, P. 1969, *Science*, **163**, 1320.
52. Fishman, G. J., Harnden, Jr., F. R., Haymes, R. C. 1969, *Astrophys. J.*, **156**, L107.
53. Boldt, E. A., Desai, U. D., Holt, S. S., Serlemitsos, P. J., Silverberg, R. F. 1969, *Nature*, **223**, 280.
54. Conklin, E. K., Howard, H. T., Miller, J. S., Wampler, E. J. 1969, *Nature*, **222**, 552.
55. Bradt, H., Rappaport, S., Mayer, W., Nather, R. E., Warner, B., Macfarlane, M., Kristian, J. 1969, *Nature*, **222**, 728.

56. Moffet, A. T. 1968, Fourth Texas Symposium on Relativistic Astrophysics, Dallas, 16–20 Dec.
57. Lyne, A. G., Smith, F. G. 1968, *Nature*, **218**, 124.
58. Ekers, R. D., Moffet, A. T. 1969, *Astrophys. J.*, **158**, L1.
59. Radhakrishnan, V., Cooke, D. J., Komesaroff, M. M., Morris, D. 1969, *Nature*, **221**, 443.
60. Ekers, R. D., Lequeux, J., Moffet, A. T., Seielstad, G. A. 1969, *Astrophys. J.*, **156**, L21.
61. Warner, B., Nather, R. E., Macfarlane, M. 1969, *Nature*, **222**, 233.
62. Wampler, E. J., Scargle, J. D., Miller, J. S. 1969, *Astrophys. J.*, **157**, L1.
63. Lynds, R. 1969, *Astrophys. J.*, **157**, L11.
64. Fazio, G. G., Helmken, H. F., Rieke, G. H., Weekes, T. C. 1968, *Nature*, **220**, 892.
65. Charman, W. N., Fruin, J. H., Jelley, J. V., Fegan, D. J., Jennings, D. M., O Mongain, E. P., Porter, N. A., White, G. M. 1969, Eleventh Internat. Conf. Cosmic Rays, Budapest, 25 Aug.–4 Sept.
66. Ogelman, H., Sobieski, S. 1969, *Nature*, **223**, 47.
67. Hegyi, D., Novick, R., Thaddeus, P. 1969, *Astrophys. J.*, **158**, L11.
68. Smith, F. G. 1968, *Nature*, **220**, 891.
69. Staelin, D. H., Reifenstein, III, E. C. 1969, *Astrophys. J.*, **156**, L121.
70. Mills, B. Y. 1969, *Proc. astr. Soc. Austr.*, **1**, 176.
71. Wielebinski, R., Vaughan, A. E., Large, M. I. 1969, *Nature*, **221**, 47.
72. Pacini, F. 1967, *Nature*, **216**, 567.
73. Gold, T. 1968, *Nature*, **218**, 731.
74. Richards, D. W., Comella, J. M. 1969, *Nature*, **222**, 551.
75. Finzi, A., Wolf, R. A. 1969, *Astrophys. J.*, **155**, L107.
76. Gold, T. 1969, *Nature*, **221**, 25.
77. Pacini, F. 1968, *Nature*, **219**, 145.
78. Gunn, J. E., Ostriker, J. P. 1969, *Nature*, **221**, 454.
79. Goldreich, P., Julian, W. H. 1969, *Astrophys. J.*, **157**, 869.
80. Chiu, H. Y., Occhionero, F. 1969, *Nature*, **223**, 1113.
81. Radhakrishnan, V., Cooke, D. J. 1969, *Astrophys. Lett.*, **3**, 225.
82. Gunn, J. E., Ostriker, J. P. 1969, *Phys. Rev. Lett.*, **22**, 728.
83. Michel, F. C. 1969, *Phys. Rev. Lett.*, **23**, 247.
84. Grewing, M., Priester, W. 1969, *Forsch. Astronom. Inst. Bonn*, No. 69–04.
85. Hirth, W., Reinhardt, M. v. 1969, *Forsch. Astronom. Inst. Bonn*, No. 69–05.
86. Lingenfelter, R. E. 1969, Eleventh Internat. Conf. Cosmic Rays, Budapest, 25 Aug.–4 Sept.
87. Cameron, A. G. W., Cohen, J. M. 1969, *Astrophys. Lett.*, **3**, 3.
88. Greenstein, G. S., Cameron, A. G. W. 1969, *Nature*, **222**, 862.
89. Ruderman, M. 1969, *Nature*, **223**, 597.
90. Dyson, F. J. 1969, *Nature*, **223**, 486.
91. Chiu, H. Y., Fassio-Canuto, L. 1969, *Phys. Rev.*, in press.
92. Eastlund, B. J. 1968, *Nature*, **220**, 1293.
93. Ginzburg, V. L., Zheleznyakov, V. V., Zaitsev, V. V. 1969, *Astrophys. Space Sci.*, **4**, 464.
94. Michel, F. C., Tucker, W. H. 1969, *Nature*, **223**, 277.
95. Canuto, V. 1969, Second Cambridge Conf. Relativity, New York, 20 May.
96. Occhionero, F. 1969, preprint, Brandeis Univ.
97. Cohen, J. M. *Astrophys. Space Sci.*, in press.
98. Cohen, J. M., Cameron, A. G. W. 1969, *Nature*, **224**, 566.
99. Gunn, J. E., Ostriker, J. P. 1969, *Nature*, **223**, 813.
100. Pacini, F. 1969, *Nature*, **224**, 160.

F. EXTRAGALACTIC RADIO ASTRONOMY: COSMOLOGY

(prepared by J. Lequeux)

1. *General remarks*

At the time of writing, it seems that progress in extragalactic radio astronomy since the last IAU meeting has been less spectacular than in the preceding period, which will be remembered as the great era of quasars. The last three years have rather seen considerable improvements in the quality and accuracy of radio data (positions, diameters and structures, polarization, spectra and flux variations, etc.); advances in angular resolution are especially remarkable, thanks to the novel technique of interferometry with independent local oscillators.

Little decisive progress has been made in the theory of extragalactic radio sources. Even the problem of the nature of quasar red-shifts is still unsettled, and the fundamental mechanisms for energy production and particle acceleration are badly understood. However decisive advances exist in certain limited fields.

Given the small space available and the general trends of extragalactic radio astronomy during the last three years, emphasis will be put on observational work; no attempts are made to give an exhaustive bibliography, as it would have to include more than 600 references.

2. *Surveys, accurate positions, identifications*

The Parkes survey has been completed in the declination zone $+20+27°$ (**1**), and the rest of the survey is edited in a useful single volume (**2**). Other surveys at relatively low frequencies include Cambridge aperture-synthesis deep surveys (**3, 4, 5, 6**). Much work has been done and is in progress at shorter wavelengths: near 21 cm at NRAO (**7**) Dwingeloo (**8**), Penticton (**9**), Ohio (**10, 11, 12**), at 11 cm at Parkes and at 6 cm at NRAO (**13, 14**). These surveys contain many sources with flat or inverted spectra which seem to constitute a major fraction of the total at high frequencies. Current positional accuracies in the surveys are about 10″, but source positions have been measured with better accuracies, notably at Parkes, Cal Tech and Green Bank.

Many new identifications have been given, but a large fraction of the possible identifications proposed in the previous period have not been confirmed. As a result, the number of certain quasars in the 3C revised catalogue is still rather low, and a large percentage of sources in this catalogue do not correspond to any visible object.

3. *Spectra, flux density, variations*

Two review articles on this subject (**15, 16**) cover the literature until the end of 1967.

Considerable progress has been made in the accuracy and number of flux measurements, both at low and high frequencies. New absolute measurements have been made, mostly at decameter wavelengths. The flux-density scales are sometimes correct to better than $\pm 5\%$ at high frequencies, but systematic errors may still reach $\pm 10\%$ at 178 MHz, 15% at 38 MHz and more at lower frequencies (see a recent discussion in ref. **17**). Emphasis has been put on measurements on complete or nearly complete samples of radio sources (**17** to **21**). A number of source fluxes are available at frequencies as low as 10 MHz (**22**), and at millimeter wavelengths.

Although a large proportion of sources have curved spectra, 40% of the revised 3C sources have perfectly straight power-law spectra between 38 and 5000 MHz at least. Intrinsically stronger radio galaxies appear to have steeper spectra. Many spectra show a downward curvature at low frequencies, perhaps connected with synchrotron self-absorption; others have concave or complex spectra, and are often variable quasars or Seyfert galaxies (**23, 24, 25**).

Flux variations of radio sources have been extensively studied and variations with time scales of weeks exist. A large number of cases can be explained by the well-known Kellermann-Van der Laan model of expanding optically thick clouds of relativistic electrons, but some require a different or less simplified model (**26**).

4. *Brightness distribution*

Major advances have been made in this field, especially towards very high resolution (see a review in ref. **27**). 2-antenna interferometry has been done for a large number of sources (**28, 29, 30**). There are similar observations without phase (**31**), some reaching resolutions of $0\overset{''}{.}1 - 0\overset{''}{.}2$ (**32**).

Aperture synthesis observations have been made at Cambridge on most of the revised 3C sources (**33**; in preparation), and more are in progress at the NRAO with resolving powers of a few arc seconds.

Lunar occultation observations have been pursued. A large number of scintillation observations have given information on structures in the range $0\overset{''}{.}01 - 1''$ (see **27**), and more are being done at Cambridge.

Very long baseline observations have been organized by American and Canadian observers, using many radiotelescopes scattered around the Earth (see **27**). Nearly 100 sources have been observed, mostly QSS and Seyfert galaxies, with baselines as large as several 10^8 wavelengths, corresponding to resolving powers of the order of 10^{-3} arc second. A variety of information, including very accurate positions, are foreseen as by-products of these observations.

Our understanding of source structure is much better than 3 years ago. The classical double source structure is obviously an oversimplification in many cases; radio galaxies frequently contain small-size components, but quasars may not contain such components. Unidentified sources tend to have small diameters, but this gives no clue as to their nature. Very small diameter quasars or Seyfert galaxies may be interpreted without difficulty as compact synchrotron sources with magnetic fields of the order of 10^{-4} to 10^{-2} G. It seems that inverse Compton losses are smaller than synchrotron losses, and that expansion is not relativistic at least in the case of 3C 84 (**25, 34**).

5. *Polarization*

Linear polarization of many sources has been observed up to wavelengths as short as 1 cm (see e.g. **35**). Major progress has been achieved in our knowledge of the structure of the polarized flux of radio sources, which bears little or no resemblance to the distribution of total flux (**36**). Under these conditions it is no surprise that no clear relationship has been found between polarization and other parameters of radio sources, except that low-brightness sources are likely to be more polarized (**37**). Linear polarization has been found to vary with time at short wavelengths in several quasi-stellar sources (**38, 39**).

Several authors have attempted to detect circular polarization in extragalactic radio sources, which would indicate large magnetic fields (**40**). Most of these attempts have been unsuccessful (see e.g. **41**), but variable circular polarization may exist in several compact sources (**42**).

6. *Spiral and weak elliptical galaxies*

Extensive studies of the continuum emission of many spiral galaxies have been made in many observatories and more are in progress (**43** to **49**). These observations show that the dimensions of the radio source are equal to, or smaller than, those of the optical galaxy, and that the emission is distributed in a disk and in a nucleus with dimensions sometimes as small as 30 parsecs. No definite evidence for radio haloes has been found. The spectra of these objects are non-thermal and seem to show a flattening at low frequencies like the spectrum of our galaxy (**49**); the spectrum of the central nucleus is not known. No clearly established relationship exists between radio and optical properties of spiral galaxies. An advance in this field has been the production of detailed supersynthesis maps for M 31 (**50**) and NGC 4631 (**51**); the former shows a correlation between radio emission and the spiral arms, and the latter galaxy has definitely no radio halo.

A radio survey of nearby elliptical galaxies at 11 cm (**52**) and work in progress at Cal Tech and NRAO have shown that half of the detected nearby ellipticals are point sources with flat or even inverted spectra, like NGC 1052 and 4278 (**53, 54**).

Hydrogen-line work on spiral galaxies has been partly summarized in ref. **55**: the 21-cm line has been observed in more than 77 objects. This has confirmed with a larger sample the relations already known between galaxy type and the (hydrogen mass)/(optical luminosity) and (hydrogen mass)/(total mass) ratios. Much work has been done or is in progress on the neutral hydrogen structure of nearby galaxies (**56** to **62**). It seems that only Sb galaxies have the ring structure in neutral hydrogen discovered in M 31; Ir galaxies have a central condensation and Sc galaxies are of both types. Asymmetries in the velocity field or hydrogen distributions are fairly common, and a bridge of neutral hydrogen has been studied between two interacting galaxies (**63, 64**).

7. *The nature of radio sources*

Although much effort has been devoted to theoretical studies of radio sources, the origin of the very large energies involved is still very uncertain and the evolution of sources is still poorly known. One has the general feeling that the physical processes taking place in quasars, radio galaxies and even the nuclei of spiral galaxies are basically the same (see e.g. **65**, **66**), and there may even be similarities between quasars and pulsars (**67**); ref. **68** and **69** are review papers on the problems of the nature of radio sources, mostly concerned with quasars.

A number of new models for the evolution of radio galaxies have been presented (see e.g; **70**, **71**, **72**); recent data on spectra of radio sources has put severe limits on their evolutional properties (**73**).

Much work has been done on physical mechanisms in compact radio sources (see e.g. **74**, **75**), including relativistic effects (**76**); as we have already said, their radio properties can often be accounted for by relatively simple models (**25**, **34**).

Not much new is known on the luminosity function of radio galaxies; for quasars, a luminosity function with time evolution has been presented (**77**).

8. *Cosmology*

No really decisive progress has been made during the 3 years by radio astronomers in the field of observational cosmology, except on the cosmic microwave background:

a. *Intergalactic matter* (see the review article ref. **78**).

More attempts have been made to detect intergalactic hydrogen, and only negative results have been obtained so far, even in the Virgo Cluster (**79**, **80**).

b. *Cosmic microwave radiation*

There are now ground-based observations of this radiation up to 3-mm wavelength (see e.g. **81**), and upper limits up to 0·36 mm are derived from observations of interstellar molecules (**82**). All agree on a blackbody radiation of about 2·7 K temperature. However, two direct rocket measurements around 1 mm give a much higher flux (**83**). The origin of the discrepancy is not clear. Careful measurements of the isotropy of this radiation have finally detected a slight anisotropy, which is interpreted as due to a relatively slow motion of the Galaxy with respect to the local Universe (**84**).

Several attempts have been made to explain the microwave background as due to unresolved discrete sources (**85**, **86**); however, there are difficulties in this interpretation (**87**).

c. *Radio-source counts*

The present status of radio-source counts at relatively low frequencies has been summarized in (**88**); the slope of the $\log N/\log S$ curve goes from about $-1{\cdot}85$ for bright sources to $-0{\cdot}8$ at the smallest observed flux densities (corresponding to 10^5 sources/steradian). At higher frequencies, there have been claims for a different $\log N/\log S$ curve (**89**); the discrepancy is discussed in (**90**).

A large number of papers have been published on the interpretation of source counts (see e.g. **91**). It seems that introduction of an evolution with cosmic time is necessary even in Lemaître-type cosmologies, but it is still extremely difficult to give more than qualitative statements because of the composite nature of the $\log N/\log S$ curve: radio galaxies, quasars and the many unidentified sources may behave differently (see e.g. the discussion in **92**), and other little-known types of sources may give an important contribution at certain frequencies and flux levels (weak radio elliptical galaxies at high frequencies, for example). There are however some hopes of using angular diameters, spectra or other parameters of radio sources to gain insight into their nature and of being able to derive separate $\log N/\log S$ relations for different groups of sources.

BIBLIOGRAPHY

1. Shimmins, A. J., Day, G. A. 1968, *Austr. J. Phys.*, **21**, 377.
2. Ekers, Jennifer A. 1969, *Austr. J. Phys., Suppl.*, **7**.
3. Branson, N. J. B. A. 1967, *Mon. Not. R. astr. Soc.*, **135**, 149.
4. Grueff, G., Vigotti, M. 1968, *Astrophys. Lett.*, **2**, 113.
5. Pooley, G. G., Kenderdine, S. 1968, *Mon. Not. R. astr. Soc.*, **139**, 529.
6. Pooley, G. G. 1969, *Mon. Not. R. astr. Soc.*, **144**, 101.
7. Höglund, B. 1967, *Astrophys. J., Suppl.*, **15**, 61.
8. Davis, M. M. 1967, *Bull. astr. Inst. Netherl.*, **19**, 201.
9. Galt, J. A., Kennedy, J. E. D. 1968, *Astr. J.*, **73**, 135.
10. Scheer, D. J., Kraus, J. D. 1967, *Astr. J.*, **72**, 536.
11. Dixon, R. S., Kraus, J. D. 1968, *Astr. J.*, **73**, 381.
12. Fitch, L. T., Dixon, R. S., Kraus, J. D. 1969, *Astr. J.*, **74**, 612.
13. Blum, E. J., Davis, M. M. 1968, *Astrophys. Lett.*, **2**, 41.
14. Kellermann, K. I., Pauliny-Toth, I. I. K., Davis, M. M. 1968, *Astrophys. Lett.*, **2**, 105.
15. Scheuer, P. A. G., Williams, P. J. S. 1968, *A. Rev. Astr. Astrophys.*, **6**, 321.
16. Kellermann, K. I., Pauliny-Toth, I. I. K. 1968, *A. Rev. Astr. Astrophys.*, **6**, 417.
17. Kellermann, K. I., Pauliny-Toth, I. I. K., Williams, P. J. S. 1969, *Astrophys. J.*, **157**, 1.
18. Andriewski, A. E., *et al.* 1969, *Astr. Cirk*, no. 494.
19. Doherty, L. H., MacLeod, J. M., Purton, C. R. 1969, *Astr. J.*, **74**, 827.
20. Shimmins, A. J., Manchester, R. M., Harris, Beverly, J. 1969, *Austr. J. Phys., Suppl.*, **8**.
21. Fanti, C., Fanti, R., Londrillo, P., Padrielli, L. 1969, *Astr. Astrophys.*, **2**, 477.
22. Bridle, A. H., Purton, C. R. 1968, *Astr. J.*, **73**, 717.
23. Kraus, J. D., Scheer, D. J., Dixon, R. S., Fitch, L. T., Andrew, B. H. 1968, *Astrophys. J.*, **152**, L35.
24. Bridle, A. H. 1969, *Nature*, **221**, 348.
25. Kellermann, K. I., Pauliny-Toth, I. I. K. 1969, *Astrophys. J.*, **155**, L71.
26. Locke, J. L., Andrew, B. A., Medd, W. J. 1969, *Astrophys. J.*, **157**, L81.
27. Cohen, M. H. 1969, *A. Rev. Astr. Astrophys.*, **7**, 619.
28. Fomalont, E. B. 1968, *Astrophys. J., Suppl.*, **15**, 203, 1969, *Astrophys. J.*, **157**, 1027.
29. Bash, F. N. 1968, *Astrophys. J., Suppl.*, **16**, 373.
30. Hogg, D. E. 1969, *Astrophys. J.*, **155**, 1099.
31. Ekers, R. D. 1969, *Austr. J. Phys., Suppl.*, **6**.
32. Palmer, H. P., *et al.* 1967, *Nature*, **213**, 789.
33. Macdonald, G. H., Kenderdine, S., Neville, Ann C. 1968, *Mon. Not. R. astr. Soc.*, **138**, 259.
34. Kellermann, K. I., *et al.* 1968, *Astrophys. J.*, **153**, L209.
35. Gardner, F. F., Morris, D., Whiteoak, J. D. 1969, *Austr. J. Phys.*, **22**, 79.
36. Seielstad, G. A., Weiler, K. W. 1969, *Astrophys. J., Suppl.*, **18**, 85.
37. Gardner, F. F., Morris, D., Whiteoak, J. B. 1969, *Austr. J. Phys.*, **22**, 107.
38. Aller, H. D., Haddock, F. T. 1967, *Astrophys. J.*, **147**, 833.
39. Hobbs, R. W., Hollinger, J. P., Marandins, G. E. 1968, *Astrophys. J.*, **154**, L49.
40. Legg, M. P. C., Westfold, K. C. 1968, *Astrophys. J.*, **154**, 499.
41. Seielstad, G. A. 1969, *Astr. Astrophys.*, **2**, 372.
42. Biraud, F. 1969, *Astr. Astrophys.*, **1**, 156.
43. Braccesi, A., Fanti-Giovannini, C., Fanti, R., Formiggini, L., Vespagnani, G. 1967, *Nuovo Cim.*, **B52**, 254.
44. De Jong, M. L. 1967, *Astrophys. J.*, **150**, 1.
45. De la Beaujardière, O., Kazès, I., Le Squeren, A. M., Nguyen Quang Rieu 1968, *Ann. Astrophys.*, **31**, 387.
46. Cameron, M. J., Glanfield, J. R. 1968, *Mon. Not. R. astr. Soc.*, **141**, 57.
47. Tovmassian, H. M. 1968, *Austr. J. Phys.*, **21**, 193.
48. Wade, C. M. 1968, *Astr. J.*, **73**, 876.
49. Lang, K. R., Terzian, Y. 1969, *Astrophys. Lett.*, **3**, 29.
50. Pooley, G. G. 1969, *Mon. Not. R. astr. Soc.*, **144**, 101.
51. Pooley, G. G. 1969, *Mon. Not. R. astr. Soc.*, **144**, 143.
52. Rogstad, D. H., Ekers, R. D. 1969, *Astrophys. J.*, **157**, 481.
53. Heeschen, D. S. 1968, *Astrophys. J.*, **151**, L135.
54. Wills, D. 1968, *Astrophys. Lett.*, **2**, 187.

55. Roberts, M. 1969, *Astr. J.*, **74**, 859.
56. McGee, R. X., Milton, Janice A. 1966, *Austr. J. Phys.*, **19**, 343.
57. Hindman, J. V. 1967, *Austr. J. Phys.*, **20**, 147.
58. Rogstad, D. H., Rougoor, G. W., Whiteoak, J. B. 1967, *Astrophys. J.*, **150**, 9.
59. Guelin, M., Weliachew, L. 1969, *Astr. Astrophys.*, **1**, 10.
60. Shobbrook, R. R., Robinson, B. J. 1967, *Austr. J. Phys.*, **20**, 131.
61. Gordon, K. J., Remage, Nancy H., Roberts, M. S. 1968, *Astrophys. J.*, **154**, 845.
62. Lewis, B. M. 1968, *Proc. astr. Soc. Austr.*, **1**, 104.
63. Roberts, M. S. 1968, *Astrophys. J.*, **151**, 117.
64. Weliachew, L. 1969, *Astr. Astrophys.*, **3**, 402.
65. Burbidge, G. R., Hoyle, F. 1968, *Astr. J.*, **73**, 907.
66. Lynden-Bell, D. 1969, *Nature*, **223**, 690.
67. Morrison, P. 1969, *Astrophys. J.*, **157**, L73.
68. Schmidt, M. 1969, *A. Rev. Astr. Astrophys.*, **7**, 527, see also *Stars and Stellar Systems*, Vol. 9, to be published.
69. Burbidge, G. R., Burbidge, E. M. 1969, *Nature*, **224**, 21.
70. Burbidge, G. R. 1967, *Nature*, **216**, 1287.
71. Ryle, M., Longair, M. S. 1967, *Mon. Not. R. astr. Soc.*, **136**, 123.
72. Rees, M. J., Setti, G. 1968, *Nature*, **219**, 127.
73. Van der Laan, H., Perola, G. G. 1969, *Astr. Astrophys.*, **3**.
74. Rees, M. J. 1967, *Mon. Not. R. astr. Soc.*, **135**, 345; **136**, 279; **137**, 429.
75. Rees, M. J. 1968, *Astrophys. Lett.*, **2**, 1.
76. Ozernoy, L. M., Sazonov, V. N. 1969, *Astrophys. Space Sci.*, **3**, 395.
77. Schmidt, M. 1968, *Astrophys. J.*, **151**, 393.
78. Gould, R. J. 1968, *A. Rev. Astr. Astrophys.*, **6**, 195.
79. Penzias, A. A., Wilson, R. W. 1969, *Astrophys. J.*, **156**, 799.
80. Allen, R. J. 1969, *Astr. Astrophys.*, **3**, 382.
81. Boyton, P. E., Stokes, R. A., Wilkinson, D. T. 1968, *Phys. Rev. Lett.*, **21**, 462.
82. Bortolot, V. J., Jr., Clauser, J. F., Thaddeus, P. 1969, *Phys. Rev. Lett.*, **22**, 307 and 806.
83. Houck, J. R., Harwit, M. 1969, *Astrophys. J.*, **157**, L45.
84. Conklin, E. K. 1969, *Nature*, **222**, 971.
85. Wolfe, A. M., Burbidge, G. R. 1969, *Astrophys. J.*, **156**, 345.
86. Pariiskii, Yu. N. 1968, *Soviet Astr.*, **12**, 219.
87. Penzias, A. A., Schraml, J., Wilson, R. W. 1969, *Astrophys. J.*, **157**, L49.
88. Ryle, M. 1968, *A. Rev. Astr. Astrophys.*, **6**, 249.
89. Shimmins, A. J., Bolton, J. G., Wall, J. V. 1968, *Nature*, **217**, 818.
90. Van der Laan, H. 1969, *Bull. astr. Inst. Netherl.*, **20**, 171.
91. Petrosian, V. 1969, *Astrophys. J.*, **155**, 1029.
92. Bolton, J. G. 1969, *Astr. J.*, **74**, 131.

G. INSTRUMENTS

(prepared by J. W. Findlay)

1. *Radio Telescopes*

a. *Filled aperture*

There have been relatively few new filled-aperture instruments to come into use during the last three years. The first new 130-ft telescope for the Owens Valley array has started observing (**1**); a new 22-m telescope good for mm waves has been built in the Crimea (**2, 3**); the 210-ft telescope at the Goldstone station of the Jet Propulsion Laboratory (**4**) has also been used for radio astronomy, and the 36-ft mm-wave telescope of the U.S. National Radio Astronomy Observatory is operating on Kitt Peak, Arizona (**5**). A 25-m dish has been completed at Chilbolton, England (**6**). The parabolic cylinder antenna in south-central India is nearing completion (**7, 8**), and the largest fully-steerable dish in the world, the 100-m Bonn telescope (**9**) should start observing by mid-1970. More information about existing large telescopes has been published (**10, 11, 12, 13, 14, 15, 16, 17**). New design ideas for minimizing the effects of structural deformations have been developed (**18, 19, 20**). There has been a growth of interest in developing means of feeding spherical reflectors (**21, 22, 23, 24, 25**) and improvements in aperture efficiency and antenna temperature have also been achieved for paraboloids (**25, 26, 27, 28**) and for the Nançay telescope (**29**).

b. *Unfilled apertures*

The variety of unfilled aperture telescopes continues to grow. For solar work, a notable event was the start of the Culgoora radioheliograph in 1967 (see reference **30** and successive papers). Other new solar instruments were built in India (**31**) and Canada (**32**). The solar array at Fleurs, near Sydney, has been converted to a synthetic array (**33**). The two-element interferometer in Haute-Provence has been described (**34**). Several groups of experimenters have used the techniques of very-long baseline interferometry first described in (**35**) and (**36**). A new synthetic antenna is almost ready to observe at Westerbork, in The Netherlands, and work has started on the Cambridge, England three-mile telescope (**37**). The one-mile Mills cross (**38**) has, among other observations, distignuished itself as a pulsar discoverer. A low-frequency array has been operating in Canada (**39**). Designs for very large arrays for the Owens Valley Observatory (**40**) and the National Radio Astronomy Observatory (**41**) are in an advanced stage.

2. *Radiometers*

There has been considerable progress in the development of low-noise stable radiometer systems. Two examples of low total system-noise may be quoted: the maser system on the Onsala, Sweden 84-ft telescope (**42**) and that used at the Jet Propulsion Laboratory (**27**). An experimental travelling-wave maser has been built at 36 GHz (**43**), and recombination line observations have been made using masers at this frequency in the U.S.S.R. Noise in travelling wave masers has been studied (**44**) and good phase stability achieved (**45**).

Many observatories now use parametric amplifiers either cooled or at room temperature. The development of closed-cycle helium cooling systems (**27**) has enabled continuous operation of masers and parametric amplifiers at temperatures down to that of liquid helium (**46**); many descriptions have been given of various types of parametric amplifiers with low noise and other desirable characteristics (**46, 47, 48, 49, 50**).

The use of low-noise transistor amplifiers has remained valuable at the lower frequencies (**51, 52, 53, 54**) as the noise characteristics of the transistor have been improved.

Methods of stabilizing radiometers have not changed very markedly. Special radiometers for interferometry (**55**), for extra stability (**56**) or using correlation (**57**) have been built.

Although considerable progress has been made in many observatories in data processing at the radiometer output, not very many details have been published. Spectral line receivers using filter techniques (**58**) and auto-correlation techniques (**59, 60**) are now widely used.

3. *Noise Sources*

The development of well-engineered sources of noise at microwave frequencies suitable for use as noise standards has proceeded in a very satisfactory way (**61, 62, 63, 64, 65**).

BIBLIOGRAPHY

1. Rule, B., Gayer, G. F. 1968, A New Radio Telescope for the Owens Valley Observatory, *Westinghouse Engr.*, **28** (5), 162–167 (Nov.).

2. Laffineur, M., Koutchmy, S. 1969, Le radiotélescope sur ondes millimétriques de l'Observatoire de Simés (U.R.S.S.), *Onde élect.* **49** (503), 246–249.

3. Ivanov, V. N., Moiseev, I. G., Monin, Yu. G. 1968, The New Radio Telescope of the Crimean Astrophysical Observatory with the 22-m Parabolic Reflector, (in Russian with summary in English) *Izv. Krym. Astrofiz. Obs.* **38**, 141–148.

4. Rocci, S. A. 1966, The 210-ft Parabolic Fully Steerable Tracking Antenna for a Deep Space Instrumentation Facility (JPL), pp. 50–70 in Deep Space and Missile Tracking Antennas: An Aviation and Space Symposium held in conjunction with the ASME winter annual meeting Nov. 27–Dec. 1, 1966, New York, N.Y. pp. 100, New York, ASME, 1966.

5. Heeschen, D. S. 1967, Annual Report of the National Radio Astronomy Observatory, *Astr. J.*, **72**, p. 1121.

6. Samuel, I. J. 1967, The Chilbolton Steerable Aerial, *AEI Engng.* **7** (1), 6–15.

7. Swarup, G. 1968, A Large Cylindrical Radio Telescope at Ootacamund for Radio Astronomy Observations, pp. I–III in Suppl. 1 to *Proceedings of the Symposium on Antennas, Radio and Telecommunication Research Committee, C.S.I.R., New Delhi held at Bangalore (Aug. 1968)*.

8. Kapahi, V. K., Damle, S. H., Balasubramanian, V. 1968, A Steerable Array of 968 Dipoles for Illuminating a Parabolic Cylindrical Antenna, pp. IX–XII in Suppl. 3 to *Proceedings of the Symposium on Antennas, Radio and Telecommunication Research Committee, C.S.I.R., New Delhi held at Bangalore (Aug. 1968)*.

9. Hachenberg, O. 1968, Studien zur Konstruktion des 100 M Teleskops, *Beitr. Radioastr.*, **1** (2), 31–61.

10. Meeks, M. L., Ball, J. A., Hull, A. B. 1968, The Pointing Calibration of the Haystack Antenna, *IEEE Trans. Ant. Prop.*, **16** (6), 746–751.

11. Fanning, W. R. 1966, The Haystack Antenna-Mechanical Concept, Design and Testing, pp. 28–49 in *Deep Space and Missile Tracking Antennas*: an Aviation and Space Symposium held in conjunction with the ASME winter annual meeting Nov. 27–Dec. 1, 1966, New York, N.Y. pp. 100, New York, ASME.

12. Heart, F. E., Mathiasen, A. A. 1966, Computer Control of the Haystack Antenna, *P.IEEE*, **54** (12), 1742–1751.

13. Puttock, M. J., Minnett, H. C. 1966, Instrument for Rapid Measurement of Surface Deformations of a 210-ft Radio Telescope, *Proc. Instn. elect. Eng.*, **113** (11), 1723–1730.

14. Higgs, L. A. 1967, Antenna Characteristics at 820 MHz of the Dwingeloo Radio Telescope, *Bull. astr. Inst. Netherl., Suppl.*, **2** (2), 59–76.

15. Palmer, H. P., Rowson, B. 1968, The Jodrell Bank Mark III Radio Telescope, *Nature*, **217** (5123), 21–22.

16. Ginat, M. 1967, Le grand radiotélescope à deux miroirs de Nançay: étude de la structure, *J. Observateurs*, **50** (2), 133–178.

17. Yang, K. S., Swenson, G. W., Jr. 1967, The Illinois 400-foot Radio Telescope; Performance and Electronic Equipment, *Radio Sci.*, **2** (2), 147–159.

18. von Hoerner, S. 1967, Homologous Deformations of Tiltable Telescopes, *J. Struct. Div. Proc. Amer. Soc. Civil Engrs.* **93** (ST5), 461–485.

19. von Hoerner, S. 1967, Design of Large Steerable Antennas, *Astr. J.*, **72** (1), 35–47.

20. Burr, D. W. 1966, Aerial Profile Automatically Controlled, *Wireless World*, **72** (7), 346.

21. Hyde, G. 1968, Studies of the Focal Region of a Spherical Reflector: Stationary Phase Evaluation, *IEEE Trans. Ant. Prop.*, **16** (6), 646–656.

22. Cohen, M. H., Perona, G. E. 1967, A Correcting Feed at 611 MHz for the AIO Reflector, *IEEE Trans. Ant. Prop.*, **AP-15** (3), 482–483.

23. Burrows, M. L., Ricardi, L. J. 1967, Aperture Feed for a Spherical Reflector, *IEEE Trans. Ant. Prop.*, **AP-15** (2), 227–230.

24. LaLonde, L. M., Harris, D. E. 1969, A High Performance Line Source Feed for the AIO Spherical Reflector, Cornell University Center for Radiophysics and Space Research Report CRSR 329 (Feb.).

25. Wielebinski, R. 1967, Spherical Reflecting Antenna as an Alternative to the Paraboloidal Dish, *Austr. elect. Eng.*, **1** (1), 4–7.

26. Herbison-Evans, D. 1968, Optimum Paraboloid Aerial and Feed Design, *Proc. Instn. elect. eng.*, **115** (1), 87–90.

27. Levy, G. S., Bathker, D. A., Higa, W., Stelzried, C. T. 1968, The Ultra Cone: An Ultra-Low-Noise Space Communication Ground Radio-Frequency System, *IEEE Trans. Microw. Theory Tech.*, **MTT-16** (9), 596–602.

28. Minnett, H. C., Thomas, B. M. 1968, Fields in the Image Space of Symmetrical Focusing Reflectors, *Proc. Instn. elect. eng.*, **115** (10), 1419–1430.

29. Biraud, F. 1967, Sources primaires pour le grand radiotélescope de Nançay, *Onde élect.*, **47** (479), 302–305.

30. Wild, J. P. 1967, The Culgoora Radioheliograph, *Proc. Instn. Radio elect. eng. Austr.*, **28** (9), 277, 278–291.

31. Kapahi, V. K., Isloor, J. D. 1968, A Radio Interferometer Consisting of 32 6-ft Dishes for Solar Radio Astronomy, pp. V–VII in Suppl. 2 to: *Proceedings of the Symposium on Antennas, Radio and Telecommunication Research Committee, C.S.I.R., New Delhi held at Bangalore (August 1968).*

32. Covington, A. E., Legg, T. H., Bell, M. B. 1967, A High-Resolution 2800 MHz Multi-Element Interferometer, *Solar Phys.*, **1** (3–4), 465–473.

33. Gore, R., Scholes, W. 1968, Australians Ready a Mighty New Radiotelescope, *Sci. Res.*, **3** (23), 50–51, 54, 57.

34. Laffineur, M., Coupiac, P. 1967, Un radio interféromètre méridien à liaisons radioélectriques: techniques et observations, *Bull. Soc. r. Sci. Liège*, **36** (7–8), 393–417. – *Contr. Inst. Astrophys. Paris Ser. A*, no. 364.

35. Broten, N. W., Legg, T. H., Locke, J. L., McLeish, C. W., Richards, R. S., Chisholm, R. M., Gush, H. P., Yen, J. L., Galt, J. A. 1967, Long Baseline Interferometry: A New Technique, *Science*, **156** (3782), 1592-1593.

36. Bare, C., Clark, B. G., Kellermann, K. I., Cohen, M. H., Jauncey, D. L. 1967, Interferometer Experiment with Independent Local Oscillators, *Science*, **157** (3785), 189–191.

37. Another Telescope for Cambridge, *Nature*, **221** (5180), 515.

38. Mills, B. Y. 1968, The One-Mile Cross, *Austr. Physicist*, **5** (4), 57.

39. Costain, C. H., Lacey, J. D. 1969, Large 22 MHz Array for Radio Astronomy, *IEEE Trans. Ant. Prop.*, **AP-17** (2), 162–169.

40. Construction of a Multi-channel Interferometer at the Owens Valley Radio Observatory. Proposal submitted to the National Science Foundation, Washington, D.C.

41. Heeschen, D. S. 1967, Radio Astronomy: A Large Antenna Array, *Science*, **158**, 75–78.

42. Rydbeck, O. E. H., Kollberg, E. 1968, Rutile Traveling-Wave Maser System for the Onsala 84-foot Radio Telescope, *IEEE Trans. Microw. Theory Techn.*, **MTT-16** (9), 799–803.

43. Swanson, P. N. 1969, A 36 GHz Traveling Wave Maser for use in Radio Astronomy, Penn. State Univ. Coll. Sci. Dept. Astron. Radio Astron. Obs. Sci. Rep. No. 017, pp. 89 (Jan. 27).

44. De Gruyl, J. A., Okwit, S., Smith, J. G. 1968, Noise Considerations in Broad-Band Traveling-Wave Masers, *IEEE Trans. Microw. Theory Tech.*, **MIT-16** (9), 586–595.

45. Smith, F. W., Booth, P. L., Hentley, E. L. 1966, Masers for a Radio Astronomy Interferometer, *Philips Tech. Rev.*, **27** (12), 313–321.

46. Josenhans, J. G. 1968, A Liquid Helium Cooled 4-GHz Broad-Band Parametric Amplifier, *IEEE Trans. Microw. Theory Techn.*, **MTT-16** (9), 791–793.

47. Aitchison, C. S., Davies, R., Gibson, P. J. 1967, A Simple Diode Parametric Amplifier Design for Use at S, C, and X Band, *IEEE Trans. Microw. Theory Techn.*, **MTT-15** (1), 22–31.

48. Delannoy, J., Ribes, J. C. 1967, Amplificateurs paramétriques à grande stabilité de gain pour la radioastronomie, *Onde élect.*, **47** (479), 192–202.

49. Kögler, W. 1966, Parametrischer Gleichlagemischer für Radioastronomische Messungen, *Telefunken ZTG*, **39** (2), 242–244, 282.

50. Etkin, V. S. 1969, Development of the Theory of Technique of Parametric Amplifier Systems with Concentrated Reactive Non-Linear Elements in the U.S.S.R. (in Russian) *Radiotekh. Elektron*, **14** (3), 486–509.

51. Cuccia, C. L. 1968, Low Noise Microwave Transistors, *Microwave J.*, **11** (2), 77–90.

52. Baechtold, W., Strutt, M. J. O. 1968, Noise in Microwave Transistors, *IEEE Trans. Microw. Theory Techn.*, **MTT-16** (9), 578–585.

53. Batchelor, R. A. 1969, Transistor R.F. and I.F. Amplifiers for Radio Astronomy Applications, *Proc. Inst. Radio elect. Eng. Austr.*, **30** (4), 99–105.

54. Michael, R., Rosatelli, C., Tomassetti, G. 1967, Low Noise Transistorized Amplifier at 408 MHz for Synthesis Radiotelescope, *Alta Frequenza*, **36** (5), 452–455 (122E-125E).

55. Wesseling, K. H. 1967, A Single-Sideband-Double-Sideband Interferometer Receiver for Radio Astronomy, *IEEE Trans. Ant. Prop.*, **AP-15** (2), 332–333.

56. Hach, J. P. 1968, A Very Sensitive Airborne Microwave Radiometer Using Two Reference Temperatures, *IEEE Trans. Microw. Theory Techn.*, **MTT-16** (9), 629–636.

57. Hubbard, J. R., Erickson, W. C. 1967, A Stabilized Cross-Correlation Radiometer, *IEEE Trans. Ant. Prop.*, **AP-15** (2), 291–294.

58. Batchelor, R. A., Brooks, J. W., Sinclair, M. W. 1969, A Multi-Channel Spectral Line Receiver for Radio Astronomy, *Proc. Instn. Radio elect. Eng. Austr.* **30** (2), 39–46.

59. Shalloway, A. M. 1964, A 100-channel Digital Autocorrelation Receiver for Observation of Spectral Lines, Northeast Electronics Research and Engineering Meeting (NEREM) Record, p. 98–99.

60. Davies, R. D., Ponsonby, J. E. B., Pointon, L., de Jager, G. 1969, The Jodrell Bank Radio Frequency Digital Autocorrelation Spectrometer, *Nature*, **222** (5197), 933–937.

61. Stelzried, C. T. 1968, Microwave Thermal Noise Standards, *IEEE Trans. Microw. Theory Techn.*, **MIT-16** (9), 646–655.

62. Denson, C. I., Halford, G. J. 1968, Plasma Noise Sources of Improved Accuracy, *IEEE Trans. Microw. Theory Techn.* **MIT-16** (9), 655–663.

63. Somlo, P. I., Hollway, D. L. 1968, The Australian National Standards Laboratory X-Band Radiometer for the Calibration of Noise Sources, *IEEE Trans. Microw. Theory Techn.*, **MIT-16** (9), 664–669.

64. Trembath, C. L., Wait, D. F., Engen, G. F., Fotte, W. J. 1968, A Low-Temperature Microwave Noise Standard, *IEEE Trans. Microw. Theory Techn.*, **MIT-16** (9), 709–714.

65. Engen, G. F. 1968, A Method of Calibrating Coaxial Noise Sources in Terms of a Waveguide Standard, *IEEE Trans. Microw. Theory Techn.*, **MIT-16** (9), 636–639.

41. HISTORY OF ASTRONOMY
(HISTOIRE DE L'ASTRONOMIE)

PRESIDENT: E. Rybka.
VICE-PRESIDENT: O. J. Gingerich.
ORGANIZING COMMITTEE: J. O. Fleckenstein, C. D. Hellman, Z. Horský, P. G. Kulikovsky, W. Petri, V. L. Tchenakal.

I. INTRODUCTION

The investigations on the history of astronomy proceed rapidly and concern all branches, viz. the history of astronomy in particular countries, the historical development of the astronomical problems, biographies of astronomers, history of observatories and of astronomical instruments etc. They are being made by professional astronomers and historians of science at observatories and institutions destined for the history of science. The president of Commission 41 cannot give a comprehensive review of all directions of investigations on the history of astronomy, because their results frequently have been published in publications not accessible to the author of this report. The present draft report is based mainly on information received by the president from members and consulting members. Besides, the present report contains short information on the results of international conferences, which were held independently of IAU.

II. MEMBERSHIP

A characteristic feature of Commission 41 is that many eminent scientists, who are not astronomers but are working actively on particular problems of the history of astronomy, are consulting members of Commission 41. At present the Commission is composed of 44 members and 34 consulting members. We suffered a loss by the death on 30 Sept. 1967 of Prof. A. Birkenmajer (Poland) an eminent scientist in Copernican problems and medieval history of exact sciences. Two scientists have been coopted as consulting members: Dr. J. K. Ravetz (U.K.) and Dr. E. Rosen (U.S.A.).

III. GENERAL HISTORY OF ASTRONOMY

At the Thirteenth General Assembly of IAU in Prague (1967) a resolution was adopted which recognized the importance and usefulness of preparing an international history of astronomy, based on original materials. The president reported the plan of the General History of Astronomy, as printed in the Information Circulars of Commission 41, Nos. 10 and 14, at the Twelfth Congress of the History of Sciences in Paris in August 1968. Further remarks to the plan have been received by the president from members and consulting members. Detailed remarks to the plan of all three volumes have been sent by N. I. Nevskaja from Leningrad. Preliminary negotations have been held with the aim to find a publisher and to organize the editorial board. The whole problem will be discussed during the General Assembly of IAU in Brighton.

IV. CONGRESSES AND SYMPOSIA

(*a*) Within a month after the Thirteenth General Assembly of IAU an international symposium "La révolution scientifique du 17e siècle et les sciences mathématiques et physiques" was held in Prague (25–29 September 1967). Four general reports were delivered by: (1) J. R. Ravetz – The evolution of science and its history, (2) I. B. Cohen – Dynamics: the key to the new science in the 17th century, (3) P. Costabel – Matière et lumière au 17e siècle, (4) A. P. Youschkevitch – Sur la révolution en mathématiques des temps modernes.

Some topics related to history of astronomy were touched in discussion. More than 10 members of Commission 41 attended the symposium.

(*b*) The Twelfth International Congress of History of Sciences was held in Paris, 25–31 August 1968. There was no special section of the history of astronomy, but reports concerning this field of science were delivered in five different sections. In all 22 contributions from different branches of the history of astronomy were delivered at the meetings of the Congress. Their titles were given in the Information Circular No. 16. Ten contributions were delivered by members of Commission 41. At the Paris Congress a Commission for the celebrations of the 400th anniversary of the birth of J. Kepler (in 1971) was appointed.

(*c*) The symposium entitled: "Oriente e Occidente – nel Medioevo: Filosofia e Scienze" (XIII Convegno Volta) was organized by the "Accademia dei Lincei", 9–15 April 1969 in Rome and Florence. 32 invited lectures were held at the "Convegno", six of them concerned the history of islamic medieval astronomy. Their lecturers were: E. Rybka, E. S. Kennedy, W. Hartner, W. Petri, M. Cimino and E. Millas Vendrell. The titles of the lecturers have been published in the Information Circular No. 17.

V. BIBLIOGRAPHY

As in previous years, P. G. Kulikovsky (Sternberg Institute, Moscow), former President of Commission 41, with the assistance of N. B. Lavrova, continued the preparation and publication of yearly bibliographies of books, papers and articles on the history of astronomy. Three issues of the bibliography were published in the years 1967–69, containing bibliographical notes on publications issued in 1966, 1967 and 1968. Almost 1000 titles are contained in the three issues 1966–68. Together 10 issues of the bibliography on the history of astronomy were prepared and published in Moscow by P. G. Kulikovsky.

VI. INFORMATION CIRCULAR

The president continued the publication of the Information Circular which was printed in Cracow under the auspices of the Committee of the History of Sciences and Techniques of the Polish Academy of Sciences. The Circular was distributed to all members and consulting members of Commission 41. It contained various information concerning congresses, symposia, anniversaries, personal notes and others. Four circulars were published since the Thirteenth General Assembly of IAU. The fifth one is planned to appear in the spring 1970.

VII. KEPLER AND COPERNICUS ANNIVERSARIES

(*a*) The Kepler Committee, appointed at the Paris Congress for History of Sciences met for the first time at Munich (DBR) 1–2 July 1969 under the presidency of J. O. Fleckenstein. A. A. Mikhailov (U.S.S.R.) has been elected as honorary President of the Committee. J. O. Fleckenstein (Switzerland) acts as President with the assistance of E. Rosen (U.S.A.) – the vice-president – and Z. Horský (Czechoslovakia) – the secretary. There are planned symposia in 1971 at Tübingen (D.B.R.), Linz (Austria) and Leningrad-Pulkovo (U.S.S.R.). Besides a scientific exhibition is to be organized in Prague. The participation of Commission 41 has been provided mainly in the organization of the Symposium in Leningrad.

(*b*) The Copernican Committee, which was appointed at the Eleventh Congress for the History of Sciences in 1965, met twice: in Warsaw (17–18 April 1968) and in Paris (26 August 1968). At the Warsaw meeting the plan of collective editions designed in Poland for the 500th Anniversary of the birth of Copernicus was adopted and approved. It is planned to publish *Opera Omnia* of Copernicus, the facsimile of the manuscript of *De Revolutionibus* and other papers relating to the Copernicus anniversary. After the death of Prof. A. Birkenmajer, Chairman of the Committee, the Committee was presided provisionally by Prof. J. Bukowski, Chairman of the Committee for the History of Sciences and Techniques of the Polish Academy of Sciences. At the Paris meeting of the Copernicus Committee J. Bukowski has been appointed Chairman of the Committee in question.

VIII. REPORTS FROM MEMBERS

Though the president requested twice all members and consulting members (Information Circular, Nos. 16, 17) to convey him information on their work in the years 1967–69, less than 20 answers have been received. The report which follows, represents therefore only partly the progress and trends of investigations, which are carried out by the members and consulting members of Commission 41. The information received by the president has been supplemented from papers being at his disposal. The summary of individual reports has been arranged according to countries in alphabetical order. The numbers in brackets refer to publications cited in the bibliography.

1. *Czechoslovakia*

Zd. Horský (Prague) was interested in general problems, which are on the border line of philosophy of sciences, as cosmology of Marsilio Ficino (**1**), aspects of changes in the scientific life (**2**), part of the platonism in the origin of modern cosmology (**3**) and others. He edited the facsimile of *Dialexis de novae et prius incognitae stellae* by Thaddaeus Hagecius ab Hayck (**4**). Together with O. Škopova he elaborated the catalogue of instruments of XV–XIXth centuries preserved in the National Technical Museum in Prague (see below Section XI.4).

2. *Denmark*

A. V. Nielsen (Aarhus) described the sundial from Vestervig, dating from about 1200 A.D.

3. *German Federal Republic*

H. C. Freiesleben (Stuttgart) wrote two booklets, each comprising about 80 pages, concerning Galilei and Kepler as scientists. He compiled some details concerning Max Wolf (1863–1932), not contained in the biography published by him in 1962. He is writing eight biographies to the *Dictionary of Scientific Biography* (Gillispie, Scribner's Son, U.S.A.). H. Waltenberg and W. Gleissberg (Frankfurt am Main) have been engaged in the problem of astronomical orientation of antique temples. They suppose that one of the conditions of the orientation of the archaic and hellenistic temples of Didyma (Asia Minor) was to render a possibility of observing Castor and Pollux at the moments when the azimuths of both stars were equal (**5**).

4. *Greece*

C. Antonacopoulos and C. Fragakis (Athens) have studied the Andronicos Horologium at the Northern foot of Acropolis in Athens. The study concerned the eight sundials which exist on the corresponding sides of the octagonal tower. The sides might be used as solar calendars (**6**).

C. S. Chassapis (Pentele) has studied the Greek astronomy of the second millennium according to the Orphic Hymns, which according to the author were formulated between XIXth and XIVth centuries B.C. The author alleges that the Greeks had conceived the heliocentric idea in the orphic times and that they were aware of the sphericity of the earth and of its rotation. D. Kotsakis (Athens) studied two sundials made in the first century B.C. by Andronicos from Kyrrhos in Macedonia, one situated on the island Tenos, the second one in Athens. S. Plakidis (Athens) published a historical account on the foundation of the National Observatory of Athens and its activity 1842–1965 (**7**).

5. *Italy*

M. Cimino, C. Chełkowska and M. A. Giannuzzi (Rome) described the exhibition of Copernican mementoes which belong to the Museum of the Rome Astronomical Observatory, as sculptures, paintings, medals, books. The exhibition was opened December 1966–January 1967.

A. Fresa (Naples) has analyzed the rules which were applied in antiquity at the astronomical orientation of temples in Egypt, Mesopotamia, Greece and Great Greece. He has aimed to explain the orientation of cardinal points in Homer's works (**8**). V. Ronchi (Florence-Arcetri) published an article on the life and versatile activity of Ruggero Boscovich (1711–87) (**9**).

6. *Poland*

B. Biliński (Rome) wrote a study on the connections of Galilei with Polish circles against a background of scientific and cultural relations between Poland and Italy in XVIth–XVIIth centuries (**10**).

J. Dobrzycki (Warsaw) and L. Hajdukiewicz (Cracow) published in *Polski Słownik Biograficzny* (Polish Biographic Dictionary) the biography of Copernicus (1968). Besides, J. Dobrzycki is preparing commentaries to a new edition of *De Revolutionibus*, which is planned for the 500-year anniversary of Copernicus. T. Przypkowski (Jędrzejów) described the Przypkowski's Museum in Jędrzejów and its rich collection of sundials (**11**).

E. Rybka (Cracow) has composed the plan of the General History of Astronomy and published it in the Information Circular of Commission 41 and in (**12**). He wrote popular articles on Copernicus and biographies of Polish astronomers (Krasowski, Kubikowski) to the Polish Biographic Dictionary. He is writing biographies of Polish astronomers for the *Dictionary of Scientific Biography* (Gillispie, Scribner's Son, U.S.A.).

J. Witkowski (Poznań) wrote biographies of T. Banachiewicz: (1) to the *Dictionary of Scientific Biography*, and (2) for the Association of Polish Geodesists. He wrote a history of astronomy in the Poznań district.

7. *Sweden*

P. Collinder (Uppsala) published a paper on the accuracy of astronomical observations in antiquity (**13**), and wrote a short history of astronomy in Sweden from the XVth to the XIXth century, which is to be published in 1970.

8. *Switzerland*

J. O. Fleckenstein (Basel) has edited Vol. I of the Collected Works of Jacob Bernoulli (Astronomy and Physics) and together with the late L. Courvoisier, Vol. 23 (ser. II) of Collected Works of L. Euler (First Lunar Theory). He wrote a short history of astronomy in Basel (1460–1960) based on Basel publications and manuscripts and prepared the historical exhibition on the occasion of IAU Symposium No. 38 on The Spiral Structure of Our Galaxy, which has been held in Basel, 29 August – 4 September 1969.

B. L. van der Waerden published a book on the beginnings of astronomy (see below Section XI.10)

9. *United Kingdom*

E. J. Aiton (Manchester) published a paper on Kepler's second Law of planetary motion (**14**).

E. G. Forbes (Edinburgh) was engaged mainly in the history of the scientific activity of the 18th-century German astronomer, Tobias Mayer (1723–1762). He published several articles on this subject. He translated: (1) from Latin into English the work of Georg Christoph Lichtenberg *Opera Inedita Tobiae Mayeri I* (Göttingen, 1775), to which he added an introduction and over 300 annotations; (2) from German into English 31 letters between Leonhard Euler and Tobias Mayer whose central theme was the lunar theory (to these letters E. Forbes wrote the introduction and over 200 annotations). He wrote a comprehensive article on the bicentenary of the *Nautical Almanac* (**15**).

At present he is engaged in editing two volumes of Tobias Mayer's unpublished papers, one of which contains his astronomical treatises, lecture notes etc.

D. J. Schove (Beckenham, Kent) has been engaged in problems of Africanistics and in the so-called "Spectrum of Time" (Meteorological and Astronomical observations in History). He published together with P. Y. Ho the Chinese records of sunspots and aurorae in the 4th century A.D. (**16**).

10. *U.S.A.*

A. Aaboe (New Haven) has been engaged in a detailed research work on cuneiform texts preserved in the British Museum. He published lunar auxiliary tables and related texts from the late Babylonian period (**17**).

S. L. Chapin (Los Angeles) published a study on the astronomical activity of the Paris Academy of Sciences during the XVIIIth century up to the eve of the Revolution (**18**).

D. S. Evans, J. Deeming, B. H. Evans and S. Goldfarb (Austin) have published the book *Herschel at the Cape* containing the diary and correspondence of John Herschel (1834–1838) (see below Section XI.2).

G. Hawkins (Cambridge) has undertaken a study of some peculiar straight lines in the desert near Nasca, Peru, and has shown that these lines do not have any obvious astronomical orientation.

C. D. Hellman (New York) has been engaged in a research work on Tycho Brahe and the Nova 1572 and she is writing a book on this subject. She wrote biographies of Tycho Brahe, G. S. Dörfel and G. Peurbach for the *Dictionary of Scientific Biography* and numerous reviews in various journals. She published a reprint with a supplement of the book *The comet of 1577*, originally printed in 1944.

O. Gingerich (Cambridge, Mass.) has continued his computer investigation of old planetary tables and ephemerides and he has calculated in cooperation with E. Poulle an extensive daily ephemeris from the Alphonsine tables. His English translation of Kepler's *Astronomia Nova* is virtually completed in an initial draft.

O. Neugebauer and R. A. Parker (Providence) published the third volume in the monumental *Egyptian Astronomical Texts*. Besides, O. Neugebauer and A. Sachs have been engaged in a detailed study of Babylonian astronomy, based primarily on cuneiform texts in the British Museum.

O. Neugebauer published the article "On the planetary theory of Copernicus" (**19**).

J. H. Reid (Houston) described the most improved planetarium, tellurian and lunarium at Dunsink Observatory (**20**).

Mr. and Mrs. R. Webster (Chicago) have been preparing an extensive *Index of Western Scientific Instrument Makers to 1850*; the first part is already available in draft form. In this work they have been particularly assisted by Derek de Solla Price.

11. *U.S.S.R.*

The following report on works concerning history of astronomy carried out in the U.S.S.R. in 1967–69 has been sent by the Astronomical Council of the U.S.S.R. Academy of Sciences.

"Soviet students of the history of astronomy have been engaged in studies of a history of the activity of individual scientists and particular institutions, as well as of a history of problems in various branches of the history of astronomy. The main results of the investigations may be summarized as follows:

(1) In 1967 considerable attention was paid to the preparation and edition of jubilee books and articles on the history of Soviet astronomy as a whole, as well as those on particular scientific institutions.

(2) The study of the history of Armenian astronomy since the beginnings of the XIXth century till 1920 has been completed and the results have been published (B. E. Tumanian).

(3) Rock engravings of astronomical contents, dated preliminarily to the period from the first to the 8th millennium B.C. have been found by B. E. Tumanian 60 km from Erevan at 2000 m above sea level. Investigations in question are being carried out in collaboration with archaeologists.

(4) A mention of the outburst of Supernova of 1006 was found in the "Matenadaran" manuscripts of Xth–XVIIth centuries (Armenia). There have been found records on meteor observations and among them those on "star showers" in 900 and 1070. The first date agrees with that one from Chinese and Japanese chronicles (B. E. Tumanian).

At present studies in the following main domains and problems of the history of astronomy are in progress:

(*a*) *History of astronomy in various epochs and countries*

I. N. Veselovsky (Moscow) is engaged in problems of Babylonian astronomy; B. E. Tumanian (Erevan) – in problems of the ancient Armenia and papers of Armenian authors who lived in Amsterdam in XIVth–XVIIIth centuries; Matvievskaya (Tashkent), Rosenfeld (Moscow) and Nevskaya (Leningrad) – in the medieval East astronomy; Kuzakov (Moscow) – in the astronomy in ancient Russia; Tchenakal, Nevskaya (Leningrad) – in the history of astronomy in Russia in XVIII–XIXth centuries; Zhelnin and Müürsepp (Tartu) – in the history of astronomy in Baltic countries (Estonia and Latvia).

(*b*) *History of particular problems.*

(1) Maistrov (Moscow) – carved calendars, Baranovskaja (Moscow) – Mongolian, Buriatian and Chinese calendars.
(2) Sokolovskaja (Moscow) – ground means for the study of the Universe.
(3) Straut (Moscow) – study of the Moon.
(4) Eremeeva (Moscow) – ideas on the structure of the Universe.

(*c*) *Personalia*

Life and activity of the following astronomers are being investigated: Eudoxos – by Veselovsky; Hevelius – by Kulikovsky; W. Herschel – by Eremeeva; directors of Tartu (Derpt) Observatory – W. Struve, Öpik, Rootsman, Maedler, Schwarz, Clausen, Levitsky, Pokrovsky – by Zhelnin; B. Schmidt – by Müürsepp and Tchenakal.

(*d*) *Preparations for publication of classical works*

al-Biruni "al-Qanun al-Mas'udi" in Russian – by Rosenfeld, Rozhanskaya and Smirnov.
Nasir al-Din al-Tusi "Tahrir al-Majisti" – exposition of "Almagest" in Azerbaijan by Mamedbeily and others (Baku).
Ptolemy "Almagest" translated with comments by Veselowsky.
(5) Reference books have been prepared and printed, among them the Album-Catalogue of scientific instruments and the bibliography of Russian astronomical literature for 1800–1900. The following two books have been translated from English into Russian: *Astronomy of the 20th Century* by O. Struve and V. Zebergs; *The exact sciences in Antiquity* by O. Neugebauer. Several books are being prepared for an edition: (1) *Astronomy and Cosmonautics*; (2) *Calendar and Chronology* (Seleshnikov, Leningrad); (3) *Scientific devices of historical importance*, Vol. II (Maistrov); (4) Album *History of Astronomy in the U.S.S.R. in Illustrations* (Kulikovsky, Tchenakal), (5) *Bibliography of literature on stellar spectroscopy* 1862–1961 (Korytnikov, Kasan); (6) *History of mechanics* (including a history of astronomy since Babylon times till D. Darwin) by Veselovsky."

In 1969 Vol. X of "Istorikoastronomitcheskije issledovania" (Investigations on history of astronomy) has been printed (Ed. P. G. Kulikovsky). It contains 14 papers written by Soviet historians of astronomy, L. Euler's letters to Wettstein, L. Euler's unknown letters, the selected bibliography of the literature on the history of astronomy 1965–68.

N. I. Nevskaya (Leningrad) advises that she is engaged in the study of ample archives of J. N. De l'Isle, L. Euler, M. V. Lomonossof and other Russian scientists of the XVIIIth century. She is studying arabic manuscripts preserved in the Institute of Asia-Peoples, viz. those of Aristotle, Ibn ash-Shatir, Ibn al-Haytham and Shichab ad-Din'Ali al-Madjdi, in connection with the problem of contacts between European and Arabic astronomy. She is engaged also in problems of Greek navigation and astronomy.

BIBLIOGRAPHY

1. *Acta historiae rerum naturalium necnon technicarum* spec. is., **2,** 57–68, Prague, 1966–67.

2. *ibid.*, spec. is., **3**, 65–78, 1967.
3. *Organon*, **4**, 47–54, Warsaw, 1967.
4. *Editio Cimelia Bohemica*, Vol. I, Prague, 1967.
5. *Sterne u. Weltraum*, Hft 8/9, 1968.
6. *Athens Annals of Archaelogy*, No. 3, 1969.
7. *Technical Annals*, No. 8, Athens, 1969.
8. *Oss. Astr. Capodimonte*, Coll. Misc., Vol. II, No. 3, Naples, 1967.
9. *Luce e Immagini*, **21**, 69–78, Florence, 1967.
10. *Accad. Polacca della Scienze, Biblioteca a Roma, Conferenze*. Fasc. 40, Wrocław-Warszawa-Kraków, 1969.
11. *Rocznik Muzeum Świętokrzyskiego*, Kraków, 1967.
12. *Kwartalnik Historii Nauki i Techniki*, **13**, 397–414, Warszawa, 1968.
13. *Astronomical Notices*, University of Göteborg, No. 10, 3–36, Göteborg, 1968.
14. *Isis*, **60**, 75–90, 1969.
15. *Brit. J. Hist. Sci.*, **3**, pt. 4, 293–394, 1967.
16. *J. Amer. Orient. Soc.*, **87**, 105–112, 1967.
17. *Danske Videnskabernes Selskab. Mat.-fys. medd.*, Bol. 36, København, 1968.
18. *French Historical Studies*, **5**, 371–404, 1968.
19. *Vistas Astr.*, **10**, 89–103, 1968, Pergamon Press.
20. *Irish Astr. J.*, **8**, 222–227.

IX. MICROFILMING OF ASTRONOMICAL MANUSCRIPTS

The working group for the description and microfilming of astronomical manuscripts, under the chairmanship of O. Gingerich has been instrumental in arranging for the microfilming of the material relating to the first four Astronomer Royals at the Greenwich Observatory, viz. J. Flamsteed, E. Halley, J. Bradley and N. Bliss. This program is now complete. Attention is called to another completed microfilm of the papers of G. E. Hale, which has been prepared by the Carnegie Institution of Washington and the California Institute of Technology. The working group has been in communication with Dr. E. W. Maddison of the Royal Astronomical Society Library, who is working on a description of the astronomical manuscripts of certain English astronomers.

More detailed information on the microfilming was printed in the Information Circular No. 16 (November, 1968).

X. JOURNAL FOR THE HISTORY OF ASTRONOMY

Since 1970 a new journal entitled *Journal for the History of Astronomy* is scheduled to be published every six months in London under the editorship of M. A. Hoskin (Cambridge University). The Journal will contain surveys of the present state of knowledge in various fields of history of astronomy, reviews of new books, regular critical bibliographies, details of manuscript collections, and information of theses completed and of how access may be obtained to this material. The subject matter of the Journal will extend to allied fields, including the history of navigation, time-keeping, and geography, and the history of relevant branches of mathematics and physics.

XI. NEW BOOKS ON THE HISTORY OF ASTRONOMY

1. *Vistas in Astronomy*, Vol. 9. New aspects in the history and philosophy of astronomy. (First Joint Symposium of the IAU and the Union Internationale d'Histoire et de Philosophie des Sciences, held at the University of Hamburg), 1967, Pergamon Press.
2. D. S. Evans, T. J. Deeming, B. H. Evans, S. Goldfarb (Eds.); *Herschel at the Cape: Diaries and Correspondence of Sir John Herschel*, 1969, University of Texas Press.
3. W. Hartner, *Oriens-Occidens, Ausgewählte Schriften zur Wissenschafts- und Kulturgeschichte. Festschrift zum 60. Geburtstag*, 1968, Georg Olms Verlagsbuchhandlung, Hildesheim.
4. Z. Horsky, O. Škopova, *Astronomy-Gnomonics, A catalogue of instruments of the 15th to the 19th centuries in the collections of the National Technical Museum Prague*, 1968, Prague.
5. T. N. Kary-Niyasov, *The Astronomical School of Ulugh Beg* (in Russian), 1967, Tachkent.

6. L. E. Maistrov (Ed.), *Scientific instruments of historical importance* (in Russian). (The description of instruments, mainly from XVIIth–XIXth centuries, as preserved in museums of U.S.S.R.) 1968, Moscow.
7. R. S. Richardson, *The Star Lovers*. (Biographies of 16 astronomers of XVIth–XXth centuries.) Macmillan, 1968, New York.
8. V. P. Shcheglov (Ed.), *Jan Hevelius. The Star Atlas* (in English, Russian and Uzbek), 1968, "Fan" Press Uzbek SSR, Tachkent.
9. B. E. Tumanian, *A History of Armenian Astronomy since beginnings of the XIXth century till 1920* (in Armenian with summaries in English and Russian), 1968, Erevan.
10. B. L. van der Waerden, *Die Anfänge der Sternkunde. Erwachende Wissenschaft II*, Birkhäuser, 1968, Basel.
11. D. Pingree, *The Thousands of Abu Ma'shar*, 1968, The Warburg Institute, London.
12. B. R. Goldstein, *Ibn al-Muthannâ's Commentary on the Astronomical Tables of al-Khwârizmî*, 1967, Yale University Press, New Haven, Conn.
13. S. Nakayama, *A History of Japanese Astronomy*, 1969, Harvard University Press, Cambridge, Mass.
14. E. Rosen, *Kepler's Somnium*, 1967, The University of Wisconsin Press, Madison, Wis.
15. C. A. Ronan, *Their Majesties Astronomers*, 1967, London (an American edition, published in 1969 bore the title *Astronomers Royal*).
16. C. A. Ronan, *Edmond Halley – Genius in Eclipse*, 1969, Garden City, N.Y.
17. K. G. Jones, *Messier's Nebulae and Star Clusters*, 1968, London.
18. H. Shapley. *Through Rugged Ways to the Stars*, 1969, Scribner's, New York.
19. O. Neugebauer and R. A. Parker, *Egyptian Astronomical Texts – III Decans, Planets, Constellations and Zodiacs*, 1969, Brown University Press, Providence, R.I.
20. E. Lessing, *Entdecker des Weltraums*, 1967, Herder, Freiburg. (An American edition has been published in 1969 with the title *Discoveries of Space*.)

XII. CONCLUDING REMARKS

The present report does not give a full representation of the actual state of research in the field of the history of astronomy, being only a short summary of main activities of the Commission and containing the information, which has been conveyed to the president by members and consulting members of Commission 41. Such a representation may be done by the bibliography of books and papers on the history of astronomy edited by P. G. Kulikovsky (Moscow). It would be unfeasible for the president to outline in this short report the general trends of investigations in the field of the history of astronomy because of their great diversity and a dispersion of printed works in various, often not easily accessible publications. It can be said, that at present there appears an increase of interest in the history of ancient and medieval astronomy, particularly in the islamic astronomy, which is not well known yet. Also in other branches of the history of astronomy, as the XVIIth-century scientific revolution, the history of modern astronomy, critical biographies of astronomers etc. we mark progress.

One of the main tasks of the IAU commissions is the coordination of scientific actions, which are needing international cooperation. For that reason the question of realization of both resolutions, adopted by the Thirteenth General Assembly of IAU comes into prominence. Especially the start of editorial work on the General History of Astronomy seems to be very urgent and desirable. The second question, which needs a strong international cooperation, is the preservation of astronomical instruments of historical interest, as well as the protection from destruction of personal papers and other heritage of deceased astronomers.

In order to make accessible the historical sources preserved in observatories or other scientific institutions, an action of microfilming them has been started, as mentioned in Section IX and in the Information Circulars of Commission 41.

Two international unions are interested in the history of astronomy: IAU and International Union of the History and Philosophy of Science (IUHPS), and therefore a close cooperation of both unions in the field of the history of astronomy is very desirable. Such cooperation is partly realized by the participation in the works of Commission 41 of eminent scientists, who are active also in the IUHPS, but its tightening would be useful for the promotion of further progress of investigations on the history of astronomy. A proposal has been made late in 1967 that the Commissions of History of Astronomy in both unions should consist of the same members. The president of Commission 41 has not been authorized to decide in that question. After the consultation with members of the organizational Committee of Commission 41 the whole question has been postponed to the Fourteenth General Assembly of IAU, when final decisions are to be taken.

During the Fourteenth General Assembly of IAU it should also be decided where the General History of Astronomy is to be published and how its Editorial Board should be organized.

E. RYBKA
President of the Commission

42. PHOTOMETRIC DOUBLE STARS
(ÉTOILES DOUBLES PHOTOMÉTRIQUES)

PRESIDENT: F. B. Wood.
VICE-PRESIDENT: M. Plavec.
ORGANIZING COMMITTEE: A. H. Batten, T. Herczeg, M. Kitamura, G. Larsson-Leander, J. E. Merrill, V. P. Cesevič.

INTRODUCTION

This report differs from earlier ones because financial limitations require the length to be half that of the preceding one. Thus, bibliographical lists are severely limited and it has not been possible to refer to all published papers or to much of the unpublished work. It is hoped that a more extensive mimeographed report will be available to commission members and others interested by the time of the 1970 General Assembly.

For invaluable aid in preparing this report, I am deeply indebted to Drs. Batten, Herczeg, Katherine Kron, Larsson-Leander, Merrill, Plavec, and Cesevič. Most of these wrote sections of the report which appear only slightly altered to prevent duplication or to take account of very recent developments. Since it was necessary to review more than 1400 papers which appeared during the three-year interval, obviously this could not have been written without extensive collaboration. I am especially indebted to Drs. Merrill and Plavec for consultation in the difficult choice of matters to be eliminated.

The Commission continued to be active between sessions of the General Assembly. Symposia or colloquia sponsored or co-sponsored include:

Non-Periodic Phenomena in Variable Stars, Budapest, Hungary, 1968; Mass Loss from Stars, Trieste, Italy, 1968; Close Binaries and Stellar Evolution, Elsinore, Denmark, 1969.

The Bibliography and Program Notes on Eclipsing Variables continued to be issued semi-annually; the extensive amount of material contained therein substitutes for the omission of certain tables in this report. Numbers 12 through 16 were issued in the interval covered in this report. Contributors include Cester, Italy; Grygar, Central and Eastern Europe; Günther, Germany; Kitamura or Kiyokawa, Japan; Mrs. Kron, North America; Kwee, Western Europe; Sahade, Southern Hemisphere; Schulberg, U.S.S.R.; and Sinvhal, India and Indonesia. Larsson-Leander has borne the heavy burden of the compilation and distribution of these reports which are now of increased value. The heavy work of collating, editing, typing, and running stencils was done by the competent staff of the Lund Observatory.

The most notable feature of the past three years has been the tremendous increase in the number of theoretical investigations, especially those connected with evolution of close double stars. Mass exchange, once scarcely considered, is now accepted as commonplace at certain times of the post main sequence evolutionary stages. However, because of the great computational difficulties involved, detailed studies of the effects of mass loss to the system as a whole, have not yet been made. Further, the stars must somehow reach the main sequence and, while promising work is under way, the theory here is not in nearly as satisfactory a condition as are the more evolved states.

Continued efforts are being made to improve the rectification processes, both in the study of the ellipticity and the re-radiation effects. Merrill's work (*Vistas in Astronomy*, in press) in particular indicates the importance of the inclusion of higher order terms in the harmonic expansion than had previously been thought necessary. After a long period of neglect various observers are again turning their attention to the problem of apsidal motion. Details of these and other developments will be found in later sections.

The discovery that many objects previously thought single are in reality close doubles has added to the significance of our studies in many lines. The work of Mumford and others has shown that many cataclysmic variables are close doubles; it now seems clear that most if not all of the peculiar A stars are also double; R. E. Wilson's work on Cygnus X-2 supports earlier suggestions of its binary nature and describes the system in terms of a semi-detached binary in which the primary component is a white dwarf. However, it seems likely that the kind of systematic X-ray observation needed for significant results must await the establishment of a lunar station where a sustained program can replace the fragmentary results obtained from satellites.

The international collaborative programs have been continued and details are given in a later section. Another effort has been initiated to aid in keeping selected systems under continuous observation for extended intervals. Beginning in November 1968, a site testing program at the South Geographic Pole has been carried on under the joint auspices of the Bartol Foundation and the University of Florida. Preliminary indications are that there will be many successive hours of cloud-free weather during the polar night. The extremely small amount of water vapor pressure also makes infrared work desirable. There is at least the possibility that a 24-inch (61 cm) telescope of special design may be part of the new pole station planned for 1971.

Various review articles have appeared or are in the process of publication; the more specialized will be referred to in later sections. Here can be mentioned Batten's "On the Interpretation of Statistics of Double Stars" in Vol. 5 of the *Annual Review of Astronomy and Astrophysics* and Popper's "Determination of Masses of Eclipsing Binary Stars" in the same volume. Several of the papers in *Mass Loss from Stars* edited by M. Hack discussed close doubles. The Elsinore colloquium brought together a small group of specialists for general discussion; details are given in a later section. Šulberg has finished a monograph on *Close Eclipsing Variables with Spherical Components*, and a group of U.S.S.R. astronomers have completed a book which contains a complete review of methods of solution and a discussion of the statistical data from an evolutionary point of view.

The important work of cataloguing has not been neglected. The Card Catalogue of eclipsing variables formerly at the Flower and Cook Observatory has now been transferred to the University of Florida and information on individual systems is continuously supplied on request. For published work, Svečnikov and Lavrov have completed a catalogue of both relative and absolute elements of 197 systems and are preparing a more extensive catalogue of photometric elements. Batten's excellent catalogue of spectroscopic elements is soon to be supplemented by a critical catalogue of photometric solutions by Koch, Plavec, and Wood. The combination of these should make possible the preparation of a complete and critical catalogue of absolute dimensions.

Finally, attention should be called to the appearance as one of the series in *Vistas in Astronomy*, edited by Beer, of the Henry Norris Russell Memorial volume. It is appropriate that, while an effort was made to include many of Russell's varied interests, the majority of the papers are on photometric double stars, the field of Russell's earliest and latest significant contributions, and one that occupied his interest throughout his career.

OBSERVATIONAL TECHNIQUES AND NEW PHOTOMETRIC DATA

An extremely large number of photometric observations has been accumulated during the past three years. It is impossible to call attention in the alloted space even to the more important individual investigations. Fortunately, references to these are given in the Bibliography and Program Notes of Commission 42; the semi-annual distribution of this makes it essential to workers in the field. Astronomers not on the mailing list may apply to Larsson-Leander at the Lund Observatory. Detailed information concerning any individual system can be obtained by application to any of the three cataloguing centers at the Cracow Observatory, the Sternberg Institute, or the University of Florida.

Work has continued at an increasing rate at the observatories long active in the photometry of eclipsing variables. It is possible here to call attention only to newly established centers or to older institutions which are for the first time (or again after a lapse of many years) beginning the photometry of eclipsing variables; even here the list is probably not complete. Among those to be mentioned are the centers in Romania, especially those at the Universities of Timisoara and of Cluj which are now making photoelectric observations and carrying on theoretical work as well in the field of eclipsing binary stars, the Louisiana State University which now has its own telescope to supplement the observations its astronomers have been obtaining at Kitt Peak and Cerro Tololo, the joint program between Nürnberg (Germany) and the Ege University, Izmir, Turkey, the new Rosemary Hill Observatory at Florida, the apsidal motion program initiated at Yale University, and the Mount John Observatory in New Zealand (operated jointly by the Universities of Pennsylvania and Canterbury) where a new 24-inch (61 cm) telescope is being installed. Attention also should be called to the large number of photoelectric light curves observed at Catania, which has now become one of the leading contributors in this field. On the negative side, nothing whatever has been heard from the Purple Mountain Observatory which contributed to the 1959 β-Lyrae campaign and which until 1965 published in *Acta Astronomica Sinica*.

In instrumentation and techniques, the chief factor has been the increasing amount of two-color or three-color photoelectric photometry on standard color systems. Now, except for discovery purposes and timing of minima, almost all of the observations are made in this manner. An increased care in making extinction corrections and color standards is noted in the work of most observers. A small but increasing amount of polarization studies are being made. Attention should be called in particular to the work of Rucinski on the changes in polarization of β Lyrae; in the Southern Hemisphere, a polarizing photometer is being installed at Mt. John. Cristaldi and Rodono at Catania have constructed an ingenious photometer which permits simultaneous photometry in three colors and thus extremely precise color measures. Cester reports from Trieste that a new 40-inch (101 cm) reflector is under construction for spectrographic and photoelectric purposes and that a simultaneous three color photometer is in operation on existing telescopes. A prototype of an infrared photometer has also been tested.

NEW SPECTROGRAPHIC DATA

Considerable activity has been reported in this field during the past three years from several observatories. Much interest has again been shown in the supergiant systems of the ζ Aurigae and VV Cephei types. The former are the subject of a review by Wright (1970, *Vistas in Astronomy*, **12**, Ed. A. Beer, in press) and the latter of one by Cowley (1969, *Publ. astr. Soc. Pacific*, **81**, 297). Popper has continued his work on improving our knowledge of masses and absolute dimensions of many eclipsing binaries.

The *Sixth Catalogue of the Orbital Elements of Spectroscopic Binary Systems* has been published (1967, Batten, *Publ. Dom. astrophys. Obs., Victoria*, **13**, 119). Batten and Ovenden (1968, *Publ. astr. Soc. Pacific*, **80**, 85) have discussed the distribution of the element ω for the systems in that catalogue. They confirm the Barr effect and point out other anomalies. Batten has also discussed observational evidence for very large stellar masses (1968, *Astr. J.*, **73**, 551) and errors affecting the determination of orbital elements (1968, *J. R. astr. Soc. Can.*, **62**, 344).

Abt and his associates have continued their work on the incidence of close binaries in different stellar classes. Abt and Levy (1969, *Astr. J.*, **74**, 908) find that the incidence is probably low among high-velocity stars, although Eggen (1967, *Astrophys. J.*, **150**, L111) reports a possible high-velocity contact system. Abt and Bidelman (1969, *Astrophys. J.*, **158**, in press) have reclassified many known spectroscopic binaries of late A types. Their results seem to support Abt's earlier conclusions that short-period binaries in this spectral range have metallic-line spectra and few of them have normal A-type spectra.

Olson has published a list of spectral classifications and surface gravities, for bright early type eclipsing binaries (1968, *Astrophys. J.*, **153**, 187), and for Am stars (including many eclipsing systems (1969, *Publ. astr. Soc. Pacific*, **81**, 97). He has also discussed the rotation of components of eclipsing binaries (1968, *Publ. astr. Soc. Pacific*, **80**, 185).

Review articles by Kraft (1967, *Publ. astr. Soc. Pacific*, **79**, 395) on W Ursae Majoris systems and by Mumford (1967, *Publ. astr. Soc. Pacific*, **79**, 283) on Novae and Nova-like variables are also of importance. The suggestion has been made that the X-ray source Cyg X-2 is a spectroscopic and possibly eclipsing binary (1967, Burbidge *et al.*, *Astrophys. J.*, **150**, L95; 1967, Kristian *et al.*,

Astrophys. J., **150**, L99). Later work, however, has shown that the interpretation of this spectrum is far from simple (1967, Kraft and Demoulin, *Astrophys. J.*, **150**, L183).

A list is given below of all those systems known either to be under spectroscopic observation or for which unpublished analyses are well in hand. They are classified according to the observatory at which the observations were made and the investigator. Fuller information on most of this work can be found in the Bibliographies issued by the Commission since the last report.

WW Aur, δ Cap, 32 Cyg, RR Lyn, TU Mon (Kiyokawa, Kondo, Kitamura, and Saito; Okayama). BF Aur, IU Aur, RS CVn, BV 267, 447, 449, 544, HD 35652, Boss 5481 (Mammano, Margoni, Catalano, Bartolini, Perinotto, and Stagni; Asiago). 32 Cyg, VV Cep, W UMa (Faraggiana, Hack, Cester; Trieste). AZ Cas (Tempesti; Teramo). R CMa, VV Ori (Beltrami and Galeotti; Milano). δ Lib, η Ori, ϕ Phe, δ Pic, V Pup (Buscombe; Mt. Stromlo). HD 159441 (Chambliss; Siding Spring). V 1143 Cyg (Avery; Kitt Peak). u Her (Sobieski and Koch; Kitt Peak). BV 268 (Hilditch; Haute Provence). BV 143, XY UMa, PV Cas (Geyer; Bonn). β Per, BD 53° 323, 32 Cyg, AR Cas, TU Cam (Herczeg, Yavuz; Hamburg). BV 447 (Wachmann; Hamburg). 31 Cyg (Mayer; München). AR Cas, SX Cas, RX Cas, U Cep, VV Cep, 31 Cyg, 32 Cyg, β Lyrae, HD 35652, HD 35921 (Batten, Wright; Victoria). GF Car, BF Cen, V346 Cen (Landolt; Cerro Tololo). CG Cyg (Milone and McClurken; Kitt Peak).

The following studies of orbits and/or absolute dimensions of eclipsing binaries have been published since the *Sixth Catalogue*:

β Aur (1968, Galeotti and Guerrero, *Mem. Soc. astr. ital.*, **39**, 269). AE Aqu (1969, Payne-Gaposchkin, *Astrophys. J.*, **158**, 429), AO Cas (1967, Galkina, *Izv. Krym. astrofiz. Obs.*, **36**, 175), 9RS Cha (1969, Chambliss, *M.N.R.A.S.*, **142**, 113; 1969, Jones, *Mon. Notes astr. Soc. Sth. Afr.*, **28**, 5), TV Ceti (1968, Popper, *Astrophys. J.*, **154**, 191), 31 Cyg (1968, Wright and Huffman, *Publ. Dominion astrophys. Obs. Victoria*, **13**, 275), V448 Cyg (1969, Cohen, *Publ. astr. Soc. Pacific*, **81**, 665), V 477 Cyg; CM Lac (1968, Popper, *Astrophys. J.*, **154**, 191), CM Lac (1968, Barnes *et al.*, *Publ. astr. Soc. Pacific*, **80**, 69), CO Lac (1967, Smak, *Acta Astr.*, **17**, 245), RR Lyn (1967, Botsula, *Astr. Zu.* **44**, 1253), TZ Men (1969, Jones, *Mon. Notes astr. Soc. Sth. Afr.*, **28**, 5), RW Mon (1969, Heard and Newton, *J. R. astr. Soc. Can.*, **63**, 208), TU Mon (1967, Popper, *Publ. astr. Soc. Pacific*, **79**, 493), IZ Per (1969, Yavuz, *Astr. Astrophys.*, **2**, 388), λ Tau (1968, Casini *et al.*, Comm. 27, *IAU Inf. Bull.*, Var. Stars, No. 257), TX UMa (1968, Swensen and McNamara, *Publ. astr. Soc. Pacific*, **80**, 192), BV 344 (HD 178001) (1968, Yavuz, *Astr. Abh. Hamburg*, **8**, No. 5), HR 7484 (1969, Snowden and Koch, *Astrophys. J.*, **156**, 667), HD 175514 (1967, Vitrichenko, *Astr. Circ. Izdav. bjuro astr. Soobsc. Kazan*, No. 448), HD 193576 (1967, Ganesh *et al.*, *Kodaikanal Bull.*, Ser. A, No. 184).

Other spectrographic investigations of eclipsing binaries are presented in Table 1.

Table 1. Recent spectrographic work

Star	References	Remarks
o And	Passinetti, L. E., *Mem. Soc. astr. ital.*, **39**, 73, 1968.	Spectrophotometry of variable shell.
ε Aur	Ivanova, N. L., *Astrofiz.*, **3**, 223, 1967.	Spectrophotometry: atmospheric conditions.
ζ Aur	Kitamura, M., *Publ. astr. Soc. Japan*, **19**, 194, 1967.	K-line obs., 1963–64 eclipse.
	Farragiana, R., Hack, M., *Mass Loss from Stars* Ed. M. Hack, D. Reidel, Holland, 1969.	
RCMa	Kitamura, M., *Astrophys. Space Sci.*, **3**, 163, 1969.	Spectrophotometry: variation of line intensities during eclipse.
AO Cas	Galkina, T. S., *Izv. Krym. astrofiz. Obs.*, **37**, 205, 1967.	Spectrophotometry Δ*m*, spectral-type variations.

Table 1 (continued)

Star	References	Remarks
RZ Cas	Ricard-Roux, M., Van 't Veer, F., *C.r.*, **266**, Ser. B, 1345, 1968.	Spectrophotometric measures of limb-darkening.
MN Cas	Lavrov, M. I., Lavrova, N. V., *Trudy Kazan astr. Obs.*, **35**.	
U Cep	Batten, A. H., Laskarides, P. G., *Publ. astr. Soc. Pacific*, **81**, 677, 1969.	Possible H emission outside eclipse.
	Batten, A. H. (in press).	Detection of H emission during eclipse.
VV Cep	Wright, K. O., Larson, S. J., *Mass Loss from Stars*, Ed. M. Hack, D. Reidel, Holland, p. 198, 1969.	Evidence from Hα for mass motions. Wright continuing work.
	Glebocki, R., Keenan, P. C., *Astrophys. J.*, **150**, 529, 1967.	Var. of O I lines at time of secondary eclipse.
31 Cyg	Faraggianna, R., Hack, M., *Mass Loss from Stars*, Ed. M. Hack, D. Reidel, Holland, 1969.	
	Wright, K. O., Morbey, C. L., *J. R. astr. Soc. Canada*, **63**, 213, 1969.	Duration of totality 1968: luminosity.
32 Cyg	Faraggianna, R., Hack, M., *Mass Loss from Stars*, Ed. M. Hack, D. Reidel, Holland, 1969.	
	Wright, K. O., Hesse, K. H., *Publ. Dom. astrophys. Obs. Victoria*, **13**, 301, 1969.	Measures of K-line at 1965 eclipse.
V444 Cyg	Guseinzade, A. A., *Astrofiz.*, **2**, 325, 1966; *Perem. Zvezdy*, **16**, 500, 1969.	Scanning spectrophotometry of emission lines.
	Kuhi, L. V., *Astrophys. J.*, **152**, 89, 1968.	Scanning of emission lines and continuum.
V453 Cyg	Cohen, H. L., Dissertation, unpublished.	
WY Gem	Martini, A., *Mem. Soc. astr. ital.*, **40**, 25, 1969.	Description of Spectrum VV Cephei-type system.
β Lyr	Svolopoulos, S. N., *Astr. Nachr.*, **290**, 155, 1967.	Spectrophotometry of line-profile variations.
	Knappenberger, P. H., Fredrick, L. W., *Publ. astr. Soc. Pacific*, **80**, 96, 1968.	Observations of He I λ10830.
AX Mon	Cowley, C. R., Marlborough, J. M., Cowley, A. P., *Publ. astr. Soc. Pacific*, **79**, 21, 1967.	Suppression of Hε component.
AG Peg	Boyarchuk, A. A., *Astr. Zu.*, **44**, 12, 1967.	Spectrophotometry, radial velocities, model.
β Persei	Glushneva, I. N., *Trudy gos. astr. Inst. Sternberga*, **34**, 81, 1966.	
	Glushneva, I. N., Esipov, V. F., *Astr. Zu.*, **44**, 1028, 1967.	Infrared spectrophotometry.
	Va'tts, I. E., Glushneva, I. N., *ibid.*, 728.	Spectrophotometry of K-line.
ν Sgr	Nariai, K., *Publ. astr. Soc. Japan*, **19**, 564, 1967.	Model for mass flow.
	Coyne, G. V., Kruszewski, A., *Astr. J.*, **74**, 528, 1969.	Polarization measurements.
RZ Scu	Karetnikov, V. G., *Astr. Zu.*, **44**, 22, 1967.	Spectrophotometry.
RW Tau	Plavec, M., *Bull. astr. Inst. Csl.*, **19**, 11, 1968.	Structure of emission ring.
BL Tel	Feast, M. W., *Mon. Not. R. astr. Soc.*, **135**, 287, 1967.	Thorough discussion: model.
HD 47129	Galkina, T. S., *Izv. Krym. astrofiz. Obs.*, **36**, 175, 1967.	Spectrophotometry.
HD 211853	Bracher, K., *Publ. astr. Soc. Pacific*, **80**, 165, 1968.	Spectrum variation.

TIMES OF MINIMA AND PERIOD VARIATIONS

The triennium 1967–69 brought again a body of observations and a number of published papers comparable with those reported in the Transactions of the previous General Assembly. One outstanding feature of the period is the extensive use of photoelectric techniques; apart from the studies based on archives of patrol plates and observations by amateurs, photo-electrically observed times of minima became far superior in number to photographic or visual determinations.

In addition to existing centers of period studies, new series of regular photoelectric observations have been started at several places with considerable effort devoted to timing of minima: University of Florida; Ege University (Izmir), in collaboration with Nürnberg; Bamberg Observatory; Cluj University; Tartu Observatory; Ashchabad Observatory. Southern hemisphere work can be supplemented now by observations in New Zealand, by close cooperation between the University of Pennsylvania and the University of Canterbury. While the Bonn Observatory seems to have stopped its program on WUMa-stars, it is to be hoped that the Vienna group remains active and continues its systematic work on short period eclipsing binaries. Some important observations came from the Haute-Provence Observatory, mainly carried out by guest investigators.

There is a steep increase of systematic photoelectric work on times of minima, supplemented by discussion of the periods, in Poland and, especially, in Italy. In Poland, observational work is mainly centered at Cracow, theoretical studies at Warsaw, results being published almost exclusively in the *Acta Astronomica*. Italian observatories (Asiago, Bologna, Catania, Merate, Teramo, Torino, Trieste), benefited by relatively favourable weather conditions, keep some 60 bright eclipsing variables (mainly BD-stars) on their photoelectric observing schedule; a list of these stars was prepared by Cester and perhaps will be circulated privately. Observations and discussions concerning about one-third of these variables, have been already published, mostly in the *Memorie della Società Astronomica Italiana*. To the three groups of amateur observers mentioned in the previous Draft Report, there is now added a Swiss group (Locher *et al.*), regularly publishing visually determined minima in the periodical *Orion*.

Although the work of non-professional observers should be clearly encouraged, we do have to notice that these observations are of very uneven quality. Looking into the $O-C$ diagrams reveals immediately that the usually high standard of reliability, represented for instance by the group around *Sky and Telescope*, is not always matched by other observers. Besides, there are cases, stars under regular photoelectric surveillance, where visual estimates of minimum epochs convey virtually no additional information, owing to the mean errors 20 or 30 times higher than those of the photoelectric measurements. It could be of considerable value to find some form of coordinating the efforts of the amateur groups or, at least, to support them by professional advice while choosing the objects to be observed. It is to be welcomed that more and more amateur observers shift to photoelectric techniques.

As to the photographic observations, surveys aiming to discover new binaries and to derive periods for them, are still the main task. Some extended series of photographic observations of selected objects or discussions of individual binaries based on patrol plates, deserve mentioning: Whitney's investigations at the University of Oklahoma or studies of numerous less known variables at the Sonneberg Observatory, dealing with about 40 published light curves and periods during the time interval reported here. Other such programs, including the comprehensive Southern hemisphere survey co-ordinated by the Bamberg Observatory, are discussed in the report of Commission 27.

A very important contribution to this field is the systematic investigation into the periods and possible earlier period changes of recently discovered binaries. These studies are carried out by the Bamberg Observatory under the direction of Strohmeier and are based on the great plate collections at the Harvard Observatory, in Bamberg and in Sonneberg. The latest issue of the publications (1969, *Veröff. Remeis-Sternw. Bamberg*, **8**, Nr. 81) deals mainly with Harvard material.

The accuracy of individual minimum epochs is indicated by a scatter of perhaps $\pm 0\overset{d}{.}05$, as a mean. This prevents discussion of finer details in the $O-C$ diagram, but is sufficient to check the period and its constancy. Out of 48 systems, 16 show with certainty changes of the period.

A list of especially important binaries, selected by Plavec, was given in the last Draft Report. At least, four-fifths of these crucial objects were observed photoelectrically during the triennium 1967–69, this indicating the major role this list of desiderata has played in planning observing programs. The following gives a certainly incomplete list of unpublished observations, collected of these systems in various observatories: RT And; WW Aur; i Boo (Teramo, Victoria, Hamburg); RS CVn (Catania); R CMa; RZ Cas (Catania, Teramo); TV Cas (Cracow); U Cep; VW Cep (Bamberg, Teramo, Victoria); U CrB (Catania); Y Cyg (Teramo, Vatican; earlier obs. Bonn); V 477 Cyg (Bologna, Catania, Teramo); W Del (Catania); TW Dra; TX Her (Hamburg, Teramo); AR Lac (Bamberg, Teramo); βLyr (Hamburg; IAU campaign 1959); U Oph (Johannesburg, Teramo); V 451 Oph (Teramo); β Per; U Sge (Teramo, Trieste); RW Tau; X Tri (Teramo); TX UMa (Bologna, Hamburg); VV UMa (Merate). Very few of these binaries can be still considered photometrically neglected. Some proposed additions to this list, of more or less *ad hoc* character, will be added at the end of this chapter.

For a thorough discussion of possible mechanisms causing period changes in close binary systems, Kruszewski's review in Vol. 4 of *Advances in Astronomy and Astrophysics* is still the most comprehensive document to consult; cf. also his recent paper in Vol. 17 of *Acta Astronomica*. Though less directly addressed to the problem of period changes, review articles by Hadjidemetriou (1967, *Adv. Astr. Astrophys.*, **5**, 131), Plavec (1968, *Adv. Astr. Astrophys.*, **6**, 202), Piotrowski (1967, *Commun. Obs. r. Belgique*, Ser. B, No. 17, 133) contain additional information.

Calculations of the evolution of close binary systems, based on the assumption of extensive mass exchange between the components, as carried out with remarkable success in Göttingen, Munich, Warsaw or Ondřejov, have important bearings on the problem of period changes. The calculated traces of evolution require, at some evolutionary phases, substantial changes of the orbital period; these again could serve as support for the theories. There is, however, up to now no clear case known where observed period changes and theoretical predictions would lend themselves to a detailed comparison. It is to be hoped that series of accurate timings of minima, extended to perhaps the next 15–20 years, may clarify this question.

Close binaries attracted much attention as possible sources of gravitational radiation. Among others, Paczynski discussed this problem (1967, *Acta Astr.*, **17**, 287) and concluded that for binaries of short and very short period (like WZ Sge), gravitational radiation may turn out, indeed, to be an evolutionary factor.

In the following some results are summarized very briefly; among the systems considered, well above a hundred in number, we had to select – somewhat arbitrarily – those investigations which can possibly contribute to future understanding the general pattern of period changes.

In many cases, when checked against recent minimum epochs, the cubic ephemeris formulae derived by D. B. Wood and Forbes do not hold. The simple assumption of "accelerated" period changes probably does not do justice to the complicated phenomenon of changing periods.

Purgathofer and Prochazka discussed the ten short period binaries observed in Vienna (1966, *Wien, Mathemat.-nat. Klasse*: *Sber. öst. Akad. Wiss.*, Abt. 2, 175, 301). In more than half of the cases they found good representations of the $O-C$ values by sudden period changes; a highly convincing object of this type is AH Vir. Other authors likewise emphasize the importance of sudden, erratic changes of the period (bot positive and negative), amounting, to the order of magnitude, to a few 10^{-5} times the period. An intriguing case is W UMa, where possibly two such changes in rapid succession shifted the phase of minimum light by $+0\overset{d}{.}0045$ (Kristenson, 1967, *J. Observateurs*, **50**, 71); the event seems to coincide with the flare of the star observed by Kuhi in 1964.

Algol type binaries, like AR Lac show a similar behaviour, $\Delta P/P$ being of the same order of magnitude as in the case of W UMa-stars. An analysis of all photoelectrically observed minima of Algol strongly suggests two such "jumps" during the last decades (Herczeg, Frieboes-Conde, 1968, *Sky Telesc.*, **35**, 288; also *Adv. Astr. Astrophys.*, in press). Two WR eclipsing systems have been discussed by Semeniuk (1968, *Acta Astr.*, **18**, 313), without establishing definite evidence of a changing period.

Mumford has been observing several old novae and nova-like variables photoelectrically (1969,

Mass Loss from Stars, Ed. M. Hack, pp. 202ff.). He found evidence of an increasing period for U Gem and DQ Her; changing period was also indicated in the case of V Sge, as proposed previously by Smak. The very shortest periods (EX Hye, WZ Sge) seem to be constant. The well-known short period binary UX UMa was observed by Kukarkin *et al.* (1969, *Perem. Zviozdy*, **16**, 599); changes of the period suggest cyclic, if not periodic nature, with a total amplitude of $0\overset{d}{.}003$.

A radical modification of the period derived for HZ 22 was proposed by Smak (1969, *Acta Astr.*, **19**, 165), from $3\overset{d}{.}6$ to possibly $0\overset{d}{.}43$. This would put the star, hitherto considered as a normal B-type binary, into the class of the close binaries with a hot subdwarf component.

As to the ultra-short periods, autocorrelation analysis revealed a period of 18 minutes for the white dwarf HZ 29 (Smak, 1967, *Acta Astr.*, **17**, 255). The light curve can be interpreted as due to orbital motion in a very close system, but other possible explanations (pulsation?) could not be dismissed entirely.

Concerning the early phases of stellar evolution, an interesting contribution by Meinunger (1966, *Mitt. Sternw. Sonneberg*), based on Sonneberg plates and dealing with eclipsing variables in T-associations, should be mentioned. He found numerous eclipsing binaries (with one exception, all Algol-type stars), part of them may turn out members and not merely background objects.

Finally, in the case of the much observed supergiant system ζ Aur, the favorable eclipse in 1963–64 made a new determination of the period possible. Kiyokawa (1967, *Publ. astr. Soc. Japan*, **19**, 209) discussed all the observed durations beginning with 1934 and found that they indicate a secular increase; a similar suggestion is also mentioned by Hardorp *et al.* (1966, *Z. Astrophys.*, **64**, 97), and was first suggested by Roach and Wood (1952, *Ann. Astrophys.*, **15**, 21).

Proposed supplement to Plavec's list

OO Aql	W UMa system with unusual deep eclipses. Sudden period changes?
SV Cam	Third body? Recent phe min scatter widely.
SW CMa	Displaced secondary, $e \approx 0{\cdot}5$.
CQ Cep	WR binary with possibly changing period.
CO Lac	Apsidal motion with $P' \approx 35^{a}$.
V 566 Oph	W UMa system showing constant period.
AG Per	Apsidal motion with $P' \approx 76^{a}$.
V 1647 Sgr	Displaced secondary, $P = 3\overset{d}{.}3$, apsidal motion possibly detectable.

Table 2. Investigations on apsidal motion

V 889 Aql	Long per. aps. motion indicated	Semeniuk **(1)**
HH Car	Long per. aps. motion indicated	O'Connell **(2)**
Y Cyg	More accurate data requested	Schneller **(3)**
V 380 Cyg	Long per. aps. motion indicated	Semeniuk **(1)**
V 453 Cyg	Evidence of aps. motion	Wachmann **(4)**
V 477 Cyg	Long per. aps. motion evident $P' \approx 349^{a}$, O'Connell, $P' \approx 349^{a}$ **(5)**,	Semeniuk **(1)**
	$P' \approx 200^{a}$	Rodono **(6)**
DI Her	Relativistic aps. motion still undetectable	Semeniuk **(1)**
CO Lac	Rapid aps. motion evident	Semeniuk **(7)**
AG Per	Rapid aps. motion evident	Semeniuk **(1)**
	Jones **(8)**	Morley **(9)**
V 526 Sgr	Evidence of aps. motion	O'Connell **(10)**
DR Vul	Rapid aps. motion evident	Semeniuk **(1)**
CW Cep	Apparently rapid apsidal motion	Nha **(11)**

(1) *loc. cit.*; **(2)** 1968, *Ric. astr. Specola astr. Vatic.*, **7** (13) 399; **(3)** 1966, *Comm. 27 IAU, Inf. Bull. Var. Stars*, No. 145; **(4)** private communication; **(5)** in press, *Vistas in Astronomy*; **(6)** 1967, *Publ. Oss. astr. Catania*; **(7)** 1967, *Acta astr.*, **17**, 223; **(8)** 1969, *Acta astr.*, **19**, 53; **(9)** 1966, *Comm. 27 IAU, Inf. Bull. Var. Stars*, No. 163; **(10)** 1968, *Ric. astr. Specola astr. Vatic.*, **7** (11) 339; **(11)** private communication.

APSIDAL MOTION

A paper by Semeniuk examining six systems with apsidal motion and quoted in *Transactions*, **13B** as being in press, appeared in 1968 (*Acta Astr.*, **18**, 1). There is a concise summary of present-day observational evidence in this paper, too, giving lists of stars with (*a*) reliable apsidal coefficients; (*b*) determined apsidal period; (*c*) evidence of apsidal motion; (*d*) suspected in apsidal motion; (*e*) displaced secondary minima. Model computations related to apsidal motion have been carried out by Semeniuk and Paczynski (1968, *Acta Astr.* **18**, 33).

CO-ORDINATED OBSERVING PROGRAMS

At the 1967 meetings the committee on co-ordinated programs agreed to concentrate on two systems: GL Car in the south and AR Cas in the north.

Dr. Serkowski, who served as co-ordinator for GL Car, reports (as of September 1969): Photometric observations were made on several nights of 1968 by Quast (São Jose de Campos) in *BV* and by Tapia (Santiago de Chile) in *UBV*. *UBV* observations were also planned by Feinstein and Snowden. The main reason for including GL Car in the co-ordinated program was the lack of spectroscopic orbit for this star with the most pronounced and shortest period apsidal motion of all eclipsing binaries. Unfortunately the only spectroscopic observations reported were by Thackeray; they suggest a value of K greater than 200 km s^{-1} for both stars and a spectral class certainly earlier than the HDE value of B3. Miss Semeniuk of Warsaw University has offered to make any needed reductions and analyses.

Dr. Herczeg acted as co-ordinator for the AR Cas program and published four circulars, the last of which appeared in August 1969 at the start of the second observing season. AR Cas was chosen because the potential yield was so great. This B3V binary has total eclipses of relatively long duration ($D \simeq 10$ h), shallow minima, a displaced secondary minimum, and inconsistencies between photometric and spectroscopic observations. Third-body and apsidal-motion questions need to be settled. Specifically reported observations are: from Hamburg: BV obs of sec min by Wachmann on July 23, 1968 and 28 spectra obtained August 22–November 4, 1968 at 72Å mm^{-1} by Herczeg, Prölss, and Wehmeyer; from Kitt Peak, 4 spectra obtained by West (University of Florida) at 13 Å mm^{-1} November 29–December 1, 1968; from Cracow, pe obs. on 4 nights, July–August 1968, by Kreiner; from Catania, pe obs. on 33 nights up to December 1968, and a few pe observations by Chen and Gleim at Florida. The Catania results by Catalano and Rodono are particularly encouraging; they consist of 460 photoelectric observations in two colors, cover about 80 % of the light curve, and define both minima reasonably well. Geyer (Bonn) reports 30 medium dispersion spectra obtained at the Hoher List Observatory. Circular 4 presents Wachmann's individual observations and estimates of 3rd and 4th contact times.

Stars of earlier co-ordinated programs continued to be studied, SX Cas by Koch and epsilon and zeta Aurigae, 31 and 32 Cygni by Gyldenkerne. Gyldenkerne has called attention in particular to the importance of observing 32 Cygni between now and the favorable 1971 eclipse; observers of any of these four systems are urged to communicate with him. Various papers resulting from these co-ordinated programs are now appearing in the literature.

An excellent summary of the photometric results of the 1959 international campaign on β Lyrae has been published by Larsson-Leander (1969, *Ark. Astr.*, **5**, 253).

DETERMINATION OF PHOTOMETRIC ORBITS

These three years have seen a gratifying surge of interest in improving our knowledge of proximity effects in close pairs, probably the weakest part of our conventional solutions of light curves. Ruciński (*Acta Astr.*, in press) has derived albedoes (directly comparable with the reflection coefficient C_1, in the Russell Model) for stars having deep adiabatic convective envelopes. He finds the "reflected" light completely limb-darkened, as in the case of radiative equilibrium, and his numerical results follow much along the lines of Sobieski's earlier ones, in these not-too-close pairs. The known need for closer definition of temperature distributions across the facing sides of close spherical pairs has occupied the attention of Chen and Rhein (*Publ. astr. Soc. Pac.*, **81**, 387, 1969), using a method valid both for the regions of full illumination and for the penumbras. They trace the somewhat higher temperatures at large polar angles, yielded by their theory as compared to Sobieski's, to more effective inclusion in their geometry of the attenuated penumbral radiation, more an effect of difference of geometrical model than of physical theory.

Departures from spherical shape and from classical von Zeipel gravity darkening have received attention, with results helpful in establishing firmer rectification procedures. Pustyl'nik (Tartu) has found a new expression for the flux distribution over the surface of a non-spherical star, more rigorous than that from von Zeipel's theorem simply applied, and has formulated the radiation transfer in an explicit form adapted to a new approach to the light curve rectification problem. Ruciński has also presented (*Acta Astr.*, **19**, 125, 1969) a method of computing the light changes in cases of moderate distortions of the binary components; his results are given as the coefficients of a Fourier cosine series. In both these "rectification" studies just described, reflection has been neglected.

Reuning (*Am. phys. Soc.*, S. E. Section, Nov. 1969; abstract) has continued his valuable computerized study of the distorted polytrope model, with special attention to polytropes of index 3·5 in contradistinction to the algebraically simpler Roche model $n=5$. He finds not only that the distorted component will be some five percent or more larger than the Roche limiting surface before instability sets in, but also that the true limiting surface for the more realistic case $n=3{\cdot}5$ approximates a triaxial ellipsoid far more closely than it does the commonly used Roche surface. For his models leakage occurs from the essentially rounded vertex of the ellipsoid, not from a Rochelike "tip". Also working from the Chandrasekhar equations for distorted polytropes, Merrill (*Vistas in Astr.*, **12**, Ed. A. Beer, p. 43) finds that the distortion may in some very close systems produce a $\cos 3\theta$ term of significant size, and that neglect of a significant $\cos 3\theta$ term during the analysis (whatever its source) can completely vitiate the deduced coefficient of $\cos\theta$ as far as its use in rectifying for reflection is concerned.

It should be noticed particularly that all the foregoing items represent improvements in our understanding of the existent models and the concomitant desirable rectification procedures; they in no way represent "new" models. In view of this substantial theoretical and procedural progress, and in view of the vast improvements in light curves resulting from increased emphasis in recent years on "epochal" rather than "mean" curves to which to apply our more refined procedures, the "new W UMa – system model" being called for by Kopal (*Astrophys. Space Sci.*, **2**, 23, 1968) and Lucy (*Ap. J.*, **153**, 877, 1968) hardly seems as much a need for W UMa systems in general as it may have 15 years ago. A few years yet of patient comparison of really good modern curves with existing refined models should at least narrow down the range of systems for which the rather considerable labor of new model-building would be likely to be justified.

Tcherepaschuk (Sternberg) is well along on working out a new method for the determination of the elements of eclipsing binaries with ellipsoidal components, with arbitrary law of limb-darkening, while Šulberg (Odessa) has nearly completed work on a method specially aimed at early-type systems and with the assumption of non-linear darkening. Cohen (*Publ. astr. Soc. Pac.*, **81**, 60, 1969) and Cesevič (*Astr. Zu.*, **44**, 139) have both published "short-cut" methods for tabular solutions of rectified light curves on assumed linear limb-darkening. Computerized solutions have multiplied; West's program (*Publ. København*, No. 184, 1965), sketchily mentioned in our 1967 report, follows the Russell Model for rectification, then Kopal's for elements from the rectified curve. Tabachnik (*Astr. Zu.*, **42**, 3, 1965; **45**, 5, 1968; **45**, 6, 1968; *Problems of Cosmic Physics*, **2**, 1968) has an extended computer program to take into account the weights of the observations and equations for semi-detached systems.

Horak (*B.A.C.*, **17**, 27, 1966; **17**, 273, 1966) continued work on his "sphere-ellipsoid" model and followed it (*B.A.C.*, **19**, 149, 1967; **19**, 241, 1968) with an "ellipsoid-ellipsoid" one. In the first the rectification process implicity assumed similar and similarly brightened ellipsoids; the second is sharply restricted in applicability to cases of very small distortions. In both models sine terms are arbitrarily "added out" and the outer faces of the components are not brightened but instead only the difference in reflection on the two inner faces removed. A computerized integration over the eclipsed area (assumed bounded by elliptical segments) is then iterated to reduce the sum of the squares of residuals to a minimum. Publication of a much more detailed exposition of the model and the associated procedures would seem to be next in order if the differences he achieves in orbital parameters from different colors are to be accepted.

Mauder (*Kl. Veröff. Bamberg*, Nr. 39, 1966) has followed his paper on solutions from the Fourier transform of the light curve, with the tables requisite for the case of uniform disks and spherical components. The method is promising and we would hope for tables covering partial limb-darkening soon.

Two books should be mentioned here. Šulberg (Odessa) has a monograph in press on "Close Eclipsing Binaries with Spherical Components". Second, a team of investigators – Martynov and Tcherepaschuk (Sternberg); Šulberg, Tabachnik and Cesevič (Odessa); Svechnikov and Snejko (Sverdlovsk') have completed the manuscript of a treatise on eclipsing binaries containing, besides other matters, a complete and full review of all methods proposed for solution of light curves! The book is scheduled for publication in 1971. Further, the new photometric catalogue by Koch, Plavec, and Wood should be available in 1970.

We close this section with three lists of new solutions of photometric light curves.

First, for the following 62 systems new orbital solutions have been published based on new photometric observations which were in practically all cases made by the respective authors and published in the orbital-solution paper:

XZ And, AD And, BL And, V535 Ara, RX Ari, OO Aql, V599 Aql, AR Aur, HD 35652 Aur, TZ Boo, TU Cam, AW Cam = BV 412, VZ CVn = BV 332, HH Car, RZ Cas, SX Cas, GG Cas, V78 in ω Cen, VW Cep, WX Cep, EK Cep, RS Cha, W Crv, BR Cyg, CV Cyg, MR Cyg (2), V448 Cyg, V463 Cyg, V470 Cyg, V477 Cyg, V836 Cyg = BV 143, UX Eri, S Equ, RY Gem, AD Her, DQ Her, V338 Her (2), RT Lac, AM Leo, AP Leo, RR Lyn, SW Lyn, TT Lyr, TY Men, TV Mon, DI Peg, EE Peg, AU Pup, AG Vir, BH Vir, BV 241, BV 267, BV 357, BV 374, BV 366, BV 412 (2) BV 419, BV 430, BV 722, BD + 34°1051, S8 279 in NGC 188, S8 280 in NGC 188.

Several of these observational series merit attention by investigators who are specialists in intermediary or definitive solutions, and several by investigators interested in tracing time-dependent deviations from some fixed model.

For the following 28 systems new orbital solutions have been published which are based on photometric observations previously available and not in general made by the authors of the respective solutions:

AB And, SX Aur, TT Aur, YZ Cas, RR Cen, 9 Cha, TZ CrA, U CrB, KR Cyg, V453 Cyg, V548 Cyg, V809 Cyg, TW Dra, TX Her, AD Her, RR Lyn, TT Lyr, β Per, V505 Sgr (2), V499 Sco, RT Scl, RZ Sct (2), RW Tau, VV UMa, W UMa, XY UMa, HD 161783.

In several cases orbital solutions from these same observations are already in the literature; new solutions therefore provide either confirmation of deduced elements or an opportunity to compare philosophies and procedures of different investigators. This latter is valuable because orbital solutions in so many cases are still an art rather than a mere technique.

For the following 27 systems, the persons named have informed the Commission (either through Program Notes to Dr Larsson-Leander or by a letter to Dr Wood) that new orbital solutions, based on new photometric data, are essentially completed:

RT And (Dean, Nelson, San Diego); V 337 Aql (Catalano, Catania); ST Aqr (Gleim, Rosemary Hill); GG Cas (Kandpal, Srivastava, Uttar Pradesh); V 701 Cen (Chen, Rosemary Hill); WX Cep (Kandpal, Srivastava, Uttar Pradesh); XX Cep (Angione, Nelson, San Diego); CQ Cep (Kartashova, U.S.S.R.); SW Cyg (Hall, Dyer); V 541 Cyg (Vetešnik, Brno); V 729 Cyg (Hall, Dyer); Z Dra (Devinney, Flower & Cook); RX Gem (Hall, Dyer); RY Gem (Hall, Dyer); AR Lac (Nelson, San Diego); V 1010 Oph (Guinan, Mt. John); TT Ori (Cristaldi, Catania); VV Ori (Marasso, Nelson, San Diego); BM Ori (Hall, Dyer); KZ Pav (Shaw, Mt. John); RW Per (Hall, Dyer); ST Per (Srivastava, Uttar Pradesh); β Per (Wilson, So. Florida); Y Psc (Hall, Dyer); AU Pup (Chambliss, Mt. Stromlo and Mt. John); BP Vul (Illès, Konkoly); BV 332 (Harris, Flower and Cook).

ABSOLUTE DIMENSIONS OF ECLIPSING SYSTEMS

While not quite as many determinations of absolute dimensions have been published in this triennium as in the previous one, the 49 determinations listed below include somewhat more systems for which no absolute dimensions at all were available before. This then is gratifying to report since such determinations are ultimately a principal reason for the existence of our field.

Three matters deserve particular mention in this connection. First, attention has been called before to Popper's long-term program; this continues to yield rich fruit both on the positive side in new determinations and on the negative side in his carefully reasoned decisions that certain stars originally on his program cannot be expected to yield, for one or another stated reason, reliable determinations of absolute dimensions. Second, special mention should be made of the study of 31 Cygni by Wright and Huffman; the depth of insight and the meticulous care employed in this investigation have provided a truly definitive work.

Third, one interesting though minor point resulting from Reuning's study, mentioned earlier, of highly distorted components of polytropic index 3·5. It appears that the geometrical center of a highly distorted component will be significantly (perhaps as much as 5%) nearer the companion than its center of mass. Since the photometric solution deals essentially with the geometrical surfaces but the spectroscopic one with (point) masses, direct combination of the two sets of elements in the conventional fashion can in cases of highly distorted components, result in a slight overestimate of the real sizes of the stars.

For the following 7 systems new determinations of absolute dimensions have been published which are based on photometric and spectroscopic observations previously available:

TV Cam, 31 Cyg, u Her, RR Lyn, AG Per, RZ Sct, BD $+53°$323.

It is an interesting commentary on the coordination of effort (or lack of it) that sometimes occurs, that it should be possible to find data for deriving such badly needed material as absolute dimensions, hitherto unused for the purpose, as in a couple of the cases listed above.

For the following 24 systems new determinations of absolute dimensions have been published which are based on new photometric observations but spectroscopic material already available:

Y Aql, ZZ Cep, EI Cep, MR Cyg, V 448 Cyg, V 453 Cyg, V 477 Cyg, χ^2 Hya, CM Lac, CO Lac, TX Leo, UV Leo, V 566 Oph, EE Peg, ST Per, V 526 Sgr, W UMa, ER Vul, BV 241, BV 342, BV 374, BV 382.

It is very satisfying to see evidence such as the above that photometrists are (presumably) taking the spectroscopic situation into account in planning their observational programs.

New determinations of absolute dimensions, from earlier photometric solutions but new radial velocity material, have been published for the following 13 systems:

SS Boo, 44 i Boo, VW Cep, 9 Cha, V 477 Cyg, 31 Cyg, WW Dra, AS Eri, CM Lac, CO Lac, TX UMa, BV 267, BV 412.

It is clear that spectroscopists have been paying heed to the progress of photometric orbital solutions.

For 7 systems, determinations of absolute dimensions have been made from new photometric and new spectroscopic observations, by the observers themselves:

V448 Cyg, S Equ, TV Mon, W UMa, BV 344, HD197406, HR 7484.

ASTROPHYSICAL AND DYNAMICAL INVESTIGATIONS

During the past triennium our ideas on the origin of binary stars have not developed very much. In particular, Roxburgh's version of the fission theory has not met with general approval. Mammano (1967, *Mem. Soc. astr. ital.*, **38**, 425) pointed out that there are contact systems that violate the upper mass limit of $4M_{\odot}$, predicted by Roxburgh, and he found from the statistics of angular momenta that the fission origin cannot be of much relevance. Hall (1968, *Publ. astr. Soc. Pacific*, **80**, 477) remarked that in Kopal's list of eclipsing binaries there are several detached Algol systems similar to KO Aql, which is considered by Roxburgh to be a system in pre-main-sequence contraction. From the galactic z-distribution of these systems, as compared with the distributions of semi-detached Algol systems and stellar associations, Hall concluded that the detached and semi-detached systems are of about the same age and certainly much older than the associations.

The various theories were reviewed by Huang (1967, *Sky Telesc.*, **34**, 368), and reference was made to a theory of contact binaries that he has developed in detail in another paper (1966, *Ann. Astrophys.*, **29**, 331). It is assumed that the components originate with larger separation and that orbital angular momentum is dissipated through mass ejection along magnetic lines of force, as suggested by Schatzman. In this way W UMa systems may be formed with subsequent mass transfer to the less massive component. Huang found the observational data in favor of the proposed theory.

Worrall (1967, *Mon. Not. R. astr. Soc.*, **133**, 83) considered the few-body problem, arising after formation of a small group of proto-stars from a cloud. Stars are ejected from the group, and the remaining objects form close binaries. Numerical integrations were made for the three-body case with various initial conditions.

At the Colloquium on the Evolution of Double Stars in Uccle 1966, one session was devoted to the genesis of binaries. The introductory review was given by Lippincott (1967, *Commun. Obs. R. Belgique*, Ser. B, No. 17, 55). Among the contributions we mention a paper by Roxburgh (1967, *Commun. Obs. R. Belgique*, Ser. B, No. 17, 60), describing his theory for double star formation, including effects of a magnetic field, a paper by van Albada (1967, *Commun. Obs. R. Belgique*, Ser. B, No. 17, 75), on the N-body problem and the formation of binaries by the escape process.

On the other hand, the theory of evolution of double stars has advanced in the most remarkable way. The detailed theoretical computations of the evolution of close binaries are summarized in the subsequent section. Here we mainly deal with the more general aspects of the theory and its confrontation with observations. General reviews were published by several authors, including Cester (1967, *Nuovo Cim. Suppl.*, **5**, 1089), Paczyński (1967, *Commun. Obs. R. Belgique*, Ser. B, No. 17, 111, 122), Plavec (1967, *Commun. Obs. R. Belgique*, Ser. B, No. 17, 83), Snezhko (1967, *Perem. Zvezdy*, **16**, 253), and Weigert (1968, *Mitt. astr. Ges.*, No. 25, 19).

Intimately connected with the evolution of close binaries are the effects of transfer of mass and angular momentum between the components and the losses occurring to the systems. The mass loss from stars in general was chosen as the topic of the second Trieste colloquium in astrophysics in 1968. As a sequel to this colloquium a smaller group of theoreticians and observers of close binaries met in 1969 at Elsinore, Denmark, for a more informal discussion. At this meeting introductory talks were given by Popper, on the knowledge of masses and their accuracy, by Koch, on observational facts, and by Paczynski, on the theoretical computations. Sahade gave a summary of the papers and the discussion. The proceedings of this meeting including the taped discussions are expected to be published in the near future as a Copenhagen publication.

Popper (1967, *Ann. Rev. Astr. Astrophys.*, **5**, 85) has given a comprehensive review on mass determinations. Batten and Ovenden (1968, *Mon. Not. R. astr. Soc.*, **140**, 81) computed masses for the objects in Sixth Catalogue of spectroscopic binary stars, for which the inclination can be found from visual or eclipsing orbits, and for which absolute magnitudes can be derived from visual orbits, trigonometric parallaxes or Hγ measurements. For $-6 < M_v < +2$ the relation between absolute magnitude and log mass is linear, and no systematic differences appear between Hγ magnitudes and dynamical or trigonometrical ones. For $M_v > +4$ there is a slight evidence for a dual relation. The observed relation is in good agreement with main-sequence models with $X = 0{\cdot}68$, $Y = 0{\cdot}30$. Křiž (1969, *Bull. astr. Inst. Csl.*, **20**, 202) summarized spectroscopic and photometric data for 26 main-sequence eclipsing binaries with two spectra and re-computed absolute dimensions and masses. Comparisons with main-sequence models show good agreement for masses smaller than $5M_\odot$, but the theoretical models are overluminous for larger masses. Popov (1968, *Astr. Zu.*, **45**, 804 = 1969, *Soviet Astr.*, **12**, 640) computed absolute dimensions for 34 eclipsing-binary systems having both components near the main sequence. In the H-R diagram the components with reliable data fall along lines of equal age. The derived mass-luminosity, mass-spectrum, and other relations show good agreement with corresponding theoretical relations for homogeneous models. The properties of the primary components of double-line Algol-like binaries were discussed by Plavec (1967, *Bull. astr. Inst. Csl.*, **18**, 334). The mass-luminosity and mass-spectrum relations agree with those for single stars. The latter relation may be used for determining masses and absolute dimensions when only one spectrum is measurable, but a certain range of masses is obtained, rather than a single definite value. In some cases the range may be narrowed because of the existence of the Roche limit. All available masses for Wolf-Rayet stars were collected by Smith (1967, *Publ. astr. Soc. Pacific*, **79**, 347). The presence of severe anomalies was emphasized, and the conslusion is that the value $10M_\odot$, commonly assumed for W-R stars, should be regarded with considerable suspicion. Batten (1968, *Astr. J.*, **73**, 551) discussed binaries with large masses. The largest reasonably well-determined masses are 30 to 35 $M_\odot$, but systematic errors have the effect to make the observed mass values too small. Considering this, it appears that no reliable mass values exceed 65 $M_\odot$, the limit set for a stable, normal star by the theory of stellar structure. A large number of visual binaries and 29 eclipsing binaries were used by Stephenson and Sawal (1969, *Astr. J.*, **74**, 689) to determine mean masses of stars above the main sequence as a function of the position in the H-R diagram. It was concluded that for non-interacting stars from B0 to K0, post-main-sequence evolution proceeds without dramatic mass loss. Absolute dimensions and masses for 18 interesting eclipsing binaries were discussed by Popov (1968, *Astr. Zu.*, **45**, 1309 = 1968, *Soviet Astr.*, **12**, 1033). Supergiants, like 31 Cyg, were found to be underluminous with respect to the evolutionary tracks corresponding to their masses. In his re-discussion of eclipsing binaries Popper (1968, *Astrophys. J.*, **154**, 191) has recently dealt with CM Lac, V477 Cyg and TV Cet. The first two systems appear to be unevolved, with components on the Sun-Sirius mass-luminosity relation. Popper (1967, *Publ. astr. Soc. Pacific*, **79**, 493) also announced that new spectrograms of TU Mon, considered to be of the R CMa class, does not show the supposed R CMa anomaly: the masses of the components are not small compared with their position in the H-R diagram. For the system RZ Sct, Kitamura and Sato (1967, *Publ. astr. Soc. Japan*, **19**, 575) found the luminosity of the primary component normal for its mass as a super-giant, while the secondary component is overluminous. Heard and Bakos (1968, *J. R. astr. Soc. Can.*, **62**, 67) reported that SZ Psc appears to be a unique example of a binary of the semi-detached type with the sub-giant component being the more massive. Burnashov and Vitrichenko (1969, *Non-periodic Phenomena in Variable Stars* (Ed. Detre), p. 427, Academic Press, Budapest) stated that the recently discovered young system HD 17514 has the large masses 48 $M_\odot + 13{\cdot}5\ M_\odot$. The components almost fill their Roche lobes, but yet the system appears to be remarkably free of complications.

The angular momenta of 126 close binaries were calculated by Gianuzzi (1967, *Commun. Obs. R. Belgique*, Ser. B, 17, 152 = 1968, *Oss. astr. Roma Mte. Mario Contr. scient.*, No. 63), according to formulae given by Eggen for visual binaries. It was found that under certain conditions detached systems of early spectral type may evolve into contact binaries through mass exchange. A general review concerning the problems of close binary systems that involve transfer of angular momentum was given by Huang (1966, *Ann. Rev. Astr. Astrophys.*, **4**, 35).

The problems of mass transfer were reviewed in great detail by Hadjidemetriou (1967, *Adv. Astr.*

Astrophys., **5**, 131) and in abridged form by Piotrowski (1967, *Astr. Zu.*, **44**, 241 =1967, *Soviet Astr.*, **11**, 191; 1967, *Commun. Obs. R. Belgique*, Ser. B, No. 17, 133). A classification of particle orbits through L_1 was made by Hadjidemetriou (1968, *Astr. J.*, **73**, 104) according to their first minimum distance from either primary. Hadjidemetriou (1968, *Astrophys. Space Sci.*, **1**, 336) studied the effects on the orbital elements of repeated mass ejections from either component of a close binary. More frequent ejections at periastron will in most cases give larger eccentricity. In another paper Hadjidemetriou (1969, *Astrophys. Space Sci.*, **3**, 31) investigated the effects of non-isotropic ejections of mass, when all ejected mass is falling on the other component. In still another paper Hadjidemetriou (1969, *Astrophys. Space Sci.*, **3**, 330) considered the case occurring when one component has expanded to fill its Roche lobe and is losing mass to the other component. If mass is transferred from the more massive component, the semi-major axis will increase in most cases.

Plavec (1968, *Adv. Astr. Astrophys.*, **6**, 201) gave an extensive review of the mass exchange in close binaries considered in connection with the computations of stellar models. Detailed comparisons of the mass-exchange theory and the model computations with the observational data were made by Plavec (1967, *Bull. astr. Inst. Csl.*, **18**, 253), by Plavec and Horn (1969, *Mass Loss from Stars*, (Ed. Hack, p. 242, D. Reidel Publ. Co.) and by Ziołkowski (1969, *Astrophys. Space Sci.*, **3**, 14; 1969, *Mass Loss from Stars*, Ed. Hack, p. 231, D. Reidel Publ. Co).

The evolutionary effects due to emission of gravitational waves were considered by Paczyński (1967, *Acta Astr.*, **17**, 287). For W UMa stars the time scale of collapse is of the same order of magnitude as the time scale of nuclear evolution. The influence of gravitational radiation on the evolution of novae, U Gem stars and pair of white dwarfs was discussed. Mironovskii (1965, *Astr. Zu.*, **42**, 977 =1966, *Soviet Astr.*, **9**, 752) gave a lower limit to the density of gravitational radiation from the space density of the W UMa stars, which were considered the most efficient source.

The properties of tides in close binaries and the importance of tides for the evolution of the systems were studied extensively by Zahn (1966, *Ann. Astrophys.*, **29**, 313, 489, 565; 1967, *Commun. Obs. R. Belgique*, Ser. B, No. 17, 108, 124; 1969, *Mass Loss from Stars*, Ed. Hack, p. 267, D. Reidel Publ. Co). The adiabatic tide, in phase with the perturbing potential, and the dissipative tide, in quadrature, are defined. These two dissipative tides produce a torque on each component, and one effect of this is a change of orbital eccentricity. One of the predictions made is that orbits for systems, having at least one component with an external convective zone, should be rapidly transformed to circular form. Another extensive study of the dynamical tides was made by Kopal (1968, *Astrophys. Space Sci.*, **1**, 179, 284, 411; **2**, 48) in a series of papers. Quantitative aspects of the tidal lag were investigated by Roach (1968, *Astrophys. Space Sci.*, **1**, 32) for components in non-synchronous rotation. The equilibrium structure was assumed polytropic, and the viscosity was taken as the 2·5th power of local temperature. Tidal friction in radiative envelopes of massive main-sequence stars was studied by Dziembowski (1967, *Commun. Obs. R. Belgique*, Ser. B, No. 17, 105).

The rotating emission rings in binary systems were discussed by Huang (1967, *Astrophys. J.*, **148**, 793). Periodic orbits around the more massive component in the restricted three-body problem were taken to represent the rings. Orbits were obtained and further improved by series expansion. A radius-velocity relation is derived and mass-ratios of the components may be deduced from this. Kruszewski (1967, *Acta Astr.*, **17**, 297; 1967, *Commun. R. Obs. Belgique*, Ser. B, No. 17, 146) considered the rings to originate from the angular momentum carried by matter travelling between the components. The smaller component may gain a large amount of angular momentum, resulting in ring formation. The theoretical ratio K/v_{em} is compared with observations. The observations show the predicted shape of velocity dependence on mass-ratio, but observed velocities are systematically smaller. It was, however, shown by Smak (1969, *Acta Astr.*, **19**, 155) that the observed velocities, derived from the emission lines, are smaller than the true velocities of ring rotation. The K/v_{em} relation was re-discussed. The discrepancy between theory and observations may be due to non-Keplerian motion in large rings and underestimated rotational velocity of rings of finite widths.

At the Trieste colloquium Wood (1969, *Mass Loss from Stars*, Ed. Hack, p. 149, D. Reidel Publ. Co.) summarized how our ideas concerning mass loss from close binaries have developed since 1941, and Sahade (1969, *ibid.*, p. 156) reviewed the spectroscopic evidence for mass loss. A new interpretation of the gaseous streams was given by Van Houten (1969, *ibid.*), who considered the streams caused by the tidal bulges. Milone (1969, *Non-periodic Phenomena in Variable Stars*, Ed. Detre, p. 457, Academic Press, Budapest) advanced the idea that unequal maxima, the so-called O'Connell effect, is partly due to gas streams, and not a periastron effect as formerly thought. Batten (1968, *J. R. astr. Soc. Can.*, **62**, 344) reviewed the statistics of ω, and its connection with gas streams. Gorbatsky (1967, *Astrofiz.*, **3**, 245; 1968, *Astrofiz.*, **4**, 504; 1969, *Astrophys. Space Sci.*, **3**, 179 = *Mass Loss from Stars*, Ed. Hack, p. 271, D. Reidel Publ. Co.; 1969, *Non-periodic phenomena in Variable Stars*, Ed. Detre, p. 391, Academic Press, Budapest) investigated the effects of gas streams from the secondary on the disk-like envelope of the primary. The envelope is supposed to be shock-heated, and the spectral index of the emitted radiation is determined, as well as $U-B$ and $B-V$ colours. It is implied that "hot spots" may be explained by this shock-heating.

The gaseous ring of RW Tau was studied by Plavec (1968, *Bull. astr. Inst. Csl.*, **19**, 11). A geometrical model is constructed and Sobolev's theory of moving envelopes is applied to derive the population of the higher energy levels. Swenssen and McNamara (1968, *Publ. astr. Soc. Pacific*, **80**, 192) found variable hydrogen absorption cores in high-dispersion spectrograms of TX UMa, indicating gas streams similar to those of U Cep and U Sge. Kitamura (1969, *Astrophys. Space Sci.*, **3**, 163 = *Mass Loss from Stars*, Ed. Hack, p. 159, D. Reidel Publ. Co.) from a spectrophotometric study of R CMa found evidence for gaseous matter surrounding the components. Nariai (1967, *Publ. astr. Soc. Japan*, **19**, 564) explained the displaced Hα absorption, which appears periodically in the spectrum of υ Sgr, as due to supersonic flow from a stellar corona, starting around L_1. Andrews (1967, *Astrophys. J.*, **147**, 1183) has measured the Hα emission strength of Algol at a variety of phases. It remains constant except during primary and secondary eclipses, when the emission region is eclipsed. The emission region is situated at L_1 and originates in a gas stream from the secondary component. Korsch and Walter (1969, *Astr. Nachr.*, **291**, 231) and Walter (1969, *Mass Loss from Stars*, Ed. Hack, p. 211, D. Reidel Publ. Co.) from photoelectric observations of AD Her found evidence for a gas stream from the secondary component. The density and mass-flow of the stream are calculated. Another star showing photometric evidence for gas streams is V448 Cyg, which, together with V453 Cyg, appears to be associated with the young cluster NGC 6871. Both stars have been observed by Cohen (1968, *Astr. J.*, **73**, S8; 1969, *Publ. astr. Soc. Pacific*, **81**, 665; 1969, *Astr. J.*, in press) and by Wachmann (1967, *Astr. Abh. Hamburg. Sternw. Bergedorf*, **8**, 89). Kumsiashvili (*Abastumani astrophys. Obs.*, unpublished) also reports photoelectric observations showing effects of gaseous streams around V448 Cyg as well as around RY Gem.

Free electrons in gaseous rings or disks are evidently responsible for intrinsic polarization of light from close double stars. This interpretation is in accord with observations of the wavelength dependence of polarization for Be-type stars, made by Coyne and Kruszewski (1969, *Astr. J.*, **74**, 528). Shakhovskoy (1969, *Astr. Zu.*, **46**, 386 = 1969, *Soviet Astr.*, **13**, 303) reported polarization measurements for 9 eclipsing variables giving the positions of the lines of nodes of their orbits and the inclinations of their orbital planes to the galactic plane. The results indicate random orientations. Shulov (1967, *Trudy astr. Obs. Leningr. gos. Univ.*, **24**, 38) published polarimetric observations of 13 eclipsing binaries. Variable polarization is found for RY Per, β Lyr, V444 Cyg, U Sge and Z Vul. Reference to further polarization observations of β Lyr are found below.

The problem of determining the type of rotation from a radial-velocity curve was treated by Porfirijev and Kalenichenko (1966, *Astr. Zu.*, **43**, 793 = 1967, *Soviet Astr.*, **10**, 631), by Porfirijev (1967, *Commun. Obs. R. Belgique*, Ser. B, No. 17, 202) and by Plavec (1967, *Bull. astr. Inst. Csl.*, **18**, 93). In an extensive study of Stellar rotation Van den Heuvel (1968, *Bull. astr. Inst. Netherl.*, **19**, 326) argued that Ap and Am stars were probably originally the less massive members of spectroscopic binaries, in which the primary components after mass-transfer have evolved into white dwarfs. Olsen (1968, *Publ. astr. Soc. Pacific*, **80**, 185) reported upon measurements of rotational velocities, $v \sin i$, for 40 components in 29 early-type binaries. Small departures from axial-orbital synchronism were revealed, which are discussed in terms of angular-momentum transfer between components and evolution away from the main sequence.

Photometric peculiarities in general were described by Fracastoro (1965, *Oss. astrofis. Catania*

Pubbl., Nuov. Ser., No. 77), and by Wood (1969, *Non-periodic Phenomena in Variable Stars*, Ed. Detre, p. 429, Academic Press, Budapest). Kozyrev (1967, *Commun. Obs. R. Belgique*, Ser. B, No. 17; 1968, *Izv. glav. astr. Obs. Pulkove*, No. 184, 108) stressed the important effects that the primary must have on the physics of the secondary component. Several photometric peculiarities of RS CVn were analyzed by Catalano and Rodono (1969, *Non-periodic Phenomena in Variable Stars*, Ed. Detre, p. 435, Academic Press, Budapest). Koch (1967, *Astr. J.*, **72**, 306) argued, with u Her as an example, that peculiarities, like transient $\cos\theta$ terms, should be removed before attempting light-curve solutions. Hall (1969, *Mass Loss from Stars*, Ed. Hack, p. 171, D. Reidel Publ. Co.) reported upon a strange secular expansion of the primary of RW Per.

Rucinski (1969, *Acta Astr.*, **19**, 125) developed a method for computing the photometric proximity effects in close binaries of early type. The brightness variation is computed by numerical integration of the emerging intensities over the visible surface, neglecting the reflection effects. Monochromatic light curves are obtained in terms of relative dimensions of components, mass ratio, inclination, average effective temperature, average gravity, frequency and mass concentration to centre. Rucinski (1969, *Non-periodic Phenomena in Variable Stars*, Ed. Detre, p. 361, Academic Press, Budapest) has also made similar computations for highly distorted white-dwarf secondaries. Another paper by Ruciński (1969, *Acta Astr.*, **19**) deals with the bolometric reflection effects for stars with deep convective envelopes. Pustylnik (1967, *Astrofiz.*, **3**, 69; 1969, *Non-periodic Phenomena in Variable Stars*, Ed. Detre, p. 423, Academic Press, Budapest) used Chandrasekhar's iterative technique to solve the transfer equation for an idealized plane-parallel atmosphere with H^- absorption. The monochromatic reflection amplitude was evaluated for the surface being that of the Roche lobe. McNapier (1968, *Astrophys. Space Sci.*, **2**, 61) developed a method for evaluating the reflection effects from light curves of high precision, except for cases when the components are in contact or nearly so. There is no restriction on the extent of the penumbra. Cester (1969, *Mem. Soc. astr. Ital.*, **40**, 169) calculated fluxes and luminous efficiency factors for main-sequence and sub-giant stars according to Chandrasekhar's model of a non-grey atmosphere, as proposed by Hosokawa, but using improved values for temperature and other parameters. The results are applied to 76 systems and a comparison is made between observed and theoretical coefficients of cosine terms. The agreement is satisfactory for all stars except for those of the earliest types. Here the choice of a lower temperature will reduce the negative trend in $O-C$, but does not eliminate it completely.

Van 't Veer (1968, *C. r. Acad. Sci. Paris*, Ser. B, **266**, 1245) studied the information that may be obtained from differential limb-darkening, deduced from multicolour observations of eclipsing binaries. Ricard-Roux and Van 't Veer (1968, *C. r. Acad. Sci. Paris*, Ser. B, **266**, 1345) specifically discussed RZ Cas using 8-band photometry. The observed differential limb-darkening is in agreement with theory if $u_{4950} = 0{\cdot}55$, which is the value predicted from stellar model atmospheres. A statistical discussion of limb- and gravity-darkening coefficients was published by Hosokawa (1968, *Sendai astr. Rap.*, No. 101). Chen and Rhein (1969, *Publ. astr. Soc. Pacific*, **81**, 387) calculated the temperature distribution on the stellar surfaces, assuming that close double stars are opaque black-bodies with limb-darkening. The increase of temperature due to radiative transfer between the components is calculated employing geometry of spheres.

Lucy (1967, *Z. Astrophys.* **65**, 89) derived a modification of von Zeipel's gravity-darkening law, valid for stars with convective envelopes. This modification is important for the theory of the light curves of W UMa stars. The gravity darkening of the W UMa systems was also considered by Kopal (1968, *Astrophys. Space Sci.*, **2**, 23). Pustylnik (*Tartu. astr. Obs.* (unpublished)) has found a new expression for the flux distribution over the surface of a non-spherical rotating star, which is more rigorous than that following from von Zeipel's theorem. The radiative transfer problem is formulated in explicit form, permitting also a new approach to the light-curve rectification problem.

We now turn to the various groups of binaries and start with the Wolf-Rayet systems. V444 Cyg has been extensively studied by several observers. Kuhi (1968, *Astrophys. J.*, **152**, 89; 1968, *Astr. Soc. Pacific*, Leafl. No. 468) made photoelectric spectrum-scanner observations of emission-line and continuum intensities, especially around secondary minimum, for the purpose of determining atmospheric stratification. Interpretation of data in terms of purely geometric effects proved impossible. Erratic fluctuations point to a random process of ejection, perhaps some type of large-scale prominence activity. Similar observations of the He II lines λλ4686 and 5411 and of the continuum were made by Guseinzade (1966, *Astrofiz.*, **2**, 325; 1969, *Perem. Zvezdy*, **16**, 500). Line variability was found rare compared to the changes of the continuum. It is supposed that the main mechanism of the formation of emission lines is not a recombination, and that the upper layers of the envelope are transparent for radiation from lower layers. Tcherepashchuk (1967, *Astr. Circ.*, No. 413; 1967, *Perem. Zvezdy*, **16**, 226) discussed *U, B, V* light curves, obtained from narrow-band photometry in the continuum, and the intensity of He II 4686. The physical fluctuations in the continuum are almost independent of the fluctuations of the He II 4686 emission. The mean light curves in the continuum are of the Algol type, while in the λ4686 region they are almost of the β Lyr type. The size of the envelope emitting He II 4686 is at least twice that of the region responsible for the continuous spectrum. Bracher (1967, *Astr. J.*, **72**, 787) investigated the spectrum variation of HD 211853. There is an absorption feature at λ3870, which is almost absent at phases 0·4 to 0·6 but increases to a very prominent absorption at phases 0·85 to 0·19, when the W-R star is in front. Tcherepashchuk (*Gos. astr. Inst. Sternberga* (unpublished)) has made narrow-band photometry of CV Ser and CQ Cep, both in emission lines and in the continuum. Narrow-band emission-line observations of some spectroscopic W-R binaries are planned to check the possibility of eclipses.

The physical conditions in the atmospheres of both components of the early type binary HD 47129 (Plaskett's star) were analyzed by Galkina (1967, *Izv. Krym. astrofiz. Obs.*, **36**, 175). The line intensities of the secondary vary appreciably with phase and wavelength. The behaviour of He II 4686 and N III 4641–4634 emissions as well as the Hβ, Hγ line profiles were studied. Galkina (1967, *Izv. Krym. astrofiz. Obs.*, **37**, 205) also made a similar analysis of the components of AO Cas. The equivalent line widths of both components show considerable variations at elongations.The primary varies from O9·1 to O8·0, the secondary is estimated as O7·2. The components are situated on the lower boundary of the main sequence. Olson (1968, *Astrophys. J.*, **153**, 187) derived line profiles for 20 eclipsing binaries between O9 and A2. Surface gravities were determined by fitting these to theoretical profiles, calculated by Strom and Peterson using the latest Griem broadening theory and a modification by Edmonds, Schluter and Wells. Comparisons with gravities found from the binary solutions indicate that the latter broadening theory is the more nearly correct one.

Knappenberger and Fredrick (1968, *Publ. astr. Soc. Pacific*, **80**, 96) reported upon the first successful observations of the He I 10830 emission line in the spectrum of β Lyr. This is the first member of the principal series of triplets, the second member of which is λ3889. The line is a broad emission, cut by a sharp absorption. Alduseva (1969, *Astr. Zu.*, **46**, 363 = 1969, *Soviet Astr.*, **13**, 286) obtained 1961–63 absolute spectrographic gradients for β Lyr in the photographic region and monochromatic light curves at λλ4000 and 4500. Differences between the 1961–62 and 1963 results are found in the gradients as a function of phase, in the shape of the light curve, and in the integrated brightness. Larsson-Leander (1969, *Ark. Astr.*, **5**, 253; 1969, *Non-periodic Phenomena in Variable Stars*, Ed. Detre, p. 443, Academic Press, Budapest) published the photometric results from the 1959 international campaign on β Lyr. A comparison with the 1958 Lick observations reveals changes in integrated light, colour, and asymmetry of minima. The central parts of the primary minima show changing asymmetry, and there appears to be a gradual decline of total light during the campaign. The color curve shows evidence of a "blue top" at zero phase and a "red dip" at phase 0·37. The Huang disk-model is regarded as most promising for explaining the various peculiarities. Changes in the polarization were reported by Ruciński (1966, *Acta Astr.*, **16**, 127; 1967, *Acta Astr.*, **17**, 207), and Shulov (1966, *Astrofiz.*, **2**, 339; 1967, *Astrofiz.*, **3**, 233), and by Appenzeller and Hiltner (1967, *Astrophys. J.*, **149**, 353). All the authors explain the variable polarization component as due to scattering by free electrons in a flat disc in the orbital plane. Ruciński finds that the shift of secondary maxima confirm that the orbital inclination can be about 75°. Shulov determines the electron density and finds that the mass of the disc is $1{\cdot}5 \times 10^{-7}\ M_\odot$. Sculsky (*Lwow Univ.* and *Krym. astrofiz. Obs.* (unpublished)) has investigated the physical conditions in the envelope by means of curves-of-growth. He concludes from high-dispersion spectrograms that there is a third body with a period of about 30 years.

The measurements of radial velocities of Algol-type stars were discussed by Plavec (1967, *Determination of Radial Velocities and their Application, IAU Symposium, No. 30*, Eds. Batten and Heard, p. 229, Academic Press), using S Equ as an example. Yavuz (1968, *Astr. Abh. Hamburg. Sternw. Bergedorf*, **8**, No. 5) made a detailed spectrographic and photometric investigation of BV 344. The Roche model indicates that it is a detached system. The spectrally invisible secondary is probably an F-type sub-giant. Azimov (*Shemaha Obs.* (unpublished)) determined absolute spectrophotometric gradients for RS Ari, S Cnc, RS CVn, SX Cas, GG Cas, U Cep, AR Lac, UX Mon, RY Per, and U Sge. The variation of the gradient with phase was studied in the case of UY Vir. A similar study of AW Peg by Azimov and Babaev (*Shemaha Obs.* (unpublished)) included also the variation of hydrogen-line intensities. Jones (1969, *Acta Astr.*, **19**, 53) determined the effective temperature of AG Per from $(B-V)_0$ using the Q-method. Svolopoulos (1968, *Z. Astrophys.*, **69**, 296) derived an empirical relation between $B-V$ colour and radiation temperature from variables with total eclipses. The relation is compared with a similar empirical relation obtained by Becker (1948). Hall (1967, *Astr. J.*, **72**, 301) used narrow-band photometry on sub-giant components of Algol-type binaries to search for a possible anomalous abundance of heavy elements. The measurements revealed an apparent under-abundance of CN. McNamara (1967, *Astrophys. J.*, **149**, 723) commented upon the apparent metal deficiency of the secondary component of U Cep. Sistero (1968, *Publ. astr. Soc. Pacific*, **80**, 474) reported that *UBV* observations of S Vel during primary eclipse show an ultraviolet excess, similar to those found for some other sub-giant components. Surkova and Skatova (unpublished) investigated the period variations of 13 binaries and evaluated the rate of mass loss among binaries with sub-giant components.

A spectrophotometric investigation of Algol in the region $\lambda\lambda$4700–3000 was made by Glushneva (1966, *Izv. gos. astr. Inst. Sternberga*, **34**, 81). Equivalent widths and depths of Balmer lines were determined at primary minimum and at maxima. The influence of Algol B and Algol C on the line shape was investigated. The Algol A Balmer lines increase during primary minimum, which can be connected with the limb-darkening. Valts and Glushneva (1967, *Astr. Zu.*, **44**, 728 = 1968, *Soviet Astr.*, **11**, 585) studied the behaviour of Ca II K from 195 spectrograms of Algol. The equivalent width is 2·4 times as large at minimum as at other phases, which confirms that the darker component is a K0 star. A doubling of the K line was observed in three spectrograms at primary minimum. Fracassini and Pasinetti (1968, *Mem. Soc. astr. ital.*, **39**, 653) determined sizes of the non-eclipsed area of Algol by means of the Balmer discontinuity, slightly modifying Chalonge and Divan's method. If the blue spectral gradient is assumed constant during eclipse the results are in agreement with McLaughlin's elements. Herczeg (1969, *Non-periodic Phenomena in Variable Stars*, Ed. Detre, p. 465, Academic Press, Budapest) and Friboes Conde, Herczeg and Høg (*Astr. Astrophys.* (in press)) discussed the Algol system, combining available astrometric, spectrographic, and photometric observations. It is concluded that the so-called great inequality, of "period" about 190 years, is in reality well represented by two sudden period-changes in the system AB, in about 1944 and 1952. The 32-year variation is probably due to apsidal motion. No further bodies beside *A*, *B*, *C* are present. The astrometric observations at primary minimum show effects of the light of Algol C.

The very interesting system BL Tel has been studied by Feast (1967, *Mon. Not. R. astr. Soc.*, **135**, 287). High-dispersion spectrograms show that the *F* super-giant primary has normal (solar) abundances. The secondary fills its Roche lobe and is slightly smaller and much fainter than the primary. It may be a cool super-giant or a hot sub-dwarf, ionizing a small and dense H II region, which produces the eclipse. High velocity and large z-coordinate (= 3·3 kpc) makes it a run-away system. The present state may be explained by the assumption that the secondary component was once a very massive star (about 400 $M_\odot$), which at some 10^7 years ago lost most of its mass very rapidly, probably as a result of a type II supernova outburst.

A general review on the W UMa stars was given by Kraft (1967, *Publ. astr. Soc. Pacific*, **79**, 395). He concludes, from statistical evidence, that the W UMa systems cannot now be transferring mass from one component to the other and cannot have descended from close-but-not-contact binaries, satisfying the mass luminosity relation. According to Lucy (1968, *Astrophys. J.*, **151**, 1123) the W

UMa systems are contact binaries, in the sense that the components have a common convective envelope. On the basis of this theory, Lucy (1968, *Astrophys. J.*, **153**, 877) proposed a model for calculating the light curves. It explains the anomalous luminosity ratios and correctly predicts the major characteristics of the light curves. A maximum of 1·28 mag. for the depth of primary minimum in integrated light is predicted and shown to be consistent with observational data. However, the asymmetries of the light curves cannot be explained, and the primary minimum for the theoretical light curves is due to the eclipse of the more massive component, which is not true for the majority of observed systems. Mauder (1968, *Mitt. astr. Ges.*, No. 25, 139; 1969, *Non-periodic Phenomena in Variable Stars*, Ed. Detre, p. 419, Academic Press, Budapest) developed a method for deriving the elements of W UMa systems by a Fourier transform of the whole light curve. Three parameters are basic: inclination, luminosity-ratio and mass-ratio, which fix the geometrical situation completely. Numerical work along these lines is in progress. Koch (1968, *Astr. J.*, **73**, S102) pointed out that theoretical models of W UMa stars give contact systems, while photometric analysis of light curves give semi-detached systems. A program with narrow-band filters has been started at Flower and Cook Observatory, to investigate if this is a band-pass effect, and if the W UMa stars show the CN-absorption strength expected for main-sequence stars. Herczeg (1969, *Mass Loss from Stars*, Ed. Hack, p. 213, D. Reidel Publ. Co.) summarized our knowledge on the period changes of the W UMa stars. It was emphasized that almost all changes are of abrupt nature, and that both positive and negative changes occur.

A monograph on the binary systems of the ζ Aur type was written by Botsula (1968, *Binary Systems with Atmospheric Eclipses* (in Russian), "Nauka", Moscow). Magnan (1965, *Ann. Astrophys.*, **28**, 512) applied the theory of deviations from local thermodynamic equilibrium for a study of the chromosphere of the K component of 31 Cyg. The case of Ca II was examined with a three-level atom and a continuum. A model atmosphere with homogeneous density-distribution was obtained. Saito (1965, *Publ. astr. Soc. Japan*, **17**, 107) used the theory of shock-pulse propagation on the K components of ζ Aur and 31 Cyg. The inferred chromospheric condensations are taken to represent shock-pulses and the turbulent velocity field originates from these pulses. Tatum (1966, *Publ. Dom. astrophys. Obs. Victoria*, **12**, 425) discussed the problem of interpreting the equivalent widths of absorption lines produced during the atmospheric eclipse. A computer programme for obtaining the radial density-distribution was applied to 32 Cyg. Wright (1968, *J. R. astr. Soc. Can.*, **62**, 67) summarized recent Victoria spectrographic results on ζ Aur and 31 Cyg. Kitamura (1967, *Publ. astr. Soc. Japan*, **19**, 194) published a study of the chromospheric K line of ζ Aur in the 1963–64 eclipse and made comparisons with theoretical profiles. Doherty (1967, *Astr. J.*, **72**, 296) analyzed photoelectric spectral scans from the 1965 eclipse 32 Cyg. The spectral type of the secondary was determined as B3. Comparisons with previous eclipses give some evidence for a decrease of the radius of the K4 I component. Mayer (*Sternw. München* (unpublished)) is reported to be working on spectrograms of 31 Cyg obtained during the 1961 eclipse.

Cowley (1969, *Publ. astr. Soc. Pacific*, **81**, 297) has given a concise summary of the photometric and spectrographic studies of the VV Cep stars. The number of objects in this group is now 14, including the related object α Sco. Besides VV Cep and AZ Cas, showing photometric eclipses, several others are suspected to show atmospheric eclipses. Spectrograms of VV Cep from the interval 1956–68 were analyzed by Wright and Larson (1969, *Mass Loss from Stars*, Ed Hack, p. 198, D. Reidel Publ. Co.) for mass motions within the system. The mass ratio of the components is found closer to 1 than to 2. Additional absorption lines may be interpreted as due to gas moving from the M-type star towards the secondary. Faraggiana and Hack (1969, *Mem. Soc. R. Sci. Liège*, Ser. 5, **17**, 317) discussed intensities and radial velocities of the forbidden lines of VV Cep. Glebocki and Keenan (1967, *Astrophys. J.*, **150**, 529) reported upon observations of the O I lines λλ7774 and 8446. These lines, which were seen in the spectrum of VV Cas in 1944, but had disappeared in 1945, reappeared in 1964 and persisted until spring or summer 1965. Both apparitions coincide with predicted times of secondary minima, and it is probable that they originate in gas surrounding the B-component. Mammano and Martini (1969, *Mass Loss from Stars*, Ed. Hack, p. 184, D. Reidel Publ. Co.) described the spectrum of Boss 5481 during the 1965–67 shell episode, which may be an eclipse phenomenon. In another paper Mammano and Martini (1969, *Non-periodic Phenomena in Variable Stars*, Ed. Detre, p. 415, Academic Press, Budapest) described changes in the spectra of three other VV Cep stars, including WY Gem. Cowley (1969, *Publ. astr. Soc. Pacific*, **81**, 297) reported that this star exhibited a shell spectrum 1968–69, probably due to an atmospheric eclipse.

The related object ε Aur was studied by Ivanova (1967, *Astrofiz.*, **3**, 223) with spectrophotometric methods during the 1955–57 minimum and also out of minimum. A comprehensive and critical review of observations and theories concerning this star has been completed by Jesse (*Sternw. München* (unpublished)).

The spectroscopic and photoelectric observations of novae and nova-like variables, pertaining to the binary nature of these stars, were reviewed by Mumford (1967, *Publ. astr. Soc. Pacific*, **79**, 283, 500). The outstanding questions are: Where did the short periods originate, and how can the varying degree of explosive activity be explained among systems constructed on approximately the same model? In a later article Mumford (1969, *Mass Loss from Stars*, Ed. Hack, p. 204, D. Reidel Publ. Co.) discussed the periods of T Aur, EM Cyg, U Gem, DQ Her, EX Hya, V Sge, and WZ Sge. Period changes are indicated for U Gem, DQ Her, and V Sge. Another review of the problems of the eruptive binaries was given by Smak (1969, *Non-periodic Phenomena in Variable Stars*, Ed. Detre, p. 345, Academic Press, Budapest). He dealt especially with period changes, as indicating mass loss or mass transfer, with colours and "hot spots", and with the origin of outbursts. Previous photometric evidence from U Gem seemed to indicate that the eruptions occur in the late type component. Recent spectroscopic observations of the non-eclipsing SS Cyg by Walker and Chincarini (1968, *Astrophys. J.*, **154**, 157) indicate, however, that the outbursts occur in the blue component. Paczyński, Ziółkowski, and Zytkow (1969, *Mass Loss from Stars*, Ed. Hack, p. 237, D. Reidel Publ. Co.) suggested that the outbursts of U Gem systems might be related to mass outflow from the late type components. These components are the more massive ones, and they fill their Roche lobes. The authors show that under these circumstances the mass outflow will occur on a dynamic time-scale, if there is a deep convective envelope. A model for nova outbursts has been developed by Rose (1968, *Astrophys. J.*, **152**, 245; 1969, *Mass Loss from Stars*, Ed. Hack, p. 77, D. Reidel Publ. Co.) on the basis of the binary nature of novae. The hydrogen-exhausted hot component is supposed to accrete small amounts of hydrogen-rich mass. A non-degenerate hydrogen-burning shell is formed, which may become thermally unstable, and pulsational instability is found to develop.

New important observations were reported on DQ Her. Dibai and Shakhovskoi (1966, *Astr. Zu.*, **43**, 1319 = 1967, *Soviet Astr.*, **10**, 1059) found that the polarization parameters vary substantially during eclipse. The polarization measurements give for the orbital plane the same position angle as that of the major axis of the envelope visible after the 1934 outburst. Nather and Warner (1969, *Mon. Not. R. astr. Soc.*, **143**, 145) have used their synchronous photometer, constructed for pulsar work, to obtain an accurate light curve for the 71·1 second-pulsation of the white dwarf component. This component does not contribute more than 2 or 3 % of the total light of the system, thus the true amplitude of the pulsation is above 1 mag. The shoulders on the light curve, preceding and following eclipse, cannot be due to a "hot spot" on the white dwarf. The nebula surrounding this component is suggested to be optically thick, and the eclipse of a bright region in the nebulosity by the larger component is considered responsible for the shoulders.

Krzeminski and Smak (1969, *Non-periodic Phenomena in Variable Stars*, Ed. Detre, p. 371, Academic Press, Budapest) proposed a new model for WZ Sge, in which the secondary contributes 20 % of the total light. The W UMa-type light curve is, except for the primary eclipse, explained as a result of the aspherical shape of the secondary. Both stars are degenerate, with effective temperatures of 20000 K and 8000 K respectively. Smak and Stepien (1969, *Non-periodic Phenomena in Variable Stars*, Ed. Detre, p. 355, Academic Press, Budapest) reported extensive photoelectric observations of TT Ari, which indicate that this star is a hot, sub-luminous close binary with $P = 0{\cdot}2658$ days. It also shows quasi-periodic fluctuations with periods between 14 and 20 min, and flickering with a time scale of about 1 min. Smak (1969, *Acta astr.*, **19**, 165) presented arguments in favour of the view that HZ 22 is not a normal B-type binary with $P = 3{\cdot}58$ days but rather a hot sub-dwarf binary with $P < 1$ day.

THEORETICAL AND STATISTICAL INVESTIGATIONS

A new feature in the past triennium was the introduction of the technique of computing time-dependent evolutionary model sequences. The integration method devised by Henyey and collaborators was adapted (first by Kippenhahn and Weigert, *Z. Astrophys.*, **65**, 251, 1967) to include mass loss due to thermal instability of the more massive component when it reaches its Roche limit. Rapid progress in this field can best be surveyed from the proceedings of four recent conferences: (1) Uccle Colloquium in 1966 (On the Evolution of Double Stars, *Comm. Obs. R. Belg. Uccle*, **B17**), (2) Joint Discussion in Prague, 1967 (section F of *Highlights of Astronomy*, Ed. L. Perek), (3) Trieste Colloquium in 1968 (*Mass Loss from Stars*, Ed. M. Hack), (4) Copenhagen Colloquium in 1969, Ed. K. Gyldenkerne. Original papers were published mainly in *Z. Astrophys.*, **65–67** by Kippenhahn, Weigert, Kohl, Thomas, Giannone, Refsdal etc., in *Acta astr.*, **17** by Paczyński and Ziółkowski, and in *Bull. astr. Csl.* (*B.A.C.*), **18–20** by Plavec, Kriz, Harmanec and Horn. See also the review papers by Plavec (*Adv. Astron. Astrophys.*, **6**, 201, 1968) and Paczyński (*Ann. Rev. Astron. Astrophys.*, in press).

Originally the purpose was to explain the evolutionary paradox of the Algol-like semidetached binaries, where the less massive component appears more advanced in evolution. It appears that the above-described theory can explain not only this but also other properties of these systems like overluminosity of the subgiant secondaries, their contact nature, and presence of gas streams. It appears that semidetached binaries are generated by mass exchange of type A (the initially more massive component reaches its Roche limit when hydrogen is still burning in its core) in case of more massive systems, while case B (hydrogen shell burning) applies to low-mass binaries, although a mixed case AB can be important, too. (Kippenhahn *et al.*, *Z. Astrophys.*, **65**, 251, **66**, 58; Refsdal, Weigert, *Astron. Astrophys.*, **1**, 167; Plavec, *Astrophys. Space Sci.*, **1**, 239, 1968; Plavec *et al.*, *B.A.C.*, **20**, 41, 1969; Ziólkowski, *Astrophys. Space Sci.*, **3**, 14, 1969).

Quantitative comparison with observed systems will probably have to take into account deviations from sphericity, possible escape of material from the system, and conversion of orbital angular momentum into that of axial rotation. The latter may explain the excess rotation of some primaries like U Cep or RZ Sct (Plavec, Van den Heuvel, 1969, Columbus Colloquium on Stellar Rotation).

Connected with the problem of angular momentum is the formation of gaseous rings around primaries (Kruszewski, *Acta astr.*, **17**, 297, 1967; Huang, *Ap. J.*, **148**, 793, 1967; Plavec, *B.A.C.*, **19**, 11, 1968; Smak, *Acta astr.*, **19**, 155, 1969; Gorbackij, *Astrofiz.*, **3**, 245).

Huang suggested that massive and opaque rings may be formed in systems like β Lyr (*Ap. J.*, **138**, 342, 1963) or ε Aur (*Ap. J.*, **141**, 976, 1965). β Lyr is likely to be in the phase of rapid mass transfer of type B. But in the supergiant systems like ε Aur or VV Cep the components apparently do not fill critical lobes and yet they lose mass. This superficial instability may in fact not be restricted to supergiants, as indicated by the erratic period changes and intermittent ring formation in some systems with giants (SX Cas, UX Mon) or subgiants (U Cep, RS CVn).

Case B of mass exchange is generally of greater importance, in particular practically all spectroscopic binaries will at one time be subject to it. The transformation is even more drastic, and mass losses as high as 80 % occur (Kippenhahn, Weigert, *Z. Astrophys.*, **65**, 251, **66**, 58; Kriz, *B.A.C.*, **20**, 127, 1969). Suggestions have been put forward that this process (or case C and higher) could explain various objects like: the Sirius-like visual binaries (Lauterborn, Trieste and Copenhagen), Wolf-Rayet stars (Paczyński, *Acta Astr.*, **17**, 355, 1967), some shell stars (Křiž, Plavec, Horn, Trieste), blue stragglers and even Am and Ap stars (Van den Heuvel, *B.A.N.*, **19**, 326, 1968 and Prague meeting), helium-rich and hydrogen-poor stars, etc. Many more detailed studies are needed, including the problem whether and how the mass-gaining component accommodates the in-flowing material. Certainly much depends on the star's nature. If it is already a degenerate star, thermal instability and violent pulses are expected (Kippenhahn *et al.*, *Z. Astrophys.*, **69**, 258, 1968; Giannone, Weigert, *ibid.*, **67**, 41, Rose, *Ap. J.*, **150**, 193, 1967), the study of which may throw new light on the problem of novae and U Gem stars.

The evolution of the W UMa systems appears to be more complex, and Huang (*Ann. Astrophys.*,

29, 331, 1966) invokes magnetic fields besides mass exchange. However, we do not even understand the character of these systems fully, since the interpretation by Lucy (*Ap. J.*, **151**, 1123, 1968) in terms of a contact binary is not in agreement with the observational evidence that they are actually semi-detached systems.

Another open problem is whether we can and do observe some binaries still at the phase of pre-main-sequence contraction (cf. Wood, Prague and Trieste). A single system of this kind would be very interesting.

Relatively little progress can be reported in the problem of the origin of binary stars. From a statistical study, Huang (*Ann. Astrophys.*, **31**, 379, 1968, and Prague meeting) found support for the contention about star formation in groups; it is hoped that intensive work on the origin of close binaries may be a feature of the next three years.

FRANK BRADSHAW WOOD
President of the Commission

43. PLASMAS AND MAGNETOHYDRODYNAMICS IN ASTROPHYSICS (PLASMAS ET MAGNÉTO-HYDRODYNAMIQUE EN ASTROPHYSIQUE)

PRESIDENT: L. Davis, Jr.
VICE-PRESIDENT: R. Lüst.
ORGANIZING COMMITTEE: T. G. Cowling, S. A. Kaplan, B. Lehnert.

A. INTRODUCTION

This report emphasizes selected problems in plasma physics and magnetohydrodynamics that appear to be of substantial future interest in astrophysics. Because of space limitations, complete coverage is impossible and bibliographical references are drastically curtailed. More complete references to the papers in Sections B and C are available on application to the undersigned. Additional references should be easily available in the well known abstracting and review journals. I am grateful to the members of Commission 43 who supplied the information on which this report is based. I am particularly grateful to Professors Mestel, Lüst, and Kaplan for writing the three main sections of this report.

LEVERETT DAVIS, JR.
President of the Commission

B. PLASMA IN THE SUN AND STARS
(by L. Mestel)

1. *Origin of stellar magnetism*

Masses condensing from magnetic gas clouds should retain some primeval flux. Models of rotating magnetic stars (Davies, 1968; Wright, 1969) suggest that even those with the strongest surface fields have a total flux less than one percent of the virial theorem limit. This may be due to field leakage during formation, or possibly to hydromagnetic instability: an analogue of the "sausage" mode in laboratory plasma theory may cause a more rapid contraction of poloidal flux-tubes into a neutral line than does decay in a stationary medium. Non-adiabatic modes, analogous to those discussed for rotating stars (see Commission 35 report) may yield a slow but steady motion of flux-tubes to the surface. Completely stable fields may require toroidal flux linking poloidal loops.

The Biermann "battery" process uses the non-irrotational partial pressure gradient per electron in a non-uniformly rotating star to drive poloidal currents which maintain a toroidal field. The process may stop if the centrifugal field relaxes rapidly into an irrotational state, but this is still under debate (see report for Commission 35). A weak primeval poloidal field constrains the rotation field sufficiently to kill the process except in stars where radiation pressure is significant (Kato and Nakagawa, 1969).

A slowly-decaying field in an Ap star or a pulsar need not be a primeval "fossil" but could be the relic of a dynamo-built field from e.g. the Hayashi convective phase. There remains uncertainty as to whether strictly isotropic turbulence can generate a large-scale field (Kraichnan and Nagarajan, 1967; Parker, 1969). New work by Moffatt (1969) and Leorat (1969) confirms the earlier result of Steenbeck, Krause and Rädler that turbulence which lacks reflection symmetry with respect to a plane generates a large-scale field with the same helicity as the turbulence. Steenbeck and Krause (1969) argue that their model is applicable to the solar and terrestrial dynamos. Numerical work by Roberts (1970) suggests that a *non*-axisymmetric field (without Cowling-type singular lines) may be maintainable against Ohmic decay by axisymmetric motions.

2. *The strongly magnetic stars*

The low rotations of the Ap stars may be due to magnetic braking, especially through coupling with a stellar wind (Mestel, 1966, 1968). The oblique rotator model of magnetic variables requires a large mutual inclination of the rotation and magnetic axes (Preston, 1967). The same coupling with a stellar wind rotates the rotation axis through the star (Mestel, 1968), but whether so as to increase or decrease the inclination depends critically on the detailed flux distribution over the stellar surface (Mestel and Selley, 1970) which in turn depends on internal magnetohydrodynamics. Sufficiently powerful laminar circulation in a rapidly rotating star can prevent flux from appearing above the surface (Mestel, 1967; Maheswaran, 1969; Wright, 1969). More work could be done to see which distributions of flux and which angles of inclination are consistent with observed Zeeman shifts and cross-over effects (Böhm-Vitense, 1967; Preston, 1967; Steinitz, 1969; Landstreet, 1969).

3. *Magnetism and stellar evolution*

A primeval field would need to be improbably strong in order to suppress convection over the bulk of a star in e.g. the Hayashi phase (Moss and Tayler, 1969), though convective efficiency may be affected. Numerical studies (Moss, 1970) confirm that even without turbulent dynamo action a field of energy much less than that of the turbulence will not be completely expelled from the zone; some of the flux will remain, concentrated into filaments. There could be a moderately strong primeval field in the solar radiative core, which would probably couple together the rotations of the core and the convective envelope. Gurm and Wentzel (1967) and Finzi and Wolf (1969) have suggested that magnetic blobs generated in convective cores may become buoyant and cause mixing with the radiative layers.

4. *Sunspots and the solar cycle*

The observed concentration of magnetic flux at the supergranular junctions is a confirmation of theoretical predictions. The clustering of active magnetic regions into complexes with a scale ten times the supergranular scale (Bumba, 1967) is a possible clue to the "giant cells" of Simon and Weiss (1968). Local magnetic interference with convective heat transport accounts qualitatively for the formation of pores and sunspots, but most sunspot models (e.g. Chitre and Shaviv, 1967) require radiative transport to be supplemented by a reduced but finite convective transport. Alfvén waves and also overstable sound waves (Syrovatskii and Zhugzhda, 1968) can develop, but it is not clear how the consequent energy transport is related to the observed umbral dots and the umbral flashes (Beckers and Tallant, 1969). Wilson (1968) has emphasized the theoretical importance of the penumbra in the energy deficit problem.

Cowling's objections to sunspot decay through Ohmic effects appear to remain valid (Altschuter, 1967), though Kopecký (1969) has emphasized the excess dissipation in the fine structure of the field. The computation by Altschuter *et al.* (1968) of an Evershed flow with a component perpendicular to the field lines suggests a dynamic model of sunspot destruction.

Following Babcock, Leighton (1969) has constructed a mathematical model of the solar cycle as a periodic dynamo, its main features being the twisting of a poloidal field by the non-uniform rotation, eruption of the consequent strong toroidal field to form sunspots, and Leighton's random-walk dispersal process. With an eye on the solar dynamo, the H.A.O. group (e.g. Nakagawa and Swarztrauber, 1969) have studied the hydromagnetics of axisymmetric, incompressible, differentially rotating systems, with and without dissipation. This work underlines the need for a dynamical theory of the differential rotation (see report to Commission 35): one theory (Gilman, 1968) simultaneously develops a solar dynamo model.

The solar cycle as the prototype of stellar magnetic variability (e.g. Tuominen and Tuominen, 1967; Godoli, 1967) is a less plausible rival to the oblique rotator model, now that the Ap stars are thought to be intrinsically slow rotators.

No theoretical model seems able yet to cope with the observed complex solar field, especially the sector structure corresponding to the interplanetary field (Wilcox and Howard, 1968). Viewed as a magnetic star, the sun has a general field currently varying with half the rotation period (Severny, 1969).

5. *Solar and stellar atmospheres*

Besides the gravitationally-modified sound waves, the solar convection zone generates coupled gravity and Alfvèn waves with no frequency limits on propagation (Lighthill, 1967). However, Kuperus (1969) concludes that most of the upward energy flux responsible for the heating of the solar corona is acoustic, though uncertainties in the turbulence spectrum in the convection zone can cause errors by an order of magnitude (Stein, 1968). More work on wave propagation and damping is needed to decide the range of stellar surface temperature and gravity for which hot coronas are formed. With only acoustic input, de Loore (1969) found no corona until the contracting sun had reached the bottom of the Hayashi track, but this may be altered if the star has a dynamo-built magnetic field, especially as the field can lead to dipole or monopole emission of acoustic waves.

Accretion heating – once suggested for the solar corona – has been revived as the energy supply for X-ray sources (Hayakawa and Sugimoto, 1968; Sakashita, 1968). Zeldovich and Shakura (1969) predict a non-Planckian spectrum.

The magnetic field channels heat conduction from the corona into the strongly magnetic chromospheric regions above the supergranular junctions. Kuperus and Athay explain spicules as due to the consequent local over-heating and hydrostatic breakdown. Pikel'ner (1969) argues that the Sweet-Petschek-Sturrock picture of the pinching-off of field lines is inadequate to explain flares, but will yield spicules (see also Uchida, 1969). Kuperus emphasizes that ultraviolet spectra and magnetic data should enable us to decide between different models of the chromosphere-coronal interface.

A strong local field with closed lines restricts solar wind expansion. If the heat input exceeds the radiation loss, a domed "helmet" structure evolves into a "cusped helmet", becoming ultimately a coronal streamer (Pneuman, 1968; Sturrock and Smith, 1968).

Most authors agree that the energy of solar flares comes from the convective zone, to be stored as the energy of non-potential magnetic fields until explosively released. Filaments modelled as cylindrical force-free fields are hydromagnetically unstable to formation of helical structures, which may therefore be more plausible pre-flare states (Anzer, 1968). The Kippenhahn-Schlüter filament model is unstable if the current density increases with height (Anzer, 1969). The tearing-mode resistive instability and its non-linear development e.g. into Petschek's mode needs further study. Yeh and Axford (1969) argue for a reconnection rate determined completely by external flow, and not even logarithmically dependent on resistivity. Syrovatskii (1969) emphasizes collisionless dissipation and the associated acceleration of runaway particles as the origin of the X-ray, optical and radio emission. The Moreton wave emitted by flares has been interpreted by Uchida (1968) as a fast-mode m.h.d. wave.

The gravitationally-distorted magnetic fields that arise naturally during condensation from the interstellar or intergalactic medium should cause large-scale flares as they relax to a lower-energy state, possibly yielding radio galaxies or quasars (Sturrock, 1969).

Zaitsev (1969) has studied shock-wave generation of type-II radio bursts. The C.S.I.R.O. group (Smerd, 1968) emphasize the continuing difficulties of the high brightness temperatures, and the need for a non-linear theory of the stabilization of radiative instabilities before meaningful comparisons with observation can be made.

6. *Pulsars*

Goldreich and Julian (1969) have followed up Gold's suggestion (1968, 1969) and studied the hydromagnetics of an axisymmetric rotating magnetic neutron star. In their model, the approximately corotating magnetosphere (of negligible thermal scale-height) is held up by the electrostatic forces associated with almost complete charge separation. The field is quasi-dipolar near the star, but field-lines from near the poles extend beyond the light-cylinder, enabling charges to be accelerated outwards to high energies in an "electrostatic wind" (although synchrotron losses are a difficulty (Chiu, 1969)).

Following Pacini (1968), Ostriker and Gunn (1969) adopt an oblique rotator model. Beyond the light cylinder the time-varying magnetic field becomes a low-frequency electromagnetic wave, as the conduction current is limited by the finite electron velocity. The wave is very efficient at accelerating particles arriving from the Goldreich-Julian magnetosphere. Emphasis is laid on the rotational braking associated with the radiation, and (Davis and Goldstein, 1970) a possible change in the mutual inclination of the rotation and dipole axes.

The work of Chiu *et al.* (Chiu, 1969) concentrates on the emission mechanism rather than on cosmic ray production. They explain the enormous brightness temperatures as laser radiation from electron bremsstrahlung in a magnetic field, with the population inversion achieved by the magnetospheric electric field. They also argue that neutron stars may be permanently magnetized, due to the moments of electrons moving in Landau orbits.

Ginzburg *et al.* (1969) discuss both the laser process and the conversion of part of the energy of amplified plasma waves into electromagnetic waves. They suggest that the second period observed in the radio pulses is a neutron star pulsation period, though this appears to conflict with computed periods of neutron star models.

C. INTERPLANETARY PLASMA

(by R. Lüst)

Remarkable progress has been made recently in our understanding of the interplanetary plasma, the solar wind. The great extent of new observations has stimulated theoretical investigations of a variety of problems which can be compared with the measurements. But at the same time observations and theories have opened new areas not yet well understood. In most cases further progress requires both theoretical and observational work. The basic theory of the dynamical behaviour of the plasma and the magnetic field as developed by Parker in 1958 has been confirmed by the observations. The currently important problems of the interplanetary plasma may be categorised under six headings:

1. *Description of the large-scale properties*

As Parker has shown the large-scale properties of the solar wind can be described by a fluid model. A number of different one and two fluid models have been solved numerically in order to improve the comparison between theoretical predictions and observed quantities like velocity, density and temperature. These calculations have shown in particular that thermal conduction alone is not sufficient for supplying the necessary energy (Hartle and Sturrock, 1968). They lead to too high an electron temperature and too low an ion temperature. Furthermore, when in such one and two fluid hydrodynamic models the only energy transport from the base of the corona is conduction and convection, the calculated density of the wind at 1 A.U. tends to be too high and the velocity too low to agree with observations.

As a consequence, other forms of energy transport must be operative too; during quiet conditions they should be effective out to at least $r = 3\ R_{\odot}$. During active periods which are indicated by higher temperatures but also by high density and wind velocity, the energy input should be enhanced out to the orbit of earth. The additional energy is presumably delivered by waves or turbulence. Near to the sun the mechanism responsible for heating the solar corona could be effective also in the outer corona, since not all energy in the shock waves will be absorbed in the inner corona. But in addition other waves might be responsible. For heating the solar wind near the sun during quiet times the fast mode of magnetoacoustic waves could be effective (Barnes, 1968, 1969).

These waves with long wavelength (about 2×10^{11} cmo r 10^{-2} A.U.) which are generated in the solar corona as remnants of processes occurring in the solar photosphere, such as the granulation or spicules, will have damped out before they reach the orbit of earth. Therefore the waves we are observing at the orbit of earth – in particular the ones with wavelengths less than 10×10^6 km – are generated in the solar wind and do not directly reflect processes occurring at the surface of the sun. The wavelength spectrum at the orbit of the earth may reflect some sort of equilibrium between generation and decay of hydromagnetic waves. The strong damping of waves provides a significant heat source for the solar wind.

However, waves generated by internal plasma instabilities cannot heat the wind, since they contribute only to the equalization of the anisotropies, and probably do not produce the long wavelength waves of interest. However, the interaction between streams, or sectors, of differing solar wind velocity distributed in solar longitude would provide a dominant source for long wavelength waves heating of the plasma (Jokipii and Davis, 1969). The velocity differences observed in such streams represent an energy source in the frame of the average solar wind corresponding to a thermal velocity of a few hundred km/s which is certainly adequate to maintain the solar wind temperature of the order of some 10^5 K and which has the required property of being variable. In addition, this mechanism would heat the particles proportionally to their mass, since the interaction produces equal velocity spreads. Therefore we would expect that during perturbed times the ion temperatures are proportional to their mass. This would indeed explain the fact that the temperature of the α-particles is about 4 times that of the protons during active times while the temperatures are more nearly equal during the times when the solar wind is cold and quiet.

Important problems are: (a) the explanation of the non-Maxwellian velocity distribution functions of the ions and the electrons and, in particular, the reason why they are different; (b) the dependence of these functions on distance from the sun and ecliptic latitude; (c) the important instabilities in the wind and the mechanisms for wave particle interactions; (d) knowledge of the abundances of the different ions in the solar wind and an understanding of any differences from the elemental abundances of the solar atmosphere. In particular, observations reported by Hundhausen (1968) and Hundhausen *et al.* (1968) of He^{+}, He^{++}, O^{+5}, O^{+6}, and O^{+7} give direct evidence of the temperature of the coronal gas expanding to produce the wind, a key parameter in most models.

2. *The interaction with the solar magnetic field and the transport of angular momentum*

The basic theoretical picture (as developed by Parker) of the spiral pattern of the magnetic field extending from the sun through interplanetary space is well substantiated by the observations of the field in space. Weber and Davis (1967) have taken into account the influence of the magnetic field on the motion of the solar wind. They evaluated also the transport of angular momentum by the magnetic field and by the azimuthal motion of the solar wind caused by the interaction with the magnetic field. The predicted value of the azimuthal velocity seems to be smaller by almost one order of magnitude than the observed value at certain periods. The difference might be due to viscosity and anisotropic pressure (Meyer and Pfirsch, 1969; Weber and Davis, 1970), or due to the interaction of fast and slow streams. The torque exerted by the magnetic field and the azimuthal motion may be sufficient to halve the rotation of the entire sun in about 10^{10} yr and of the outer part of the convection zone in about 10^9 yr.

3. *Fluctuations, discontinuities and waves*

Many more detailed observations are now available about the fluctuations in the plasma and in the magnetic fields, in particular the power spectra (Coleman, 1966; Holzer and McLeod, 1966; Burlaga *et al.*, 1969). Jokipii and Hollweg (1970) have studied the connection of the correlation length observed for scintillations of point radio sources with the power spectra of the magnetic field. The longer wavelengths (lower frequencies) may be of solar origin, but the rest of the spectrum observed near the orbit of earth must originate in space. The basis for proposed generation mechanisms is the observed variation in the wind velocity. Coleman (1968) suggests that the large-scale shear in the wind leads to something like hydrodynamic turbulence. Jokipii and Davis (1969) propose that the fluctuations are produced more directly by the collision of fast and slow streams (see Subsection 1). Another mechanism (only for higher frequencies) is based on the thermal anisotropies in the wind (Parker, 1958). The calculations and a summary of Kennel and Scarf (1968) show that the strength and nature of the dominant instabilities in the solar wind depend on a number of parameters, including the pressures of both the electrons and ions relative to the magnetic field pressures and on the mode which one considers.

Individual fluctuations can now be compared with theory. Shock transitions have been studied in some detail by Razdan *et al.* (1967), Schubert and Cummings (1967), and Nishida (1968). Dryer and Jones (1968) investigated the problem of heating the solar wind by shock waves generated in a solar flare disturbance. Hundhausen and Gentry (1969) have studied the propagation of shock waves through the interplanetary space.

Besides shocks, many individual fluctuations are of the nature of shear planes (McCracken and Ness, 1966; McCracken *et al.* (1968), Burlaga, 1968, Meyer and Pfirsch, 1969). The shear planes are separated by distances of a few times 10^{11} cm, suggesting that they may be related to the super granulation and related solar motions (Gosling *et al.*, 1968).

4. *Modulation of galactic cosmic rays and propagation of energetic solar particles*

Progress has been made too in the understanding of the modulation of galactic cosmic rays by the solar wind and of the propagation of energetic solar particles (Gosling *et al.*, 1968; Jokipii and Parker, 1967, 1968 and 1969; Parker, 1967 and 1968; Axford, 1967; Jokipii, 1967 and 1968; Jokipii and Coleman, 1968; Noerdlinger, 1968). A theoretical connection between the cosmic-ray diffusion tensor and the power spectrum of the small-scale irregularities in the interplanetary magnetic field has been established.

A comparison of theory with observation indicates that the fluctuations in the interplanetary field die out with increasing distance from the sun, so that cosmic rays propagate more freely beyond 5–10 A.U. than at the orbit of earth (Simpson and Wang, 1967; Bachlet *et al.*, 1967; O'Gallagher, 1967; O'Gallagher and Simpson, 1967). The conditions inside the orbit of earth and out of the ecliptic plane have also been theoretically investigated (Gosling *et al.*, 1968). Three distinct effects combine to produce the anisotropy of cosmic rays. One is the adiabatic deceleration of cosmic-ray particles in the field carried in the expanding solar wind gas, another is Fermi acceleration and solar production, and the third is the variations with solar latitude of the ratio of the solar wind velocity to the diffusion coefficient. A comparison of theory and observations of cosmic ray anisotropy suggests tentatively, that, compared to conditions observed at the orbit of earth, the wind is less turbulent nearer to the sun and/or faster or more turbulent at higher solar latitudes, particularly at sunspot latitudes.

5. *The transition into interstellar space*

The supersonic solar wind will reach interstellar space after undergoing a shock transition. This shock transition should occur about 30 to 300 A.U. from the sun, the uncertainty being due to our uncertain knowledge of the interstellar pressure and of the effects of charge neutralisation of the solar wind particles due to charge exchange with the interstellar hydrogen atoms. Also the relative motion of the sun with respect to the neighboring interstellar gas must be considered.

More detailed studies of this transition should be of great interest. The charge exchange interaction between solar wind ions and interstellar neutral atoms followed by the sweeping up of all newly created ions by the solar wind magnetic field must contribute to the deceleration and heating of the solar wind. Within a transition region of radius of order 10^3 A.U. the electrons and ions may largely recombine and the lines of force of the solar magnetic field merge with those of the interstellar field. Presumably plasma instabilities are important here.

6. *The interaction of the solar wind with the earth, moon, and planets*

The main features of the interaction of the solar wind with the geomagnetic field to form a bow shock, magnetosheath, magnetosphere, and tail have become reasonably well understood but many features require further study. It is expected that the interaction with Jupiter will be similar. The moon appears to have so little intrinsic magnetic field and such a low conductivity that the solar wind impinges directly on its surface and leaves an empty wake, occupied by a largely undisturbed interplanetary magnetic field just behind. Venus and Mars appear to have no significant intrinsic magnetic fields but to have an ionosphere with high enough pressure and conductivity that the solar wind cannot easily penetrate it. Thus there is a bow shock and an as yet very poorly understood wake and interface between solar wind and atmosphere.

REVIEW PAPERS

Dessler, A. J. 1967, *Rev. Geophys.*, **5**, 1.
Axford, W. I. 1968, *Space Sci. Rev.*, **8**, 331.
Ness, N. F. 1968, *A. Rev. Astr. Astrophys.*, **6**, 79.
Wilcox, J. M. 1968, *Space Sci. Rev.*, **8**, 258.
Hundhausen, A. J. 1968, *Space Sci. Rev.*, **8**, 690.
Parker, E. N. 1969, *Space Sci. Rev.*, **9**, 325.
Lüst, R. 1969, *IAU Symposium* no. 39.
Colburn, D. S., Sonett, C. P. 1966, *Space Sci. Rev.*, **5**, 439.

D. MAJOR PROBLEMS IN PLASMA ASTROPHYSICS AND MAGNETOHYDRODYNAMICS

(by S. A. Kaplan)

This report reviews the main directions of work currently done in the U.S.S.R. on the astronomically significant problems of plasma astrophysics and magnetohydrodynamics. The only references are to books or review papers; in addition, names of authors are given for some cases where essential results were obtained.

1. *Theory of plasma turbulence*

Much attention is paid to the general theory of wave turbulence in collisionless plasma. Different plasma instabilities excite different modes of plasma wave turbulence. Computations of increments of plasma instabilities, the various coefficients of interaction between the same and different waves and the coefficients of decay of turbulence are now in progress (B. B. Kadomtsev, R. S. Sagdeev, V. I. Karpman, V. N. Tsytovich, S. A. Kaplan, V. Yu. Trakhtengertz, V. V. Zaitsev, A. S. Chikhachev etc.).

Among the problems of plasma turbulence which are most important for plasma astrophysics are:

a. *Interactions between plasma waves and electromagnetic waves*

(i) Plasma waves of different modes are converted into electromagnetic waves both at the plasma frequencies and at higher frequencies. This is due to the nonlinear and Compton scattering of plasma waves on thermal, superthermal and relativistic electrons and ions, making possible the observation of plasma turbulence.

(ii) Coalescence of plasma and electromagnetic waves gives a number of effects which also have made possible the observation of plasma turbulence. These are: angular scattering of electromagnetic waves (increasing, in particular, the angular sizes of sources); changes of electromagnetic frequency (increasing, in particular, the width of emission spectral radio-lines). Coalescence of the same and different modes of plasma waves can also lead to the generation of electromagnetic waves with combination frequencies.

(iii) Induced conversion of plasma waves into electromagnetic waves due to nonlinear and Compton scattering on anisotropically distributed fast charged particles made possible the occurrence under cosmic conditions of maser-effects. It was shown that this process may lead to the very fast transformation of nearly all the energy of relativistic electrons into electromagnetic waves.

A summary of theoretical investigations of the interaction between plasma and electromagnetic waves is given in the review article "Plasma mechanism of radiation in astrophysics" (S. A. Kaplan, V. N. Tsytovich, *Usp. fiz. Nauk*, **97**, 77, 1969) and in the book under the preparation by the same authors.

Knowledge of different probabilities of interaction between waves of one or different modes makes it possible to calculate the spectra of plasma turbulence (distribution of energy between various modes and between waves with different wave numbers of the same mode). Only a few data of this kind are now known and further investigation is needed for the classification of other physical aspects of the effect of plasma turbulence (B. B. Kadomtsev, S. B. Pikel'ner, V. N. Tsytovich).

b. *Interaction between plasma wave turbulence and charged particles*

(i) Streams of fast particles with anisotropic velocity distributions are the most common source of plasma turbulence (especially on high frequency modes) in cosmic conditions. Many of these effects are observable now, e.g. sporadic solar radiobursts of type III. In exciting the plasma turbulence, the streams usually decay and anisotropy vanishes but in some cases there is a stabilization of the streams.

(ii) Plasma waves of different modes accelerate charged particles and even neutrinos. It has been shown that in many cases that the plasma acceleration is much more effective than the well known Fermi mechanism and can also lead to power-law spectra.

(iii) The high degree of isotropy of cosmic rays may be attributed to plasma instabilities. Any significant anisotropy is removed by the generation of Alfvèn and magnetosound waves with a time scale of some hundred years.

A review of plasma acceleration was given by B. B. Kadomtsev and V. N. Tsytovich at the VIth Joint IAU/IUTAM Symposium (*IAU Symposium* no. 39) held in Crimea 8–19 September 1969 (to be published soon).

(iv) The effect of synchrotron instability due to cold plasma in radiosources with a high concentration of relativistic particles is proposed and investigated (V. V. Zheleznyakov, E. V. Suvorov).

(v) Many papers are devoted to the plasma instability and interaction of plasma and different wave modes inside and outside the magnetosphere of the Earth (V. Tu. Trakhtengerz). Here the most important are the plasma waves connected with magnetic fields.

c. *Influence of plasma turbulence on the physical and dynamical state of the cosmic medium*

The occurrence of plasma turbulence in the cosmic medium changes the gas motion and leads to new effects in the usual hydromagnetic flow.

(i) Plasma turbulence makes possible the formation of shock waves in collisionless plasma, both in plasma with high magnetic pressure and in plasma with no magnetic fields. The stability conditions for tangential discontinuity are also changed by the presence of plasma turbulence.

(ii) There are many problems which must be solved by using hydromagnetic and plasma turbulence equations together; such problems must be formulated in a self-consistent way. These are (besides the shock structure): flow of gas around bodies with magnetic fields, annihilation of magnetic field in the regions near neutral lines, instability of electric current sheets, etc. In these investigations, it is necessary to use the tensors of viscosity, electric and thermal conductivity, stress and so on in the hydromagnetic equations for plasma with strong turbulence. This has not been done yet in part because of the complexity of the nonlinear equations for plasma turbulence.

2. *Hydromagnetics of solar and stellar winds*

One of the main problems of modern Hydromagnetics and Plasma Astrophysics is the study of gas motions around stars and the Sun. The possibility of observational and experimental investigations using cosmic probes makes such work particularly significant.

a. *The collisionless shock*

The possibility of a shock having a scale much less than the collision mean free path, particularly where a magnetic field provides a mechanism for particle interaction or where plasma turbulence and the interaction of waves substitutes for particle collisions, has been studied ever since it was realized that many of the very low density gas flows in astrophysics were supersonic. Such shocks very probably exist in interstellar space where they may maintain the isotropy of galactic cosmic rays and may heat the gas in H I regions. They definitely exist in interplanetary space, where they can be studied with space probes. Space probe observations give us much information on the structure of the magnetic field, the gas density and velocity, etc. on both sides of the Earth's bow shock. Both theory and observation show that the shock thickness is of the order of the ion Larmor radius. The magnetic fields and other features of the shock are very irregular. All ions are heated more than electrons and fluctuating electric fields are found.

It is believed that the plasma turbulence excited in the fronts may serve as a mechanism of increasing entropy. Such turbulence also accelerates charged particles.

The theory of collisionless shocks may also explain some features of solar radio activity. No review is available on the theory of collisionless shocks but many papers have been published by R. Z. Sagdeev, V. V. Zaitsev and others.

b. *The flow of solar wind around the Earth*

(i) This flow, including the stability of the gas motion, the position of the shock, and the directions of stream lines, may be studied using phenomenological hydromagnetic equations. Such studies have been made by M. V. Samokhin and, for the flow around a comet, by Z. M. Jaffe.

(ii) Alternatively, the kinetic equations for the state of plasma turbulence in and around the magnetosphere may be studied, some progress having been achieved by R. Z. Sagdeev, A. I. Beresin, and V. Yu. Trakhtengerz.

c. *Modelling of plasma flow in interstellar space*

Laboratory modelling of cosmic plasma phenomena gives information which is difficult to obtain by any other means. The principle of similitude points out phenomena which may be reproduced in the laboratory with different parameters. For instance, it is possible to modulate the collisionless shocks and the tail (or wake) of the Earth's magnetosphere. The experiments have shown the micropulsations and other features in accordance with theory. The modelling of plasma flow around the Moon and planets is planned. A review article is "Physics of interplanetary plasma and laboratory experiments" (R. Z. Sagdeev, I. M. Podgorny, *Usp. fiz. Nauk.*, **98**, 411, 1969).

3. *Hydromagnetics of stellar interiors and surfaces*

a. *Theory of hydromagnetic convection, generation of stellar and solar magnetic fields*

Convective transport of energy inside stars plays an important role in the determination of stellar structure and stellar (solar) activity. The presence of magnetic fields can drastically change the whole character of convection and turbulence. Thus the problem of thermal convection and turbulence in hydromagnetics is one of the most important.

(i) Generalization of the Schwarzschild criteria for convective instability for the hydromagnetic case shows that external magnetic fields usually impede the convection but that it is possible to find conditions in which the magnetic field relieves the convection. Special cases are the vibrational convection in strong magnetic fields with transfer of energy by waves, and the heating of the solar corona by the transport of energy from the convective zone by Alfvén and magnetoacoustic waves. No review is available but there are many papers of S. A. Kaplan, N. S. Petruchin, S. B. Pikelner, S. I. Syrovatsky, Yu. I. Zhugzhda and others.

(ii) It is now believed that the magnetic fields of stars are created by internal motions (dynamo-model). Many modes of convective, circular or meridional motions of gas inside stars are investigated and it is shown that it is possible to explain the occurrence of stellar magnetic fields. The general method of calculation of the dynamo-model was developed by S. I. Braginsky. Some papers are published by E. M. Drobyshevsky, A. I. Kipper, R. S. Iroshnikov and others.

(iii) In usual stars, magnetic fields are irrelevant for the equilibrium or stability. But the magnetic field may have an important role in supermassive configurations which now are attributed to the quasars. For instance, the magnetohydrodynamics turbulence and pulsations can prevent collapse of such supermassive objects (L. M. Ozernoy, Ya. B. Zeldovich, V. V. Vandakurov).

b. *Behaviour of the magnetic fields in neutron stars and pulsars*

It is now widely believed that pulsars are rotating and pulsating neutron stars with very strong magnetic fields.

(i) Possibly superconductivity may occur during the collapse thus influencing the formation of a pulsar magnetic field (V. L. Ginzburg, D. A. Kirznitz).

(ii) The rotation of a pulsar with a strong magnetic field must lead to the generation of longwave electromagnetic radiation. If the velocity of motion of the poles of the magnetic field is relativistic, this longwave radiation is concentrated into a comparatively narrow beam in which the acceleration of electrons causes the generation of high-frequency radiation (S. A. Kaplan, V. Ya. Eidman).

(iii) The magnetospheres of pulsars have a great influence on the motion of nearby charged particles. Concentration of relativistic particles to the rotating equatorial plane of a pulsar magnetosphere can explain the pulsed nature of its radiation (review article on magnetic models of pulsars by V. I. Ginzburg, V. V. Zheleznyakov, V. V. Zaitsev "Coherent mechanisms of cosmic radiation and magnetic model of pulsars", *Usp. fiz. Nauk*, **98**, 201, 1969).

(iv) It is now assumed that the main sources of cosmic rays are pulsars (I. S. Shklovsky).

E. COSMIC RAYS

(by L. Davis)

A number of significant astrophysical problems are related to cosmic rays. The mechanisms by which cosmic rays are accelerated, an understanding of their isotopic abundances, a determination of the regions in which galactic cosmic rays are stored, their relationship to the various recently discovered highly energetic objects, the way they diffuse in the solar system are all important both as cosmic ray problems and for their relation to basic problems in other areas of astronomy. Surveys of, and extensive references for, these topics may be found in recent reviews listed below. A detailed account of recent work is given in the Proceedings of the 11th International Conference on Cosmic Rays, Budapest, 1969 (in press).

In some of these problems the plasma properties of the cosmic ray gas are significant and there are important interactions between cosmic rays and lower energy plasmas. Plasma instabilities appear to hold the net flux of cosmic rays throughout the Galaxy to a very low value and hence require near isotropy of cosmic rays. The balance between cosmic ray pressure, gravitation on gas, and magnetic forces determines the large scale structure of gas and field; instabilities in this balance should be involved in many of the irregularities and condensations of the structure. Energy derived from very low energy cosmic rays may be significant in the temperature balance of H I regions (Spitzer and Scott, 1969; Goldsmith, Habing, and Field, 1969).

REVIEWS

Meyer, P. 1969, *A. Rev. Astr. Astrophys.*, **7**, 1.
Ginzburg, V. L. 1969, *Comments Astrophys. Space Sci.*, **1**, 207.
Ginzburg, V. L. 1970, *Comments Astrophys. Space Sci.*, **2** (in press).
Parker, E. N. 1969, *Space Sci. Rev.*, **9**, 651.

44. ASTRONOMICAL OBSERVATIONS FROM OUTSIDE THE TERRESTRIAL ATMOSPHERE (OBSERVATIONS ASTRONOMIQUES AU-DEHORS DE L'ATMOSPHERE TERRESTRE)

PRESIDENT: R. Wilson.
VICE-PRESIDENT: V. K. Prokof'ev.
ORGANIZING COMMITTEE: J. E. Blamont, C. de Jager, L. Goldberg, R. Lüst, F. G. Smith, R. Tousey.

INTRODUCTION

In view of the limitation on length set by the cost of producing the IAU Reports, the present report can only be a selection of the most important advances during the past three years and not a complete review of the field.

Attempting to pick out a few points of interest, since the last report improvements have been made in recording solar X-ray spectra and in high spatial resolution imaging; the OSO satellite programme continues to produce valuable data. The eclipse experiments planned for March 1970 will, if successful, provide a considerable amount of new information. The solar iron abundance is still a subject of interest in view of the revision of oscillator strengths used in the photospheric analyses. Further interplanetary observations have shown that the general nature of the solar wind at the Earth does not depend on the phase of the solar activity cycle, and have revealed large increases in the abundance of helium in the solar wind following solar flares.

There have been considerable advances in UV Astronomy. Good quality spectra of coarse resolution, obtained from pointing rockets, are now available for a number of bright early type stars, whereas the successful operation of the OAO-AS satellite is providing a large amount of new data, largely photometric but including a coarse spectral scan. Mass loss has been detected from the early type supergiants and the interstellar extinction curve has been extended to 1150 Å and reveals a fine structure near 2200 Å. Observations have been made of some of the brighter external galaxies and the nucleus of Andromeda shows an unexpected UV emission component.

One of the most exciting events in X-ray astronomy has been the discovery that the pulsar in the Crab Nebula has very strong X-ray emission pulsating with the same period as the radio and optical radiation. Observations of the soft X-ray cosmic background have shown that the intensity is larger than expected from a simple extrapolation of the power law spectrum in the 1–10 keV region, and the excess is probably of galactic origin.

The Report is divided according to subject matter into six sections. These sections have been prepared by R. Tousey, M. Neugebauer, H. E. Butler, L. Gratton, W. Kraushaar and F. G. Smith and deal respectively with solar astronomy, the interplanetary medium, stellar astronomy, X-ray sources, the X- and γ-ray cosmic background and radio-astronomy. I am extremely grateful to the above authors for their contributions, and to Dr Carole Jordan, who as Scientific Secretary to the Commission carried out much of the organization of this Report.

R. WILSON
President of the Commission

SOLAR ASTRONOMY

(by R. Tousey)

In the period since the 1967 IAU General Assembly there has been great activity in solar research, particularly in X-rays, imaging, and monitoring. Vehicles launched carrying solar experiments include the following: *Balloons:* Employed principally by France and the U.S.A.; equipped with solar pointing systems. *Rockets:* The Aerobee-150 and 170 (U.S.A.); Skylark (U.K.); Veronique (France); three-axis stabilization and solar pointing is accomplished with SPARCS (U.S.A.); and Stage-3 ACU (U.K.) with a stability of about 2 arc sec, peak-to-peak; biaxial pointing-controls continue in use. *Satellites:* Orbiting Solar Observatories, OSO III–VI (NASA); OGO IV–VI (NASA) and OV 1–10 and 17 (USOAR and Aerospace), all with solar-pointed sections; SOLRAD 9 (NRL) for monitoring; Cosmos-166, 230, 262, with solar-pointed sections (U.S.S.R.).

Vehicles soon to become available for solar research are: The Aerobee-350 and Black Brant V-B with greatly increased load-volume-altitude capability; both can be equipped with type II SPARCS solar pointing and stabilization to 0·5 arc sec peak-to-peak. Water recovery is now practicable. OSO is being improved; OSO-H by more than doubled size and power; and OSO-I will have telemetry increased up to 6400 bits/s, increased stability and additional commands. In the manned program, the Apollo Telescope Mount, ATM, (**1**), is to be launched in 1972 and operated in combination with the Orbital Workshop during three visits for a total of at least twenty weeks. Solar experiments proposed for automatic operation or operation by astronauts from a lunar base are under consideration by NASA.

Solar observations from rockets and balloons

The XUV spectrum

Advances include: a 30mÅ resolution 2000–2200Å echellogram (**2**); line spectra, 12–120Å and 140–800Å with intensities and $\Delta\lambda$ to $\pm$40mÅ (**3**); photoelectric line and continuum intensities at 70mÅ resolution, 360–1340Å (**4**); photographic line and continuum intensities at 0·2Å, resolution, 1400–2200Å (**5**), and at 0·5Å, 20 arc sec resolution, 1984–2839Å (**6**); a continuing program of photoelectric spectral intensity measurements, 300–1300Å (**7**). Line lists expected during 1970 are 3000–2100Å (NRL), and 2200–2000Å (Culham). Line profile work includes: spectra of $\pm$7 arc sec, 35mÅ resolution, 2770–2845Å, including MgII, H, and K, from a balloon (**8, 9**); MgII, H, and K with 30mÅ resolution and 6 arc sec spatial resolution along a solar diameter using a Fabry-Pérot and echelle (**10**); echelle-photographed profiles, at 15mÅ resolution and with partial spatial resolution, of H-$L\alpha$ (**11**), SiIII 1206Å and OI 1302, 5,6Å (**12**), and CII 1334,6Å (**13**).

XUV imaging from rockets and balloons

Images in MgII 2795·5Å, obtained with a double-Solč 2·1Å filter and resolution approaching 8 arc sec, are similar to but show more contrast than CaK at 0·1Å bandpass (**14**). Broad-band (80Å), images at 1975, 2190, and 2235Å, with as much as 15 arc sec resolution, obtained from a rocket and two balloons show decreased limb-darkening and greater similarity to CaK at 1975Å, which is below the 2090Å absorption edge and is chromospheric (**6**). With a scanning photoelectric technique the H-$L\alpha$ disc was shown to contain structural detail as small as 2 arc sec, the instrumental resolution (**15**). Spectroheliograms from 171 to 650Å were obtained on several rocket flights (**16**); on 4 November 1969 a spatial resolution near 5 arc sec was achieved; spicules are prominent in HeII 304Å; the network is conspicuous in HeI 584Å, HeII 304Å, and OV 630Å, but is absent or masked in all lines from ions of higher excitation. Unpublished data show that plages retain their fine structure through NeVII, lose it partially in MgIX and completely in FeXV and FeXVI. Coronal emission is intense from MgIX and higher excitation ions; the emission above the HeII 303·78Å limb is from SiXI 303·31Å (λ from (**3**)), and is resolved in several spectroheliograms. Broad-band images, 171–500Å, continue to show emission far above the limb, reaching 30 arc min on some occasions (**16, 17**). X-ray images obtained by pinhole and zoneplate techniques (**18–20**), although of value, are less powerful than the Wolter lens (**21**), with which images having a few arc seconds resolution have been obtained (**22**). On two occasions these instruments were flown during a solar flare (**23**). Combined with a transmission grating, the X-ray lens has produced X-ray spectroheliograms of small active centers and flares (**23**).

White light and infrared corona

White light coronal photographs from 3 to $10 R_s$ were obtained together with XUV spectroheliograms and heliograms during the total eclipse on 22 September 1968, on 27 and 29 April 1968, and on 16 and 17 April 1969 (**24**). The principal findings are that the streamers are straight and nearly always approximately radial, and that they change greatly, even in 24 hours. The coronal streamers have also been photographed from balloons from 1·7 to $5 R_s$ (**25**). At 2·2μ, with a coronagraph on a balloon at 93000 ft, measurement of the brightness from 3·5 to 9·2 R_s confirmed a peak at 4 R_s (**26**), whose origin had been proposed to be thermal emission from refractory interplanetary dust (**27**); the earlier observations were made from 13000 ft in Bolivia during the 1966 eclipse (**26, 27**).

Eclipse experiments

During the 12 November 1966 eclipse two rockets carrying solar-pointed photometers for 1050–1250 Å and 1250–1350 Å, were flown into the eclipse cone (**28**). After correction for a plage, the results from the better flight indicate that H-$L\alpha$ for the quiet sun is brighter by approximately 30 % at the center of a ring of 15 arc sec width, centered 6·4 arc sec inside the near-infrared limb. During the same eclipse, three rockets, each carrying eight curved-crystal monochromators for selected emission lines between 16 and 40 Å were launched into the cone. Less than 1 % of full-sun emission was recorded during totality (**29**). On 7 March 1970 the eclipse cone can be reached by rockets launched from Wallops Island, Virginia. Experiments are being prepared under NASA sponsorship for two launches; one by a consortium of Harvard and York Universities, Imperial College and Culham, the other by NRL. The principal experiments will photograph the flash spectrum and the corona in various wavelength ranges. NRL will fly also a photoelectric scanning spectrometer for the range 1300–2000 Å.

Orbiting Solar Observatory (OSO)

Highly successful launches took place with OSO-III on 8 March 1967; OSO-IV, 18 October 1967; OSO-V, 22 January 1969; and OSO-VI, 9 August 1969. Results from OSO-III are published (**30**), but only scattered reports from OSO-IV–VI have appeared. OSO-H is scheduled for 1971; this and later OSO's will have a pointed section of size up to 15 × 15 × 57 inches, and will carry up to 250 pounds. The wheel diameter will be 56 inches, and power 75 watts for OSO-H. Like OSO-VI, the follow-on spacecraft will have offset pointing and small rastering (5 × 5 arc min for OSO-H), and roll will be command-controlled by magnetic torquing. In OSO-H the pointed section will contain a group of Bragg spectrometers (6–25 Å), and a grazing incidence spectrometer (170–400 Å) equipped with an image-forming collector so that spectroheliograms in selected lines can be recorded by rastering (Neupert, GSFC); the other part of the pointed section will be instrumented to record in alternation the white light corona from 3–10 R_s using an SEC Vidicon and the sun-centered mode, and the XUV (171–500 Å) corona using a mirror-aluminium filter-channel photomultiplier instrument in the raster-mode, in order to observe coronal changes and transients and correlate them in white light and in the XUV (Tousey, NRL). In the wheel are soft and hard solar X-ray monitors by Peterson, University of California at San Diego, and Chupp, University of New Hampshire.

XUV imaging from satellites

The Harvard scanning spectrometers on OSO-IV and VI and the NRL spectroheliograph on OSO-V have been used in the spacecraft raster-mode to record many monochromatic solar images. On OSO-IV the entire sun was scanned once per minute in single, ground-commanded wavelengths between 300 and 1400 Å; on OSO-V with a similar raster-scan, images were obtained simultaneously in He II 304 Å and Ne VII 465 Å, and in Si XII 499 Å and H I 1216 Å. For both instruments the resolution was 1 × 1 arc minute. The OSO-VI experiment was like the OSO-IV, but with significant improvements: Spatial resolution 35 × 35 arc sec; scanning over a 46 × 46 arc min sun-centered area once per 8 min, or over a 7·5 × 7 arc min area once per 30 s; the latter area could be selected anywhere over the 46 × 46 arc min area by ground-command. From OSO-IV more than 4000 images were recorded in fifty-two wavelengths (**31**). Limb brightening was observed in various lines and used to derive a new model for the equatorial chromospheric-coronal transition region (**32**), with results for T_e vs height in good agreement with (**33**). Unpublished center to limb analyses by Noyes and Kalkofen, in different parts of the Lyman continuum show limb-darkening near the head. Withbroe and Noyes point out that the contrast between active and quiet regions appears to go through a minimum for lines in the chromospheric-coronal transition region. Also flown in OSO-IV was an X-ray spectroheliometer of the AS & E Corporation that produced one solar image per 5 min with several arc minutes spatial resolution for 3·5–12 Å (**34**). Both impulsive and gradual variations in plage X-ray emission were found (**35**). An X-ray imaging system built by Leicester and University College London is operating in OSO-IV.

X-ray spectra

Spectra of plages and of flares have been obtained by several groups using Bragg crystal spectrometers in satellites:

OSO-III	1·3–3 Å and 8–20 Å ,	GSFC (**36**);
OSO-IV	0·5–3·9 Å, 1·4–8·5 Å ,	NRL (**37**);
OSO-V	1 –3·8, 3·6–8·4, 8·2–25 Å,	GSFC (Neupert);
OSO-VI	0·5–3·9 Å, 1·4–8·5, 5 –16 Å,	NRL (Kreplin);
OV 1–10	8 –25 Å ,	Aerospace (**38**);
OV 1–17	1·5–25 Å ,	Aerospace (**39**).

The results, although differing in details, show the following for the 1·6–25 Å range: Plages emit both continuum and lines; great intensity increases occur for all spectral features during a flare; many new lines are emitted during a flare; the lines include both optical and inner shell transitions from highly-stripped atoms. The broad 1·9 Å feature, prominent during flares, is associated with K-line emission from Fe in all stages between helium-like and 16-electron, and perhaps lower; a small peak at 1·86 Å, reported during the most intense flares, may be $L\alpha$ of one-electron Fe. Similar but less intense emission from 1·60–1·66 Å is the Ni counterpart. The changes with time of the lines during a flare show that the lines of highest ionization stage reach maximum intensity and decay soonest, in accordance with cooling of the flare plasma and recombinations. Several lines of previously doubtful identification have been explained as the helium-like ion resonance lines caused by single-photon decay of the $1s2s\,^3S_1$ level (**40**). The intensity ratio of the ^1S–^3P intercombination to the ^1S–^3S forbidden line allows derivation of the coronal density (**41, 39**). Within the continuum an edge observed at 3·59 Å is identified as recombination of bare *S* ions; edges at 2·06, 2·8, and 4·46 Å are not easily explained; dielectronic recombination has been suggested by Doschek and Meekins. The production of the continuum appears to be complex, perhaps involving more than one mechanism.

Solar monitoring

Experiments for monitoring the total solar flux in various spectral bands from the near ultraviolet all the way to 250 KeV X-rays have been flown by many groups and in many spacecraft. Data from the NRL Solrad 9 and the University of Iowa detectors on Explorers 33 and 35 are now published by ESSA in *Solar-geophysical Data*. Worldwide reception of solar data is available from Solrad and has been used by various observatories, e.g. by Arcetri (**42**). Solar X-ray maps (9·1–10·5 Å) produced by the University College London-Leicester experiment on OSO-V are being sent to ESSA on a regular basis. One or more monitoring experiments were placed in OSO-I–VI, OGO-IV–VI, and several Explorers of NASA; several OV satellites of the AFOAR; Vela of the AEC; ESRO IA, IB, and II. The principal groups providing instrumentation are: in soft X-rays, NRL, Aerospace Corporation, Lockheed, Los Alamas, the Universities of Michigan and Iowa, University College London, and Utrecht; in the XUV, AFCRL, the University of Colorado, and University College London; in hard X-rays, the Universities of Minnesota, California at San Diego and at Berkeley, and GSFC, together with many groups using balloons. The U.S.S.R. has a continuing program of

XUV and X-ray monitoring with "Cosmos" (**43**). At present reports of results are widely scattered (**44, 45**). Nevertheless the following facts seem to be established (**46**): High energy X-ray emission and radio noise increase together with rising solar activity, but the physics involved is still not well understood. Flares are usually preceded by a slow rise in X-ray flux that is greatest for <3Å; during this period there are often precursors in Hα and in microwave emission. In the main phase the X-ray flux increases rapidly and is coincident with a microwave burst, whose structure is related to that in the accompanying burst of >20 keV X-rays. The X-ray flare peak is reached first in 0–3 Å and last in 8–20 Å or longer. X-ray flares always accompany an Hα flare but are sometimes present by themselves, presumably when the flare is located over the limb (**47**). Emission of X-rays extends high into the corona during a large flare. Quasi-periodic structure has been found in the output from 9-channel spectra of solar X-ray bursts (14–250 keV) recorded with 1·85 s time-resolution by a GSFC experiment in OSO-V (**48**).

Flare observations

Flare history data with 30-s time and 30-arc sec spatial resolution are being obtained with the Harvard spectrometer on OSO-VI; flares were recorded on OSO-IV with 5-min and 1-arc-min resolution. Flare histories without spatial resolution but with 0·16 s time resolution were recorded with the AFCRL spectrometer on OSO-III for lines in the range 260–1300 Å (**49**); these flare produced enhancements in total emission have been found to correlate well with ionospheric Sudden Frequency Deviations (**50**). Structure in X-ray images of flares has been found with scanning slit heliographs on Cosmos-230; these instruments produced 20 arc sec $\times$ 2 arc min resolution for 2–8Å and 20 $\times$ 15 arc sec for 8–14 Å (**51, 52**). A grazing incidence X-ray telescope, and also a photoelectric scanning spectrometer (20–300 Å) are in orbit on Cosmos-262 (**43**). On a rocket flight on 8 June 1968 AS & E photographed fine structure in a Class 1N flare with a few arc seconds resolution (**23**). On near-simultaneous flights on 4 November 1969 AS & E obtained a series of X-ray images with great time resolution of a 1B flare near the limb, and NRL recorded the spectrum of the same flare, both with about 5 arc sec resolution (**23**).

Radio observations

Radio noise burst observations have been made by the Research Telecommunications Institute (Canada) from Alouette I and II (**53**), and by GSFC from ATS-I and -II, and RAE-I (**54**), and by the U.S.S.R. from Venus 2 (**55**). The frequencies covered are in the range 0·1–15 MHz. The results for type-III bursts indicate a high density, low-temperature model for coronal streamers to $50R_s$ but alternative models are possible.

Infrared

In the infrared ($\approx$ 50–300μ), the sun's brightness gives directly the value of temperature at the minimum between the chromosphere and corona, because of the high opacity of H^-. Measurements made from an aircraft near the tropopause using high resolution Fourier spectroscopy, after correction for residual absorption, gave values from 4338 to 4438 K in four windows between 238 and 312μ (**56**). From a balloon, 4600 K was found near 60μ and 4000 K at 180μ (**57**). In this difficult spectral region accuracies are less than desired. The Bilderberg Model value of minimum temperature (**58**), 4600 K, lies almost within the error bars of the infrared data.

Interpretation

The controversy over the ten times greater coronal than photospheric abundance of Fe, first noted by Pottasch from XUV spectra and confirmed by others, appears to be nearly resolved. New *f*-values of Fe I, lower than those of King and King, have been found by (**59, 60**), and by the beam foil method by (**61**). These results have been extended over a wide range of excitation potentials by means of a special arc by (**62**). According to (**63**) the new *f*-values imply an abundance increase by just the factor of 10 required to bring the photospheric abundance into agreement with the coronal. A somewhat similar conclusion has been reached by (**64**). Further consequences of this revision are expected. Dielectronic recombination (e.g., **65**), has been established as essential for interpreting XUV line intensities; with its use, temperature and abundance values from the XUV resonance lines and those from the visible forbidden lines come into reasonable agreement. For example for the quiet corona, Fe XI and Si X are the most abundant species and their resonance lines lead to $T_e = 1{\cdot}7 \times 10^6$ K, and $N_{Fe}/N_{Si} = 1{\cdot}5$ (**66**). For extensive calculations of ionization equilibrium for the most abundant solar atoms from carbon through nickel, see (**67**). Much work on abundances, and on the prediction of XUV line intensities and comparison with observation has been done by (**68**). The Bilderberg Model of the photosphere and low chromosphere, (**58**), has been subjected to some uncertainty in the neighborhood of the temperature minimum, for which 4600 K was assumed, as a result of a disagreement in the value of observed brightness temperature near 1650 Å (**69**): The NRL photographic spectra lead to 4650 K (**70**), while the Harvard photoelectric spectra give 4400 K (**4**). The higher temperature is supported by data for 2080–1950 Å (**71**), and the lower by results in the infrared (**56**). Additional work is required. XUV spectroheliograms have been analyzed and interpreted in terms of chromospheric structure (**72, 73**).

REFERENCES

1. *Trans. IAU*, **XIIIA** (Reports 1967), Commission 44, Solar Astronomy.
2. Jones, B. B., Boland, B. C., Wilson, R., Engstrom, S. T. F. 1970, *IAU Symposium* no. 36, Lunteren.
3. Jones, B. B., Freeman, F. F., Wilson, R. 1968, *Nature*, **219**, 252.
4. Parkinson, W. H., Reeves, E. M. *Solar Phys.* (in press).
5. Sandlin, G. D., Widing, K. G. 1967, *Astrophys. J.*, **149**, L129.
6. Bonnet, R. M., Blamont, J. E. 1968, *Solar Phys.*, **3**, 64.
7. Hinteregger, H. E. 1969, Ann. IQSY, *Solar-Terr. Phys.*, **5**, 305.
8. Lemaire, P., Blamont, J. E. 1967, *Astrophys. J.*, **150**, L129.
9. Lemaire, P. *IAU Symposium* no. 36, Lunteren.
10. Bates, B., Bradley, D. J., McKeith, C. D., McKeith, N. E., Burton, W. M., Paxton, H. J. B., Shenton, D. B., Wilson, R. 1969, *Nature*, **224**, 161.
11. Bruner, E. C., Jr, Parker, R. W. 1969, *J. geophys. Res.*, **74**, 107.
12. Bruner, E. C., Jr, Rense, W. A. 1969, *Astrophys. J.*, **157**, 417.
13. Berger, R. A., Bruner, E. C., Jr. 1969, *Astrophys. J.*, **155**, L115.
14. Fredga, K. 1969, *Solar Phys.*, **9**, 358.
15. Sloan, W. A. 1968, *Solar Phys.*, **4**, 196; **5**, 329.
16. Purcell, J. D., Tousey, R., Koomen, M. J. 1968, *Space Res.*, **8**, p. 450.
17. Tousey, R., Sandlin, G. D., Purcell, J. D. 1968, *IAU Symposium* no. 35, Budapest, p. 411.
18. Bol Raap, B. E., Le Poole, J. B., Dijkstra, J. H., De Graaff, W., Lantwaard, L. J. 1969, *Small Rocket Instrumentation Techniques*, North-Holland Publishing Co., Amsterdam, p. 203.
19. Burton, W. M. 1969, *Solar Phys.*, **8**, 53.
20. Unzicker, A. E., Meekins, J. F., Conrads, H. Submitted to *Appl. Opt.*
21. Mangus, J. D., Underwood, J. H. 1969, *Appl. Opt.*, **8**, 95.
22. Reidy, W. P., Vaiana, G. S., Zehnpfennig, T., Giacconi, R. 1968, *Astrophys. J.*, **151**, 333.
23. Vaiana, G. S., Reidy, W. P., Zehnpfennig, T., Van Speybroeck, L., Giacconi, R. 1968, *Science*, **161**, 564.
24. Koomen, M. J., Seal, R. T., Purcell, J. D., Tousey, R. 1969, *Sky Telesc.*, **37**, 356.
25. Dolfuss, A., Fort, B., Morel, C. 1968, *C. r. Acad. Sc. Paris*, **266**, 1537.
26. MacQueen, R. M. 1968, *Astrophys. J.*, **154**, 1059; *Appl. Opt.*, **7**, 1149.
27. Peterson, A. W. 1967, *Astrophys. J.*, **148**, L37.
28. Blamont, J. E., Malique, C. 1969, *Astr. Astrophys.*, **3**, 135.
29. Argo, H. V., Bergey, J. A., Evans, W. D., Singer, S. 1968, *Solar Phys.*, **5**, 551.
30. Brandt, J. C., Hinteregger, H. E., Hall, L. A., Neupert, W. M. *et al.*; Teske, R. G.; Hudson, H. S. *et al.*; Kaplon, M. F., Valentine, D.; Kraushaar, W. L. *et al.*; Neel, C. B. *et al.*; Papers. 1969, *Solar Phys.*, **6**, 171–240.
31. Goldberg, L., Noyes, R. W., Parkinson, W. H., Reeves, E. M., Withbroe, G. L. 1968, *Science*, **162**, 95.
32. Withbroe, G. L. 1969, To appear in *Bull. AAS*.
33. Dupree, A. K., Goldberg, L. 1967, *Solar Phys.*, **1**, 229.

34. Paolini, F. R., Giacconi, R., Manley, O., Reidy, W. P., Vaiana, G. S. 1968, *Astr. J. Suppl.*, **73.**
35. Krieger, A. S., Vaiana, G. S., Special Meeting AAS. 1969, *Bull. AAS*, **1**, 283.
36. Neupert, W. M., White, W. A., Swartz, M., Young, R. M. 1969, *Solar Phys.*, **6**, 183.
37. Meekins, J. F., Kreplin, R. W., Chubb, T. A., Friedman, H. 1968, *Science*, **162**, 891.
38. Walker, A. B. C., Rugge, H. R. 1969, *Solar Flares and Space Research*, North-Holland, Amsterdam, p. 102.
39. Walker, A. B. C., Rugge, H. R. 1969, Submitted to *Astr. Astrophys.*
40. Gabriel, A. H., Jordan, C. 1969, *Nature*, **221**, 947.
41. Gabriel, A. H., Jordan, C. 1969, *Mon. Not. R. astr. Soc.*, **145**, 241.
42. Landini, M., Fossi, B. C. M., Poletto, G., Tagliaferri, G. L. 1968, *Solar Phys.*, **5**, 546.
43. Prokof'ev, V. K. Communicated to Commission 44.
44. Friedman, H. 1969, Energetic Solar Radiations, *Solar-Terrestrial Physics*, **4**, 36. MIT Press.
45. Hudson, H. S., Peterson, L. E., Schwartz, D. A. 1969, *Solar Phys.*, **6**, 205; *Astrophys. J.*, **157**, 389.
46. Kreplin, R. W., Moser, P. J., Castelli, J. P. To appear in *Space Res.*, Vol. 10.
47. Teske, R. F. 1969, *Solar Phys.*, **6**, 193.
48. Frost, K. J. *Astrophys. J. Lett.*, in press.
49. Hinteregger, H. E., Hall, L. A. 1969, *Solar Phys.*, **6**, 175.
50. Donnelly, R. F. 1969, *J. geophys. Res.*, **74**, 1873.
51. Vasil'ev, B. N., Žitnik, I. A., Corneev, V. V., Crutov, V. V., Mandel'stam, S. L., Tindo, I. P., Tsheremuchin, G. S., Shurigin, A. I. 1968, *Kosm. Issled.*, **6**, 420.
52. Beigman, I. S., Grineva, Yu. I., Mandel'stam, S. L., Vainstein, L. A., Žitnik, I. A. 1969, *Solar Phys.*, **9**, 160.
53. Hartz, T. R. 1969, *Planet. Space Sci.*, **17**, 267.
54. Alexander, J. K., Malitson, H. H., Stone, R. G. 1969, *Solar Phys.*, **8**, 338.
55. Slysh, V. I. 1967, *Soviet Astr.*, **11**, 389.
56. Eddy, J. A., Lena, P. J., MacQueen, R. M. 1969, *Solar Phys.*, **10**, 330.
57. Mankin, W. G., Strong, J. 1970, *Solar Phys.*, (in press).
58. Bilderberg Model: 1968, *Solar Phys.*, **3**, 5–239.
59. Huber, M., Tobey, F. L. 1968, *Astrophys. J.*, **152**, 609.
60. Grasdalen, G. L., Huber, M., Parkinson, W. H. 1969, *Astrophys. J.*, **156**, 1153.
61. Whaling, W., King, R. B., Martinez-Garcia, M. 1969, Report OAP-166, Cal. Inst. Tech.
62. Garz, T., Kock, M. 1969, *Astr. Astrophys.*, **2**, 274.
63. Garz, T., Holweger, H., Kock, M., Richter, J. 1969, *Astr. Astrophys.*, **2**, 446.
64. Withbroe, G. L. 1969, *Solar Phys.*, **9**, 19.
65. Burgess, A., Summers, H. P. 1969, *Astrophys. J.*, **157**, 1007.
66. Widing, K. G., Sandlin, G. D. 1968, *Astrophys. J.*, **152**, 545.
67. Jordan, C. 1969, *Mon. Not. R. astr. Soc.*, **142**, 499.
68. Nikolsky, G. M. 1969, *Solar Phys.*, **6**, 399. (See also Ivanov-Holodny, G. S., Nikolsky, G. M. 1969, *The Sun and The Ionosphere*, p. 455.)
69. Gingerich, O., Cuny, Y., Noyes, R. W. 1969, To appear in *Bull. AAS.*
70. Sandlin, G. D., Widing, K. G. 1967, *Astrophys. J.*, **149**, L129.
71. Bonnet, R. 1968, *Ann. Astrophys.*, **31**, 597.
72. Allen, C. W. 1969, *Solar Phys.*, **8**, 72.
73. Zirin, H. 1969, *Solar Phys.*, **9**, 77.

THE INTERPLANETARY MEDIUM

(by M. Neugebauer)

It is the purpose of this report to summarize briefly the most significant advances in our understanding of the interplanetary medium derived from satellite and space-probe experiments during the past three years and to suggest possible major future developments in this area of research.

Since 1965, the conditions in interplanetary space have been monitored not only nearly continuously, but often by more than one spacecraft simultaneously. One of the most important results is that the general nature of the solar wind near 1 A.U. does not depend in any spectacular way on the phase of the solar activity cycle, even though the average values of plasma density, velocity, etc., may vary due to changes in the number or intensity of high-velocity plasma streams.

At the time of the previous report, essentially nothing was known about the electrons in the solar wind. We now know that, as predicted, the electron convective (or flow) velocity is equal to that for the positive ions and that the solar wind is electrically neutral. The Vela 4 measurements (**1**) showed that in the solar-wind reference frame the electrons have an anisotropic velocity distribution function with the maximum velocity away from the sun and parallel to the magnetic field. The magnitude of the thermal anisotropy is smaller than that for protons; the electron anisotropy is probably collision limited. Vela 4 observed the electron temperature to range from 7×10^4 to 2×10^5 K when averaged over all directions; it varies less with time than does the proton temperature. The electron temperature was 1·5 to 5 times greater than the proton temperature; comparison of this temperature ratio with the value of ~ 16 predicted by a two-fluid model (**2**) of the solar wind indicates that more energy is transferred from the electrons to the positive ions than can be accounted for by ion-electron collisions.

Early measurements indicated a large temporal variation in the composition of the solar wind; the ratio of alpha-particle to proton number density varied from near 0 to 0·20, with rapid variations. Recent observations (**3, 4**) indicate that flare-associated plasmas are sometimes unusually helium rich. The alpha-particle thermal distribution is anisotropic and similar to that observed for protons. The fact that the ratio of alpha-particle temperature to proton temperature is often ~ 4 (corresponding to equal proton and alpha thermal velocities) is another indication of the collisionless transfer of energy from the electrons to the positive ions. The alpha-particle temperature is nearly equal to the proton temperature only when the protons are unusually cool; i.e., when little collisionless heating occurs (**5**).

When the proton and alpha-particle temperatures are unusually low, it is possible to resolve other helium ions and oxygen ions in the electrostatic-analyzer spectra. Vela measurements showed the following ranges of ion number density ratios (**5**); $^3He/^4He$ from $\leqslant 4 \times 10^{-4}$ to $1{\cdot}3 \times 10^{-3}$; $^4He^+/^4He^{++}$ from $\leqslant 10^{-3}$ to $\sim 3 \times 10^{-3}$; He/O from 25 to 80; and $O^{+7}:O^{+6}:O^{+5} = (\leqslant 0{\cdot}2 \text{ to } 3):1:\leqslant 0{\cdot}1$. The ratio of O^{+6} to O^{+7} is thought to indicate the temperature of the coronal region from which the solar wind emanated (**6**). Several different hypotheses have been proposed to explain the relatively large amounts of He^+ and O^{+5}.

Conflicting results have been obtained on the direction of the solar wind: the Vela data gave an average of 1·5° in the direction of corotation with the sun (**7**) for the period July 1964 through July 1965, while the IMP 1 data indicated 1·5° anticorotation for the period November 27, 1963, through February 24, 1964 (**8**).

The sector structure of the interplanetary magnetic field has shown a large amount of evolution with the increase of solar activity (**9**). The polarity of the field still remains steady for extended periods of time, but the very long-lived sectors of solar minimum are no longer evident (**10**). A study of the periodicities of the field direction has shown an apparent lengthening of the recurrence period which has been interpreted as an increase in the latitude of the solar source from 10–15 °N for IMP 1 to 20–25 °N for IMP 3 (**11**). There is some evidence for a correlation of the component of the field perpendicular to the ecliptic with the solar latitude of the point of observation (**12**).

Power spectra of variations in the interplanetary field have been determined over the frequency range $f = 3 \times 10^{-6}$ to $f = 5 \times 10^{-1}$ Hz (**13–16**). The spectral power density is generally proportional to f^{-x}, where $1 \leqslant x \leqslant 2$. This power density probably consists of contributions from both a large number of hydromagnetic discontinuities (**15**) and from large amplitude, aperiodic Alfvén (transverse) waves propagating outward from the sun along the field lines (**14**).

In addition to the by-now common plasma, energetic-particle and magnetic-field experiments, several recent space probes (Zond 3, Venus 2, Luna 11 and 12, Pioneers 8 and 9, and OGO 5) have carried apparatus for the measurement of the electric fields in space (**17–19**). Low-frequency (hundreds of Hz) electric-field oscillations are essentially always detectable; these waves have large amplitude variations which correlate with other changes in the solar wind. Large-amplitude high-frequency (tens of kHz) waves are also occasionally detected.

More data are slowly being acquired relevant to the radial gradients of properties of the interplanetary medium. Mariner 4 (**20**) observed a less rapid change in the pitch of the spiral angle between 1·0 and 1·5 A.U. than would be expected for a constant solar wind velocity whereas the field magnitude varied approximately as expected. The relative amplitude of the transverse fluctuations in the field (from 10^{-5} to 10^{-2} Hz) was independent of heliocentric distance and the amplitude of compressional fluctuations increased as $r^{0.69}$. Experiments to determine the gradient of cosmic-ray intensity between 0·7 and 1·5 A.U. have given contradictory results (**21–26**) ranging from -14 to $+500$ %/AU.

The increased data rates of recent spacecraft have allowed the study of some of the smaller-scale hydromagnetic structures. Observations of interplanetary shock waves have demonstrated that (i) the conditions across the shocks are not inconsistent with the Rankine Hugoniot relations (**27, 28**), (ii) the shock is greatly decelerated as it moves away from the sun (**28**), and (iii) the shock front can appreciably affect the spectrum of energetic solar particles incident upon it (**29**).

Discontinuities other than shocks have also been observed and studied in detail. That these surfaces are convected by, but do not propagate through, the solar wind is demonstrated both by the abrupt change in composition, as shown by the relative number of alpha particles (**30**), and by the velocity, as determined by sequential observations of the surface by more than one space probe (**31, 32**). The observed jump conditions across the surfaces (**30, 32, 33**) have all been consistent with theoretical predictions. The preferential alignment of the discontinuities along the spiral of the interplanetary field (**16, 32, 33**) has been suggested to cause a correlation between solar-wind velocity and direction with the faster streams coming from a more westerly direction (**33**). The thickness of these shocks and other "discontinuities" ranges from $\lesssim 500$ to ~ 20000 km (**16, 32**); current interest is focused on their internal structure, which is often fairly complicated.

Observations have also been made of what is believed to be the merging or reconnection of magnetic-field lines with accompanying plasma heating and high-frequency electrostatic waves (**34, 35**).

Much of the future progress expected in this field of research will be extensions of the best recent experiments; more information will be obtained on the details of the wave-particle interactions in the solar wind which transfer thermal energy from the electrons to the heavier ions and which limit the growth of the proton thermal anisotropy. More complete data will be obtained on the chemical (and isotopic) composition of the solar wind. Better and more complete measurements will be made on the long-term averages and variations in the flow direction of the solar wind. More detailed spectra will be obtained of magnetic- and electric-field oscillations. The cosmic-ray gradient will be better understood. The internal structure of collisionless shock waves, other discontinuities, and regions of field-line reconnection will be more accurately mapped to give a better understanding of the physical processes occurring within them.

Significant results can be expected from the spacecraft which go much closer to or much farther from the sun than the 0·7 to 1·5 A.U. studied to date. Data acquired close to the sun might reveal whether the solar wind is accelerated only in certain regions or whether the entire corona is involved. Outward from the sun, some of the major questions are: How far does the solar wind extend? What are the termination mechanisms? How and where is the galactic cosmic radiation modulated and what is the interstellar cosmic-ray spectrum and intensity?

Finally, the study of materials returned from the moon may determine the flux of many different elements in the solar wind over both short (but accurately measured) and very long (but not accurately determined) time intervals.

REFERENCES

1. Montgomery, M. D., Bame, S. J., Hundhausen, A. J. 1968, *J. geophys. Res.*, **73**, 4999.
2. Sturrock, P. A., Hartle, R. E. 1966, *Phys. Rev. Lett.*, **16**, 628.
3. Ogilvie, K. W., Burlaga, L. F., Wilkerson, T. D. 1968, *J. geophys. Res.*, **73**, 6809.
4. Hirshberg, J., Bame, S. J., Hundhausen, A. J. 1968, *Trans. Amer. geophys. Un.*, **49**, 728.
5. Bame, S. J., Hundhausen, A. J., Ashbridge, J. R., Strong, I. B. 1968, *Phys. Rev. Lett.*, **20**, 393.
6. Hundhausen, A. J., Gilbert, H. E., Bame, S. J. 1968, *J. geophys. Res.*, **73**, 5485.
7. Hundhausen, A. J., Ashbridge, J. R., Bame, S. J., Strong, I. B. 1967, *J. geophys. Res.*, **72**, 1979.
8. Egidi, A., Pizzella, G., Signorini, C. 1969, *J. geophys. Res.*, **74**, 2807.

9. Wilcox, J. M., Colburn, D. S. 1969, *J. geophys. Res.*, **74**, 2388.
10. Coleman, P. J., Jr., Davis, L., Jr., Smith, E. J., Jones, D. E. 1966, *J. geophys. Res.*, **71**, 2831.
11. Wilcox, J. M., Ness, N. R. 1967, *Solar Phys.*, **1**, 437.
12. Rosenberg, R. L., Coleman, P. J., Jr. 1969, UCLA Preprint no. 737.
13. Coleman, P. J., Jr. 1966, *J. geophys. Res.*, **71**, 5509.
14. Belcher, J. W., Davis, L., Jr., Smith, E. G. 1969, *J. geophys. Res.*, **74**, 2302.
15. Sari, J. M., Ness, N. F. to appear in *Solar Phys.*
16. Siscoe, G. L., Davis, L., Jr., Coleman, P. J., Smith, E. J., Jones, D. E. 1968, *J. geophys. Res.*, **73**, 61.
17. Slysh, V. I. 1967, *Soviet Astr.*, **11**, 72.
18. Slysh, V. I. 1967, *Soviet Astr.*, **11**, 389.
19. Scarf, F. L., Crook, G. M., Green, I. M., Virobik, P. F. 1968, *J. geophys. Res.*, **73**, 6665.
20. Coleman, P. J., Jr., Smith, E. J., Davis, L., Jr., Jones, D. E. 1969, *J. geophys. Res.*, **74**, 2826.
21. Anderson, H. R. 1968, *J. geophys. Res.*, **73**, 2897.
22. Krimigis, S. M. 1968, *Can. J. Phys.*, **46**, S976.
23. Krimigis, S. M., Venkatesan, D. 1969, *J. geophys. Res.*, **74**, 4129.
24. O'Gallagher, J. J. 1967, *Astrophys. J.*, **150**, 675.
25. O'Gallagher, J. J., Simpson, J. A. 1967, *Astrophys. J.*, **147**, 819.
26. Vernov, S. N., Chudakov, A. E., Vakulov, P. V., Logatchev, Yu. I., Lyubimov, G. P., Nikolaev, G. P., Pereslegina, N. V. 1967, *Spaceflight*, **9**, 168.
27. Sonett, C. P., Colburn, D. S., Davis, L., Jr., Smith, E. J., Coleman, P. J., Jr. 1964, *Phys. Rev. Lett.*, **13**, 153.
28. Gosling, J. T., Ashbridge, J. R., Bame, S. J., Hundhausen, A. J., Strong, I. B. 1968, *J. geophys. Res.*, **73**, 43.
29. Van Allen, J. A., Ness, N. F. 1967, *J. geophys. Res.*, **72**, 935.
30. Gosling, J. T., Ashbridge, J. R., Bame, S. J., Hundhausen, A. J., Strong, I. B. 1967, *J. geophys. Res.*, **72**, 3357.
31. Fairfield, D. H. 1968, *J. geophys. Res.*, **73**, 6179.
32. Burlaga, L. F., Ness, N. F. to appear in *Solar Phys.*
33. Siscoe, G. L., Turner, J. M., Lazarus, A. J. to appear in *Solar Phys.*
34. Burlaga, L. F. 1968, *Solar Phys.*, **4**, 67.
35. Scarf, F. L. 1969, talk presented at 16th General Assembly of URSI, Ottawa, Canada.

APPENDIX I: RESULTS FROM INTERPLANETARY "VENERA" STATIONS

(by V. G. Kurt)

Automatic interplanetary stations "Venera-4", "Venera-5" and "Venera-6" were used to carry out observations far from the Earth in three spectral regions: 1050–1340, 1225–1340 and 1050–1180 Å. Radiation in the spectral region including the H-$L\alpha$-line was discovered; it is concentrated to the plane of the Milky Way and has a maximum intensity of about 100 Rayleigh. Investigations of the scattered ultraviolet radiation in the vicinity of the Earth and of Venus were carried out with these stations. An extended hydrogen corona around Venus was discovered and an upper limit for the atomic oxygen density in the upper atmosphere of Venus was estimated.

ULTRAVIOLET STELLAR ASTRONOMY

(by H. E. Butler)

General

The past three years have seen appreciable advances in uv astronomy, the following being particularly noteworthy:

(a) The Orbiting Astronomical Observatory (OAO-A2), launched in December, 1968, has produced large amounts of high quality observational photometric and spectrophotometric data that must be most gratifying to the two astronomical groups concerned – the Smithsonian Astrophysical Observatory in Harvard and the Washburn Observatory, Wisconsin – as well as to all working in this field.

(b) Medium resolution (~ 1 Å) spectra of early type stars are now being obtained regularly from stabilised rockets, the uv region having been covered down to 928 Å. The spectra have shown a complex structure of absorption lines shortward of about 1800 Å. Early type supergiants have shown P-Cygni type profiles for several resonance lines, with ejection velocities of up to 3800 km s^{-1}: the consequent rate of loss of mass from such stars is appreciable.

(c) New spectrophotometric data have confirmed and extended the first uv interstellar extinction curve. Considerable theoretical efforts have been made to interpret the curve in terms of model absorbing grains of graphite and silicates with, and without, ice or solid H_2 mantles.

(d) Interpretation of the strength of the interstellar Lyman α absorption has led to a figure for the average density of neutral hydrogen along the line of sight to the stars being observed. In the direction of Orion the figure is 0·1 atoms cm^{-3}, approximately one tenth of the figure derived from 21-cm radio observations. The recent data that have been obtained for stars in other directions in the sky show larger figures, even exceeding the corresponding 21-cm values.

(e) Photometric and spectrophotometric absolute intensity observations have continued but with increasing awareness of the pitfalls associated with inadequate instrumental calibrations. Stellar model-atmosphere work has continued simultaneously and a series of line blanketed models is now in good agreement with observation for most of the UV region.

(f) At least four new groups (one in the U.S., two in Europe and one in Japan) have recorded their first stellar observations from rockets and there is the preliminary report of successful observations from a Russian orbiting astronomical satellite.

Stellar observations

Photometric observations of many thousands of stars have been made in four passbands within the spectral range 1050–3000 Å by Project Celescope, the Smithsonian Astrophysical Observatory cluster of uv television cameras on board the Orbiting Astronomical Observatory (OAO-A2). The faintest stars observed were of visual magnitude ten at spectral type A0 early in the life time of the instruments but the sensitivity has considerably decreased. The instrumentation is described by Davis (**1**). Observational results will be circulated to institutions and individual scientists in a series of "Celescope Observational Data Reports" (**2**). Only a small part of the data has as yet been completely reduced: Davis reports that the relative accuracy cannot be relied on to better than $0^{m}.25$ in any channel. One consistent picture emerging from the data is that the observed stars are about equally divided between those that fall within $0^{m}.5$ of the predicted uv brightness and those that are significantly fainter than predicted. Most of the giant stars exhibit these deficiencies, including stars for which no interstellar reddening is predicted.

The Wisconsin package on OAO-A2 has obtained medium resolution spectrophotometric data between 1050 and 3800 Å and photometric data in nine different passbands between $L\alpha$ and the visual blue region. None of these data is in print at the time of writing although four papers detailing them were read at the IAU Symposium no. 36 on *Ultraviolet Stellar Spectra and Related Ground-based Observations* in June 1969 (**3–6**). The spectrum scanning instruments (15 Å slit width; scanning in 10 Å steps) have been very successful in providing several hundreds of observations and, in particular, details were given of 50 stellar spectra having "a wide range of spectral type, luminosity class, reddening and peculiarities". It was stated that the calibration remained satisfactorily consistent to better than $0^{m}.02$ throughout several months.

Comparison of observations with model atmosphere theories

The early photometric observations indicated that the continuum intensities of early type stars at wavelengths shorter than about 2000 Å were much less than contemporary models indicated.

Since the work of Morton, Mihalas and Guillaume prior to 1967, the following studies have been made. The line blocking to be encountered for each 100 Å of a spectrum between 1900 and 3000 Å has been calculated (**7**). Comparisons between models and observations have been made for a

B4V star (**8**), for O5V and B0V stars (**9**), and for O type stars (**10**). New scales of effective temperatures and bolometric corrections for main sequence stars hotter than the Sun have been suggested (**11**).

Wavelength lists of 283 lines observed within 1100–1950 Å in Orion stars (**12**), and of 100 lines from ζ Pup (**13**) have been published. Identifications are suggested for over half of the lines. Other wavelength lists include: 50 lines in the region 1050–1300 Å, also for Orion stars (**14**), 128 lines in the region 928–1350 Å for α Vir with identifications offered for most (**15**), and in particular identifications of all except one of the lines predicted for a B IV star (**16**). A low resolution study of ζ UMa (B3V), (**17**), has shown that line blanketing is still greater than recent models indicate by about $0^{m}.5$ and it is suggested that an additional source of continuous opacity is needed.

The results of Stuart (**18**) differed from those of other observers by as much as $1^{m}.2$ at 1376 Å and he generally obtained larger figures than others, being in agreement with Mihalas and Morton's B1·5 model. Intercomparison of many observations have shown (**19**), first, that about $0^{m}.25$ was a reasonable absolute accuracy for current photometric observations at that time, and, secondly, that blanketed models of Morton and colleagues, together with the Morton-Adams (**11**) temperature scale, adequately represented most of their observations. The inaccuracies associated with the large correction for uv interstellar absorption that has to be obtained by a considerable extrapolation from (U–B) observations are emphasised.

Carruthers (**20**) agrees that the position for main sequence stars is satisfactory but that the intensities of giants and supergiants are up to one magnitude less than those of main sequence stars at 1115 Å. Stecher (**21**) examining the spectra of 29 stars of types between O 5 and F 0 in the range 1700–3000 Å, shows that line blanketing is evident in the region 1800–2100 Å for the hotter stars but that the region moves to about 2300 Å for A and F types. Maran *et al.* (**22**) computed a synthetic spectrum for A stars between 2000 and 3000 Å, for given temperature and surface gravity. Bless (**3**) states that his recent observations suggest that earlier deficiencies were probably the result of calibration errors. However, late B and A types may well emit less in the uv than the models, which do not yet allow for Si, Mg or C opacities. Gingerich and Latham (**23**) agree for Si and C at type A0: Strom and Strom (**24**) consider the effect of Si opacity in B and A star atmospheres. Code and Bless (**5**) displayed 50 stars with 10 Å resolution within 1050–2000 Å to give a qualitative description of spectral variations with type. They compared the equivalent widths of Si IV (1403 Å) and C IV (1550 Å) with calculated values and suggested a revision of effective temperatures for stars earlier than B 3. Bless and Savage (**4**) made an important contribution by showing for a small sample that if two early type stars of luminosity classes II-V have the same visual spectral type, then they have essentially the same continuum in the ultraviolet.

Stellar mass loss

Morton (**25**) found that the uv spectra of certain hot stars in Orion showed not only that the resonance lines of Si IV (1403 Å) and C IV (1550 Å) were present strongly in emission, but that each was accompanied on its shortwave side by an absorption line shifted by up to 10 Å – i.e. a P Cygni-like profile – showing matter streaming outwards with velocities of up to 3800 km s^{-1}. The lines are considered to be formed in the expanding shell and since the ion states present in these hot atmospheres have all their resonance lines in the uv, the effect will not generally be present in the visible region. Later observations by the Princeton Group, (**12, 13**) and by Carruthers (**14**) have extended the observations to 8 stars and to 8 lines (5 of which are resonance lines) of the ions He II, C III and IV, N IV and V and Si III and IV, showing various emission speeds up to 2140 km s^{-1}. Morton (**26**) reviews the data and concludes that probably all supergiants and some giants of type B 0·5 and earlier show the phenomenon. He does not anticipate that the rate of loss of mass – estimated to be of the order of 10^{-6} $M_{\odot}$ y^{-1} – is sufficient to affect the evolution of such a star. Lucy and Solomon (**27**) proposed that the outward velocity can be produced by the absorption of photons by the ions: for this case they compute a figure for the rate of mass loss smaller than Morton's by a factor of about ten. The effect of the phenomenon on stellar evolution is considered by Hartwick (**28**).

General interstellar absorption

A careful photoelectric comparison of the uv spectra of ε and ζ Per from a rocket (**29**), has produced an improved observational curve: further far-ultraviolet data have been given (**30**). Coincident with these observations, silicates were reported in space and Greenberg and Stoeckly (**31**) examined the new possibility of silicate grains, possibly with ice mantles. Hoyle and Wickramasinghe (**32**) in a review, and Wickramasinghe (**33**) have considered mixtures of graphite and silicates with and without ice or H_2 mantles. The first OAO has now made an appreciable contribution to our observational data on interstellar absorption and Bless and Savage (**4**), presenting it, conclude from spectra of about 50 stars that

(a) all extinction curves show a maximum at about 2200 Å
(b) most curves show a minimum at about 1750 Å
(c) all curves continue to rise towards shorter wavelengths
(d) there appear to be real differences between the curves for different stars.

Interstellar absorption by atomic hydrogen

Spectra showing the interstellar Lyman α absorption line were first reported by Jenkins (**34**) and Jenkins and Morton (**35**). Stecher (**36**) mentioned similar observations. The observations are interpreted as indicating an average density along the line of sight to Orion of about 0·1 atoms cm^{-3}, appreciably less than the 1 atom cm^{-3} computed from 21-cm radio observations. In general the density in other directions in the sky is more in accord with 21-cm results – see, for example, (**37**) for β^1, δ and π Sco. However, Carruthers (**30**) discusses anomalous results for θ Ori. The subject is discussed in detail by Jenkins (**34**) who considers the possibility of the hydrogen being in small, dense, hot clouds well separated from each other. Savage and Code (**6**) give probably the most consistent set of data for 48 stars measured with the OAO. They generally confirm the above conclusions and favour as explanation that the 21-cm emission mostly originates beyond Orion.

Interstellar absorption by molecular hydrogen

The first deliberate attempt to observe the interstellar molecular hydrogen bands shortward of 1108 Å was reported by Carruthers (**38, 14**). He interpreted the fact that he failed to see them superimposed on the spectra of γ Vel and ζ Pup as indicating that the number of hydrogen molecules in the line of sight to these stars was less than 10^{19} cm^{-2}. Likewise Smith (**15**) has an upper limit for the numbers of molecules in the line of sight to α Vir of $6{\cdot}5 \times 10^{16}$ cm^{-2} or a mean number density of less than $3{\cdot}1 \times 10^{-4}$ cm^{-3}.

Radiation from the celestial background in the region 1050–1350 Å

This radiation is of two different types: general continuum radiation due to the integrated light from hot stars and $L\alpha$ emission. The latter can be either (i) diffuse emission from the Galaxy or (ii) sunlight scattered by the geocorona, by interplanetary hydrogen or by hydrogen at the extreme edge of the solar system. Photometric observations have been made by both Russian and U.S. spacecrafts at considerable distances from the Earth where any geocoronal light can be discounted. Kurt and Dostovalov (**39**) report data from the satellite Venera 4 and Dimov *et al.* (**40**) give data for satellites Kosmos 51 and 213. Kurt and Sunyaev (**41**) review the observations where they quote that the $L\alpha$ intensity varies round the Milky Way in the wide range $2{\cdot}5 \times 10^{-5}$ to $2{\cdot}5 \times 10^{-4}$ erg cm^{-2} s^{-1} $ster^{-1}$. The integrated continuum intensity from hot stars in the Milky Way (i.e. omitting $L\alpha$) was found to be 3×10^{-7} erg cm^{-2} s^{-1} $ster^{-1}$ from the spectral band 1225–1340 Å, a figure which is considerably greater than that which simple summation of observations of individual stars would suggest. They examine this problem in some detail. Barth (**42**) reports similar data for the $L\alpha$ and oxygen (1304 Å) intensities from the interplanetary flight of Mariner 5. He found 5×10^{-4}

erg cm^{-2} s^{-1} $ster^{-1}$ to the south of the galactic plane and 7×10^{-4} erg cm^{-2} s^{-1} $ster^{-1}$ to the north. There was an enhancement when crossing the galactic plane in Vela but none in Cygnus. Interpretation of the data in terms of atomic hydrogen within, and on the outskirts of, the solar system is given.

Extragalactic observations

There are two observations to date. Goldberg (**42**) reports an OAO spectral scan, supplied to him by Code, of the nucleus of the Andromeda nebula. The intensity is known from ground observations to decrease as we move to shorter wavelengths through the blue and near uv regions. The OAO data show this decrease as continuing until about 2700 Å where the trend reverses and the intensity is still steadily rising at 2000 Å, the shortest wavelength observed. Cruvellier *et al.* (**44**) report these data as being in agreement with two photometric observations made by them at 3450 Å and 2710 Å with rocket equipment. Goldberg comments on the data indicating the presence of hot stars in the nucleus and also on the possible cosmological importance of the observation should it be common to all galaxies.

Future work

Unmanned satellites

(i) OAO III will carry the Goddard 36-inch telescope and spectrograph. The spectral region is 1000–3000 Å, the resolutions are 2 Å, 8 Å and 64 Å and launch is expected in 1970.

(ii) OAO IV will carry the Princeton 80-cm telescope which is capable of being pointed to an accuracy of 0·1 second of arc on O and B stars as faint as $7\overset{m}{.}0$. The spectral resolution will be 0·05 Å from 800 to 1600 Å and 0·10 Å from 1600 to 3200 Å. It will be used primarily to study the composition and physical state of the interstellar gas.

(iii) TD 1 is a European Space Research Organisation satellite with a stabilised scan and which will cover the whole sky in a period of 6 months: launch is expected in 1972. It will carry two UV astrophysical instruments. The first is a 290 cm^2 aperture telescope of Utrecht Observatory which will examine spectral regions centred on 2085, 2500, 2800 Å at a resolution of about 1 Å. Its magnitude limit is $V = 5$ for a B 0 star. The second instrument uses a 27-cm telescope which is being jointly provided by the Astrophysical Institute at Liège and the Royal Observatory at Edinburgh. It will examine four uv regions within the range 1450–3000 Å to give photometric observations and low resolution spectra. The limiting magnitude will be about 9 for photometric measures of early type stars.

Stabilised rockets

The work using these rockets is scheduled to continue in the U.S. with steadily increasing numbers of flights. The Culham and Edinburgh groups in Britain have plans for flights beginning in 1970 using Skylark rockets in the British National programme. Within the ESRO programme the Liège and Munich groups, as well as the Culham and University College London groups, have plans for flights in 1970 and 1971. All are for medium and high resolution spectroscopy of single stars except Edinburgh, which will fly wide angle, objective prism cameras.

Balloons

The group at the Geneva Observatory has further plans to obtain data in the 2000–3000 Å region from balloons.

Calibration

The importance of absolute calibration in stellar photometric and spectrophotometric work has led to increased efforts to ensure the highest possible individual accuracy: the interchange of absolute standards amongst groups is proposed.

REFERENCES

1. Davis, R. J. 1968, *Smithsonian Astrophys. Obs. Special Reports* no. 282 and 1969, *ibid.* no. 310.
2. Davis, R. J. 1969, *Celescope Observational Data Reports, Smithsonian Astrophys. Obs.*
3. Bless, R. C. 1970, *IAU Symposium* no. 36, Lunteren.
4. Bless, R. C., Savage, B. D. 1970, *IAU Symposium* no. 36, Lunteren.
5. Code, A. D., Bless, R. C. 1970, *IAU Symposium* no. 36, Lunteren.
6. Savage, B. D., Code, A. D. 1970, *IAU Symposium* no. 36, Lunteren.
7. Elst, E. W. 1967, *Bull. astr. Inst. Netherl.*, **19**, 20.
8. Adams, T. F., Morton, D. C. 1968, *Astrophys. J.*, **152**, 195.
9. Hickok, F. R., Morton, D. C. 1968, *Astrophys. J.*, **152**, 203.
10. Bradley, P. T., Morton, D. C. 1969, *Astrophys. J.*, **156**, 687.
11. Morton, D. C., Adams, T. F. 1968, *Astrophys. J.*, **151**, 611.
12. Morton, D. C., Jenkins, E. B., Bohlin, R. C. 1968, *Astrophys. J.*, **154**, 661.
13. Morton, D. C., Jenkins, E. B., Brooks, H. H. 1969, *Astrophys. J.*, **155**, 875.
14. Carruthers, G. R. 1968, *Astrophys. J.*, **151**, 269.
15. Smith, A. M. 1969, *Astrophys. J.*, **156**, 93.
16. Mihalas, D. M., Morton, D. C. 1965, *Astrophys. J.*, **142**, 253.
17. Opal, C. B., Moos, H. W., Fastie, W. G., Bottema, M., Henry, R. C. 1968, *Astrophys. J.*, **153**, L179.
18. Stuart, F. E. 1969, *Astrophys. J.*, **157**, 1255.
19. Bless, R. C., Code, A. D., Houck, T. E. 1968, *Astrophys. J.*, **153**, 561.
20. Carruthers, G. R. 1970, *IAU Symposium* no. 36, Lunteren.
21. Stecher, T. P. 1969, *Astr. J.*, **74**, 98.
22. Maran, S. P., Kurucz, R. L., Strom, K. M., Strom, S. E. 1968, *Astrophys. J.*, **153**, 147.
23. Gingerich, O., Latham, D. 1970, *IAU Symposium* no. 36, Lunteren.
24. Strom, S. E., Strom, K. M. 1969, *Astrophys. J.*, **155**, 17.
25. Morton, D. C. 1967, *Astrophys. J.*, **150**, 535.
26. Morton, D. C. 1969, *Astrophys. Space Sci.*, **3**, 117.
27. Lucy, L. B., Solomon, P. M. 1967, *Astr. J.*, **72**, 310.
28. Hartwick, F. D. A. 1967, *Astrophys. J.*, **150**, 953.
29. Stecher, T. P. 1969, *Astrophys. J.*, **157**, L125.
30. Carruthers, G. R. 1969, *Astrophys. J.*, **156**, L97.
31. Greenberg, J. M., Stoeckly, R. 1970, *IAU Symposium* no. 36, Lunteren.
32. Hoyle, F., Wickramasinghe, N. C. 1969, *Nature*, **223**, 459.
33. Wickramasinghe, N. C. 1970, *IAU Symposium* no. 36, Lunteren.
34. Jenkins, E. B. 1967, *Astr. J.*, **72**, 304.
35. Jenkins, E. B., Morton, D. C. 1967, *Nature*, **215**, 1257.
36. Stecher, T. P. 1968, *Bull. astr. Inst. Csl.*, **19**, 279.
37. Jenkins, E. B., Morton, D. C., Matilsky, T. A. 1969, *Astrophys. J.*, **158**, in press.
38. Carruthers, G. R. 1967, *Astrophys. J.*, **148**, L141.
39. Kurt, V. G., Dostovalov, S. B. 1968, *Nature*, **218**, 258.
40. Dimov, N. 1970, *IAU Symposium* no. 36, Lunteren.
41. Kurt, V. G., Sunyaev, R. A. 1970, *IAU Symposium* no. 36, Lunteren.
42. Barth, C. A. 1970, *IAU Symposium* no. 36, Lunteren.
43. Goldberg, L. 1969, *Sci. Am.*, **220**, 92.
44. Cruvellier, P., Roussin, A., Valerio, Y. 1970, *IAU Symposium* no. 36, Lunteren.

X-RAY SOURCES

(by L. Gratton)

Since the last Report of Commission 44 there has been a great advance in the study of X-ray sources. There are at present a score of groups engaged in active work on this field and already some general conclusions are beginning to take shape.

At the General Assembly of the IAU in Prague a joint discussion of Commissions 28, 33, 34 and 44 (Chairman: S. B. Pikelner) was held (**1**) on the subject of X-ray astronomy; in November 1968 a one-day conference was organized by the Royal Society in London (**2**) and a three-day Symposium cosponsored by IAU and COSPAR was held in May 1969 in Rome (**3**) on non-solar gamma and X-ray Astronomy. The papers presented in this last Symposium (see especially the invited papers by H. Gursky, G. W. Clark, L. E. Peterson, R. Giacconi, H. M. Johnson, L. Gratton, L. Woltjer, J. E. Felten) cover the field very thoroughly, so that the present report, and especially the bibliography, can be made quite short. Among other review papers there may be mentioned those by Gould (**4**), Friedman (**5**), Rossi (**6**) and by Giacconi *et al.* (**7**) and also L. Gratton's lecture at the 1969 NATO summer school in Lagonissi (Greece) (**8**).

The individual X-ray sources known at present number about 50, but there is still considerable disagreement in several cases concerning whether a particular source noted by two observers is really the same source with a large error in the given positions. For six sources there is a definite identification with optical objects and two more may be accepted with reasonable confidence, although some observers are still expressing reserve.

A point which perhaps should be discussed by Commission 44 is that of the naming of the sources; the present system is very unsatisfactory, each group naming the sources according to its own system. This adds considerably to the confusion which naturally arises from the poor determination of the positions; a procedure similar to that in use for naming variable stars or minor planets might be desirable.

The eight identified objects fall into (at least) three categories:

(a) *supernova remnants*, which include Tau X-1 (Crab nebula), Cas XR-1 (Cas A) and an unnamed source coincident with Tycho SN 1572;

(b) *peculiar starlike objects*, including 4 objects, two of which are still doubted by some observers: Sco X-1, Cyg X-2, Cen XR-2 (WX Cen) and GX 3 + 1;

(c) *radiogalaxies*, including at present only Vir XR-1 identified with M 87 (Virgo A).

Long series of photometric observations of Sco X-1 (**9**) and spectral and color data (**10–13**) will have been published for objects of class (b) showing light variations and spectral features reminiscent of those of old novae, but no known old nova has been yet discovered to possess a detectable emission in the X-ray domain. It was also noted that the galactic sources – which appear in their majority to belong to class (b) – follow a galactic distribution closely resembling that of known novae. But the proper motions of Sco X-1 and Cen XR-2 (**14, 15**), point towards a membership of these objects in the Scorpio-Centaurus group. This however contradicts Wallerstein's earlier results from the interstellar Ca II lines (**16**). The whole situation concerning the physical nature and statistical properties of the objects of this class (e.g. their stellar population type) is very obscure and conclusions concerning the theoretical interpretation – except perhaps very rough and qualitative ones – might be at present premature and misleading. The binary nature of Sco X-1 and Cyg X-2 suggested by the radial velocity variations is also doubtful (**17**). Thermal bremstrahlung from an optically thin plasma in the X-ray domain and optically thick in the visual and radio spectrum seems to be a likely mechanism of radiation (**18–21**) the temperatures required are of the order of 10^6 K. Apparently the X-ray flux from these objects is subject to large variations not correlated with similar variations in the visual spectrum; the large decrease of the X-ray flux from Cen XR-2 in 1967 and the outburst – or "birth" – of a source between Lupus and Centaurus in 1969 are especially noteworthy.

The only object of class (c) discovered so far is M 87 and this might well be an exceptional case, since as a radio-source it belongs to the relatively rare "core-halo" class of objects. The physical interpretation of M 87 has been discussed by Felten (**22**). A large number of faint and distant X-ray sources is postulated in some theories of the diffuse X-ray background radiation.

One of the most exciting discoveries in the field of the X-ray Astronomy was made almost simultaneously by the NRL (**23**), the MIT (**24**) and the Rice University (**25**) groups early in 1969. They found that the pulsar, NP0582 in the Crab nebula has very strong X-ray emission pulsating with the same period as the radio and optical radiation. Indeed the power emitted in the X-ray domain far exceeds that in the rest of the electromagnetic spectrum. If, as many now believe, the pulsar mechanism is the source of the energy radiated by the entire Crab nebula, these observations and those of the X-ray emission from the nebula may be of enormous importance in providing experimental evidence for the mechanism itself.

All X-ray observations of individual sources have been obtained by means of rocket experiments; there can be little doubt that great advances will be made when it will be possible to fly a satellite borne experiment. This will permit detection of fainter sources and a great improvement of the position of the stronger ones. An Explorer satellite is planned by the ASE group in 1970, which will probably detect sources of an intensity $10^{-4} \times$ Crab; Gursky estimates that this experiment will detect as many as 10^3 sources, compared with the 50 reported so far. A satellite experiment will also permit long series of simultaneous X-ray and optical observations of some objects. A grazing incidence telescope giving images of the Crab, M 87 and other extended or multiple sources with a resolution of a few arc sec may bring decisive results for the physical interpretation of these objects.

Another field in which decisive advances may be expected from a satellite borne experiment is that of X-ray spectroscopy; at present spectral data of sources are very rough and may be compared to those obtained in broad band multicolor photometry of stars. Perhaps before the 1973 General Assembly various experiments of this sort will have been made.

REFERENCES

1. *Highlights of Astronomy* presented at the XIIIth General Assembly of the IAU. 1968, Ed. L. Perek, p. 177.
2. *Proc. R. Soc. London* (in press).
3. 1969, *IAU Symposium* no. 37, Rome, in press.
4. Gould, R. J. 1967, *Am. J. Phys.*, **35**, 376.
5. Friedman, H. 1967, *Ann. Rev. Nucl. Sci.*, **17**, 317.
6. Rossi, B. 1969, *Astronomia in raggi X* – Lecture held at the Accademia Nazionale dei Lincei, published in *Annuario della ES & T*, Mondadori, Milano, p. 272.
7. Giacconi, R., Gursky, H., Van Speybroeck, L. R. 1968, *A. Rev. Astr. Astrophys.*, **6**, 373.
8. Gratton, L. 1969, NATO International Summer School at Lagonissi (Greece), in press.
9. Among others Hiltner, W. A., Mook, D. E. 1967, *Astrophys. J.*, **150**, 851.
10. Westphal, J. A., Sandage, A. R., Kristian, J. 1968, *Astrophys. J.*, **154**, 139.
11. Eggen, O. J., Freeman, K. C., Sandage, A. R. 1968, *Astrophys. J.*, **154**, L27.
12. Lynds, C. R. 1967, *Astrophys. J.*, **149**, L41; Burbidge, E. M., Lynds, C. R., Stockton, A. 1967, *Astrophys. J.*, **150**, L95; Kraft, R. P., Demoulin, M. H. 1967, *Astrophys. J.*, **150**, L183.
13. Peimbert, M., Spinrad, H., Taylor, B. F., Johnson, H. M. 1968, *Astrophys. J.*, **151**, L93.
14. Gatewood, G., Sofia, S. 1968, *Astrophys. J.*, **154**, L69.
15. Eggen, O. J. 1969, *Astrophys. J.*, **158**, L31.
16. Wallerstein, G. 1967, *Astrophys. Lett.*, **1**, 31.
17. Sofia, S., Wilson, R. E. 1968, *Nature*, **218**, 73.
18. Meekins, J. F., Henry, R. C., Fritz, G., Friedman, H., Byram, T. E. 1969, *Astrophys. J.*, **157**, 197.
19. Shklovsky, I. S. 1967, *Astrophys. J.*, **148**, L1.
20. Prendergast, K. H., Burbidge, G. R. 1968, *Astrophys. J.*, **151**, L83.
21. Neugebauer, G., Oke, J. B., Becklin, E., Garmire, G. 1969, *Astrophys. J.*, **155**, 1.
22. Felten, J. E. 1968, *Astrophys. J.*, **151**, 861.
23. Fritz, G., Henry, R. C., Meekins, J. F., Chubb, T. A., Friedman, H. 1969, *Science*, **164**, 709.
24. Bradt, H., Rappaport, S., Mayer, W., Nather, F. H., Warner, B., McFarlane, M., Kristian, J. 1969, *Nature*, **222**, 728.
25. Haymes, R. C., Freeman, K. C., Harnden, F. R., Jr. 1969, *Astrophys. J.*, **156**, L107.

THE X- AND GAMMA-RAY BACKGROUND RADIATION

(by W. Kraushaar)

Observational status

Prior to 1966, there was good evidence for cosmic background radiation (or at least spatially unresolved radiation) in the 1 keV to 1 MeV energy region. The region below 1 keV was not yet seriously examined and only tentative evidence existed for cosmic gamma radiation at energies larger than a few MeV. Spectral measurements could be fitted satisfactorily to a single power law of the from $I \sim E^{-\gamma}$ with γ near 2·3 and with I at 5 keV about 1 photon cm^{-2} sec^{-1} $sterad^{-1}$ keV^{-1}. Little attention had been given to a possible anisotropic nature of the radiation. The earlier observations and their interpretation have been summarized by Gould (**1**), and the following paragraphs review briefly the highlights of extended observations made since late 1966.

Rocket-borne (**2–10**), satellite-borne (**11, 12**), and balloon-borne (**8, 13, 14**) detectors have all contributed to a vast amount of new data in the 1 keV to 1 MeV energy region. Much of the new data were discussed at the Rome IAU Symposium no. 37 and will be published in the forthcoming proceedings of that meeting. An overall view suggests that no single power law can be made to fit the accumulated data satisfactorily and that certainly a single power law with index as large as 2·45 is ruled out. Various authors have suggested the existence of "breaks" in the photon number spectrum. It seems likely that positioning of these breaks is at present largely subjective. There is fairly good evidence, however, that γ, the logarithmic slope of the differential photon number spectrum, is about 1·5 or 1·6 near 5 keV, has become about 1·8 near 50 keV and has become 2 or more in the region between 100 keV and 1 MeV. Only better quality data can determine whether these changes are gradual or do indeed occur at reasonably well-defined energies.

Few groups have given serious attention to the important question of just how isotropic the background radiation in the 1 keV to 1 MeV region really is. The LRL group (**2**) noted an approximately 10 % enhancement of the background intensity near the galactic plane over the intensity at high galactic latitudes. The statistical significance of the measurement was marginal, however. The Nagoya-Tokyo group (**10**) divided the parts of the sky their detector scanned into cells and examined the data for fluctuations beyond those to be expected from counting statistics. No unexpected fluctuations were found. The A.S.E. group (**3**) reported the background radiation to be uniform in intensity to within 8 % over the narrow strip of sky scanned by their detector. The Leicester group (**15**) has reported a statistically significant increase in the intensity of the diffuse radiation near the galactic plane in a region ($l_{II} = 220°$ to $320°$) where there are no reported discrete sources. They suggest that the enhanced intensity may arise from unresolved galactic plane sources.

Measurements of the background radiation in the energy region just below the carbon K absorption edge (0·28 keV) have been carried out by several groups (**16–20**). The measurements and data reduction procedures are less straight-forward here than in the higher X-ray energy regions and there exist significant differences in the reported intensities. At the time of writing (October 1969) there is fair evidence that the soft X-ray background intensity is larger than one would expect on the basis of a simple extrapolation of the power law spectrum that seems most appropriate in the 1–10 keV region. There is good evidence that at energies near the carbon K edge (280 eV) the intensity does not fall off with decreasing galactic latitude in the manner one would expect were the radiation all extragalactic. Involved in this conclusion is the calculation of X-ray absorption coefficients from radio astronomy 21-cm emission measurements of the columnar hydrogen density and the assumed universal abundance of the elements. At energies near 1 keV, on the other hand, there is some evidence, though of poor statistical quality, that the apparent absorption based on the above reasoning is about as to be expected for extragalactic origin (**20**). (Effective X-ray absorption cross-sections per hydrogen atom have been assembled (**21**), and revised somewhat (**22**) to take into account improved partial cross sections for helium and molecular hydrogen and an assumed lower universal abundance of neon.)

Until very recently, the especially important energy region from 1 to 100 MeV has remained unexplored. A scintillation detector sensitive in the 0·25 to 6 MeV energy region has been flown aboard a small satellite (**11**). The preliminary results indicate a quite unexpected (positive) deviation from the nominal E^{-2} power law spectrum extrapolated from lower energies. Up to 1 MeV the data are consistent with this extrapolated power law. At 6 MeV the reported intensity is about a factor of 4 higher. The physics of gamma-ray interactions in the MeV region makes directional detection especially difficult and background effects are hard to evaluate.

In the 100 MeV and greater energy region, a large flux of gamma rays coming preferentially from the galactic plane and particularly from the galactic center region has been observed (**23**). The intensity variation with galactic longitude matches rather well the galactic longitude variations of non-thermal galactic radio noise. Structure in galactic latitude is unresolved by the broad (25° FWHM) angular response of the orbiting instrument. There is apparently also a diffuse or isotropic flux with an intensity in approximate agreement with an extrapolation of the power law evident in the 100 keV to 1 MeV region. A number of other groups (**24–28**) have provided supporting evidence, though of low statistical significance, for enhanced ~100 MeV gamma-ray emission from the galactic plane. On the other hand, the results of balloon-borne spark chamber investigations (**29**) indicate a discrete source of gamma rays near $\alpha = 288°$ $\delta = -35°$ and provide no evidence for the galactic plane emission so evident in the data reported by (**23**).

Interpretive status

Initial discussions of the possible origin of the diffuse X-radiation included synchrotron radiation (**30**), bremsstrahlung from otherwise unsuspected galactic electrons in the kilovolt region (**31**), a hot intergalactic medium (**31**), production in intergalactic space through electron-photon collisions (**32**), and production in distant unresolved galaxies (**33, 34**). Gould's review paper (**1**), quoted earlier, and that of Ginzburg and Syrovatsky (**35**) summarize the early interpretative as well as observational results. Mechanisms that appear plausible in the light of more recent and more refined observations are discussed briefly below. A review of X-ray mechanisms has been prepared recently (**36**).

Inverse Compton collisions of energetic *galactic* electrons with either stellar or the universal 3 K photons fall far short of explaining the observed X- and gamma-ray intensity (**37–40**). (The exact factor depends on how the galactic electron intensity and extent is estimated.) In addition, it is difficult to reconcile a galactic origin with the observed X-ray isotropy. These facts have led to the suggestion (**37**) that remote electrons in radio galaxies or possibly in intergalactic space are responsible, through Compton collisions with photons of the universal radiation field, for the diffuse X-ray intensity. These first considerations, made without the inclusion of detailed general relativistic and evolutionary effects, indicated that the required remote electron flux was very large. General relativistic and evolutionary effects pertaining particularly to the inverse Compton universal radiation field phenomena have been discussed by a number of authors. Inclusion of these effects tends generally to aggravate rather than relieve the remote electron flux problem (**41**). If one considers only those electrons confined in strong radio-emitting objects (radio galaxies and quasars) magnetic fields must be taken to be more than an order of magnitude less than the equipartition or minimum energy values in order to reconcile radio and X-ray background data (**42, 43**). If instead one imagines the electrons to have had their origin in strong radio galaxies but to have escaped into intergalactic space, one is forced to adopt relatively small leakage times and, therefore, (probably) unrealistic values for the efficiency (per unit galactic mass) of energetic particle production. Brecher and Morrison (**44**) have suggested that electrons that have escaped from normal galaxies are sufficient to provide the required intergalactic electron flux. This suggestion relieves somewhat the difficulty of the high electron acceleration efficiency, basically because a larger total number (and therefore larger total mass) of galaxies is assumed to be contributing.

Bremsstrahlung by known non-thermal electrons (e.g., electrons in radio galaxies) is not capable of explaining the X-ray background radiation. Silk and McCray (**45**) have suggested that the remnant particles of an *ad hoc* burst of electrons at a remote epoch ($z \approx 10$) could, through bremsstrahlung in a low density intergalactic gas, account for the background flux.

Protons in the tens of MeV region are capable of producing bremsstrahlung X-rays through collisions with ambient electrons (**46–48**). If protons in this energy region exist at the levels needed to account for the heating of interstellar H I regions (**48**), a marginally detectable X-ray flux would be expected from the galactic plane. While this may conceivably be the mechanism responsible for the slight intensity enhancement detected by the Leicester group (**15**), it does not seem capable of explaining the general X-ray background unless a very large and otherwise unsuspected intergalactic flux of protons is assumed.

As mentioned earlier, observations of the diffuse X-radiation at energies below 1 keV are on a considerably less firm basis than are the observations at higher energies.

Numerous causes of the apparently anomalous soft X-ray absorption by galactic gas have been suggested. There may be unsuspected local (solar or terrestrial) sources of soft X-rays (**20**). There may be numerous unresolved soft X-ray emitters within a few hundred parsecs of the solar system (**18, 20**). The interstellar gas may be non-uniformly distributed to the extent that to soft X-rays some parts of the sky are essentially opaque while other parts are relatively open (**20, 49**). There may be less helium in the interstellar medium than has been supposed (**16, 18, 22**). None of these possibilities can at present be easily dismissed and so it is by no means clear that the soft X-rays detected are of extragalactic origin. Because an excess soft X-ray flux (e.g., a flux larger than that to be expected from an extrapolation of the power law appropriate at energies greater than 1 keV) can be taken as providing evidence for a hot extragalactic ionized gas, the interpretation has received considerable attention (**16, 19, 20**). Several authors (**50–52**) have made predictions of incident radiation intensities for various intergalactic gas temperatures, densities and various cosmological models, some (**51, 52**) including detailed models for the heating (and cooling) of the supposed gas. Sunyaev (**53**) has suggested that the ultraviolet radiation from any hot intergalactic plasma, if it exceeds 10^{-23} erg cm^{-2} sterad^{-1} Hz^{-1} in intensity, would ionize the H I regions on the edges of galaxies. While this argument would apparently exclude extremely hot and dense intergalactic media (**19**), a recent re-examination of the question (**54**) relaxes the Sunyaev condition considerably. Shklovsky (**55**) has suggested an entirely different origin of the excess soft-X-ray intensity – the cosmologically superposed line radiation from the O VII, O VIII (20 Å) transitions in the remnants of type-II supernovae in remote galaxies.

Most mechanisms suggested to explain the isotropic X-radiation in the 1 to 1000 keV region do not predict large departures from the power law spectrum extrapolated to 100 MeV or greater. The isotropic observation (**23**) at ~ 100 MeV may, therefore, require no special explanation. This is not so of any excess radiation, such as that reported (**11**) in the region of 6 MeV. It has been pointed out (**56, 57**) that π° mesons produced by cosmic rays at some early epoch would decay into gamma rays which, when detected now, would have a maximum flux at an energy near $70/(1+z)$ MeV where z is the red shift parameter.

Clayton and Silk (**58**) have suggested that if ^{56}Fe is synthesized as ^{56}Ni, the superposed decay gamma rays from remote galactic matter could produce detectable intensities near 1 MeV.

The galactic or non-isotropic component of the 100 MeV gamma ray intensity was at first thought to be too large to be accounted for by π° meson production in the galaxy through cosmic ray interstellar gas collisions (**23**) and a number of alternative mechanisms have been suggested (**59–62**). However, a re-evaluation of the effective meson production cross-sections (**63**), based on recent high energy accelerator data, has the effect of increasing the predicted intensities over the values based on the earlier cross section evaluations by a factor of nearly three. It is possible that the remaining apparent discrepancy is within the uncertainty of the columnar hydrogen and gamma-ray intensity measurements.

REFERENCES

1. Gould, R. J. 1967, *Am. J. Phys.*, **35**, 376.
2. Seward, F., Chodil, F., Mark, H., Swift, C., Toor, A. 1967, *Astrophys. J.*, **150**, 845.
3. Gorenstein, P., Kellogg, E. M., Gursky, H. 1969, *Astrophys. J.*, **156**, 315.
4. Cooke, B. A., Pounds, K. A., Stewardson, E. A., Adams, D. 1969, *Astrophys. J.*, **150**, L189.
5. Green, D. W., Wilson, B. G., Baxter, A. J. 1969, *Space Res.*, vol. 9, p. 222.
6. Ducros, G., Ducros, R., Rocchia, R., Tarrius, A. 1969, preprint.
7. Boldt, E. A., Desai, U. D., Holt, S. S., Serlemitsos, P. J. 1969, *Nature*, **224**, 677.
8. Bleeker, J. A. M., Deerenberg, A. J. M. 1969, submitted to *Astrophys. J.*
9. Brini, D., Fuligni, E., Horstman-Moretti, E. 1969, *IAU Symposium* no. 37, Rome.
10. Matsuoka, M., Oda, M., Ogawara, Y., Hayakawa, S., Kato, T. 1969, *Astrophys. Space Sci.*, **4**, 44.
11. Vette, J., Matteson, J. L., Gruber, D., Peterson, L. E. 1969, *IAU Symposium* no. 37, Rome.
12. Mandel'stam, S. L., Tindo, I. P. 1967, *Soviet Phys. JETP Lett.*, **6**, 251.

13. Rocchia, R., Rothenflug, R., Boclet, D., Ducros, G., Labeyrie, J. 1967, *Space Res.*, vol. 7, p. 1327.

14. Agrawal, P. C., Biswas, S., Gokhale, G. S., Iyengar, V. S., Kunte, P. K., Manchanda, R. K., Sreekantan, B. U. 1969, *IAU Symposium* no. 37, Rome.

15. Cooke, B. A., Griffiths, R. E., Pounds, K. A. 1969, *Nature*, **224**, 134.

16. Bowyer, C. S., Field, G. B., Mack, J. E. 1968, *Nature*, **217**, 32.

17. Baxter, A. J., Wilson, B. G., Green, D. W. *Can. J. Phys.* (in press).

18. Hayakawa, S., Kato, T., Makino, F., Ogawa, H., Tanakya, Y., Yamashita, K., Matsuoka, S., Miyamoto, S., Oda, M., Ogawara, Y. 1969, *IAU Symposium* no. 37, Rome.

19. Henry, R. C., Fritz, G., Meekins, J. F., Friedman, H., Bryman, E. T. 1969, *Astrophys. J.*, **153**, L11.

20. Bunner, A. N., Coleman, P. L., Kraushaar, W. L., McCammon, D., Palmieri, T. M., Shilepsky, A., Ulmer, M. 1969, *Nature*, **223**, 1222.

21. Bell, K. L., Kingston, A. E. 1967, *Mon. Not. R. astr. Soc.*, **136**, 241.

22. Brown, R. L., Gould, R. J. *Astrophys. J.* (in press).

23. Clark, G. W., Garmire, G. P., Kraushaar, W. L. 1968, *Astrophys. J.*, **153**, L203.

24. Delvaille, J. P., Albats, P., Greisen, K. I., Ogelman, H. B. 1968, *Can. J. Phys.*, **46**, S425.

25. Fichtel, C. E., Kniffen, D. A., Ogelman, H. B. 1969, *IAU Symposium* no. 37, Rome.

26. Hutchinson, G. W., Ramsden, D., Wills, R. D. 1969, *IAU Symposium* no. 37, Rome.

27. Sood, R. K. 1969, *Nature*, **222**, 650.

28. Valdez, J. V., Waddington, C. J. 1969, *Astrophys. J.*, **156**, L85.

29. Frye, G. M., Staib, J. A., Zych, A. D., Hooper, V. D., Rawlinson, W. R., Thomas, J. A. 1969, *Nature*, **223**, 1320.

30. Giacconi, R., Gursky, H., Paolini, F. R., Rossi, B. B. 1962, *Phys. Rev. Lett.*, **9**, 439.

31. Hoyle, F. 1963, *Astrophys. J.*, **137**, 993.

32. Felten, J. E., Morrison, P. 1963, *Phys. Rev. Lett.*, **10**, 453.

33. Gould, R. J., Burbidge, G. R. 1963, *Astrophys. J.*, **138**, 969.

34. Oda, M. 1965, *Proc. Intern. Conf. Cosmic Rays*, (London).

35. Ginzburg, V. L., Syrovatski, J. 1965, *Space Sci. Rev.*, **4**, 267.

36. Setti, G., Rees, M. J. 1969, *IAU Symposium* no. 37, Rome.

37. Felten, J. E., Morrison, P. 1966, *Astrophys. J.*, **146**, 686.

38. Fazio, G. G., Stecker, F. W., Wright, J. P. 1966, *Astrophys. J.*, **144**, 611.

39. Maraschi, L., Perola, G. C., Schwarz, S. 1968, *Nuovo Cim.*, **53B**, 443.

40. Shen, C. S., Berkey, G. 1968, *Astrophys. J.*, **151**, 895.

41. Brecher, K., Morrison, P. 1967, *Astrophys. J.*, **150**, L61.

42. Bergamini, R., Londrillo, P., Setti, G. 1967, *Nuovo Cim.*, **52B**, 495.

43. Fukui, M., Hayakawa, S. *Prog. Theor. Phys.* (in press).

44. Brecher, K., Morrison, P. 1969, *Phys. Rev. Lett.*, **23**, 802.

45. Silk, J., McCray, R. 1969, *Astrophys. Lett.*, **3**, 59.

46. Hayakawa, S., Matsuoka, M. 1964, *Prog. Theor. Phys. Suppl.*, **30**, 204.

47. Hayakawa, S. 1969, *IAU Symposium* no. 37, Rome.

48. Boldt, E., Serlemitsos, P. 1969, *Astrophys. J.*, **157**, 557.

49. Bowyer, C. S., Field, G. B. 1969, *Nature*, **223**, 573.

50. Field, G. B., Henry, R. C. 1964, *Astrophys. J.*, **140**, 1002.

51. Weymann, R. 1967, *Astrophys. J.*, **147**, 887.

52. Bergeron, J. 1969, *Astr. Astrophys.*, **3**, 42.

53. Sunyaev, R. A. 1969, *Astrophys. Lett.*, **3**, 42.

54. Felten, J. E., Bergeron, J. *Astrophys. Lett.* (in press).

55. Shklovsky, I. 1969, *Astrophys. Lett.*, **3**, 1.

56. Ginzburg, V. L. 1968, *Astrophys. Space Sci.*, **1**, 125.

57. Stecker, F. W. 1968, *Nature*, **220**, 675; 1969, *Astrophys. J.*, **157**, 507.

58. Clayton, D. C., Silk, J. 1969, *Astrophys. J.*, **158**, L43.

59. Ogelman, H. 1969, *Nature*, **221**, 753.

60. Shen, C. S. 1969, *Phys. Rev. Lett.*, **22**, 568.

61. Cowsik, R., Pal, Y. 1969, *Phys. Rev. Lett.*, **22**, 550.

62. O'Connell, R. F., Verma, S. D. 1969, *Phys. Rev. Lett.*, **22**, 1443.

63. Stecker, F. W. *Astrophys. Space Sci.* (in press).

RADIO ASTRONOMY

(by F. G. Smith)

Observations of radio noise from space vehicles are directed towards four objectives; solar radio waves, Jovian emissions, the background from galactic and extragalactic sources, and the noise which is generated locally in the magnetosphere by streams of high energy particles. Although it has been difficult to calibrate the sensitivity of even simple receiving systems, because of the effects of the ionosphere on antenna impedance, the general shape of the background spectrum is becoming known. The sharp drop at frequencies below 1 MHz is attributed to absorption in Galactic ionised hydrogen (**1**, **2**).

Solar outbursts extending to low radio frequencies have been recorded by the Alouette satellites, the U.S.S.R. satellites (**3**), and by the U.S.A. satellite RAE I. Most of these are outbursts of type 3, sweeping downwards in frequency as a disturbance travels out from the Sun. They can be followed to frequencies below 250 KHz, at which frequency they are apparently generated at distances of the order of 1 A.U. from the Sun.

Noise from particle streams occurs in frequency bands defined by the electron density and magnetic field strength in the ionosphere. Observations from the U.K./U.S.A. satellite Ariel 2 suggested that part of this noise was associated with the inner trapped radiation belt (**4**). Observations from satellite Ariel 3 showed that part of this noise was due to auroral particles (**5**).

Accurate measurement of noise power in space depends on the theory of the behaviour of antennae immersed in plasmas. Theory and measurements directed to this problem are to be found in papers referred to in a recent review by T. R. Kaiser and J. K. E. Tunaley (**6**) and in papers by D. Walsh and H. Weil (**7**) and G. J. Daniell (**8**).

Further results are to be expected shortly from the RAE I satellite, whose long antennae have considerable directional properties, and from the satellites OGO 3 and OGO 5.

REFERENCES

1. Weber, R. R., Stone, R. G., Somerlock, C. R. 1969, *Astr. Astrophys.*, **1**, 47.
2. Alexander, J. K., Brown, L. W., Clark, T. A., Stone, R. G., Weber, R. R. 1969, *Astrophys. J.*, **157**, L163.
3. Slysh, V. I. 1967, *Kosm. Issled.*, **5**, 897.
4. Harvey, C. C. 1969, *Nature*, **217**, 50.
5. Gregory, P. C. 1969, *Nature*, **221**, 350.
6. Kaiser, T. R., Tunaley, J. K. E. 1968, *Space Sci. Rev.*, **8**, 32.
7. Walsh, D., Weil, H. 1968, *Radio Science*, **3**, (New Series), 936.
8. Daniell, G. J. 1968, *Proc. R. Soc. London*, **A303**, 157.

45. CLASSIFICATIONS SPECTRALES ET INDICES DE COULEUR A PLUSIEURS BANDES (SPECTRAL CLASSIFICATIONS AND MULTI-BAND COLOUR INDICES)

PRÉSIDENT: Ch. Fehrenbach.
VICE-PRÉSIDENT: B. Westerlund.
COMITÉ D'ORGANISATION: W. P. Bidelman, E. K. Haradze, C. O. C. Jaschek, H. L. Johnson, W. W. Morgan.

INTRODUCTION

Des classifications spectrales et des mesures photométriques ont été faites dans de nombreux Observatoires.

La classification à un seul paramètre, comme celle de Harvard, a été remplacée dans la plupart des cas par une classification bidimensionelle. La classification simplifiée est réservée aux étoiles faibles. Les classifications dans le système MK ont été faites dans de nombreux Observatoires et nous signalons les importantes listes d'Abastumani.

De nombreux astronomes ont recherché les étoiles riches ou pauvres en métaux, ce qui est l'amorce des classifications à trois paramètres. Des listes d'étoiles Am et Ap ont été publiées.

De nombreux Observatoires se sont attachés à déterminer directement les magnitudes absolues M_v par des mesures de largeur de raies (H γ, He I, etc.). La classification de Chalonge qui est à trois paramètres a été poursuivie.

Une situation analogue existe pour la classification par photométrie photoélectrique. Le système de Lick est de moins en moins utilisé. Le système de Johnson et Morgan est encore très employé mais diverses variantes sont utilisées notamment à Genève. Ces classifications sont en principe bidimensionnelles mais divers auteurs montrent qu'il est possible d'atteindre un troisième paramètre. C'est le système *uvby* Hβ de Strömgren et Crawford qui est l'approche la plus efficace d'une classification photométrique à trois ou quatre paramètres. Diverses variantes à ce système, notamment celles de Walraven et de van den Bergh, permettent des mesures analogues. Il est regrettable qu'une entente pour le choix d'un système ne puisse se faire. On constate que peu de progrès ont été faits dans les mesures des systèmes à bandes fines ou par exploration spectrale (scanning). Ceci s'explique par les difficultés inhérentes à ces méthodes.

Nous donnons ici un résumé des mesures qui ont été faites dans la classification spectrale et les mesures photométriques.

1. CLASSIFICATION SPECTRALE PAR LA PHOTOGRAPHIE

1.2.2.0. Petrie
1.2.2.1. Vatican – Torun – Herman
1.3. Les classifications par mesure de gradient
1.3.0. Chalonge
1.4. Listes de spectres et atlas
1.4.0. Atlas spectraux
1.4.1. Catalogues
1.4.1.0. Étoiles OB
1.4.1.1. Étoiles Ap, Am

2. CLASSIFICATIONS PAR PHOTOMÉTRIE PHOTOÉLECTRIQUE

2.1. Système à bandes larges
2.1.0. Système Johnson *UBV*, *RI*
2.1.1. Système de Stebbins Whitford (Lick)
2.1.2. Système de Genève
2.2. Systèmes à bandes intermédiaires
2.2.1. Strömgren – Crawford *u*, *b*, *v*, *y*
2.2.2. Walraven – *ULBV*
2.2.3. Van den Bergh
2.2.4. Système de Vilnius
2.2.5. Divers
2.3. Système à bandes fines
2.3.0. Sinnerstad
2.3.1. Nissen
2.3.2. SCI
2.4. Mesures par monochromateur (scanning)
2.4.0. Uttar Pradesh
2.5. Catalogues

1.1.1. *Classification MK*

De nombreux observatoires ont continué la classification de types spectraux dans le système MK.

1.1.1.0. *Étoiles standards*

A Perkins Observatory, P. C. Keenan a étudié un certain nombre d'étoiles G0-M5 géantes et supergéantes pour compléter les listes de W. W. Morgan et P. Keenan un peu insuffisantes dans ce domaine (dispersion 40–80 Å mm^{-1}, $6 < V < 10$).

A Cordoba, J. L. Dessy a fait en liaison avec P. Keenan un travail analogue pour les étoiles brillantes K2-M6 avec une dispersion de 42 Å mm^{-1}. Une liste analogue pour les types F-K a été publiée, une liste F0-G2 est en préparation. Un atlas de tous les types spectraux sera publié en collaboration avec M. et C. Jaschek.

1.1.1.1. *Nouvelles classifications*

Nous indiquons les classifications spectrales effectuées dans les divers Observatoires. Souvent les clichés ont été pris dans l'hémisphère austral.

U.S.A. A Northwestern University, Evanston, W. Buscombe a indiqué les classes MK d'un certain nombre d'étoiles pour lesquelles il a aussi mesuré les largeurs équivalentes des raies (121 étoiles australes, 30 dans M 7).

A la NASA, N. G. Roman a publié les classifications de 679 étoiles $8 < V < 12$, ainsi que de 30 étoiles dans les SA 164, 166, 167 et 185.

A Kitt Peak, Ch. L. Perry et H. E. Bond ont déterminé le type spectral pour 20 étoiles près de IC 2391 observées à Tololo (Chili).

A Yerkes Observatory, A. Cowley, W. A. Hiltner et A. Witt ont publié les spectres des étoiles $m < 10$ des étoiles de Luyten à grand mouvement propre.

A Michigan University, une classification de 1700 étoiles brillantes de type spectral entre B9 et A9 ($\delta > -21°$) a été exécutée par A. et C. Cowley et M. et C. Jaschek.

W. P. Bidelman et F. D. Miller projettent la classification de toutes les étoiles entre 7 et 9·5 avec le télescope de Schmidt de l'Université de Michigan installé à Tololo. F. Miller demande si on peut trouver une aide pour cet important programme; 593 excellents clichés sont déjà pris et représentent 50 % des clichés nécessaires à la couverture du Ciel austral.

A Naval Observatory, Harry Guetter a classé 182 étoiles de types B et A dans les associations Per OB1, Ori OB1, Lac OB1.

A Perkins Observatory, A. Slettebak, R. R. Wright et J. A. Graham ont classé 77 étoiles de type A $70 < m < 9{\cdot}5$.

S. C. Simonson a classé 75 étoiles dans Cep OB2.

En Argentine. A La Plata, la classification de toutes les étoiles plus brillantes que 6·5 est prévue par M. et C. Jaschek.

En Europe. A. Lund, B. Karlsson: 39 étoiles OB distantes, $l_{II} = 203°$, $b_{II} = 2°$; G. Lyngå: 17 étoiles, $l_{II} = 300°$.

A l'Observatoire de Marseille, sur des clichés pris à l'Observatoire de Haute Provence ou à l'Observatoire Européen Austral du Chili à la Silla les classifications suivantes ont été faites.

Les étoiles dont les vitesses radiales sont mesurées sont aussi classées dans le système MK. Avec le petit prisme objectif ($m < 10$), la classification est très précise mais avec le grand prisme objectif ($m < 12{\cdot}5$), très sensible à la turbulence, les classes de luminosité sont plus difficiles à déterminer.

Les listes suivantes ont été publiées ou sont sous presse:

322 étoiles dans un champ de $4 \times 4°$, $l_{II} = 124{\overset{\circ}{.}}6$, $b_{II} = -1{\overset{\circ}{.}}40$ – M. Barbier;
275 étoiles dans un champ de $4 \times 4°$, $l_{II} = 139°$, $b_{II} = 2{\overset{\circ}{.}}3$ – D. Jonas;
236 étoiles de l'amas NGC 752 – E. Rebeirot.

Les classifications des champs suivants sont très avancées:

6 champs de $4 \times 4°$ aux longitudes (hémisphère austral); $l_{II} = 283°$, 302°, 320°, 327°, 331°, 352° – Y. Georgelin;
4 champs de $2 \times 2°$ à $l_{II} = 313°$, 318°, 324°, 333° – A. Laval;
1 champ de $4 \times 3°$ à $l_{II} = 188°$, $b_{II} = 3°$;
1 champ de $4 \times 5°$ à $l_{II} = 101°$, $b_{II} = -1{\overset{\circ}{.}}25$ – M. Barbier, N. Martin *et al.*;
2 champs galactiques de $2 \times 2°$, $l_{II} = 245°$, 338° – D. Jonas;
champ de $4 \times 4°$ près de NGC 6231 – A. Laval;
champ de $4 \times 4°$ près de IC 4628, $l_{II} = 343°$.

Au spectrographe à fente, avec une dispersion de 73 Å mm^{-1}, Ch. Fehrenbach et E. Maurice ont classé 200 étoiles supergéantes du Grand Nuage de Magellan.

Y. Georgelin a entrepris la classification des étoiles excitatrices de diverses nébuleuses gazeuses entre $l_{II} = 200°$ et $l_{II} = 360°$ (dispersion 73 Å mm^{-1}).

L. Prévot exécute la classification d'étoiles faibles bleues de population II aux grandes latitudes galactiques.

En U.R.S.S. A l'Observatoire d'Abastumani, de très nombreuses classifications spectrales ont été faites par E. K. Kharadze et R. A. Bartaya *et al.* Les listes très importantes publiées sont les suivantes:

6000 étoiles dans les SA 7, 8, 16, 17, 18, 19, 20, 21, 37, 38, 39, 40, 41, 42 et 43;
3000 étoiles dans les associations O: Cam I, Cas III, Cas IV, Cas VII;
6500 étoiles dans NGC 2129, 6802, 6823, 6834, 6913, 7086, 7654 (N. B. Kalandadze et M. D. Metreveli);
4000 étoiles dans les SA 25, 48, 50, 73, 75, 97, 99 et 122 (A. D. Chuadze);
2000 étoiles ont été classées près du pôle galactique nord (D. G. Chipashvili et L. I. Zaitseva);
4000 étoiles dans association T And T1, Oph T1, Lyr T1 (M. A. Shiukashvili et L. Skhirtladze).

T. A. Uranova finit la classification des étoiles O-F5 dans 7 champs de la région 1 du plan P. P. Parenago (environ 7000 étoiles).

1.1.2. *Classifications simplifiées*

Les classifications spectrales sur des spectres à petite dispersion sont faites par simple inspection par divers observatoires.

U.S.A. N. Sanduleak, C. B. Stephenson ont commencé l'extension au ciel austral des étoiles à haute luminosité. Ce travail complètera ceux de Cleveland et Hambourg.

A Perkins Observatory, A. Slettebak *et al.* classent 900 étoiles moins avancées que F0 et plus brillantes que 14 près du pôle galactique sud (580 Å mm^{-1} vers $H\gamma$) spectres obtenus à Tololo.

B. E. Westerlund a classé des étoiles OB australes autour des restes des Supernovae RCW 86 et RCW 103 ainsi que dans Norma (dispersion 140 Å mm^{-1}).

Europe. A Bonn, Dr. W. C. Seitter a publié un Atlas spectral pour les dispersions de 645 et 1280 Å mm^{-1} vers $H\gamma$. Cet atlas sera très utile pour la classification des étoiles faibles. Au même observatoire, une classification des étoiles < A5 du ciel austral ($l_{II} = 233°$ à $l_{II} = 20°$ $|b_{II}| < 12.^{\circ}5$) est en cours.

A Lund, B. Karlsson a classé 800 étoiles situées vers $l_{II} = 203°$ $b_{II} = +2°$ et B. Virdefors met au point une classification bidimensionnelle pour les étoiles B à F (480 Å mm^{-1} vers $H\gamma$).

A Stockholm, 446 étoiles de la SA M3 ont été classées (206 Å mm^{-1} vers $H\gamma$).

1.1.3. *Groupes naturels*

M. F. McCarthy et P. J. Treanor étudient la distribution des étoiles dA et gK d'après les critères des groupes naturels de W. W. Morgan (SA 28, 54, 106 et 107).

1.1.4. *Étoiles particulières*

1.1.4.0. *Étoiles froides*

La classification précise des étoiles froides a été étudiée dans divers observatoires.

A Abastumani, G. N. Dzhimsheleishvili a mis au point une méthode pour classer les étoiles carbonées; une séquence de 24 étoiles a été établie et 40 autres étoiles ont été classées. A Perkins Observatory, A. Slettebak, P. C. Keenan et R. K. Brundage ont aussi classé 7 étoiles carbonées du pôle galactique nord. P. Keenan essaye la plus petite dispersion qui permette une détection des étoiles C.

De nombreux essais de classification d'étoiles particulières ont été tentés. Les variables du type Mira Ceti ont été étudiées dans ce but à Perkins, Mt Wilson et Toronto par P. Kennan, R. Garrison et A. Deutsch (1000 clichés).

A l'Observatoire de Marseille, M. Barbier a dressé une nouvelle liste d'étoiles M, S, C découvertes au cours des inspections de clichés de prisme objectif.

B. E. Westerlund a terminé un catalogue de 1200 étoiles C de l'hémisphère austral (clichés de Mont Stromlo).

Un catalogue des étoiles C est en préparation (C. B. Stephenson à Warner Swasey Observatory).

1.1.4.1. *Classifications diverses*

C. Jaschek, M. Jaschek et M. Arnal ont commencé une recherche d'étoiles trop bleues pour leurs types spectraux dans toutes les aires qu'ils ont étudiées.

D. S. Evans, S. V. M. Clube et D. H. P. Jones ont observé les spectres d'un nombre d'étoiles RR Lyrae australes et D. Evans avec P. A. T. Wild a recherché des sous-naines.

D. D. Meisel a déterminé les types spectraux, les vitesses de rotation et Δm pour 50 composantes d'étoiles binaires visuelles. 10 de ces étoiles sont des étoiles Be.

A. Markowitz a fait la classification de toutes les étoiles à spectres composites $V < 70$, $b > -25°$, le type spectral, la luminosité et la rotation ont été estimés pour toutes les étoiles.

1.1.4.2.

Des groupes d'étoiles O et B particulières ont été étudiés par W. W. Morgan (Yerkes Observatory).

(1) Les étoiles à hydrogène très fort (membres des amas de la Nébuleuse d'Orion). Ces étoiles de type O à B2 permettent une discrimination des étoiles se trouvant à un certain état d'évolution. Ces étoiles sont moins lumineuses que les étoiles normales.

(2) Les étoiles à He I de type B2, membres d'amas.

(3) Les étoiles "quasi géantes B8" sur des spectres à petite dispersion. Elles ont la classe de luminosité III mais sur des grandes dispersions, C II 4267 a l'intensité correspondante aux classes B2-B3. Ces étoiles apparaissent dans certains amas mais elles peuvent aussi exister isolées.

(4) Des étoiles B avancées avec des raies K anormalement intenses.

1.1.5. *Spectres infrarouges*

R. F. Wing et H. Spinrad ont fait un inventaire des spectres connus dans la région $\lambda > 9000$Å. Pour les étoiles M les bandes de CN sont particulièrement intenses dans la région 1–2·5 μ. L'identification de ces bandes avec H_2O est rejetée.

1.2.0. *Classification par mesure photométrique*

1.2.0.1. *Classification de Stockholm-Uppsala*

Les astronomes suédois ont continué les mesures des spectres en vue de leur classification. L'état de ce travail est le suivant:

(1) Dans le programme des étoiles galactiques $|b| < 3°$:
- (*a*) résultats publiés $73° < l_{II} < 133°$;
- (*b*) résultats pratiquement achevés $43° < l_{II} < 73°$;
- (*c*) plaques prises $133° < l_{II} < 153°$.

(2) Les résultats dans les SA sont analysés par ailleurs, 19 champs publiés.

(3) Des études sur le pôle galactique sud sont en cours: région de 40 degrés carrés $V < 15$.

(4) K. G. Malmquist a étudié une région de 120 degrés carrés 3000 étoiles $m_{pg} < 13{\cdot}5$. Cette étude a permis de mettre en évidence des nuages absorbants à 200 et 400 pc.

(5) Häggkvist et T. Oja ont mis en route un programme d'observations qui comprendra la mesure des étoiles brillantes par photoélectricité.

Les étoiles faibles seront mesurées photographiquement.

Le programme comprend les étoiles plus brillantes que 11 (magnitude *B*) pour $b > 70°$ et les étoiles plus brillantes que 14 pour $b > 75°$. Les clichés sont pris.

(6) L. O. Lodén et Mlle B. Nordström font une classification préliminaire pour 15000 étoiles de types spectraux très chauds ou très froids dans le Centaure entre $280° < l_{II} < 320°$.

1.2.0.2.

A. Broerfelde, J. Andersen et J. V. Clausen ont commencé la classification de spectres de 105 Å mm^{-1} vers $H\gamma$.

1.2.0.3.

A l'Université de Bochum, Th. Schmidt-Kaler a entrepris la classification ($B < 9$) dans 107 SA (étendues) de l'hémisphère austral. Les classifications seront basées sur des mesures photoélectriques.

1.2.2. *Mesures de largeurs équivalentes*

La mesure de la largeur équivalente des raies permet de déterminer indépendemment la magnitude absolue, la composition chimique et la turbulence lorsque les raies sont choisies spécialement.

1.2.2.0. *Méthode de Petrie*

Poursuivant les mesures de R. M. Petrie, W. Buscombe à Evanston a classé 121 étoiles australes de type O5 à A3 et a déterminé les magnitudes absolues de 380 autres étoiles australes; il a aussi pu déterminer les vitesses de rotation $v \sin i$ pour une centaine de ces étoiles.

B. E. Westerlund a déterminé les M_v pour 25 étoiles.

A l'Observatoire de Marseille, N. Martin a indiqué une méthode très expéditive, dérivée d'une méthode d'Öhman, pour déterminer une grandeur qui est en relation simple avec W_λ.

En fait, toutes les étoiles OB-A2 des champs de P.O. indiquées dans les listes de spectres MK sont ainsi mesurées. Ce qui permet de déterminer M_v en se servant des étoiles de Petrie. N. Martin détermine aussi le rayonnement de ces étoiles sur les mêmes clichés. Une vingtaine de champs galactiques sont ainsi mesurés par N. Martin, M. Barbier, Y. Georgelin et A. Laval.

A l'Observatoire de Toulouse, M. Azzopardi et R. Bouigue poursuivent des travaux sur la mesure des magnitudes absolues à partir de spectres obtenus au prisme objectif de 25 cm.

Les mesures de W_λ sont digitalisées. Les raies de Balmer Hγ et Hδ ainsi que les raies de l'hélium sont mesurées. Cette méthode doit permettre de préciser les critères de luminosité en établissant conjointement les critères de classification spectrale.

L'application de ces résultats à des champs situés sur l'anticentre et pôle galactique nord est en cours pour les étoiles de types O, B, A et F.

G. A. H. Walker, à l'Université de British Columbia a effectué des mesures des raies Hγ et 4388 et 4471 de He I (largeur équivalente et profondeur); il en déduit la relation entre l'abondance de He et la vitesse de rotation des étoiles: 425 étoiles O, B étudiées. Il trouve que ces résultats suppriment une difficulté de la théorie de la diffusion de He I de J. L. Greenstein.

A Meudon, Mme R. Herman et N. Yilmas font des mesures de l'intensité des raies de l'hydrogène pour des étoiles Be.

1.2.2.1.

M. F. McCarthy, P. J. Treanor et W. K. Ford ont classé les étoiles K4 à M5 à l'aide de l'intensité des raies de Na I (228 Å mm^{-1}) et des bandes de CaH (6365 et 6850 Å) et MgH (5622 et 5211).

A Ambastumani, V. V. Lenshin a fait une classification quantitative de 29 étoiles normales B3-F0 et 15 étoiles Ap avec des spectres de 15 Å mm^{-1}.

R. M. West à Copenhague a classé 194 spectres de P.O. en ultraviolet, obtenus à Abastumani (110 Å mm^{-1} à 3900 Å). Il a mesuré 1031 raies et calculé 47 rapports d'intensité de raie. Il en déduit les classes spectrales, la luminosité et le rapport Fe/H. La région 4000–4550 est étudiée pour 250 spectres.

A Torun, J. Smolinski, A. Strobel *et al.* font aussi des classifications multiparamétriques par la mesure de la profondeur et de l'intensité des raies dans des spectres de P.O. Ils calibrent leurs clichés à l'aide de spectres de grande dispersion pour déterminer en plus de Te, g, Fe/H la vitesse de microturbulence des étoiles.

1.3.0. *Classification BCD* (Barbier, Chalonge et Divan)

A Paris, L. Divan continue ce travail. Les résultats donnent les trois gradients φ_{rb}, φ_b, φ_{uv}, la valeur de la discontinuité de Balmer D et sa position λ a été déterminée pour 380 étoiles; les 30 étoiles standards sont publiées et 350 sont prêtes pour la publication. Cette classification est très précise et permet une bonne détermination du rougissement interstellaire.

Une étude particulière des étoiles O et B est entreprise avec M. L. Burnichon. Environ 1000 spectres sont déjà photographiés. Ces auteurs montrent que les magnitudes des étoiles O varient avec les valeurs de λ et D.

J. Berger effectue, à l'Institut d'Astrophysique de Paris, sur des clichés obtenus à l'Observatoire de Haute Provence, la classification BCD des étoiles bleues faibles du halo galactique. 130 étoiles ($7{\cdot}5 < m < 13$) sont classées. Un effort spécial est fait pour établir des séquences d'étoiles de diverses populations.

Dans une étude sous presse, A. Ardeberg étudie les gradients des étoiles O9 à G0 par classe de luminosité séparées. Ces résultats sont comparés aux théories des atmosphères stellaires.

1.4.0.

An Atlas of Low-dispersion Grating Stellar Spectra a été publié par Abt, Meinel, Morgan et Tapscott. Des exemplaires peuvent être obtenus à Kitt Peak National Observatory. Des suppléments sont en préparation. Des améliorations nettes au système MK sont indiquées dans cet atlas.

Listes diverses

Nous avons signalé au fur et à mesure les listes de classifications spectrales de diverses étoiles. Ajoutons les catalogues actuellement en préparation pour des étoiles particulières.

Étoiles à émission. Le catalogue des étoiles à émission préparé par F. C. Bertiau et M. F. McCarthy contient 3216 étoiles à émission $<F$ et reprend notamment les listes de Mont Wilson, de Tonantzintla et de Case-Hambourg.

K. G. Henize publie une liste de 1929 étoiles présentant Hα en émission, la plupart de ces étoiles sont des premiers types spectraux et situées dans le ciel austral.

Le catalogue L. R. Wackerling contient 5290 étoiles à émission des premiers types spectraux dont l'auteur a eu connaissance.

C. Jaschek, M. Jaschek et L. Ferrer ont préparé une bibliographie contenant 990 références pour 1900 étoiles Be.

Étoiles Ap. A. Cowley *et al.* ont publié une nouvelle liste d'étoiles Ap. Ch. Bertaud a publié un catalogue d'étoiles Ap.

2.1. SYSTÈMES À BANDES LARGES

2.1.0. *Système UBV RI de Johnson-Morgan*

De nombreuses classifications spectrales ont été accompagnées par la mesure des magnitudes dans le système de Johnson et Morgan. Ceci est le cas pour les travaux déjà signalés:

(1) 25 étoiles OB près de RCW 86 et RCW 103 étudiées par B. E. Westerlund.

(2) Pour 764 étoiles, classées par N. Roman, la publication n'est pas encore faite à cause d'une difficulté de raccordement.

(3) Les amas galactiques de l'hémisphère sud étudiés par E. R. Hill et Ch. L. Perry.

(4) L'Association Céphée OB2 étudiée par S. C. Simonson III.

(5) Les étoiles brillantes du programme d'Uppsala pour $V<5{\cdot}05$ et $b_{\text{II}}>-10°$, pour $V<6{\cdot}05$ et $b_{\text{II}}>60°$.

(6) Les étoiles des agrégats stellaires faites par E. E. Mendoza comportent des mesures dans le système *UBV RI.JHKLM.*

(7) A Lyon, J. Bigay *et al.* mesurent les magnitudes des *U*, *B*, *V* des étoiles excitatrices de région H II, étoiles Be Ap Am, les amas jeunes NGC 6823, 6871, 6910, IC 1805–1848 et des amas et associations dans M33, M51-LMC.

Une très importante série de mesures dans le système étendu (*UBVRIJKL*) a été faite par E. E. Mendoza à Mexico. 500 étoiles ont été mesurées.

D'après son analyse, la meilleure valeur de la température effective est obtenue à partir de *V-R* et *R-I*.

Parmi ces étoiles, 50 ont des excès de couleur infrarouge, qui ne peuvent pas être expliqués par l'extinction interstellaire; diverses causes sont étudiées: T Tau, enveloppes de gaz ou de poussières, atmosphères anormales, etc.

C. et M. Jaschek et L. Ferrer montrent que l'effet de serre (blanketing effect) pour les étoiles Ap et Am peut être déterminé par la photométrie *UBVRI.* Il est possible de déterminer la métallicité des étoiles Am.

2.1.1. *Système de Lick* (J. Stebbins et A. E. Whitford)

Ce système n'a pas été utilisé par de nombreux auteurs. J. Rousseau a mesuré et étudié les étoiles de O. C. Wilson de type spectral G5 à K3 dans ce système. Son travail a surtout consisté à discuter les deux thèses en présence, sur la raison de la dispersion des indices $P-V$ pour ces étoiles pour un type spectral donné (O. C. Wilson et B. E. J. Pagel). Il existe une contradiction formelle entre les conclusions de ces deux auteurs: pour O. Wilson, les étoiles à petit $P-V$ sont des étoiles chaudes et riches en métaux alors qu'elles sont pauvres en métaux pour B. Pagel. Cet auteur tient compte de l'effet de serre (blanketing). Les indices bleus étant plus sensibles à la composition chimique et les indices rouges à la température, J. Rousseau montre, par des observations en six couleurs, que l'effet Pagel est prédominant pour les étoiles G8-K0 alors que les deux effets existent pour les étoiles plus avancées. Ceci montre que les indices $P-V$ seuls ne permettent pas une classification distinguant ces deux effets.

2.1.2. *Système de Genève*

Le système à 7 couleurs utilisé par les astronomes de Genève dans diverses stations d'observation est le système *UBV* complété par les couleurs B_1B_2 qui coupent l'intervalle *B* en deux parties et V_1G l'intervalle *V*.

Cinq paramètres sont définis: l'un [*d*] est un paramètre de luminosité, [Δ] permet d'estimer la rotation des étoiles B0-B7; pour les étoiles Am, il permet de déterminer la métallicité et pour les étoiles F, la composition chimique.

Le programme comprend surtout l'observation des étoiles Ap.

2.2. LES SYSTÈMES À BANDES INTERMÉDIAIRES

2.2.0.

Le système photométrique de Strömgren u, v, b, y quelquefois complété par des mesures de Hβ a été utilisé par divers observateurs.

Ch. L. Perry a publié un catalogue *ubvy* de la plupart des étoiles F5 à G0 entre 6·5 et 7·5 et $-10° < \delta < +60°$.

Ces observations doivent servir de base pour des études plus spécialisées. Des mesures dans ce système complétées par des mesures Hβ ont été faites à Cerro Tololo par Ch. Perry et E. R. Hill pour les amas IC 2391, IC 2602, NGC 6231.

L'appartenance des étoiles aux amas et la détermination des âges des amas ont été faites.

H. E. Bond a mesuré à Kitt Peak plus de 250 étoiles de magnitude 7 à 9 suspectées comme étoiles à raies métalliques faibles (weak lines). La fréquence de ces étoiles est rare; elles seraient des géantes évoluées. H. Bond trouve que la relation entre composition chimique et excentricité des orbites galactiques est douteuse.

E. H. Olsen a comparé les calibrations dans le système *u, v, b, y, β* pour les étoiles A-F de la série principale avec les calibrations dans le système *k, n, m, g* pour les géantes G-K, en étudiant des étoiles doubles. L'accord est bon particulièrement pour les rapports [Fe/H].

Pour les géantes, les résultats sont moins clairs et semblent indiquer une dépendance de $M_v(K)$ avec [Fe/H]. Les mesures continuent.

A. G. Davis Philip a mesuré un certain nombre d'étoiles dans le système *uvby*. 40 étoiles de champ de la branche horizontale ont été trouvées et leur couleur normale déterminée. Des étoiles bleues des branches horizontales dans les amas globulaires M 4, NGC 362 et 6809 ont été mesurées et comparées aux étoiles normales.

La distribution des étoiles de Pop. I et II dans la région du pôle galactique nord est étudiée pour les étoiles A. (mesure *uvby* et *UBV* pour 57 étoiles $5 < m < 14·5$). 11 sont de Pop. II, 10 de Pop. I pour $m > 12$ les étoiles de Pop. II prédominent.

A l'Observatoire National de Kitt Peak, D. L. Crawford *et al.* ont étudié les étoiles brillantes et des étoiles d'amas pour calibrer le système $u, b, v, y\ \beta$ (couleurs intrinsèques et magnitudes absolues) pour les étoiles B A et F. En général, les couleurs intrinsèques $(b-y)$, sont prédites avec une dispersion cosmique de $\pm 0{\cdot}010$, et les magnitudes absolues avec des précisions de $\pm 0{\cdot}6$ ou $\pm 0{\cdot}2$ suivant le type spectral pour les étoiles B et $\pm 0{\cdot}2$ pour les étoiles A. Les étoiles F des Hyades sont anormales. Liste des mesures faites:

1. Sp B5 toutes les étoiles $V \leqslant 6{\cdot}5$
2. Sp G0 toutes les étoiles $V \leqslant 5{\cdot}0$
3. Sp B0 $\delta > 0°$ $V \leqslant 6{\cdot}5$

Amas: 1. Hyades; 2. Coma; 3. UMa; 4. Praesepe; 5. α Per; 6. Pléiades; 7. IC 2391; 8. IC 2602; 9. h et χ Per; 10. NGC 6231 et Sco OB1; 11. IC 4665; 12. NGC 752.

L'étude d'autres amas est en cours.

2.2.1.

D'autres systèmes intermédiaires sont à l'étude. L. Hansen et P. U. Jacobsen ont étudié 215 étoiles G-K (géantes et naines) pour les longueurs d'onde 3440, 4060, 4590, 5110, 5400 et 6390 Å et comparé ce système avec le système k, n, m, g. Un catalogue dans le système k, n, m, g de 1160 étoiles de type avancé (K. Gyldenkerne *et al.*) et un catalogue d'étoiles A (K. T. Johansen et K. Gyldenkerne) sont sous presse.

P. B. Boyce *et al.* ont exécuté une classification des étoiles M géantes à l'aide d'un photomètre avec la bande 7050 de TiO.

La calibration des quantités k, n, m, g pour déterminer les indices de couleur, [Fe/H] et M_v sont en cours – en partie sous presse. (5 P. K. Rasmuseen).

2.2.2. *Système de Walraven*

De nombreuses mesures dans le système à bandes intermédiaires de Th. et J. Walraven ont été faites par les astronomes de Leyde.

Ce système permet de définir des indices de couleur corrigés de l'absorption interstellaire. Les quantités calculées sont des mesures photoélectriques de la discontinuité de Balmer, de gradient φ_{uv} et de la magnitude absolue M_v.

Les listes publiées sont: 469 étoiles OB dont 391 de la liste de Lyngå dans l'association Centaure-Norma (Van Houten); 69 supergéantes du G. N. M. (Van Genderen); 30 étoiles dans la partie nord de Sco OB1; 169 étoiles dans NGC 5662 (Pel et Velthuijse); 20 étoiles dans NGC 4755; 24 étoiles dans NGC 6421 (Wamsteker); 36 étoiles dans NGC 6475; 500 étoiles bleues entre le PNM et le GNM.

2.2.3. *Le système de van den Bergh*

S. van den Bergh a utilisé un système photométrique à cinq couleurs à bandes intermédiaires avec des largeurs d'ondes effectives.

	λ_{eff}	$\Delta\lambda$
35	3490	± 370
38	3795	176
41	4165	84
42	4255	83
45	4513	84

L'indice C(35–38) est sensible à la fois à T_e et à M_V; C(38–41) permet de mesurer l'effet "line blanketing"; C(41–42) est une mesure de l'absorption de CN (Lindblad) et est sensible à M_V; C(42–45) mesure l'influence de CH et est sensible à T_e et M_V.

Van den Bergh a mesuré 210 étoiles isolées, 70 amas stellaires et 56 galaxies. La détermination des paramètres essentiels, type spectral, m_v, abondances et rougissements peuvent être obtenus en combinaison avec les mesures *UBV*.

La séparation des étoiles riches et pauvres en métal pour les étoiles naines est particulièrement nette dans un diagramme $C(38\text{–}41)\text{-}C(42\text{–}45)$ corrigé du rougissement. La mesure globale des galaxies et amas a été faite et est recommandée par van den Bergh.

2.2.4. *Système de Vilnius*

V. Straizys de l'Université de Vilnius a défini un nouveau système photométrique *UPXYZVS* avec les longueurs d'ondes effectives 3500, 3750, 4050, 4600, 5100, 5400 et 6500 Å. Des filtres en verre ont remplacé, sauf pour 6500, les filtres interférentiels. Sept jeux ont été réalisés et distribués à cinq observatoires. 400 étoiles ont été mesurées.

Des indices corrigés de l'absorption interstellaire sont définis et permettent de classer les étoiles de B à M. L'auteur pense que son système donne des résultats comparables au système de B. Strömgren même pour des étoiles rougies et permet de faire dans tous les cas une classification à deux dimensions et quelquefois à trois.

Il serait important de comparer ces systèmes.

2.2.5.

A l'Université d'Arizona, R. I. Mitchell et H. L. Johnson ont défini un nouveau système photométrique à bandes intermédiaires (narrow band photometry dans leur publication) centrées sur les longueurs d'ondes 3370, 3530, 3750, 4020, 4590, 5180, 5830, 6350, 7240, 8000, 8580, 9850 et 11080 Å dont les largeurs sont variables allant de 110 Å pour −3370 à 580 Å vers 7240. Les flux transmis pour les filtres pour des étoiles A0V sont indiqués.

Le système est défini par la mesure de 1000 étoiles brillantes pour lesquelles on donne la magnitude 52 et 12 couleurs. Ces mesures sont complétées par une photométrie dans le rouge (8 et 6 couleurs) pour un certain nombre d'étoiles. Ces données sont surtout utiles pour un rattachement au système rouge de Johnson *JHKL*. Cette photométrie sera étendue à l'hémisphère austral.

R. F. Wing utilise deux systèmes photoélectriques à bandes multiples. Un système à 27 couleurs pour la classification des étoiles de type M avancé. Il y a accord avec les types spectraux de P. C. Keenan.

Une différence sensible de la température effective T_e existe entre les étoiles Mira et M non variables pour le même type spectral.

R. F. Wing retrouve pour certaines étoiles géantes M une anomalie d'abondance de CN comme pour certaines étoiles K.

Un système à 13 couleurs (3000–5000 Å) permet l'étude des étoiles avancées (mesure des raies d'émission et différence des répartitions spectrales de l'énergie).

Signalons que Th. Walraven construit un photomètre à canaux multiples. Le spectre est décomposé en 22 bandes alternativement brillantes et obscures à contours rectangulaires. Un dispositif optique simple permet de changer les bandes brillantes en bandes obscures et vice versa. De sorte que 11 photomultiplicateurs permettent de déterminer en deux mesures les 22 bandes. La mesure de polarisation est aussi possible.

2.3. SYSTÈMES À BANDES FINES

2.3.0.

La photométrie à bandes très étroites a été faite dans divers Observatoires.

A Lyon et à l'Observatoire de Haute Provence, J. Bigay *et al.* utilisent un système à bandes passantes étroites.

λ	3600	4035	4102	4220	4340	4500	4700	4861	4900Å
$\Delta\lambda$	120	60	30	98	35	85	92	35	98

Ce qui permet de déterminer la valeur de la discontinuité de Balmer et les largeurs équivalentes de Hβ, Hγ, Hδ ainsi que des grandeurs analogues à φ_b et $B-V$. Il s'agit d'une classification à trois paramètres.

Le programme comporte la détermination de la distance des étoiles O-B, une étude des étoiles Be ainsi qu'un indice de métallicité.

A Paris et à l'Observatoire de Haute Provence, N. Morguleff et M. P. Véron utilisent la classification décrite par D. Barbier (1960). Ce systéme comprend 11 bandes centrées sur NaI, MgI, Hβ, CaI, CN, CaII et 5 bandes dans des régions peu perturbées par des raies vers 5950, 5325, 4960, 4040 et une vers 3612Å. Ceci permet de définir; trois gradients, une discontinuité de Balmer et six intensités de raies. Une étude de la correlation entre ces grandeurs et les grandeurs $B-V$, $U-B$, classe de luminosité, est déterminée. Cette classification faite pour 100 étoiles G et K donne des résultats très intéressants.

U. Sinnerstad et J. Arkling ont mesuré les largeurs équivalentes des raies Hβ, Hγ et 4471 et 4922 de HeI. La combinaison d'un filtre de 6Å avec un autre de 28Å pour Hβ permet de déterminer $v \sin i$ avec une précision de ± 45 km s^{-1} pour les étoiles de classe V.

Les étoiles Be sont très faciles à mettre en évidence.

A l'Observatoire de Copenhague, de nombreux travaux ont été exécutés à l'aide de divers systèmes photométriques à bandes étroites notamment pour la classification des géantes G et K.

2.3.1.

P. E. Nissen a mesuré avec le télescope solaire McMath l'intensité d'une bande très étroite centrée sur 4800Å avec une largeur totale de 3·51Å.

Le paramètre $A = 2{\cdot}5 \log(I_{\text{cont}}/I_{\text{raie}})$ est déterminé.

P. E. Nissen montre qu'on peut déterminer à l'aide de ce paramètre l'abondance [Fe/H] si on connaît l'indice Hβ déterminé par D. L. Crawford *et al.*

Une photométrie à bandes étroites faite par M. Rudkjøbing et P. Nissen à l'Observatoire de Haute Provence et à Kitt Peak montre que la microturbulence n'est pas un paramètre indépendant de la structure stellaire. L'indice m de B. Strömgren a été d'autre part calibré.

2.3.2.

R. F. Griffin à Cambridge (U.K.) a étudié l'intensité de la raie Sc I 6305 pour les étoiles avancées.

2.4.0.

La classification spectrale par la mesure directe des répartitions spectrales à l'aide de monochromateur (spectral scanning) ne semble guère avoir fait de grands progrès.

J. P. Chaturvedi à Uttar Pradesh State Observatory a commencé des mesures des étoiles K et M avec une définition de 50Å. G. S. D. Babu étudie avec le même instrument les étoiles magnétiques.

2.5.

Le problème de la constitution de centres de données pour l'établissement d'un catalogue et d'un fichier des données spectrales et photométriques pour toutes les étoiles mesurées est devenu d'une importance capitale et une coordination s'impose.

L'U.S. Naval Observatory a publié un catalogue photométrique (V. M. Blanco, S. Demers, G. C. Douglas et M. P. Fitzgerald) contenant les magnitudes V et les couleurs $(B-V)$, $(U-B)$ et $(U-B)$ contenant ces mesures pour environ 17000 étoiles. Ce catalogue présente un intérêt considérable.

"Un catalogue photométrique" est préparé pour la publication à La Plata; il contient toutes les mesures photoélectriques de magnitudes et couleurs publiées avant 1968·0. Il contient 20000 étoiles et 1800 références bibliographiques (C. Jaschek).

A l'Université de Lausanne, B. Hauck a mis en route la compilation d'un catalogue analogue à celui publié par l'U.S. Naval Observatory pour les principaux systèmes actuellement utilisés avec une homogénéisation et la corrélation entre les divers systèmes.

CONCLUSION

De nombreuses classifications spectrales et des mesures photométriques ont été exécutées. Elles sont souvent des données accessoires pour l'étude de nombreux problèmes sur la Structure galactique, les Nuages de Magellan, l'étude de l'absorption interstellaire. De nombreux résultats obtenus sont signalés dans les rapports des Commissions correspondantes.

Indiquons que pour la classification, par spectrographie ou photoélectricité, deux problèmes se posent: la standardisation des systèmes photoélectriques et la collection des données. Il s'agit là de deux problèmes d'entente entre les astronomes que notre Commission doit essayer de résoudre.

CH. FEHRENBACH
Président de la Commission

46. TEACHING OF ASTRONOMY
(ENSEIGNEMENT DE L'ASTRONOMIE)
(Committee of the Executive Committee)

PRESIDENT: E. A. Müller.
VICE-PRESIDENT: J. Kleczek.
ORGANIZING COMMITTEE: G. Abell, E. K. Kharadze, E. Kononovich, V. Kourganoff, M. Minnaert, P. Routly, E. Schatzman, T. Swihart.

MEMBERSHIP

During the 13th IAU General Assembly in Prague it was decided that each nation adhering to the IAU should be represented in Commission 46 by one member interested in the teaching of Astronomy. Six nations were already represented by the members of the Organizing Committee. For all other nations the Commission President contacted the head of the National Delegations to the IAU with the request to designate one IAU member of the particular country as member of Commission 46. With only a few exceptions every IAU member nation designated a representative in this Commission which now contains representatives of 33 nations. It is hoped that the missing nations in our Commission will also soon be represented by an IAU member active in the promotion of Astronomy education. Each member is responsible to (a) promote the teaching of Astronomy in his own country, (b) to keep in touch with the Astronomy teachers of his country, and (c) to communicate to the Commission President any programs and developments, problems and needs concerned with Astronomy education in his country. It is also hoped that institutions in countries not yet adhering to the IAU but which might wish to start some Astronomy teaching, could contact the Commission and, in some specific cases, could designate a Consulting Member to the Commission.

WORK OF THE COMMISSION

1. *Astronomy educational material*

A. *Bibliography*

With the kind collaboration of the Commission members from 22 nations a world wide list of *Astronomy Educational Material* has been prepared which collects the available teaching material recommended by Astronomy teachers of IAU member nations. It contains the following Tables: I. Books; II. Atlases, Catalogues, Encyclopedias, Dictionaries; III. Magazines, Journals, Yearbooks; IV. Lecture Notes, Exercises, Examination Problems; V. Films; VI. Filmstrips, Photographs, Slides. For each item information is given as to the educational level it refers to, where it can be ordered, what its price is, and any other remark which might be of interest. In order to keep its cost low, the A.E.M. list is being reproduced in mimeographed form and will be distributed free of charge to members of the Commission and to the Division of Science Teaching of UNESCO. It is intended to make it available to any astronomer or science teacher involved in the teaching of astronomy. The Commission members will be responsible to give the list of A.E.M. a wide distribution in their respective countries. How this can best be done will be discussed during one of the Commission sessions in Brighton. Every three years, at successive IAU General Assemblies an Addendum of the A.E.M. should be prepared which will contain the recent material useful for the teaching of Astronomy.

B. *Didactics of astronomy*

So far only little has been published concerning the didactics and methods of Astronomy teaching which is essential for the education in Astronomy at elementary and secondary Schools and College level as well as for the education of Astronomers. It will be one of the topics to be discussed during one of the Commission sessions in Brighton. We should clearly distinguish between "Education *of* Astronomers" and "Education *in* Astronomy" both being of equal importance for our Commission. The education *in* astronomy should already start at the school level. It also should be done in public lectures and in courses for non-scientists, as Taylor (1969) puts it: "We should try to encourage people who will not become astronomers to take an intelligent interest in astronomy". The few books on didactics of Astronomy teaching which have been published in recent years are included in the A.E.M. list. Some additional useful references to recent articles on the subject may be mentioned here. A special issue of the journal *Astronomie in der Schule* (1969) is devoted to the development of the teaching of Astronomy at the secondary school level in the German Democratic Republic and in Czechoslovakia. In the United Kingdom a special meeting was held in 1968 on the Education of Astronomers (1969) during which a number of British Astronomers involved in the teaching of Astronomy presented their views on the subject. R. C. Roeder reported the following: "The Dominion Observatory in Ottawa (Ontario) publishes pamphlets in English or French for the elementary school level on the following subjects: Moon, Sun, Eclipses, Planets, Stars, the Solar System, Jupiter, Mars, Saturn, the Orbits of the Planets, the Space Age, Astronomy in Canada". R. Berendzen sent the following information: "Articles on teaching astronomy to children appear from time to time in several English-language journals, e.g. *Science and Children*, *Science Teacher* (both of which are published by the National Science Teachers' Association in the U.S.), and *School Science Review* (which is published in England). The entire issue of *Science and Children*, **5**, no. 4, December 1967, was devoted to teaching about the Moon to elementary school children." R. Berendzen is editor of the *Journal of College Science Teaching*, a journal "which is, hopefully to become an organ for distinguished papers on astronomy education". In addition, R. Berendzen mentions that M. Hoskin (Cambridge University) is editor-in-chief of another newly formed publication: the *Journal of History of Astronomy* which is to include the educational use of history of astronomy. S. S. Ross (1969) describes the session on Laboratory Exercises in Astronomy which was held during the session devoted to methods and needs in astronomical education after the August 1969 meeting of the American Astronomical Society. An Informal Resource Letter on astronomy laboratory exercises was prepared by Wentzel and Smith (1969) which contains published manuals and articles on laboratory exercises as well as sources for photographs and other material for teaching aid. An interesting outline of an "Astronomy" course given to non-science students has been prepared by Wentzel (1969) in which he "demonstrates that science is a human endeavour, that it involves social judgments and philosophical implications as well as established facts".

C. *Astronomy as a career*

It is important to inform the young people with some interest in astronomy what the general requirements are for becoming an astronomer and what possibilities they have when choosing astronomy as a career. The following recent publications on this subject were communicated to the President; C. Schalén published a booklet in Swedish on the study of astronomy. McNally (1968) summarizes what is needed in an astronomer, gives advise on what and where to study, and lists Universities and Observatories in the United Kingdom where an astronomical education can be obtained. R. C. Roeder mentions a pamphlet on Astronomy as a Career published by the Guidance Centre, Ontario College of Education, Univ. of Toronto. Berendzen (1968) wrote a thesis on The Career Development and Education of Astronomers in The United States in which he studied the education, vocational developments, pedagogical viewpoints, and employment of U.S. astronomers. He compiled information from astronomers themselves, astronomy departments, governmental and private organizations and the literature, and he has a large compilation of the literature on the subject. A brochure on How to Become an Astronomer was published by Miller (1969) which contains among others a section on "qualifications and expectations", and one on "education of an astronomer", and a very useful appendix on Graduate Departments of Astronomy in the United States. The American Institute of Physics has published two handbooks concerning Astronomy studies: (1) *Graduate Programs in Physics and Astronomy in the United States (1968)*, this being a handbook for advisors of prospective doctoral and master's students; and (2) *Directory of Physics and Astronomy Faculties in 1968–96 in the United States, Canada, and Mexico (1969)*.

D. *Material most needed*

All Commission members were invited to list the material for Astronomy teaching which they considered most needed in their own countries. Many members expressed the need for good slides and photographs as teaching aids. The Table VI of the A.E.M. list will give detailed information about photographs, slides and film-strips available and where they can be purchased. It is suggested that in each individual country a rental service is organized either through the astronomical society or the astronomy department of the principal university, such as is done for example in Canada (Roy. Astr. Soc. of Canada), in France (Soc. Astr. France), Great Britain (Roy. Astr. Soc.), Switzerland (S. A. G.), U.S.S.R. (Soc. Moscow Planetarium), etc. Some members mentioned the need of a planetarium; others felt the lack of a real good modern textbook on Astronomy and Astrophysics – the latter being possibly a language problem and also the quite different educational level at which Astronomy is taught in different countries. An interesting comment on "needs" was communicated by J. Hers who wrote: "J. Wolterbeek feels that there is a real need to make star catalogues available to students, so that they will have an opportunity to learn how to use them. Although many catalogues are today almost antique, and have been shown to contain many systematic errors, they are still of outstanding educational value, and some might well be reprinted. Catalogues are required of accurate positions, approximate positions, proper motions, radial velocities, trigonometrical and other parallaxes, magnitudes, spectral types, nebulae and clusters, variable stars, zodiacal stars, observational results of *E*-regions and of Kapteyn Selected Areas, visual and other double stars, minor planets, comets, etc. etc.". This question on Catalogues should be discussed by the Commission members present at the Brighton meeting.

E. *Miscellaneous teaching material offered*

Upon the President's encouragement Commission members kindly made a survey of possible exercise material that might be available in various Observatories of the different countries, and that could be placed at the disposal of astronomy teachers in other universities. The result of this enquiry allowed the listing of the following material. Colleagues interested in any of the following items are invited to contact directly the person or Observatory offering the material.

i. *The Sun*

The Göttingen Observatory offers reproductions of used observational material (for example spectra of sunspots). Address: Prof. H. H. Voigt, Universitäts-Sternwarte, Geismarlandstrasse 11, 34 Göttingen, Germany.

The Kandilli Observatory offers reproductions of filtergrams of the sun made in H-alpha and in the Ca II K-line. Address: Prof. M. Dizer, Astronomical Observatory, Kandilli, Istanbul, Turkey.

The Utrecht Observatory can offer spectrograms, microphotometer records, and interferometric records of radio-frequency observations, either already used for research work or rejected but still sufficiently good for teaching purposes. Address: Prof. M. Minnaert, Sterrewacht "Sonnenborgh", Zonnenburg 2, Utrecht, The Netherlands.

ii. *Planets*

The Konkoly Observatory offers copies of photographic plates on minor planets. Address: Dr. B. Szeidl, Konkoly Observatory, Szabadsaghegy, Budapest XII, Hungary.

iii. *Stars*

Mme M. Hack offers reproductions on paper of spectra of Bp and Ap stars (dispersion 9·4 Å/mm). Address: Prof. M. Hack, Osservatorio Astronomico, Via G. B. Tiepolo 11, I-34131 Trieste, Italy.

Mme R. Herman offers reproductions on paper of used spectra of Be type stars. Address: Mme

Dr. R. Herman, Observatoire de Paris-Meudon, Section d'astrophysique, 92-Meudon, France.

The Konkoly Observatory offers (1) copies of used photoelectric records and photographic plates of RR Lyrae stars; (2) copies of used photographic plates of globular clusters. Address: See above, *Planets*.

iv. *Exercise material requested*

In exchange for Exercise Material the Remeis Sternwarte can spare the following motion pictures:
(1) Solar Prominences, 16 mm, length: 119 m, duration 15 min;
(2) Solar Prominences, 16 mm, length: 80 m, duration 10 min;
(3) The Moon, 16 mm, length: 198 m, duration 18 min.
Address: Prof. W. Strohmeier, Remeis Sternwarte, Sternwartestr. 7, Bamberg, Germany.

F. *Atlas of Stellar Spectra*

At the IAU General Assembly in Prague Dr. Anne Underhill reported on the possibility and cost of preparing an atlas of tracings of high dispersion stellar spectra. The atlas could either be a compilation of tracings or be available in digitized form, the latter being preferred from the standpoint of low cost of production. Since one or several atlases of high-dispersion stellar spectra are of very great help for teaching purposes, the Commission members were asked to give an estimated number of copies that might be ordered by Observatories and Astronomical Institutes in their own countries. Obviously, the number of copies ordered of such an atlas will largely depend on the cost of a single copy. The result of the enquiry gave a total of about 100 copies which might be ordered, and more if the cost is low.

2. *Visiting professors project*

During the IAU meetings in Prague both Commissions 38 and 46 stressed the importance of the Visiting Professors Project which is to help the needs of countries remote from the main centers of astronomical research. As President of Commission 38 and member of the Organizing Committee of this Commission, Prof. M. Minnaert prepared a questionnaire for countries wishing to participate in the project and mailed it to various countries together with an explanatory letter in which he points out that "it is not sufficient to enable a few young astronomers from such countries to work for some time in more developed parts of the world, but that in some respects it might be more useful and economic if experienced astronomers were prepared to stay for some time in countries wishing a closer participation in modern astronomical research". So far, two positive answers have been received, one from Hyderabad (India) and one from Bandung (Indonesia). In the next IAU Information Bulletin an announcement will be published inviting colleagues to contact either Prof. M. Minnaert or Prof. E. A. Müller if they are interested to teach for some time at one or the other of the Observatories desiring to have a visiting professor.

3. *List of Observatories*

For many years Commission 38 has prepared a list of Observatories which offer facilities for visiting astronomers. The most recent list published by M. Minnaert may be found in the *IAU Information Bulletin* no. 23, p. 47, 1970. Furthermore many Observatories prepare pamphlets listing their main astronomical instrumentation and their various research programs. Undoubtedly, all solar physicists know that the journal *Solar Physics* publishes regularly brief descriptions of the Solar Observatories of the world. A pamphlet on Observatories of the World was prepared by T. L. Page for the last IAU General Assembly which, however, needed to be corrected and brought up-to-date. It is now understood that Commission 9 will be in charge of this project.

BIBLIOGRAPHY

Astronomie in der Schule, **6**, 113, 1969.

Berendzen, R. 1968, *The Career Development and Education of Astronomers in the United States,* thesis Harvard University.
Education of Astronomers 1969, *Q. J. R. astr. Soc.*, **10**, 21.
McNally, D. 1968, *Education for Astronomy,* Careers Bulletin autumn 1968, p. 6 (Central Youth Employment Executive, London).
Miller, F. D. 1969, *How to Become an Astronomer,* Sky Publishing Corp., Cambridge, Mass.
Ross, S. S. 1969, Conference on Laboratory Exercises, *Sky Telesc.*, **38**, no. 5, 304.
Taylor, R. J. 1969, *Q. J. R. astr. Soc.*, **10**, 44.
Schalén, C. 1965, *Att Studera Astronomi,* Svenska Bokförlaget (Scandinavian Univer. Books).
Wentzel, D. G. 1969, *Science Education: Case for Astronomy,* Astronomy Program, Univ. of Maryland, College Park, Md. 20742, U.S.A.
Wentzel, D. G., Smith, E. v. P. 1969, *Informal Resource Letter on Astronomy Laboratory Exercises,* Astronomy Program, University of Maryland, College Park, Md. 20742, U.S.A.

THE INTERNATIONAL SCHOOLS FOR YOUNG ASTRONOMERS

With the financial support of UNESCO the IAU organized three International Schools for Young Astronomers (ISYA) so far, the first in Manchester, United Kingdom (1967), the second in Arcetri, Italy (1968), and the third in Hyderabad, India (1969). The organization was in the hands of Dr. J. Kleczek, secretary of the ISYA. The report on the Manchester School is published in the *IAU Trans.* **XIIIB**.

The *Arcetri* ISYA was devoted to modern Solar Physics. Ten students from seven different countries attended the school (Bolivia: 1, Brazil: 2, Czechoslovakia: 3, Hungary: 1, Poland: 1, Sweden: 1, and Turkey: 1). The Arcetri astronomers were in charge of most of the 70 hours lectures, the 142 hours of practical training and the series of seminars held during the school of $8^1/_2$ weeks duration. In addition, two European professors were invited to give a couple of lectures. The school was financed in part by UNESCO and in part by the Arcetri Observatory and the Italian Government. M. Rigutti prepared a detailed report on the basis of (a) his and his colleagues' experiences during the ISYA, and (b) the answers to a questionnaire he had mailed after the end of the school to all attending students. Prof. Rigutti's excellent report is a very useful document and should be used as guideline for the planning of future ISYA.

The subject of the ISYA at the Osmania University in *Hyderabad* was Stellar Astronomy and Astrophysics. It was attended by 23 students from 5 Eastern countries (Ceylon: 2, China: 1, India, 18, Japan: 1, New Zealand: 1). During 8 weeks the students attended a total of 200 hours of lectures and practical training which were given by 7 invited professors from abroad and 6 professors from various Indian Universities and the astronomical staff of the Hyderabad Observatory. The school was financed in part by UNESCO, in part by the New Delhi University Grant Commission, and in part by the IAU.

The following is a brief synthesis of the results of the three ISYA as reported by J. Kleczek:

"In general, both the lecturers and the young astronomers considered the ISYA a success. They appreciated the 'home-like' and friendly atmosphere. To many young astronomers the school was of great help in formulating and preparing their future research programs. This is especially true about young astronomers from countries where the possibilities of education and research in astronomy are not very advanced. The contacts between the young astronomers and the lecturers, and among the young astronomers themselves do not end with the termination of the school. These post-school relations are an important result of our schools."

"In organizing the schools in Manchester and Arcetri young astronomers from various countries were brought to an Observatory to work with local lecturers and instruments. At Hyderabad, on the contrary, a new scheme was tried, namely to bring foreign lecturers and experienced research workers to an Observatory which, with newly acquired telescopes, is developing its astronomical research programs. The scheme was effective and it has the advantage that the experienced astronomers can help the host Institute to plan future research programs. For example, a special seminar was arranged at Hyderabad dealing with the exploitation of medium-size telescopes. (In the nearest future, the three principal Indian Observatories will begin to work with new telescopes of about one meter diameter)."

"Participants of all three schools complained that the instruction was too concentrated and that they had difficulties assimilating the new material. We are facing a dilemma: either to limit the lectures or to have schools of three months' duration. To reduce the lectures would mean to cripple the whole program; nothing unimportant is taught. Prolongation of the school, on the other hand, would require the reduction of the number of participants; we are limited by the available finances. For the moment, the young astronomers must try to finish assimilating the new material at their home institutes. They receive mimeotyped texts of the lectures for this purpose."

"It is the Secretary's opinion, that the ISYA brings great benefit to young astronomers from astronomically developing countries. Preference should be given to them, while a small number of participants from astronomically more developed countries should be admitted too. We had a few very good young astronomers who, for quite different reasons, had no possibility to continue their studies and research at home. The ISYA Secretary is grateful to the Commission members for not being indifferent to such cases."

The IAU Executive Committee has re-appointed Dr J. Kleczek as Secretary of the ISYA for another period of three years.

Dr J. Kleczek hopes to organize the 1970 ISYA in Argentina for the benefit of the Latin American young astronomers, and for the 1971 ISYA he is considering an African country.

In the name of the Commission and of all students attending the ISYA, the Commission President wishes to thank the UNESCO, the IAU, and the host countries for their generous financial support without which our schools would not have been feasible. Our greatest appreciation goes to Dr J. Kleczek for all his efforts and the efficient organization of the ISYA, and to the lecturers and astronomers of the host Observatories for their kind collaboration. They all contributed to make our ISYA so successful.

COLLABORATION WITH INTERNATIONAL COMMISSIONS ON SCIENCE TEACHING

In 1968 the CIES (Commission Internationale de l'Enseignement des Sciences) – an inter-union commission of ICSU organized a congress which was held from September 11 to 19 at the international center in Droujba near Varna (Bulgaria). The IAU was represented by Prof. M. Minnaert replacing Prof. E. A. Müller who was unable to attend. The proceedings of this congress are published by the Secretary of the CIES, Prof. P. Fleury (3, Boulevard Pasteur, Paris XV). The following two papers related to astronomy were presented:

M. G. J. Minnaert: The Integration of Astronomy into a Course of Physics

E. Schatzman: Place de l'astronomie par rapport au autres sciences.

Prof. *M. Minnaert*'s report on the congress is reproduced here in abbreviated form:

"The general tendency of the congress was to have at the elementary school an integrated teaching by one and the same teacher, including mathematics as well as sciences. In the first cycle of the secondary school the sciences of nature should be integrated, while mathematics ought to have a separate teaching. In the second cycle the same could apply to pupils not wishing to follow later a career in sciences. For the others, integration seemed not desirable, except perhaps between physics and chemistry; but close coordination between the different disciplines was expected. The eastern countries and Denmark were least inclined to the integration which different groups in the U.S.A., Australia, and Great Britain propagated. – It was most interesting to see the display of excellent textbooks written for integrated courses in sciences for secondary schools. The subjects treated are essentially: air, water, the forest, the farm, food, the seasons, motion, light, energy, etc. The connection with life at large was most enjoyable."

"The subject of integration in the teaching of the different sciences and of mathematics is, of course, of considerable importance for the teaching of astronomy. Here is an opportunity to have our science taught to a great part or to almost all of the pupils. On the other hand, there might be the danger that it would lose its proper value, especially if it were taught by a teacher who is not sufficiently prepared for this part of his task. – In the higher classes there is the problem of whether astronomy should be integrated with physics or with the earth-sciences (geology, meteorology, etc.). In the present situation I prefer the combination with physics, because as it is now, the teacher of earth sciences will not have been educated in modern astronomy; moreover, earth sciences are only taught in some countries."

"In a plenary session I stressed the great significance of astronomy in secondary education. Later, in one of the working groups, I showed in a more extensive and somewhat more technical manner how astronomy may be integrated with physics, but what precautions have to be taken in this case in order to save the proper value of astronomy as such. – At the end of the congress a concluding report was read and approved to which were added 15 statements, summarizing the main recommendations of the congress to pedagogical authorities and departments. In the final *report* the following paragraph appears:

Astronomie et physique

L'astronomie fait emploi des mathématiques et de la mécanique, mais elle participe surtout de la physique sous toutes ses formes. Elle a en outre des points communs avec les sciences de la terre et avec la chimie. Des notions d'astronomie moderne, comprenant spécialement l'astrophysique, sont indispensables dans l'enseignement secondaire. L'astronomie mérite d'être enseignée comme une discipline autonome; elle peut être aussi intégrée dans l'enseignement de la physique. Dans ce dernier cas il sera nécessaire, à fin des cours, de récapituler en quelques leçons les notions acquises, en les regroupant de manière à construire un modèle synthétique de l'univers."

In the concluding statements, issued by the congress and which will be diffused on a much larger scale, Prof. Minnaert succeeded in having the following statement printed:

"The physics teacher should be responsible for teaching *mechanics* as well as *astronomy*; the introduction of some notions of astronomy in secondary schools is desirable."

On December 12 and 13, 1969, E. A. Müller, as the President of the Commission 46, was invited by ICSU to Paris to attend the first meeting of a newly formed international Committee on Science Teaching (CST) under the chairmanship of Dr M. Matyas of Czechoslovakia. Participants were some representatives of ICSU and of UNESCO, and the presidents of the Teaching Commissions of the various Scientific Unions who reported on the activities of their Teaching Commissions. The topics discussed for the future programs of the CST were the following: modernisation and coordination of science teaching; integration of sciences; cooperation with UNESCO; exchange of information and teaching experiences among the various Scientific Unions; continuous education of the science teachers; guide-lines for improving the science teaching in developing and developed countries; social responsibilities of scientists. A second meeting of the CST is scheduled for the second part of April 1970.

REPORTS ON NATIONAL ACTIVITIES IN ASTRONOMY EDUCATION

The Commission members were invited to submit a report on new developments in the teaching of astronomy in their own countries. 75 % of the members sent a report, and some of the reports are very interesting and detailed and may interest other members of the Commission. Due to lack of space in the IAU Draft Report only a brief summary of the trends in the astronomy teaching activities can be presented here. The individual complete reports will be reproduced in mimeographed form and distributed to all Commission members. It should be added, that any astronomer interested in the detailed report of a particular country may order a copy either through the Commission member representative of that country, or through the Commission President.

1. *Astronomy education in elementary and secondary schools*

Members of the following countries reported that astronomy has been – or is in the process of being – introduced in Schools: Argentina, Belgium, Bulgaria, Canada, Czechoslovakia, Denmark, German Dem. Rep., German Fed. Rep., Japan, The Netherlands, New Zealand, Poland, Roumania, Switzerland, United Kingdom, U.S.S.R., and U.S.A. The level of astronomy teaching is quite different in the different countries. In the elementary school level astronomy is usually integrated in the earth sciences. At the secondary school level some notions of astronomy are given in either mathematics or physics courses, and in only a few places astronomy is taught as a separate course (usually optional) in either one of the last two years of secondary school. Often the teaching of astronomy in schools does not go much beyond the Earth-Moon and the Earth-Sun system. The main problem of school astronomy is that the teachers themselves are not well prepared, they know very little about modern astronomy. In order to overcome this difficulty, astronomy courses for teachers are being organized in several countries (e.g.: Argentina, German Fed. Rep., The Netherlands, Poland, Switzerland) during which the school teachers get acquainted with modern astronomy and astrophysics through lectures and practical training on school telescopes. Didactic books and guide-lines in astronomy for school teachers would be of very great help. In a few countries such teaching aids have been prepared, but much more should be done about this.

2. *General astronomy education at the university level*

In several countries new programs have been or are being developed for the introductory courses in astronomy and astrophysics. These courses are given to science students who may or may not become astronomers later on. A number of these students may eventually go into secondary school and/or college level teaching and, thus, will be better prepared for the astronomy teaching in schools than the teachers have been so far. Also, in some countries (e.g.: France, German Fed. Rep. and others) one tries to coordinate and unify the introductory and basic courses of astronomy and astrophysics given at various universities.

3. *Education of astronomers*

Everyone agrees that the education of astronomers starts with a good education in physics and mathematics. The training for becoming an astronomer then requires a number of basic astronomy and astrophysics lectures and seminars, and much practical training and research work. Obviously, the education of astronomers very much depends on the facilities available in individual countries and universities. In recent years many universities have reconsidered and modernised their programs for astronomy studies. Universities which only recently have started an astronomical program may make use of the facilities of an observatory of another university in the same country or in a neighbouring country. How in individual countries the education of astronomers has been improved or recently introduced can be read in the detailed report of the national activities in astronomy education.

4. *Education in astronomy of the general public*

Various Commission members report that astronomy education of the general public is done (a) by regular public lectures given by the astronomers themselves or by amateur astronomers, (b) by visits to planetaria, (c) by series of radio or television transmissions on astronomy, and (d) by a number of popular books which have appeared on the market in recent years.

EDITH A. MULLER
President of the Commission